PHARMACOKINETICS AND TOXICOKINETICS

PHARMACOKINETICS AND TOXICOKINETICS

MEHDI BOROUJERDI
College of Health Sciences – School of Pharmacy
University of Massachusetts Lowell
Lowell, Massachusetts USA

CRC Press
Taylor & Francis Group
Boca Raton London New York

CRC Press is an imprint of the
Taylor & Francis Group, an **informa** business

CRC Press
Taylor & Francis Group
6000 Broken Sound Parkway NW, Suite 300
Boca Raton, FL 33487-2742

Printed on acid-free paper
Version Date: 20150116

International Standard Book Number-13: 978-1-4822-2134-3 (Hardback)

Visit the Taylor & Francis Web site at
http://www.taylorandfrancis.com

and the CRC Press Web site at
http://www.crcpress.com

Dedicated to my beloved family: Parvin, Mazy, Bob, Maggie, Lila and Ruby

Contents

Preface

The fields of pharmacokinetics and toxicokinetics have made great progress since the introduction of the term *pharmacokinetics* by Professor F. H. Dost in 1953. To cite tangible evidence of the growing interest in both disciplines, it would be sufficient just to consider the application of their principles in decision making by the regulatory agencies; the abundance of books, journals, and articles with the word pharmacokinetics or toxicokinetics; and the permeation of the fields in all related basic and applied disciplines. In large measure, this progress parallels the remarkable growth of computing technology.

Combining pharmacokinetics and toxicokinetics in one book stemmed from the most basic principle of science and pedagogy and the one most applicable to them, the Aristotle principle of noncontradiction, which is the second of the following three principles of "thought" or "logic":

- The principle of identity
- The principle of noncontradiction
- The principle of excluded middle

The principle of identity can simply be stated as A is equally the same as A and cannot be both A and not A at the same time and in the same sense, or "whatever is, is." Stated differently, kinetics is kinetics and cannot be non-kinetics at the same time and in the same sense.

The principle of noncontradiction can be verbalized plainly as A cannot be both B and non-B at the same time and in the same sense, that is, "nothing can both be and not be." If one assumes that pharmacokinetics is different from toxicokinetics, or vice versa, then kinetics is the principle of only one and not the other.

The third principle is the principle of excluded middle, which means that a statement is either true or false and there is no middle ground between true and false; otherwise stated, the "principles of pharmacokinetics are the same as toxicokinetics" is either true or false and there is no position in-between. This book is written on the assertion that the principles of pharmacokinetics are the same as those of toxicokinetics and the congruence of the two fields is conclusive.

Therefore, the purpose of this book is to provide systematic overlapping principles of pharmacokinetics and toxicokinetics. It covers a wide range of concepts basic to both fields and the mathematics of modeling and interconnectivity of the topics. The emphasis is on the understanding of concepts and the governing principles of kinetics, and thus isolated details, theoretical and complex mathematics of stochastic modeling and neural networks, and computer-generated curves of published experimental data have been avoided.

The publisher deserves the author's gratitude for their untiring efforts toward the goal of producing a book of maximal utility to scientists, researchers, educators, and students.

1 The Influence of Routes of Administration on ADME

1.1 INTRODUCTION

The word xenobiotic, a combination of two Greek words ("xenos" and "biōtkós"), literally means "stranger to life" and is referred mostly to an exogenous chemical or substance foreign to the human body or other organisms. Xenobiotics may act as therapeutic agents, toxins, chemical carcinogens, food additives, pollutants, etc. Pharmacokinetics (PK) and toxicokinetics (TK) are the quantitative science of physiological and biochemical processes involved in absorption, distribution, metabolism, and excretion (ADME) of xenobiotics, and their corresponding pharmacological or toxicological response *in vivo* and *in vitro* (Figure 1.1). The overlapping goals of both disciplines are to understand the behavior of xenobiotics in the body, how the body handles them, and predict their pharmacological or toxicological response. The end result of PK/TK analysis is establishing appropriate criteria and approaches to avoid undesirable outcomes, and safeguard human or animal life. The desirable outcome can be as practical as developing appropriate dosage forms or a dosage regimen for the purpose of achieving an optimum therapeutic outcome or pharmacological response, establishing regulatory guidelines/policies for consistency of the outcomes through bioequivalence and biosimilar evaluations, or as challenging as low dose extrapolation of chemical carcinogens and the understanding of safe doses in hazardous environmental pollutants.

PK and TK use the same principles of kinetics in data analysis, interpretation, and deduction or prediction. The difference is noticeable when one looks at the biological outcome: PK deals with low doses of drugs and therapeutic outcome; TK uses large doses of therapeutic agents or low doses of highly toxic substances. In both cases, regardless of the route of administration, the body has to deal with an invaded foreign substance.

1.2 PHARMACOKINETICS (PK) AND PHARMACODYNAMICS (PD)

The time course of xenobiotics in biological samples was studied as early as 1924 and 1932 (Widmark and Tandberg, 1924; Widmark, 1932) and 1937 (Teorell, 1937a,b), but the term pharmacokinetics was first introduced in 1953 (Dost, 1953). The history of the first fifty years of pharmacokinetics is well chronicled in a paper published by John Wagner, another pioneer of pharmacokinetics, in 1981 (Wagner, 1981). The following definitions have gained wide acceptance by the scientists of the field: Pharmacokinetics is the study of the time course (i.e., kinetics) of absorption, distribution, metabolism, and excretion (ADME) of drugs and their metabolites in the body (Figure 1.1).

The corresponding influence(s) on the intensity and time course of pharmacological response or therapeutic outcome is known as pharmacodynamics. Thus, pharmacodynamics (PD) can be defined as the quantitative study of pharmacological, biochemical, physiological, and therapeutic effects of a drug and their mechanism of action on living structures, *in vivo*, *in vitro*, and *in situ*. In essence, pharmacokinetics and pharmacodynamics (PK/PD) are two integrated subjects that in combination elucidate what the body does to a drug molecule and what it acquires from the drug. PK/PD modeling via different approaches has developed into an advanced mathematical field relating the kinetic behavior of a drug in the body, to its interaction with the receptor site and to the magnitude of the physiological/pharmacological response. Given that certain physiological, pathophysiological, and demographical factors can alter the therapeutic outcome of a drug in certain populations, a branch

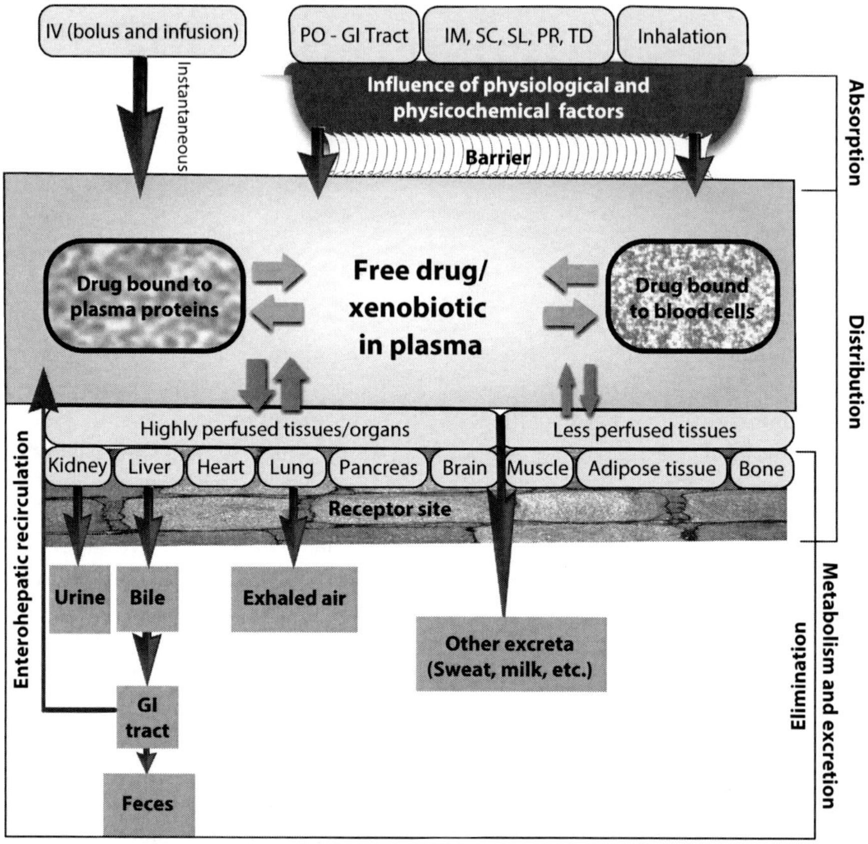

FIGURE 1.1 Schematic illustrations of physiological processes of absorption, distribution, metabolism, and excretion (ADME) of xenobiotics; the routes of entry into the body; and the main biological samples.

of pharmacokinetics known as population pharmacokinetics has received much attention in recent years. Population pharmacokinetics can be defined as the study of the variability in plasma drug concentrations between individuals when standard dosage regimens are administered (Aarons, 1991; FDA Guidance for Industry, 1999). Another key factor influencing how the individuals respond to therapeutic agent is their genetic make-up. The study of how humans respond to xenobiotics due to their genetic inheritance is called pharmacogenetics (PGt). The interindividual variability in genes (encoding protein transporters, drug metabolizing enzymes, some receptors, ion channels, and immune molecules) influences the PK/PD of a drug. In particular, genetic differences in enzymes of drug metabolism (Evans and Relling, 1999) have direct impact on metabolic clearance and variability in formation of active or toxic metabolites. Advances in PGt in combination with sophisticated PK/PD modeling and analysis will provide guidance for the practice of personalized medicine.

1.3 TOXICOKINETICS (TK) AND TOXICODYNAMICS (TD)

The focus of PK/PD analysis is mainly on optimization of therapeutic dose, dosing regimen, and therapeutic outcome. While the central focus of toxicokinetics is in safety assessment of pharmaceuticals (ICH Guideline S 3A, 1995; Baldrick, 2003; Ploemen et al., 2007), risk assessment of industrial chemicals (fertilizers, pesticides, biocides, carcinogens, nanomaterials, etc.) and safety of non-pharmaceutical ingredients added to food, cosmetics, and personal care products. The application of toxicokinetics for the safety assessment of pharmaceuticals is an integrated part of the science of drug discovery and development. Its definition according to the ICH (International Conference on Harmonization) is "the generation of pharmacokinetic data, either as an integral component in the conduct of non-clinical toxicity studies or in specially designed supportive studies, to assess systemic exposure (ICH, 1995)." Furthermore, the application of toxicokinetics in improvement of the chemical risk assessment has been recognized by organizations such as International Life Sciences Institute (ILSI), Organization for Economic Co-operation and Development (OECD), Health and Environmental Sciences Institute (HESI), and chemical industries. Thus, a more clear definition of toxicokinetics would be the study of the time course of absorption, distribution, elimination (i.e., excretion and metabolism), and uptake of potentially harmful xenobiotics leading to a toxic response.

Toxicodynamics (TD) can then be defined as the *in vivo*, *in vitro*, and *in situ* determination and quantification of the sequence of events at the cellular and molecular levels leading to apoptosis, toxic risk or lethal response after exposure to a chemical agent (Figure 1.2).

Based on the principles of population pharmacokinetics, a branch of toxicokinetics known as population toxicokinetics has emerged, which is based on the Bayesian statistical inferences and physiological modeling (Bois et al., 1996; Chiu and Bois, 2006; Amzal et al., 2009). Similar to pharmacogenetics, the interindividual variability in genes, which influences the TK/TD of a chemical, is also known as toxicogenetics. In particular, genetic differences in enzymes have direct impact on the clearance and formation of reactive metabolites.

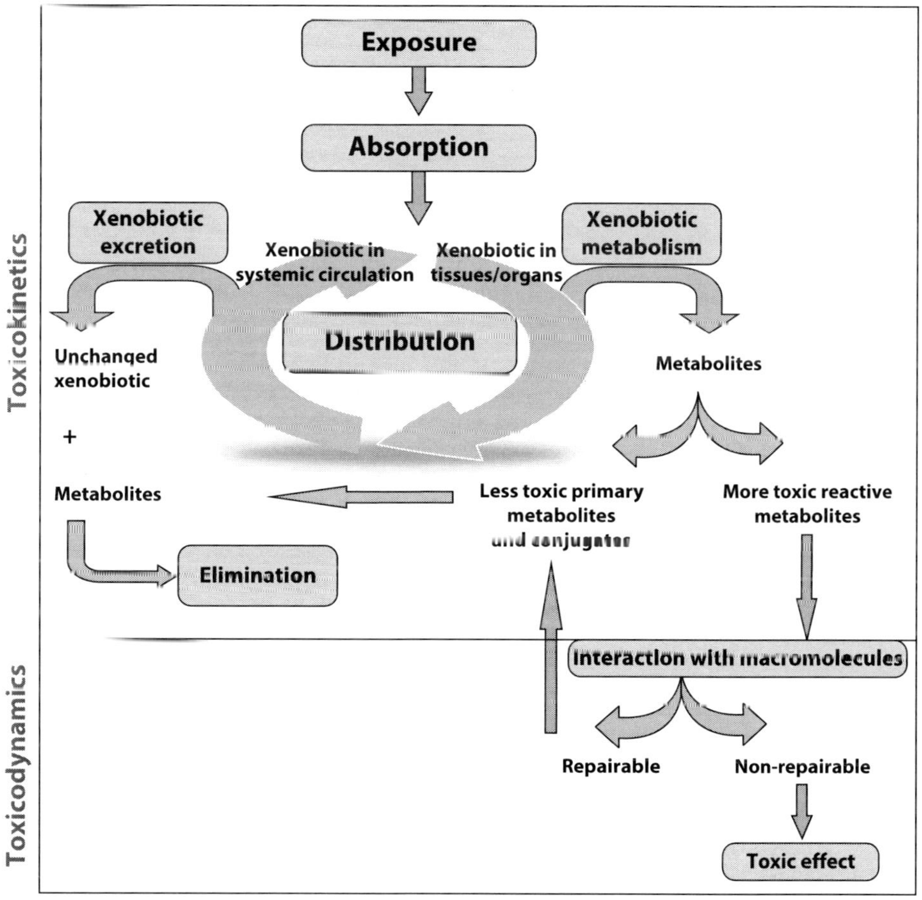

FIGURE 1.2 Conceptual depiction of ADME influences on the overall toxicokinetics and toxicodynamics of xenobiotics.

Toxicology (Tcole) is the study of toxic effects of xenobiotics on living organisms; it includes the study of mechanisms, symptoms and detection of toxic response. Similar to pharmacology, the areas of toxicology are numerous and, depending on the diverse fields of toxicology (environmental, biochemical, regulatory, reproductive, forensic, analytical, clinical, food, pharmaceutical, nutritional, behavioral, occupational, dermato-, neurto-, immune-, idiosyncratic organo-, nano-, and mechanistic), toxicokinetics have different applications ranging from complementary data to establish clinical safety to environmental low dose extrapolation and risk assessment (IUPAC, 2007).

1.4 CONCEPTS BASIC TO PK AND TK

As indicated in the definition of PK and TK, the important factors influencing an individual's response to a therapeutic agent or a toxic chemical is the uptake or absorption into the body, followed by distribution in the body via systemic circulation which reaches equilibrium rapidly with highly perfused organs and tissues and slowly with less perfused tissues. The liver, the main organ of metabolism, is one of the highly perfused tissues. Thus, the conversion of a xenobiotic to its nontoxic or toxic/reactive metabolites occurs rather rapidly following the administration or exposure to a chemical. Another highly perfused organ is the kidney, which removes the unchanged and water-soluble metabolites from the body while the distribution of the parent compound is underway. The response, either pharmacological or toxicological, is a function of the total amount of xenobiotic to which the body is exposed to; the higher the dose, the more intense the response. The dose can make all xenobiotics poison. This concept was first recognized by Paracelsus (1493–1541) who wrote:

> *"Alle Ding sind Gift und nichts ohn' Gift; allein die Dosis macht, das ein Ding kein Gift ist."* [All substances are poisons; there is none which is not a poison. The right dose differentiates a poison from a remedy.]

The dose refers to the amount of xenobiotic at a site of administration, which will be subjected to the physiological processes of ADME. Absorption is the initial process a compound encounters, and its complexity depends on the route of administration. All therapeutic agents, depending on their design, can enter the systemic circulation through a specific route of administration. They go through preclinical and clinical PK/TK evaluation during the course of drug development. The main distinction between TK and PK in drug development is that the doses used in TK studies are higher than doses used in PK studies that produce pharmacological response. Both evaluations are based on methodical approaches to yield information on drug plasma levels, relevant PK/TK parameters and constants in order make PD and TD assessments.

The xenobiotics' routes of entry into the body through environmental exposure are mainly skin (topical), lung (inhalation), eye (ophthalmic), and gastrointestinal (GI) tract (ingestion). For environmental exposure to chemicals, it is often difficult to determine the exact dose. The probability of adverse effects for this type of exposure depends on the frequency, duration, and extent of exposure, and offers challenges for defining the effective dose, which will be discussed later in this book.

1.5 ROUTES OF EXPOSURE/ADMINISTRATION

An important factor affecting ADME and thus the onset of action, duration of action, bioavailability and intensity of the pharmacologic or toxic response is the route where the body comes into contact with a xenobiotic. The route of exposure or administration determines the biological processes involved in handling the compound. For example, the parenteral routes avoid the absorption process and provide immediate onset of action. Whereas, for drugs that diffuse through a barrier, a delay to the onset action may occur, the magnitude of which depends on the complexity of the barrier and the physicochemical characteristics of xenobiotic, and, for therapeutic agents, the type of dosage form.

The recent advances in delivery technology and medical devices have extended the application of some of these routes of administration beyond the traditional delivery systems.

Classification of the routes of administration, which is helpful in PK/TK modeling and analysis, are the following four categories (Boroujerdi, 2002):

First Category: There are four groups (A–D) of routes of administration in this category; they all share a common feature of having a biological barrier. The structure of some barriers can be as complex as the gastrointestinal wall and some as basic as sublingual or rectal mucosa. Xenobiotics must pass through these barriers to reach the systemic circulation and exhibit the expected or unexpected systemic effect. Depending on the chemical reactivity of a compound at the site of administration, the systemic effect may also be associated with the local effect. Different regions of the gastrointestinal tract, various zones of the pulmonary tract, peripheral sites of injections and administration through a natural orifice are in this category (Table 1.1).

Second Category: The routes of administration in this category circumvent the barriers, and are used for immediate onset of action and direct access to the systemic circulation (Table 1.1).

Third Category: Interaction of xenobiotics with the skin can lead to local and/or systemic effect. The presence of the outermost horny layer of the skin (stratum corneum) makes the absorption more complex than the routes of the First Category. In addition to the physicochemical characteristics of xenobiotics and their vehicle, environmental factors such as humidity, temperature, and physiological condition of the skin also play important roles in the absorption process.

Fourth Category: Specialized routes of administration for achieving local and targeted therapy are included in this category. Although the intended use of these routes is mostly to achieve local or targeted effect, nonetheless a portion of administered dose may reach the systemic circulation and exhibit systemic effect (Table 1.1).

In the remaining sections of this chapter, major routes of administration/exposure and their important characteristics as well as factors that may influence the analysis of data and PK/TK parameters/constant will be highlighted.

1.5.1 First Category

1.5.1.1 Gastrointestinal Site of Absorption

The detailed discussion of this route of administration is presented in Chapters 2 and 8.

1.5.1.2 Sublingual and Buccal Absorption

The mouth cavity is used for both localized treatment of conditions such as fungal conditions, ulcers, and periodontal disease, and also achieving systemic effect. The oral mucosa, particularly buccal and sublingual sites are used to achieve the systemic effect. Both sites are highly vascular regions of the mouth; the buccal site is between the cheek and gums (gingival) and sublingual site is under the tongue between the ventral surface of the tongue and floor of the mouth.

The sublingual/buccal absorption of xenobiotics takes place in the presence of saliva in an environment with a pH of 6–7 through the mucosal barrier into the systemic circulation by passive paracellular and transcellular diffusions (Rathbone and Tucker, 1993)—see also Chapter 2. The diffusion is limited by molecular weight (Walton, 1935). There has been no report on the presence of active absorption, facilitated transport, or pinocytosis at this site. The mucosal barrier is comprised of (i) a mucus layer, (ii) a variable keratinized layer, (iii) an epithelial layer, (iv) a basement membrane, (v) connective tissue, and (vi) a submucosal region. The sublingual site is believed to be more permeable than the buccal and palatal (roof of mouth) region. The difference is attributed to the

TABLE 1.1
Routes of Administration/Exposure

FIRST CATEGORY

Group A	Group B	Group C	Group D
GI Tract	Nasopharyngeal	Subcutaneous	Rectal route
Sublingual/buccal	Tracheobronchial	Intramuscular	Vaginal route
	Pulmonary	Intraperitoneal	

SECONG CATEGORY

Intravenous Injection	Intra-Arterial Injection

THIRD CATEGORY

Percutaneous Absorption

FOURTH CATEGORY

Specialized Routes of Administration

Route of Administration	Site of Administration	Route of Administration	Site of Administration
Auricular (OTIC)	Ear	Intramedullary	Bone marrow cavity
Conjunctival	Conjunctiva	Intrameningeal	Within the meninges
Endocervical	Canal of the cervix uteri	Intraocular	Within the eye
Endosinusial	Nasal sinuses of the head	Intraovarian	Within the ovary
Endotracheal	Directly into the trachea	Intrapericardial	Within the pericardium
Enteral	Directly into the intestines	Intrapleural	Within the pleura
Epidural	Dura mater	Intraprostatic	Within the prostate gland
Extra-amniotic	Membrane cover of fetus	Intrasinal	Nasal or periorbital sinuses
Interstitial	Interstices of a tissue	Intraspinal	Vertebral column
Intra-abdominal	Abdomen	Intrasynovial	Synovial cavity of a joint
Intra-amniotic	Within the amnion	Intratendinous	Within a tendon
Intra-articular	Within a joint	Intratesticular	Within the testicle
Intrabiliary	Bile ducts or gallbladder	Intrathecal	Cerebrospinal fluid
Intrabronchial	Within a bronchus	Intrathoracic	Thorax (internal to the ribs)
Intrabursal	Within a bursa	Intratubular	Tubules of an organ
Intracardiac	Heart	Intratumor	Tumor
Intracartilaginous	Within a cartilage	Intratympanic	Within the aurus media
Intracaudal	Within the cauda equina	Intrauterine	Uterus
Intracavernous	Within a pathologic cavity	Intraventicular	Within a ventricle
Intracerebral	Within the cerebrum	Intravesical	Within the bladder
Intracisternal	Cisterna magna	Intravitreal	Vitreous body of the eye
Intracorneal	Within the cornea	Laryngeal	Larynx
Intracoronary	Coronary arteries	Nasogasteric	Through nose into stomach
Intracorporus cavernosum	Corporus cavernosa of penis	Ophthalmic	External eye
Intradermal	Within the dermis	Oropharyngeal	Mouth and pharynx
Intradiscal	Within a disc	Periarticular	Around a joint
Intraductal	Within the duct of a gland	Peridural	Outside of the dura mater
Intraduodenal	Within the duodenum	Perineural	Around nerves
Intradural	Within or beneath the dura	Retrobulbar	Behind the eyeball
Intraepidermal	Epidermis	Subarachnoid	Beneath the arachnoid
Intraesophageal	Within the esophagus	Subconjuctival	Beneath the conjunctiva
Intragasteric	Within the stomach	Transplacental	Placenta
Intragingival	Within the gingivae	Transtracheal	Wall of the trachea
Intraileal	Ileum	Transtympanic	Tympanic cavity
Intralesionalo	Into a localized lesion	Ureteral	Ureter
Intralymphatic	Within the lymph	Urethral	Urethra

Source for Specialized Routes of Administration: FDA Data Element OID: 2.16.840.1.113883.3.26.1.1.1, Version 4.

vascularity of the region, the degree of kreatinization of the barrier, and physicochemical properties of the drug (Kurosaki et al., 1991; Bartlett et al., 2012). However, in an evaluation of absorption of fentanyl tablet administered buccaly and sublingually, no significant differences were observed in human subjects (Darwish et al., 2008), which may be related to the physicochemical characteristics of fentanyl. The important physicochemical properties of xenobiotics affecting the oral mucosal absorption include lipophilicity of the compound and its partition coefficient, solubility in saliva, pH of the saliva, pKa of the compound, and binding to oral mucosa, which reduces the absorption of the compound (Katz and Barr, 1955).

The sublingual/buccal absorption is often considered an alternative route for oral absorption of compounds that are unstable at a low pH in the stomach, are metabolized by the enzymes of GI tract (e.g., CYP3A4), are substrates for efflux proteins, and are subjected to the hepatic first-pass effect. The flow of saliva, ~0.5 mL/min, however, may wash away the compound from the site of absorption into the stomach, which in turn reduces the consistency of the rate and extent of absorption. An important factor in oral mucosal absorption is the retention of the dosage at the site of absorption. Polymeric buccal patches with adhesion characteristics have been investigated, which may help the bioavailability of their active ingredient, dependent on if they don't impair the site of attachment. Other sublingual/buccal dosage forms (e.g., solutions, ointment, emulsions, and powder) will also be subjected to the parallel loss of the compound due to the dilution in saliva and involuntary swallowing. The diagram presented in Figure 1.3 represents a model for the sequence of rate processes for kinetic analysis of buccal/sublingual administration for achieving systemic effect, taking into consideration the parallel removal due to swallowing for a compound that follows three-compartment model (Figure 1.3).

A modified version of the compartmental model (Figure 1.3) in conjunction with non-compartmental analysis has been reported as a practical approach for estimation of absorption from sublingual/buccal mucosa (Wang et al., 2013):

$$\frac{(dA_{abs})_i}{dt} = (\alpha)_i \times (P_{eff})_i \times (V)_i \times (C_1(t)_i - C_2(t)_i) \quad (1.1)$$

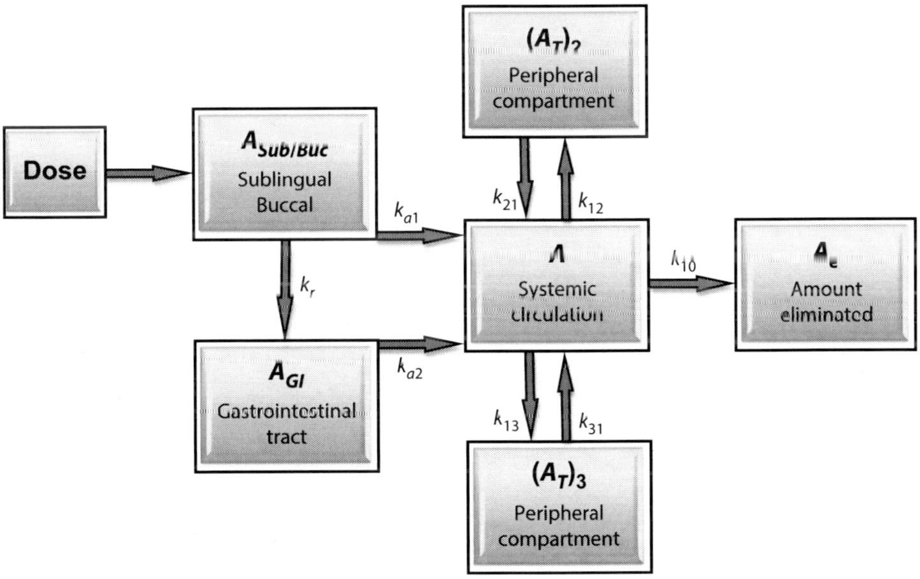

FIGURE 1.3 Diagram of a three-compartment pharmacokinetic model for buccal/sublingual absorption of a xenobiotic, taking into consideration the sequence of rate processes including the parallel removal from the site of absorption by involuntary swallowing.

where $(dA_{abs})_i/dt$ is the rate of absorption for compartment (i); $(\alpha)_i$ is identified as the absorption scale factor; $(P_{eff})_i$ is the permeability of compound in the compartment with volume of $(V)_i$ with two concentration terms $(C_1(t)_i - C_2(t)_i)$ representing the concentration at the site of administration and concentration in mucosa (Wang et al., 2013).

In addition to the removal from the site of absorption due to the involuntary swallowing, the oral mucosa contains proteases. Enzymes such as aminopeptidase, carboxypeptidase, and esterase have also been identified in homogenates of human buccal epithelial cell culture of human and porcine buccal mucosa (Nielsen and Rassing, 2000). The use of tissue homogenates does not allow one to distinguish between the membrane and cytoplasmatic enzymes. However, the presence of these enzymes may reduce the bioavailability of the peptides and protein drugs administered buccally or sublingually.

1.5.1.2.1 Saliva

An important issue associated with sublingual/buccal administration is the stability of xenobiotics in saliva. The issue is even important for the antimicrobial agents used in the prevention and treatment of plaque-related oral disease (Na et al., 2007).

The salivary excretion is a route for elimination of compounds administered through other routes of administration (Wagner, 1971). Depending upon the molecular size, lipid solubility, and pKa, the unbound fraction of administered or exposed xenobiotic is transferred from the systemic circulation through the salivary glands into saliva. In the ductal system of salivary glands, secretion and reabsorption play major roles in the flow rate of saliva and the concentration of its solutes. The concentration of a compound in saliva is much lower than the urine or plasma. The degree of protein binding in saliva is negligible and back diffusion of the compound into the plasma depends mainly on the degree of ionization of a compound at the pH of saliva. Thus, the ratio of saliva/plasma concentrations is a function of the pH of saliva and under equilibrium conditions between concentration of free drug in the plasma and saliva using the Henderson–Hasselbalch principles (Henderson, 1908; Hasselbalch, 1917) the following equations can be estimated for weakly acidic and basic compounds:

$$\begin{aligned} \textit{Acidic compounds:}\ \frac{Saliva}{Plasma} &= \frac{\left[A_{saliva}\right]\left[HA_{plasma}\right]}{\left[A_{plasma}\right]\left[HA_{saliva}\right]} \times \frac{(f_u)_{plasma}}{(f_u)_{saliva}} \\ &= \frac{1+10^{\left(pH_{saliva}-pK_a\right)}}{1+10^{\left(pH_{plasma}-pK_a\right)}} \times \frac{(f_u)_{plasma}}{(f_u)_{saliva}} \end{aligned} \tag{1.2}$$

$$\textit{Basic compounds:}\ \frac{Saliva}{Plasma} = \frac{1+10^{\left(pK_a-pH_{saliva}\right)}}{1+10^{\left(pK_a-pH_{plasma}\right)}} \times \frac{(f_u)_{plasma}}{(f_u)_{saliva}} \tag{1.3}$$

where $[A_{saliva}]$ and $[A_{plasma}]$ are the total concentration of ionized and unionized molecules in saliva and plasma, respectively, that is, $[A] = [A^-] + [HA]$ and f_u is the free fraction of the compound.

In many cases saliva can be used instead of blood in the therapeutic or toxicity monitoring of many compounds. In general, saliva testing is used when the ease of collection as compared to the other biological samples like urine or blood outweighs the cost and required sensitive methodology, like mass spectrometry (Kidwell et al., 1998). The detection of illicit compounds such as amphetamines, PCP, cocaine, marijuana, etc., in saliva has been established in forensic science and practice. Saliva screening for therapeutic agents is also used in pediatric patients, and therapeutic drug monitoring of many agents such as tolbutamide, propranolol, procainamide, etc.

1.5.1.3 Pulmonary Route of Administration and Absorption

The nasopharyngeal airways of pulmonary tracts is comprised of the nasal cavity, a highly vascular region divided into three distinct regions of superior, middle, and inferior nasal concha, sinuses, nasopharinx, oropharynx, and larynx. The inferior turbinate is the highly vascular region of the site. Approximately 10,000 L of ambient air pass through the nasal airway daily and about one liter of moisture is added to the air during this time.

The alveolar surface is covered with a layer of fluid of about 0.2 μm in thickness, which contains significant concentration of surfactants such as 1,2-dipalmitoyl phosphatidylcholine. The presence of surfactants, airflow, and moisture influence the absorption of solution, inhaled gases, aerosols, and vapor of volatile solvents as well as their rate and extent of absorption. The interaction of surfactants with xenobiotics after deposition in the mucosa of tracheobronchial airways or alveolar region facilitates the solubility of lipophilic and cationic compounds (Wiedmann et al., 2000; Liao and Wiedmann, 2003). However, the interaction of surfactants with peptides, proteins and other macromolecules has the opposite effect, causes aggregation of the molecules and prevention of their absorption (Patton, 2007). The pulmonary absorption of xenobiotics is by active absorption, passive transcellular and/or paracellular diffusion, pore formation, and vesicular transport. Nanoparticles of less than 100 nm may penetrate the epithelial membrane and absorb into systemic circulation by endocytosis (Kemp et al., 2008). The mechanism involves the transit of nanoparticles across epithelia of the respiratory tract and entering the circulation via lymphatic pathways (Medina et al., 2007). If the particles do not get deposited and remain afloat in the aerial milieu, they are removed by exhalation from the pulmonary tract by outflow. The pulmonary route of administration provides a rapid onset of action locally and systemically for xenobiotics when their absorption is thermodynamically favorable. Furthermore, inhalation and nasal routes are considered viable alternatives to the oral administration of compounds with poor bioavailability, the compounds that are sensitive to the low pH of the stomach and/or eliminated by the hepatic and/intestinal first-pass metabolism, or their absorption is reduced by GI tract efflux proteins. The shortcoming, however, is the optimization of the administered dose, residence time, and absorption variability due to the random removal of the particles by exhalation, variability of airflow for absorption of gases and particles, and inter-individual anatomical and physiological differences. The dissolution of the particles in the mucous layer of the pulmonary tract is also an important factor that contributes to the variability of the effective dose. Particles with a slow rate of dissolution will be removed by mucociliary clearance. Moreover, the highly vascular region of the nasal airway is the nasal tubinates and particles containing xenoibiotic must be deposited in the turbinate region to provide optimal systemic effects (Kimbell et al., 2004).

1.5.1.3.1 Pulmonary Protein Transporters and Enzymes

Throughout the respiratory tract, a number of protein transporters are expressed, which include the organic cation transporters (e.g., OCT1, OCT2, OCT3, OCTN1) (Lip et al., 2005; Horvath et al., 2007), organic anion transporters (e.g., OATP2B1, OATP2B1, OATP3A1, OATP4C1) (Bleasly et al., 2006), ABC transporters (e.g., MRP1, MRP3) (Young et al., 2001; Van der Deen et al., 2006; Okamura et al., 2013), and peptide transporters (e.g., PEPT2) (Gronebrg et al., 2004).

The metabolism of xenobiotics in the lung complements other respiratory tract clearances, such as mucociliary removal (Stahlhofen et al., 1990), phagocytosis by alveolar macrophages (Oberdorster, 1998), and other ancillary processes (Patton et al., 2004). The enzyme systems of the lung carry out both Phase I and Phase II metabolism (Spivack et al., 2003) (see also Chapters 6 and 8). The Phase I oxidative reactions are carried out by cytochrome P450 isozymes (Shimada et al., 1994; Raunio et al., 1999; Ding and Kaminsky, 2003) including CYP1A1, CYP3A4, CYP3A5, CYP1B1, CYP2B6, CYP2D6, CYP2E1, and CYP2J2 (Anttila et al., 1997; Kelly et al., 1997; Spivack et al., 2001; Hukkanen et al., 2002; Nebert et al., 2004; Zhang et al., 2006; Peng et al., 2013). The expression levels of these isoforms in the lung are lower than the intestinal or hepatic enzymes and their activity is also generally lower than other major sites of metabolism. The Phase II metabolic conjugation can also occur in the lungs with the help of enzymes such as UDP glucuronosyl transferases, glutathione S-transferarses, and sulfotransferases (Zheng et al., 2002). Other metabolic enzymes of the respiratory tract include peptidases, flavin mono-oxygenases, estrases, and cyclo-oxygenases.

1.5.1.3.2 Pulmonary Disposition of Particles

The deposition of particles in the environment or in aerosol depends on the size and charge of the particles. The epithelium of nasopharyngeal airway and tracheobronchial region (i.e., trachea, bronchi, and bronchiols) are the site of absorption for gases and deposited particles. However, the deposited particles in tracheobronchial region can be subjected to removal by the upward mucociliary movement of the region (Figure 1.4). If the mucous is swallowed, the absorption of particles would be partially through the GI tract. Thus, the clearance of particles from the site of absorption depends on the dissolution and absorption at the site; removal by the mucociliary transport and likely absorption at the GI tract; and pulmonary metabolism of the dissolved particles (Figures 1.4 and 1.5). These processes occur in parallel and PK/TK analyses of hybrid data may not differentiate the concentration that reaches the systemic circulation from the lungs and that of the GI tract; in particular, when one intends to define the true pulmonary absorption rate or rate constant and/or pulmonary elimination parameters and constants. For compounds with negligible GI absorption, the kinetic evaluation of pulmonary absorption can be accomplished with the standard approaches in PK/TK analysis. However, for drugs with significant GI absorption, it has been suggested that the kinetics of pulmonary absorption can be defined by blocking gastrointestinal absorption with charcoal, or by studying absorption during the first 30 minutes post inhalation before appreciable oral absorption has occurred (Derendorf et al., 2001). Another element of inhalation that may influence the pulmonary absorption of particles is their size and the pattern of airflow, which can limit their migration through the pulmonary system. The airflow depends on the physical activity of the subjects, which may also vary from person to person, or species to species. The default value for

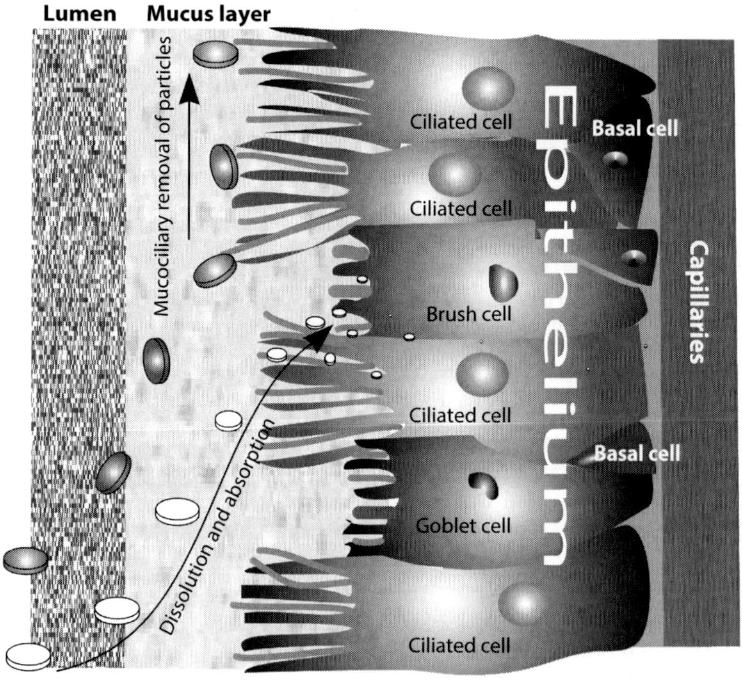

FIGURE 1.4 Handling of particles in the epithelium of the tracheobronchial and nasopharyngeal airway; subjected either to dissolution followed by absorption, or removal by the upward mucociliary transport by ciliated cells.

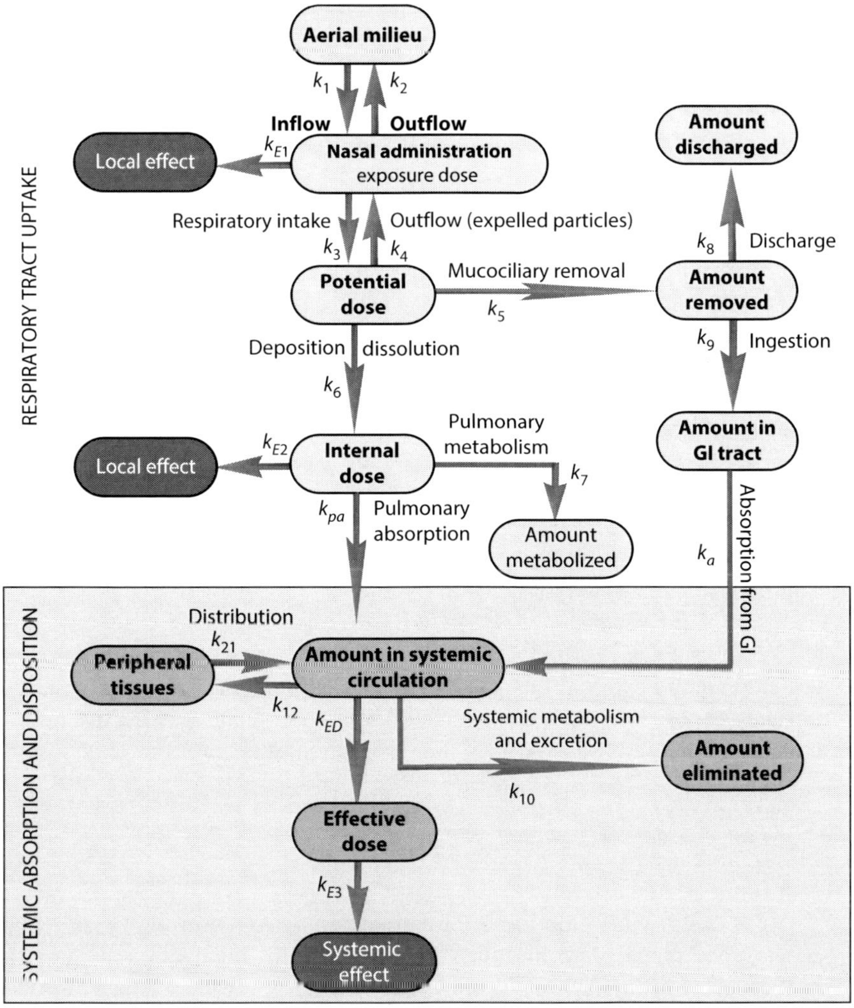

FIGURE 1.5 Schematic representation of respiratory ADME of particles and sequential steps from the exposure dose to potential dose to internal dose and ultimately to the effective dose; k_1 and k_2 are the aerial rate constants of inflow and outflow, respectively; k_{E1} is the local effect rate constant of amount absorbed from nasal site; k_3 and k_4 are the rate constants of respiratory intake and respiratory outflow, respectively; k_5 is the rate constant of mucocilіry removal; k_6 is the deposition and dissolution rate constant; k_7 is the pulmonary metabolic rate constant; k_8 is the discharged rate constant; k_9 is the rate constant of ingestion into the GI tract; k_a is the absorption rate constant from GI tract; k_{E2} is the local effect rate constant of deposited amount of pulmonary site; k_{pa} is the pulmonary absorption rate constant; k_{12} and k_{21} are the distribution rate constant; k_{10} is the overall elimination rate constant from the body; k_{ED} is the rate constant that makes up the effective dose; k_{E3} is the systemic effect rate constant.

a human is 20 m^3/day and when performing the occupational exposure assessment is 10 m^3/8 h (USEPA, 1992).

The maximum air capacity volume of the lung is about 5700 cm^3 and the total air that moves in and out of the lung in active normal breathing is approximately 4500 cm^3; therefore, the lung retains 1200 cm^3 of air. The number of breaths per minute in humans is approximately 12–20. At rest, the volume of air is reduced to 500 cm^3. The air velocity is very high in nasopharyngeal region and becomes milder as it reaches the alveolar region. Large particles (5–30 μm) are retained in the nasopharyngeal region by inertial impaction. Small particles (1–5 μm) are retained in the trachea, bronchial, and bronchiolar region by sedimentation. Smaller particles (1 μm) penetrate deep into the pulmonary tree, reach the alveolar sacs, and dissolve in the available fluid followed by passive diffusion. There is always a greater tendency for smaller particles to be exhaled. Cough or sneeze, with an air velocity of 75–100 miles per hour, can remove a significant number of particles from the site of absorption, particularly if they occur immediately after the administration of a compound.

1.5.1.3.3 Pulmonary Absorption of Gases and Vapors

The absorption of gases is from the mucous membrane of nose, pharynx, trachea, bronchi, bronchioles, alveolar sacs, and alveoli. Therefore, a large surface area is available for the absorption of gases and vapors. The solubility and reactivity of the gases determine their regional absorption. The upper pulmonary tract is the site for absorption of water-soluble and chemically reactive gases. The distal portion of the tract is the site for absorption of lipid soluble and non-reactive gases. The absorption of lipid soluble and volatile gases is very rapid and the onset of systemic action is rather instantaneous. Because of their lipophilicity, high levels of these compounds can be accumulated in systemic circulation and tissue compartments, which can be measured by the blood:air or plasma:air partition coefficient. They are assumed to be 100% bioavailable. The amount of gas entrapped in the lung, a highly perfused organ, is mainly due to their extraction in the lung blood. There are significant intra-species differences and caution should be exercised in extrapolation from experimental animals to humans (Morris, 2012).

1.5.1.3.4 Relevant Pulmonary Kinetic Parameters

Because of the internal and external factors influencing the pulmonary absorption and residence time of administered/exposed xenobiotic it is often difficult to estimate the effective dose. An exposure equation recommended by (USEPA, 1997) for estimation of the average daily dose is

$$(D_{average})_{daily} = \frac{C_{inhaled} \times Rate_{inhalation} \times t_{duration}}{BW \times t_{average}} \tag{1.4}$$

where $(D_{average})_{daily}$ is the average daily dose (often denoted as ADD) in mg/kg/day, $C_{inhaled}$ is the chemical concentration in inhaled air in μg/m^3, $Rate_{inhalation}$ is the inhalation rate in m^3/day, $t_{duration}$ is the duration of exposure in days, BW is the body weight, and $t_{average}$ is the average time with units of days. For non-carcinogenic effect $t_{average} = t_{duration}$ and for carcinogenic or chronic effect $t_{average}$ is equal to 70 years = 25,550 days, that is, lifetime average daily dose (often denoted as LADD).

The total potential dose may be estimated as

$$\left(D_{total}\right)_{potential} = C_{inhaled} \times Rate_{inhalation} \times t_{duration} \tag{1.5}$$

The approximated inhalation rate for children less than one year is 4.5 m^3/day; for twelve-year-old children it is 8.7 m^3/day; for adult females it is 11.3 m^3/day; and for adult males it is 15.2 m^3/day (USEPA, 1997).

The calculations of the inhalation rate as a function of basal metabolic rate is proposed according to the following relationship (Layton, 1993):

$$Rate_{inhalation} = Rate_{BM} \times A \times H \times V_Q \tag{1.6}$$

where $Rate_{BM}$ is the estimated basal metabolic rate in units of kcal/d (kilo-calories/day) or MJ/d (mega joules/day) determined by the empirical equations developed through regression analysis (Layton, 1993). For example:

$$\text{Human (Male) } 18 \leq \text{ Age } \leq 30 \quad Rate_{BM} = (0.063 \times bwt) + 2.896 \tag{1.7}$$

$$\text{Human (Female) } 18 \le \text{Age} \le 30 \quad Rate_{BM} = (0.062 \times bwt) + 2.036 \tag{1.8}$$

where (*bwt*) is the body weight in kg. The equations of other ages have also been reported (Layton, 1993).

The parameter *H* has units of liter oxygen intake per units of energy (L/kJ or L/kcal). It represents the volume of oxygen at STPD (standard temperature and pressure dry air) consumed for the production of one kJ of energy. It is also defined as the reciprocal of the energy yield of oxygen consumption equal to 0.0476, 0.0508, and 0.0529 L/kJ for carbohydrate, fat, and protein (McLean and Tobin, 1987; Layton, 1993). The parameter *VQ* is the ratio of minute volume (L/min) to the oxygen uptake rate (L/min), a unitless number that varies from individual to individual and represents the oxygen uptake capacity as a function of the lung physiology and metabolic processes. It is identified as a ventilatory equivalent and varies from 25 to 30 with a geometric mean of ~27.

Because $Rate_{BM}$ is estimated from empirical equations based on the characteristics of population used to generate the equations, a multiplier or correction factor is needed to normalize the calculations according to the daily food-energy intake of a different population. This multiplier (denoted *A*) is called the basal metabolic rate multiplier and is estimated by dividing the average daily food-energy intake (denoted EFD) by the basal metabolic rate. EFD has the same units as the basal metabolic rate and thus, *A* is a fraction with no units, that is,

$$A = \frac{\text{EFD}}{Rate_{BM}} \tag{1.9}$$

Equation 1.9 can also be corrected according to the intensity of physical activity during the active hours, that is,

$$A_{act} = [(24 - S)F + S]/24 \tag{1.10}$$

where A_{act} is the corrected value of *A* for activity and *S* is the number of hours the subject sleeps. *F* is the ratio of the rate of energy expenditure during active hours to the estimated basal metabolic rate and is calculated according to the following equation:

$$F = [(24A - S)]/(24 - \text{S}) \tag{1.11}$$

1.5.1.3.5 Role of the Lungs in PK/TK of Xenobiotics: Pulmonary First-Pass Metabolism

Being a highly perfused organ, the lungs play an important role in PK/TK profile of xenobiotics given by other routes of administration. Following the intravenous or oral administration of a compound, the concentration in the lung achieves equilibrium immediately with the systemic circulation. Thus, the lungs can metabolize and eliminate the compounds that are either retained or passing through. There are also compounds that are preferentially taken up by the lungs and are metabolized by the enzyme system of this organ; among them are amino compounds, for example, phenylethylamine (Gillis and Pitt, 1882; Junod, 1985) and basic drugs like imipramine, fentanyl and chlorpromazine (Philpot, 1977), estrogen (Peng et al., 2013), hormones (Junod, 1975), prostaglandins (Bakhle and Ferreira, 1985), and peptides. The pulmonary first-pass metabolism, although is not as significant as the hepatic and intestinal first-pass metabolism, nonetheless affects the bioavailability of a compound given orally or intravenously. The sequence of uptake events in the pulmonary tract is summarized in Figure 1.5. The exposure dose, which starts with nasal/mouth breathing and represents the amount of a compound that enters in the body per unit of time and body weight, is estimated by the following equation:

$$\text{Exposure Dose} = \frac{C \times IR \times F \times EF}{BW} \tag{1.12}$$

where C is the inflow concentration in mass/volume (e.g., mg/L, or parts per million), IR is the intake rate in units of volume/time (e.g., L/day), F is the bioavailability (unitless), BW is body weight in kg, and EF is called exposure factor and represents the length and frequency of an exposure, that is,

$$EF = (\textit{Frequency of Exposure})\ (\textit{Duration of Exposure})/\textit{Averaging Time} \tag{1.13}$$

For short-term exposure EF is approximately equal to 1.

1.5.1.3.6 Pulmonary Rate Equations

Derived from Figure 1.5, the rate equations of the exposure dose, potential dose and internal dose can be written as

$$\frac{dA_{Exposure}}{dt} = k_1 A_M + k_4 A_P - A_{Exposure}\left(k_2 + k_3 + k_{E1}\right) \tag{1.14}$$

$$\frac{dA_P}{dt} = k_3 A_{Exposure} - A_P\left(k_4 + k_5 + k_6\right) \tag{1.15}$$

$$\frac{dA_R}{dt} = k_5 A_P - A_R\left(k_8 + k_9\right) \tag{1.16}$$

$$\frac{dA_{GI}}{dt} = k_9 A_R - k_a A_{GI} \tag{1.17}$$

$$\frac{dA_{discharged}}{dt} = k_8 A_R \tag{1.18}$$

$$\frac{dA_{ID}}{dt} = k_6 A_P - A_{ID}\left(k_7 + k_{pa} + k_{E2}\right) \tag{1.19}$$

$$\frac{dA_1}{dt} = k_{pa} A_{ID} + k_a A_{GI} + k_{21} A_2 - A_1\left(k_{12} + k_{10}\right) - k_{ED} A_1 \tag{1.20}$$

$$\frac{dA_2}{dt} = k_{12} A_1 - k_{21} A_2 \tag{1.21}$$

where $A_{Exposure}$ is the exposure dose, k_1 and k_2 are the rate constant of inflow and outflow, respectively; A_P is the potential dose, k_3 is the input rate constant and k_4, k_5, and k_6 are the output rate constants of the potential dose, k_{E1} is the local effect rate constant; A_R is the amount removed by mucociliary transfer, k_4 is the rate constant of transfer from the potential dose, and k_8 and k_9 are the rate constants of removal due to discharge (spit/cough up) and swallowing, respectively; A_{GI} is the swallowed amount in GI tract, k_9 is the swallowing rate constant, and k_a is the absorption rate constant from GI tract into systemic circulation; A_{ID} is the internal or absorbable dose in pulmonary tract, k_6 is the hybrid rate constant of deposition and dissolution, k_7 is the rate constant of pulmonary metabolism assuming linear process and may change to the non-linear rate upon saturation of enzyme system, k_{pa} is the pulmonary absorption rate constant, and k_{E2} is the local effect rate constant; A_1 is the amount in the systemic circulation, A_2 is the amount in the peripheral tissues/organs, k_{12} and k_{21} are the distribution rate constants, and k_{10} is the rate constants of elimination representing excretion and metabolic rate constants.

The scheme presented in Figure 1.5 is mostly applicable to the pulmonary absorption of particles. For volatile gases or lipophilic vapors, the potential dose is the internal dose and Figure 1.5 will change to a simpler scheme as presented in Figure 1.6.

Among different approaches to PK/TK analysis of xenobiotics in the body after the exposure by inhalation, the physiologically-based modeling has been used more frequently. A typical model is presented in Figure 1.7 and more discussion on physiological modeling is provided in Chapter 9.

The general mass balance for the combined processes occurring in each organ/compartment of the model presented in Figure 1.7 can be described by a series of differential equations. The simplest set up of these equations defines the rate of uptake by each organ as the difference between the rate of input (inflow) and rate of output (outflow), as shown in Equation 1.22. For the organs with elimination (excretion/metabolism) process, the rate of uptake can be presented in the form of Equation 1.23.

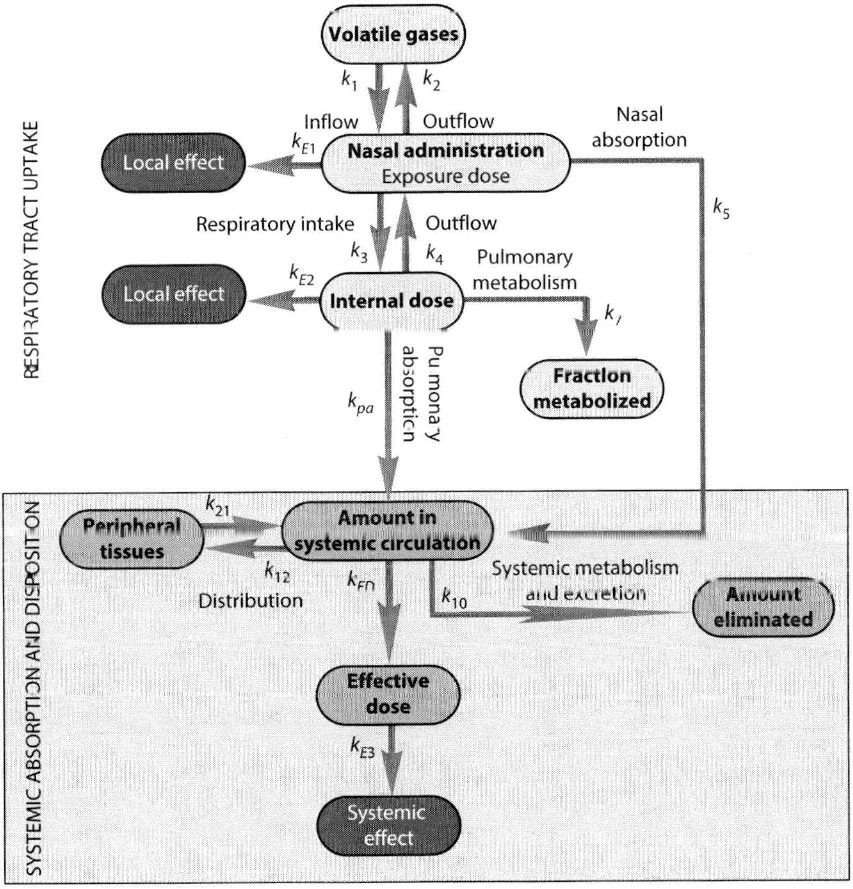

FIGURE 1.6 Schematic illustration of respiratory ADME of gases and the sequential steps from exposure dose to effective dose.

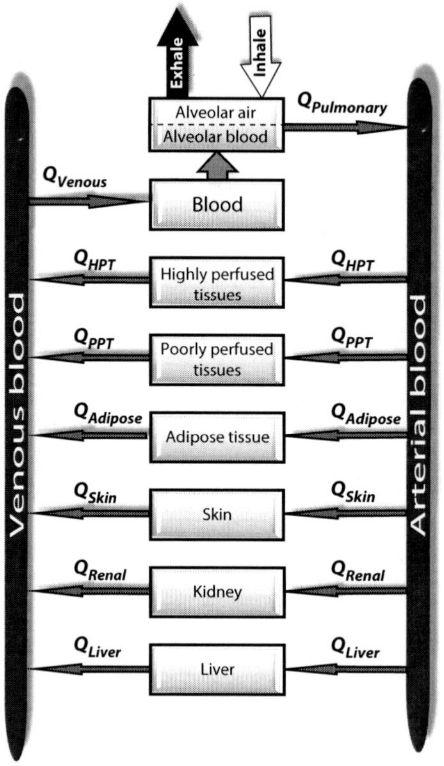

FIGURE 1.7 A typical physiologically-based pharmacokinetic model for absorption of xenobiotics from pulmonary tract and distribution to a selected group of organs and tissues, the elimination from the liver and kidney is not included in the figure.

$$\frac{dA_{organ}}{dt} = \frac{dA_{in}}{dt} - \frac{dA_{out}}{dt} \tag{1.22}$$

$$\frac{dA_{organ}}{dt} = \frac{dA_{in}}{dt} - \frac{dA_{out}}{dt} - \frac{dA_e}{dt} \tag{1.23}$$

where dA_{organ}/dt is uptake rate at time t with units of mass/time, dA_{in}/dt is the input rate, dA_{out}/dt is the output rate, and dA_e/dt is the elimination rate.

The mass balance equations can be as simple as Equations 1.22 and 1.23, or as complex as Equation 1.24 (Willems et al., 2001).

$$\frac{dA_{lung}}{dt} = \frac{A_{lung-cap}}{V_{lung-cap}} \times Q_{blood} \times P_{cap} - \frac{dA_{lung}}{V_{lung}} \times \frac{Q_{blood}}{P_{T:BCoeff}} \times P_{cap} - \frac{V_{\max} \times V_{lung} \times A_{lung}}{K_M \times V_{lung} + A_{lung}} \times MP_{lung} \tag{1.24}$$

where A_{lung} is the amount of xenobiotic in the lung (units of mass); $A_{lung-cap}$ is the amount in the lung capillary blood (units of mass); $V_{lung-cap}$ is the volume of blood in capillaries (units of volume); Q_{blood} is the total blood flow (volume/time), P_{cap} is the capillary permeability constant (unitless); V_{lung} is the volume of lungs; $P_{T:BCoeff}$ is tissue:blood partition coefficient (unitless); V_{max} is the maximum rate of metabolism (units of [mass/MP_{lung}]/time); MP_{lung} is the amount of microsomal protein (units of mass/volume of tissue); K_M is the Michaelis–Menten constant (units of mass/volume).

Although Equation 1.24 seems to be significantly different from Equation 1.23, nonetheless it is the same equation. The first term $((A_{lung-cap}/V_{lung-cap}) \times Q_{blood} \times P_{cap})$ is the input rate with units of mass/time, the second term $((dA_{lung}/V_{lung}) \times (Q_{blood}/P_{T:BCoeff}) \times P_{cap})$ is the output rate with units of mass/time, and the third term $((V_{max} \times V_{lung} \times A_{lung}/K_M \times V_{lung} + A_{lung}) \times MP_{lung})$ is the rate of elimination due to metabolism of the compound (Willems et al., 2001).

A compartmental approach, proposed for simulation purpose of inhaled particles (Weber and Hochhaus, 2013), is presented in Figure 1.8. The model considers the lung as having two compartments: the central lung (*LC*) and peripheral lung (*LP*). Each of these compartments is divided further into two compartments of LC_1, LC_2 and LP_1, LP_2. The undissolved particles are in compartments LC_1 and LP_1 and dissolved particles, that is, in solution form, are in LC_2 and LP_2. The mucociliary clearance occurs from compartment LC_1. The inhaled particles upon administration are fractionated into three fractions: deposited in the mouth, deposited in the lung and exhaled from the mouth/lung (Figure 1.8). The exhaled fraction is considered negligible.

The differential equations of the model are

$$\frac{d(LC_1)}{dt} = -\left(k_{C_1C_2} + k_{muc}\right)LC_1 \tag{1.25}$$

$$\frac{d(LP_1)}{dt} = -k_{P_1P_2} \times LP_1 \tag{1.26}$$

$$\frac{d(LC_2)}{dt} = \left(k_{C_1C_2} \times LC_1\right) - \left(ka_{LC} \times LC_2\right) \tag{1.27}$$

$$\frac{d(LP_2)}{dt} = \left(k_{P_1P_2} \times LP_1\right) - \left(ka_{LP} \times LP_2\right) \tag{1.28}$$

$$\frac{dA_{GI}}{dt} = \left(-ka_{GI} \times A_{GI}\right) + \left(k_{muc} \times LC_1\right) \tag{1.29}$$

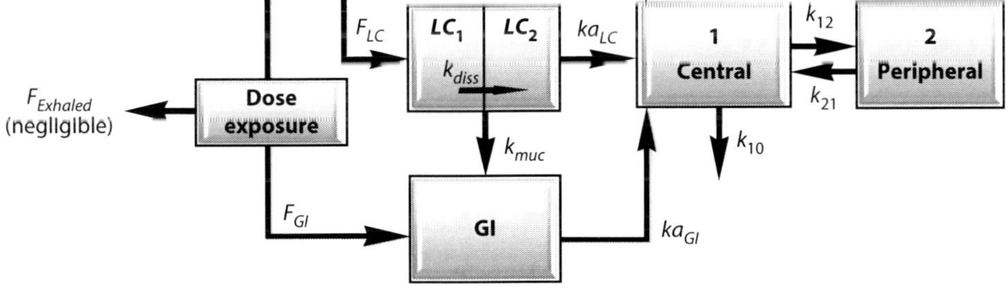

FIGURE 1.8 A proposed two-compartmental model for the pulmonary absorption of particulate matters including the dissolved and undissolved particles as separate compartments with the dissolution rate constants and the mucociliary clearance from the site of absorption.

$$\frac{dA_1}{dt} = -k_{10}A_1 - k_{12}A_1 + k_{21}A_2 + ka_{LC}LC_2 + ka_{LP}LP_2 + ka_{GI}A_{GI} \tag{1.30}$$

$$\frac{dA_2}{dt} = k_{12}A_1 - k_{21}A_2 \tag{1.31}$$

where $k_{C_1C_2}$ and $k_{P_1P_2}$ are the dissolution rate constant in the lung central and peripheral compartments, respectively. The tacit assumption of the model is that $k_{C_1C_2}$ is the same as $k_{P_1P_2}$ and represent the dissolution rate constant of the particles in the central and peripheral compartment of LC and LP compartments and is identified as (k_{diss}). (LC_1) and (LP_1) are the amount in the respective compartments; (k_{muc}) is the mucociliary rate constant of transfer from LC_1 to the GI tract; (LC_2) and (LP_2) are the amount in the lung peripheral compartments; (ka_{LC}) and (ka_{LP}) are the rate constants of absorption from lung peripheral compartments to the systemic circulation/central compartment; (A_{GI}) is the amount in GI tract; (ka_{GI}) is the absorption rate constant from GI tract into the systemic circulation/central compartment; (A_1) and (A_2) are the mount in the central and peripheral compartments of the body; (k_{12}) and (k_{21}) are the rate constants of distribution between the central and peripheral compartments of the body; and k_{10} is the elimination rate constant from the central compartment.

From the following Laplace transform of Equations 1.25 through 1.31 (Weber et al., 2013), it is feasible to generate the integrated equations representing the amount in each compartment (Appendix A, Section A.3):

$$\widetilde{LC_1} = \frac{LC_{1,0}}{(s + k)} \tag{1.32}$$

$$\widetilde{LP_1} = \frac{LP_{1,0}}{(s + k_{diss})} \tag{1.33}$$

$$\widetilde{LC_2} = \frac{k_{diss} \times LC_{1,0}}{(s + ka_{LC})(s + k)} \tag{1.34}$$

$$\widetilde{LP_2} = \frac{k_{diss} \times LP_{1,0}}{(s + ka_{LP})(s + k_{diss})} \tag{1.35}$$

$$\widetilde{A_{GI}} = F_{GI} \times \left(\frac{k_{muc} \times LC_{1,0}}{(s + ka_{GI'})(s + k)} + \frac{A_0}{(s + ka_{GI})} \right) \tag{1.36}$$

$$\widetilde{A_1} = \frac{k_{21} \times \overline{A_2} + ka_{LC} \times \overline{LC_2} + ka_{LP} \times \overline{LP_2} + ka_{GI} \times A_{GI}}{(s + k_{10} + k_{12})} \tag{1.37}$$

$$\widetilde{A_2} = \frac{k_{12} \times \overline{A_1}}{\left(s + k_{12}\right)} \tag{1.38}$$

where F_{GI} is the fraction of inhaled dose reaches the GI tract, and s is the Laplace operator.

1.5.1.4 Subcutaneous Absorption

The subcutaneous route of administration is used for injection of small volume vaccines or drugs such as insulin, heparin, local anesthetics or proteins. This route is also used to implant prolonged-release dosage forms such as polymeric rods, discs, or pellets (e.g., Oreton®, Percorten®, Norplant®, and Alzet® osmotic mini-pump). The rate and extent of absorption are strongly influenced by factors such as physicochemical characteristics of drug, the lipophilicity of the compound, the site and depth of injection, the volume of dose that may not exceed 2 mL, the dose concentration, pH–pKa relation and degree of ionization, type of vehicle, presence of surfactants, viscosity of the solution or emulsion, particle size of suspension, etc. The subcutaneous region is less vascular that the muscle, and thus has a lower rate of absorption than intramuscular administration, but their extent of absorption is somewhat similar for equally administered dose.

The mechanism of absorption in the subcutaneous region is by passive diffusion through the capillary wall into the systemic circulations. If the compound is lipophilic, it can diffuse directly through the membrane of capillaries by transcellular passive diffusion. Water-soluble drugs diffuse through cleft pores and vesicular channels of the membrane by paracellular passive diffusion. There has been no report to indicate that other mechanisms of absorption such as pinocytosis, active or facilitated transport may take place at this site. Proteins administered subcutaneously are absorbed into the systemic circulation either directly through the capillary wall, or indirectly through the lymphatic capillaries within the interstitial space (Figure 1.9). Proteins larger than 16 kDa are absorbed mainly through the lymphatic capillaries (Supersaxo et al., 1990; Charman et al., 2000; McLennan et al., 2006; Kota et al., 2007).

The PK/TK models and analyses for most compounds injected subcutaneously, other than insulin, are similar to the intramuscular approaches, which is discussed in Section 1.5.1.5. The models describing the plasma concentration–time profile of exogenous insulin are, however, diverse and depend mainly on the type of insulin; the concentration of their injected dose; the expected pharmacological response; the time to their onset of action; and their duration of action. For example, Lispro is a monomeric short-acting insulin analogue with a time to the onset of action 5–15 minutes and duration of action of 3–5 hours; the regular insulin is also another short acting-insulin with a

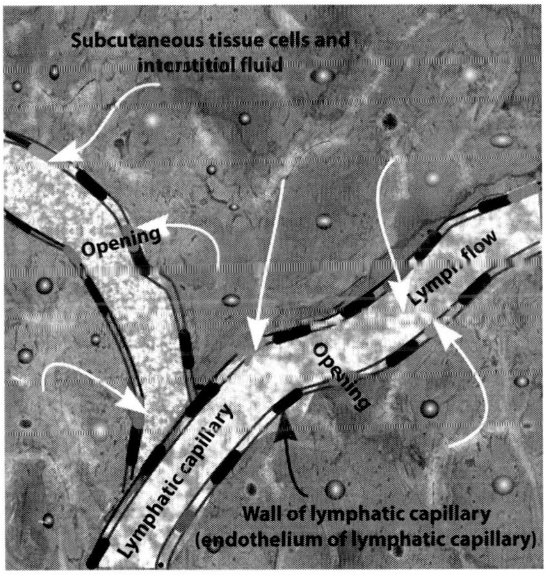

FIGURE 1.9 An illustration of passive diffusion from the subcutaneous region into the blood and lymphatic capillaries within the interstitial space.

time to the onset of action of 30–60 minutes and duration of action of 6–10 hours; whereas, insulin glargine is a long-acting insulin with a time to the onset of about 1–1.2 hours and duration of action of 24 hours. Most insulin models assume the diffusion in the subcutaneous region as an absorption compartment that is treated separately from insulin in plasma. Often the hexameric complex of the molecule has to split up to a dimeric complex and then to a monomeric structure in the subcutaneous region, which adds to the complexity of the model in the absorption compartment. A few absorption models of insulin relevant to the objectives of this section are discussed below.

1.5.1.4.1 The Diffusion Rate-Limited Model

The absorption site, that is, the subcutaneous space holding the injected insulin, is assumed compartment one and the diffused molecules available for absorption (i.e., the diffusion compartment) as compartment two. The insulin available in compartment two is the molecules absorbable into the systemic circulation. The decomposition of insulin is assumed to occur in both compartments (Kraegen and Chisholm, 1984; de Meijer et al., 1989). The equations of the model are

$$\frac{dU_0}{dt} = k_d U_0 - k_{12} U_0 \tag{1.39}$$

$$\frac{dU_t}{dt} = k_{12} U_0 - k_a U_t - k_d U_t \tag{1.40}$$

$$U_t = \frac{U_0 (k_a e^{-(k_d + k_{12})t} - k_{12} e^{-(k_d + k_a)t}}{k_a - k_{12}} \tag{1.41}$$

where U_t is the number of absorbable units of insulin in compartment two at time t, U_0 is the dose introduced in the subcutaneous region at time zero, k_a is the absorption rate constant for transfer of insulin from the absorption compartment (compartment two) into the systemic circulation, k_d is the decomposition rate constant, and k_{12} is the first-order diffusion rate constant from compartment one to two (Figure 1.10).

1.5.1.4.2 The Dissolution Rate-Limited Model

The assumptions of the model are

- The rate of delivery depends on the volume of injected dose and follows the cube root model of dissolution.
- The injected dose in the subcutaneous region distributes in a coin-shaped volume (de Meijer et al., 1989).

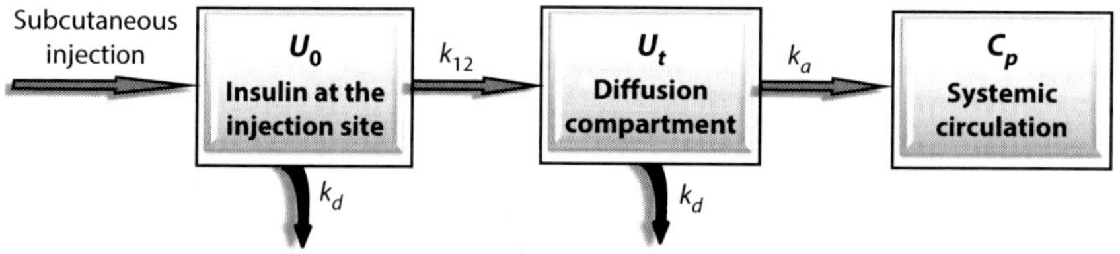

FIGURE 1.10 A diffusion rate-limited model for absorption of insulin from subcutaneous space, holding the injected dose, into the diffusion compartment followed by absorption into the systemic circulation.

The equation of the model is

$$U_t = \left(U_0 - k_{12} \times t\right)^3 + U_{rest} \tag{1.42}$$

The parameter U_{rest} was added to the equation to improve the fit (de Meijer, 1989).

1.5.1.4.3 The Capacity-Limited Model

This model is based on the assumption that the movement of the molecules in the subcutaneous region is capacity limited and the transfer rate constant at high concentrations is not governed by first-order kinetics (de Meijer, 1989)

$$\frac{dU_t}{dt} = \frac{V_{max}\left(U_t - U_{rest}\right)/V}{K_M + \left(U_t - U_{rest}\right)/V} \tag{1.43}$$

where dU_t/dt is the reduction rate of the injected dose, V_{max} and K_M are the maximum rate of transport and Michaelis–Menten constant, respectively, V is the volume of injected dose.

1.5.1.4.4 Models Based on Diffusion Equations

The compartmentalization of the process is the same as before, except the models take into consideration the hexameric and dimeric forms of insulin (Mosekilde et al., 1989; Trajanoski et al., 1993; Nucci and Cobelli, 2000; Tarín et al., 2005; Li and Kuang, 2009). These models are mostly mechanistic and conceptual relationships that are often difficult to adapt to experimental data. The following are the examples of these models (Tarín et al., 2005):

$$\frac{\partial C_d(t,\bar{p})}{\partial t} = D\nabla^2 C_d(t,\bar{p}) \tag{1.44}$$

$$\frac{\partial C_h(t,\bar{p})}{\partial t} = D\nabla^2 C_h(t,\bar{p}) \tag{1.45}$$

$$\frac{\partial C_b(t,\bar{p})}{\partial t} = d_b D\nabla^2 C_b(t,\bar{p}) \tag{1.46}$$

where C_d and C_h are dimeric and hexameric concentration, respectively; C_b is a virtual insulin state concentration, identified as bound concentration; t is time, $\bar{p}$ is the position vector and D is the diffusion constant, which is the same for both dimeric and hexameric forms. The diffusion is assumed to be homogeneous with rotational symmetry with respect to the site of subcutaneous injection with spherical coordinate, that is,

$$\frac{\partial C_d(t,r)}{\partial t} = D\nabla^2 C_d(t,r) = D\frac{1}{r^2}\frac{\partial}{\partial r}\left(r^2 \frac{\partial C_d(t,r)}{\partial t}\right) \tag{1.47}$$

$$\frac{\partial C_h(t,r)}{\partial t} = D\nabla^2 C_h(t,r) = D\frac{1}{r^2}\frac{\partial}{\partial r}\left(r^2 \frac{\partial C_h(t,r)}{\partial r}\right) \tag{1.48}$$

$$\frac{\partial C_b(t,r)}{\partial t} = d_b D\nabla^2 C_b(t,r) = d_b D\frac{1}{r^2}\frac{\partial}{\partial r}\left(r^2 \frac{\partial C_b(t,r)}{\partial r}\right) \tag{1.49}$$

where r is the distance from the subcutaneous site of injection, and the concentration terms are defined locally as a function of time and r through a theoretical spherical model.

The hexameric-dimeric dissociation at the site of absorption (Mosekilde et al., 1989; Tarín et al., 2005) is then identified as

$$Hexameric_{insulin} \underset{k_{-1}}{\overset{k_1}{\rightleftarrows}} 3Dimeric_{insulin}$$

The rate constants of dissociation and association are k_1 and k_{-1}, respectively. The partial differential equations representing the law of mass action are

$$\frac{\partial C_d(t,r)}{\partial t} = k_1 C_h(t,r) - k_{-1} C_d(t,r)^3 \tag{1.50}$$

$$\frac{\partial C_h(t,r)}{\partial t} = -k_1 C_h(t,r) + k_{-1} C_d(t,r)^3 \tag{1.51}$$

Because only the absorption of dimeric form is significant (Mosekilde et al., 1989), its absorption rate from subcutaneous region to the systemic circulation can be defined as

$$\frac{\partial C_d(t,r)}{\partial t} = -ka_{dimeric} C_d(t,r) \tag{1.52}$$

where $ka_{dimeric}$ is the absorption rate constant of dimeric insulin.

The rate of change of insulin concentration in plasma can then be defined as (Lehmann et al., 2009)

$$\left(\frac{dCp_{insulin}(\text{IU/Volume})}{dt(\text{Time} \times \text{kg})}\right) = \frac{\text{Rate of insulin absorption (IU/Time)}}{V_{insulin}(\text{volume}) \times \text{Body weight (kg)}} - KCp_{insulin}(\text{IU/Volume} \times \text{Time} \times \text{kg}) \tag{1.53}$$

The concentration is expressed in international units, IU, per volume and the rate as IU/time. $V_{insulin}$ is the relative volume of insulin distribution, K is the overall elimination rate constant, $KC_{Pinsulin}$ is the rate of elimination.

1.5.1.5 Intramuscular Route of Administration and Absorption

The primary sites of intramuscular injection (IM) are the gluteus maximus, deltoid (upper arm), triceps, pectoral and vastus lateralis (thigh) muscles, and because they have a higher blood perfusion the absorption from intramuscular site of administration is faster than the subcutaneous region. Similar to the subcutaneous region, the small molecules absorb directly into capillaries whereas large molecules access the systemic circulation indirectly through the lymphatic capillaries. The deltoid muscle has a greater blood perfusion rate compared to the other muscle injection sites, but depending on the weight of the individual, it can hold, painlessly, not more than 2 mL, versus 5–10 mL for the gluteus maximus. Muscle blood flow at rest is about 3–4 mL per minute per 100 grams of muscle. It increases to a maximum of 80–90 mL/minute per 100 grams of muscle during activity. Thus, the muscular absorption of xenobiotics is faster in active individuals than those confined to bed. In general, xenobiotics are absorbed faster from the deltoid rather than gluteal muscle and the absorption from the gluteal muscle of individuals with lower fat/muscle ratio is faster than those with a higher ratio. The local uptake of lipophillic compounds, however, is significant in individuals with higher fat/muscle ratio. Contrary to the gastrointestinal tract where the absorption is a function of GI tract transit time, compounds injected in the muscle or subcutaneous region have no time limit for residence, and the injected dosage form (solution, suspension or emulsion) forms a depot which gradually mixes with interstitial fluids and absorbs through capillaries until completely absorbed. The long-acting lipophilic intramuscular solutions, prepared by use of a suitable vegetable oil-containing vehicle and a lipophilic drug or prodrug, prolongs the duration of action and influences the PK/TK performance of the compound *in vivo* (Murdan and Florence, 2000; Larsen and Larsen, 2009).

The significant advantages of IM injection are (i) the absorption of high lipid-soluble molecules is rapid and complete; (ii) the absorption of lipid-insoluble molecules is not as rapid, but is complete due to diffusion into interstitial fluids and through the pores of capillary membrane; (iii) IM injection can be used effectively to bypass the low pH of the stomach and hepatic/intestinal first-pass metabolism; (iv) the site is ideal for the sustained–release polymeric dosage form or crystalline suspension. However, there are some uncommon side effects and complications (Greenblatt and Allen, 1978) namely sciatic nerve damage (Van Alstine and Dietrich, 1988), skin pigmentation, tissue necrosis, hemorrhage, abscesses (Rossi and Conen, 1995), cellulitis, quadricepts myofibrosis (Alvarez et al., 1980), and even gangrene. Furthermore, incomplete absorption may also occur in muscular environments because of precipitation or decomposition of drugs, and conditions such as hypotension or circulatory diseases may also reduce the rate and extent of absorption.

1.5.1.5.1 Pharmacokinetics of Intramuscularly Injected Compounds

The blood level data following intramuscular administration can be analyzed by any of the different approaches in PK/TK modeling, including non-compartmental analysis, compartmental analysis, physiologically-based pharmacokinetic models, etc. In many respects, the pharmacokinetic analysis of intramuscularly injected compounds mimics the simple models of oral administration without the complexities associated with the role of transport/efflux proteins and intestinal/hepatic first-pass metabolism in GI tract. This is also true for subcutaneously administered xenobiotics. The main assumption for IM or SC injection is that the transfer and absorption of small molecules from the site of injection toward the capillaries is linear and follows first-order kinetics, and because the molecules are removed by blood flow, the transfer essentially takes place under the sink condition.

The simplest PK/TK model is based on the assumption that the body behaves as a single homogeneous compartment and the rate of change in plasma concentration of an intramuscularly injected compound is the difference between two linear differential equations: (1) the rate of absorption from the site of injection into the systemic circulation, and (2) the rate of elimination from the body by metabolism and urinary excretion

$$\frac{dA}{dt} = k_a A_a - KA \tag{1.54}$$

where the rate of change in plasma level is dA/dt (with units of mass/time); the absorption rate constant is k_a (with units of time^{-1}), the absorbable amount at the site of injection is A_a, the rate of absorption is $k_a A_a$ (mass/time), the overall rate constant of elimination that encompasses metabolism and excretion is K, and the rate of elimination is KA.

The integration of Equation 1.54 yields the plasma concentration of the intramuscularly injected compound at time t

$$Cp = \frac{FD \times k_a}{Vd(k_a - K)}(e^{-Kt} - e^{-k_a t}) = \overleftarrow{Cp}^0(e^{-Kt} - e^{-k_a t}) \tag{1.55}$$

where F represents the fraction of the dose absorbed in most instances, D is the injected dose, Vd is the volume of distribution, k_a and K described previously, $\overleftarrow{Cp}^0$ is the y-intercept of the extrapolated line of the terminal data points (Figure 1.11).

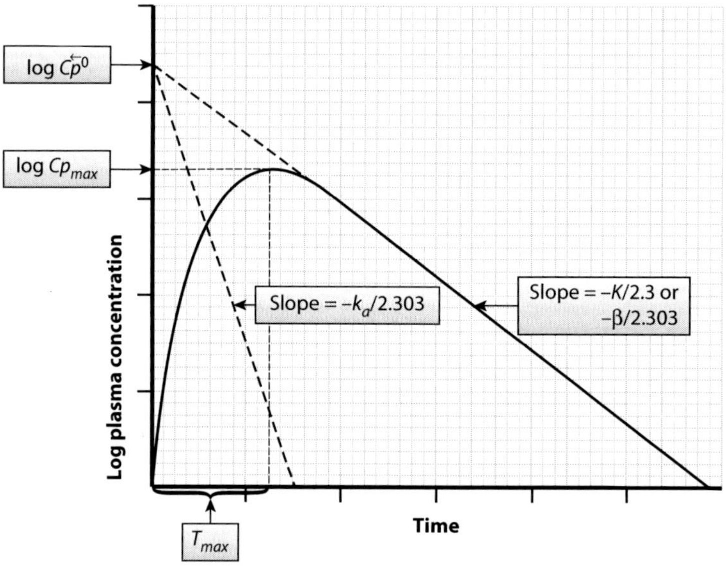

FIGURE 1.11 A plot logarithm of plasma concentration versus time after an intramuscularly injected dose representing two simultaneous linear biological processes of absorption from the site of injection into the systemic circulation and elimination from the body by metabolism and urinary excretion.

When the absorption of the intramuscularly injected dose is complete (i.e., $F = 1$) and very rapid, (i.e., $k_a \gg K$), Equation 1.55 changes to Equation 1.56, which is the same as an intravenous administration (Figure 1.11)

$$Cp = \frac{D}{Vd}\left(e^{-Kt}\right) = Cp^0 e^{-Kt} \tag{1.56}$$

In using more complex models, such as two and three compartment models, the equations are the same as the models for oral absorption. For example, for compounds that their plasma disposition curve follows two-compartment model, the related equation is

$$Cp = \frac{k_a FD}{V_1}\left(\frac{(k_{21} - k_a)}{(\alpha - k_a)(\beta - k_a)}e^{-k_a t} + \frac{(k_{21} - \alpha)}{(k_a - \alpha)(\beta - \alpha)}e^{-\alpha t} + \frac{(k_{21} - \beta)}{(k_a - \beta)(\alpha - \beta)}e^{-\beta t}\right) \tag{1.57}$$

where V_1 is the volume of the central compartment, k_{21} is the distribution rate constant from peripheral compartment into the central compartment and α and β are the hybrid rate constants. When absorption is fast and complete, $k_a > \alpha \gg \beta$ and F is equal to 1, Equation 1.57 is simplified to Equation 1.58

$$Cp = \frac{k_a D}{V_1}\left(\frac{(k_{21} - \beta)}{(k_a - \beta)(\alpha - \beta)}e^{-\beta t}\right) \tag{1.58}$$

Similar to oral absorption, the non-compartmental analyses are also employed for intramuscular absorption for determining parameters and constants such as area under the plasma concentration–time curve, mean residence time, the volume of distribution, and the mean absorption time.

The use of pharmacologically-based pharmacokinetic models for the intramuscularly injected compound is similar to the models with input from the extra-vascular region and the selection of organs or groups of organs and tissues depends on the physiochemical characteristics and pharmacokinetic profile of the compound.

1.5.1.6 Intraperitoneal Administration of Drugs

The intraperitoneal (IP) route is used both for administration and removal of xenobiotics. Peritoneum is a thin membrane, which covers the abdominal cavity, GI tract, the peritoneal muscles, and organs within the cavity. The membrane with its large surface area and blood rich capillaries is the site of absorption of compounds injected with IP, or exsorption of endogenous and exogenous compounds when filled with dilysate. Compounds absorbed into the peritoneal capillary network reach the portal circulation and pass through the liver before the distribution in the body. Therefore, the compound may undergo a hepatic first-pass effect. In humans, the IP route is used in continuous ambulatory peritoneal dialysis to remove endogenous and exogenous toxic molecules from the body. It is also used effectively to treat peritonitis caused by peritoneal dialysis (Wiggins et al., 2007). Xenobiotics introduced into the abdominal cavity are absorbed into the systemic circulation by passive transcellular and paracellular diffusion through peritoneum. They can also be removed from systemic circulation into a dialysate based on the concentration gradient between the blood and dialysis solution, which can then be removed from the body by drawing off the dialysate.

The intraperitoneal route of administration is also used for advanced ovarian cancer treatment as the single route of administration for chemotherapeutic agents, or in combination with intravenously administered anticancer drugs (Armstrong et al., 2006). A favorable advantage of the IP route in treatment of solid tumors or peritonitis is the targeting nature of treatment and flexibility in enhancing the dose and subjecting the target areas to higher concentrations of therapeutic agents. Most often, compounds with high molecular weight and/or extensive protein binding, may not diffuse in optimum amount into the peritoneal target areas to provide apt therapeutic outcome. This, in conjunction with the low vascularity of tumors and presence of their efflux proteins, will further reduce the ineffectiveness of drugs given intravenously or orally intended for the peritoneal tumors.

1.5.1.6.1 Kinetics of IP Administration

The kinetics and dynamics of peritoneal-plasma transport can be defined by a distributed model (Flessner et al., 1984) and its modified version for diffusive transport in superfused tissue applicable to cancer chemotherapy (Flessner et al., 2006; Ze Lu et al., 2010). Both approaches are based on the assumptions that peritoneum is a well perfused exchange site between the peritoneal cavity and the plasma, and drug transfer through the barrier tissue is governed by the Fick's second law of diffusion, and removal of a drug from the site of absorption is by permeation through the capillaries. The lymphatic uptake is considered significant in a distributed model. The schematic diagram of a distributed model is presented in Figure 1.12.

According to the model (Flessner et al., 1984), the transfer of solute between the plasma and distribution compartment is very rapid and collectively behave as a single distribution compartment. For IP injection, the transfer of solute occurs between the peritoneal cavity and peritoneal barrier, from the barrier into plasma and distribution compartment, and from the plasma to the body exchange compartment. The volumes of body exchange and distribution compartments are assumed constant, that is,

$$\frac{dV_I}{dt} - \frac{dV_D}{dt} - 0 \tag{1.59}$$

Diffusion through the tissue and permeation through the capillaries are assumed to be only a function of the molecular size of a compound. All compartments of the model are assumed well-mixed and peritoneal barrier is considered a distributive space. This theoretical model defines the diffusion of a compound, such as anticancer drugs, from the peritoneal cavity. It also applies to the dynamics of peritoneal dialysis when a compound is removed from the distribution compartment

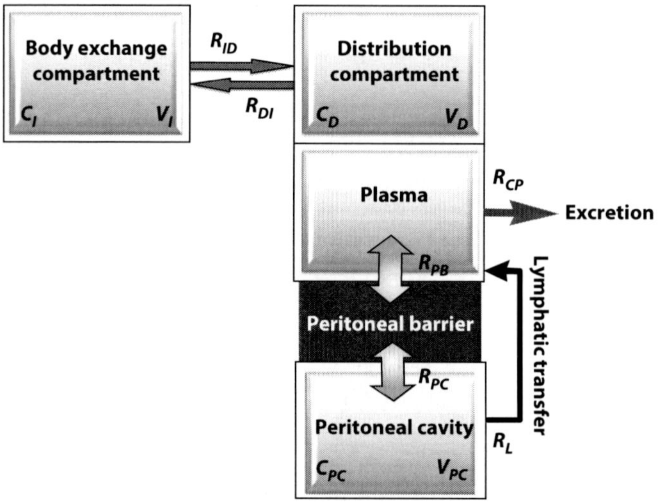

FIGURE 1.12 Diagram of a distributed model for absorption from peritoneal cavity into the plasma and related distributional compartments.

and plasma through the peritoneal barrier into the dialysate infused in peritoneal cavity. The differential equations of the absorption from peritoneal cavity for well-mixed compartments (Flessner et al., 1984) are

$$\frac{d\left(C_{PC} \times V_{PC}\right)}{dt} = -(R_L + R_{PC}) \tag{1.60}$$

$$\therefore \frac{d\left(C_{PC}\right)}{dt} = \frac{-(R_L + R_{PC})}{V_{PC}} \tag{1.61}$$

$$\frac{d\left(C_D V_D\right)}{dt} = R_{PB} + R_L + R_{ID} - (R_{DI} + R_{Cp}) \tag{1.62}$$

$$\therefore \frac{d\left(C_D\right)}{dt} = \frac{R_{PB} + R_L + R_{ID} - (R_{DI} + R_{Cp})}{V_D} \tag{1.63}$$

$$\frac{d\left(C_I V_I\right)}{dt} = R_{DI} - R_{ID} \tag{1.64}$$

$$\therefore \frac{d\left(C_I\right)}{dt} = \frac{R_{DI} - R_{ID}}{V_I} \tag{1.65}$$

where C_{PC}, C_D, and C_I are the concentrations (mass/volume) in the peritoneal cavity, distribution compartment and body exchange compartment, respectively; V_{PC} is the volume in the peritoneal cavity, V_D is the volume of distribution compartment and V_I is the volume of the body exchange compartment; the rates of mass transfer with units of mass/time are: R_L for the lymphatic rate of mass transfer, R_{PC} for the rate of mass transfer from the peritoneal cavity into the barrier, R_{PB} for the rate of mass transfer from barrier to the distribution compartment via the plasma, R_{ID} and R_{DI} for the rates of mass transfer between the body exchange and distribution compartments.

The rate of volume change in the peritoneal compartment is a function of the lymphatic volume flow rate F_L (volume/time) and peritoneal cavity to tissue volume flow rate F_{PC} (volume/time), that is,

$$\frac{dV_{PC}}{dt} = -(F_L + F_{PC}) \tag{1.66}$$

$$\therefore V_{PC} = -\int_0^t (F_L + F_{PC}) \tag{1.67}$$

Thus, the lymphatic mass transfer rate can be defined as

$$R_L = F_L \times C_{PC} \tag{1.68}$$

To solve for the rate equation of transcapillary transport, that is, the portion of the transport from peritoneal cavity that is through the tissue barrier and capillaries into the distribution compartment, Equation 1.45 can be expanded and modified by substituting Equations 1.51 and 1.53 as follows:

$$V_{PC}\frac{dC_{PC}}{dt} = C_{PC}\left(F_L + F_{PC}\right) - \left(R_L + R_{PC}\right) = F_{PC}C_{PC} - R_{PC} \tag{1.69}$$

$$\therefore \frac{dC_{PC}}{dt} = \frac{F_{PC}C_{PC} - R_{PC}}{V_{PC}} \tag{1.70}$$

The initial conditions of the model are

At $t = 0$: $C_{PC} = 0$, $C_D = 0$, $C_I = 0$, $V_{PC} = V_{PC}(0)$; $C_{PB}(x,0) = g(x) = 0$ and $\partial C_{PB}/\partial x = 0$

where C_{PB} is the concentration of the barrier and x is the thickness of the barrier.

Other mass transfer rates equations are defined as

$$R_{DI} = k_{DI}C_DV_D \tag{1.71}$$

$$R_{ID} = k_{ID}C_IV_I \tag{1.72}$$

$$R_{Cp} = k_{Cp}C_DV_D \tag{1.73}$$

To make the model more practical and applicable to the clinical application of IP cancer chemotherapy, a simpler model without the body exchange compartment and distribution compartment has been suggested that considers the exchange between plasma, the peritoneal cavity, and barrier (Flessner et al., 2006). The diagram of the model is presented in Figure 1.13 with two sampling compartments of plasma and solution in peritoneal cavity. A new hybrid coefficient that represents all of the diffusive and solute transfer characteristics through the barrier is identified in the model as the mass transfer area coefficient (*MTAC*). The rate of mass transfer from the peritoneal cavity into the plasma is defined as

$$\frac{d\left(C_{PC} \times V_{PC}\right)}{dt} = -MTAC \times (C_{PC} - Cp) \tag{1.74}$$

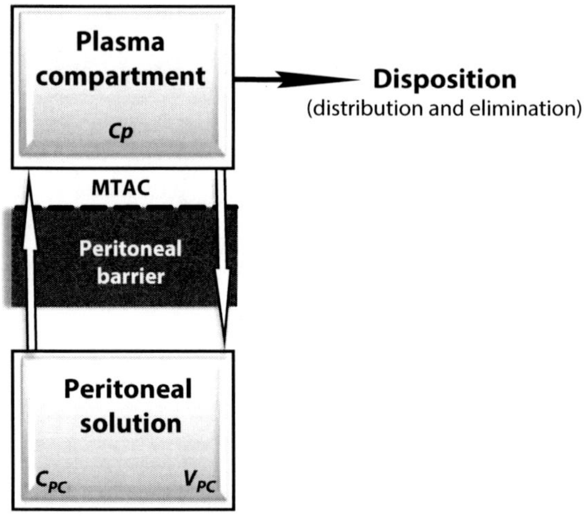

FIGURE 1.13 A simple model for intraperitoneal absorption of xenobiotics with mass transfer area coefficient (MTAC) through the barrier, and two sampling sites of systemic circulation and peritoneal cavity.

where C_{PC} and Cp are concentrations of peritoneal cavity and plasma, respectively, and V_{PC} is the volume of solution in peritoneal cavity.

MTAC is the product of mass transfer coefficient and the fluid contact area in the peritoneal cavity, that is,

$$MTAC = MTC \times Area \tag{1.75}$$

where

$$MTC = \sqrt{D_{eff} \times Pa} \tag{1.76}$$

When dealing with blood perfusion limitation

$$MTC = \sqrt{D_{eff} \times Q} \tag{1.77}$$

where the blood flow is Q, D_{eff} is the effective diffusion coefficient of the solute, P is the capillary permeability, and a is the capillary surface area per unit of volume. The mass (M) of MTC is

$$M = C_{PC} \times V_{PC} \tag{1.78}$$

Thus, the mass transfer rate in peritoneal cavity is

$$\frac{dM}{dt} = \frac{\text{Amount removed}}{\text{Period of experiment}} = -MTC \times Area(C_{PC} - Cp) \tag{1.79}$$

The mass balance on the tissue side is defined based on the Fick's second law (Flessner et al., 2006)

$$\frac{\partial(\theta_s \times C_s)}{\partial t} = \frac{\partial}{\partial x}\left[D_{eff}\frac{\partial C_s}{\partial x} + (C_s \times f \times J_v \times \theta_s)\right] + R_{cap} \tag{1.80}$$

where θ_s is the fraction of tissue accessible to the solute; C_s is the concentration of the solute in the interstitial space; x is the distance into the tissue from the cavity; f is the ratio of the solute rate of diffusion to the solvent rate of diffusion, identified as the solute retardation factor; J_v is the fluid flux through the tissue, which is dependent on the osmotic and hydrostatic pressure between the interstitial fluid and capillary network; and R_{cap} is the rate of solute exchange defined as

$$R_{cap} = -Pa\left(C_s - C_{plasma}\right) \tag{1.81}$$

In the case of blood perfusion limitation

$$R_{cap} = -Q\left(C_s - C_{plasma}\right) \tag{1.82}$$

Under steady state condition, Equation 1.66 can be written as

$$\frac{\partial(\theta_s \times C_s)}{\partial t} = 0 = D_{eff}\frac{d^2C_s}{\partial x^2} + f \times \theta_s\left[\frac{d}{dx}(C_s \times J_v)\right] - Pa(C_s - Cp) \tag{1.83}$$

For isotonic solutions, $J_v \cong 0$ and Equation 1.69 will be simplified to

$$D_{eff}\frac{d^2C_s}{\delta x^2} = Pa\left(C_s - Cp\right), \text{ or} \tag{1.84}$$

$$D_{eff}\frac{d^2C_s}{\delta x^2} = Q\left(C_s - Cp\right) \tag{1.85}$$

$$\frac{C_s - Cp}{C_{PC} - Cp} = e^{-\left(\sqrt{\frac{Pa}{D_{eff}}}\right)x} \tag{1.86}$$

$$\frac{C_s - Cp}{C_{PC} - Cp} = e^{-\left(\sqrt{\frac{Q}{D_{eff}}}\right)x} \tag{1.87}$$

As noted earlier, there is a renewed interest in the use of the intraperitoneal route of administration for targeting the cancerous tumor in intraperitoneal cavity. The flexibility of using a high dose of anticancer agents for better therapeutic outcome, averting the normal tissue encounter with high plasma concentration, the direct measurement of mass transfer at the site of absorption, direct evaluation of pharmacological effect, and the influence of osmotic and hydrodynamic pressure on the absorption of solutes after IP infusion or retrieval of compounds through peritoneal dialysis all offer exciting research opportunities in PK/PD modeling and TK/TD evaluations of compounds like anticancer agent.

1.5.1.7 Rectal Route of Administration

The rectal mucosal route of administration for the purpose of achieving systemic effect is used for many therapeutic agents. The rectum, with a length of 15–20 cm and surface area of 200–400 cm^2, is located in the terminal portion of pelvic colon. This site of absorption has a pH range of 7.4–8.0 in adults and 7.2–12.2 in children (Jantzen et al., 1989); this pH range favors the absorption of xenobiotics with pKa near the physiological range of 7.4–7.6. The rectal canal has a constant temperature of 37°C and limited buffer capacity; thus, the pH may fluctuate depending on the pH of the dosage form. It is a highly vascular region. The superior rectal artery is the main artery and its venous network comprises the upper, middle, and lower rectal vein system. The upper rectal vein is linked with the hepato-portal vein system; the middle and lower veins enter into the inferior vena cava. Therefore, drugs that are absorbed from the upper part of the rectum enter the upper rectal vein and are subject to first-pass metabolism in the liver before distribution. Drugs that are absorbed in the middle and lower rectal veins avoid the liver and distribute in the body following the absorption into the systemic circulation (Choonara, 1987; Van Hoogdalem et al., 1991). The mechanisms of absorption at this site are transcellular and paracellular passive diffusion. There is no active site for absorption and there is no proof of carrier-mediated transport. The rectal absorption is rather rapid and can be substituted for the intramuscular, oral, or even intravenous route of administration in adult and pediatric patients; however, not for irritant drugs. Other advantages are

- It is a stable environment for the absorption of drugs. The pH, viscosity, and temperature are constant; the rectal motility is very low and the residence time of drugs is long and only limited by defecation. Overall, the condition is most suitable for administration of controlled-release dosage form.
- The lymphatic circulation of the rectal region is significant; thus, the site can be used effectively for lymph-targeted drugs.
- The rectum has no enzymes and thus no metabolism takes place at this site; however, the metabolic activity of gut microbiota continues to pose a problem.
- It is an alternative route for administration of drugs that have low bioavailability following oral administration.
- It is a convenient route for patients, such as infants, who cannot take solid dosage forms orally.

However, the most important limitation of the rectal route of administration is the inconsistent absorption and bioavailability of the administered dosage forms and consequent inter- and intra-individual variability due to a number of factors such as

- Absorption into the upper rectal vein and hepatic first-pass effect
- Presence of fecal particles
- Formulation of the dosage form, suppositories, solution and micro-enemas
- Time for liquefaction and dispersion of suppository bases followed by absorption of active ingredient
- The dielectric constant of suppository base
- Presence of surfactants
- The amount of active ingredient in the dosage form
- The volume and retention of medicated enema
- pH of the environment and pKa of the drug
- Disease state
- Metabolism by microbiota
- Quick discharge of the dosage forms due to the irritant nature of active and/or inactive ingredients

The diffusion through the barrier and general pharmacokinetic approaches for rectal administration of therapeutic agents is similar to buccal and sublingual routes, where the input into the systemic circulation is through extravascular sites. One approach in defining kinetics of rectal absorption (Riegelman and Crowell, 1958) is summarized below

$$\log \frac{A_t - A_{end}}{A_0 - A_{end}} = -k_a t \tag{1.88}$$

where the amount in the body at time t is A_t; the amount remaining at the site of absorption after the end of experiment (i.e., residuum in rectum) is A_{end}; the total dose administered at time zero is A_0; and the pseudo first-order diffusion and absorption rate constant is k_a. Equation 1.88 does not take into consideration the permeation of molecules through the membrane and ignores the overall elimination of absorbed drug from the body. The kinetics of absorption through other mucous membranes, such as different segments of the GI tract, may also apply to rectal absorption (Chapter 8).

1.5.1.8 Intravaginal Route of Administration

Absorption is from the highly vascular region of the vagina. The pH of this environment in adult populations is usually acidic. Lactic acid produced from glycogen by the *Lactobacillus acidophilus* present in the vagina acts as a buffer to maintain the vaginal pH between 3.8 and 4.2. Most of the vaginal dosage forms are intended for local effects. However, this site can be used effectively for achieving the systemic effect such as progesterone-containing dosage forms. The absorption from this site is by transcellular and paracellular passive diffusion. No active absorption or carrier-mediated transport has been reported for this site. Menopausal changes decrease the regional vascularity and thinning the mucosa, which leads to a significant increase in permeability of the barrier.

The advantages of using this route for achieving systemic effect include the ease of administration, circumventing the low pH of stomach and hepatic first-pass effect, and first-uterine-pass effect. It should be pointed out that the vaginal basal cell layers exhibit proteases enzymatic activity (Lee, 1988), which is a barrier for absorption of peptides and proteins. The disadvantages include vaginal changes during menstrual cycle, inter- and intra-individual absorption variability and mucosal irritation of drugs intended for systemic effect. However, the vagina is considered an effective site for the administration of drugs important in women's health. It is used for administration of contraceptives, antimicrobials, and antifungals. The vaginal absorption of therapeutic agents is influenced by their lipophilicity, degree of ionization, chemical structure and molecular weight, and interaction with vaginal barrier and secretion. The dosage forms appropriate for this route include ointments, creams, emulsion, suspension, solution, foams, douches, sprays, gels, solid dosage forms, vaginal suppositories, and polymeric intrauterine devices such as a vaginal ring.

The pharmacokinetics of a vaginally administered drug depends to some extent on the type of dosage form. For example, the absorption of a compound dispersed in the matrix of a polymeric IUD depends on the release mechanism and rate from the delivery system. In general, the absorption through mucosal barrier of drugs and their related pharmacokinetic analysis are the same as other extravascular routes of administration.

1.5.2 Second Category: Parenteral Routes of Administration

1.5.2.1 Intravenous Injections (IV Bolus and Infusion)

The intravenous route of administration is the reference route for pharmacokinetic and toxicokinetic analyses. Given xenobiotics directly into the vein of humans or experimental animals provides immediate onset of action (~10–25 seconds after the injection), complete bioavailability (100%), minimal hepatic first pass effect, and no intestinal first-pass metabolism. Because of the complete introduction of the dose into the systemic circulation, the dose can be titrated according to its response and the limit of linearity of PK/TK behavior of a compound in the body can be established. It is the only way highly irritant xenobiotics, such as anticancer agents, can be administered into the body. The sites of intravenous injection for adults include the veins on the forearm, hand, and wrist, which connect to antecubital fossa. For infants the sites include veins on the scalp, umbilical vessels, and the legs' cursory veins. The major risks associated with intravenous administration are toxicity due to overdose, embolism, vascular injury, speed, and anaphylactic shock. Xenobiotics can be administered intravenously as a bolus injection or infusion. The intravenous infusion can be fast or slow, or a combination of fast and slow infusions, or bolus and infusion (see Chapters 10 and 11). Intravenous infusion is accomplished by the use of infusion pumps; it is called an IV drip when the dose is added to IV bags such as a normal saline or dextrose in water and gravity is the driving force for the continuous dripping and infusing over several hours; it is also called IV piggyback when the dose is mixed in a separate small volume IV bag and connected to an existing port of the IV drip.

Most, if not all, pharmacokinetic models of xenobiotics, whether compartmental, physiological or non-compartmental analysis, are first developed based on their intravenous administration. The process of assessing the risk associated with exposure to environmental chemical, extrapolation from animals to humans and low dose extrapolation relies on the predictive power of PK/TK models that are developed following intravenous administration of the chemical.

1.5.2.2 Intra-Arterial Injection

Intra-arterial injections are not recommended in clinical practice and cannot be considered a route of administration for PK/TK studies. Many therapeutic agents are toxic when administered intra-arterially. Accidental arterial injections may occur in the antecubitasl fossa where arterial branches are more accessible. The intra-arterial injections are painful, and the pain lingers for hours after the injection. It is, however, used for targeting certain medications to regions that require first brunt of the administered dose, such as injection of chemotherapeutic drugs into the tumors major feeding artery.

1.6 BIOMARKERS

Any attribute that can be objectively assessed and appraised as an indicator of normal biological processes, pathogenic manifestations, or pharmacological/toxicological responses to administration of or exposure to a xenobiotic is called a biomarker, or biological marker. Biomarkers have been used in clinical laboratory diagnostics, medical prognostics, and environmental/toxicity safety predictions. In PK/PD and TK/TD modeling, the quantitative measurement of biomarkers is used to monitor the response and quantify the effect. When a biomarker is used instead of a clinical or laboratory endpoint, it is known as surrogate endpoint. A surrogate endpoint is expected and must be validated to predict benefit, harm, or no effect based on scientific, pathophysiologic, and/or population substantiation (Biomarkers Definitions Working Group, 2001). Examples of surrogate endpoints are bone density measurement, blood pressure, hemoglobin A1C measurement, etc. In ecotoxicological terms, surrogate endpoint biomarkers would be most valuable if they could be linked to individual adverse effects as well as to population and ecosystem effects. It is not uncommon that a biomarker may represent the intermediate step between exposure and response. Biomarkers are used routinely in clinical practice and drug development for selecting the appropriate dose, achieving maximum therapeutic efficiency and minimum toxicity, evaluating toxicogenomics, and monitoring therapeutic outcome. PK/TK biomarkers are the parameters and constants associated with analysis of serum levels, which may include area under plasma concentration time curve, maximum plasma concentration and time to achieve the maximum concentration. The examples of PD/TD biomarkers are blood sugar level, pulmonary function tests and blood pressure. Biomarkers are used in pharmacology and toxicology for evaluation of dose-response curves for achieving the optimum therapeutic outcome, or toxicity dose-response curves associated with safety and refining low dose extrapolation. Examples of safety biomarkers are: enzyme CYP 2D6 for atomoxetine HCl (Strattera®—used in attention-deficit/hyperactivity disorder [ADHD]); enzyme UGT1A1 for irinotecan (Camptosar®—an antineoplastic agent); or enzyme 2D6 for warfarin. The use of biomarkers in environmental toxicology is essential as indicators of public health. For example, increased activity of enzymes like alanine aminotransferase and alkaline phosphatase is associated with liver damage; or reduction of cholinesterase activity is the biomarker for the organophosphate pesticides exposure, which causes inhibition of cholinesterase and breakdown in nervous system function; eggshell thickness as an indicator of organochlorine pesticide DDT, which interfere with calcium transport causing eggshell thinning and breaking (Eason and O'Halloran, 2002); C-reactive protein (CRP) as a marker of inflammation (Pradham et al., 2001; Ridkar et al., 2002; Shacter and

Weitzman, 2002); and also argininosuccinate synthase as a biomarker for inflammatory conditions (Cao et al., 2013). A number of chemical carcinogenicity biomarkers are reported, which in most cases are organ specific, such as ethylene oxide hemoglobin adducts or acetaldehyde-DNA and protein adducts for lung cancer (Hatsukami et al., 2006). In addition to enzymes and other validated biomarkers, promising genetic biomarkers for different population of patients are also under development, among them are MDR1 gene polymorphism as predictor of hepatocellular carcinoma in the Chinese population (Yang et al., 2013); or use of serum metabolomics, which is a method for evaluation and measurement of all small molecule metabolites due to genetic mutation in variations of lipid metabolism between coronary artery disease and congestive heart failure (Luan et al., 2013).

Many of the biomarkers utilize genomics and proteomics, and are expected to facilitate the implementation of personalized medicine or precision medicine. The progress is rather slow at this time, which is mainly due to the genomic assay development and the related technical difficulties. In developing the genomic biomarkers, in addition to the understanding of gene function at the DNA level, the transcriptional (RNA) and post-transcriptional (protein) data are also crucial. Gene expression at the RNA and protein levels is often non-linearly correlated and somewhat inconsistent due to the translation, regulation, transport, and degradation. Genomic biomarkers will continue to significantly influence the development and administration of therapeutic agents. However, their role in PK/TK modeling and analysis is yet to be determined.

Another critical limitation in the use of biomarkers is the technical difficulties in large epidemiological studies. Examples are the enzymatic induction of aryl hydrocarbon hydroxylase; or measurement of DNA repair enzyme systems, which both have been recommended by the non-profit Institute of Medicine, among a number of other biomarkers, as the lung cancer biomarkers (Lowe et al., 2013). In support of developing new and validated biomarkers, federal, state and private agencies, including the U.S. Food and Drug Administration (FDA Guidance—Qualification of drug development tools, 2010), NIH, NCI, PhRMA in collaboration with academic institutions, have established working groups, qualification programs, and consortia, which have collectively contributed to the library of validated or potential biomarkers.

REFERENCES

Aarons, L. 1991. Population pharmacokinetics: Theory and practice. *Br J Clin Pharmac* 32:669–70.

Alvarez, E. V., Munters, M., Lavine, L. S., Manes, H., Waxman, J. 1980. Quadriceps myofibrosis. A complication of intramuscular injections. *J Bone Joint Surg Am* 62:58–60.

Amzal, B., Julin, B., Vahter, M., Wolk, A., Johanson, G., Åkesson, A. 2009. Population toxicokinetic modeling of cadmium for health risk assessment. *Environ Health Perspect* 117(8):1293–301.

Anttila, S., Hukkanen, J., Hakkola, J., Stjernvall, T., Beaune, P., Edward, R. J., Boobis, A. R., Pelkonen, O., Raunio, H. 1997. Expression and localization of CYP3A4 and CYP3A5 in human lung. *Am J Resp Cell Mol* 16:242–49.

Armstrong, D. K., Bundy B., Wenzel, L., Huang, H. Q., Baergen, R., Lele, S., Copeland, L. J., Walker, J. L., Burger, R. A. 2006. Intraperitoneal cisplatin and paclitaxel in ovarian cancer. *N Engl J Med* 354(1):34–43.

Bakhle, Y. S., Ferreira, S. H. 2011. Lung metabolism of eicosanoids: prostaglandins, prostacyclin, thromboxane and leukotrienes. In: *Handbook of Physiology, The Respiratory System, Circulation and Nonrespiratory Functions,* Wiley Online Library: doi: 10.1002/cphy.cp030111.

Baldrick, P. 2003. Toxicokinetics in preclinical evaluation. *Drug Discov Today* 8(3):127–33.

Bartlett, J. A., Kees Voort Maarschalk, K. V. 2012. Understanding the oral mucosal absorption and resulting clinical pharmacokinetics of asenapine. *AAPS Pharm Sci Tech* 13(4):1110–5.

Biomarkers Definitions Working Group. 2001. Biomarkers and surrogate endpoints: Preferred definitions and conceptual framework. *Clin Pharmacol Ther* 69(3):89–95.

Bleasly, K., Castle, J. C., Roberts, C. J., Cheng, C., Baily, W. W., Sina, J. F., Kulkami, A. V. et al. 2006. Expression profiles of 50 xenobiotic transporter genes in himans and pre-clinical species: A resource for investigations into drug disposition. *Xenibiotica* 36:963–88.

Bois, F. Y., Jackson, E. Y., Pekari, K., Smith, M. T. 1996. Population toxicokinetics of benzene. *Environ Health Perspect* 104(6):1405–11.

Boroujerdi, M. 2002. *Pharmacokinetics, Principles and Applications*, 52–53. New York: McGraw Hill.

Cao, M., George, T. J., Prima, V., Nelson, D., Stanislav Svetlov, S. 2013. Argininosuccinate synthase as a novel biomarker for inflammatory conditions. *Biomarkers* 18(3):242–9.

Charman, S. A., Segrave, A. M., Edwards, G. A., Porter, C. J. 2000. Systemic availability and lymphatic transport of human growth hormone administered by subcutaneous injection. *J Pharm Sci* 89:168–77.

Chiu, W. A., Bois, F. Y. 2006. Revisiting the population toxicokinetics of tetrachloroethylene. *Arch of Toxicol* 80(6):382–5.

Choonara, I. A. 1987. Giving drugs per rectum for systemic effect. *Arch Dis Child* 62(8):771–2.

Darwish, M., Kirby, M., J. G., Tracewell, W., Robertson, P. Jr. 2008. Bioequivalence following buccal and sublingual placement of fentanyl buccal tablet 400 micrograms in healthy subjects. *Clin Drug Invest* 28(1):1–7.

de Meijer, P. H. E. M., Russel, F. G. M., Van lier, M. J. J., Van Ginneken, C. A. M. 1989. A comparison of three mathematical models to describe the disappearance curves of subcutaneously injected 125 I-labeled insulin. *Brit J Clin Pharmaco* 27:461–7.

Derendorf, H., Hochhaus, G., Möllmann, H. 2001. Evaluation of pulmonary absorption using pharmacokinetic methods. *J Aerosol Med Pulm D* 14(1):9–17.

Ding, X., Kaminsky, L. S. 2003. Human extrahepatic cytochrome P450: Function in xenobiotic metabolism and tissue-selective chemical toxicity in the respiratory and gastrointestinal tracts. *Annu Rev Pharmacol Toxicol* 43:149–73.

Dost, F. H. 1953. Der Bliitspiegel-Kinetic der Konzentrationsablaiife *in der Krieslaufüssigkeit*, 244. Leipzig: G. Thieme.

Eason, C., O'Halloran, K. 2002. Biomarkers in toxicology versus ecological risk assessment *Toxicology* 181–182:517–21.

Evans, W. E., Relling M. V. 1999. Pharmacogenomics: Translating functional genomics into rational therapeutics. *Science* 286:487–91.

FDA Guidance for Industry. 1999. Population Pharmacokinetics. www.fda.gov/downloads/Drugs/…/Guidances/UCM072137.pdf

Flessner, M. F., Dedrick, R. L., Schultz, J. S. 1984. A distributed model of peritoneal-plasma transport: Theoretical considerations. *Am J Physiol-Reg* I 246:R597–R607.

Flessner, M. F., Deverkadra, R., Smitherman, J., Li, X., Credit, K. 2006. *In vivo* determination of diffusive transport parameters in a superfused tissue. *Am J Physiol Renal* 291:F1096–F1103.

Gillis C. N., Pitt, B. R. 1982. The fate of circulating amines within the pulmonary circulation. *Annu Rev Physiol* 44:269–81.

Greenblatt, D. J., Allen, M. D. 1978. Intramuscular injection-site complications. *JAMA* 240:542–44.

Groneberg, D. A., Fischet, A., Chung, K. F., Danial, H. 2004. Molecular mechanisms of pulmonary peptidomimetic drug and peptide transport. *Am J Resp Cell Mol* 30:251–60.

Guidance for Industry and FDA Staff—Qualification Process for Drug Development Tools. 2014. U.S. Department of Health and Human Services, Food and Drug Administration, CDER.

Hasselbalch, K. A. 1917. Die berechnung der wasserstoffzahl des blutes aus der freien und gebundenen kohlensäure desselben, und die sauerstoffbindung des blutes als funktion der wasserstoffzahl. *Biochem Z* 78:112–44.

Hatsukami, D. K., Benowitz, N. L., Rennard, S. I. 2006. Biomarkers to assess the utility of potential reduced exposure tobacco products. *Nicotine Tob Res* 8:169–91.

Henderson, L. J. 1908. A diagrammatic representation of equilibria between acids and bases in solution. *J. Am. Chem. Soc.* 30(6):954–60.

Horvath, G., Schmid, N., Fragosa, M. A., Schmid, A., Conner, G. E., Salathe, M. 2007. Epithelial organic cation transporters ensure pH-dependent drug absorption in the airway. *Am J Resp Cell Mol* 36:53–60.

Hukkanen, J., Pelkonen, O., Hakkola, J., Raunio, H. 2002. Expression and regulation of xenobiotic metabolizing cytochrome P450 (CYP) enzymes in human lung. *Crit Rev Toxicol* 32:391–411.

ICH Topic S 3 A—Step 5. 1995. Note for guidance on toxicokinetics: A guidance for assessing systemic exposure in toxicology, toxicokinetics. http://www.emea.eu.int

IUPAC (International Union of Pure and Applied Chemistry). 2007. Chemistry and human health division. *Glossary of Terms Used in Toxicokinetics*. 2nd edition. 79(7):1153–1344.

Jantzen, J. P., Tzanova, I., Witton, P. K., Klein, A. M. 1989. Rectal pH in children. *Can J Anaesth* 36:665–7.

Junod, A. F. 1975. Metabolism, production and release of hormones and mediators in the lung. *Am Rev Respir Dis* 112:93–110.

Junod, A. F. 1985. 5-Hydroxytryptamine and other amines in the lungs. In: *Handbook of Physiology*; Section 3, Vol. 1. Eds. A. P. Fishman, and A. B. Fisher, 337–50. Betheseda, MD: American Physiological Society.

Katz, M., Barr, M. 1955. A study of sublingual absorption I. Several factors influencing the rate of adsorption. *J Am Pharm Assoc Am Pharm Assoc (Baltim)* 44(7):419–23.

Kelly, J. D., Eaton, D. L., Guengerich, F. P., Coulombe, R. A. Jr., 1997. Aflatoxin B1 activation in human lung. *Toxicol Appl Pharmacol* 144:88–95.

Kemp, S. J., Thorley, A. J., Gorelik, J., Seckl, M. J., O'Hare, M. J., Arcaro, A., Korchev, Y., Goldstraw, P., Tetley, T. D. 2008. Immortalization of human alveolar epithelial cells to investigate nanoparticle uptake. *Am J Resp Cell Mol* 39:591–97.

Kidwell, D. A., Holland, J. C., Athanaselis, S. 1998. Testing for drugs of abuse in saliva and sweat. *J Chromatogr B* 713(1):111–35.

Kimbell, J., Shroeter, J. D., Asgharian, B., Wong, B. A., Segal, R. A., Dickens, C. J., Southall, J. P., Millerk, F. J. 2004. Optimization of nasal delivery devices using computational models. *Res Drug Del* 9:233–38.

Kota, J., Machavaram, K. K., McLennan, D. N., Edwards, G. A., Christopher J. H. Porter, C. J. H., Susan, A., Charman, S. A. 2007. Lymphatic absorption of subcutaneously administered proteins: Influence of different injection sites on the absorption of darbepoetin alfa using a sheep model. *Drug Metab Dispo* 35(12):2211–17.

Kraegen, E. W., Chisholm, D. J. 1984. Insulin responses to varying profiles of subcutaneous insulin infusion: Kinetic modeling studies. *Diabetologia* 26:208–13.

Kurosaki, Y., Takatori, T., Nishimura, H., Nakayama, T., Kimura, T. 1991. Regional variation in oral mucosal drug absorption permeability and degree of keratinization in hamster oral cavity. *Pharm Res* 8:1297–301.

Larsen, S. W., Larsen, C. 2009. Critical factors influencing the *in vivo* performance of long-acting lipophilic solutions—Impact on *in vitro* release method design. *AAPS J* 11(4):762–70.

Layton, D. W. 1993. Metabolically consistent breathing rates for use in dose assessment. *Health Phys* 64(1):23–36.

Lee, V. H. L. 1988. Enzymatic barriers to peptide and protein absorption. *CRC Crit Rev Ther Drug Delivery Sys* 5:69–97.

Lehmann, E. D., Tarín, C., Bondia, J., Teufel, E., Deutsch, T. 2009. Incorporating a generic model of subcutaneous insulin absorption into the AIDA v4 diabetes simulator 3. early plasma insulin determinations. *J Diabetes Sci Technol* 3(1):190–201.

Li, J., Kuang, Y. 2009. Systematically modeling the dynamics of plasma in subcutaneous injection if insulin analogues for type 1 diabetes. *Math Biosci Eng* 6(1):41–58.

Liao, X., Wiedmann, T. S. 2003. Solubilization of cationic drugs in lung surfactant. *Pharm Res* 20:1858–63.

Lip, K. S., Volk, C., Schmitt, B. M., Pfeil, U., Aendt, P., Miska, D. 2005. Polyspecific cation transporters mediate luminal release of acetylcholine from bronchial epithelium. *Am J Resp Cell Mol* 33:79–88.

Lowe, F. J., Luettich, K., Gregg, E. O. 2013. Lung Cancer Biomarkers for the assessment of modified risk tobacco products: An oxidative stress perspective. *Biomarkers* 18(3):183–95.

Lu, Z., Wang, J., Wientjes, M. G., Jessie, L-S., Au, J. L-S. 2010. Intraperitoneal therapy for peritoneal cancer. *Future Oncol* 6(10):1625–41.

Luan, H., Chen, X., Zhong, S., Yuan, X. Meng, N., Zhang, J., Fu, J. et al. 2013. Serum metabolomics reveals lipid metabolism variation between coronary disease and congestive heart failure: A pilot study. *Biomarkers* 18(4):314–21.

McLean, J. A., Tobin, G. 1987. *Animal and Human Calorimetry*. Cambridge, MA. Cambridge University Press.

McLennan, D. N., Porter, C. J., Edwards, G. A., Heatherington, A. C., Martin, S. W., Charman, S. A. 2006. The absorption of darbepoetin alfa occurs predominantly via the lymphatics following subcutaneous administration to sheep. *Pharm Res* 23:2060–66.

Medina, C., Santos-Martinez, M. J., Radomski, A., Corrigan, O. I., Radomski, M. W. 2007. Nanoparticles: Pharmacological and toxicological significance. *Brit J Pharmacol* 150(5):552–58.

Morris, J. B. 2012. Biologically-based modeling insights in inhaled vapor absorption and dosimetry. *Pharmacol Ther* 136(3):401–13.

Mosekilde, E., Jensen, K. S., Binder, C., Pramming, S., Thorsteinsson, B. 1989. Modeling absorption kinetics of subcutaneous injected soluble insulin. *J Pharmacokinet Biopharm* 17:67–87.

Murdan, S., Florence, A. T. 2000. Non-aqueous solutions and suspensions as sustained-release injectable formulations. In: *Sustained-Release Injectable Products*, eds. J. Senior and M. Radomsky, 71–107. Denver, CO: Interpharm Press.

Na, D. H., Jabar Faraj, J., Capan, Y., Kai P. Leung, P., DeLuca, P. P. 2007. Stability of Antimicrobial Decapeptide (KSL) and its analogues for delivery in the oral cavity. *Pharm Res* 24(8):1544–50.

Nebert, D. W., Dalton, T. P., Okey, A. B., Gonzalez F. J. 2004. Role of aryl hydrocarbon receptor mediated induction of the CYP1 enzymes in environmental toxicity and cancer. *J Biol Chem* 279:23847–50.

Nielsen, H. M., Rassing, M. R. 2000. TR146 cells grown on filters as a model of human buccal epithelium: V. Enzyme activity of the TR146 cell culture model, human buccal epithelium and porcine buccal epithelium, and permeability of leu-enkephalin. *Int J Pharm* 200:261–70.

Nucci, G., Cobelli, C. 2000. Models of subcutaneous insulin kinetics. A critical review. *Comput Prog in Biomed* 62:249–57.

Oberdörster, G. 1998. Lung clearance of inhaled insoluble and soluble particles. *J Aerosol Med* 1:289–330.

Okamura, T., Kikuchi, T., Okada, M., Wakizaka, H., Zhang, M. R. 2013. Imaging of activity of multidrug resistance-associated protein1 in the lung. *Am J Resp Cell Mol* 49(3):335–40.

Patton, J. S., Fishburn, C. S., Weers, J. G. 2004. The lung as a portal of entry for systemic drug delivery. *Proc Am Thorac Soc* 1:338–44.

Patton, J. S. 2007. Mechanisms of macromolecule absorption by the lungs. *Drug Deliv Rev* 19:3–36.

Peng, J., Xu, X., Mace, B. E., Vanderveer, L. A., Workman, L. R., Slifker, M. J., Sullivan, P. M., Veenstra, T. D., Clapper, M. L. 2013. Estrogen metabolism within the lung and its modulation by tobacco smoke. *Carcinogenesis* 34(4):909–15.

Philpot, R. M. 1977. Uptake, accumulation, and metabolism of chemicals by the lung. In: *Metabolic Functions of the Lung*, Eds Y. S. Bahkle and J. R. Vane, 123–71. New York: Marcel Dekker.

Ploemen, J-P. H. T. M., Kramer, H., Krajnc, E., Martin, I. 2007. The use of toxicokinetic data in preclinical safety assessment: A toxicologic pathologist perspective. *Toxicol Pathol* 35:834–37.

Pradham, A. D., Manson, J. E., Rifai, N. 2001. C-reactive protein, interleukin 6, and risk of developing type 2 diabetes mellitus. *JAMA* 286:327–34.

Rathbone, M. J., Tucker, I. G. 1993. Mechanisms, barriers and pathways of oral mucosal drug permeation. *Adv Drug Del Rev* 13:1–22.

Raunio, H., Hakkola, J., Hukkanen, J., Lassila, A., Paivarina, K., Pelkonen, O., Anttila, S., Piipari, R., Boobis, A., Edwards, R. J. 1999. Expresssion of xenobiotic-metabolizing CYPs in human pulmonary tissue. *Exp Toxicol Pathol* 51:412–17.

Ridkar, P. M., Rifai, N., Rose, L. 2002. Comparison of C-reactive protein and low-density lipoprotein cholesterol levels in the prediction of first cardiovascular events. *N Engl J Med* 347:1557–65.

Riegelman, S., Crowell, W. J. 1958. The kinetics of rectal absorption III. The absorption of undissociated molecules. *J Pharm Sci* 47(2):127–33.

Rossi, L., Conen, D. 1995. Die intramuskuläre Injektioneine überholte Applikation sart? 6 Fälle von Staphylococcus-aureus-Sepsis nach intramuskulären Injektionen. *Schweiz Med Wochenschr* 125:1477–82.

Shacter, E., Weitzman, S. A. 2002. Chronic inflammation and cancer. *Oncology* 16:217–32.

Shimada, T., Yamazaki, H., Mimura, M., Inui, Y, Guengerich, F. P. 1994. Interindividual variations in human liver cytochrome P-450 enzymes involved in the oxidation of drugs, carcinogens and toxic chemicals: Studies with liver microsomes of 30 Japanese and 30 caucasians. *J Pharmacol Exp Ther* 270:414–23.

Spivack, S. D., Hurteau, G. J., Andrew, A., Reilly, A. A., Aldous, K. M., Ding, X., Kaminsky, L. S. 2001. CYP1B1 expression in human lung. *Drug Metab Dispos* 29(6):916–22.

Spivack, S. D., Hurteau, G. J., Michael, J., Fasco, M. J., Laurence, S., Kaminsky, L. S. 2003. Phase I and II carcinogen metabolism gene expression in human lung tissue and tumors. *Clin Cancer Res* 9:6002–11.

Stahlhofen, W., Köbrich, R., Rudolf, G., Scheuch, G. 1990. Short term and long term clearance of particles from the upper human respiratory tract as function of particle size. *J Aerosol Sci* 21:S407–S41.

Supersaxo, A., Hein W. R., Steffen, H. 1990. Effect of molecular weight on the lymphatic absorption of water-soluble compounds following subcutaneous administration. *Pharm Res* 7:167–9.

Teorell, T. 1937a. Kinetics of distribution of substances administered to the body I. The extravascular modes of administration. *Archs Int Pharmacodyn Ther* 57:205–55.

Teorell, T. 1937b. Kinetics of distribution of substances administered to the body II. The intravascular modes of administration. *Archs Int Pharmacodyn Ther* 57:226–40.

Trajanoski, Z., Wach, P., Kotanko, P., Ott, A., Skraba, F. 1993. Pharmacokinetic model for the absorption of subcutaneously injected soluble insulin and monomeric insulin analogues. *Biomed Technik* 38:224–31

Tarín, C., Teufel, E., Picó, J., Bondia, J., Pfleiderer, H-J. 2005. Comprehensive pharmacokinetic model of insulin glargine and other insulin formulations. *IEEE Trans Biomed. Eng* 52:1994–05.

USEPA. 1992. Guidelines for exposure assessment: Notice. *Fed. Reg.* 57(104):22888–938.

USEPA. 1997. *Exposure Factors Handbook.* National Center for Environmental Assessment Office of Research and Development U.S. Environmental Protection Agency, Washington D.C., Chapter 5.

Van Alstine, W. G., Dietrich, J. A. 1988. Porcine sciatic nerve damage after intramuscular injection. *Comp Contin Educ Pract Vet* 10:1329–32.

Van der Deen, M., Marks, H., Willemse, B. W. A., Poetma, D. S., Muller, M., Smit, E. F. 2006. Diminished expression of multidrug resistance-associated protein 1 (MRP1) in bronchial of COPD patient. *Virchows Archi* 449:682–8.

Van Hoogdalem, E. J., de Boer, A. G., Breimer, D. D. 1991. Pharmacokinetics of rectal drug administration: Part II: Clinical applications of peripherally acting drugs, and conclusions. *Clin Pharmacokinet* 21:110–28.

Wagner, J. 1971. *Biopharmaceutics and Relevant Pharmacokinetics*, 1st edition, Drug Intelligence publications, 45–46.

Wagner, J. G.1981. History of pharmacokinetics. *Pharmacol Therapeut* 12:537–62.

Walton, R. P. 1935. Absorption of drugs through the oral mucosa. III Fat-water solubility coefficient of alkaloids. *Proc Soc Exp Bio Med* 32:1488.

Wang, Y., Wang, Z., Zuo, Z., Tomlinson, B, Benjamin, T. K., Lee, B. K., Michael, B., Bolger, M. B., Chow, M. S. S. 2013. Clinical pharmacokinetics of buffered propranolol sublingual tablet (Promptol™)— Application of a new "physiologically based" model to assess absorption and disposition. *AACP J* 15(3):787–96.

Weber, B., Hochhaus, G. 2013. A pharmacokinetic simulation tool for inhaled corticosteroids. *AAPS J* 15(1):159–71.

Widmark, E., Tandberg, J. 1924. Uber die bedingungen f'tir die Akkumulation Indifferenter Narkoliken Theoretische Bereckerunger. *Biochem Z* 147:358–69.

Widmark, E. 1932. Die theoretischen Grundlagen und die praktische Verwendbzrkeit der gerichtlichmedizinischen Alkoholbestimmung, Berlin.

Wiedmann, T. S., Bhatia, R., Wattenberg, L.W. 2000. Drug solubilization in lung surfactant. *J Contr Release* 65:43–47.

Wiggins, K. J., Johnson, D.W., Craig, J. C., Strippoli, G. F. 2007. Treatment of peritoneal dialysis-associated peritonitis: A systematic review of randomized controlled trials. *Am J Kidney Dis* 50(6):967–88.

Willems, B. A. T., Melnick, R. L., Kohn, M. C., Portier, C. J. 2001. A physiologically based pharmacokinetic model for inhalation and intravenous administration of naphthalene in rats and mice. *Toxicol Appl Pharma* 176:81–91.

Yang, D., Zhou, F., Wang, X., Gao, H., Li, G., Xue, M. 2013. Association analysis between MDR1 gene polymorphisms and risk of hepatocellular carcinoma in Chinese population. *Biomarkers* 18(3):236–41.

Young, L. C., Campling, B. G., Cole, S. P., Deeley, R. G., Gerlach, J. H. 2001. Multidrug resistance proteins MRP3, MRP1 and MRP2 in lung cancer: Correlation of protein levels with drug response and messenger RNA levels. *Clin Cancer Res* 7(6):1798–804.

Zhang, J. Y., Wang, Y. F., Prakash, C. 2006. Xenobiotic metabolizing enzymes in human lung. *Curr Drug Metab* 7:939–48.

Zheng, Z., Fang J. L., Lazarus, P. 2002. Glucuronidation: An important mechanism for detoxification of benzo(a)pyrene metabolites in aerodigestive tract tissue. *Drug Metab Disp* 30:397–403.

2 Mechanisms of Absorption and Permeation of Xenobiotics

2.1 INTRODUCTION

The fundamental considerations of pharmacological activity or organ toxicity include those aspects of biochemical and physiological processes which influence the handling of xenobiotics in the body. These biological processes occur concurrently, impact the overall pharmacokinetic or toxicokinetic profile of xenobiotics and most are predictable in normal populations except when the genetic differences are the variability factors among humans.

The simplest mechanism of xenobiotic absorption is through passive diffusion and it depends on the partitioning of a compound through a barrier. The barrier can be as simple as the lipid bilayer of a cell membrane, or as complex as the wall of gastrointestinal tract, or as shielding as the multilayer barrier of the skin. The partitioning is the result of the compound interacting with the barrier. The nature of such interaction can be as simple as transcellular or paracellular diffusion, or as intricate as active or facilitated transport. The interaction was first observed from the relationship between the rate of penetration and the lipid/water partition coefficient (Overton, 1902). The concept was further investigated (Hober and Hober, 1937; Travel, 1940; Hogben et al., 1959; Hogben, 1960), refined, and presented as the concept of pH partition hypothesis. The hypothesis, however, has been shown to have certain limitations, which do not reduce the significance of the hypothesis, yet obscure the understanding of absorption process. Examples are the complete absorption of weakly acidic drugs in the alkaline environment of the small intestine, or absorption of quaternary ammonium compounds at the different pH of the GI tract. Several explanations have been suggested to define these deviations in support of the pH partition theory. For example, the absorption of weakly acidic drugs in an alkaline pH is because of the presence of an acidic microclimate adjacent to the apical surface of the intestinal epithelium (Tsuji et al., 1978). However, the observation of transportation of charged species of xenobiotics via the paracellular diffusion weaken the validity of the intestinal acidic microclimate theory (Palm et al., 1999; Neuhoff et al., 2003). Thus, the controversy regarding the exceptions to the pH partition theory will remain to be clarified.

Other mechanisms of absorption are as intricate and involved as the pH partition theory. The absorption of xenobiotics from absorption sites such as buccal, sublingual, rectal, etc., is explained by the passive diffusion. However, for the routes such as the pulmonary or GI tract, there are other mechanisms of absorption, which are as essential as the passive diffusion.

The evaluations of these mechanisms within the context of the absorption sites are the focus of this chapter.

The permeation/transport of a compound through physiological barriers is governed by one of the following mechanisms. The GI tract is the only site of absorption that utilizes all these mechanisms:

- Passive diffusion
  - Transcellular and paracellular diffusion
    - Carrier-mediated transcellular diffusion or facilitated diffusion, or passive-mediated transport
- Transcellular diffusion subject to P-glycoprotein efflux
- Active transport

Pinocytosis and receptor mediated endocytosis
Solvent drag, osmosis, and two-pore theory
Ion-pair absorption

2.2 PASSIVE DIFFUSION

2.2.1 Transcellular and Paracellular Diffusion

The permeation of xenobiotics through biological barriers is mostly governed by passive diffusion. The mechanism of transport is the passage of molecules from the aqueous or hydrophilic surroundings into the lipophilic environment of the barriers. The degree of lipophilicity of the molecules is quantified by its partition coefficient or distribution coefficient. The pH-partition theory (Shore et al., 1957; Hogben et al., 1957, 1959; Bekett et al., 1967, 1968, 1970), which assumes only unionized molecules of a xenobiotic can transfer across biological barriers, is still considered the encompassing feature of passive diffusion (Schaper et al., 2001). Transcellular and paracellular passive diffusions are mainly in the GI tract, nasal, buccal, pulmonary, rectal, and vaginal routes of administration. Small hydrophilic of hydrophobic molecules may absorb through vaginal, buccal, and rectal routes by paracellular diffusion. In the GI tract, the extent of paracellular absorption of the small molecules is as significant as trascellular diffusion (Figure 2.1). The nasal route, however, has been shown to have a nonlinear correlation between the lipophilicity of a compound and its permeation, which may be related to the complex structure of the nasal epithelium. Diffusion of xenobiotics through the cornea and its component layers (epithelium, stroma, and endothelium) is also by passive transcellular and paracellular permeability through its different layers, and depends on the molecular size of the compound and its distribution coefficient (Edwards and Prausnitz, 2001).

For most biological barriers, the transcellular diffusion is assumed as the preferred route for absorption of xenobiotics by passive diffusion. The absorption by the paracellular route is limited to the size of the paracellular opening and the presence of a tight junction barrier; despite this limitation, it is an important mechanism for the absorption of hydrophilic compounds with little membrane permeability.

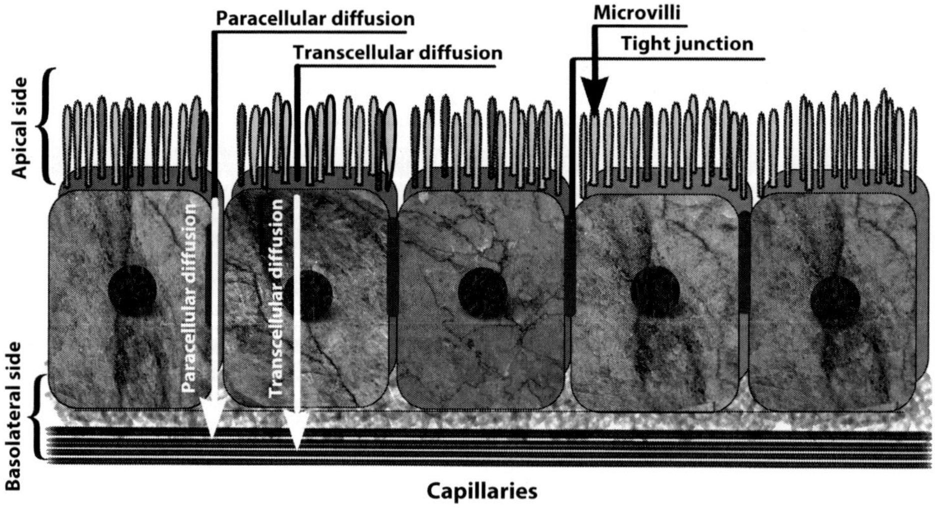

FIGURE 2.1 Illustration of transcellular diffusion by intracellular trafficking and paracellular diffusion through the disrupted tight junctions.

To differentiate kinetically between transcellular diffusion and paracellular pathway, the following three-compartment closed model is proposed for the accumulation and transport in Caco-2 cells (Bourdet et al., 2006). The combined differential equations representing the transcellular and paracellular pathways are

$$\frac{dA_1}{dt} = -(k_{12} + k_{13})A_1 + k_2A_2 \tag{2.1}$$

$$\frac{dA_2}{dt} = k_{12}A_1 - (k_{21} + k_{23})A_2 + k_{32}A_3 \tag{2.2}$$

$$\frac{dA_3}{dt} = k_{23}A_2 + k_{13}A_1 - k_{32}A_3 \tag{2.3}$$

where (A_1) is the amount in the apical compartment; (A_2) is the amount in the cells; (A_3) is the amount in the basolateral compartment; (k_{12}) is the first-order input rate constant in the cellular compartment; (k_{21}) is the first-order input rate constant from the cellular compartment into the apical compartment; (k_{23}) is the rate constant from the cellular compartment into the basolateral compartment; (k_{13}) is the input rate constant from apical into the basolateral compartment; (k_{32}) is the rate constant from the basolateral compartment into the cellular compartment; (dA_1/dt) is the rate of change in amount in the apical compartment; (dA_2/dt) is the rate of change in amount of the cellular compartment; and (dA_3/dt) is the rate of change in the basolateral compartment.

The following differential equations set apart the rate equations of paracellular from transcellular diffusion:

Paracellular rate equations:

$$\frac{dA_1}{dt} = -k_{13}A_1 \tag{2.4}$$

$$\frac{dA_3}{dt} = k_{13}A_1 \tag{2.5}$$

Transcellular rate equations:

$$\frac{dA_1}{dt} = -k_{12}A_2 + k_{21}A_2 \tag{2.6}$$

$$\frac{dA_2}{dt} = k_{12}A_1 - (k_{21} + k_{23})A_2 + k_{32}A_3 \tag{2.7}$$

$$\frac{dA_3}{dt} = k_{23}A_2 - k_{32}A_3 \tag{2.8}$$

The specific input rate constants into the cellular compartment, (k_{12}) and (k_{32}), are estimated by the following equations, taking into consideration the saturable and non-saturable components of the cellular uptake (Bourdet et al., 2006):

$$\frac{dA_2}{dt} = \frac{V_{max} \times A_{1(OR3)}}{K_M + A_{1(OR3)}} + \frac{K_dA_1}{V_{AP(OR\ BL)}} \tag{2.9}$$

The rate of input from the apical compartment or basolateral compartment into the cellular compartment is

$$\frac{dA_2}{dt} = k_{12_{(OR32)}} A_{1_{(OR3)}} \tag{2.10}$$

Setting Equation 2.9 equal to Equation 2.10 and solving for $k_{12_{(OR32)}}$ yields the following relationship:

$$k_{12_{(OR32)}} = \frac{V_{max}}{K_M + A_{1_{(OR3)}}} + \frac{K_d}{V_{AP_{(OR\,BL)}}} \tag{2.11}$$

where (V_{max}) is the maximum uptake rate; (K_M) is the Michaelis–Menten constant; (K_d) is the nonsaturable parameter of the uptake with units of (volume/time)/mg of protein; ($V_{AP_{(orBL)}}$) is the volume of the donor compartment.

The transcellular absorption is facilitated by the ability of xenobiotics to partition into the cell membrane. In other words, the molecules should have optimum lipophilicity to cross the membrane; and not being too lipophilic to be retained in the lipid environment of membrane. The lipophilicity of a compound in terms of its chemical structure is defined as

$$Lipophilicity = Hydrophobicity - Polarity$$

Hydrophobicity makes up the hydrophobic and dispersion forces and polarity comprises for hydrogen bonds, orientation, and induction forces (Liu et al., 2011).

A measure of the lipophilicity of xenobiotic molecules is the magnitude of its partition coefficient and distribution coefficient. In certain cases the distribution coefficient replaces the diffusion coefficient. As will be discussed in this section, the diffusion coefficient is inversely proportional to the molecular weight of the compound. Thus, small hydrophobic molecules have a better chance of getting absorbed by passive diffusion faster and more completely.

2.2.1.1 Partition Coefficient

The partition coefficient implies the extent a xenobiotic partitions between a hydrophilic phase and a lipophilic phase. It represents the degree of lipophilicity or hydrophobicity of a compound. There are various *in vitro* methodologies to estimate a realistic value for the partition coefficient of xenobiotics. These methodologies include (1) a classical n-octanol/water partitioning system, (2) liposome (Betageri and Rogers, 1988) and immobilized artificial membrane partitioning systems (Pauletti and Wunderli-Allenspach, 1994; Taillardat-Bertschinger et al., 2002), and (3) high-performance liquid chromatography partitioning systems using stationary phases (Pidgeon et al., 1995; Ong et al., 1995). The partition coefficient determined by the octanol/water system or ODS system corresponds only to the hydrophobic nature of a compound and not the interaction of its molecules with the structural characteristics of a barrier, for example, polar head groups of the lipid bilayer. The liposome partitioning systems provide the estimates that include both partition coefficient and the membrane interaction. The discussion on comparison of different methodology to determine the partition coefficient is beyond the scope of this chapter. The following discussion is focused on the data generated from the classical n-octanol/water system.

Using the octanol/water system, the aqueous phase (water) is a buffer containing the xenobiotic with concentrations of ($C_{aqueous}$). It represents the hydrophilic environment at the absorption site. The organic or lipid phase is n-octanol and it represents the lipophilic barrier with concentrations of (C_{lipid}). At the lipid-aqueous (octanol/water) interfaces, a discontinuity in solute concentration

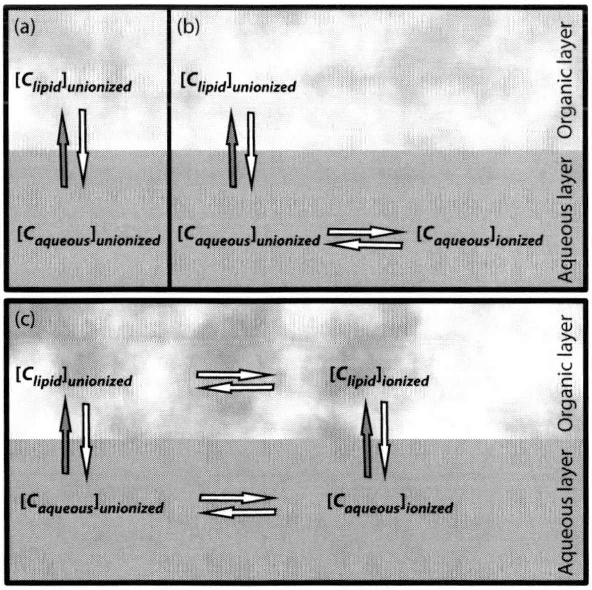

FIGURE 2.2 Partitioning diagram when dealing with neutral molecules (a), ionized and unionized molecules when ionized molecules are present only in aqueous phase (b), and when ionized and unionized molecules are present in aqueous and lipid phases (c).

exists at equilibrium, which is the result of the molecular hindrance that exists for a solute molecule migrating from the aqueous phase into the lipid phase. The discontinuity between ($C_{aqueous}$) and (C_{lipid}) is represented by the lipid-aqueous partition coefficient P_{coeff} as the ratio of concentration in the hydrophobic core, (C_{lipid}), to its concentration in the hydrophilic phase, ($C_{aqueous}$) (Equation 2.12). The ratio represents the relative affinity of a compound for the lipid environment of biological barrier (Figure 2.2a).

$$P_{coeff} = \left(\frac{C_{lipid}}{C_{aqueous}}\right) \tag{2.12}$$

Partition coefficient, P_{coeff}, is a unitless fraction and its value may vary according to the experimental conditions (pH, temperature, ionic strength, etc.) and selection of organic solvent (octanol, chloroform, dichloromethane, etc.). Thus, a validated procedure is essential in determining the partition coefficient. Its logarithm, identified as log P_{coeff}, also signifies the lipophilicity of xenobiotics and provides numbers that are more convenient to use. Similar to pKa, log P_{coeff} (Equation 2.13) is also an important descriptor in understanding the permeability behavior of xenobiotics.

$$\log P_{coeff} = \log\left(\frac{C_{lipid}}{C_{aqueous}}\right) \tag{2.13}$$

The (log P_{coeff}) is $\log_{10}$ scale, thus when it is equal to zero the ratio ($C_{lipid}/C_{aqueous}$) is equal to 1, which indicates that the compound under evaluation is equally present in both the lipid and aqueous layer. When log P_{coeff} is a positive number, for example, 2 or 5 (see also Chapter 8, Section 8.2.2.2), it means that the compound has 100 or 100,000 times more affinity for the lipid phase and when it is negative, for example, –1 or –2, the compound is 10 or 100 times more soluble in the aqueous phase.

A number of different approaches have been recommended for the theoretical calculation of log P_{coeff}. These theoretical approaches are based mostly on predicting lipophilicity of a compound from the structural parameters of the molecule or solvent accessible surface area (SASA). These approaches include: CLOGP (Leo et al., 1975; Hansch and Leo, 1979; Leo, 1987, 1991, 1993), the acronym stands for calculated log P (described below); Σ*f*-system (Nys and Rekker, 1973, 1974; Rekker, 1977; Rekker and de Kort, 1979) is used to study the aliphatic hydrocarbon/water partitioning and the system is based on the relationship of $Log\ P = \Sigma f$, where is the fragmental constant; KLOGP (Klopman et al., 1994; Zhu et al., 2005) or Klopman LOGP is based on atomic composition of the molecule; KOWWIN (Meylan and Howard, 1995)—Syracuse Research Corporation—the KOW part of the acronym stands for K_{OW} representing the partition coefficient of oil/water system, the system is based on hydrophobic contribution of atomic structure and molecular fragments of a compound; ACD/LOGP (Petrauskas and Kolovanov, 2000), or Advanced Chemistry Development Inc LOGP is also used for quantitative measure of lipophilicity of a neutral molecule; AB/LOGP (Japertas and Didziapetris, 2002), is based on Algorithm Builder (AB) for all computational steps; XLOGP (Wang et al., 1997), developed by the Institute of Physical Chemistry of Peking University, is based on classifying atoms by their hybridization states and their neighboring atoms, X is used in the acronym as the unknown LOGP; CLIP (Gaillard et al., 1994) or Calculated Lipophilicity Potential, is based on the atomic lipophilic system and molecular lipophicity potential; HINT (Kellogg et al., 1991, 1992; Abraham and Kellogg, 1994; Kellogg and Abraham, 1999, 2000) or Hydrophobic Interactions is based on hydrophobic fragment constants and could be used to assess interactions between small and large molecules; MLOGP (Moriguchi et al., 1992, 1994) or Moriguchi LOGP uses sum of hydrophilic and hydrophobic atoms (described below); VLOGP (Gombar and Enslein, 1996; Gombar, 1999), is a model developed based on the value of linear free energy relationship and uses molecular topology; TLOGP (Junghans and Pretsch, 1997) or Topological LOGP is also based on topological and substructure coding of molecules description; AUTOLOGP (Devillers et al., 1998) is derived from heterogenous organic chemicals and used especially for highly lipophilic substances; BLOGP (Bodor and Huang, 1992; Klopman and Iroff, 1981), is based on parameters such as geometric descriptos including molecular surface area, ovality, and charge density and dipole moment; QLOGP (Bodor and Buchwald, 1997) or Quantum LOGP is based on quantum mechanical semi-empirical calculations of molecular size; ClogPalk (Kenny et al., 2013) or Calculated logP of alkane/water partition is based on the relationship between alkane/water partition coefficient and molecular surface area.

The two most often used methodologies are briefly described below.

2.2.1.1.1 $CLOGP_{coeff}$

CLOGP stands for calculated partition coefficient and is determined empirically by the CLOGP program developed by Pomona Med Chem. The manual of the program is available at http://www.daylight.com/dayhtml/doc/clogp

The calculated value is based on molecular structure and values of different fragments of the molecule. The program essentially uses the following relationship in conjunction with Hansch and Leo's database (Hansch et al., 1979; Leo, 1993), in Equation 2.14:

$$\log P_{coeff} = \sum a_n f_n + \sum b_m F_m \tag{2.14}$$

where *a* is the number of occurrences of fragment *f* of type *n*, *b* is the number of occurrences of correction factor *F* of type *m*.

2.2.1.1.2 $MLOGP_{coeff}$

MLOGP or the Moriguchi partition coefficient is calculated by using the following relationship (Moriguchi et al., 1992; Moriguchi et al., 1994):

$$
\begin{aligned}
MLOGP_{coeff} &= -1.014 + 1.244(F_{CX})^{0.6} - 1.017(N_O + N_N)^{0.9} + 0.406F_{PRX} - 0.145N_{UNS}^{0.8} + 0.511I_{HB} \\
&+ 0.268N_{POL} - 2.215F_{AMP} + 0.912I_{ALK} - 0.392I_{RNG} - 3.684F_{QN} + 0.474N_{NO_2} \\
&+ 1.582F_{NCS} + 0.773I_{\beta L}
\end{aligned}
\tag{2.15}
$$

There are 13 structural parameters in the above regression equation (Equation 2.15), which are identified as independent Moriguchi based lipophilicity molecular descriptors. The definition and values of the descriptors (i.e., regression coefficients and variable of Moriguchi model) are as follows (Todeschini and Consonni, 2000).

The y-intercept, is $b_0 = -1.014$; the coefficient of F_{CX} is 1.244 and F_{CX} is the total number of carbon and halogen atoms weighted by carbon = 1.0, fluorine = 0.5, chlorine = 1.0, bromine = 1.5, and iodine = 2.0; the coefficient of $(N_O + N_N)$ is −1.017 and $(N_O + N_N)$ are the total number of oxygen and nitrogen atoms; the coefficient of F_{PRX} is 0.406 and F_{PRX} is the proximity influence of oxygen and nitrogen for N-O, or O-N: $F_{PRX} = 2$, for N-A-O where A = C, or S, or P, $F_{PRX} = 1$ and for carbonamid bond $F_{PRX} = -1$; the coefficient of N_{UNS} is −0.145 and N_{UNS} represents the total number of unsaturated bonds, except those in NO_2; the coefficient of I_{HB} is 0.511 and I_{HB} stands for the presence of intramolecular H-bond; the coefficient of N_{POL} is 0.268 and N_{POL} is the number of polar substituent; The coefficient of F_{AMP} is −2.215 and F_{AMP} represents the amphoteric property: for α-amino $F_{AMP} = 1.0$ for aminobenzoic acid or pyridinecarboxylic acid $F_{AMP} = 0.5$; The coefficient of I_{ALK} is 0.912 and I_{ALK} is a variable for alkanes, alkenes, cycloalkanes, cycloalkenes (hydrocarbons with 0 or 1 double bond); the coefficient of I_{RNG} is −0.392 and I_{RNG} is a variable for the presence of ring structures (not benzene and its condensed rings); the coefficient of F_{QN} is −3.684 and F_{QN} represents quaternary nitrogen $F_{QN} = 1$ for quaternary nitrogen, $F_{QN} = 0.5$ for nitrogen oxide; the coefficient of N_{NO_2} is 0.474 and N_{NO_2} is the number of nitro groups in the molecule; the coefficient of F_{NCS} is 1.582 and $F_{NCS} = 1$ for $-N = C = S$ group and $F_{NCS} = 0.5$ for $S - CN$ group; the coefficient of $I_{\beta L}$ is 0.773 and $I_{\beta L}$ stands for the presence of β-lactam in the molecule.

($MLOG_{coeff}$) and ($CLOG_{coeff}$), as well as other approaches are used for estimating lipophilicity of chemicals and predicting the solubility, permeability, and absorption of xenobiotics. One of these applications is the Lipinski's Rule of Five (see Chapter 8).

2.2.1.2 Distribution Coefficient

The estimation of the partition coefficient for neutral molecules is straightforward and follows Equation 2.12 (Figure 2.2a). However, for ionizable molecules, monoprotic weakly acidic or basic xenobiotics or polyprotic ampholytes, for example, peptides, the partition coefficient depends on the pH and the degree of ionization of their functional group(s). Thus, their true partition coefficient, known as distribution coefficient, is estimated based on the presence of both ionized and unionized forms in the lipid phase and aqueous phase (Stopher and McClean, 1990; Avdeef, 1993; Scott and Clymer, 2002, Davies and Flower, 2013). It is often assumed that the ionized form of a molecule may not partition into octanol or a biological barrier, but both octanol and a biological barrier maintain a certain percent of water that makes it possible for ionized molecules to partition. Octanol is known to absorb water and the solubility of water in octanol is about 26.4 mol% or 4.73% w/w (Margolis and Levenson, 2000), which is similar to the water content of biological barriers.

When the unionized molecules are present in both the aqueous and lipid phase and ionized molecules are present only in the aqueous phase, the following relationship applies (Figure 2.2b):

$$
(Dist)_{coeff} = \frac{[C_{lipid}]_{unionized}}{[C_{aqueous}]_{unionized} + [C_{aqueous}]_{ionized}} \tag{2.16}
$$

Comparing Equation 2.16 with Equation 2.12 indicates that $(Dist)_{coeff} < P_{coeff}$.

Dividing the numerator and denominator of Equation 2.16 by $[C_{aqueous}]_{unionized}$ and setting the dissociation constant as

$$[C_{aqueous}]_{unionized} \xrightleftharpoons{H_2O} [C_{aqueous}]_{ionized} + H^+ \quad (2.17)$$

$$\text{Dissociation constant} = K_d = \frac{H^+ + [C_{aqueous}]_{ionized}}{[C_{aqueous}]_{unionized}} \quad (2.18)$$

The distribution coefficient can then be estimated as

$$\text{For weak acids: } (Dist)_{coeff} = \frac{P_{coeff}}{1 + (K_d / H^+)} \quad (2.19)$$

$$\text{For weak bases: } (Dist)_{coeff} = \frac{P_{coeff}}{1 + (H^+ / K_d)} \quad (2.20)$$

When the ionized and unionized molecules are present in both aqueous and lipid phase, the distribution coefficient is estimated as (Figure 2.2c)

$$(Dist)_{coeff} = \frac{[C_{lipid}]_{unionized} + [C_{lipid}]_{ionized}}{[C_{aqueous}]_{unionized} + [C_{aqueous}]_{ionized}} \quad (2.21)$$

$$\therefore \text{ For weak acids: } (Dist)_{coeff} = \frac{P_{coeff}}{1 + (K_d / H^+)} + \frac{\overline{P_{coeff}}}{1 + (H^+ / K_d)} \quad (2.22)$$

$$\text{For weak base: } (Dist)_{coeff} = \frac{P_{coeff}}{1 + (H^+ / K_d)} + \frac{\overline{P_{coeff}}}{1 + (K_d / H^+)} \quad (2.23)$$

where $(\overline{P_{coeff}})$ is the partition coefficient of ionized molecules.

For molecules with single ionizable group, the relationship between distribution coefficient and partition coefficient can be defined as Equations 2.24 and 2.25.

$$\text{For weak monoprotic acids: } \log(Dist)_{coeff} = \log P_{coeff} - \log[1 + 10^{(pH - pKa)}] \quad (2.24)$$

$$\text{For weak monoprotic bases: } \log(Dist)_{coeff} = \log P_{coeff} - \log[1 + 10^{(pKa - pH)}] \quad (2.25)$$

For molecules with more than one ionizable group the relationships become more complex, such as amphoteric compounds with both acidic and basic functional groups.

To avoid the pH-related variability for estimating the distribution coefficient, buffers with a physiological pH of 7.4 or 7.2 are often selected in PK/TK inferences.

2.2.1.3 Diffusion Coefficient

Diffusion coefficient (D_{coeff}) is another constant required for defining the passive diffusion/absorption. This coefficient is not expressed in terms of amount or concentration, and may have multiple definitions in different fields of physics or engineering. For biological systems, it is defined as a constant that reflects diffusivity of a compound with a unit of distance²/time, or area/time. A better understanding of diffusion coefficient is based on the solution of Fick's second law of diffusion

$$\frac{\partial C}{\partial t} = D_{coeff}\frac{\partial^2 C}{\partial x^2} \tag{2.26}$$

Equation 2.26 is also known as a one-dimensional diffusion equation. The three-dimensional equation (Equation 2.27) is known as a diffusion equation

$$\frac{\partial C}{\partial t} = \frac{\partial C}{\partial x^2} + \frac{\partial C}{\partial y^2} + \frac{\partial C}{\partial z^2} = D_{coeff}\Delta C \tag{2.27}$$

$$\text{Where } \Delta = \frac{\partial}{\partial x^2} + \frac{\partial}{\partial y^2} + \frac{\partial}{\partial z^2} \tag{2.28}$$

For one-dimensional diffusion in the x direction, the assumption is that the barrier is uniform in the y and z direction and $\partial/\partial y^2 = \partial/\partial z^2 = 0$.

A solution for Equation 2.26 is

$$C = \frac{1}{(4\pi D_{coeff}t)^{1/2}} e^{-x^2/4D_{coeff}t} \tag{2.29}$$

The mean square displacement of a diffusing molecule is defined by the indefinite integral of Equation 2.29 (Rubinow, 1975; Michalet, 2010)

$$\overline{x^2} = \frac{1}{(4\pi D_{coeff}t)^{1/2}} \int_{-\infty}^{\infty} [e^{-x^2/4D_{coeff}t} x^2 dx] \tag{2.30}$$

Setting $\alpha = \frac{1}{4D_{coeff}t}$ and definite integration between $0 \rightarrow \infty$, Equation 2.30 yields

$$\overline{x^2} = 2D_{coeff}t \tag{2.31}$$

$$\therefore D_{coeff} = \frac{\overline{x^2}}{2t} \tag{2.32}$$

Thus, the diffusion coefficient with units of distance²/time, or area/time is equal to the mean square displacement of a diffusing molecule during the time interval divided by 2.

The diffusion coefficient was also defined and interpreted by the Brownian movement theory (Einstein, 1905) and is known as the Stokes–Einstein relation. The Stokes–Einstein relation is a form of the fluctuation-dissipation theorem and relates the diffusion coefficient of colloidal particles to the viscosity via the thermal energy.

$$D_{coeff} = \frac{k_B T}{6\pi\eta r} \tag{2.33}$$

The equation defines the relationship between D_{coeff} the diffusion coefficient, η, the solvent viscosity, k_B the Boltzmann's constant (1.38×10^{-23} J/K), T the absolute temperature of the solution, and r the radius of the solute molecule. The radius is related to the molecular weight according to the following equation:

$$MW = N\rho V = N\rho\left(\frac{4}{3}\pi r^3\right) \tag{2.34}$$

Solving for r:

$$r = \left[\frac{3(MW)}{4\pi N\rho}\right]^{1/3} \tag{2.35}$$

where N is the Avogadro's number, V is the molal volume of solute, r is the hydrodynamic radius, that is, solvent bound to solute, and ρ is the density.

Equation 2.33 applies to the systems where the solute molecule is large compared to the solvent molecule and is assumed to have spherical shape. Although it is developed with the diluted systems in mind, the relationship works well at higher densities (Bonn and Kegel, 2003).

2.2.1.4 Permeation and Permeability Constant

Based on the definition of diffusion coefficient and partition/distribution coefficient, one can describe the permeation of a compound (P_m) by passive diffusion through a biological barrier as a process that is directly proportional to the partition coefficient and diffusion coefficient and inversely proportional to the thickness of the barrier, or distance it permeates through

$$P_m = \frac{D_{coeff} \times P_{coeff}}{x} \tag{2.36}$$

where (P_m) is the permeation with units of distance/time. Standardizing the permeation (Equation 2.36) with respect to the surface area of permeation yields the permeability constant (Figure 2.3)

$$P = \frac{D_{coeff} \times P_{coeff} \times Area}{x} = \frac{P_m \times Area}{x} \tag{2.37}$$

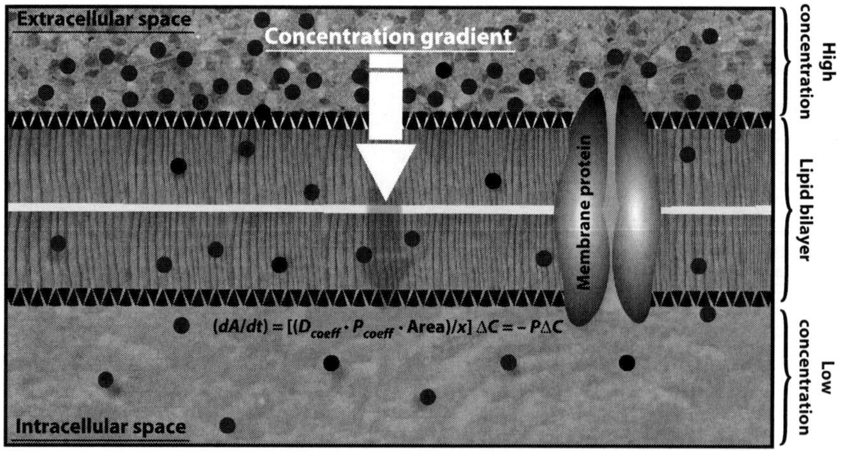

FIGURE 2.3 Depiction of Fick's first law of diffusion when the mass flux is related to the concentration gradient over the diffusion distance.

The units of permeability are (distance^3/time), which is the three-dimensional consideration of the barrier per a unit of time (Equation 2.37).

According to the Fick's first law of diffusion, the driving force of passive diffusion is the concentration gradient, which is the movement of molecules, or mass transfer, from high concentration to the lower concentration region. The first law defines the mass flux as

$$J = -D_{coeff} \frac{dC}{dx} \tag{2.38}$$

or

$$J = -D_{coeff} \frac{\partial C}{\partial x} \tag{2.39}$$

or

$$J = -D_{coeff} \frac{\Delta C}{\Delta x} \tag{2.40}$$

where D_{coeff} has units of area/time; dC or ΔC represents change in concentration, ∂C is partial change in concentration and all three terms have units of mass/volume; dx or Δx or ∂x is the distance that the molecules permeate with respect to their initial location. The negative sign states that the molecules permeate in the direction of decreasing concentration (i.e., gradient is negative).

Rearranging Equation 2.38 and integrating with respect to the thickness of membrane ($0 \rightarrow x$) and concentration ($C_{outside} \rightarrow C_{inside}$) yield

$$\int_0^x J dx = -D_{coeff} \int_{C_{outside}}^{C_{inside}} dC \tag{2.41}$$

$$Jx = -D_{coeff}[C_{inside} - C_{outside}] = D_{coeff}[C_{outside} - C_{inside}] \tag{2.42}$$

$$\therefore J = \frac{D_{coeff}}{x}[C_{outside} - C_{inside}] \tag{2.43}$$

Equation 2.43 is a straight line equation with slope of (D_{coeff}/x), which approximates the permeability constant.

The mass flux has units of mass/time and surface area

$$\frac{x^2}{time} \times \frac{mass}{x^3} \times \frac{1}{x} = \frac{mass}{x^2 \times time} \tag{2.44}$$

Thus, according to the Fick's first law of diffusion, the mass flux is related to the concentration gradient over the diffusion distance. The mass flux may also be defined as the "rate of diffusion/area" or "rate of absorption/area." Thus, a higher concentration gradient leads to a faster rate of absorption and mass flux (Figure 2.3).

The rate of absorption of a compound by passive diffusion across a barrier with a thickness of L and known surface area according to the Fick's law of diffusion (Figure 2.3) is

$$\frac{dA}{dt} = -\left(\frac{D_{coeff} \times P_{coeff} \times Area}{x}\right)\Delta C = -P\Delta C \tag{2.45}$$

where *dA/dt* is the amount absorbed per unit of time that is, the rate of absorption, and ΔC is the concentration gradient of the compound with partition coefficient of P_{coeff} across a barrier with thickness of *x* and a defined surface *Area*. The permeability constant *P*, as presented by Equation 2.37 is equal to

$$P = \frac{D_{coeff} \times P_{coeff} \times Area}{x} \tag{2.46}$$

The unit analysis of the absorption rate is

$$\frac{dA}{dt} = -\left(\frac{D_{coeff}(x^2/Time) \times P_{coeff} \times Area(x^2)}{x}\right)\Delta C(Mass/x^3) = Mass/Time \tag{2.47}$$

Equation 2.46 can be applied to a number of absorptive processes through biological barriers. However, it is applicable only for the cases where the rate-limiting step is diffusion across the barrier. Equation 2.45 is also used to estimate the diffusion coefficient or diffusivity of a compound through a biological barrier

$$D_{coeff} = \frac{\left(\frac{dA}{dt}\right)(x)}{(P_{coeff})(Area)(\Delta C)} = \frac{(Mass/Time)(x)}{(x^2)(Mass/x^3)} = Area/Time \tag{2.48}$$

It is also important to recognize that not all biological barriers are the same. Using a gastrointestinal tract (GI) as an example, the driving force of diffusion is the concentration gradient between the GI tract and the systemic circulation. Considering the large volume of systemic circulation and the small volume of GI fluid, the concentration of drug (*mass/volume*) at the site of absorption will always be greater than the concentration of free drug in the systemic circulation that is, $C_{GI} \gg C_{plasma}$ and $\Delta C = C_{GI}$. Thus, the rate of absorption (Equation 2.45) for passive diffusion under this sink condition across the GI tract can be written as

$$\frac{dA}{dt} = -P\Delta C = -PC_{GI} \tag{2.49}$$

The sink condition exists for most, if not all, routes of administration with a biological barrier between the site of absorption and systemic circulation, that is, the concentration of xenobiotics at the site of absorption is always higher than its concentration in the systemic circulation. Under this condition, the rate of absorption (*dA/dt*) is equal to the permeability constant multiplied by the concentration at the absorption site, for example, Equation 2.49. The rate of absorption is then equal to a constant (permeability constant) multiplied by a variable (concentration), which is indicative of first-order kinetics.

2.2.1.5 Estimation of Apparent Permeability Constant Using Caco-2 Cell

The cultured Caco-2 cells, the human colon adenocarcinoma cell line (HTB-37), is used extensively for prediction of xenobiotic absorption in the intestine. They undergo spontaneous differentiation and polarization in culture to have an apical and basolateral surface with tight junctions and many normal functions of enterocytes. The culture is used in evaluation of paracellular and transcellular absorption (Hidalgo et al., 1989; Artursson, 1991; Balimane et al., 2000; Le Ferrec et al., 2001; Balimane and Chong, 2005; Sambuy et al., 2005; Farrell et al., 2012) and other features including CYP3A4 and P-glycoprotein expressions, and absorption of zwitterions (Pieksaritanont et al., 1998). Several theoretical relationships for permeation through *in vitro* biological barriers have been developed (Camenisch et al., 1998; Hubatsch et al., 2007; Farrell et al., 2012), which are based on Caco-2 cells and the assumption that permeation is mostly through paracellular and transcellular diffusion. The apparent permeability coefficient according to these observations is described as

$$P_{app} = \frac{V_{apical}}{Area \times (A_{apical})_{t=0}} \times \frac{dA_{basal}}{dt} \tag{2.50}$$

(P_{app}) is the apparent permeability constant (distance/time) for an apical donor compartment with volume of V_{apical} (cm^3), surface area of *Area* (cm^2), and an initial amount of $(A_{apical})_{t=0}$ in the donor compartment multiplied by (dA_{basal}/dt), which is the rate of change of amount in the basal receiver compartment. To estimate a realistic value for P_{app}, the amount in the basal compartment should exceed 10% of the apical compartment (Hubatsch et al., 2007), which is indicative of a constant concentration gradient.

The relationship between the apparent permeability constant and partition coefficient, or distribution coefficient, is often defined by the following sigmoidal relationship:

$$\log P_{app} = a \log D_{coeff} - b \log(1 + \beta D_{coeff}) + c \tag{2.51}$$

where a, b, β, and c are the data-related curve-fitting coefficients, D_{coeff} is the diffusion coefficient, which is often replaced by the distribution coefficient.

2.2.2 Carrier-Mediated Transcellular Diffusion

The facilitated diffusion, also known as passive mediated transport or carrier-mediated transcellular diffusion, refers to the absorption mechanism that is facilitated by transport proteins also known as carrier proteins, or absorbed through ion channels of biological barriers. The absorption is still a passive diffusion and the role of transport proteins is just to provide a passageway for the molecules to permeate across a barrier without requiring energy. This means that the diffusion is not due to the binding of xenobiotics to the membrane protein and subsequent energy requiring conformational change of the protein, but rather related to the kinetic energy of moving molecules through the passageway. Compounds that are quite insoluble in lipids and/or are larger than the barrier's pores may use this mechanism of absorption (Figure 2.4).

In contrast to passive diffusion, mass flux, J, for facilitated diffusion is not just directly proportional to the concentration gradient of the compound but rather to quantity of the protein (i.e., carrier) present for the absorption, the diffusion coefficient, and the rate of reversible interaction and splitting between the carrier and the compound, if such an interaction exists between the carrier and the compound. Any of these factors can act as a rate-limiting step for the absorption of compounds by carrier-mediated diffusion. This type of diffusion and its complex parameters mathematically require a boundary condition that deals with proteins that are stationary and do not move from the membrane with the compound. This scenario can only be defined by a nonlinear second-order differential equation that has no exact solution, and the numerical analysis of this type of equations mostly provides an approximation of actual diffusion.

Although the facilitated transport is a non-energy requiring process, and the xenobiotic molecules do not cross the barrier against the concentration gradient, there is a marked contrast with

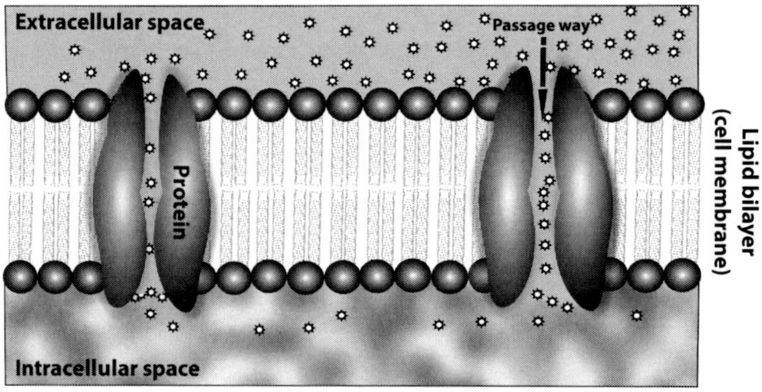

FIGURE 2.4 The facilitated transport, a non-energy requiring process, with dependency on the amount of carrier present in the barrier.

passive diffusion in that the process is dependent on the amount of carrier present in the barrier and, thus, a saturation phenomenon may occur. Therefore, the kinetics of absorption may become nonlinear at high concentrations of the compounds and influence the overall PK/TK of a compound.

The following relationship may clarify the distinction between the passive and carrier-mediated permeability (Usansky and Sinko, 2002). The relationship is based on the *in vitro* apical to basolateral transport in the brain microvessel endothelial cell (BMEC) monolayers. It is used to experimentally estimate the combined processes of permeability by passive diffusion and carrier-mediated transport (Figure 2.4)

$$P_{combined} = \frac{J_{max}}{K_M + C_{apical}} + P_{passive} \tag{2.52}$$

($P_{combined}$) is the combined permeability of passive diffusion and carrier-mediated transport; (J_{max}) and (K_M) are the maximum flux and the Michaelis–Menten constant for the carrier-mediated transport and ($P_{passive}$) is the permeability for the passive diffusion. When $K_M \gg C_{apical}$, the relationship is expressed as

$$P_{combined} = \frac{J_{max}}{K_M} + P_{passive} = (P_{intrinsic})_{carrier-mediated} + P_{passive} \tag{2.53}$$

It is worth noting that exogenous compounds have been shown to use a facilitated transport. Examples are the receptor-facilitated endocrine steroid cellular absorption and glycoprotein-facilitated intestinal transport of vitamin B_{12}.

2.3 TRANSCELLULAR DIFFUSION SUBJECT TO P-GLYCOPROTEIN EFFLUX

P-Glycoprotein (Pgp) is a member of the ATP-Binding Cassette (ABC) transporter superfamily, capable of transporting a wide range of structurally unrelated xenobiotics from the intracellular environment into the extracellular space (Gatmaitan and Arias, 1993; Schinkel, 1997). The ABC superfamily represents the largest family of transmembrane proteins. Pgp was first isolated from Pgp-overexpressing MDR tumor cells. However, detectable amounts of this protein have also been identified in the normal tissues and organs of several mammalian species including rats and humans (Thiebaut et al., 1987; Cordon-Cardo et al., 1990). Subsequent immunochemical studies on Pgp has confirmed the existence of highly expressed Pgp family members in the adrenal cortical cells, the brush border of the renal proximal tubule epithelium, the luminal surface of biliary hepatocytes, small and large intestine mucosal cells, and pancreatic ductules (Thiebaut et al., 1987, 1989; Cordon-Cardo et al., 1990). The capillary endothelial cells of the brain, testis, heart, placenta, lung, prostate, and stomach have been shown to express Pgp, although to lower extent (Matsuoka et al., 1999; Regina et al., 2002; Breedveld et al., 2005; Leslie et al., 2005; Kodaira et al., 2010). In the GI tract, in particular the small intestine, Pgp is expressed on the brush-border membrane of entrocytes where it pumps xenobiotics out of the cytosol into the lumen environment of the intestine. Thus, the efflux role of Pgp counteracts the absorption of xenobiotics in GI tract and reduces the rate and extent of absorption. This section focuses only on the influence of Pgp on the absorption and permeability models of xenobiotics and highlights the role it plays in PK/TK evaluation and bioavailability studies, which will be discussed later in this book.

2.3.1 Pgp Structure and Function

The following is a brief review of structural characteristics and function of Pgp to elucidate the influence of this protein on the rate and extent of absorption of xenobiotics. Pgp is a 170 kDa glycosylated transmembrane protein, composed of about 1200 amino acids (1280 in human). Studies on human Pgp have shown that this protein is synthesized as a 150 kDa glycosylated intermediate in the endoplasmic reticulum. Modification of the carbohydrates of this intermediate molecule in Golgi apparatus results in the mature protein, and glycosylation occurs at 91, 94, and 99 positions of the first transmembrane loop (Loo and Clarke, 1998, 1999, 2001; Loo et al., 2003). Reversible phosphorylation has been postulated to have an effect on the function of Pgp. A linker region between the first nucleotide binding site and the second transmembrane domain of the protein has been identified as a potential phosphorylation site. The linker region contains a number of serine residues that are phosphorylated by protein kinase C and/or protein kinase A in human (Chambers, 1998; Dong et al., 1998; Cordon-Cardo et al., 1990). Primary structure studies using electron microscopy and image analysis techniques suggest four domains for the polypeptide. There are two hydrophobic transmembrane domains and two nucleotide-binding domains located at the cytoplasmic site of membrane (Figures 2.5 and 2.6).

Each of the hydrophobic domains consists of six membrane-spanning segments separated by hydrophilic loops and each contains either the amino or the carboxylic terminal of the polypeptide chain (Rosenberg et al., 1997, 2003; Jones and George, 1998; Loo and Clarke, 1999). The proposed structural studies identified 14 anti-parallel transmembrane β-strands and 5 intracellular α-helices for each half of the hydrophobic domains as well as two β strands and one α helix for each nucleotide binding domain (Jones et al., 1998) According to this model, each half of the molecule can be

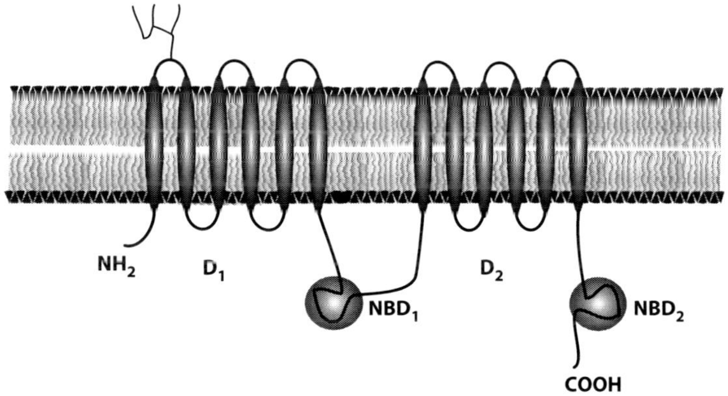

FIGURE 2.5 Portrayal of Pgp structure with two hydrophobic domains (D1 and D2) and two nucleotide-binding domains NBD1 and NBD2.

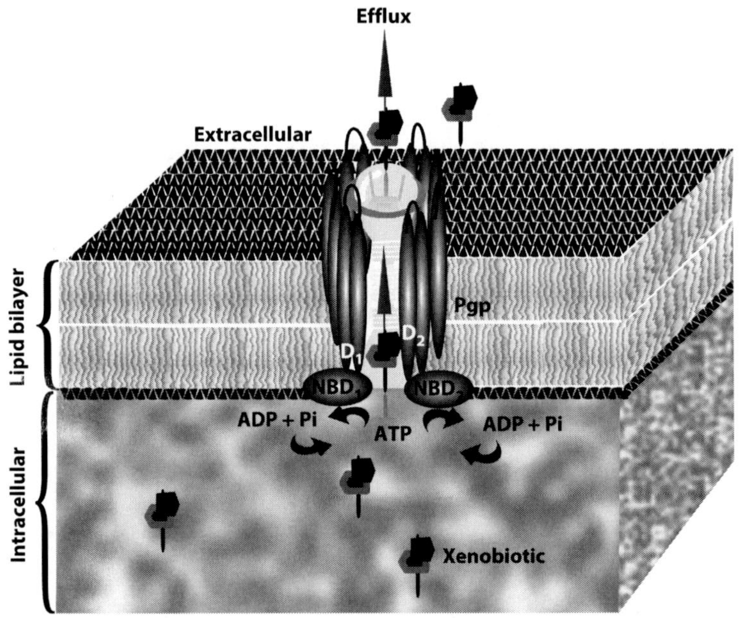

FIGURE 2.6 Schematic configuration of Pgp within the lipid bilayer of biological membrane, the two nucleotide binding domains (NBD1, NBD2) are partially embedded in the lipid bilayer.

assembled as a 16 β-sheet transmembrane barrel connected to a 6 α-helix bundle by short loops. Charged residues of the hydrophobic domains are often located immediately at the extracellular site of the membrane, which enables them to interact with the negative phospholipid head groups in the bilayer or with the cations capable of establishing an association with them.

Using an electron microscopy (Rosenberg et al., 1997, 2001, 2003) the overall shape of Pgp is reported to be similar to a cylinder of about 10 nm in diameter with a maximum height of about 8 nm. In comparison to the thickness of the lipid bilayer, which is about 4 nm, it was suggested that about one-half of the Pgp molecule is embedded in the membrane. Studies on Pgp structure using fluorescent resonance energy transfer (FRET) also suggested a cylindrical structure of 10 nm wide and 8 nm deep for the with two L-shaped nucleoside binding domains (NBD1, NBD2). This model also identifies two non-identical putative drug binding sites (D1, D2) within the transmembrane region of Pgp, which interact with Pgp substrates and modulators (Sharom, 1995; Sharom et al., 1998, 1999; Wang et al., 2000; Marcoux et al., 2013). The two nucleotide binding domains (NBD1, NBD2) demonstrate ATPase activity, which is considered essential in the efflux activity of Pgp. Studies on the Pgp structure using FRET have also shown that the two nucleotide binding domains (NBD1, NBD2) are partially embedded in the lipid bilayer in its cytosolic side (Figure 2.6).

The two non-identical putative drug binding sites (D1 and D2) within the transmembrane region are large enough to accommodate at least two structurally different drug substrates simultaneously. The interaction of xenobiotics with D1 and D2 is not fully clarified yet. Using a cysteine-free human Pgp it has been shown that there are two conformational changes: (1) when there is a central cavity between the N- and C-terminal halves of the molecule, and (2) when the two halves have moved sideways closing the cavity (Lee et al., 2008). The consensus is that dimerization of NBDs driven by nucleotide binding is an essential step in the efflux cycle and the details of how this is linked to ATP hydrolysis and substrate transport investigated remains yet unclear (Siarheyeva et al., 2010). Although the two NBDs appear to be located close to one another, they do not appear to be tightly associated and, thus, this structure does not provide any additional information on their mode of interaction during catalysis (Aller et al., 2009).

Two types of Pgp have been identified in humans. Type I, termed multidrug resistance type 1 Pgp (MDR1Pgp, encoded by the MDR1 gene) and type II, MDR2Pgp (encoded by MDR2 gene). In rodents, the genes are identified as *mdr1a* and *mdr1b*, which together are expressed roughly as the single MDR1 gene in humans and perform the same set of function as MDR1Pgp.

Pgp at the absorption site having binding sites for interaction with a variety of compounds, including its inhibitors/modulators, co-administration of multiple drugs, or exposure to various compounds may enhance or reduce the absorption or bioavailability of compound of interest. For example, it is well documented that the oral bioavailability of fexofenadine increases significantly when ketoconazole or erythromycin, known inhibitors of Pgp, is co-administered in humans. In addition to absorption, Pgp has significant influence on the disposition of drugs in the body, for example, reduction of biliary elimination of vincristine when it is co-administered with verapamil (another known inhibitor of Pgp). However, several *in vitro* and *in vivo* studies have shown that coadministration of doxorubicin and Pgp modulators such as verapamil, cyclosporine A and amiodarone results in an increase in the cardiotoxicity of doxorubicin. There are numerous xenobiotics that can act as modulator/inhibitor of Pgp; among them are flavonoids and other polyphenols, vinca alkaloids and taxenes. As it was indicated earlier, Pgp is involved in secretory transportation of its substrates through its strategic location in the apical membrane of the proximal tubule of nephrons (Ernest and Bello-Reuss, 1998). Thus, the inhibitors of this transporter can significantly decrease renal excretion of its substrates, and consequently have an impact on pharmacokinetics and pharmacodynamics of xenobiotics and are considered an important factor in drug-drug interaction. The synergistic role of metabolic enzyme (CYP 3A4) and Pgp in the small intestine is well documented and recognized as a major factor in limiting oral absorption and bioavailability of drugs.

2.3.2 Pgp Computational Equations

The Pgp-mediated *in vitro* efflux rate in cell culture is governed by the following equations, which represent the saturable nature of the efflux (Kuh et al., 2000; Jang et al., 2001):

$$J_{Pgp-mediated} = \frac{J_{max} \times C_{f_{cell}}}{K_{M_{Pgp}} + C_{f_{cell}}} \tag{2.54}$$

where the efflux rate per cell is $J_{Pgp-mediated}$, the maximum efflux rate per cell is J_{max}; $K_{M_{Pgp}}$ represents the dissociation constant of Pgp-mediated efflux; and $C_{f_{ree}}$ is the free concentration of xenobiotic in cells. The mass balance equations for the amount inside the cells and in medium as a function of time are

$$\frac{dA_{total_{cells}}}{dt} = V_{cells} \times \frac{dC_{total_{cells}}}{dt} + C_{total_{cells}} \times \frac{dV_{cells}}{dt} \tag{2.55}$$

where the rate of change in total amount (i.e., bound and free) as a function of time is $(dA_{total_{cells}}/dt)$, the total volume of cells is (V_{cells}) and the total concentration of free and bound is $(C_{total_{cells}})$. Equation 2.55 assumes the volume of cells changes with time.

Given that *Rate* = *Clearance* × *Concentration*, the equation can also be written as

$$\frac{dA_{total_{cells}}}{dt} = \left[Cl_f \times C_{f_{medium}} - Cl_f \times C_{f_{cell}} - \frac{J_{max} \times C_{f_{cell}}}{K_{M_{Pgp}} + C_{f_{cell}}} \right] \times N_{cells} \tag{2.56}$$

where the clearance of free xenobiotic by passive diffusion is (Cl_f), the free concentration of the compound in the medium and inside the cell is ($C_{f_{medium}}$) and ($C_{f_{cell}}$), respectively, and N_{cells} is the cell number.

The rate of change in the total amount of compound in the medium is

$$\frac{dA_{total_{medium}}}{dt} = V_{medium} \times \frac{dC_{total_{medium}}}{dt} \tag{2.57}$$

where ($C_{total_{medium}}$) represents the total concentration (free and bound) in the medium and (V_{medium}) is the volume of the medium.

In comparison with Equations 2.55 and 2.56, the change in the volume of the medium is assumed negligible, and in terms of clearance Equation 2.57 can be written as

$$\frac{dA_{total_{medium}}}{dt} = \left[-Cl_f \times C_{f_{medium}} + Cl_f \times C_{f_{cell}} + \frac{J_{max} \times C_{f_{cell}}}{K_{M_{Pgp}} + C_{f_{cell}}} \right] \times N_{cells} \tag{2.58}$$

The rate equations were further expanded to define the time-dependent changes in intracellular and extracellular drug concentration as a function of cell volume, binding affinity and capacity, and Pgp-mediated efflux (Kuh et al., 2000; Jang et al., 2001) as follows.

The total concentration (bound and free) in the cells, $C_{total_{cells}}$, and in the medium, ($C_{total_{medium}}$), as functions of free concentration in the cells, ($C_{f_{cell}}$), and the medium, ($C_{f_{medium}}$), can be defined as

$$C_{total_{cells}} = C_{f_{cells}} + \frac{B_{max_{cells}} \times C_{f_{cells}}}{K_{M_{cells}} + C_{f_{cells}}} + B_{NS} \times C_{f_{cells}} \tag{2.59}$$

$$C_{total_{medium}} = C_{f_{medium}} + \frac{B_{max_{medium}} \times C_{f_{medium}}}{K_{M_{medium}} + C_{f_{medium}}} \tag{2.60}$$

($B_{max_{cells}}$) is the maximum binding to the cellular components; ($K_{M_{cells}}$) is the corresponding Michaelis–Menten constant; (B_{NS}) stands for binding to the cells' nonsaturable binding sites. ($B_{max_{medium}}$) is the maximum binding to the medium and $K_{M_{medium}}$ is the medium Michaelis–Menten constant.

The total volume of cells (V_{cells}) is estimated by Equation 2.61

$$V_{cells} = V_{cell} \times N_{cells} \tag{2.61}$$

where (V_{cell}) is the volume of one cell and (N_{cells}) is the number of cells at time t. Equation 2.61 is also defined as

$$V_{cells} = (V_{cell} \times N_{t=0})e^{(k_{cell} \times t)} \tag{2.62}$$

where the initial number of cells is ($N_{t=0}$) and, depending on the type of xenobiotic and its influence on the cells, the rate constant for changes in cell number, k_{cell}, can be positive when the xenobiotic mediates cell proliferation, and negative when the compound reduces the number of cells. For compounds when there is a linear enhancement in the number of cells with respect to time, the relationship between $B_{max_{cells}}$ at time t and zero can be defined as (Kuh et al., 2000)

$$(B_{max_{cells}})_t = (B_{max_{cells}})_0 \times (1 + k_{B_{max_{cells}}} \times t) \tag{2.63}$$

where the cellular maximum binding sites at time t and 0 are $(B_{max_{cells}})_t$ and $(B_{max_{cells}})_0$, respectively. The rate constant for increases in the cellular binding site is $(k_{B_{max_{cells}}})$, which is a concentration-dependent rate constant.

Substitution of Equations 2.59–2.60 and 2.62–2.63 in Equations 2.56 and 2.58 yields the following relationships, which define the time-course of intracellular and extracellular concentrations of a xenobiotic as a function of Pgp-mediated efflux binding capacity and affinity, and total cellular and medium volumes:

$$\frac{dC_{total_{cells}}}{dt} = \left[Cl_f \times X - Cl_f \times Y - \frac{J_{max} \times Y}{K_{M_{Pgp}} + Y}\right] \times \frac{1}{V_{cell}} - (k_{cell} \times C_{total_{cells}}) \tag{2.64}$$

$$\frac{dC_{total_{medium}}}{dt} = \left[-Cl_f \times X + Cl_f \times Y + \frac{J_{max} \times Y}{K_{M_{Pgp}} + Y}\right] \times \frac{N_{t=0}e^{k_{cell} \times t}}{V_{medium}} \tag{2.65}$$

where

$$X = \left[-(K_{M_{medium}} + B_{max_{medium}} - C_{total_{medium}}) + \sqrt{\begin{array}{l}(K_{M_{medium}} + B_{max_{medium}} - C_{total_{medium}})^2 + 4 \\ \times K_{M_{medium}} \times C_{total_{medium}}\end{array}}\right] \bigg/ 2 \tag{2.66}$$

$$Y = \frac{-Z + \sqrt{Z^2 + 4(1 + B_{NS}) \times K_{M_{cells}} \times C_{total_{cells}}}}{2(1 + B_{NS})} \tag{2.67}$$

where

$$Z = (1 + B_{NS}) \times K_{M_{cells}} + (B_{max_{cells}})_0 \times (1 + K_{M_{cells}} \times t) - C_{total_{cells}} \tag{2.68}$$

The following are Pgp-independent parameters of the above relationships:

$$B_{max_{medium}}, B_{max_{cells}}, K_{M_{medium}}, K_{M_{cells}}, B_{NS} \text{ and } Cl_f \tag{2.69}$$

Thus, the Pgp-mediated efflux rate can be defined as

$$Rate_{Pgp-efflux} = Cl_f \times (C_{free_{medium}} - C_{free_{cells}}) = \frac{J_{max} \times C_{f_{cells}}}{K_{M_{Pgp}} + C_{f_{cells}}} \tag{2.70}$$

The efflux clearance of xenobiotic from intracellular environment through Pgp efflux is

$$Cl_{Pgp} = \frac{J_{max}}{K_{M_{Pgp}} + C_{f_{cells}}} \tag{2.71}$$

The overall influence of Pgp efflux on *in vivo* PK/TK will considered in Chapters 12 and 15.

2.4 ACTIVE TRANSPORT

Active transport generally refers to the transport of molecules against the concentration gradient. It requires interaction with protein carriers and the expenditure of energy. Thus, the process is an endergonic one that must be joined with an exergonic process, namely the hydrolysis of ATP (Figure 2.7). Both processes occur in the presence of enzymes and are inhibited with toxins such as dinitrophenol and fluoride. The active transport occurs only at certain sites such as the intestines, liver, lung, and kidneys. Due to the energy-requiring characteristic of the process and limited concentration of protein carriers, the active transport is apt to produce saturation occurrence. Most often, the kinetics of active absorption follow the Michaelis–Menten equation and exhibit nonlinear characteristics. Furthermore, the protein carriers are subject to competitive inhibition by compounds with similar structure.

$$\frac{dC}{dt} = \frac{(dC/dt)_{max} \times C}{K_M + C} \tag{2.72}$$

where (dC/dt) is the rate of absorption, $(dC/dt)_{max}$ is the maximum rate of absorption, K_M is the Michaelis–Menten constant, and C is the concentration at the site of absorption.

Depending upon the formation and use of ATP, the active absorption is identified as either primary or secondary active transport. The primary active transport is referred to the movement of ions across biological barrier by protein transporters that combine the transport of the substrate with hydrolysis of ATP (Figure 2.7). Calcium pumps, sodium pumps, and proton pumps are examples of primary active transport.

The secondary active transport occurs when the transport of ions by primary active transport stores and releases enough energy that forms ion gradients and transports other ions across the

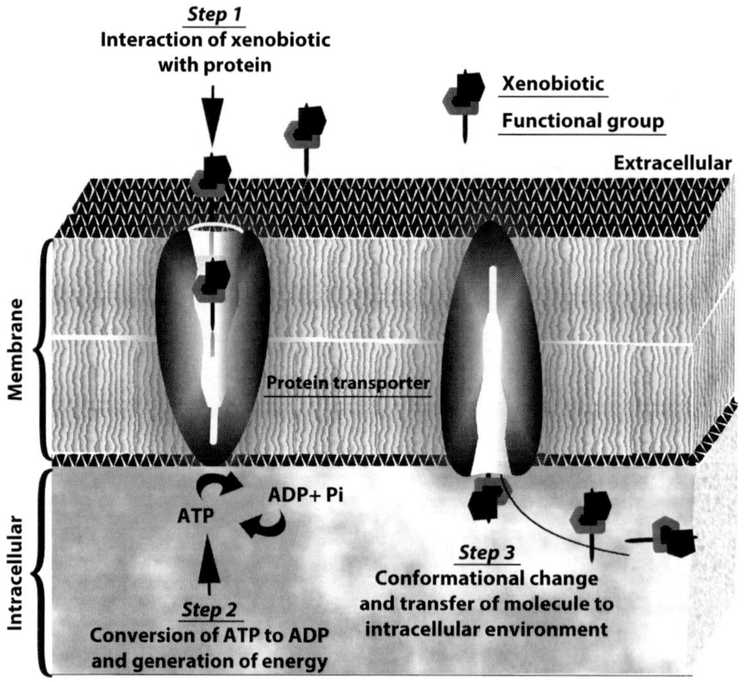

FIGURE 2.7 The stepwise energy requiring active transport of molecules against the concentration gradient through the biological barriers.

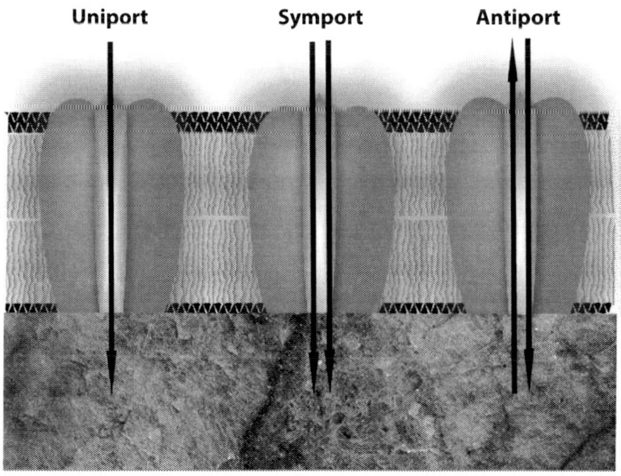

FIGURE 2.8 The uniport, symport, or antiport direction of secondary active absorption.

barrier. This is also called the co-transport process. The direction of the secondary transport can be the same or opposite that of the primary transport and is identified as uniport, symport, or antiport (Figure 2.8). A well-known example of secondary transport is the co-transport of glucose and sodium in the small intestine.

The GI tract is the main site of active absorption in the body. The pathways of active transport in the intestine are structure-specific and exist for the purpose of transferring certain nutrients such as uracil, choline, folate derivatives, bile salts, monosaccharides, l-amino acids, certain vitamins, etc. Xenobiotics that are structurally similar to these nutrients use the same pathways for absorption. For example, the anticancer drug 5FU (5-Fluorouracil) uses the uracil pathway and methotrexate, another anticancer drug, uses the folate derivative pathway, examples include iron salts, levodopa, propylthiouracil, and fluorouracil. Active transport is more common for the disposition of endogenous compounds such as glucose, amino acids, and certain ions.

The intestinal peptide transport, known as PepT1, has a wide range of substrate specificity and transports, not only di- and tri-peptide of food products (Mathews and Adibi, 1976; Furst et al., 1990, Gilbert et al., 2008), but also a number of xenobiotics with the structure similar to peptides, for example, ACE (angiotensin converting enzyme) inhibitors (Kim et al., 1994; Covitz et al., 1996; Moore et al., 2000) or beta-lactam antibiotics (Snyder et al., 1997). Substrate mimicry, that is, development of a therapeutic agent with a 3 D structure similar to the substrate of PepT1, is an approach for the development of new therapeutic peptides.

The apical sodium-dependent bile acid transporter, ASBT, is another transporter involved in the re-absorption of bile acid. It is worth noting that from the total amount of bile acids in a human (about 3–5 g), which circulates 6–10 times a day with a turnover of 20–30 g, only 0.2–0.5 g eliminates through fecal elimination (Zhang et al., 2002). The amount eliminated is replenished by the synthesis of new molecules. ASBT plays an important role in maintaining the balance of bile acids and governs the enterohepatic re-circulation of bile acids (Hofmann, 1976, 1989; Erlinger, 1987). Compounds that are structurally similar to bile acids are also absorbed via this transporter.

The molecular mechanism of primary active transport at the protein level has not been fully elucidated. There are, however, certain postulations that have been accepted as plausible scenarios. For example, a protein transporter may interact with a xenobiotic molecule, undergo the energy-requiring conformational change, rotate 180°, and release the molecule at the opposite side (Figure 2.7). Another scenario is that when the protein is a large molecule with multiple binding sites and a functional group across its surface, it is capable of shuttling the molecule between the functional group to the opposite side.

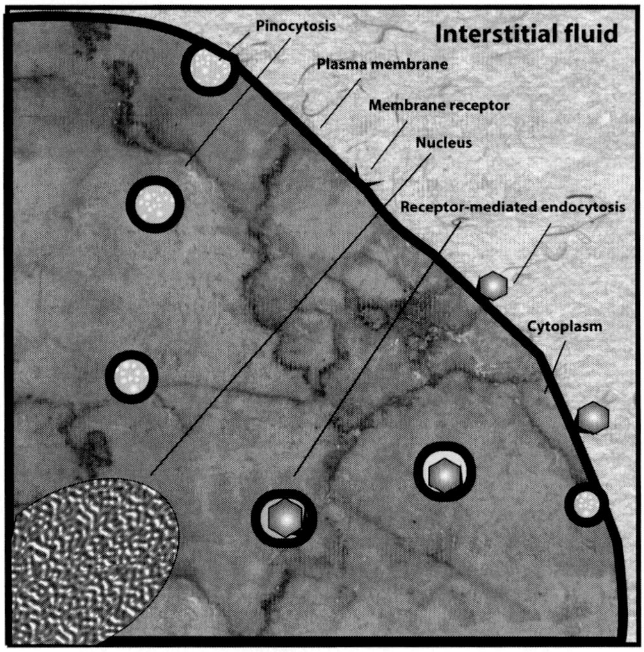

FIGURE 2.9 Illustration of differences between pinocytosis, a non-selective transport, and receptor-mediated endocytosis, a receptor selective process.

2.5 ENDOCYTOSIS AND PINOCYTOSIS

In general, both of these processes are initiated by the cell barrier engulfing the compound/particles, then separating from the barrier, forming a vesicle, and transfering toward lysosomes and lytic vacuoles for assimilation (Figure 2.9). The endocystosis of large particles/entities, known as phagosome or vacuole, is referred to as phagocytosis. The endocytosis of small particles in solution forming small vesicles is known as pinocytosis. Pinocytosis, known as "cell-drinking," is a non-selective physical absorption that forms small vesicles inside the cell from extracellular fluids before interacting with lysosomal acidic hydrolases and breaking down into smaller particles. Pinocytosis occurs continuously in nearly all cells. For small particles to be absorbed by pinocytosis, they must be associated with the extracellular fluid. The process, from forming the vesicles to breaking down by lysosomes, is an energy requiring absorption. Compounds such as proteins/polypeptides and particulate entities like therapeutic or non-therapeutic nanoparticles, lipophilic carriers such as small unilamellar liposomes, or droplet of micro-emulsion, etc., may cross the membrane by pinocytosis.

Distinction should be made between pinocytosis, which is a non-selective transport, and the receptor-mediated endocytosis, which is a selective process and depends on the specific role of receptor on the membrane (Figure 2.9).

2.6 SOLVENT DRAG, OSMOSIS, AND TWO-PORE THEORY

The transfer of appropriately sized xenobiotic molecules by solvent through the pores and loosely connected biological barrier is referred to as solvent drag. The basic principles of solvent drag are somewhat intertwined with the principles of osmosis and the two-pore theory (Kedem and Katchalsky, 1958; Rippe and Haraldsson, 1994; Hammel and Schlegel, 2005). The following equation (Kedem and Katchalsky, 1985) shows the link between the solvent drag and osmosis:

$$J_S = L_p\Delta_p - \sigma L_p\Delta\pi \tag{2.73}$$

where J_S is the solvent flux; L_p if the filtration coefficient of the solvent; Δ_p and $\Delta\pi$ are the hydrostatic and osmotic pressure differences, respectively; σ is a coefficient of reflection used in estimation the solvent drag coefficient as described below.

The flux of solute (J_{solute}) by the solvent drag is defined as

$$J_{solute} = J_S(1-\sigma)(C_{ave})_{solute} + P_{solute}\Delta\pi \tag{2.74}$$

The coefficient of solvent drag is $(1 - \sigma)$; the average concentration of solute is $(C_{ave})_{solute}$; the solute permeability is P_{solute} estimated by the ratio of the apparent transcapillary diffusion coefficient to the diffusion distance; and $\Delta\pi$ is the osmotic pressure differences across the barrier, which also represents the solute concentration gradient, that is,

$$\begin{aligned} J_{solute} &= J_S(1-\sigma)(C_{ave})_{solute} + P_{solute}\Delta C \\ \Delta C &= (C_{outside} - C_{inside}) \end{aligned} \tag{2.75}$$

Integration of Equation 2.75 across the membrane according to the boundary conditions yields the following equation:

$$J_{solute} = J_S(1-\sigma)\frac{C_{outside} - C_{inside}e^{Pe}}{1-e^{-Pe}} \tag{2.76}$$

where (Pe) is a modified Peclet number equal to

$$Pe = \frac{J_S(1-\sigma)}{P_{solute}} \tag{2.77}$$

Therefore, the solvent flux is approximately equal to

$$J_S = \frac{Pe \times P_{solute}}{1-\sigma} \tag{2.78}$$

When Pe is larger than 3, the transport of the solute is mainly by solvent convection.

Thus, one can conclude that the flux of a solute absorbed by passive paracellular diffusion and solvent drag is the sum of both fluxes. The microvascular endothelium permeability has been suggested to have small pores, represented by clefts in the intracellular junctions. In contrast, macromolecules use a group of large pores, which are either real openings between cells or intracellular vesicular systems (Pappenheimer et al., 1951; Eleni et al., 2011). In fact, most of the macrovascular walls have two types of pores: (1) the small functional pores that are abundant, but restrictive toward large molecules such as proteins; and (2) non-size selective pathways that are scarce, but permit the passage of large molecules (Rippe and Haraldsson, 1994). Large pores are predicted to make up 1 part per 10,000 to 30,000 of the total population of pores. The convection through the large pore size is considered an essential element of the permeation of proteins by solvent drag.

In general, the role of solvent drag is considered more significant for the flux of large molecules than the small ones. The reason is the significant diffusive flux of small molecules compared to the solvent drag movement and the related velocity of solute in one direction.

An interesting but unresolved aspect of the solvent drag is whether the back flux of an absorbed drug from blood into intestinal lumen can be ignored while the back flux of water cannot be ignored.

2.7 ION-PAIR ABSORPTION

The pH-partition theory postulates that only unionized lipophilic compounds are able to penetrate through the biological barriers and ionized forms of drugs cannot readily diffuse through the barrier. However, it can occur that a drug is completely ionized at the site of absorption in the GI tract, yet it is significantly bioavailable. This observation cannot be explained by the pH-partition theory. Thus, it has been postulated that such absorption can be a consequence of the ion-pair formation at the site of absorption (Jonkman and Hunt, 1983; Neubert et al., 1987; Neubert, 1989; Härtl et al., 1990). This hypothesis has been a highly debated subject in the literature. According to this hypothesis, oppositely charged molecules may interact and form a neutral complex. The neutral complex will then cross the biological barrier by passive diffusion. Since the formation of a neutral complex is a simple chemical equilibrium, an excess of one ion can increase the formation of the complex. Thus, the formation and absorption of the complex depends on the concentration of one or both ions and this dependency contributes to an erratic absorption of drug (Neubert et al., 1988). This means that the formation of the complex is the rate-limiting step in the absorption process. The breakdown of the complex is also based on the chemical equilibrium. Once the neutral complex is inside the membrane a new equilibrium will be established and all three species of positive, negative, and neutral will be present in the membrane at the same time (Irwin et al., 1969; Lippold and Schneider, 1974; Lippold and Lettenbauer, 1980; van der Giesen and Janssen, 1982; Boroujerdi, 1987; Neubert et al., 1988). As the charged molecules leave the barrier, more of the neutral complex breaks down into the charged molecule. The absorption of the neutral complex is by passive transcellular or paracellular diffusion.

Thus, one can assume that the ion-pair formation increases the lipophilicity of hydrophilic ionized compounds and enhances their partition coefficient. It has been shown that the partition coefficient of hydrophilic compounds like buformine, bretylium, and pholedrine is markedly increased by lipophilic ions such as hexylsalicylate or, the lipophilicity of doxorubicin is enhanced in the presence of the counter ions diotylsulfosuccinate and dodecylsulfonate in the low pH range. There are a number of other counter ions, which have shown to increase the lipophilicity of ionized molecules. However, the argument against use of these compounds is that they are too harsh for *in vivo* applications. Among the counter ion compounds are: alkylcarboxylase, cholate, *n*-alkylsulfates, *n*-alkylcarbonates, and trichloracetate, which are considered absorption enhancers of hydrophilic quaternary ammonium compounds. The application of this hypothesis in absorption of peptides from different sites (Quintanar-Guerrero et al., 1997; Soares et al., 2007), enhancement of prodrugs absorption (Samiei et al., 2013) and antiviral drugs (Miller et al., 2010) has renewed the interest in ion-pair formation as a means of enhancing absorption of ionized molecules.

The following are the relationships of lipophilic behavior of ion-pair complex species according to pH-partition theory in a simple model of the aqueous-organic phase system.

The dissociation constant of reversible interaction of an ionized drug with the ion pair agent to form an ion-pair complex in the aqueous phase is (van der Giesen and Janssen, 1982)

$$K_{[AB]_w} = \frac{[A^-]_w[B^+]_w}{[AB]_w} \tag{2.79}$$

The dissociation constant in the organic phase is

$$K_{[AB]_o} = \frac{[A^-]_o[B^+]_o}{[AB]_o} \tag{2.80}$$

where $[A^-]_w$ and $[A^-]_o$ are the concentration of the ionized drug in aqueous phase (w) and organic (o), respectively.

$[B^+]_w$ and $[B^+]_o$ are the positively charged ion-pair agents; $[AB]_w$ and $[AB]_o$ are the ion-pair complexes.

The partition coefficient of the ion-pair complex between the two layers can be defined as

$$P_{AB} = \frac{[AB]_o}{[AB]_w} \tag{2.81}$$

When the ion-pair agent is dissociated completely in the aqueous phase into charged ions, the apparent partition coefficient, $(P_{coeff})_{app}$, is

$$(P_{coeff})_{app} = \frac{[AB]_o + [A^-]_o}{[A^-]_w} \tag{2.82}$$

To estimate $(P_{coeff})_{app}$ in terms of measurable quantities $[B^+]_w$ and $[A^-]_w$, it would be convenient to use a quantity, known as extraction constant, K_{ext}, which is the ratio of $(P_{coeff})_{app}/K_{[AB]_w}$ (Equations 2.64 and 2.67)

$$K_{ext} = \frac{(P_{coeff})_{app}}{K_{[AB]_w}} = \frac{[AB]_o}{[B^+]_w[A^-]_w} \tag{2.83}$$

Combining Equations 2.65 and 2.68 gives rise to

$$[A^-]_o[B^+]_o = K_{ext}K_{[AB]_o}[B^+]_o[A^-]_w \tag{2.84}$$

When the ion-pair complex is the only group of molecules that can pass through the barrier, then $[B^+]_o = [A^-]_o$ and Equation 2.70 can be written as

$$[A^-]_o^2 = (K_{ext}K_{[AB]_o}[B^+]_w[A^-]_w) \tag{2.85}$$

$$\therefore [A^-]_o = \sqrt{K_{ext}K_{[AB]_o}[B^+][A^-]_w} \tag{2.86}$$

Combining Equations 2.82 (P_{app}), 2.83 (K_{ext}), and 2.86 ($[A^-]_o$) and solving for P_{app} gives rise to the following relationships:

$$P_{app} = [B^+]_w\left(K_{ext} + \left(K_{ext} \times K_{[AB]_o} \times [B^+]^{-1} \times [A^-]_w^{-1}\right)^{1/2}\right) \tag{2.87}$$

If $K_{[AB]_o} = 0$, then

$$P_{app} = [B^+]_w K_{ext} \tag{2.88}$$

When $K_{[AB]_o} \neq 0$, the equation can be written in terms of known quantities of the system

$$P_{app}[B^+]_w^{-1} = K_{ext} + \left(K_{ext} \times K_{[AB]_o} \times [B^+]_w^{-1} \times [A^-]_w^{-1}\right)^{1/2} \tag{2.89}$$

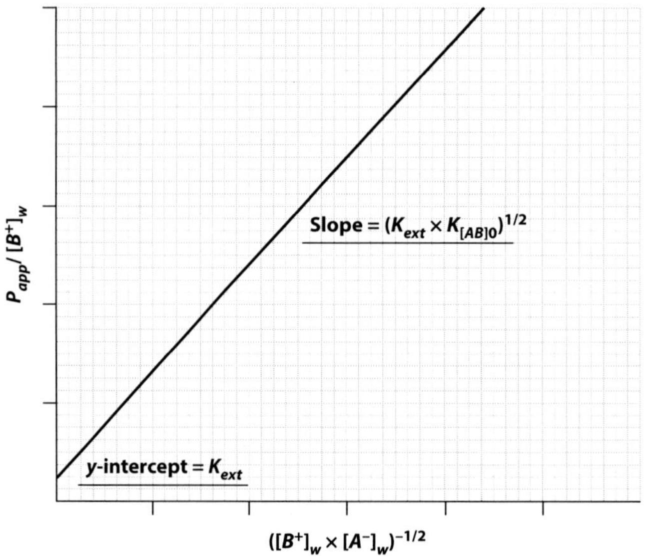

FIGURE 2.10 Plot of normalized apparent partition coefficient with respect to positively charged ion-pair agent against reciprocal of square root of drug molecule and ion-pair agent for the purpose of estimating the ion-pair complex in aqueous and lipid layers from known values of the experiment.

A plot of $P_{app}[B^+]_w^{-1}$ against $([B^+]_w \times [A^-]_w)^{-1/2}$ should generate a straight line with the y-intercept of K_{ext} and slope of $(K_{ext} \times K_{[AB]_0})^{1/2}$ (Figure 2.10). This approach has been used successfully to estimate the important parameters of ion-pair complex lipophilicity for compounds such as 4-hydroxycumarin.

REFERENCES

Abraham, D. J., Kellogg, G. E. 1994. The effect of physical organic properties on hydrophobic fields. *J Comput Aided Mol Des* 8:41–49.

Aller, S. G., Yu, J., Ward, A., Weng, Y., Chittaboina1, S., Zhuo, R., Patina M. Harrell, P. M., Trinh, Y. T., Zhang, Q., Urbatsch, I. L., Chang, G. 2009. Structure of P-glycoprotein reveals a molecular basis for poly-specific drug binding. *Science* 323(5922):1718–22.

Artursson, P. 1991. Cell cultures as models for drug absorption across the intestinal mucosa. *Crit Rev Ther Drug Carrier Syst* 8:305–330.

Avdeef, A. 1993. Part II: Refinement of partition coefficients and ionization constants of multiprotic substances. *J Pharm Sci* 82(2):183–90.

Balimane, P. V., Chong, S. 2005. Cell culture-based models for intestinal permeability: A critique. *Drug Discov Today* 10:335–343.

Balimane, P. V., Chong, S., Morrison, R. A. 2000. Current methodologies used for evaluation of intestinal permeability and absorption. *J Pharmacol Toxicol Methods* 44:301–12.

Beckett, A. H., Triggs, E. J. 1967. Buccal absorption of basic drugs and its application as an *in vivo* model of passive drug transfer through lipid membranes. *J Pharm Pharmacol* 19:31S–41S.

Beckett, A. H., Boyes, R. N., Triggs, E. J. 1968. The kinetics of buccal absorption of amphetamines. *J Pharm Pharmacol* 20:92–97.

Beckett, A. H., Moffatt, A. C. 1970. Kinetics of buccal absorption of some carboxylic acids and correlation of the rate constants and n-heptane aqueous phase partition coefficient. *J Pharm Pharmacol* 22:15–19.

Betageri, G. V., Rogers, J. A. 1988. The liposome as a distribution model in QSAR studies. *Int J Pharm* 46:95–102.

Bodor, N., Buchwald, P. 1997. Molecular size based approach to estimate partition properties for organic solutes. *J Phys Chem* B101:3404–12.

Bodor, N., Huang, M. J. 1992. An extended version of a novel method for the estimation of partition coefficient. *J Pharm Sci* 81(3):272–81.

Bonn, D., Kegel, W. K. 2003. Stokes–Einstein relations and the fluctuation-dissipation theoremin a supercooled colloidal fluid. *J Chem Phys* 118(4):2005–09.

Boroujerdi, M. 1987. Kinetics of ion-pair absorption. *Drug Dev Ind Pharm* 13(1):181–19.

Bourdet, D. L., Pollack, G. M., Thakker, D. R. 2006. Intestinal absorptive transport of the hydrophilic cation ranitidine: A kinetic modeling approach to elucidate the role of uptake and efflux transporters and paracellular vs. transcellular transport in Caco-2 cells. *Pharm Res* 23:1178–87.

Breedveld, P., Pluim, D., Cipriani, G., Wielinga, P., van Tellingen, O., Alfred, H., Schinkel, A. H., Schellens, J. H. M. 2005. The effect of Bcrp1 (Abcg2) on the *in vivo* pharmacokinetics and brain penetration of imatinib mesylate (gleevec): Implications for the use of breast cancer resistance protein and P-glycoprotein inhibitors to enable the brain penetration of imatinib in patients. *Cancer Res* 65:2577–87.

Camenisch, G., Alsenz, J., van de Waterbeemd, H., Folkers, G. 1998. Estimation of permeability by passive diffusion through Caco-2 cell monolayers using the drugs' lipophilicity and molecular weight. *Eur J Pharm Sci* 6:313–319.

Chambers, T. C. 1998. Identification of phosphorylation sites in human MDR1 P-glycoprotein. *Method Enzymol* 292:328–442.

Cordon-Cardo, C., O'Brien, J. P., Boccia, J., Casals, D., Bertino, R., Melamed, M. R. 1990. Expression of multidrug resistance gene product (P-glycoprotein) in human normal and tumor tissues. *J Histochem Cytochem* 38:1277–87.

Covitz, K. M., Amidon, G. L., Sadee, W. 1996. Human dipeptide transporter, hPEPT1, stably transfected into Chinese hamster ovary cells. *Pharm Res* 13:1631–34.

Dong, M., Ladaviere, L., Penin, F., Deleage, G., Baggetto, L. G. 1998. Secondary structure of P-glycoprotein investigated by circular dichroism and amino acid sequence analysis. *Biochim Biophys Acta* 1371:317–34.

Davies, M. N., Flower, D. R. 2013. A Benchmark Dataset Comprising Partition and Distribution Coefficients of Linear Peptides. http://www.hindawi.com/dpis/biology/2013/976758/http://dx.doi.org/10.7167/2013/976758

Devillers, J., Domine, D., Guillon, C., Karcher, W. 1998. Simulating lipophilicity of organic molecules with a back-propagation neural network. *J Pharm Sci* 87:1086–90.

Einstein, A. 1905. On the movement of small particles suspended in stationary liquid required by the molecular-kinetic theory of heat. *Annalen der Physic* 17:549–60.

Eleni, A., Irini, T., Chrissi, H., Paschalis-Adam, M., Konstantinos, G. I. 2011. Marker Measurements in Transudates. *Open Resp Med* 5:70–72.

Erlinger, R. H. 1987. Physiology of bile secretion and enterohepatic circulation in *Physiology of the Gastrointestinal Tract* (2nd Edition), ed. L. R Johnson, 1557–80. New York: Raven Press.

Ernest, S., Bello-Reuss, E. 1998. P-glycoprotein functions and substrates: Possible roles of MDR1 gene in the kidney. *Kidney Int* 53(65):S11–S17.

Edwards, A., Prausnitz, M. R. 2001. Predicted permeability of the cornea to topical drugs. *Pharm Res* 18(11):1497–1508.

Farrell, T. L., Poquet, L., Dew, T. P., Barber, S., Williamson, G. 2012. Predicting phenolic acid absorption in Caco-2 cells: A theoretical permeability model and mechanistic study. *Drug Metab Dispos* 40(2):397–406.

Furst, P., Albers, S., Stehle, P. 1990. Dipeptides in clinical nutrition. *Proc Nutr Soc* 49:343–59.

Gaillard, P., Carrupt, P. A., Testa, B, Boudon, A. 1994. Molecular lipophilicity potential, a tool in 3D QSAR: Method and applications. *J Comput Aided Mol Des* 8:83–96.

Gatmaitan, Z., Arias, I. 1993. Structure and function of P-glycoprotein in normal liver and small intestine. *Adv Pharmacol* 24:77–97.

Gilbert, E. R., Wong, E. A., Webb, K. E. Jr. 2008. Board-invited review: Peptide absorption and utilization: Implications for animal nutrition and health. *Anim Sci J* 86(9):2135–55.

Gombar, V. K., Enslein, K. 1996. Assessment of n-octanol/water partition coefficient: When is the assessment reliable? *J Chem Inf Comput Sci* 36:1127–34.

Gombar, V. K. 1999. Reliable assessment of logP of compounds of pharmaceutical relevance. SAR QSAR *Environ Res* 10:371–80.

Hammel, H. T., Schlegel, W. M. 2005. Osmosis and solute-solvent drag—Fluid transport and fluid exchange in animals and plants. *Cell Biochem Biophy* 42(3):277–45.

Hansch, C., Leo, A. J. 1979. *Substituent Constants for Correlation Analysis in Chemistry and Biology*. New York: Wiley.

Härtl, A., Amlacher, R., Neubert, R., Hause, C. 1990. Influence of the ion-pair-formation of bretylium and hexylsalicylic acid on their blood plasma levels in dogs. *Pharmazie* 45(4):295.

Hidalgo, I. J., Raub, T. J., Borchardt, R. T. 1989. Characterization of the human colon carcinoma cell line (Caco-2) as a model system for intestinal epithelial permeability. *Gastroenterology* 96:736–49.

Hober, R., Hober J. 1937. Experiments on the absorption of organic solutes in the small intestine of rats. *J Cell Comp Physiol* 10:401–422.

Hofmann, A. F. 1989. Enterohepatic circulation of bile acids. In *The Gastrointestinal System*, 3rd Edition, eds. S. Schultz and J. Forte, 567–96. Bethesda, MD: American Physiological Society.

Hofmann, A. F. 1976. The enterohepatic circulation of bile acids in man. *Adv Intern Med* 21:501–34.

Hogben, C. A. M. 1960. The first common pathway. *Fed Proc* 19:864–69.

Hogben, C. A. M., Schanker, L. A., Tocco, D. J., Brodie, B. B. 1957. Absorption of drugs from the stomach II. The Human. *J Pharmacol* 120:540–45.

Hogben, C. A. M., Tocco, D. J., Brodie, B. B., Schanker, L.S. 1959. On the mechanism of intestinal absorption of drugs. *J Pharmacol Exp Ther* 125:275–82.

Hubatsch, I., Ragnarsson, E. G., Artursson, P. 2007. Determination of drug permeability and prediction of drug absorption in Caco-2 monolayers. *Nat Protoc* 2:2111–19.

Irwin, G. M., Kostenbauder, H. B., Dittert, L. W., Staples, R., Misher, A., Swintosky, J. V. 1969. Enhancement of gasterointestinal absorption of a quaternary ammonium compound by trichloroacatate. *J Pharm Sci* 58(3):313–15.

Jang, S. H., Wientjes, M. G., Au, J. A-S. 2001. Kinetics of P-glycoprotein-mediated efflux of paclitaxel. *J Pharmacol Exp Ther* 298(3):1236–42.

Japertas, P., Didziapetris, A. A. 2002. Fragmental methods in the design of new compounds applications of the advanced algorithm builder. *Quant Struct-Act Relat* 21:23–27.

Jones, P. M., George, A. M. 1998. A new structural model for P-glycoprotein. *J. Membrane Biol.* 166:133–47.

Jonkman, J. H. G., Hunt, C. A. 1983. Ion pair absorption of ionized drugs—Fact or fiction? *Pharm Weekblad Sci* 5:41–48.

Junghans, M., Pretsch, E. 1997. Estimation of partition coefficients of organic compounds: Local data base modeling with uniform-length structure descriptors. *Fresenius J Anal Chem* 359:88–92.

Kedem, O., Katchalsky, A. 1958. Thermodynamic analysis of the permeability of biological memberanes to non-electrolytes. *Biochim Biophys Acta* 27(2):229–46.

Kellogg, G. E., Abraham, D. J. 1999. Development of empirical molecular interaction models that incorporate hydrophobicity and hydropathy. The HINT paradigm. *Analysis* 2:19–23.

Kellogg, G. E., Abraham, D. J. 2000. Hydrophobicity: Is LogP(o/w) more than the sum of its parts? *Eur J Med Chem* 35:651–61.

Kellogg, G. E., Joshi, G. E., Abraham, D. J. 1992. New Tools for modeling and understanding hydrophobicity and hydrophobic interaction. *Med Chem Res* 1:444–53.

Kellogg, G. E., Semus, S. F., Abraham, D. J. 1991. HINT: A new method of empirical hydrophobic field calculation for CoMFA. *J Comput Aided Mol Des* 5:545–52.

Kenny, P. W., Montanari, C. A., Prokopczyk, M. 2013. ClogPalk: A metod for predicting alkane/water partition coefficient. *J Comput Aided Mol Des* 27(5):389–402.

Kim, J. S., Oberle, R. L., Krummel, D. A., Dressman, J. B., Fleisher, D. 1994. Absorption of ACE inhibitors from small intestine and colon. *J Pharm Sci* 83:1350–56.

Klopman, G., Iroff, L. D. 1981. Calculation of partition coefficients by the charge density method. *J Comp Chem* 2:157–60.

Klopman, G., Li, J. W., Wang, S., Dimayuga, M. 1994. Computer automated lop P calculations based on an extended group contribution approach. *J Chem Inf Comput Sci* 34:752–81.

Kodaira, H., Kusuhara, H., Ushiki, J., Fuse, E., Yuichi Sugiyama, Y. 2010. Kinetic analysis of the cooperation of P-glycoprotein (P-gp/Abcb1) and breast cancer resistance protein (Bcrp/Abcg2) in limiting the brain and testis penetration of erlotinib, flavopiridol, and mitoxantrone. *JPET* 333(3):788–96.

Kuh, H. J., Jang, S. H., Wientjes, M. G., Au, J. L. S. 2000. Computational model of intracellular pharmacokinetics of paclitaxel. *J Pharmacol Exp Ther* 293:761–70.

Lee, J-Y, Urbatsch, I. L., Senior, A. E., Stephan Wilkens, S. 2008. Nucleotide-induced structural changes in P-glycoprotein observed by electron microscopy. *J Biol Chem* 283:5769–79.

Le Ferrec, E., Chesne, C., Artrusson, P., Brayden, D., Fabre, G., Gires, P., Guillou, F., Rousset, M., Rubas, W., Scarino, M. L. 2001. *In vitro* models of the intestinal barrier. The report and recommendations of ECVAM Workshop 46. European Center for the Validation of Alternative Methods. *Altern Lab Anim* 29:649–68.

Leo, A. J. 1987. Some advantages of calculating octanol-water partition coefficients. *J Pharm Sci* 76:166–68.

Leo, A. J. 1991. Hydrophobic parameter: Measurement and calculation. *Methods Enzymol* 202:544–91.

Leo, A. J. 1993. Calculating logPoct from structure. *Chem Rev* 93:1281–1306.

Leo, A., Jow, P. Y., Silipo, C. 1975. Calculation of hydrophobic constant from pi and f constants. *J Med Chem* 18:865–868.

Leslie, E. M., Deeley, R. G., Cole, S. P. C. 2005. Multidrug resistance proteins: Role of P-glycoprotein, MRP1, MRP2, and BCRP (ABCG2) in tissue defense. *Toxicol Appl Pharmacol* 204(3):216–37.

Lippold, B. C., Lettenbauer, W. A. 1980. The absorption of homologous alkyl sulphonates of tridihexethyl. *Pharmazie* 35(10):612–20.

Lippold, B. C., Schneider, G. F. 1974. The optimization of availability of homologous quaternary ammonium compounds. 1. Physico-chemical properties and ion-pair dissociation of homologous benzilic acid esters of dimethyl (2-hydroxyethyl)alkylammonium bromides. *Arzneimittel-forsch* 24(12):1952–56.

Liu, X., Testa, B., Alfred Fahr, A. 2011. Lipophilicity and its relationship with passive drug permeation. *Pharm Res* 28:962–977.

Loo, T. W., Bartlett, M. C., Clarke, D. M. 2003. Simultaneous binding of two different drugs in the binding pocket of the human multidrug resistance P-glycoprotein. *J Biol Chem* 278:39706–10.

Loo, T. W., Clarke, D. M. 2001. Determining the dimensions of the drug-binding domain of human P-glycoprotein using thiol cross-linkers as molecular rulers. *J Biol Chem* 276:36877–80.

Loo, T. W., Clarke, D. M. 1998. Quality control by proteases in the endoplasmic reticulum. *J Biol Chem* 273(32):373–76.

Loo, T. W., Clarke, D. M. 1999. The transmembrane domains of the human multidrug resistance P-glycoprotein are sufficient to mediate drug binding and trafficking to the cell surfaces. *J Biol Chem* 274:24759–65.

Loo, T. W., Clarke, D. M. 1999. The glycosylation and orientation in the membrane of the third cytoplasmaic loop of human P-glycoprotein is affected by mutations and substrates. *Biochem* 38:5124–29.

Marcoux, J., Wang, S. C., Politis, A., Reading, E., Ma, J., Biggin, P. C., Zhou, M. et al. 2013. Mass spectrometry reveals synergistic effects of nucleotides, lipids, and drugs binding to a multidrug resistance efflux pump. *Proc Natl Acad Sci USA* 110(24):9704–09.

Margolis, S. A., Levenson, M. 2000. Certification by the Karl Fischer method of the water content in SRM 2890, water saturated 1-octanol, and the analysis of associated interlaboratory bias in the measurement process. *J Anal Chem* 367:1–7.

Mathews, D. M., Adibi, S. A. 1976. Peptide absorption. *Gastroenterology* 71:151–61.

Matsuoka, Y., Okazaki, M., Kitamura, Y., Taniguchi, T. 1999. Developmental expression of P-glycoprotein (multidrug resistance gene product) in the rat brain. *J Neurol* 39:383–392.

Meylan, W. M., Howard, P. H. 1995. Atom/fragment contribution method for estimating octanol-water partition coefficients. *J Pharm Sci* 84:83–92.

Michalet, X. 2010. Mean square displacement analysis of single-particle trajectories with localization error: Brownian motion in isotropic medium. *Phys Rev E Stat Nonlin Soft Matter Phys* 82(4 Pt 1): 1–26. http://www.ncbi.nlm.nih.gov/pmc/articles/PMC3055791/pdf/nihms273605.pdf

Miller, J. M., Dahan, A., Gupta, D., Varghese, S., Amidon, G. L. 2010. Enabling the intestinal absorption of highly polar antiviral agents: Ion-pair facilitated membrane permeation of zanamivir heptyl ester and guanidino oseltamivir. *Mol Pharm* 7(4):1223–34.

Moriguchi, I., Hirono, S., Liu, Q. 1992. Simple method of calculating octanol/water partition coefficient. *Chem Pharm Bull* 40:127–30.

Moriguchi, I., Hirono, S., Nakagome, I., Hirano, H. 1994. Comparison of log P values for drugs calculated by several methods. *Chem Pharm Bull* 42:976–78.

Moore, V. A., Irwin, W. J., Timmins, P., Lambert, P. A., Chong, S., Dando, S. A., Morrison, R. A. 2000. A rapid screening system to determine drug affinities for the intestinal dipeptide transporter 2: Affinities of ACE inhibitors. *Int J Pharm* 210:29–44.

Neubert, R. 1989. Ion pair transport across membranes. *Pharm Res* 6(9):743–47.

Neubert, R., Albrecht, G., Taube, C., Weiss, M., Fürst, W. 1988. Influence of ion-pair-formation on the pharmacokinetic properties of drugs. Part 3: Influence of hexylsalicylic acid on the pharmacokinetics of pholedrine. *Pharmazie* 43(9):632–33.

Neubert, R., Fürst, W., Schulze, P., Loh, H. J., Jirka, M., Wenzel, U. 1987. The effect of ion pair formation on the pharmacokinetics of drugs. 2. The effect of hexylsalicylic acid on the pharmacokinetics of quinine. *Pharmazie* 42(6):393–94.

Neubert, R., Ritter, A., Stolte, E., Albrecht, G., Loh, H. J., Jirka, M., Fürst, W. 1988. Influence of ion-pair-formation on the pharmacokinetic properties of drugs. Part 4: Influence of hexylsalicylic acid on the pharmacokinetics of bretylium. *Pharmazie* 43(12):848–50.

Neuhoff, S., Ungell, A-L., Zamora, I., Per Artursson, P. 2003. pH-dependent bidirectional transport of weakly basic drugs across Caco-2 monolayers: Implications for drug–drug Interactions. *Pharm Res* 20(8):1141–48.

Nys, G. G., Rekker. R. F. 1973. Statistical analysis of a series of partition coefficients with special reference to the predictability of folding of drug molecules: Introduction of hydrophobic fragmental constants (f-values). *Chim Therap* 8:521–35.

Nys, G. G., Rekker, R. F. 1974. The concept of hydrophobic fragmental constants (f-values): II. Extension of its applicability to the calculation of lipophilicities orf aromatic and hetero-aromatic structures. *Chim Therap* 9:361–75.

Ong, S., Liu, H., Qiu, X., Bhat, G., Pidgeon, C. 1995. Membrane partition coefficients chromatographically measured using immobilized artificial membrane surfaces. *Anal Chem* 67:755–62.

Overton, E. 1902. Beigrage zur allgemeinen muskel—und nerve physiologie. *Pflugers Archiv fur die gesamte physiologie des Menschen und der Tiere* 92:115–20.

Palm, K., Luthman, K., Ros, J., Grasjo, J., Artursson, P. 1999. Effect of molecular charge on intestinal epithelial drug transport: pH-dependent transport of cationic drugs. *J Pharmacol Exp Ther* 291:435–43.

Pappenheimer, J. R., Renkin, E. M., Borrero, L. M. 1951. Filtration, diffusion and molecular sieving through peripheral capillary membranes; a contribution to the pore theory of capillary permeability. *Am J Physiol* 167:13–46.

Pauletti, G. M., Wunderli-Allenspach, H. 1994. Partition coefficients in vitro: Artificial membranes as a standardized distribution model. *Eur J Pharm Sci* 1:273–82.

Petrauskas, A. A., Kolovanov, E. A. 2000. ACD/LogP method description. Perspect. *Drug Discov Des* 19:99–116.

Pidgeon, C., Ong, S., Liu, H., Qiu, X., Pidgeon, M., Dantzig, M., Dantzig, A.H. et al. 1995. IAM chromatography: An *in vitro* screen for predicting drug membrane permeability. *J Med Chem* 38:590–94.

Pieksaritanont, T., Deluca, P., Gorham, L. M., Ma, B., Cohn, D., Pang, J., Xu, X., Leung, K., and Lin, J.H. 1998. *In vitro* and *in vivo* evaluations of intestinal barriers for the zwitterions L-767, 679 and its carboxyl ester prodrug L-775, 318: Roles of efflux and metabolism. *Drug Metabo Dispos* 26(6):520–27.

Quintanar-Guerrero, M., Eric Allémann, E., Fessi, H., Erie Doelker, E. 1997. Applications of the ion-pair concept to hydrophilic substances with special emphasis on peptides. *Pharmaceutical Research* 14(2):119–27.

Regina, A., Koman, A., Piciotti, M., El Hafny, B., Center, M. S., Bergmann, R., Couraud, P. O., Roux, F. 1998 (Online 2002). Mrp1 multidrug resistance-associated protein and P-glycoprotein expression in rat brain microvessel endothelial cells. *J Neurochem* 71(2):705–15.

Rekker, R. F. 1977. *The Hydrophobic Fragmental Constant. Its Derivation and Application*, Amsterdam: Elsevier.

Rekker, R. F., de Kort, H. M. 1979. The hydrophobic fragmental constants; an extension to a 1000 data point set. *Eur J Med Chem* 14:479–88.

Rippe, B., Haraldsson, B. 1994. Transport of macromolecules across microvascular walls: The two-pore theory. *Physiol Rev* 74(1):163–219.

Rosenberg, M. F., Callaghan, R., Ford, R. C., Higgins, C. F. 1997. Structure of the multidrug resistance P-glycoprotein to 2.5 nm resolution determined by electron microscopy and image analysis. *J Biol Chem* 272:10685–94.

Rosenberg, M. F., Kamis, A. H., Callaghan, R., Higgins, C. F., Ford, R. C. 2003. Three-dimensional structures of the mammalian multidrug resistance P-glycoprotein demonstrate major conformational changes in the transmembrane domains upon nucleotide binding. *J Biol Chem* 278:8294–99.

Rosenberg, M. F., Velarde, G., Robert C. Ford, R. C., Martin, C., Georgina Berridge, G., Kerr, I. D., Callaghan, R. et al. 2001. Repacking of the transmembrane domains of P-glycoprotein during the transport ATPase cycle. *EMBO J* 20(20):5615–25.

Rubinow, S. I. 1975. *Introduction to Mathematical Biology*. 207–209. New York: John Wiley & Sons.

Sambuy, Y., DeAngelis, I., Ranaldi, G., Scarino, M. L., Stammati, A., Zucco, F. 2005. The Caco-2 cell line as a model of the intestinal barrier: Influence of cell and culture-related factors on Caco-2 cell functional characteristics. *Cell Biol Toxicol* 21:1–26.

Samiei, N., Mangas-Sanjuan, V., González-Álvarez, I., Foroutan, M., Shafaati, A., Zarghi, A., Bermejo, M. 2013. Ion-pair strategy for enabling amifostine oral absorption: Rat *in situ* and *in vivo* experiments. *Eur J Pharm Sci* 49(4):499–504.

Schaper, K-J, Zhang, H., Raivsky, O. A. 2001. pH-dependent partitioning of acidic and basic drug into liposomes—A quantitative structure-activity relationship analysis. *Mol Inform* 20(1):46–54.

Schinkel, A. H. 1997. The physiological function of drug-transporting P-glycoproteins. *Semin Cancer Biol* 8:161–70.

Scott, D. C., Clymer, J. W. 2002. Estimation of distribution coefficients from the partition coefficient and pKa. *Pharm Techn*: http://www.pharmtech.com/pharmtech/data/articlestandard/pharmtech/462002/38266/article.pdf

Sharom, F. J. 1995. Characterization and functional reconstitution of the multidrug transporter. *J Bioenerg Biomembr* 27:15–22.

Sharom, F. J., Liu, R., Romsicki, Y. 1998. Spectroscopic and biophysical approaches for studying the structure and function of the P-glycoprotein multidrug transporter. *Biochem Cell Biol* 76:695–708.

Sharom, F. J., Liu, R., Romsicki, Y., Lu, P. 1999. Insights into structure and substrate interactions of the P-glycoprotein multidrug transporter from spectroscopic studies. *Biochim Biophys Acta* 1461:327–45.

Shore, P. A., Brodie, B. B., Hogben, C. A. M. 1957. The gastric secretion of drugs: A pH partition hypothesis. *J Pharmacol Exp Ther* 119: 361–69.

Siarheyeva, A., Liu, R., Frances J., Sharom, F. J. 2010. Characterization of an asymmetric occluded state of P-glycoprotein with two bound nucleotides: Implications for catalysis. *J Bio Chem* 285:7575–86.

Soares, A. F., Carvalho, R. A., Veiga, F. 2007. Oral administration of peptides and proteins: Nanoparticles and cyclodextrins as biocompatible delivery systems. *Nanomedicine* 2(2):183–202.

Snyder, N. J., Tabas, L. B., Berry, D. M., Duckworth, D. C., Spry, D. O., Dantzig, A. H. 1997. Structure–activity relationship of carbacephalosporins and cephalosporins: Antibacterial activity and interaction with the intestinal proton-dependent dipeptide transport carrier of Caco-2 cells. *Antimicrob Agents Chemother* (41):1649–57.

Stopher, D., McClean, S. 1990. An improved method for the determination of distribution coefficients. *J Pharm Pharmacol* 42:144.

Taillardat-Bertschinger, A., Martinet, C. A. M., Carrupt, P-A., Reist, M., Caron, G., Roberta Fruttero, R., Bernard Testa, B. 2002. Molecular factors influencing retention on immobilized artifical membranes (IAM) compared to partitioning in liposomes and n-Octanol. *Pharm Res* 19(6):729–37.

Thiebaut, F., Tsuruo, T., Hamada, H., Gottesman, M., Pastan, I., Willingham, C. 1989. Immunochemical localization in normal tissues of different epitopes in the multidrug transport protein P170: Evidence for localization in brain capillaries and crossreactivity of one antibody with a muscle protein. *J Histochem Cytochem* 37:159–64.

Thiebaut, F., Tsuruo, T., Hamada, H., Gottsman, M. M., Pastan, I., Willingham, M. C. 1987. Cellular localization of the multidrug resistance gene product P-glycoprotein in normal human tissues. *Proceedings of the National Academy of Sciences of the United States of America* 84:7735–38.

Todeschini, R., Consonni, V. 2000. Handbook of molecular descriptors. In *Methods and Principles in Medicinal Chemistry*, Vol. 11, eds. R. Mannold, H. Kubinyi, and H. Timmerman, 276 pp. Weinheim, Germany: Wiley-VCH Verlag GmbH.

Travel, J. 1940. The influence of the hydrogen ion and concentration on the absorption of alkaloids from the stomach. *J Pharmacol Exp Ther* 69:21–33.

Tsuji, A., Miyamoto, E., Hashimoto, N., Yamana, T. 1978. GI absorption of β-lactam antibiotics II: Deviation from pH-partition hypothesis in penicillin absorption through *in situ* and *in vitro* lipoidal barriers. *J Pharm Sci* 67:1705–11.

Usansky, H. H., Sinko, P. J. 2003. Computation of log BB values for compounds transported through carrier-mediated mechanisms using in vitro permeability data from brain microvessel endothelial cell (BMEC) monolayers. *Pharm Res* 20:390–96.

van der Giesen, W. F., Janssen, L. H. M. 1982. Influence of ionization and ion-pair formation on lipophilicity of some 4-hydroxycoumarin derivatives in the octanol-water system. *Inter J Pharmaceutics* 12:231–49.

Wang, E. J., Casciano, C. N., Clement, R. P., Johnson, W. W. 2000. Two transport binding sites of P-glycoprotein are unequal yet contingent: Initial rate kinetic analysis by ATP hydrolysis demonstrates intersite dependence. *Biochim Biophys Acta* 1481:63–74.

Wang, R., Fu, Y., Lai, L. 1997. A new atom-additive method for calculating partition coefficient. *J Chem Inf Comp Sci* 37:615–21.

Zhang, E. Y., Phelps, M. A., Cheng, C., Ekins, S., Swaan, P. W. 2002. Modeling of active transport systems. *Adv Drug Deliv Rev* 54(3):329–54.

Zhu, H., Sedykh, A., Suman K., Chakravarti, S. K., Klopman, G. 2005. A new group contribution approach to the calculation of LogP. *Current Computer-Aided Drug Design* 1:3–9.

3 *In Vivo* Distribution of Xenobiotics

3.1 INTRODUCTION

After gaining access to the systemic circulation through one of the routes of administration, xenobiotics distribute in different tissues and organs of the body. Depending on their physicochemical characteristics, a series of physical and physiological processes occur simultaneously that shape the distinctive pattern of their distribution in the body. An example of physical processes is the simple dilution of drug or toxin in the intracellular and extracellular fluids and the examples of physiological processes are protein binding, tissue uptake, permeation of the compound through different biological barriers in the body, interaction with transport proteins in conjunction with the processes of metabolism, and excretion. The distribution processes transfer the compound to various regions of the body including the receptor sites, target organs, excretory organs, and other tissues. Thus, if the compound is not actively or passively targeted to a receptor site in the body, all organs and tissues will have to endure the presence of the xenobiotic. The subject of targeting the therapeutic agents is beyond the scope of this book.

3.2 FACTORS INFLUENCING THE DISTRIBUTION OF XENOBIOTICS IN THE BODY

The distribution of a xenobiotic in the body is influenced by factors such as

- Total body water (TBW)
- Blood flow and organ/tissue perfusion
  - Transcapillary exchange of xenobiotics
  - Perfusion-limited distribution
  - Permeability-limited distribution
- Binding to plasma proteins
  - Estimation of parameters of protein binding
- Physicochemical characteristics of the compounds
- Extent of penetration through physiological barriers and parallel elimination processes Physiological barriers
  - Blood–brain barrier
  - Blood–lymph barrier
  - Placental barrier
  - Blood–testis barrier
  - Blood–aqueous humor barrier
- Body weight and composition
- Disease states

3.2.1 Total Body Water (TBW)

Under normal conditions, the total volume of water in the body responsible for the distribution of xenobiotics and nutrients is about 57%–60% of body weight. It is often estimated by one of the following empirical equations:

Watson Formula (Watson et al., 1980):

$$(♂): TBW(L) = 2.447 - (0.09156 \times age) + (0.1074 \times Height\ (\text{cm})) + (0.3362 \times Weight\ (\text{kg})) \tag{3.1}$$

$$(♀): TBW(L) = -2.097 + (0.1069 \times Height\ (\text{cm})) + (0.296785 \times Weight\ (\text{kg})) \tag{3.2}$$

Hume Formula (Hume and Weyers, 1971):

$$(♂): TBW(\text{L}) = 0.194786 \times Height\ (\text{cm})) + (0.296785 \times Weight\ (\text{kg})) - 14.012934 \tag{3.3}$$

$$(♀): TBW(\text{L}) = 0.34454 \times Height\ (\text{cm})) + (0.183809 \times Weight\ (\text{kg})) - 35.27 \tag{3.4}$$

The calculated total body water for a 70 kg male subject with standard height is about ≈40 L. In newborns it constitutes 75% of the body weight and it reduces to the normal value within the first 10–12 years of life. Approximately 62.5% of the calculated TBW or 36% of total body weight is the intracellular fluid (≈25 L). The remaining volume, which is 37.5% of total body water, or 21.5% body weight, is equal to the extracellular fluid (≈15 L). The extracellular fluid consists of interstitial fluid, systemic circulation, cerebrospinal fluid, and other compartments. The total blood volume, which is a part of extracellular fluid, in normal subjects is about ≈12.5% of total body water (≈5 L) and changes in adults depending on their sex, weight, and disease states. The five liters of blood consists of approximately three liters of plasma and two liters of blood cells. An equation that is often used to determine blood volume is

$$(Volume)_{blood} = \frac{(Volume)_{plasma}}{1 - Hematocrit} \tag{3.5}$$

Thus, the blood volume for the 70 kg subject is approximately 71.5 mL/kg body weight. This value is close to 79 mL/kg for mouse, 65 mL/kg for rat and 85 mL/min for dogs. The total blood volume is an essential physiological factor in determining the PK/TK of xenobiotics. When it is necessary to remove multiple large volumes of sample blood, the total reduction in the blood volume may influence the distributional profile of the compound. Thus, the gradual replenishment of the volume may often be necessary.

3.2.2 Blood Flow and Organ/Tissue Perfusion

Blood flow is the volume of blood that circulates in the cardiovascular system and reaches different organs and tissues in the body per unit of time. It is expressed in volume/time and its magnitude differs in body's different regions. The total blood flow is equal to 5000 mL/min that correspond to the cardiac output at rest. The cardiac output is the volume of blood pumped by the heart per minute. In addition to the cardiac output, an important factor in blood flow is the variable volume of blood in different network of systemic circulation. The heart usually holds 7% of total blood; the pulmonary system maintain 9%; arteries capacity is about 13%; the arterioles and capillaries contain 7%; the remaining 64% is in the veins, venules, and venous system. The capillaries with their permeable walls allow the exchanges of drug, nutrients, toxins, etc., with the interstitial fluids of the organ/tissue and merge with the venules network that gradually converges into larger veins. During this transcapillary exchange xenobiotics are transported across the capillary wall into the tissue. This transport may result from the difference in pressures on the inside and outside of the capillary wall, such as osmotic pressure or hydrostatic pressure, or may result mainly from a concentration gradient across capillary wall. The distribution of a compound in a region of the body depends on the blood flow to that region and how fast the drug is transferred from the site of administration to the region. Table 3.1 is a representation of the differences in volume and blood flow of different organs/tissues of the human body for a 30–39 year-old man with 70 kg body weight and 1.73 m^2 body surface area.

The blood flow essentially determines how fast a compound is delivered to an organ/tissue and reflects the relative rate a compound in the organ/tissue is expected to achieve at equilibrium with blood.

The amount of xenobiotic that can be stored or distributed in a tissue depends on the mass of tissue or the size of organ as well as physicochemical characteristics of xenobiotic such as partition coefficient or distribution coefficient. Figure 3.1 exhibits the influence of blood flow on the distribution time course of a compound injected intravenously and disseminated to various tissues. The accumulation in various regions is a function of blood flow, as well as other physiological factors and physicochemical characteristics of the compound.

Physiologically, the blood flow is determined by the Ohm's law, that is,

$$Q = \frac{\Delta P}{R} \tag{3.6}$$

where the flow rate (volume/time) is (Q), the pressure difference between the two ends of a vessel (cylinder) in mmHg is ΔP and the resistance (mmHg × time/volume) is R. According to Equation 3.6, the flow rate is directly proportional to pressure differences and inversely proportional to the resistance. This principle can be applied to a single cylinder (vessel) or a network of cylinders (e.g., the vascular network of an organ). Resistance to blood flow influences not only the cardiovascular function, but also the distribution and transfer to organs/tissues. It has been suggested that in small microvessels, in addition to the rheological variability of blood, the interaction between the inner vessel surface and blood components also causes high flow resistance (Pries et al., 1994, 1997).

TABLE 3.1
Blood Flow to Different Organs/Tissue for a 30–39 Year-Old Man at Rest with 70 kg Body Weight and 1.73 m^2 Body Surface Area

Organ/Tissue	Volume (L)	Blood Flow (mL/min)	Blood Flow (mL/min)/ 100 mL Tissue	Volume of Blood in Equilibrium with Tissue (mL)
Adrenals	0.02	100	300	62
Kidneys	0.30	1240	410	765
Thyroid	0.02	80	400	49
Gray Matter	0.75	600	80	371
Heart	0.30	240	80	148
Liver + Portal System	3.90	1760	45	976
Red Marrow	1.40	120	9	74
Muscle	30.00	600	2	370
Fat	10.00	200	2	123
Fatty Marrow	2.20	60	2.70	37
Bone Cortex	6.40	0	0	0
Total		5000		

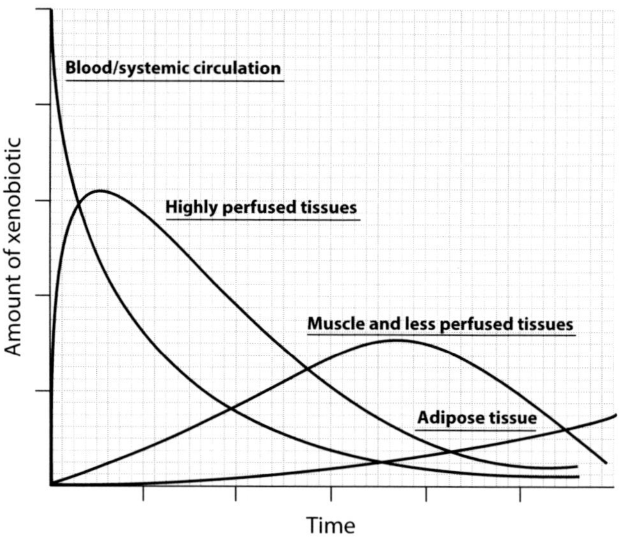

FIGURE 3.1 The influence of blood flow on the distribution profile of xenobiotics in various regions of the body as a function of time following intravenous administration.

In general, the resistance to blood flow depends on the radius and length of vessels and viscosity of blood according to the Hagen–Poiseuille equation (Hagen, 1839; Poiseuille, 1838; Sutera and Skalak, 1993)

$$R = \frac{8 \times Length \times \eta}{\pi \times r^4} \tag{3.7}$$

where η is blood viscosity and r is the inner radius of the vessel. Thus, a small change in radius, raised to the fourth power, will increase the resistance significantly. Substitution of Equation 3.7 into Equation 3.6 yields the following equation applicable to blood flow:

$$Q = \frac{\Delta P \times \pi \times r^4}{8 \times Length \times \eta} \tag{3.8}$$

Thus, the blood flow is directly proportional to the radii of vessels and pressure differences and inversely proportional to the viscosity of blood and length of the vessels.

Depending upon the physicochemical characteristics of a xenobiotic, the transcapillary exchange and storage in tissue is either a blood flow-limited phenomenon known as perfusion-limited distribution or a transfer-limited phenomenon known as permeability-limited distribution.

3.2.2.1 Perfusion-Limited Distribution and Permeability-Limited Distribution (Transcapillary Exchange of Xenobiotics)

To differentiate the dynamics of perfusion-limited distribution from permeability-limited distribution, the blood capillary can be assumed as a cylinder of length (L) and radius of (r) in which blood flows with a velocity of (v) in the positive direction of (x). The blood flow carries the xenobiotic with a concentration of (C_{blood}), which transfers the compound to the tissue bed surrounding the capillary with a concentration of (C_{tissue}). The compound diffuses across the capillary membrane by passive diffusion due to the concentration gradient of the compound between the blood and tissue. If we consider a segment of the capillary between x and $x + \Delta x$, the difference in mass flux of

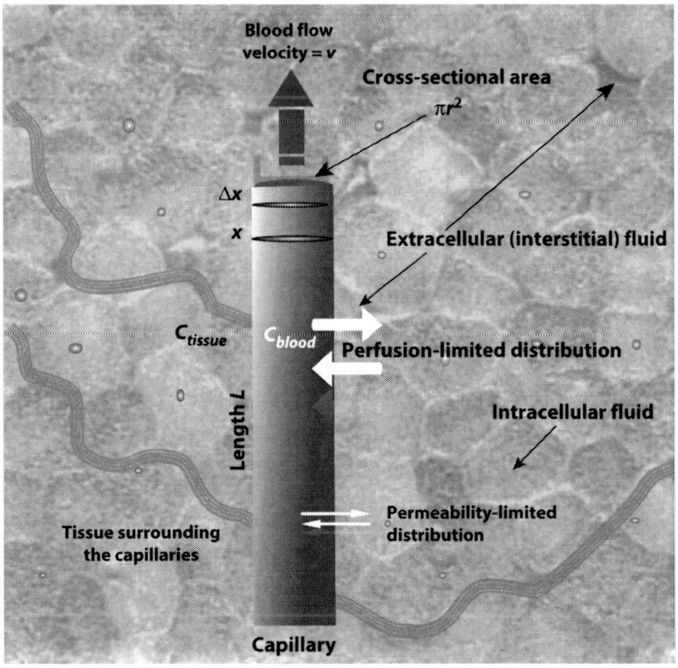

FIGURE 3.2 Theoretical distinction between perfusion-limited distribution and permeability-limited distribution by assuming the blood capillary as a cylinder of the length (L), radius of (r) with blood velocity (v) in the positive direction of (x) transporting xenobiotic concentration to the tissue.

the compound between the beginning and the end of Δx would be equal to the mass flux across the capillary wall into the surrounding tissue (Figure 3.2).

This can be written in the following form where the cross-sectional area of the capillary is πr^2, the surface area of diffusion is $2r\pi\Delta x$, and the permeability constant for the compound in the capillary membrane is P:

$$v\left[C_{blood}(x+\Delta x) - C_{blood}(x)\right]\pi r^2 = -P\left[C_{tissue} - C_{blood}(x)\right]2r\pi\Delta x \tag{3.9}$$

As the limit of $\Delta x \to 0$, Equation 3.9 changes to

$$\pi r^2 v(dC_{blood}/dx) = -2r\pi P[C_{blood} - C_{tissue}] \tag{3.10}$$

The cross-sectional area multiplied by the velocity ($\pi r^2 v$) is equal to the blood flow Q.

The surface area of the capillary under observation (S_{Area}) is equal to $2r\pi L$. Substitutions of Q and S_{Area} in Equation 3.10 yields

$$\frac{dC}{dx} = -\frac{PS_{Area}}{QL}(C_{blood} - C_{tissue}) \tag{3.11}$$

where ($C_{blood} - C_{tissue}$) is the concentration gradient between the capillary and surrounding tissue.

Integration of Equation 3.11 yields the following relationship:

$$C_{blood}(x) - C_{tissue} = (C_{arterial} - C_{tissue})e^{-PSx/QL} \tag{3.12}$$

Setting $x = L$ and $C_{blood}(L) = C_{venous}$, Equation 3.12 changes to 3.13

$$C_{venous} = C_{tissue} + (C_{arterial} - C_{tissue})\, e^{-PS/Q} \quad (3.13)$$

Also, the mass flux across the capillary wall into the tissue (J_{tissue}) in terms of net mass-current leaving the capillary per length of L, is

$$J_{tissue} = 2r\pi P \int_0^L \left[C_{blood}(x) - C_{tissue}\right] dx = QC_{arterial} - QC_{venous} \quad (3.14)$$

Substituting Equation 3.13 into 3.14 yields

$$J_{tissue} = (C_{arterial} - C_{tissue})Q(1 - e^{-PS/Q}) \quad (3.15)$$

Thus, the mass flux is directly proportional to the blood flow and concentration gradient of the compound across the capillary wall.

Dividing the mass flux across the capillary wall by the concentration gradient yields the capillary clearance, which represents the volume of blood from which the compound is extracted from the blood and distributed into the tissue per a unit of time

$$Clearance = \frac{J_{tissue}}{C_{arterial} - C_{tissue}} = Q(1 - e^{-PS/Q}) \quad (3.16)$$

Where the extraction ratio of the distribution is

$$Extraction\ Ratio = 1 - e^{-PS/Q} \quad (3.17)$$

Thus, the capillary clearance is equal to the blood flow times the extraction ratio.

The extraction ratio is an equilibration fraction between the xenobiotic concentration in the tissue, arterial capillaries and the venous side of the capillaries

$$E = \frac{C_{arterial} - C_{venous}}{C_{arterial} - C_{tissue}} \quad (3.18)$$

Equations 3.16 through 3.18 present the following two scenarios:

SCENARIO 1: WHEN $Q > PS$

The blood flow is greater than the permeability-surface area product. The permeability-surface area product, *PS*, characterizes the diffusion of the compound from the blood vessels into the tissue environment. Under this scenario, the extraction ratio is less than one and the distribution of the compound is limited by how fast it permeates from the blood, through the capillary wall into the surrounding tissue during its transit by the blood flow (Figure 3.2). Under this condition the distribution is permeability-limited (or diffusion-limited) distribution and the driving force for the transfer of compound into tissue is the concentration gradient; the diffusion is slow and limited by the membrane permeability and physicochemical characteristics of the compound; and the mass flux into the tissue is defined as

$$J_{tissue} = PS\,(C_{arterial} - C_{tissue}) \quad (3.19)$$

SCENARIO 2: WHEN $Q < PS$

In this situation, the extraction ratio is equal or close to one, the uptake of xenobiotic by the tissue is thermodynamically more favorable than its residence in the blood and the limit of distribution is how fast the compound can be delivered to the tissue. When the compound reaches the tissue its uptake is immediate. This scenario most often applies to highly lipophilic compounds (Figure 3.2). This type of distribution is perfusion-limited distribution or blood flow-limited distribution and the mass transfer into the tissue can be defined as

$$J_{tissue} = Q(C_{arterial} - C_{tissue}) \tag{3.20}$$

Thus, the blood flow is the principal factor in distribution, and concentration gradient plays little or no role in mass transfer of the compound.

The distribution of a compound, whether it is perfusion-limited or permeability-limited, includes diffusion into the extracellular fluid, followed by diffusion through the cell membrane, which may include reversible interaction with cell membrane, followed by internalization and intracellular handling of the compound by the cells, which may include metabolism and exocytosis.

The overall distribution profile of xenobiotics, after reaching the systemic circulation and before diffusion into the interstitial fluid, starts with interaction with blood components, namely plasma proteins and blood cells.

3.2.3 Binding to Plasma Proteins

Most xenobiotics, if not all, bind reversibly to plasma proteins, particularly albumin and α-acid glycoprotein. Binding to plasma proteins influences not only the distribution of xenobiotics in the body, but also their pharmacologic and/or toxic response as well as their elimination. The binding creates a xenobiotic-protein complex with a molecular size equal to the sum of the molecular weights of the compound and protein. In general, it is the unbound or free molecule of the compound that diffuses through the capillaries wall to the extravascular environment where the interactions with cellular receptors and passage through the membrane takes place. The xenobiotic-protein complex, because of its large molecular weight, does not permeate easily through the barriers; thus its distribution is different from the free compound and remains mainly in the systemic circulations. Even if the complex crosses the capillaries and enters into interstitial fluid its diffusion into the intracellular fluid is doubtful, unless the complex dissociation constant favors the formation of free molecule at the site of cellular diffusion. A protein-drug complex is pharmacologically inactive, its formation affects the distribution of xenobiotics in the body and influences the therapeutic/toxic response. For this reason, the concentration of an unbound drug in plasma is used mainly for PK/TK analysis.

The plasma protein binding of one compound given individually is often different from the time they are administered concurrently with other compounds. The interaction involves the affinity of two or more compounds for binding to the same protein and the same binding site (Mattos and Ring, 1996). This type of interaction causes displacement of one compound by the other at the binding site of protein, which increases the free concentration of the compound and its unbound fraction in the systemic circulation. The outcomes of such an increase are the changes in most *in vivo* PK/TK parameters, such as increase in the area under plasma concentration–time curve, changes in the apparent volume of distribution and clearance, and may influence the pharmacological or toxicological response. The displacement occurs when the interaction of one compound is thermodynamically more favorable than the other. Similar displacement can also occur at the cellular levels with other proteins/enzymes. Theoretically the displacement causes an increase in free fraction of xenobiotic in plasma and a proportionate elevation of its concentration at receptor sites. It is important, however, to validate whether the displacement observed *in vitro* reflects the corresponding occurrence *in vivo* (Aarons and Rowland, 1981).

Human serum albumin (HSA) is the major protein of plasma and consists of 585 amino acids forming a single polypeptide of a known sequence with a molecular weight of 66 kDa with three homologous helical domains with each domain having at least two sub-domains (Ghuman et al.,

2005) and a catabolic half-life of approximately 17–18 days. Albumin is produced exclusively by the hepatocytes of the liver and is the major protein of plasma and most tissues. It binds to a wide range of exogenous (Herve et al., 1994; Kratochwil et al., 2002) and endogenous compounds (Curry et al., 1998; Bhattacharya et al., 2000; Petitpas et al., 2001) both covalently and reversibly. Overcoming its capacity to bind a lead compound presents a major challenge in drug discovery and development (Ghuman et al., 2005). In spite of its large molecular size it can diffuse into the extravascular compartment. The interaction of small molecule xenobiotics with HSA has been extensively investigated (Bowmer and Lindup, 1978; Ekman et al., 1980; Aarons et al., 1983; Nowak and Shaw, 1995). Albumin possesses both negatively and positively charged binding sites. Most acidic, neutral, and basic compounds bind to HSA. The strength and extent of the binding depends on the structure of the compound. The interaction of drugs with HSA occurs by hydrogen binding, hydrophobic binding, and van der Waals forces.

The extent of protein binding significantly influences not only the distribution of a compound, but also its renal elimination as well. The kidneys do not filter albumin and therefore drugs that are bound to albumin also remain unfiltered. Furthermore, since only free molecules of xenobiotic can be taken up by the hepatocyte of the liver and interact with metabolic enzyme systems, the extent of binding also influences the rate and extent of metabolism. In general, plasma protein binding influences the PK/TK profile of xenobiotics significantly, and has been widely studied for decades (Koch-Weser and Sellers, 1976; Wagner, 1976; Yacobi et al., 1976; Shary et al., 1978; McNamara et al., 1979; Rowland, 1980; Greenblatt et al., 1982; Buur et al., 2009; Smith et al., 2010).

Other proteins of plasma are fibrinogen, globulins such as γ-globulins that are mainly antibodies (IgA, IgM, IgG, IgE, and IgD), α_1 globulin (α_1-antitrypsin, orosomucoid [α_1 acid glycoprotein]) (Edwards et al., 1982), α_2 globulin (ceruloplasmin, haptoglobulin), α_2 macroglobulin, β_1 and β_2 globulins (transferrin hemopexin plasminogen fibrinogen), and α and β lipoproteins.

The normal plasma protein level is 63–83 g/L; albumin constitutes 31.5–41.5 g/L; α_1 globulin $\approx$3 g/L; α_2 globulin $\approx$1.6 g/L; α_2 macroglobulin $\approx$3 g/L; β_1 and β_2 globulins $\approx$8.2 g/L; and the remaining balance is γ-globulins and lipoproteins (Vallner and Chen, 1977).

The physicochemical characteristic of a compound influences its binding to a specific protein in plasma. Neutral molecules that are highly lipophilic tend to bind mostly to lipoproteins. Acidic and weakly acidic compounds in ionized form with moderate or market lipophilicity bind mainly to albumin. Ionized basic compounds with moderate lipophilicity bind mostly to α_1 acid glycoprotein, and those with market lipophilicity bind not only to α_1 acid glycoprotein, but also to albumin and lipoproteins. Globulins usually interact negligibly with the majority of drugs. The hydrophobic nature of compounds is the key factor on the interaction between compounds and proteins. Cationic drugs are of particular interest because they constitute 70%–80% of all drugs and they constitute the majority of drugs showing a high first pass effect. It is worth noting here that a wide variety of factors such as pregnancy, surgery, age, sex, and interethnic and racial differences affect the interaction of drugs with proteins. Also, metabolites of certain compounds, either the reactive metabolites of Phase I metabolism or conjugates of Phase II metabolism, interact significantly with plasma proteins. In certain cases, depending on the affinity of metabolites for interaction with a protein, the metabolites may interact competitively with the protein and displace the parent compound.

3.2.3.1 Estimation of Parameters of Protein Binding

The binding of xenobiotics to plasma proteins is usually determined *in vitro* under physiological pH and temperature. The methodology includes equilibrium dialysis, dynamic dialysis, ultrafiltration, gel filtration chromatography, ultracentrifugation, microdialysis, electrophoresis, and the methodologies for high throughput measurements. The objective of all these techniques is to estimate the concentration of the free xenobiotic that is in equilibrium with its protein complex. The free concentration is then used to determine the free fraction, or fraction unbound, (f_u)

$$f_u = \frac{\text{Free xenobiotic concentration}}{\text{Total xenobiotic concentration}} \tag{3.21}$$

or,

$$f_u = \frac{\text{Free xenobiotic concentration}}{\text{(Bound + Free) xenobiotic concentration}} \tag{3.22}$$

$$\therefore \text{ Free concentration} = f_u \times \text{Total concentration} \tag{3.23}$$

As mentioned earlier, most of the protein complexes, in particular HSA, are held together by the weak chemical interactions such as electrostatic interactions, van der Waal's forces and hydrogen binding, which tend to separate under non-physiological pH, temperature, and osmotic pressure. The selected methodology and conditions of the experiment should be such that the stability of the complex and equilibrium is persevered and the concentration of free molecules is not overestimated by break down of the complex during the measurement.

To estimate the parameters of protein binding, a brief description of *in vitro* methodology is necessary. They all share the same objective of determining the bound and unbound fractions. Choosing the conventional and classical method of equilibrium dialysis, plasma, or HSA at a physiological pH (7.2–7.4) is spiked with different concentrations of xenobiotic. The mixture is then dialyzed at 37°C through a dialysis membrane with molecular cutoffs ≈12,000–14,000 Dalton against an equivalent volume of a phosphate buffer (≈67 mM, pH 7.2–7.4), made isotonic with NaCl. The dialysis membrane is used in the form of a bag that holds protein and drug, and the bag is placed in the buffer solution. A modified version of the bag is a prefabricated two-compartment unit that is separated by a dialysis membrane. The equilibrium of free molecules across the membrane is usually achieved in about 2 to 3 hours. The concentration of free molecules is measured in the buffer side, that is, outside of the bag or compartment separated by the membrane, which is assumed to be equal to the concentration of free molecules inside the bag or compartment. The concentration of free molecules inside the bag or compartment is assumed to be in equilibrium with the bound molecules to protein.

The binding parameters, such as free fraction or association constant, can then be determined by the following application of the Law of Mass Action:

$$Xenobiotic + Protein \underset{k_2}{\overset{k_1}{\rightleftharpoons}} [Xenobiotic - Protein] \tag{3.24}$$

that is,

$$C_{free} + C_{protein} \underset{k_2}{\overset{k_1}{\rightleftharpoons}} C_{complex} \tag{3.25}$$

The association constant of the interaction (Equation 3.25) is

$$K_a = \frac{C_{complex}}{C_{free} \times C_{protein}} = \frac{k_1}{k_2} \tag{3.26}$$

(C_{free}) is the concentration of free molecules in moles, ($C_{protein}$) is the concentration of protein with free binding sites, ($C_{complex}$) is the concentration of substrate-protein complex, and k_1 and k_2 are the rate constants of forward and reverse reactions, respectively. The reciprocal of the association constant is known as the dissociation constant

$$K_d = \frac{C_D \times C_P}{C_{DP}} = \frac{k_2}{k_1} = \frac{1}{K_a} \tag{3.27}$$

The magnitude of the association constant, K_a, represents the extent of protein binding. Compounds that bind extensively to plasma proteins usually have a very large association constant.

From Equation 3.26 the concentration of complex is

$$C_{complex} = K_a C_{free} C_{protein} \tag{3.28}$$

The concentration of total protein ($C_{Tprotein}$), known at the start of experiment, and the estimated value of the complex ($C_{complex}$), are used to determine the concentration of protein at the equilibrium ($C_{protein}$)

$$C_{protein} = C_{Tprotein} - C_{complex} \tag{3.29}$$

Substitution of Equation 3.29 in 3.28 yields

$$C_{complex} = K_a C_{free}(C_{Tprotein} - C_{complex}) \tag{3.30}$$

$$\therefore C_{free}(1 + K_a C_{free}) = K_a \times C_{free} \times C_{Tprotein} \tag{3.31}$$

Therefore, the number of moles of drug bound per moles of protein at equilibrium (r) is

$$r = \frac{C_{complex}}{C_{Tprotein}} = \frac{K_a \times C_{free}}{1 + K_a C_{free}} \tag{3.32}$$

When there are n identical binding sites per mole of protein, the equation is multiplied by n

$$r = \frac{n(K_a \times C_{free})}{1 + K_a C_{free}} \tag{3.33}$$

Equations 3.32 and 3.33 are known as Langmuir equations (Langmuir, 1916) and plot r versus C_{free} yields the Langmuir hyperbolic isotherm (Figure 3.3). The straight-line modifications of Equation 3.33 are

Double reciprocal plot (Figure 3.4)

$$\frac{1}{r} = \frac{1}{nK_a C_{free}} + \frac{1}{n} \tag{3.34}$$

Scatchard plot (Figure 3.5)

$$\frac{r}{C_{free}} = nK_a - rK_a = K_a(n - r) \tag{3.35}$$

Plot of ($C_{complex}/C_{free}$) versus $C_{complex}$ (Figure 3.6)

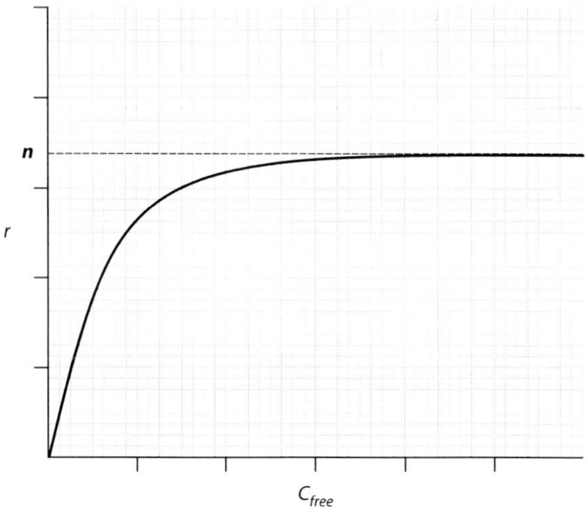

FIGURE 3.3 Langmuir isotherm with dependent variable (y) representing the number of moles of xenobiotic bound per moles of protein versus (x) the free concentration of the compound.

From Equation 3.32

$$C_{complex} + C_{complex} K_a C_{free} = nC_{Tprotein} K_a C_{free} \tag{3.36}$$

$$\frac{C_{complex}}{C_{free}} = nK_a C_{Tprotein} - K_a C_{complex} \tag{3.37}$$

When the protein has multiple binding sites (e.g., albumin has two main positively charged binding sites with high association constants of 10^4–10^6 M^{-1}, which binds largely anionic drugs) and the

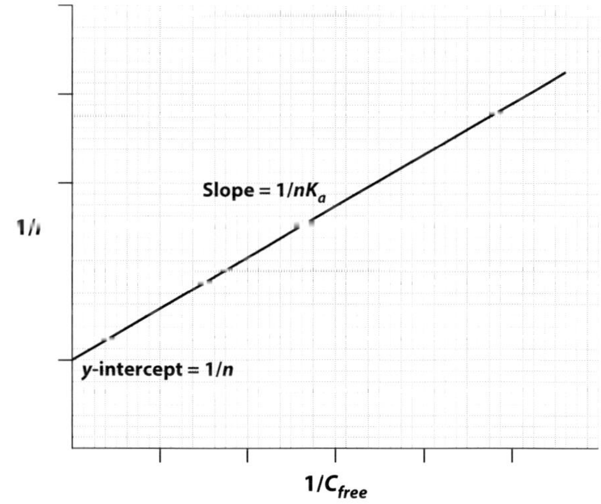

FIGURE 3.4 The relationship between the reciprocal of moles of xenobiotic bound per mole of protein versus the reciprocal of free concentration for estimating the number of binding sites per mole of protein (reciprocal of the y-intercept) and the association constant (slope/y-intercept) according to Equation 3.34.

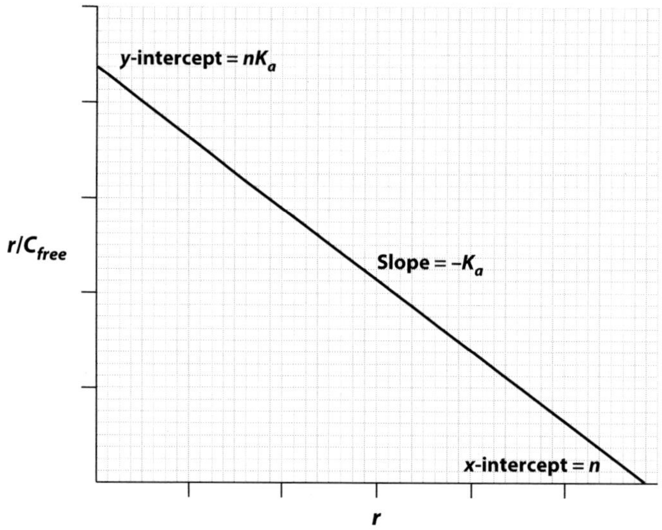

FIGURE 3.5 Scatchard plot based on Equation 3.35 with a slope equal to ($-K_a$), y-intercept of nK_a, and x-intercept of n.

interaction of xenobiotic is with different functional groups, the following Langmuir equation is preferred:

$$r = \frac{n_1 K_{a_1} C_{free}}{1 + K_{a_1} C_{free}} + \frac{n_2 K_{a_2} C_{free}}{1 + K_{a_2} C_{free}} + \cdots + \frac{n_n K_{a_n} C_{free}}{1 + K_{a_n} C_{free}} \tag{3.38}$$

where n_1 and K_{a_1} are the parameters of one type of identical binding sites, n_2 and K_{a_2} are the parameters of the second type of identical binding sites and so on. For example, the $-COO^-$ of aspartic acid or glutamic acid residue can be one type of binding site and $-S^-$ of cysteine or $-NH^{2+}$ of histidine

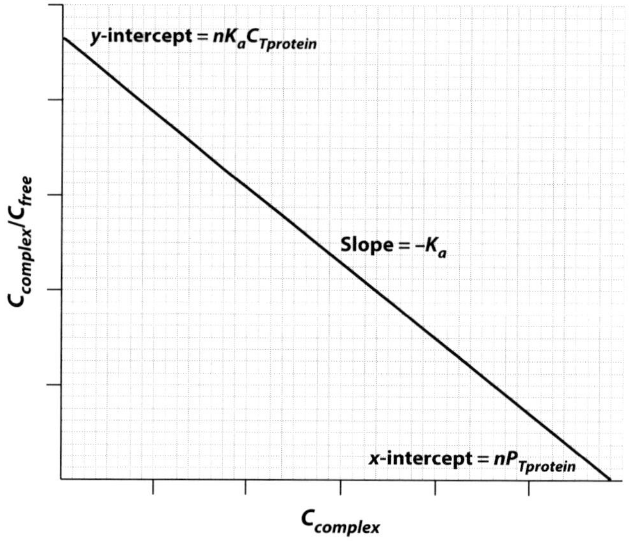

FIGURE 3.6 Plot of bound to free concentration ratio versus concentration of bound according to Equation 3.6 for estimation of K_a from slope and n from y-intercept or x-intercept.

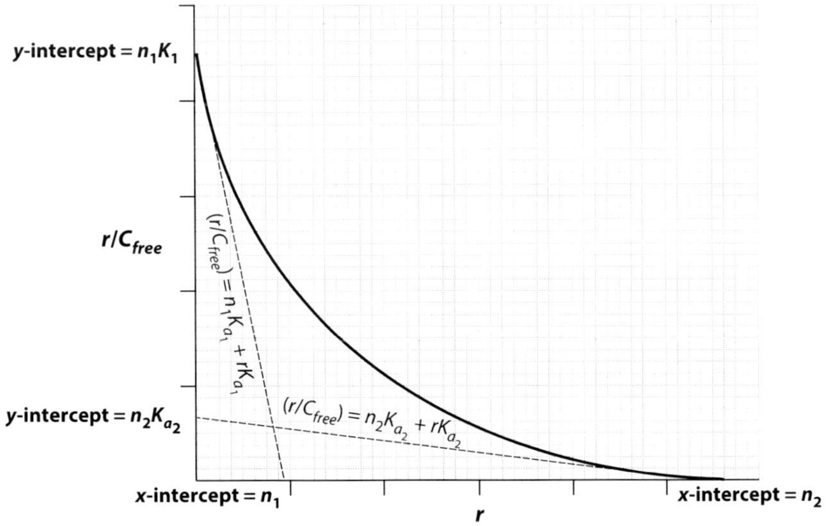

FIGURE 3.7 Scatchard plot of binding to a protein with two different classes of binding sites; the line with y-intercept of $n_1K_{a_1}$ represents the "high-affinity, low-capacity" class of binding sites, and the line with y-intercept of $n_2K_{a_2}$ represents the "low-affinity, high-capacity" class of binding sites.

residue, the second type of binding site. When the drug molecule has affinity for two types of binding sites, the Scatchard plot of r/C_{free} versus r is not a straight line (Figure 3.7) and extrapolation of the initial and terminal linear segments of the curve yields two straight lines with equations

$$\frac{r}{C_{free}} = n_1K_{a_1} + rK_{a_1} \tag{3.39}$$

$$\frac{r}{C_{free}} = n_2K_{a_2} + rK_{a_2} \tag{3.40}$$

As the affinity and capacity of the two binding sites are different, the line with greater y-intercept and smaller x-intercept is identified as "high-affinity, low-capacity" sites, whereas the line with smaller y-intercept and greater x-intercept represents the "low-affinity, high-capacity" binding sites.

3.2.4 Physicochemical Characteristics of Xenobiotics

Lipophilicity, among the physicochemical attributes of a compound, plays a principal role in its distribution in the body. The permeation of lipophilic compounds in the highly perfused organs, in particular, the liver and kidney (perfusion-limited distribution) and adipose tissue (permeability-limited distribution) is significant. When the plasma concentration declines, the amount stored in the adipose tissue is released into the systemic circulation for redistribution, elimination and also re-uptake by the adipose tissue, although it may not be significant. The redistribution often creates a lingering pharmacological response such as long-term drowsiness in the case of certain barbiturates or anesthetics. As discussed in Chapter 2, Section 2.2.1.1, the ability of a lipophilic compound to permeate through the capillary wall or partition into an organ/tissue is expressed in terms of its partition coefficient.

The degree of ionization also plays an important role in the distribution of xenobiotics. Most compounds are salts of weak acids or bases and, according to pH–partition theory, only unionized form of drugs can penetrate through the capillary wall and cellular barrier. Therefore, the pKa of xenobiotics and pH of the environment that they are introduced in are also important in their permeation and distribution in the tissue (see also Chapter 2).

3.2.5 Extent of Penetration through Physiological Barriers and Parallel Removal Processes

The biological mechanisms discussed in Chapter 2 not only govern the absorption and permeation of xenobiotics through biological barriers, but also influence their distribution and elimination. For example, efflux and transport protein in organs of elimination, mainly the liver and kidneys, not only influence the elimination, but also the distribution of xenobiotics in the body. To signify these shared characteristics among ADME processes, terms such as disposition are used to imply complementary and collective processes of distribution and elimination. Elimination is also terminology used for the combined processes of metabolism and excretion. Thus, disposition is a term used to refer to the distribution metabolism and excretion of xenobiotics.

The expression and activity of efflux proteins (P-glycoproteins) and their distribution throughout the body play a critical role in the disposition of xenobiotics (Ayrton and Morgan, 2001; Fromm, 2003; Fenneteau et al., 2009). Those distributed in the kidneys maintain the total body homeostasis by selective elimination of endogenous and detoxification of exogenous compounds, and those in the liver coordinate the detoxification processes with the role of metabolizing enzymes. In addition to the important role efflux proteins play in the disposition of xenobiotics (Silverman, 1999; Fromm, 2000; Litman et al., 2001; Schinkel and Jonker, 2003), their presence in all physiological barriers, which morphologically are different from each other, are also significant and influence the overall disposition profile of a compound.

3.2.6 Physiological Barriers

The function of physiological barriers in the body is essentially to protect different organs/regions from xenobiotics. There are libraries of literature on each of these barriers. The main physiological barriers, which influence the distribution and PK/PD or TK/TD parameters, are briefly discussed here.

3.2.6.1 Blood–Brain Barrier

The distribution of xenobiotic into the brain and the permeation from the blood into the extracellular space is highly regulated and is limited by the blood–brain barrier. The tight junctions of endothelial cells of cerebral capillaries create this barrier. Nutrients are selectively transported from the blood to the brain through specific receptors or transporters. These transporters are located on the luminal and abluminal of the endothelium. They regulate the passage of nutrients like glucose or amino acids and prevent the permeation of xenobiotics. Examples are the glucose transporter (GLUT-1), large neutral amino acid transporter, organic anion transporters (OATP), and the monocarboxylic acid transporter (MCT1). The transfer of xenobiotics by pinocytosis through this tight barrier is very rare and the transport of water-soluble drugs similar to nutrients requires a specific carrier or receptor. Viruses and large molecules like peptides and protein can be transferred by fluid phase endocytosis (Hawkins and Egleton, 2007), receptor-mediated transcytosis and adsorptive mediated transcytosis (Abbott, 2004). The permeation of lipophilic compounds occurs by passive transcellular diffusion (Abbott et al., 2006) and therefore their passage is dependent on their lipophilicity and molecular size (Alavijeh et al., 2005; Chancellor et al., 2012; Waring, 2009). The higher the partition coefficient, the easier is the passage through the barrier. However, this permeation is not absolute even for highly lipophilic compounds if their molecular weight is above 500–600 Da (Pardridge, 1998). Because of the presence of tight junctions, the paracellular diffusion practically does not occur. However, under certain pathological conditions leakage in the junctions may occur, which may allow the permeation of proteins or xenobiotics (Cirrito et al., 2005; Abbott et al., 2006, 2010; Bennett et al., 2010; Pittet et al., 2011).

Brain capillary endothelial cells express Pgp (Sun et al., 2003; Fromm, 2003; Löscher and Potschka, 2005; Bauer et al., 2012); thus, the low permeation may also be explained by the occurrence of this efflux protein at the luminal face of the blood–brain barrier. It has been shown that the brain uptake of the xenobiotics can be enhanced when the body is subjected to the Pgp inhibitors (Sadeque et al., 2000; Bauer et al., 2005). MDR1Pgp and MDR2Pgp are the two types of Pgp in humans. MDR1Pgp is responsible for xenobiotics efflux at the blood–brain barrier. In rodents both *mdr1a*Pgp and *mdr1b*Pgp are present in the brain, but in two different locations: *mdr1a* is localized in brain capillaries whereas *mdr1b* is only in brain parenchyma. It is believed that Pgp is expressed by the gene encoding the protein (MDR1) at the apical membrane of brain capillary endothelial cells in humans and other mammals. There are other transport proteins in the brain, including multidrug resistance-associated protein (MRP) family, which has several members and they transport a wide range of anticancer drugs out of the cells, including the tumor cells. MRP2 has been identified as an important MRP in the barrier. Contrary to Pgp, which interacts with diverse groups of xenobiotics, MRPs interact and transport anionic and neutral compounds conjugated to acidic ligands. A more recent efflux protein identified in the brain is the breast cancer resistance protein (BCRP).

All the influx and efflux proteins are expressed in the blood–brain barrier and the blood-CSF-barrier (BCSFB), and become over-expressed in certain diseases or conditions (Figure 3.8), which adds to the challenges of overcoming the multiple drug resistance (Löscher and Potschka, 2005). It should also be mentioned that continuous production of CSF (cerebrospinal fluid) is an important factor complementing the role of the barrier by transporting xenobiotics across arachnoid granulations into the systemic circulation. The function of efflux and most influx proteins is an active and energy demanding process. However, this active transport is not uniform for all xenobiotics and it depends on whether the distribution into the brain is a permeability- or flow-limited distribution.

In summary, Pgp, MRPs, and BCRP in association with the tightly joined capillary endothelial cells severely restrict the distribution of xenobiotics in the brain (Figure 3.8).

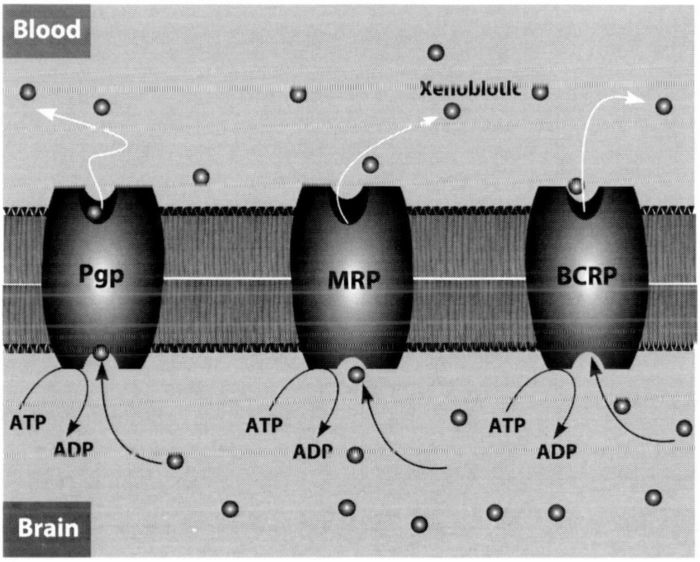

FIGURE 3.8 Restriction of distribution of xenobiotic in the brain due to the presence of ABC efflux transporters P-glycoprotein (Pgp), multidrug resistance protein (MRP) family and breast cancer resistance protein (BCRP) in association with the tightly joined capillary endothelial cells.

The two important parameters of PK/TK analysis of brain uptake are identified as the extent of brain equilibrium with plasma concentration and the time to achieve this equilibrium (Liu et al., 2005). The extent of uptake depends largely on the partition coefficient of the xenobiotic, which in turn depends on the free fraction of the compound in plasma, the fraction removed by efflux proteins, the extent of metabolism in the brain and removal by the cerebrospinal fluid. It has been suggested that the ratio of brain to plasma concentration at the steady state is a more relevant partition coefficient, reflecting the brain uptake.

Among the different approaches to PK/TK analysis and estimation of parameters and constants of the brain uptake and the influence of blood–brain barrier (Tunblad et al., 2003), the following is a practical one (Liu et al., 2005; Peng et al., 2001). The assumptions of the model include: (1) the brain intracellular and extracellular compartments are considered two separate and interconnected compartments, where the intravascular space is in exchange with the central compartment of a two compartment model, (2) only free concentration in plasma may cross the blood–brain barrier and equilibrate in each compartment, (3) the uptake and efflux clearance is equal to a parameter identified and quantified as a permeability-surface area product (*PS*) representing the distribution clearance across the barrier by passive diffusion, (4) CSF flow does not significantly influence the compound's disposition in the brain. The model can be perceived as a four-compartment model where the central compartment (plasma and highly perfused organs/tissues) is in exchange with the peripheral compartment (less perfused tissues) and brain intravascular space (Figure 3.9).

When the drug is injected through an administration route with complete absorption, the mass balance equations are

Central compartment:

$$\frac{dA_1}{dt} = V_1\frac{dCp}{dt} = (k_a \times Dose) - (Cl \times Cp) - Q(Cp - (C_{brain})_{\text{int}}) - Cl_{12}(Cp - C_2) \quad (3.41)$$

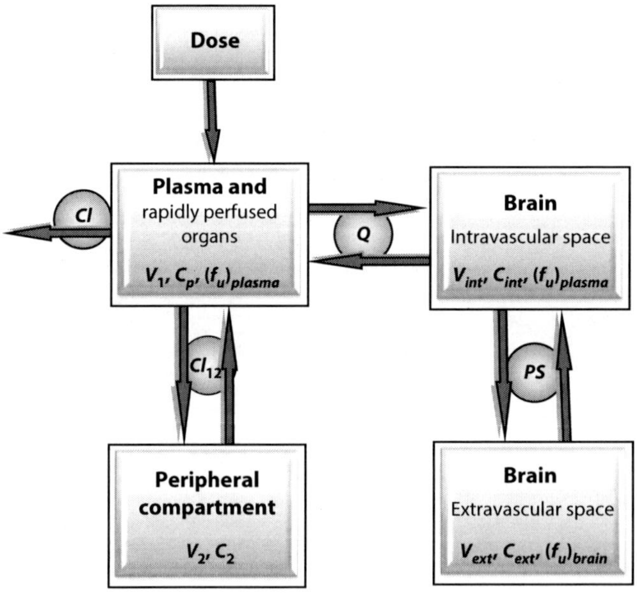

FIGURE 3.9 Diagram of a four-compartment model for estimation of parameters and constants of the brain uptake and influence of blood–brain barrier; plasma and highly perfused organs/tissues are the central compartment in exchange with the peripheral compartment (less perfused tissues) and the intravascular space of the brain in exchange with the extravascular space of the brain.

where the rate of change in the amount of the central compartment is $V_1(dCp/dt)$ (mass/time); the absorption rate constant is k_a (time^{-1}); the plasma concentration is Cp (mass/volume); the systemic clearance is Cl (volume/time); the rate of elimination from the body is $ClCp$ (mass/time) the cerebral blood flow is Q (volume/time); the concentration in the intravascular compartment of the brain is $(C_{brain})_{int}$; The concentration of the peripheral compartment is C_2; and the distribution clearance between the central and peripheral compartment is Cl_{12}.

Peripheral compartment:

$$\frac{dA_2}{dt} = V_2\left(\frac{dC_2}{dt}\right) = k_{12}A_1 - k_{21}A_2 = Cl_{12}(Cp - C_2) \tag{3.42}$$

where the rate of change in the peripheral compartment is dA_2/dt; the volume of the peripheral compartment is V_2; the concentration of the peripheral compartment is C_2; and the distribution rate constants are k_{12} and k_{21}, respectively, and represent $k_{12} = Cl_{12}/V_1$ and $k_{21} = Cl_{12}/V_2$.

The intravascular compartment of the brain:

$$\frac{dA_{int}}{dt} = V_{int}\frac{dC_{int}}{dt} = Q(Cp - C_{int}) - PS(f_u)_{plasma}\left(C_{int} - \frac{C_{ext}}{K_p}\right) \tag{3.43}$$

where the amount in the brain intravascular compartment is A_{int}; the volume of the intravascular compartment is V_{int}; the rate of change in the intravascular amount is $V_{int}(dC_{int}/dt)$; the unbound fraction in the plasma is $(f_u)_{plasma}$; the xenbiotic concentration in the intravascular and extravascular compartments are C_{int} and C_{ext}, respectively; the equilibrium brain/plasma partition coefficient is $K_p = (f_u)_{plasma}/(f_u)_{brain}$; and the uptake-efflux clearance between intravascular and extravascular compartments is PS.

The extravascular compartment of the brain:

$$\frac{dA_{ext}}{dt} = V_{ext}\frac{dC_{ext}}{dt} = \left(PS \times (f_u)_{plasma}\right)\left(C_{int} - \frac{C_{ext}}{K_p}\right) \tag{3.44}$$

where the volume and concentration of the extravascular compartment are V_{ext} and C_{ext}, respectively.

The initial condition of the differential Equations 3.41 through 3.44 is: all concentration terms at time zero are equal to zero, unless the dose is injected intravenously where at time zero, the plasma concentration is $CP_{t=0} = Dose/V_1$. If the dose is administered intravenously, the rate equations are

$$V_1\frac{dCp}{dt} = -Cl \times Cp \tag{3.45}$$

$$V_b\frac{dC_b}{dt} = \left(PS \times (f_u)_{plasma} \times Cp\right) - \left(PS \times (f_u)_{brain} \times C_b\right) \tag{3.46}$$

where the volume of brain issue is V_b and the concentration of compound in the brain is C_b.

The integration of Equations 3.45 and 3.46 yields equations that define the time course of xenobiotic concentration in plasma and the brain

$$Cp = \frac{Dose}{V_1}e^{-(Cl/V_1)t} \tag{3.47}$$

$$C_b = \frac{Dose \times PS \times (f_u)_{plasma}}{V_b \times V_1\left((PS \times (f_u)_{brain})/V_b - (Cl/V_1)\right)}\left(e^{-(Cl/V_1)t} - e^{-\left((PS\times(f_u)_{brain})/(V_b)\right)t}\right) \quad (3.48)$$

Setting $(Cl/V_1) = k_{10}$ the rate constant of elimination from the central compartment and $(PS \times (f_u)_{brain})/(V_b) = k_{out}$ the rate constant of elimination from the brain and dividing Equation 3.48 by Equation 3.4 yields the ratio of brain concentration to plasma concentration as follows:

$$\frac{C_b}{Cp} = \frac{PS \times (f_u)_{plasma}}{V_b(k_{out} - k_{10})}(1 - e^{-(k_{out}-k_{10})t}) \quad (3.49)$$

Therefore, when $k_{out} < k_{10}$ the ratio increases over time, the compound resides longer in the brain with a half-life longer than plasma, and never reaches a plateau. When $k_{out} > k_{10}$ the ratio increases initially, reaches a plateau with half-life equal to the plasma concentration.

3.2.6.2 Blood–Lymph Barrier

Physiologically, the balance of body water is maintained by the lymphatic system. This system drains the extra volume of interstitial fluids with all their waste products and proteins, known as lymph, into the blood circulation. The lymphatic system, similar to the blood system, is formed from lymphatic capillaries that join together and form lymphatic vessels. The vessels usually follow the same route as the blood vessels. They go through successive sets of lymph nodes to filter the lymph and finally join blood vessels at the junction of the jugular and left subclavian veins. The exchange of fluids containing electrolytes, hormones, gasses, nutrients and xenobiotics between blood, tissue and blood–lymph barrier are coordinated effectively to maintain interstitial fluid and cellular homeostasis. For example, in the liver, water and solutes are transported into the sinusoidal space and interstitial area where they drain into lymphatic collecting vessels (Henriksen et al., 1984). Therefore, the liver blood–lymph barrier consists of a series of areas, barriers, and pathways connected sequentially, which are: lining wall of sinusoids, sinusoidal space, and interstitial fluid ending with terminal lymphatic capillaries.

Because of its extensive network, the lymphatic system plays an important role in the distribution of lipophilic compounds. Small polar molecules are usually excluded from distribution by the lymphatic system, which is mainly due to the anatomical structure of the barrier and lymphatic capillaries. Proteins with an effective radius of about 120 Å with a large molecular weight are able to cross this barrier. In large experimental animals (e.g., dogs) solid spherical particles with a radius of up to 700 Å are able to cross the blood–lymph barrier.

3.2.6.3 Placental Barrier

The placenta receives nutrients and oxygen from maternal arteries for distribution through fetal capillaries and disposes fetal waste products and carbon dioxide into maternal veins. The placenta barrier is essentially a dynamic fetal:maternal interface for regulating the exchange of essential nutrients, hormonal levels, gases, metabolic wastes, toxins, and drugs. The barrier is between fetal and maternal systemic circulations and is made up of various cell layers consisting of an endothelial cell layer, a thin layer of connective tissue, and a continuous syncytiotrophoblast with some individual trophoblasts underneath, which are thicker at the early phase of pregnancy and gradually become thinner near term, such that the distance of over 50 μm at the late second month may get reduced to less than 5 μm by the 37th week of pregnancy. Nutrients such as glucose, amino acids, minerals, some vitamins, purines, and pyrimidines actively transported across the placenta against the concentration gradient. The barrier, although considered remarkable for its function, often acts as a leaky filter for certain xenobiotics and teratogenic compounds. Xenobiotics may actively or passively cross the barrier and enter the fetal circulation. Several prominent efflux transporters have been identified in the placenta suggesting a more active role for the barrier in regulating exposure of the fetus to xenobiotics circulating in the maternal circulation. The efflux proteins and enzyme systems of the developing fetus may not be able to adequately extrude or metabolize the compound and may cause embryotoxicity or teratogenicity. The distribution of xeobiotics within the fetus follows essentially the same pattern as in the maternal body, with some differences related to the volume of drug distribution, plasma-protein binding, blood circulation, and greater permeability of other membranous barriers.

In addition to the presence of multilayer cells that form the physical fetal:maternal barrier, the placenta and the fetus are equipped with enzymes of phase I and phase II metabolism. The metabolism of xenobiotics complement the role of the placental barrier in protecting the fetus (Ganapathy et al., 2000; Syme et al., 2004). Cytochrome P450 (CYP) enzymes are characterized in the placenta at the level of mRNA, protein, and enzyme activity and CYP1A1, 2E1, 3A4, 3A5, 3A7, and 4B1 have been detected in the term placenta (Syme et al., 2004). Furthermore, the influx and efflux proteins are abundant in the placenta (Allikmets et al., 1998; Myllynen et al., 2007; Vähäkangas and Myllynen, 2009) and include: ATP-binding cassette transporters (ABC transporters: ABCB1/MDR1/Pgp, ABCC1-3/MRP1-3, ABCG2/BCRP [breast cancer resistance protein]) (Hitzl et al., 2004; Ceckova-Novotna et al., 2006; Behravan and Piquette-Miller, 2007); organic anion transporters (OAT: SLC21A8/OATP-8, SLC21A12/OATPA-E, SLC22A11/OAT4) (Babu et al., 2002); organic cation transporters (OCT: SLC22A3/OCT3, SLC22A4/OCTN1, SLC22A5/OCTN2); 5-HT (serotonin) transporter (SLC6A4/SERT), as well as a noradrenaline transporter (SLC6A2/NET) (Bottalico et al., 2004; Ganapathy and Prasad, 2005). The efflux proteins are mostly expressed on the apical membrane of syncytiotrophoblast facing maternal blood, and thus their role is to protect the fetus against xenobiotics. There are significant interspecie differences in the expression of transport proteins, which should be taken into consideration in investigative teratogenicity of therapeutic agents or chemicals intended for use in the environment.

3.2.6.4 Blood–Testis Barrier

The capillary endothelium and the Sertoli cells mainly create this barrier against the uptake of xenobiotics and it is often called Sertoli cell barrier (Russell, 1977). The Sertoli cells, which play an important role in spermatogenesis, are tightly joined together and form an additional layer after the endothelium of capillaries, making this barrier similar to the blood–brain barrier. The barrier has a complex pattern of transport protein expressions, where several proteins are highly expressed, whereas others are expressed at lower levels or are absent in Sertoli cells. A differential expression of protein transporters in Sertoli cells of adult Sprague Dawley rats has been observed (Augustine et al., 2005), which may play an important role in the distribution of xenobiotic in the male reproductive system. The over expressed efflux and influx protein in Sertoli cells include multidrug resistant protein 1, 7, and 8 (Mrp1, Mrp7, Mrp8); testis specific transporter 1 and 2 (Tst1, Tst2); organic cation/carnitine transporter (OCTN2); Wilson's disease zinc transporter 1 (Znt1); and equilibrative nucleoside transporter 1 and 2 (Ent1 and Ent2). Testicular toxicity due to heavy metal (lead, cadmium, arsenic, etc.) poisoning occurs often in various species, but with variable sensitivity (Liu et al., 2001).

The testis, due to the presence of metabolizing enzyme systems such as isozymes of Cytochrome P450, is capable of metabolizing the molecules that succeed in crossing the barrier. The testis microsomal incubation of various experimental animals has shown interspecies differences in CYP1A1, CYP1A2, and CYP1B1 isoforms (Smitha et al., 2007). The barrier itself and the gonadal metabolism limit the distribution of xenobiotics into the testis.

Similar to the blood–brain barrier, the transfer of drugs by pinocytosis through this tight barrier is very rare and the permeation of lipophilic compounds is by passive diffusion. In addition to heavy metal poisoning, the testis is susceptible to mutagens and cancerous cells.

3.2.6.5 Blood–Aqueous Humor Barrier

The ocular barrier may not have a major impact on the overall distribution of xenobiotics in the body, but it limits the distribution to a vital region of the body. For delivery systems of therapeutic agents intended for the retina, it poses a major challenge particularly for retinal disorders and related visual impairment. Even the systemic administration of the compounds is strictly regulated by ocular barrier system. Upon access of a xenobiotic to the systemic circulation, a fraction of a dose may diffuse into the anterior chamber of the eye; however, some compounds do not appear there at all. This restriction, which is mostly for small hydrophilic compounds and large molecular weight proteins and polymers, is analogous to the function of the blood–brain barrier. Similar to the blood–brain barrier, the presence of tight junctions of capillary endothelium and transport proteins impede the free exchange of drugs between blood and the barriers. The ATP-binding cassette (ABC) protein such as Pgp, and solute carrier (SLC) transporters, such as ABCB, ABCC, ABCG, SLC7, SLC16, SLC19, SLCO/SLC21A, SLC22A, and SLC29 transporters are all present and influence the permeability of drugs across the blood–ocular barriers (Tomi and Hosoya, 2010).

3.2.7 Body Weight and Composition

The distribution of xenobiotics in the body is influenced by the body weight. A lower lean body mass and higher adipose tissue would increase the physiological volume of distribution and uptake of lipophilic compounds. Doses of therapeutic agents are usually adjusted according to the Ideal Body Weight (IBW), or Body Surface Area (BSA). Body Mass Index (BMI) and Lean Body Mass (LBM) are also used. The following are the equations used to determine these quantities:

3.2.7.1 Ideal Body Weight (IBW in kg)

$$(\text{♂}): \text{IBW (kg)} = 50\text{ kg} + 2.3\text{ kg for each inch over 5 feet} \tag{3.50}$$

$$(\text{♀}): \text{IBW (kg)} = 45 + 2.3\text{ kg for each inch over 5 feet} \tag{3.51}$$

3.2.7.2 Body Surface Area (BSA in m²)

DuBois Formula (Du Bois et al., 1916):

$$BSA = 0.007184 \times Weight(\text{kg})^{0.425} \times Height(\text{cm})^{0.725} \tag{3.52}$$

Mosteller Formula (Mosteller, 1987):

$$BSA = \left(\frac{Height(\text{cm}) \times Weight(\text{kg})}{3600}\right)^{1/2} \tag{3.53}$$

It is also common to use the values reported in Table 3.2 for children of different age groups.

3.2.7.3 Body Mass Index (BMI in kg/m²)

$$BMI(\text{kg/m}^2) = \frac{Weight(\text{kg})}{Height(\text{m})^2} \tag{3.54}$$

$$BMI(\text{kg/m}^2) = \frac{Weight(\text{lb})}{Height(\text{inch})^2} \times 703.07 \tag{3.55}$$

TABLE 3.2
Body Surface Area of Different Age Groups

Children		Adults	
Age (year)	**BSA (m^2)**	**♂ (m^2)**	**♀ (m^2)**
New Born	0.25	1.6	1.9
2	0.5	Average	
9	1.07	1.73 (m^2)	
10	1.14		
12–13	1.33		

3.2.7.4 Lean Body Mass (LBM in kg)

Lean body mass, or lean body weight is the estimation of body weight without adipose tissue, and is different from Ideal Body Weight (IBW). The IBW takes into consideration the normal proportion of fat required for an ideal body weight.

The formulas for estimation of LBM are (Hume, 1966)

$$(♂)\text{: } LBM(\text{kg}) = (1.10 \times Weight(\text{kg})) - 128 \times (Weight^2/(100 \times Height(\text{m}))^2 \tag{3.56}$$

$$(♀)\text{: } LBM(\text{kg}) = (1.07 \times Weight(\text{kg})) - 148 \times (Weight^2/100 \times Height(\text{m}))^2 \tag{3.57}$$

An alternative equation is

$$(♂)\text{: } LBM(\text{kg}) = (0.32810 \times Weight(\text{kg})) + (0.33929 \times Height(\text{cm})) - 29.5336 \tag{3.58}$$

$$(♀)\text{: } LBM(\text{kg}) = (0.29569 \times Weight(\text{kg})) + (0.41813 \times Height(\text{m}))^2 - 43.2933 \tag{3.59}$$

3.2.8 Disease States

Certain disease states can modify the normal distribution of xenobiotics in the body, and the PK/TK parameters can change to values different from the normal states. Briefly they are

3.2.8.1 Congestive Heart Failure (CHF)

CHF is associated with hypoperfusion to various organs and tissues, particularly highly perfused organs such as the liver and kidneys. Thus, CHF not only reduces the physiological volume of distribution significantly, it can reduce the overall elimination and clearance of xenobiotics.

3.2.8.2 Acute and Chronic Kidney Disease (CKD)

Depending on the degree of severity of renal failure, the plasma protein binding can become impaired, and thus influence not only the distribution of xenobiotics, but also the reduction of the excretion of unchanged compound and elimination of water-soluble metabolites. It may even modify the linear profile of a compound to capacity-limited and nonlinearity of PK/TK parameters.

3.2.8.3 Hepatic Diseases

Hepatic diseases such as hepatitis, alcohol liver, hepatic cancers, liver cirrhosis, hepatic steatosis and many other genetic or non-genetic liver diseases can modify the metabolic profile of xenobiotics significantly and reduce the rate of formation of metabolites, thus, increasing the free fraction in plasma, modifying the equilibrium of protein binding, and changing not only the profile of elimination, but also their distribution in the body.

3.2.8.4 Cystic Fibrosis (CF)

CF decreases the deposition of particulate compounds administered through inhalation, for example, aerosols and other pulmonary delivery systems. It also modifies the distribution and free fraction of xenobiotics in the body, in particular aminoglycosides and other antibiotics. The disposition profile of xenobiotics in CF patients is usually different than with healthy lung individuals.

3.2.8.5 Other Conditions

Dehydration, edema caused by sunburn, pregnancy, etc., and malnutrition can modify the disposition of xenobiotics in the body significantly.

REFERENCES

Aarons, L. J., Rowland, M. 1981. Kinetics of drug displacement interactions. *J Pharmacokin Biopharm* 9(2):181–90.

Aarons, L., Grennan, D. M., Siddiqui, M. 1983. The binding of ibuprofen to plasma proteins. *Eur J Clin Pharmacol* 25:815–19.

Abbott, N. J. 2004. Prediction of blood–brain barrier permeation in drug discovery from *in vivo*, *in vitro* and in silico models. *Drug Discov Today Technol* 1:407–16.

Abbott, N. J., Patabendige, A. A. K., Dolman, D. E. M., Yusof, S. R., Begley, D. J. 2010. Structure and function of the blood–brain barrier. *Neurobiol Dis* 37:13–25.

Abbott, N. J., Ronnback, L., Hansson, E. 2006. Astrocyte-endothelial interactions at the blood–brain barrier. *Nat Rev Neurosci* 7:41–53.

Alavijeh, M. S., Chishty, M., Qaiser, M. Z., Palmer, A. M. 2005. Drug metabolism and pharmacokinetics, the blood–brain barrier, and central nervous system drug discovery. *Neurotherapeutics* 2:554–71.

Allikmets, R., Schriml, L. M., Hutchinson, A., Romano-Spica, V., Dean, M. 1998. A human placenta-specific ATP-binding cassette gene (ABCP) on chromosome 4q22 that is involved in multidrug resistance. *Cancer Res* 58:5337–39.

Augustine, L. M., Markelewicz, R. J. Jr., Boekelheide, K., Cherrington, N. J. 2005. Xenobiotic and endobiotic transporter mRNA expression in the blood–testis barrier. *Drug Metab Dispo* 33(1):182–89.

Ayrton, A., Morgan, P. 2001. Role of transport proteins in drug absorption, distribution and excretion. *Xenobiotica* 31:469–97.

Babu, E., Takeda, M., Narikawa, S., Kobayashi, Y., Enomoto, A., Tojo, A. 2002. Role of human organic anion transporter 4 in the transport of ochratoxin A. *Biochim Biophys Acta* 1590:64–75.

Bauer, M., Zeitlinger, M., Karch, R., Matzneller, P., Stanek, J., W Jäger, W., Böhmdorfer, M. et al. 2012. Pgp-mediated interaction between (R)-[11C]verapamil and tariquidar at the human blood–brain barrier: A comparison with rat data. *Clin Pharmacol Ther* 91(2):227–33.

Bauer, B., Hartz, A. M. S. Fricker, G., Miller, D. S. 2005. Modulation of P-glycoprotein transport function at the blood–brain barrier *Exp Biol Med* 230(2):118–27.

Behravan, J., Piquette-Miller, M. 2007. Drug transport across the placenta, role of the ABC drug efflux transporters. Expert. Opin. *Drug Metab Toxicol* 3:819–80.

Bennett, J, Basivireddy, J., Kollar, A., Biron, K. E., Reickmann, P., Jefferies, W. A., McQuaid, S. 2010. Blood–brain barrier disruption and enhanced vascular permeability in the multiple sclerosis model EAE. *J Neuroimmunol* 229:180–91.

Bhattacharya, A. A., Grüne, T., Curry, S. 2000. Crystallographic analysis reveals common modes of binding of medium and long-chain fatty acids to human serum albumin. *J Mol Biol* 303:721–32.

Bottalico, B., Larsson, I., Brodszki, J., Hernandez-Andrade, E., Casslén, B., Marsál, K. 2004. Norepinephrine transporter (NET), serotonin transporter (SERT), vesicular monoamine transporter (VMAT2) and organic cation transporters (OCT1, 2 and EMT) in human placenta from pre-eclamptic and normotensive pregnancies. *Placenta* 25:518–29.

Bowmer, C. J., Lindup, W. E. 1978. Binding of phenytoin, L-tryptophan and o-methyl red to albumin. Unexpected effect of albumin concentration on the binding of phenytoin and L-tryptophan. *Biochem Pharmacol* 27:937–42.

Buur, J. L., Baynes, R. E., Smith, G. W., Riviere, J. E. 2009. A physiologically based pharmacokinetic model linking plasma protein binding interactions with drug disposition. *Res Vet Sci* 86(2):293–301.

Ceckova-Novotna, M., Pavek, P., Staud, F. 2006. P-glycoprotein in the placenta: Expression, localization, regulation and function. *Reproductive Toxicol* 22(3):400–10.

Chancellor, M. B., Staskin, D. R., Kay, G. G., Sandage, B. W., Oefelein, M. G., Tsao, J. W. 2012. Blood–brain barrier permeation and efflux exclusion of anticholinergics used in the treatment of overactive bladder. *Drugs Aging* 29:259–73.

Cirrito, J. R., Deane, R., Fagan, A. M., Spinner, M. L., Parsadanian, M., Finn, M. B. Jiang, H. et al. 2005. P-glycoprotein deficiency at the blood–brain P-glycoprotein deficiency at the blood–brain barrier increases amyloid-beta deposition in an Alzheimer disease mouse model. *J Clin Invest* 115(11): 3285–90.

Curry, S., Mandelkow, H., Brick, P., Franks, N. 1998. Crystal structure of human serum albumin complexed with fatty acid reveals an asymmetric distribution of binding sites. *Nature Struct Biol* 5:827–35.

Du Bois, D., Du Bois, E. F. 1916. A formula to estimate the approximate surface area if height and weight be known. *Arch Internal Med* 17(6): 863–71.

Edwards, D. J., Lalka, D., Cerra, F., Slaughter, R. L. 1982. Alpha acid glycoprotein concentration and protein binding in trauma. *Clin Pharmacol Ther* 31(1):62–7.

Ekman, B., Sjodin, T., Sjoholm, I. 1980. Binding of drugs to human serum albumin. XV. Characterization and identification of the binding sites of indomethacin. *Biochem Pharmacol* 29:1759–65.

Fenneteau, F., Turgeon, J., Couture, L., Michaud, V. Li, J., Nekka, F. 2009. Assessing drug distribution in tissues expressing P-glycoprotein through physiologically based pharmacokinetic modeling: Model structure and parameters determination. *Theor Biol Med Model* 6:2. http://www.ncbi.nlm.nih.gov/pmc/articles/PMC2661039/

Fromm, M. F. 2000. P-glycoprotein: A defense mechanism limiting oral bioavailability and CNS accumulation of drugs. *Int J Clin Pharmacol Ther* 38:69–74.

Fromm, M. F. 2003. Importance of P-glycoprotein for drug disposition in humans. *Eur J Clin Invest* 33[Suppl 2]: 6–9.

Ganapathy, V., Prasad, P. D. 2005. Role of transporters in placental transfer of drugs. *Toxicol Appl Pharmacol* 207:381–87.

Ganapathy, V., Prasad, P. D., Ganapathy, M. E., Leibach, F. H. 2000. Placental transporters relevant to drug distribution across the maternal–fetal interface. *J Pharmacol Exp Ther* 294 (2):413–20.

Ghuman, J., Zunszaina, P. A., Petitpasa, I., Bhattacharya, A. A., Otagirib, M., Curry, S. 2005. Structural basis of the drug-binding specificity of human serum albumin. *J Mol Biol* 353(1):38–52.

Greenblatt, D. J., Sellers, E. M., Koch-Weser, J. 1982. Importance of protein binding for the interpretation of serum or plasma drug concentration. *J Clin Pharmacol* 22(5–6):259–63.

Hagen, C. H. L. 1839. Uber die Bewegung des Wassers in engen cylindrischen Rohren. *Annalen der Physik und Chemie* 42:423–42.

Hawkins, B. T., Egleton, R. D. 2007. Pathophysiology of the blood–brain barrier: Animal models and methods. In: *Current Topics in Developmental Biology,* eds. Gerald, P. S. New York: Academic Press, 277–309.

Henriksen, J. H., Horn, T., Christoffersen, P. 1984. The blood–lymph barrier in the liver. A review based on morphological and functional concepts of normal and cirrhosis liver. *Liver* 4:221–32.

Herve, F., Urien, S., Albengres, E., Duche, J. C., Tillement, J. P. 1994. Drug binding in plasma. A summary of recent trends in the study of drug and hormone binding. *Clin Pharmacokinet* 26:44–58.

Hitzl, M., Schaeffeler, E., Hocher, B., Slowinski, T., Halle, H., Eichelbaum, M. 2004. Variable expression of P-glycoprotein in the human placenta and its association with mutations of the multidrug resistance 1 gene (MDR1, ABCB1). *Pharmacogenetics* 14:309–18.

Hume, R., Weyers, E. 1971. Relationship between total body water and surface area in normal and obese subjects. *J Clin Pathol* 24:234–38.

Hume, R. 1966. Prediction of lean body mass from height and weight. *J Clin Path* 19(4):389–91.

Koch-Weser, J., Sellers, E. M. 1976. Binding of drugs to serum albumin. *N Engl J Med* 294:526–31.

Kratochwil, N. A., Huber, W., Muller, F., Kansy, M., Gerber, P. R. 2002. Predicting plasma protein binding of drugs: A new approach. *Biochem. Pharmacol* 64:1355–74.

Langmuir, I. 1916. The constitution and fundamental properties of solids and liquids. Part I. solids. *J Am Chem Soc* 38:2221–95.

Litman, T., Druley, T. E., Stein, W. D., Bates, S. E. 2001. From MDR to MXR: New understanding of multidrug resistance systems, their properties and clinical significance. *Cell Mol Life Sci* 58:931–59.

Liu, X., Smith, W. J., Chen, C., Callegari, E., Becker, S. L., Chen, X. Julie Cianfrogna, J. et al. 2005. Use of a physiologically based pharmacokinetic model to study the time to reach brain equilibrium: An experimental analysis of the role of blood–brain barrier permeability, plasma protein binding, and brain tissue binding. *J Pharmacol Exp Ther* 313(3):1254–62.

Liu, J., Corton, C., Dix, D. J., Liu, Y., Waalkes, M. P., Klaassen, C. D. 2001. Genetic background but not metallothionein phenotype dictates sensitivity to cadmium-induced testicular injury in mice. *Toxicol Appl Pharmacol* 176:1–9.

Löscher, W., Potschka, H. 2005. Blood–brain barrier active efflux transporters: ATP-binding cassette gene family. *J Am Soc Exper Neurother* 2(1):86–98.

Mattos, C., Ringe, D. 1996. Locating and characterizing binding sites on proteins. *Nature Biotech* 14:595–99.

McNamara, P. J., Levy, G., Gibaldi, M. 1979. Effect of plasma protein and tissue binding on the time course of drug concentration in plasma. *J Pharmacokin Biopharm* 7:195–206, 1979.

Mosteller, R. D. 1987. Simplified calculation of body-surface area. *N Engl J Med* 317:1098.

Myllynen, P., Pasanen, M., Vähäkangas, K. 2007. The fate and effects of xenobiotics in human placenta. *Expert Opin Drug Metab Toxicol* 3:331–46.

Nowak, I., Shaw, L. M. 1995. Mycophenolic acid binding to human serum albumin: Characterization and relation to pharmacodynamics. *Clin Chem* 41(7):1011–17.

Pardridge, W. M. 1998. CNS drug design based on principles of blood–brain barrier transport. *J Neurochem* 5:1781–92.

Peng, B., Andrews, J., Nestorov, I., Brennan, B., Nicklin, P., Rowland, M. 2001. Tissue distribution and physiologically based pharmacokinetics of antisensephosphorothioate oligonucleotide ISIS 1082 in rat. *Antisense Nucleic Acid Drug Dev* 11:15–27.

Petitpas, I., Grüne, T., Bhattacharya, A. A., Curry, S. 2001. Crystal structures of human serum albumin complexed with monounsaturated and polyunsaturated fatty acids. *J Mol Biol* 314:955–60.

Pittet, C. L., Newcombe, J., Prat, A., Arbour, N. 2011. Human brain endothelial cells endeavor to immunoregulate CD8 T cells via PD-1ligand expression in multiple sclerosis. *J Neuroinflam* 8:155.

Poiseuille, J. L. M. 1838. Ecoulement des Liquides: Societe Philomatique de Paris. Extraits des Procés-Verbaux des Séances Pendant I'Année 1838. Paris: René et Cie, pp. l–3, 77–81.

Pries, A. R., Secomb, T. W., Gessner, T., Sperandio, M. B., Gross, J. F., Gaehtgens, P. 1994. Resistance to blood flow in microvessels *in vivo*. *Circulation Res* 75:904–15.

Pries, A. R., Secomb, T. W., Jacobs, H., Speradio, M., Osterloh, K., Gaehtgens, P. 1997. Microvascular blood flow resistance: Role of endothelial surface layer. *Am J Physiol* 273:H2272–9.

Rowland, M. 1980. Plasma protein binding and therapeutic monitoring. *Ther Drug Monitor* 2:29–37.

Russell, L. 1977. Movement of spermatocytes from the basal to the adluminal compartment of the rat testis. *Am J Anat* 148:313–28.

Sadeque, A. J. M., Wandel, C., He, H., Shah, S., Wood, A. J. J. 2000. Increased drug delivery to the brain by P-glycoprotein inhibition. *Clin Pharmacol Ther* 68:231–37.

Schinkel, A. H., Jonker, J. W. 2003. Mammalian drug efflux transporters of the ATP binding cassette (ABC) family: An overview. *Adv Drug Deliv Rev* 55:3–29.

Shary, W. L., Aarons, L. J., Rowland, M. 1978. Representation and interpretation of drug displacement interactions. *Biochem. Pharmacol* 27:139–44.

Silverman, J. A. 1999. Multidrug-resistance transporters. *Pharm Biotechnol* 12:353–86.

Smith, D. A., Di, L., Kerns, E. H. 2010. The effect of plasma protein binding on *in vivo* efficacy: Misconceptions in drug discovery. *Nat Rev Drug Discov* 9(12):929–39.

Smitha, T. L., Merry, S. T., Harris, D. L., Ford, J. J., Ike, J., Archibong, A. E., Ramesh, A. 2007. Species-specific testicular and hepatic microsomal metabolism of benzo(a)pyrene, an ubiquitous toxicant and endocrine disruptor. *Toxicology in vitro* 21(4):753–58.

Sun, H., Dai, H., Shaik, N., Elmquist, W. F. 2003. Drug efflux transporters in the CNS. *Adv Drug Deliv Rev* 55:83–105.

Sutera, S. P., Skalak, R. 1993. The history of Poiseuille's law. *Ann Rev Fluid Mechanics* 25:1–19.

Syme, M. R., Paxton, J. W., Keelan, J. A. 2004. Drug transfer and metabolism by the human placenta. *Clin Pharmacokinet* 43(8):487–514.

Tomi, M., Hosoya, K. 2010. The role of blood–ocular barrier transporters in retinal drug disposition: An overview. *Expert Opin Drug Metab Toxicol* 6(9):1111–24.

Tunblad, K., Jonsson, E. N., Hammarlund-Udenaes, M. 2003. Morphine blood–brain transport is influenced by probenecid co-administration. *Pharm Res* 20:618–623.

Vähäkangas, K., Myllynen, P. 2009. Transporters in the human blood–placental barrier. *Br J Pharmacol* 158(3):665–67.

Vallner, J. J., Chen, L. 1977. β-Lipoproteins, possible plasma transport proteins for basic drugs. *J Pharm Sci* 66(3):420–1.

Wagner, J. G. 1976. Simple model to explain effects of plasma protein binding and tissue binding on calculated volumes of distribution, apparent elimination rate constants and clearances. *Eur J Clin Pharmacol* 10:425–32.

Waring, M. J. 2009. Defining optimum lipophilicity and molecular weight ranges for drug candidates–molecular weight dependent lower logD limits based on permeability. *Bioorg Med Chem Lett* 19(10):2844–51.

Watson, P. E., Watson, I. D., Batt, R. 1980. Total body water volumes for adult males and females estimated from simple anthropometric measurements. *Am J Clin Nutr* 33:27–39.

Yacobi, A., Stoll, R. G., DiSante, A. R., Levy, G. 1976. Intersubject variation of warfarin binding to protein in serum of normal subjects. *Res Comm Chem Pathol Pharmacol* 14:743–47.

4 Elimination Rate, Clearance, and the Theoretical Models

4.1 INTRODUCTION

Xenobiotics eliminate from the body predominantly by renal excretion and hepatic metabolism. Both processes are discussed in more detail in Chapters 5 and 6. The major route of excretion of unchanged xenobiotics from the body is the kidney. The removal of unchanged compounds may also occur through other routes including biliary, pulmonary, salivary, mammary, and skin/dermal/sweat glands. Liver, the major site of metabolism, is responsible for biotransformation of xenobiotics, in particular lipid-soluble compounds to pharmacologically active and/or inactive metabolites by the enzymatic reactions classified as Phase I and II metabolism (Chapter 6). Phase I metabolism is responsible for the formation of primary metabolites through oxidation, reduction, hydrolysis and hydroxylation, which may or may not be active pharmacologically, or reactive toxicologically. Phase II metabolism is responsible for the conjugation of unchanged and primary metabolites through acetylation, glucuronidation, sulfation, glutathione formation, etc. (see Chapter 6) leading to the formation of water-soluble conjugates. Although the major site of metabolism is the liver, other organs like the kidney, lung, skin, and gastrointestinal tract contribute significantly to the overall metabolism of xenobiotics in the body. The ultimate outcome of metabolism is to modify xenobiotics to water-soluble and less toxic compounds and facilitate their removal from the body.

Thus, the overall elimination from the body is the combination of excretion and metabolism of xenobiotics by the major and minor routes of excretion and metabolism. Through appropriate PK/PD modeling and data analysis, the combined and individual rate constants of excretion, formation of metabolites and removal of metabolites can be estimated. Kinetically, the process of elimination (i.e., combined excretion and metabolism) is either dose independent and linear (i.e., governed by first-order kinetics) or dose dependent and nonlinear (i.e., governed by Michaelis–Menten type equations), or a combination of both linear and nonlinear processes. Linear PK/TK constants and parameters are predictable, consistent, and can be extrapolated to the population of the subject. Whereas, nonlinear PK/TK parameters and constants estimated from a given dose of a compound may not be reproducible from a different dose of the same compound. Therapeutically, to maintain the optimum response, the effective concentration should be monitored continually during the therapy. The detection of nonlinearity for a therapeutic xenobiotic often occurs during the pre-clinical phase of drug discovery and development, and during the toxicity testing, when high doses of the lead compound are used to determine the margin of safety. PK/TK nonlinearity may also be related to nonlinear absorption, or nonlinear protein binding, or tissue uptake, etc. A reason for nonlinearity of elimination is the capacity-limited metabolism by Phase I/II enzyme systems. Often a single dose data of a xenobiotic may predict linear PK/TK, whereas multiple dosing or chronic exposures of the same compound generate dose-disproportionate changes in kinetic constants and parameters. If the chronic exposure follows a constant zero-order kinetics and the elimination is governed by a saturable process, the rate of change of amount in the body can be defined as

$$\frac{dA}{dt} = k_0 - \frac{V_{max} A}{K_M + A} \tag{4.1}$$

where dA/dt is the rate of amount change in the body, k_0 is the zero-order chronic rate of exposure and $V_{max}A/K_M + A$ is the rate of elimination from the body. In terms of concentration and population of patients, the equation changes to (Sedman and Wagner, 1974; Jonsson et al., 2000)

$$\frac{dCp_{ij}}{dt} = \frac{k_0}{V_j} - \frac{(V_{max})_j \, Cp_{ij}}{(K_M)_j + Cp_{ij}} \tag{4.2}$$

where (Cp_{ij}) is the plasma concentration for jth individual at ith time and V_j is the volume of distribution of jth individual. If, in addition to nonlinear removal from the body, a linear first-order excretion occurs simultaneously, that is, is a combination of a zero-order exposure/input, linear excretion output and the the nonlinear metabolism, the equation representing the rate of change of concentration in the body is expressed as

$$\frac{dCp_{ij}}{dt} = \frac{k_0}{V_j} - \left(\frac{(V_{max})_j \, Cp_{ij}}{(K_M)_j + Cp_{ij}} + k_e Cp_{ij} \right) \tag{4.3}$$

where (k_e) is the excretion rate constant; $(k_e Cp_{ij})$ is the excretion rate per units of volume; dCp_{ij}/dt is the rate of concentration change in the body; k_0/V_j is the rate of input and the overall rate output is $\{((V_{max})_j \times Cp_{ij})/((K_M)_j + Cp_{ij})\} + k_e Cp_{ij}$.

4.2 RATE OF ELIMINATION

In PK/TK analysis, several parameters and constants are used to define quantitatively the elimination from an organ. The rate of elimination is one of the parameters described by the following abstract relationship:

$$\begin{aligned} \textit{Rate of elimination (mass/time)} &= (\textit{Rate of input into the organ}) \\ &\quad - (\textit{Rate of output from the organ}) \end{aligned} \tag{4.4}$$

Equation 4.4 indicates that the amount of xenobiotic removed by an organ of elimination per units of time is equal to the amount of compound that permeates from arterial inflow into the organ of elimination per units of time, minus the rate of the remaining amount of xenobiotic in the blood leaving the organ through venous outflow per units of time. Thus, the rates of input and output with units of mass/time are the product of blood flow with units of volume/time and arterial or venous concentration of xenobiotic with units of mass/volume. The general relationship (Equation 4.4) is used to define various rates of elimination. For example, the rate of renal excretion or hepatic metabolism using the above relationship is

$$\begin{aligned} \textit{Rate of excretion} &= (\textit{Blood flow} \times \textit{Arterial concentration}) \\ &\quad - (\textit{Blood flow} \times \textit{Venous concentration}) \end{aligned} \tag{4.5}$$

that is,

$$\frac{dA_e}{dt} = QC_{Arterial-Kidney} - QC_{Venous-Kidney} = Q(C_{Arterial-Kidney} - C_{Venous-Kidney}) \tag{4.6}$$

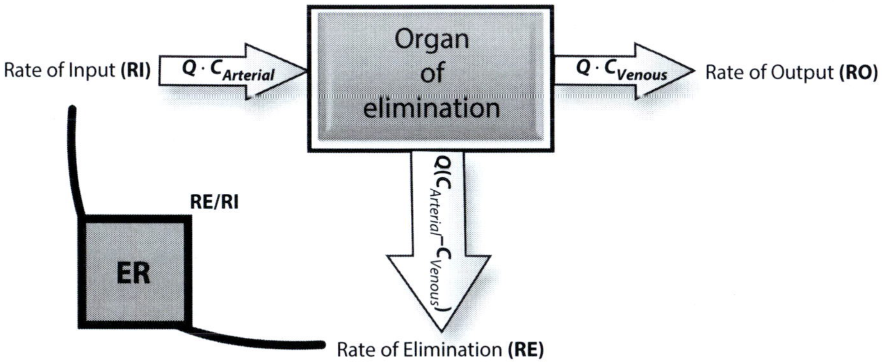

FIGURE 4.1 Schematic diagram of an organ of elimination with input, output and elimination rates and representation of extraction ratio (*ER*).

where A_e is the amount of xenobiotic excreted unchanged in urine, dA_e/dt is the rate of excretion, Q is the blood flow and $C_{Arterial-Kidney}$ and $C_{Venous-Kidney}$ are the arterial and venous concentrations of xenobiotic, respectively (Figure 4.1).

For hepatic metabolism:

$$\begin{aligned} \text{Rate of metabolism} &= (\text{Blood flow} \times \text{Arterial concentration}) \\ &\quad - (\text{Blood flow} \times \text{Venous drug concentration}) \end{aligned} \tag{4.7}$$

$$\frac{dA_m}{dt} = QC_{Arterial-Liver} - QC_{Venous-Liver} = Q(C_{Arterial-Liver} - C_{Venous-Liver}) \tag{4.8}$$

where (dA_m/dt) is the rate of metabolism.

4.3 EXTRACTION RATIO

The extraction ratio (*ER*) is the fraction of a compound that is extracted in a single pass through an organ of elimination and is considered the measure of relative efficiency of the organ. Being a ratio, its magnitude is between zero and one ($0 \le ER \le 1$). An extraction ratio of one indicates the compound eliminates rather completely (i.e., 100%) in a single pass through the organ, and the organ is efficient in removing the entire compound presented by the input. An extraction ratio of zero indicates that the compound evades the elimination completely and its venous concentration is identical to its arterial concentration. The magnitude of extraction ratio depends on the physicochemical characteristics of the xenobiotic and the performance of the elimination organ. Xenobiotics often are categorized as compounds with high extraction ratio when $ER \ge 0.7$, medium extraction ratio when $0.7 > ER > 0.3$ and low extraction ratio when $ER \le 0.3$. A compound may have a high extraction ratio in the kidney, but low extraction ratio in the liver, or vice versa. Thus, *ER* can also be considered an organ-specific number, which represents the relative ability of an organ to extract a compound from the systemic circulation. Dividing the elimination rate of an organ by the rate of input into the organ yields the *ER* for that organ, that is,

$$\text{Extraction ratio} = \frac{\text{Rate of elimination}}{\text{Rate of input}} \tag{4.9}$$

$$ER = \frac{Q(C_{Arterial} - C_{Venous})}{QC_{Arterial}} = \frac{C_{Arterial} - C_{Venous}}{C_{Arterial}} \tag{4.10}$$

The blood flow in and out of the organ is assumed constant equal to blood flow to the organ.

According to Equation 4.10, when ($C_{Venous} \cong C_{Arterial}$) the extraction ratio is equal to zero, which is lack of extraction/elimination by the organ. The overall *ER* that represents the efficiency of the body to eliminate is the summation average of all extraction ratios:

$$ER = \frac{(ER)_{hepatic} + (ER)_{renal} + (ER)_{pulmonary} + \cdots}{n} \tag{4.11}$$

Theoretically, the extraction ratio of a compound in a given population should remain constant as long as the elimination processes follow first-order kinetics (linear pharmacokinetics). The fraction of xenobiotic that escapes the extraction enters into the general systemic circulation and will be available for distribution in the body; this fraction is considered the bioavailability of the compound and is expressed by Equation 4.12

$$F = 1 - ER \tag{4.12}$$

For example, the bioavailability of compound administered orally is estimated as (Lee et al., 2001):

$$Bioavailability = F_{abs} \times (1 - ER)_{GI} \times (1 - ER)_{hepatic} \tag{4.13}$$

where (F_{abs}) is the fraction of the administered dose absorbed into the GI tract wall, $(1 - ER)_{GI}$ is the fraction that escapes gut metabolism and $(1 - ER)_{hepatic}$ is the fraction that escapes hepatic metabolism.

4.4 CLEARANCE

The effectiveness of an elimination process by an organ of elimination is expressed in terms of the organ's clearance. The distinction should be made between clearance and the rate of elimination (Equations 4.4 through 4.8). The rate of elimination by an organ or the overall rate of elimination from the body expresses the elimination in terms of amount eliminated per units of time. Its magnitude depends on the administered dose or plasma/blood concentration. This dependency makes the rate of elimination a variable for linear PK/TK process. In contrast to the rate of elimination, the clearance of an organ of elimination in linear pharmacokinetics is a constant representing the normalized rate of elimination with respect to plasma concentration. Therefore, it is independent of the amount or concentration, and it is defined as the volume of plasma or blood from which the compound is removed per unit of time by an organ of elimination. As a constant, clearance is a measure of an organ's ability to eliminate a xenobiotic. In terms of body clearance, that is, the overall clearance of a compound, it is identified as plasma clearance or blood clearance of free drug.

The clearance is estimated as

$$Clearance = \frac{Rate\ of\ elimination}{Concentration\ of\ input} \tag{4.14}$$

In terms of related variables, Equation 4.14 is defined as

$$Clearance = \frac{Q(C_{input} - C_{output})}{C_{input}} = Q(ER) \tag{4.15}$$

where the rate $Q(C_{input} - C_{output})$ has units of mass/time, and the concentration C_{input} has units of mass/volume. Thus, the clearance is expressed in units of volume/time, that is,

$$\frac{Mass/time}{Mass/volume} = Volume/time$$

According to Equation 4.15, the clearance can also be defined as the product of blood flow (with units of volume/time) and extraction ratio (unitless). Thus, it represents irreversible extraction of a compound by an organ of elimination in a unidirectional input and output. The relationship between extraction ratio, blood flow, and clearance is depicted in Figure 4.2.

Dividing a xenobiotic's clearance by its volume of distribution yields the first-order rate constant of elimination of the compound by that organ, for example, excretion rate constant, metabolic rate constant, etc. The quotient of the body clearance (i.e., plasma or blood clearance) compound and its volume of distribution is the overall first-order rate constant of elimination, that is,

$$Overall\ rate\ constant\ of\ elimination = \frac{Plasma\ clearance}{Volume\ of\ distribution} \quad (4.16)$$

Therefore, the total body clearance is the sum of all clearances and represents a measure of the body's ability to eliminate a xenobiotic. Renal clearance (Cl_r) and metabolic clearance (Cl_m) are considered the most significant ones, but depending on the route of administration, physicochemical properties of xenobitic, and disposition profile of the compound, other clearances such as pulmonary clearance (Cl_p), sweat gland clearance (Cl_{sg}), mammary clearance (Cl_{milk}), and metabolism at sites other than liver, etc., are also parts of the total body clearance. Depending upon their physiological functions, these clearances are mutually independent, and collectively constitute the total body clearance (Cl_T)

$$Cl_T = Cl_r + Cl_m + Cl_{other} \quad (4.17)$$

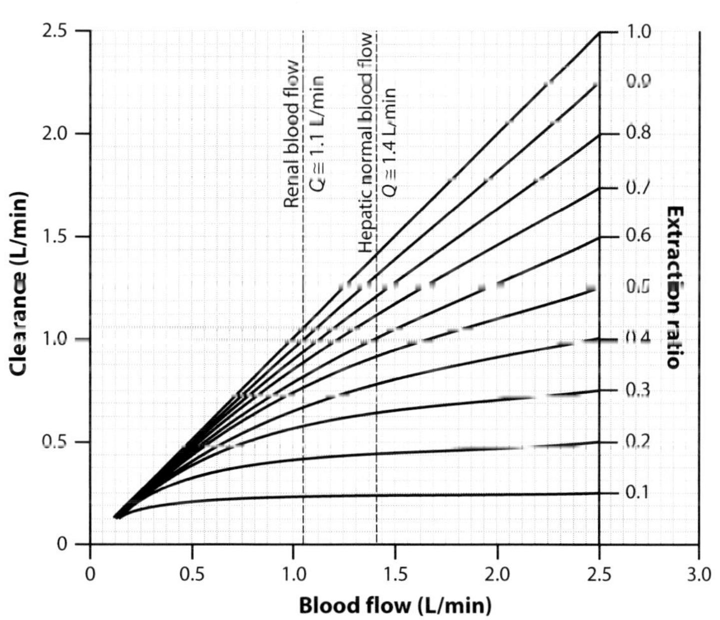

FIGURE 4.2 Plot of clearance as a function of blood flow and extraction ratio according to Equation 4.15.

4.4.1 Estimation of Clearance Using Theoretical Models

To define clearance more clearly and consider the influence of protein binding on the elimination of xenobiotics, various theoretical models have been proposed that take into consideration the magnitude of protein binding and its influence on the clearance (Bass et al., 1976). These models are listed next.

4.4.1.1 Well-Stirred Model

The well-stirred model (Gillette, 1971; Rowland et al., 1973; Wilkinson and Shand, 1975; Pang and Rowland, 1977; Benet and Hoener, 2002; Benet, 2010), also known as the venous equilibration model (Bass, 1979) assumes the organ of elimination, in particular the liver, is a single well-stirred compartment. This means that the compound distributes instantaneously and homogenously throughout the organ (e.g., liver) and the free concentration of fully mixed xenobiotic in the organ is in equilibrium with and identical to the free concentration in the plasma/blood. According to this model, the net clearance of the organ is a function of blood flow (Q), the free fraction of drug in plasma or blood (f_u), and the intrinsic clearance of the organ (Cl_{int})

$$(Clearance)_{organ} = \frac{Q \times (f_u \times Cl_{int})}{Q + (f_u \times Cl_{int})} \tag{4.18}$$

The intrinsic clearance is the ability of the organ to remove free molecules with no limitations due to blood flow or binding to erythrocytes or other binding sites (see also Chapter 6). The relationship (Equation 4.18) essentially means the elimination is a function of blood flow, and governed by perfusion-limited distribution.

When the elimination capacity of an organ ($f_u \times Cl_{int}$) is very high and larger than the blood flow, that is, $f_u \times Cl_{int} \gg Q$, in the denominator of Equation 4.18, Q becomes negligible compared to ($f_u \times Cl_{int}$), the two ($f_u \times Cl_{int}$) terms will be cancelled, and clearance of the organ will be nearly equal to the blood flow

$$(Clearance)_{organ} = \frac{Q \times \overbrace{(f_u \times Cl_{int})}^{Cancel}}{\underbrace{Q}_{negligible} + \underbrace{(f_u \times Cl_{int})}_{Cancel}} \cong Q \tag{4.19}$$

This means the perfusion of the organ limits the elimination of the compounds with high extraction ratio (ER ≥ 0.7).

Conversely, if the elimination capacity is low in relation to the rate of input, that is, $f_u \times Cl_{int} \ll Q$, the ($f_u \times Cl_{int}$) term in the denominator is negligible and Q terms will be cancelled, and the clearance will be equal to the elimination capacity of the organ

$$(Clearance)_{organ} = \frac{\not{Q} \times (f_u \times Cl_{int})}{\not{Q} + \underbrace{(f_u \times Cl_{int})}_{negligible}} \cong f_u \times Cl_{int} \tag{4.20}$$

Most compounds with a low extraction ratio ($ER \leq 0.3$) fall into this category.

The initial model was based on the free fraction of xenobiotics in plasma with the tacit assumption that there is no binding to blood cells. However, for many xenobiotics the magnitude of clearance exceeds the blood flow. One plausible explanation is that the uptake of drugs by blood cells is significant and the organ of elimination extracts the drug not only from the plasma but also from blood cells. Under this condition it is more desirable to express the blood clearance rather than plasma clearance. The analytical determination of the concentration of a compound in whole blood is not an easy task. There are limitations that hinder an accurate assessment of the concentration in blood. One approach in overcoming this problem is to adjust the plasma levels by a correction factor (CF) that is based on the blood hematocrit (Hc) and drug concentration in plasma (C_p), and blood cells (C_{BC})

$$CF = (1 - Hc) + \frac{C_{BC}}{C_p} \tag{4.21}$$

There have been a number of suggested modifications to the well-stirred model that are essentially based on the definition of free fraction and the role of blood cells in defining f_u (Jansen, 1981; Masimirembwa et al., 2003; Yang et al., 2007). For example, if the free fraction in the blood is defined by Equation 4.22, the blood clearance is equal to (Masimirembwa et al., 2003)

$$f_{u\,blood} = f_u\left(\frac{C_p}{C_{blood}}\right)(1 - Hc) \tag{4.22}$$

$$Cl_{blood} = \frac{Q \times \left(f_{u_{blood}} Cl_{int}\right)}{Q(1 - Hc) + \left(f_{u_{blood}} \times Cl_{int}\right)} \tag{4.23}$$

The magnitude of binding to blood cells can be estimated from *in vitro* measurements, where total blood and isotonic phosphate buffer pH 7–7.2 are incubated with equal and known concentrations of xenobiotic simultaneously. After incubation, the concentrations of drug in the plasma portion of the blood and in the buffer are determined. Based on the assumption that the concentration of drug in the buffer is equal to the concentration of whole blood, the ratio of buffer concentration to plasma concentration is used as the correction factor to convert the plasma clearance to blood clearance as follows:

$$\text{Blood clearance} = \text{Plasma clearance} \times \left(\frac{\text{Buffer concentration}}{\text{Plasma concentration}}\right) \tag{4.24}$$

$$Cl_{blood} = Cl_{plasma} \times \frac{[Buffer]}{[Plasma]} \tag{4.25}$$

Hepatic microsomes, or hepatocytes, are used routinely to measure the *in vitro* intrinsic metabolic clearance ($Cl_{int\,invitro}$) of xenobiotics. The estimated values are then used to predict the *in vivo* metabolism profile of a compound and the elimination capacity of an organ (Houston, 1994; McGinnity and Riley, 2001). For the well-stirred model the following equation is proposed to predict the *in vivo* hepatic clearance from *in vitro* data:

$$Cl_m = \frac{Q \times \left(Cl_{int\,invitro} \times SF/f_{u\,inc}\right) \times f_u}{Q + \left(Cl_{int\,invitro}/f_{u_{inc}}\right) \times f_u} \tag{4.26}$$

where Q is the hepatic blood flow, SF is the number of hepatocytes or weight of microsomal protein, $f_{u_{inc}}$ is the free fraction of the compound added to the incubation and f_u is the free fraction in plasma. Under linear conditions and with respect to microsomal protein concentration, or hepatocytes density, the initial rate of metabolism is measured for different concentrations of xenobiotic at different time interval.

There are other theoretical models similar to the well-stirred model that are used to explore and define clearance, in particular hepatic clearance. Among them are the parallel-tube model and the dispersion model.

4.4.1.2 Parallel Model

The parallel model, also known as sinusoidal perfusion model, has been fully investigated, theoretically and quantitatively (Pang and Rowland, 1977; Bass et al., 1977; Pang and Gillette, 1978; Bracken and Bass, 1979; Lau et al., 2002; Ahmad et al., 1983; Ito and Houston, 2004, 2005). The model assumes the organ, notably the liver, as a succession of a parallel tube, with evenly distributed metabolic enzymes around the tubes and declining concentration along the length of each tube. The equation of the parallel-tube model is

$$Cl = Q\left[1 - \exp\left(-\frac{f_u \times Cl_{int}}{Q}\right)\right] \tag{4.27}$$

When the model is used for the hepatic metabolism, (Q) is the hepatic blood flow and Cl_{int} represents the liver intrinsic metabolic clearance (see Chapter 6).

4.4.1.3 Dispersion Model

The dispersion model is based on the residence distribution of a compound in the liver according to the following differential equation, which reflects the dimensions of the liver viewed as a cylinder (Roberts and Rowland, 1986a,b):

$$\frac{\partial C_N}{\partial t_N} = D_N \frac{\partial^2 C_N}{\partial d_N{}^2} - \frac{\partial C_N}{\partial d_N} - R_N C_N \tag{4.28}$$

where (C_N) is the normalized concentration term with respect to the input concentration; (d_N) is the normalized distance along the cylinder with respect to the length of the cylinder (L); (t_N) is the time normalized to the mean residence time of substrate within the cylinder (i.e., liver); (D_N) is defined as the axial dispersion number representing the degree of dispersion as the compound migrates through the liver and is defined as

$$D_N = \frac{D_{coeff}}{vL} = \frac{D_{coeff} A}{QL} \tag{4.29}$$

(D_{coeff}) is the axial diffusion coefficient, (A) is the cross sectional area of blood in the liver, (v) is the average velocity of blood in the liver (Q/A), and (R_N) is calculated as

$$R_N = (f_u)_{blood} \times Cl_{\text{int}} \times \frac{\rho}{Q} \tag{4.30}$$

where ($(f_u)_{blood}$) is the unbound fraction of the compound in the blood, (Q) is the hepatic blood flow and (ρ) is estimated as

$$\rho = \frac{P_{coeff}}{P_{coeff} + Cl_{int}} \tag{4.31}$$

(P_{coeff}) is the permeability coefficient.

The integration of Equation 4.28 at steady state (i.e., $\partial C_N/\partial t_N = 0$) assuming first-order process yields the following equation (Wehner and Wilhelm, 1956):

$$Cl_H = Q_H\left[1 - \frac{4a}{(1+a)^2 e^{[(a-1)/2D_N]} - (1-a)^2 e^{[-(a-1)/2D_N]}}\right] \tag{4.32}$$

where

$$a = \left(1 + 4R_N D_N\right)^{1/2} \tag{4.33}$$

The dispersion model is further discussed in Chapter 6.

4.4.2 Clearance Scale Up in Mammalian Species

4.4.2.1 Extrapolation of Clearance from Animal to Human

The conventional approach for scaling up the inter-mammalian-species extrapolation of PK/TK parameters and constants is using the allometric methodology based on differences in body weight (also discussed in detail in Chapter 18):

$$Y = aW^b \tag{4.34}$$

where (Y) is the parameter or constant under evaluation, (W) is the body weight, "a" is the allometric coefficient and "b" is the allometric exponent. Both a and b are species-related coefficients for (Y).

For example, the extrapolation of *in vivo* clearance from animal to human is

$$Cl = aW^b \tag{4.35}$$

In logarithmic form

$$\log Cl = \log a + b \log W \tag{4.36}$$

For the extrapolation of metabolic clearance from *in vitro* data, the ratio of estimated *in vitro* clearances (i.e., ($Cl_{int\ invitro-human}/Cl_{int\ invitro-animal}$)) is used as a correction factor in the following relationship:

$$Cl_{m_{human}} = Cl_{m_{animal}} \times \left(\frac{Cl_{int_{invitro-human}}}{Cl_{int_{invitro-animal}}}\right) \times \left(\frac{W_{human}}{W_{animal}}\right)^b \tag{4.37}$$

Depending on the species, the value of "b" is between 0.08 and 1.31. The allometric coefficients (a) and (b) are estimated by linear least square analysis of log Cl versus log W based on Equation 4.36 (Figure 4.3).

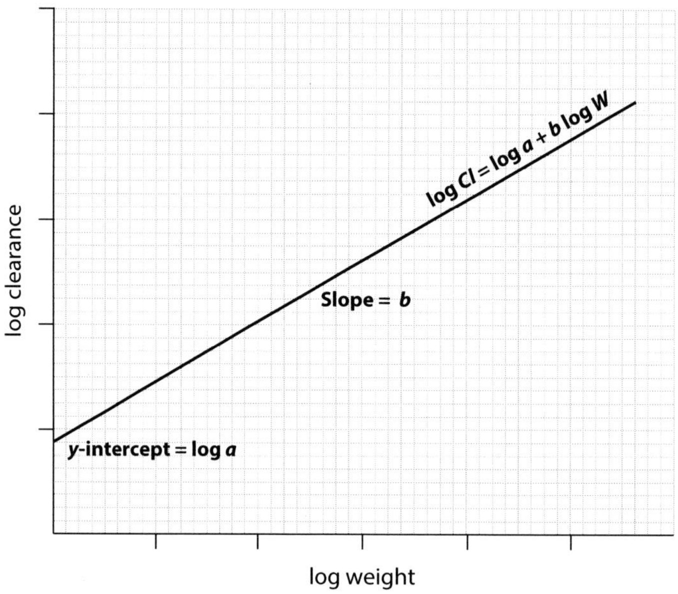

FIGURE 4.3 Plot of logarithm of clearance as a function of logarithm of body weight according to the allometric Equation 4.36.

Equation 4.38 is the "*b*" value for extrapolation from experimental animals to humans (Chiou and Hsu, 1988; Chiou et al., 1998)

$$b = \frac{\log\left(Cl_{human}/Cl_{rat}\right)}{\log\left(W_{human}/W_{rat}\right)} \tag{4.38}$$

Other extrapolation approaches involve using parameters such as maximum life span (MLP), or brain weight (BRW) for incorporation into Equation 4.35 (Boxenbaum, 1980, 1984, 1986; Davidson et al., 1986; Boxenbaum and Dilea, 1995; Feng et al., 1998, 2000; Mahmood and Bailian, 1996)

$$Cl = \frac{a \cdot W^b}{MLP} \tag{4.39}$$

$$Cl = \left(aW^b\right)BRW^c \tag{4.40}$$

The parameter (*MLP*) has been shown to correlate well with *BRW* and *W* (Boxenbaum, 1986; Feng, 2000)

$$MLP = 10.839 \times BRW^{0.636} \times W^{-0.225} \tag{4.41}$$

The inclusion of *MLP* or *BRW* in Equation 4.35 is often based on the value of exponent (*b*) and discretion of the investigator. For example for $b > 0.85$, BRW is included and for $b \le 0.85$ BRW is not included (Feng et al., 2000).

4.4.2.2 Body-Weight Dependent Extrapolation of Clearance in Humans

For scaling the clearance value across a human life span, the b value is used as $b = 2/3$, or $b = 3/4$ (Kleiber, 1932; Brody et al., 1934; Stahl, 1967; Peters, 1983; Lavé et al., 1997). The values are estimated from the slope of logarithm linear plot of the basal metabolic rate against the logarithm of body weight (also discussed in detail in Chapter 18). The allometric scaling of $b = 3/4$ has gained more acceptance for scaling PK/TK parameters and constants like clearance. The 3/4 allometric scaling is used not only in species extrapolation but also for extrapolation in the human life span and changes in body size. Equation 4.42 is a recommended equation for human body-weight scaling of clearance across the human lifespan is (Wang et al., 2012):

$$Cl_i = Cl_{std} \times \left(\frac{W_i}{70}\right)^b \tag{4.42}$$

where Cl_i is the estimated clearance of ith individual with body weight of W_i and (Cl_{std}) is the standard clearance with body weight of 70 kg. Using four different approaches to estimate (b), it has been reported that, for example, for propofol clearance in populations of humans at different ages, the 3/4 allometric scaling was adequate for ages of 1 to 81 years old with a body weight of 9–123 kg, but using the value underestimated the clearance in infants and overestimated in term and preterm neonate (Wang et al., 2012).

4.4.3 Clearance Estimation in Linear Pharmacokinetics

As long as the elimination of a compound follows first-order kinetics, the plasma clearance remains constant and it is used as a proportionality constant to relate the rate of elimination of a drug to its unbound plasma concentration

$$Rate\ of\ elimination = Clearance \times Concentration \tag{4.43}$$

Depending on the type of compound, analytical methodology and relevance of the calculated value to the objectives of the investigation, the term "*Concentration*" refers to plasma or blood concentration of the administered compound, that is,

$$\frac{dA}{dt} = Cl_T \times Cp \quad or \quad \frac{dA}{dt} = Cl_T \times C_{blood} \tag{4.44}$$

where (dA/dt) is the rate of elimination; Cl_T is the total body clearance; Cp and C_{blood} are plasma concentration and blood concentration, respectively.

Based on Equations 4.43 and 4.44, the clearance is estimated as

$$Clearance = Rate\ of\ elimination \div Concentration \tag{4.45}$$

$$Cl_T = \frac{dA/dt}{Cp} \quad \text{or} \quad Cl_T = \frac{dA/dt}{C_{blood}} \tag{4.46}$$

$$Concentration = Rate\ of\ elimination \div Clearance \tag{4.47}$$

$$Cp = \frac{dA/dt}{Cl_T} \tag{4.48}$$

The total body clearance (Cl_T), or plasma clearance in non-compartmental analysis is estimated by the following model independent equation that presumes first-order elimination:

$$Cl_T = \frac{F \times Dose}{AUC} \tag{4.49}$$

where (F) is the bioavailability with a value between zero to one; for intravenous bolus injection $F = 1$ and for the other routes $F \leq 1$. AUC is the area under the plasma concentration–time curve estimated by the trapezoidal or log trapezoidal rules (Appendix A).

In linear compartmental analysis, depending on the number of compartments and the mode of administration, various equations are used that essentially estimate the same constant with model dependent differences.

One-compartment model—IV bolus:

$$Cl_T = K \times Vd \tag{4.50}$$

$$Cl_T = \frac{\int_0^{\infty} dA/dt}{\int_0^{\infty} Cp} \tag{4.51}$$

where K is the overall elimination rate constant; Vd is the apparent volume of distribution; $\int_0^{\infty} dA/dt$ is the total amount of the administered dose that ultimately eliminates from the body, which is equal to the dose for intravenous administration and equal to FD for other routes of administration; $\int_0^{\infty} Cp$ is the area under the plasma concentration–time curve with units of mass time/volume.

One-compartment model—Zero-order input after achieving steady-state plasma concentration (e.g., intravenous infusion):

$$Cl_T = \frac{k_0}{Cp_{ss}} \tag{4.52}$$

where k_0 is the zero-order rate of input and Cp_{ss} is the steady-state plasma concentration.

One-compartment model—First-order input (e.g., oral administration, rectal, sublingual, vaginal, etc.):

$$Cl_T = \frac{FD}{\int_0^{\infty} Cpdt} \tag{4.53}$$

Two-compartment open model—IV bolus:

$$Cl_T = (Vd)_{\beta} \times \beta \tag{4.54}$$

$$Cl_T = V_1 \times k_{10} \tag{4.55}$$

$$Cl_T = \frac{Dose}{AUC} = \frac{Dose}{\frac{A}{\alpha} + \frac{B}{\beta}} \tag{4.56}$$

where β is the first-order hybrid rate constant of disposition; k_{10} is the first-order overall exit rate constant from the central compartment; V_1 and $(Vd)_\beta$ are the volumes of distribution of the central compartment and the overall volume of distribution, respectively, and are estimated as

$$V_1 = \frac{Dose}{A + B} = \frac{Dose}{Cp^0} \tag{4.57}$$

$$(Vd)_\beta = \frac{Dose}{\beta \times AUC} \tag{4.58}$$

where A and B are the y-intercepts of the terminal and extrapolated lines of the bi-exponential curve, which in combination represent the initial plasma concentration that is, $A + B = Cp^0$; and $A/\alpha + B/\beta$ is equal to the area under the plasma concentration–time curve (see also Chapter 10).

Two-compartment open model—Zero-order input: The equation is the same as the one-compartment model that is, equal to the rate of input divided by the steady-state plasma concentration.

One-compartment model—First-order input: The equation is similar to the previous equations of dose divided by the area under the plasma concentration–time curve

$$Cl_T = \frac{F \times Dose}{\int_0^\infty Cpdt} \tag{4.59}$$

4.4.4 Clearance Estimation in Nonlinear Pharmacokinetics

The kinetics of the compounds that are nonlinear is referred to as dose-dependent PK/TK, a scenario that may occur in multiple dosing, intravenous infusions, or any chronic exposures. The dose-dependent PK/TK data deviate from linearity, which reflects the involvement of one or more capacity-limited processes. This often occurs when the compound is extensively metabolized or is subjected to renal tubular active re-absorption and/or secretion. The nonlinear dose-dependent PK/TK is usually described by the Michaelis–Menten equation, where the rate of elimination and clearance are defined as follows:

$$\frac{dA}{dt} = \frac{V_{max} \times Cp}{K_M + Cp} \tag{4.60}$$

$$Cl_T = \frac{dA/dt}{Cp} = \frac{V_{max}}{K_M + Cp} \tag{4.61}$$

The clearance in dose-dependent PK/TK is a variable, directly proportional to the maximum rate of the nonlinear biological process and inversely proportional to the plasma concentration and therefore to dose. Therefore, the calculation of clearance based on the assumption that it is a constant is no longer applicable (see Chapter 9, Section 9.3.2).

REFERENCES

Ahmad, A. B., Bennett, P. N., Rowland, M. 1983. Models of hepatic drug clearance: Discrimination between the well stirred and parallel-tube models. *J Pharm Pharmacol* 35(4):219–24.

Bass, L.1979. Current models of hepatic elimination. *Gastroenterology* 76:1504–5.

Bass, L., Kieding, S., Winkler, K., Tygstrup, N. 1976. Enzymatic elimination of substrates flowing through the intact liver. *J Theor Biol* 61:393–409.

Bass, L., Bracken, A. J. 1977. Time-dependent elimination of substrates flowing through the liver or kidney. *J Theor Biol* 67:637–52.

Benet, L. Z. 2010. Clearance (née Rowland) concept: A downdate and an update. *J Pharmacokinet Phar* 37:529–39.

Benet, L. Z., Hoener, B. A. 2002. Changes in plasma protein binding have little clinical relevance. *Clin Pharmacol Ther* 71: 115–21.

Boxenbaum, H. 1980. Interspecies variation I liver weight, hepatic blood flow, and antipyrene intrinsic clearance: Extrapolation of data to benzodiazepines and phenytoin. *J Pharmacokinet Biop* 2:165–76.

Boxenbaum, H. 1984. Interspecies pharmacokinetic scaling and the evolutionary-comparative paradigm. *Drug Metab Rev* 15:1071–21.

Boxenbaum, H. 1986. Time concepts in physics, biology, and pharmacokinetics. *J Pharm Sci* 75:1053–62.

Boxenbaum, H., Dilea, C. 1995. First-time-in-human dose selection: Allometric thoughts and perspectives. *J Clin Pharmacol* 35:957–66.

Bracken, A. J., Bass, L. 1979. Statistical mechanics of hepatic elimination. *Mathl Biol Sci* 44:97–120.

Brody, S., Proctor, R. C., Ashworth, U. S. 1934. Basal metabolism, endogenous nitrogen, creatinine, and sulphur excretions as functions of body weight. *Univ Mo Agric Exp Sta Res Bull* 220:1–40.

Chiou, W. L., Robie, G., Chug, S. M., Wu, T-C, Ma, C. 1998. Correlation of plasma clearance of 54 extensively metabolized drugs between humans and rats: Mean allometric coefficiet of 0.66. *Pharm Res* 15(9):1474–79.

Chiou, W. L., Hsu, F. H. 1988. Correlation of unbound clearance of fifteen extensively metabolized drugs between humans and rats. *Pharm Res* 5:668–72.

Davidson, I. W. F., Parker, J. C., Beliles, R. P. 1986. Biological basis for extrapolation across mammalian species. *Regul Toxicol Pharmacol* 6:211–37.

Feng, M. R., Rossi, D., Strenkoski, C., Black, A., DeHart, P., Lovdahl, M., McNally, W. 1998. Disposition of cobalt mesoporphyrin in mice, rat, monkey, and dog. *Xenobiotica* 4:413–426.

Feng, M. R., Lou, X., Brown, R. R., Hutchaleelaha, A. 2000. Allometric pharmacokinetic scaling toward the prediction of human oral pharmacokinetics. *Pharm Res* 17(4):410–18.

Gillette J. R., 1971. Factors affecting drug metabolism. *Ann NY Acad Sci* 179:43–66.

Houston, J. B. 1994. Utility of *in vitro* drug metabolism data in predicting *in vivo* metabolic clearance. *Biochem Pharmacol* 47:1469–79.

Ito, K., Houston, J. B. 2004. Comparison of the use of liver models for predicting drug clearance using *in vitro* kinetic data from hepatic microsomes and isolated hepatocytes. *Pharm Res* 21(5):785–792.

Ito, K., Houston, J. B., 2005. Prediction of human drug clearance from *in vitro* and preclinical data using physiologically based and empirical approaches. *Pharm Res* 22(1):103–12.

Jansen, J. A. 1981. Influence of plasma protein binding kinetics on hepatic clearance assessed from a Tube model and a Well-Stirred model. *J Pharmacokinet Biop* 9(1):15–26.

Jonsson, E. N., Karlsson, M. O., Wade, J. R. 2000. Nonlinearity detection: Advantages of nonlinear mixed-effects modeling. *AAPS Pharm Sci* 2(3):114–23.

Kleiber, M. 1932. Body size and metabolism. *Hilgardia* 6:315–33.

Lavé, T., Dupin, S., Schmitt, C. 1997. Integration of *in vitro* data into allometric scaling to predict hepatic metabolic clearance in man: Application to 10 extensively metabolized drugs. *J Pharm Sci* 86:584–90.

Lau, Y. Y., Krishna G., Yumibe, N. P., Grotz, D. E., Sapidou, E., Norton, L., Chu, I., Chen, C., Soares, A.D., Lin, C. C. 2002. The use of *in vitro* metabolic stability for rapid selection of compounds in early discovery based on their expected hepatic extraction ratios. *Pharm Res* 19(11):1606–10.

Lee, Y-H., Perry, B. A., Lee, H-S., Kunta, J. R., Sutyak, J. P., Sinko, P. J. 2001. Differentiation of gut and hepatic first-pass effect of drugs: 1. Studies of verapamil in ported dogs. *Pharm Res* 18:1721–28.

Mahmood, I., Bailian, J. D. 1996. Interspecies scaling: Predicting clearance of drugs in humans. Three different approaches. *Xenobiotica* 26(9):887–95.

Masimirembwa, C. M., Bredberg, U., Andersson, T. B. 2003. Metabolic stability for drug discovery and development: Pharmacokinetic and biochemical challenges. *Clin Pharmacokinet* 42:515–28.

McGinnity, D. F., Riley, R. J. 2001. Predicting drug pharmacokinetics in humans from *in vitro* metabolism studies. *Biochem Soc Trans* 29:135–39.

Pang, K. S., Rowland, M. 1977. Hepatic clearance of drugs I. Theoretical consideration of a well stirred model and a parallel tube model. Influence of hepatic blood flow, plasma and blood cell binding, and the hepatocellular enzymatic activity on hepatic drug clearance. *J Pharmacokinet Biop* 5:625–53.

Pang, K. S., Gillette, J. R. 1978. A theoretical examination of the effects of gut wall metabolism, hepatic elimination, and enterohepatic recycling on estimates of bioavailability and hepatic blood flow. *J Pharmacokinet Biop* 6:355–67.

Peters, H. P. 1983. Physiological correlates of size. In *The Ecological Implications of Body Size*, eds. E. Beck, H. J. B. Birks and E. F. Conner, 48–53pp. Cambridge: Cambridge University Press.

Roberts, M. S., Rowland, M. 1986a. A dispersion model of hepatic elimination; 1. Formulation of the model and bolus considerations. *J Pharmacokinet Biop* 14(3):227–60.

Roberts, M. S., Rowland, M. 1986b. A dispersion model of hepatic elimination; 1. Formulation of the model and bolus considerations. *J Pharmacokinet Biop* 14(3):261–88.

Rowland, M., Benet, L. Z., Graham, G. G. 1973. Clearance concepts in pharmacokinetics. *J Pharmacokinet Biop* 1:123–35.

Sedman, A. J., Wagner, J. G. 1974. Importance of the use of the appropriate pharmacokinetic model to analyze *in vivo* enzyme constants. *J Pharmacokinet Biop* 2:161–73.

Stahl, W. R. 1967. Scaling of respiratory variables in mammals. *J App Physiol* 22:453–600.

Wang, C., Peeters, M. Y. M., Allegaert, K., van Oud-Alblas, H. J. B., Krekels, E. H. J., Tibboel, D., Meindert Danhof, M., Knibbe, C. A. J. 2012. A bodyweight-dependent allometric exponent for scaling clearance across the human life-span. *Pharm Res* 29:1570–81.

Wehner, J. F., Wilhelm, R. M. 1956. Boundary conditions of flow reactor. *Chem Eng Sci* 65:89–93.

Wilkinson, G. R., Shand, D. G. 1975. A physiological approach to hepatic drug clearance. *Clin Pharmacol Ther* 18:377–90.

Yang, J., Jamie, M., Yeo, K. R., Rostami-Hodjegan, A., Tucker, G. T. 2007. Misuse of the well-stirred model of hepatic drug clearance. *Drug Metab Dispos* 35(3):501–2.

5 Renal Excretion of Xenobiotics and Estimation of Related PK/TK Parameters and Constants

5.1 INTRODUCTION

The kidneys are the major organs of excretion for water-soluble endogenous and exogenous compounds. In addition to the excretion, kidneys have the capacity to metabolize xenobiotics, play a central role in the clearance and PK/TK of xenobiotics and their metabolites, and act as the principle organs in management of metabolic waste of endobiotics (Hall et al., 1983; Masereeuw et al., 2000; Shitara et al., 2005). The physiological processes of renal elimination, which contribute to the magnitude of its clearance, consist of four different processes: glomerular filtration, active tubular secretion, active reabsorption, and passive reabsorption (Figure 5.1). In PK/TK analysis, urinary data is considered important in mass balance studies, simultaneous curve fitting in compartmental analysis, and provides information on the metabolic profile and clearances that are not readily attainable from other biological samples. The urinary data since 1977 is considered by the United States Food and Drug Administration as an alternative approach if the bioanalytical method lacked the appropriate sensitivity to characterize a pharmacokinetic profile of a compound from plasma concentration–time data. Furthermore, as it will be discussed in this chapter, the integrity of the renal function is essential for achieving an optimum therapeutic outcome from normal dosage regimen of therapeutic agents.

5.2 GLOMERULAR FILTRATION

The rate at which the glomeruli of the kidney removes catabolic products, xenobiotics, and other impurities in the blood is referred to as the glomerular filtration rate (GFR). GFR is defined as the volume of blood that is filtered per units of time. The permeability of the capillary wall of the glomerulus is due to the large pore size, which is approximately 25 times more permeable than regular capillary walls. The large capillary pore sizes in association with the cardiac contraction force facilitate the transcapillary passage of water, nutrients, ions, and ionized and unionized xenobiotics. The normal GFR is about 125 mL/min or 7.5 L/h or 180 L/day. GFR varies among healthy individuals with a range between 110 and 125 mL/min and it represents how much filtrate is removed from the blood per minute. The normal rate of plasma flow (RPF) through both kidneys is approximately 650 mL/min, and the normal rate of blood flow (RBF) is about 1100 1200 mL/min, thus about 10% of renal blood flow is filtered by the glomerular filtration (Leggett and Williams, 1991, 1995; Walton et al., 2004). The glomerular filtration of compounds that follow linear pharmacokinetics is equal to GFR multiplied by the free fraction of the compounds in plasma. The composition of the filtrate is similar to plasma, but without the large molecular weight proteins. The extent of filtration of a compound depends on the molecular weight and protein binding. All compounds of low molecular weight (<60,000 Dalton) are filtered out of the blood. The filtrate travels from the Bowman capsule toward the collecting duct and eventually the bladder during which the body reabsorbs water, nutrients and ions, and secrets other solutes. Ultimately, from the 180 L/day original filtrate, only about one liter per day reaches the bladder and leaves the body as urine.

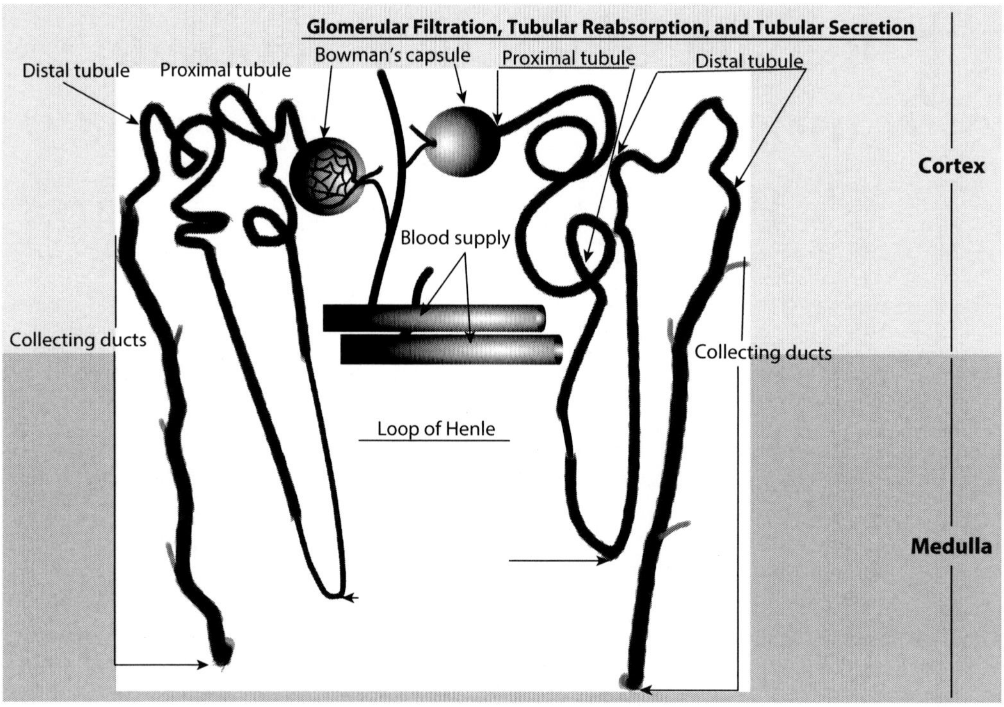

FIGURE 5.1 Nephron, the kidney's primary functional unit, with capillary network (glomerulus) engulf by Bowman's capsule attached to the renal tubule positioned in the cortex and collecting duct and loop of Henle positioned partially in the renal medulla.

5.3 TUBULAR REABSORPTION AND SECRETION

Filtering substances into the tubules and then reabsorbing nearly 100% of them requires energy and may seem to be a very wasteful process, but it allows the body to quickly remove many toxic substances from the blood without being reabsorbed. Many endogenous and exogenous compounds are reabsorbed by active or passive reabsorption in the proximal tubule. Sodium and glucose transported by the proximal tubules occurs with participation of various transport proteins such as the SGLT family, present in the luminal membrane. The majority of organic anions and cations reabsorb by passive diffusion with the concentration gradient being the driving force, and the rate of reabsorption is proportional to the concentration gradient of the compound across the barrier. The passive diffusion is influenced by the urine (filtrate) pH, flow rate, and whether the compound is a weak acid or base. The pH of the filtrate can greatly influence the extent of tubular passive reabsorption for many compounds. When urine is acidic, weak acid drugs tend to be reabsorbed. Alternatively when urine is more alkaline, weak bases are more extensively reabsorbed. Making the urine more acidic reduces reabsorption of weak bases and enhances their excretion. In the case of drug toxicity it is possible to increase the excretion of some xenobiotics by the suitable adjustment of urine pH. For example, excretion of a pentobarbital (a weak acid) overdose may increase by making the urine more alkaline with sodium bicarbonate injection. Studies have indicated that a variety of proteins filtered at the glomerulus are absorbed by the proximal tubule by luminal endocytosis and are hydrolyzed by lysosomal enzymes (Cortney et al., 1970). The transport of intact protein molecules across tubular cells inhuman is inconclusive.

In addition to the active and passive reabsorption, a number of compounds such as radiocontrast agents, aminoglycosides, low molecular weight proteins like peptide hormones (growth hormone, insulin, and glucagon), certain enzymes, and immunoprotein fragments are reabsorbed by endocytosis. The beginning of the proximal tubules is the major sites for endocytosis. The energy required for the active secretion pathway in the proximal tubules is used by many weak acids and electrolytes. Active tubular reabsorption and secretion occur in the proximal tubule and are mediated by transporters, which are expressed at the apical and basolateral membranes of proximal tubules. Thus, the extent of the active processes depends on the affinity of the compound to interact with the transporters. The interaction may become competitive or uncompetitive inhibition in the presence of the other substrates of the proteins, a scenario that may lead to drug–drug interaction. A classical example among therapeutic agents is the inhibition of penicillin by probenecid, which results in a reduction of penicillin excretion and increases penicillin residence time in the body.

In addition to glomerular filtration, the kidney through carrier-mediated pathways eliminates ionized xenobiotics via anion and cation transporters (Pritchard and Miller, 1993; Cha et al., 2001) in proximal tubules. The anion transports play a significant role in the elimination of xenobiotics from the body through the kidneys. The tubular epithelial cell membranes possess both organic cation and organic anion ATP-requiring transport proteins that mediate transport of ions across the cell membrane. Xenobiotics that are actively secreted by the cation system are structurally heterogeneous and include quaternary ammonium salts and primary, secondary, and tertiary amines. The secretion of anionic organic compounds is driven by an active transport mechanism in which the sodium gradient is the driving force. The gradient is governed by Na^+–K^+–ATPase, the organic anion enters the proximal cell in exchange for a dicarboxylate ion. The organic anion-decarboxylate exchange (OAT1) has wide substrate specificity with a large group of anionic xenobiotics. Other examples of transporters involved in tubular secretion are OAT1/3, OCT2, OCTNs, and MATE1, which transport endogenous waste products and xenobiotics and have a wide-ranging substrate specificity (Hediger et al., 1995; Ohashi et al., 1999; Sekine et al., 2000; Hasegawa et al., 2003; Tamai et al., 2004; Koepsell et al., 2007; Tanihara et al., 2007; Muller et al., 2011). There are other transporters that are responsible for the transport of peptides and nutrients, but may also participate in the transport of xenobiotics. Examples are MCTs, SMCTs, PEPT1, SGLT2 (Leibach et al., 1996; Launay-Vacher et al., 2006; Ganapathy et al., 2008; Morris et al., 2008). These transporters have expansive substrate specificity and most have a high capacity for transport, which makes it difficult to predict renal elimination of actively secreted compounds.

Several PK/TK mechanistic models have been proposed to describe the influence of passive and saturable processes of the kidneys on the elimination of xenobiotics (Felmlee et al., 2010, 2011, 2013). The diagram of one of the comprehensive models is presented in Figure 5.2.

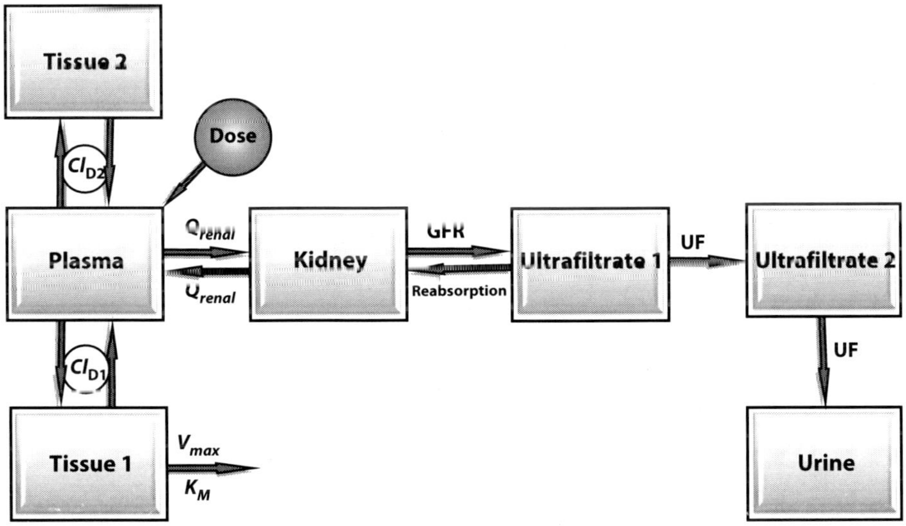

FIGURE 5.2 Diagram of a PK/TK mechanistic model describing the influence of passive and saturable processes on the elimination of xenobiotics (Equations 5.1 through 5.8).

The rate equations of different compartments of the model are as follows (Felmlee et al., 2010):

$$\frac{dA_{plasma}}{dt} = -\left((Q_{renal} + Cl_{D1} + Cl_{D2}) \times \frac{A_{plasma}}{V_{plasma}}\right) + \left(Q_{renal} \times \frac{A_{kidney}}{V_{kidney}}\right) + \left(Cl_{D1} \times \frac{A_{tissue1}}{V_{tissue1}}\right) + \left(Cl_{D2} \times \frac{A_{tissue2}}{V_{tissue2}}\right) \quad (5.1)$$

$$\frac{dA_{tissue1}}{dt} = Cl_{D1} \times \frac{A_{plasma}}{V_{plasma}} - \left(Cl_{D1} + \frac{(V_{max})_{metabolism}}{(K_M)_{metabolism} + (A_{tissue1}/V_{tissue1})}\right) \times \frac{A_{tissue1}}{V_{tissue1}} \quad (5.2)$$

$$\frac{dA_{tissue2}}{dt} = Cl_{D2} \times \frac{A_{plasma}}{V_{plasma}} - Cl_{D2} \times \frac{A_{tissue2}}{V_{tissue2}} \quad (5.3)$$

$$\text{Reabsorption} = \frac{(V_{max})_{Reabsorption}}{(K_M)_{Reabsorption} + (A_{ultrafil-1}/V_{ultrafil-1})} \quad (5.4)$$

$$\frac{dA_{kidney}}{dt} = Q_{renal} \times \frac{A_{plasma}}{V_{plasma}} - (Q_{renal} + GFR) \times \frac{A_{kidney}}{V_{kidney}} + \text{Reabsorption} \times \frac{A_{ultrafil-1}}{V_{ultrafil-1}} \quad (5.5)$$

$$\frac{dA_{ultrafil-1}}{dt} = GFR \times \frac{A_{kidney}}{V_{kidney}} - UF \times \frac{A_{ultrafil-1}}{V_{ultrafil-1}} - \text{Reabsorption} \times \frac{A_{ultrafil-1}}{V_{ultrafil-1}} \quad (5.6)$$

$$\frac{dA_{ultrafil-2}}{dt} = UF \times \frac{A_{ultrafil-1}}{V_{ultrafil-1}} - UF \times \frac{A_{ultrafil-2}}{V_{ultrafil-2}} \quad (5.7)$$

$$\frac{dA_{urine}}{dt} = UF \times \frac{A_{ultrafil-2}}{V_{ultrafil-2}} \quad (5.8)$$

where V_i represents the volume of a different compartment; A_i is the amount in the ith compartment, for example, $A_{ultrafil-2}$ represent the amount in the ultrafiltrated-2 compartment; Cl_{ij} represents the distributional clearance; Q_{renal} and UF are renal blood flow and urine flow, respectively.

The model and related differential equations (Equations 5.1 through 5.8) is proposed for the γ-hydroxybutyric acid and is identified as a mechanistic toxicokinetic model representing a semi-physiological model when the elimination is governed by glomerular filtration, saturable reabsorption, and capacity-limited metabolism (Felmlee et al., 2010).

The overall renal elimination of a compound that follows linear pharmacokinetics, that is, the processes of filtration, reabsorption, and secretion follow first-order kinetics, the renal clearance is a constant of proportionality that relates the rate of excretion to the plasma concentration of unbound xenobiotic:

$$Cl_{renal} = \frac{(dA_e/dt)}{Cp} \tag{5.9}$$

(dA_e/dt) is the rate of excretion and Cp is the plasma concentration of free compound in plasma.

5.4 ESTIMATION OF GLOMERULAR FILTRATION RATE (GFR)

The exogenous or endogenous compounds that do not bind to plasma proteins, filter through the glomerular filtration and leave the body without tubular secretion or reabsorption, have renal clearance identical to the glomerular filtration rate, and are used to estimate GFR. Among the exogenous markers are inulin and a number of radiolabeled chemicals. Among the biological markers, creatinine and cystatin C provide a relatively accurate estimate of GFR. Creatinine is used routinely and urea nitrogen is often mentioned as a biomarker of renal function. However, urea undergoes substantial re-absorption and its clearance depends on urine flow and various pathological and medication factors that modify blood urea nitrogen (BUN). Thus, the clearance of urea might not be a reliable marker of GFR.

The estimated GFR (eGFR) is traditionally used as the best overall indicator of kidney function. The National Kidney Disease Education Program (NKDEP) within the National Institute of Diabetes and Digestive and Kidney Diseases reported the mean of eGFR for normal adult age of 20–29 as 116 mL/min/1.73 m^2; 30–39 as 107 mL/min/1.73 m^2; 40–49 as 99 mL/min/1.73 m^2; 50–59 as 93 mL/min/1.73 m^2; for 60–69 as 85 mL/min/1.73 m^2; for 70 and above as 75 mL/min/1.73 m^2. GFR values of <60 mL/min/1.73 m^2 for 3 months or more, irrespective of cause, are considered chronic kidney disease states. The value of 60 mL/min/1.73 m^2 represents about 50% of an adult's normal kidney function (NKDEP, 2012).

5.4.1 Exogenous Markers of GFR

5.4.1.1 Radioisotope-Labeled Compounds

The exogenous markers have a radioisotope element in their chemical structure and are excreted entirely by glomerular filtration, examples are: ^{14}C-carboxyl-, ^{131}I- and ^{125}I-allyl-inulin; ^{131}I- (also ^{125}I-) dia-trizoate; ^{131}I- (also ^{125}I-) iothalamate; ^{57}Co-vitamin B_{12}; ^{51}Cr-ethylenediamine tetra-acetate, and ^{99m}Technetium (V) Citrate Complex (Kountz et al., 1967). The radiolabeled markers used to determine renal blood flow is ^{125}I-, ^{131}I-Human Serum Albumin (HSA) and the markers for renal plasma flow is radiolabeled para-aminohippuric acid (^{3}H-, ^{14}C-PAH). It is worth noting that markers for plasma and blood flow do not estimate GFR.

5.4.1.2 Inulin

Inulin is a non-radioactive marker that does not bind to plasma proteins; filters through the glomerular filtration; does not secret, re-absorb, or metabolize; and is removed from the body solely by glomerular filtration. Inulin and its derivatives are a group of naturally occurring polysaccharides (polymer of fructose, MW: 5200) produced by plants. They mix evenly with blood components and their plasma clearance is equal to GFR. Its molecular weight is low enough to pass through glomerulus freely. It is used for the determination of both GFR and renal plasma flow (RPF). A derivative of inulin known as Inuset is also used to determine GFR and RPF. To estimate GFR, inulin is infused over a period of time long enough to attain a steady-state level. The half-life of excretion is less than 2 hours. Using Equation 5.11, inulin clearance can be estimated and set equal to GFR. For example, if the steady-state plasma concentration of inulin is 100 mg/dL and the volume of urine collected over three hours is 270 mL with a concentration of 80 mg/L, the GFR can then be estimated as follows:

$$(Cp_{ss})_{inulin} = \frac{k_0}{(Cl_T)_{inulin}} \tag{5.10}$$

$$(Cl_T)_{inulin} = \frac{k_0}{(Cp_{ss})_{inulin}} \tag{5.11}$$

where k_0 is the zero-order rate of infusion, which at steady state is equal to the rate of excretion of inulin, that is, $k_0 = k_e A_e$, and since inulin does not go under metabolism, its rate of excretion = rate of elimination, $k_e A_e = KA$

$$\begin{aligned} KA &= k_e A_e = (270 \text{ mL/3 h}) \times 80 \text{ mg/mL} = 7200 \text{ mg/h} \\ Cl_{inulin} &= GFR = \frac{7200 \text{ mg/h}}{100 \text{ mg/dL}} = 72 \text{ dL/h} = 7200 \text{ mL/h} = 120 \text{ mL/min} \end{aligned} \tag{5.12}$$

Inulin is a good exogenous marker for GFR estimation. However, its drawback is the inconvenience of application, which is considered somewhat cumbersome for outpatients.

Inulin clearance is used occasionally to determine whether a compound is being secreted or reabsorbed in the tubules. If the ratio of clearance of the compound to the clearance of inulin is greater than one, $(Cl_{compound}/Cl_{inulin}) > 1$, it is indicative of tubular secretion and if the ratio is less than one it is indicative of tubular reabsorption $(Cl_{compound}/Cl_{inulin}) < 1$.

5.4.1.3 Iohexol

An alternative to inulin for estimation of GFR is iohexol, a nonionic low osmolar contrast agent used in radiologic procedures. Iohexol excretes completely unmetabolized in the urine by glomerular filtration (Schwartz et al., 2006). When it is used for the estimation of GFR, the dose of iohexol is 10–50 times lower than its dose for radiographic procedures. It has a very low toxicity even at high doses.

5.4.2 GFR Biomarkers

5.4.2.1 Creatinine Clearance

The serum level of creatinine and the clearance of creatinine by the kidney have been used universally in conventional clinical practice as indicators of renal function, roughly reflecting the glomerular filtration rate. Creatinine is a secondary amino acid and anhydride of creatine. It has long been clear that the body is capable of synthesizing whatever creatine, a primary amino acid, it needs since its excretory product, creatinine, appear in constant amounts in the urine even when the diet contains no creatine. Creatinine biosynthesis in mammalian systems requires three enzymes: amidinotransferase (L-arginine: glycin amidinotransferase), *N*-methyltransferase (*S*-adenosylmethionine: guanidoacetate *N*-methyl transferase), and creatine phosphokinase. The overall synthesis of creatinine in the body is presented in the diagram of Figure 5.3.

The loss of creatine is limited. It is assumed that creatine following glomerular filtration is reabsorbed and secreted in the tubules by active transport. Approximately 98% of all creatine in the body is normally accumulated within skeletal muscle. The major loss of creatine from the body occurs through the nonenzymatic formation of creatinine and subsequent urinary excretion. Hence, creatinine excretion is a function of lean body mass in normal individuals and shows little or no response to dietary changes or alteration in electrolyte balance. Creatinine leaves the body after

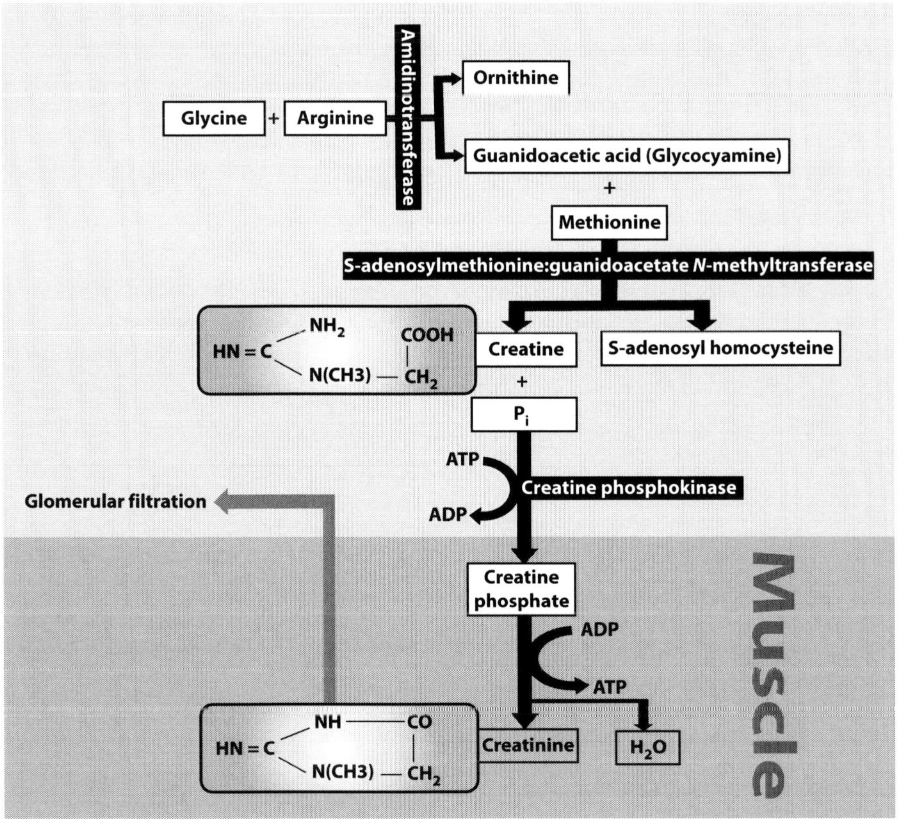

FIGURE 5.3 The overall biosynthesis of creatinine in the body.

the biosynthesis and apparently has no physiological or biochemical role in the body. Due to the zero-order rate of formation in the body, the serum concentration and daily urinary excretion of creatinine are relatively constant, and its clearance unquestionably provides an informative approximation of GFR. Creatinine in the filtrate is not reabsorbed; however, it is weakly secreted. Cimetidine, a histamine receptor antagonist, has been found to inhibit tubular creatinine secretion in patients with renal disease, leading to a creatinine clearance value more representative of the true GFR (Hellerstein et al., 1998). The kinetic analysis of creatinine in the body follows a two-compartment flow model (Boroujerdi and Mattocks, 1979) and since it is considered a non-threshold compound, the creatinin clearance of experimental animals correlates well with human data (Boroujerdi, 1982).

Creatinine clearance, C_{Cr}, can be measured directly by collecting a 24-h urine and a serum sample, or it can be estimated from the serum concentration by using various empirical equations or nomograms.

5.4.2.1.1 Estimation of Creatinine Clearance by Direct Measurement of 24-h Urine Sample

This method requires the complete collection of 24-h urine. After measuring the total amount of creatinine excreted in 24 hours, the rate of excretion is determined by dividing the amount of creatinine by 1440 minutes or 24 hours. By dividing this rate of excretion by serum concentration of creatinine (S_{cr}) with units of mg% (i.e., mg/dL), the creatinine clearance can be estimated as follows:

$$C_{cr}(\text{mL/min}) = \frac{(C_U(\text{mg\%}) \times V_U(\text{mL}))_{24\text{h}}}{S_{cr}(\text{mg\%}) \times 1440} \tag{5.13}$$

where V_U is the volume of urine collected in 24 h, C_U is the urine creatinine concentration of the collected volume, and 1440 is the collection time in minutes.

Equation 5.13 assumes a zero-order production of creatinine and steady-state serum levels. The equation is based on the concept of *Rate/Concentration* = *Clearance*. There are several disadvantages to measuring creatinine clearance using a 24-h urine sample. The complete collection of urine is laborious, sometimes interferes with other diagnostic studies, and the inconvenience involved for outpatients is substantial.

5.4.2.1.2 Estimation of Creatinine Clearance (C_{cr}) from Serum Creatinine (S_{cr})

Numerous investigators and clinicians have sought appropriate mathematical transformation of the data that would result in a linear relationship between the creatinine clearance and serum creatinine so that the value of clearance could be predicted from the plasma concentration (Finney et al., 2000). The relationships are developed for the data from a population of patients that the investigator used for the study. The following have been selected for their historical and practical significance:

Equations with gender and age:

(Jelliffe, 1973):

$$\text{For Men: } C_{cr} = \frac{98 - (16(Age - 20)/20)}{S_{cr}(\text{mg\%})} = \frac{98 - 0.8(Age - 20)}{S_{cr}(\text{mg\%})} \tag{5.14}$$

$$\text{For Women: } C_{cr} = 0.9\left[\frac{98 - 0.8(Age - 20)}{S_{cr}(\text{mg\%})}\right] \tag{5.15}$$

Equations with gender, age, and weight:

(Mawer et al., 1972):

$$\text{For Men: } C_{cr} = \frac{Weight(\text{kg})(29.3 - (0.203 \times Age))}{S_{cr} \times 1.44} \times (1 - 0.03S_{cr}) \tag{5.16}$$

$$\text{For Women: } C_{cr} = \frac{Weight(\text{kg})(25.3 - (0.175 \times Age))}{S_{cr} \times 14.4} - (1 - 0.03S_{cr}) \tag{5.17}$$

(Cockcroft and Gault, 1976):

$$\text{For Men: } C_{cr} = \frac{(140 - Age)(Weight(\text{kg}))}{72 \times S_{cr}(\text{mg\%})} \tag{5.18}$$

$$\text{For Women: } C_{cr} = 0.9\left[\frac{(140 - Age)(Weight(\text{kg}))}{72 \times S_{cr}(\text{mg\%})}\right] \tag{5.19}$$

(Hull et al., 1981):

$$\text{For Men: } C_{cr}(70\,\text{kg}) = \frac{88 \times (145 - Age)}{S_{cr}(\mu\text{mol/L})} - 3 \tag{5.20}$$

$$\text{For Women: } C_{cr}(70\,\text{kg}) = \frac{75 \times (145 - Age)}{S_{cr}(\mu\text{mol/L})} - 3 \tag{5.21}$$

(Agarwal and Nicar, 1994):

$$\text{For Men: } C_{cr} = \frac{Weight(\text{kg}) \times (28.2 - (0.172 \times Age))}{S_{cr}(\text{mg\%}) \times 14.4} \tag{5.22}$$

$$\text{For Women: } C_{cr} = \frac{Weight(\text{kg}) \times (21.9 - (0.115 \times Age))}{S_{cr}(\text{mg\%}) \times 14.4} \tag{5.23}$$

(Toto and Kirk, 1995):

$$\text{For Men: } C_{cr} = -0.3 \times (Age - 52) + \frac{105}{S_{cr}(\text{mg\%})} + (Weight(\text{kg}) - 86) \tag{5.24}$$

$$\text{For Women: } C_{cr} = -0.29 \times (Age - 52) + \frac{88}{S_{cr}(\text{mg\%})} - 0.77(BMI - 30) \tag{5.25}$$

Body mass index (BMI) is calculated according to Equation 3.54 or 3.55 (Chapter 3).

Equations that take gender, weight, age, and body surface area (BSA) into consideration are

$$\text{For Men: } C_{cr} = \frac{(140 - Age) \times 2.12 \times Weight(\text{kg})}{S_{cr}(\text{mg\%}) \times BSA} \tag{5.26}$$

$$\text{For Women: } C_{cr} = 0.85\left[\frac{(140 - Age) \times 2.12 \times Weight(\text{kg})}{S_{cr}(\text{mg\%}) \times BSA}\right] \tag{5.27}$$

The BSA is calculated by Equations 3.52 and 3.53 (Chapter 3) and the ideal body weight (IBW) according to Equations 3.50 or 3.51 (Chapter 3).

Obese patients usually excrete less creatinine per kg body weight; the following equation is proposed for the measurement of GFR in obese patients:

(Salazar and Corcoran, 1988):

$$\text{Men: } C_{cr} = \frac{(137 - age) \times \left[(0.285 \times Weight(\text{kg})) + (12.1 \times (Height(\text{m}))^2)\right]}{51 \times S_{cr}(\text{mg\%})} \tag{5.28}$$

$$\text{Women: } C_{cr} = \frac{(146 - age) \times \left[(0.287 \times Weight(\text{kg})) + (9.74 \times (Height(\text{m}))^2)\right]}{60 \times S_{cr}(\text{mg\%})} \tag{5.29}$$

Weight = Actual body weight in kg.

The advantages of using the empirical equations are the ease and the convenience of estimating GFR by taking one blood sample to determine the serum creatinine. These equations are based mostly on stepwise multiple regression analysis of serum creatinine and creatinine clearance measured in a selected population of patients in the clinic/hospital of the investigator. As the ages of participants in most of these investigations were in the range of 18–70, use of these equations in young children is not recommended. Equations used for pediatric patients are (Schwartz et al., 1976)

$$C_{cr} = \frac{AF \times Height(\text{cm})}{S_{cr}(\text{mg\%})}\left(\frac{BSA}{1.73\ \text{m}^2}\right) \tag{5.30}$$

BSA is estimated according to Equations 3.52 and 3.53 (Chapter 3).

AF is the patient *Age Factor* for patient populations of ages over 1 week old through adolescence (18 years old)

$AF = 0.45$ for infants 1–52 weeks old

$AF = 0.55$ for children 1–13 years old

$AF = 0.55$ for adolescent females 13–18 years old

$AF = 0.7$ for adolescent males 13–18 years old

(Shull et al., 1978):

$$C_{cr} = \frac{100 \times ((0.035 \times Age) + 0.236)}{S_{cr}(\text{mg\%}) \times 1.73\ \text{m}^2} \tag{5.31}$$

Prediction of the creatinine clearance from serum creatinine is not accurate unless renal function and serum creatinine are at the steady state. For example, the prediction of creatinine clearance can be greatly overestimated from serum creatinine in the early phase of acute renal failure before serum creatinine has risen significantly. Furthermore, errors in the laboratory measurement of serum creatinine and biological factors such as blood pressure, pregnancy, and taking antihypertensive medications may mar the true measurement of serum creatinine and GFR. A plan to standardize and enhance the accuracy of serum creatinine measurement has been recommended by the National Kidney Disease Education Program (NKDEP), which is based on creatinine calibration to be traceable to an isotope dilution mass spectrometry (IDMS) reference measurement procedure (Gary et al., 2006). Based on the IDMS data the NKDEP has recommended the following equation with elements of sex, age, BSA, and race:

$$\begin{aligned}\text{GFR}(\text{mL/min/1.73 m}^2) = 175 \times (S_{Cr})^{-1.154} \times (Age)^{-0.203} \times (0.742 \text{ if female})\\ \times (1.212 \text{ if African American})\end{aligned} \tag{5.32}$$

5.4.2.2 Cystatin C

Another biomarker of GFR is Cystatin C (CysC), a non-glycosylated low-molecular-weight (13.3 kDa) basic protein. It is a member of the cystatin superfamily of cysteine protease inhibitors, consisting of 120 amino acids created by all nucleated cells (Grubb, 1992). CysC has been described as the product of a "house-keeping gene," which is suited for a stable production rate by most cells, even during inflammation, and its production remains unmodulated. CysC is freely filtered by the glomerulus (Grubb, 2010; Grubb et al., 2012). It is neither secreted nor reabsorbed in the tubules, and does not metabolize in the body (Simonsen et al., 1985). Thus, its clearance from the body approximates the GFR. Examples of equations used to estimate GFR from Cystatin C concentration are

(Grubb et al., 2005):

Age ≥ 18-year-old men:

$$\text{GFR}(\text{mL/ min /1.73 m}^2) = 86.49 \times \text{Cystatin C(mg/L)}^{1.686} \tag{5.33}$$

Age ≥ 18-year-old women:

$$\text{GFR}(\text{mL/ min /1.73 m}^2) = (86.49 \times \text{Cystatin C(mg/L)}^{1.686}) \times 0.948 \tag{5.34}$$

Children:

$$\text{GFR}(\text{mL/ min /1.73 m}^2) = 84.69 \times \text{Cystatin C(mg/L)}^{1.68} \tag{5.35}$$

For children younger than 14 years of age, Equation 5.35 is multiplied by 1.384.

5.5 ANALYSIS OF URINARY DATA

The total amount of unchanged xenobiotic excreted in urine is the combined amount of unchanged molecules that are filtered plus the amount secreted, if any, minus the amount reabsorbed:

$$(A_e)_{total} = (A_e)_{filtered} + (A_e)_{secreted} - (A_e)_{reabsorbed} \tag{5.36}$$

Thus, the excretion rate is the collective rates of filtration, secretion, and reabsorption of the unchanged compound:

$$\left(\frac{dA_e}{dt}\right) = \left(\frac{dA_e}{dt}\right)_{filtration} + \left(\frac{dA_e}{dt}\right)_{secretion} - \left(\frac{dA_e}{dt}\right)_{reabsorption} \tag{5.37}$$

The corresponding equation for renal clearance is

$$Cl_r = \frac{(dA_e/dt)_t}{Cp_t} \tag{5.38}$$

The renal clearance (Equation 5.38) is defined as the rate of excretion, expressed as differential equation dA_e/dt or difference equation $\Delta A_e/\Delta t$, divided by the plasma concentration or xenobiotic. Renal clearance values can range from 0 mL/min, the normal value for glucose, which is usually completely reabsorbed in healthy humans, to a value equal to the renal plasma flow of about 650 mL/min for compounds like *p*-aminohippuric acid.

For compounds that are fully or partially eliminated in urine, the analysis of the cumulative amount excreted unchanged or eliminated as metabolites provides information that can only be obtained from urinary data. For example, the percent of dose excreted unchanged, or percent of dose eliminated as metabolites, may not be easily estimated from plasma data. In addition, urine is considered a convenient biological sample; it can be collected easily and non-invasively and due to the lack of proteins, it is a cleaner sample than plasma. PK/TK analysis of urinary data have a practical use in the evaluation of therapeutic agents in pediatrics, assessment of bioavailability in humans and experimental animals, and understanding of mass balance of the agents in the body. The volume of urine is a variable that changes irregularly and is a function of several factors, some related to the consumption of fluid and diet and some related to the state of mind and other physiological factors. To exclude the effect of variability of the volume from analysis of urinary data, the appearance of xenobiotics in urine is expressed in terms of amount and not concentration. The amount in the urine is estimated as the amount excreted per interval of urine collection, or as the cumulative amount excreted between the time of dose administration and any time t thereafter. The method of calculation for the amount of drugs in urine is presented in Table 5.1.

The first and second columns of Table 5.1 are the time of urine collection and the collected volume. The third column is the concentration of the compound in the collected sample and the fourth column is the total amount excreted unchanged per each interval of urine collection. The fifth column represents the cumulative amount of xenobiotic excreted in the urine from time zero to any time t. If the time of urine collection from $0 \rightarrow t_n$ were equal or greater than $7T_{1/2}$ of the administered compound, the last value of the cumulative column should represent the total amount excreted unchanged in the urine and the plot of cumulative amount versus time yields a hyperbolic curve as is shown in Figure 5.4 with a plateau level equal to A_e^∞.

5.5.1 PK/TK Analysis of Urinary Data—Bolus Input

The model presented in Figure 5.5 is a one-compartment model with instantaneous (bolus) input and first-order excretion and metabolism.

TABLE 5.1
Stepwise Calculations of Cumulative Amount of Drug Excreted Unchanged in Urine

Time (Δt)	Volume of Urine (V)	Concentration (C_e)	Amount Excreted/Δt (ΔA_e)	Cumulative Amount (A_e)
$t_1 - 0$	V_1	C_{e_1}	ΔA_{e_1}	$A_{e_1} = \Delta A_{e_1}$
$t_2 - t_1$	V_2	C_{e_2}	ΔA_{e_2}	$A_{e_2} = A_{e_1} + \Delta A_{e_2}$
$t_3 - t_2$	V_3	C_{e_3}	ΔA_{e_3}	$A_{e_3} = A_{e_2} + \Delta A_{e_3}$
$t_4 - t_3$	V_4	C_{e_4}	ΔA_{e_4}	$A_{e_4} = A_{e_3} + \Delta A_{e_4}$
: :	: :	:	:	:
$t_n - t_n{-}1$	V_n	C_{e_n}	ΔA_{e_n}	A_{e_n}
$t_n = \infty$				A_e^{∞}

The first-order rate of excretion, that is, the rate of emergence of unchanged compound in the urine, is directly proportional to the concentration or amount of unchanged compound in the body, that is,

$$\frac{dA_e}{dt} = k_e \times A \tag{5.39}$$

$$\frac{dA_e}{dt} = k_e \times (Cp \times Vd) \tag{5.40}$$

$$\frac{dA_e}{dt} = Cl_r \times Cp \tag{5.41}$$

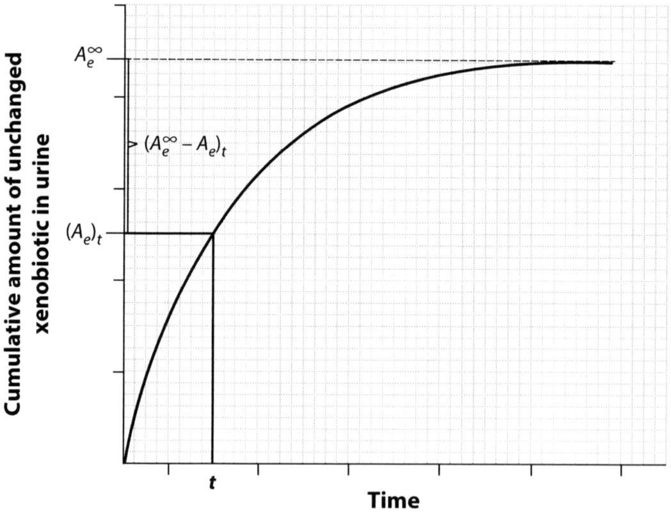

FIGURE 5.4 Plot of cumulative amount of xenobiotic excreted unchanged in urine versus time with a plateau level that corresponds to the total amount excreted unchanged in urine.

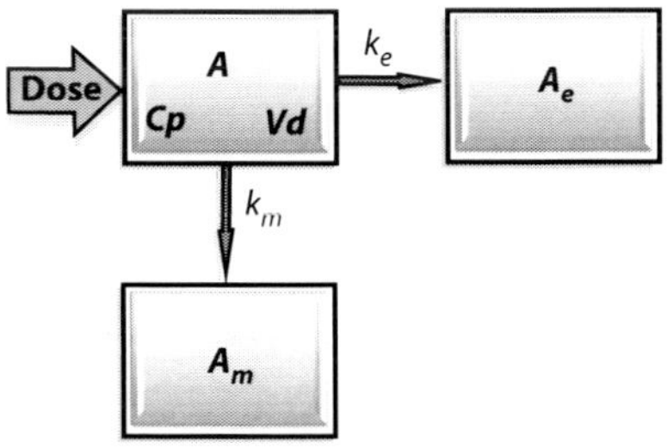

FIGURE 5.5 Diagram of one-compartment open model with intravenous bolus input and first-order excretion and metabolism.

where (dA_e/dt) is the rate of excretion, (k_e) is the first-order excretion rate constant, A and Cp are the amount and concentration of the compound in the body at time t, Vd is the volume of distribution, and Cl_r is the renal clearance.

The rate of excretion after the injection of a drug is high and gradually declines as the amount or concentration of the drug in the body decreases. This change in the rate of excretion is the reason for the initial rise and plateau level of cumulative amount of the unchanged drug in the urine (Figure 5.4). The computational approaches for analysis of urinary data based on Equation 5.39 include the Rate Plot and ARE (Amount Remaining to be Excreted) Plot, also known as Sigma-Minus Plot (Gibaldi and Perrier, 1982).

5.5.1.1 Rate Plot—Bolus Input

For the PK/TK analysis of urinary data, it is necessary to convert the differential equation of the excretion rate (Equation 5.39) into the difference equation to fit the scale of urine collections and the related time intervals. The excretion rate dA_e/dt represents the infinitesimal rate of excretion and it is hardly a measurable quantity in laboratory. Thus, the conversion to $(\Delta A_e/\Delta t)$ reflects the summation of quantities of dA_e that forms ΔA_e over summation of periods of dt that forms Δt (i.e., $\Delta A_e = \Sigma_{i=1}^{n} dA_{e_i}$ and $\Delta t = \Sigma_{i=1}^{n} dt_i$). Dividing (ΔA_e) by (Δt) provides an average value for the excretion rate that corresponds to the midpoint of interval of urine collection

$$\frac{\Delta A_e}{\Delta t} = k_e A_t = k_e \times Vd \times Cp = Cl_r \times Cp_t \tag{5.42}$$

where (ΔA_e) is the amount of the unchanged drug excreted in urine in interval of Δt and A_t is the amount of the unchanged drug in the body. Assuming the plasma concentration–time profile follows the one compartmental model of $Cp_t = Cp^0 e^{-Kt}$ (Figure 5.5); its substitution in Equation 5.42 yields the following rate equations:

$$\frac{\Delta A_e}{\Delta t} = Cl_r \times Cp^0 e^{-Kt_{midpoint}} \tag{5.43}$$

$$\therefore \quad \log\frac{\Delta A_e}{\Delta t} = \log(Cl_r \times Cp^0) - \frac{Kt_{midpoint}}{2.303} \tag{5.44}$$

$$\frac{\Delta A_e}{\Delta t} = k_e A^0 e^{-Kt_{midpoint}} \tag{5.45}$$

$$\therefore \log\frac{\Delta A_e}{\Delta t} = \log(k_e \times Dose) - \frac{Kt_{midpoint}}{2.303} \tag{5.46}$$

TABLE 5.2
Stepwise Calculations of Rate Plot

Δt	$t_{midpoint}$	ΔA_e	$\Delta A_e/\Delta t$
$\Delta t_1 = t_1 - 0$	$t_{midpoint_1} = (0 + t_1)/2$	ΔA_{e_1}	$\Delta A_{e_1}/\Delta t_1$
$\Delta t_2 = t_2 - t_1$	$t_{midpoint_2} = (t_1 + t_2)/2$	ΔA_{e_2}	$\Delta A_{e_2}/\Delta t_2$
$\Delta t_3 = t_3 - t_2$	$t_{midpoint_3} = (t_2 + t_3)/2$	ΔA_{e_3}	$\Delta A_{e_3}/\Delta t_3$
$\vdots \quad \vdots$	$\vdots \quad \vdots \quad \vdots$	$\vdots$	$\vdots$
$\Delta t_n = t_{n-1} - t_n$	$t_{midpoint_n} = (t_{n-1} + t_n)/2$	ΔA_{e_n}	$\Delta A_{e_n}/\Delta t_n$

The method of Rate Plot calculations are presented in Table 5.2 and the linear plot of $\log(\Delta A_e/\Delta t)$ against $t_{midpoint}$ is shown in Figure 5.6.

The Rate Plot, according to the Equations 5.43 through 5.46, is used to estimate the excretion rate constant, k_e, from the y-intercept and the overall elimination rate constant, K, from the slope of the log-linear plot of $\log(\Delta A_e/\Delta t)$ versus $t_{midpoint}$

$$k_e = \frac{\text{antilog}(y\text{-intercept})}{Dose} \tag{5.47}$$

If the excretion of the unchanged drug occurs mainly in the urine, the first-order metabolic rate constant can be estimated as

$$k_m = K - k_e \tag{5.48}$$

The knowledge of excretion and metabolic rate constants enables one to estimate the following parameters and constants:

Fraction of dose excreted unchanged:

$$f_e = \frac{k_e}{K} = \frac{A_e^\infty}{Dose} \tag{5.49}$$

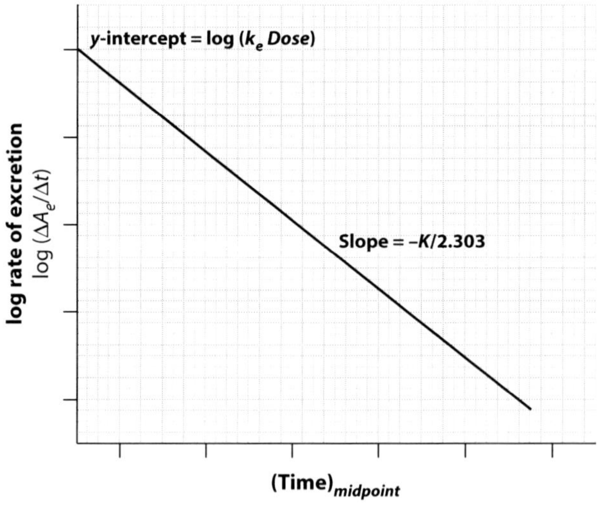

FIGURE 5.6 Plot of logarithm of rate of excretion ($\log(\Delta A_e/\Delta t)$) versus $t_{midpoint}$.

The total amount excreted unchanged:

$$A_e^\infty = f_e \times Dose \tag{5.50}$$

which can also be defined as

$$A_e^\infty = \frac{\text{antilog}(y\text{-intercept})}{K}$$

Fraction of dose eliminated as metabolite(s):

$$f_m = \frac{k_m}{K} = \frac{Dose - A_e^\infty}{Dose} \tag{5.51}$$

$$f_m = 1 - f_e \tag{5.52}$$

Total amount eliminated as metabolite(s):

$$A_m^\infty = f_m \times Dose \tag{5.53}$$

$$A_m^\infty = Dose - A_e^\infty \tag{5.54}$$

By knowing the apparent volume of distribution from plasma data and the rate constants of excretion and metabolism, the clearance terms can be estimated as

$$\text{Plasma clearance} = Cl_T = Vd \times K \tag{5.55}$$

$$\text{Renal clearance} = Cl_r = Vd \times k_e \tag{5.56}$$

$$Cl_r = Cl_T \times f_e \tag{5.57}$$

$$\text{Metabolic clearance} = Cl_m = Vd \times k_m \tag{5.58}$$

$$Cl_m = Cl_T \times f_m \tag{5.59}$$

When a compound is eliminated by metabolism and excretion:

$$f_e + f_m = 1 \quad \text{and} \quad Cl_T = Cl_r + Cl_m \tag{5.60}$$

When the compound is eliminated entirely by metabolism:

$$f_m = 1 \Rightarrow f_e = 0 \Rightarrow K = k_m \quad \text{and} \quad Cl_T = Cl_m \tag{5.61}$$

When the compound is eliminated entirely unchanged:

$$f_e = 1 \Rightarrow f_m = 0 \Rightarrow K = k_e \quad \text{and} \quad Cl_T = Cl_r \tag{5.62}$$

5.5.1.2 ARE Plot (Amount Remaining to Be Excreted), or Sigma-Minus Plot—Bolus Input

This approach is based on the integration of Equation 5.39 between the injection time and the time of last urine collection

$$\int_0^t dA_e = k_e \int_0^t A dt = k_e A^0 \int_0^t e^{-Kt} dt \tag{5.63}$$

$$\therefore A_e = \frac{k_e A^0}{K}(1 - e^{-Kt}) = f_e \ Dose(1 - e^{-Kt}) = A_e^\infty (1 - e^{-Kt}) \tag{5.64}$$

where A_e is the cumulative amount excreted up to time t (Table 5.3) and A_e^∞ is the total amount that ultimately will be excreted unchanged and corresponds to the plateau level of Figure 5.4.

Equation 5.64 can also be presented as

$$(A_e^\infty - A_e) = A_e^\infty e^{-Kt} \tag{5.65}$$

The quantity $(A_e^\infty - A_e)$ is the amount remaining in the body to be excreted unchanged (ARE) at time t. The semi-logarithmic form of Equation 5.65 is

$$\log(A_e^\infty - A_e) = \log A_e^\infty - \frac{Kt}{2.303} \tag{5.66}$$

The graph of Equation 5.66 is presented in Figure 5.7 and the method of ARE (Sigma-Minus) calculations is shown in Table 5.3.

The calculations of ARE data are dependent on a correct estimate of A_e^∞, that is, urine samples should be collected long enough to obtain an accurate assessment of this parameter. The Rate Plot, however, is independent of A_e^∞, although the data points are usually more scattered than those of the ARE Plot.

The information obtained from ARE Plot is somewhat similar to that obtained from the Rate Plot. The slope of the line (Equation 5.66) is equal to $-(K/2.303)$, but the y-intercept is log A_e^∞ that is, log ($f_e \times Dose$):

TABLE 5.3
Calculations of ARE Plot

Time (h)	Cumulative Amount	ARE
0	0	A_e^∞
t_1	$A_{e_1} = \Delta A_{e_1}$	$ARE_1 = A_e^\infty - A_{e_1}$
t_2	$A_{e_2} = A_{e_1} + \Delta A_{e_2}$	$ARE_2 = A_e^\infty - A_{e_2}$
t_3	$A_{e_3} = A_{e_2} + \Delta A_{e_3}$	$ARE_3 = A_e^\infty - A_{e_3}$
t_4	$A_{e_4} = A_{e_3} + \Delta A_{e_4}$	$ARE_4 = A_e^\infty - A_{e_4}$
⋮	⋮	⋮
t_n	A_{e_m}	$ARE_n = A_e^\infty - A_{e_m}$
∞	A_e^∞	0

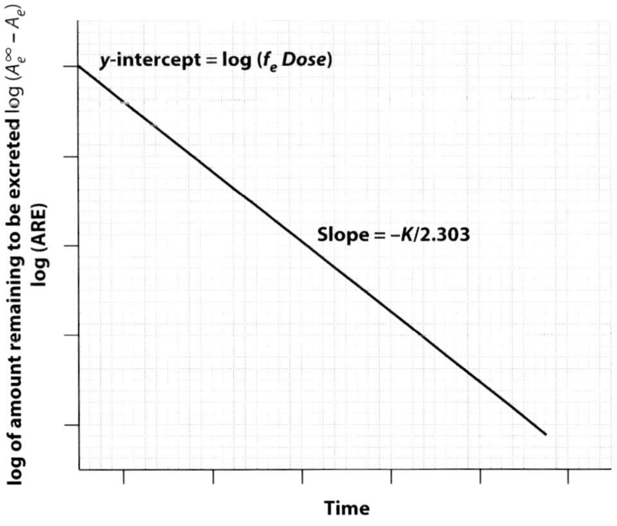

FIGURE 5.7 Plot of logarithm of amount remaining to be excreted unchanged versus time.

$$\therefore \text{antilog}(y\text{-intercept})_{ARE} = \frac{\text{antilog}(y\text{-intercept})_{Rate}}{K} \tag{5.67}$$

$$k_e = \frac{\text{antilog}(y\text{-intercept})_{ARE}}{Dose} \times K \tag{5.68}$$

$$K = 2.303 \times slope \tag{5.69}$$

$$A_e^\infty = \text{total amount excreted unchanged} = \text{antilog}\,(y\text{-intercept})_{ARE} \tag{5.70}$$

Other relationships are the same as Equations 5.49 through 5.62.

5.5.2 PK/TK Analysis of Urinary Elimination of Metabolites—Bolus Input

The compartmental model used in this section is the same as the one discussed in Section 5.5.1, with the proviso that the rate constant of metabolite formation is different from the rate constant of elimination of metabolite, and the amount of metabolite in the body is a separate pool from the amount eliminated in urine. The diagram of the model is presented in Figure 5.8. Compartment A represents the homogeneous compartment of systemic circulation and highly perfused tissues, which receives the compound as a bolus dose; compartment A_e is the cumulative amount of injected compound that is excreted unchanged in urine; compartment A_m is the amount of metabolite in the body; and compartment A_{me} is the cumulative amount of metabolite eliminated in urine. The time course of amount for compartments A and A_m and cumulative amount for A_e and A_{me} are presented in Figure 5.9.

The differential equations of the model are

$$\frac{dA}{dt} = -KA, \quad \text{where} \quad K = k_e + k_m \tag{5.71}$$

$$\frac{dA_e}{dt} = k_e A \tag{5.72}$$

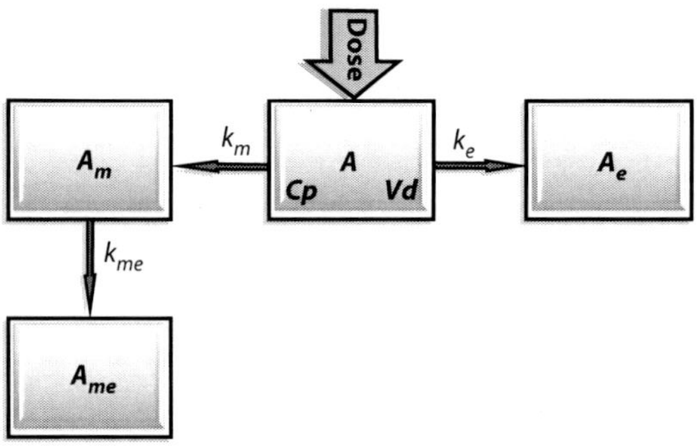

FIGURE 5.8 Diagram of a compartmental model with distinction between the rate constant of metabolite formation in the body and the rate constant of metabolite elimination.

$$\frac{dA_m}{dt} = k_m A - k_{me} A_m \tag{5.73}$$

$$\frac{dA_{me}}{dt} = k_{me} A_m \tag{5.74}$$

The integrated equations of the model, representing the amount in each compartment at time t, using the Laplace transform (Appendix A), are

$$A = A^0 e^{-Kt} \tag{5.75}$$

$$A_e = \frac{k_e A^0}{K}(1 - e^{-Kt}) \tag{5.76}$$

$$A_m = \frac{k_m A^0}{K - k_{me}}(e^{-k_{me}t} - e^{-Kt}) \tag{5.77}$$

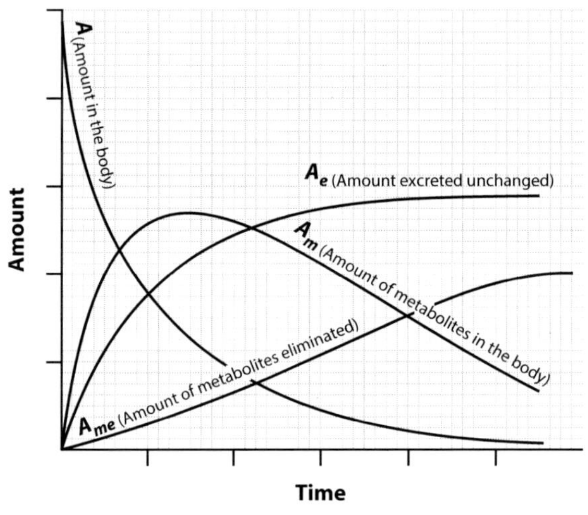

FIGURE 5.9 Theoretical time course of unchanged amount in the body (A) and in the urine (A_e), amount of metabolite in the body (A_m) and the amount of metabolite eliminated in the urine (A_{me}).

$$A_{me} = \frac{k_m A^0}{K}\left(1 + \frac{1}{k_{me} - K}(Ke^{-k_{me}t} - k_{me}e^{-Kt})\right) \tag{5.78}$$

The first-order rate constants k_m and k_{me} represent the formation and elimination of metabolite, respectively, and A^0 is the amount in the body at time zero and it is equal to the injected dose. As $t \Rightarrow \infty$ the exponential terms of Equation 5.78 become negligible and the total amount eliminated as metabolite is expressed as

$$A_{me}^{\infty} = \frac{k_m \times A^0}{K} = f_m Dose \tag{5.79}$$

Substitution of Equation 5.79 in 5.78 gives rise to Equation 5.81 that describes the amount of metabolite remaining to be eliminated from the body, that is, the ARE equation of the metabolite

$$A_{me} = A_{me}^{\infty}\left(1 + \frac{1}{k_{me} - K}(Ke^{-k_{me}t} - k_{me}e^{-Kt})\right) \tag{5.80}$$

$$(A_{me}^{\infty} - A_{me}) = \frac{A_{me}^{\infty}}{k_{me} - K}(k_{me}e^{-Kt} - Ke^{-k_{me}t}) \tag{5.81}$$

The Rate equation of metabolite elimination is developed by the substitution of Equation 5.77 in Equation 5.74

$$\frac{dA_{me}}{dt} = k_{me}A_m = \frac{k_{me}k_m A^0}{K - k_{me}}(e^{-k_{me}t} - e^{-Kt}) \tag{5.82}$$

$$\frac{\Delta A_{me}}{\Delta t} = \frac{k_{me}k_m \times A^0}{K - k_{me}}(e^{-k_{me}(t_{midpoint})} - e^{-K(t_{midpoint})}) \tag{5.83}$$

The plot of $\log(A_{me}^{\infty} - A_{me})$ versus t or $\log(\Delta A_{me}/\Delta t)$ versus $t_{midpoint}$ has a linear terminal portion with a slope of $(-k_{me}/2.303)$. Subtraction of the linear segment of the curve from the Rate or ARE equation provides the equation of the residual line with a slope of $-K/2.303$. A typical Rate Plot of Equation 5.83 is presented in Figure 5.10. An estimate of the metabolic rate constant, k_m, can be determined from the y-intercept of the extrapolated line of the Rate Plot of metabolite in urine, that is,

$$k_m = \frac{(y\text{-intercept}) \times (K - k_{me})}{k_{me} \times Dose} \tag{5.84}$$

5.5.3 PK/TK Analysis of Urinary Excretion Data—Zero-Order Input

The Rate and ARE plots for compounds that are introduced in the body by zero-order input are also based on the methodology applied for bolus injection. The rate equations of the model based on the diagram presented in Figure 5.11 are

$$\frac{dA}{dt} = k_0 - k_e A \tag{5.85}$$

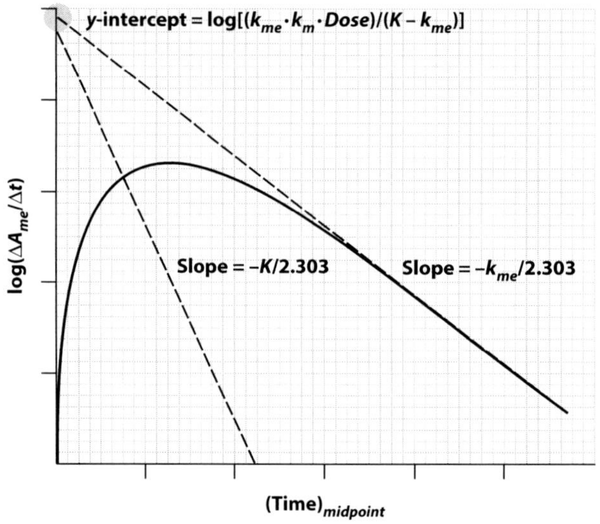

FIGURE 5.10 Rate Plot of elimination of metabolite in urine as a function of midpoint of urine collection; the slope of extrapolated and residual lines provide the estimate of the rate constant of metabolite elimination and the overall elimination rate constant, respectively.

$$A = \frac{k_0}{K}(1 - e^{-Kt}) \tag{5.86}$$

$$\frac{dA_e}{dt} = k_e A = \frac{k_e k_0}{K}(1 - e^{-Kt_{midpoint}}) \tag{5.87}$$

$$\frac{\Delta A_e}{\Delta t} = \frac{k_e k_0}{K}(1 - e^{-Kt_{midpoint}}) = k_e A_{ss}(1 - e^{-Kt_{midpoint}}) \tag{5.88}$$

where (A_{ss}) is the steady-state amount of the compound if the zero-order input continues at the same rate for a time equal to about seven half-lives of the compound; (k_0) is the zero-order input rate with units of mass/time; (k_e) and K are the first-order rate constants of excretion and overall elimination rate constant, respectively; and amount of the compound at steady state is $A_{ss} = (k_0/K)$.

Equation 5.88 indicates the rate of excretion, and as long as the rate of infusion remains constant it is a function of the fraction of steady-state level ($f_{ss} = 1 - e^{-Kt_{midpoint}}$). During the infusion and before achieving the steady-state level, the rate of excretion changes with changing f_{ss} and

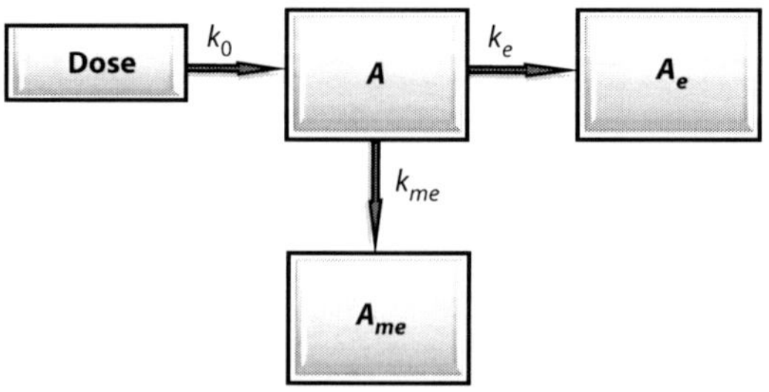

FIGURE 5.11 Diagram of a one-compartment model with zero-order input and first-order excretion and metabolism.

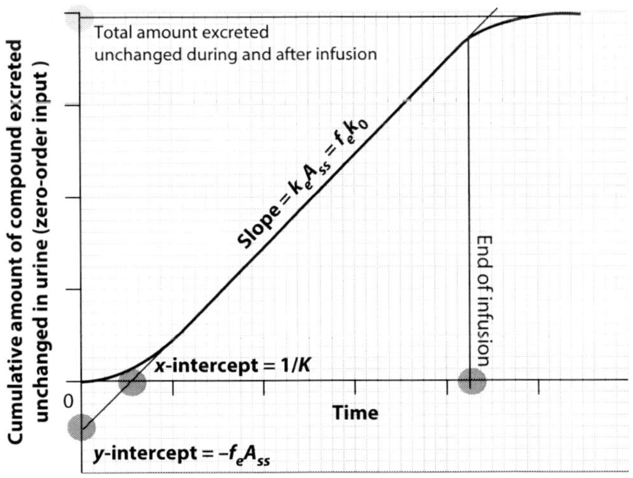

FIGURE 5.12 Plot of amount excreted unchanged in urine during infusion (Equation 5.90) with the linear segment of the curve (with y-intercept of $-f_eA_{ss}$) according to Equations 5.91 and 5.92; the plateau level of the plot represents the total amount excreted unchanged during and after stopping the infusion.

time. At steady state ($f_{ss} = 1$), the amount of compound in the body is maintained constant, the rate of infusion is equal to the rate of elimination and thus, the first-order rate of excretion, k_eA_{ss}, is invariable and equal to $f_e \times k_0$.

The urinary excretion data of a zero-order input can also be analyzed by integrating a differential form of Equation 5.87 between zero and time t

$$\int_0^t \frac{dA_e}{dt} = \int_0^t \frac{k_ek_0}{K} - \int_0^t \frac{k_ek_0}{K} e^{-Kt} \quad (5.89)$$

$$\therefore A_e = \frac{k_ek_0}{K} t - \frac{k_ek_0}{K^2}(1 - e^{-Kt}) \quad (5.90)$$

As $t \Rightarrow \infty, \quad e^{-Kt} \Rightarrow 0$

$$A_e^{\infty} = \frac{k_ek_0}{K} t - \frac{k_0k_e}{K^2} \quad (5.91)$$

Equation 5.91 indicates that the total amount excreted unchanged in urine, A_e^{∞}, is a time-dependent variable during the continuous input (e.g., intravenous infusion). When the infusion is discontinued the amount in the body declines gradually and A_e^{∞} reaches a plateau that represents the total amount excreted during and after infusion. Equation 5.91 is a straight-line relationship with a slope of (f_ek_0) or (k_eA_{ss}) and y-intercept of $-f_eA_{ss}$ (Figure 5.12). Equation 5.91 may also be written as

$$A_e^{\infty} = A_{ss}(k_et - f_e) \quad (5.92)$$

5.5.4 PK/TK Analysis of Urinary Excretion Data—First-Order Input

The development of Rate and ARE equations of xenobiotic excreted unchanged in urine following a first-order input (e.g., compounds given orally, nasally, rectally, sublingually, etc.) is similar to

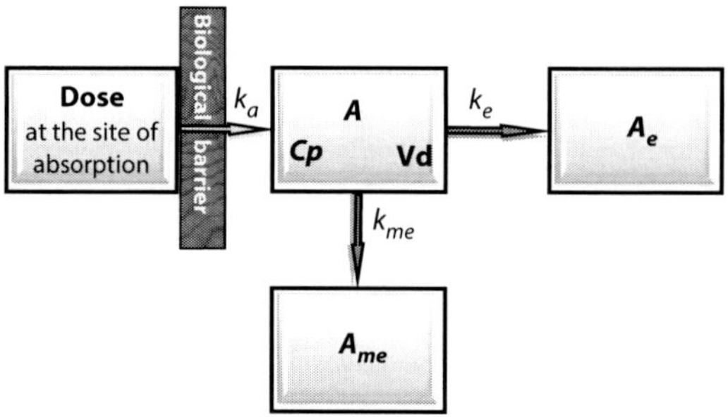

FIGURE 5.13 Diagram of a one-compartment model with first-order input and first-order excretion and metabolism.

those previously described for intravenous injections. The diagram of the model is presented in Figure 5.13.

The differential equations of the model are

$$\frac{dA}{dt} = k_a F Dose - KA \tag{5.93}$$

$$\frac{dA_e}{dt} = k_e A \tag{5.94}$$

The integrated equation of the central compartment is

$$A = \frac{k_a FD}{k_a - K}(e^{-Kt} - e^{-k_a t}) \tag{5.95}$$

The excretion rate of the compound given by a route of administration that requires first-order absorption is

$$\frac{dA_e}{dt} = \frac{k_e k_a FD}{k_a - K}(e^{-Kt} - e^{-k_a t}) \tag{5.96}$$

The corresponding equation for the Rate Plot is expressed as

$$\frac{\Delta A_e}{\Delta t} = \frac{k_e k_a FD}{k_a - K}(e^{-Kt_{midpoint}} - e^{-k_a t_{midpoint}}) \tag{5.97}$$

where k_a is the first-order absorption rate constant and F is the fraction of dose absorbed (absolute bioavailability). Equation 5.97, similar to its plasma concentration profile (Equation 5.95), has a maximum and a linear terminal portion that has a slope of $-K/2.303$ (Figure 5.14). Depending on how fast the administered compound is absorbed, the curve may rise sharply or less sharply and more softly. When the absorption rate constant is greater than the overall elimination rate constant ($k_a \gg K$) and F is equal to one, the y-intercept would be equal to $k_e Dose$.

The extrapolated line (Figure 5.14) corresponds to the terminal section of the curve and post absorption phase, and represents the overall elimination of the compound from the body. Its equation is

$$\left(\frac{\Delta A_e}{\Delta t}\right)_{extrapolated} = \frac{k_e k_a FD}{k_a - K}(e^{-Kt_{midpoint}}) \tag{5.98}$$

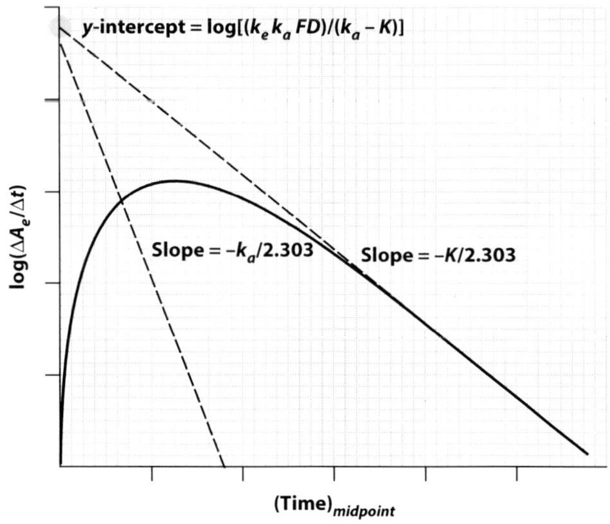

FIGURE 5.14 Theoretical plot of logarithm of excretion rate versus the midpoint of urine collection for single dose of a xenobiotic administered through an extravascular route of administration.

Or,

$$\log\left(\frac{\Delta A_e}{\Delta t}\right)_{extrapolated} = \log\frac{k_e k_a FD}{k_a - K} - \frac{K \times t_{midpoint}}{2.303} \tag{5.99}$$

The residual line, which results from subtraction of Equation 5.98 from Equation 5.97, represents the overall absorption process with a slope of $-k_a/2.303$ that provides an estimate of the absorption rate constant. Therefore, by knowing K, k_a, and FD the excretion rate constant, k_e, can be estimated from the y-intercept of the extrapolated line (Figure 5.14).

The equation of the ARE Plot is developed by integrating Equation 5.96 using the Laplace transform (Appendix A). The integrated equation for cumulative amount excreted unchanged at time t is

$$Ae = \frac{keFD}{K}\left[1 + \frac{k_a}{K - k_a}e^{-Kt} - \frac{K}{K - k_a}e^{-k_a t}\right] \tag{5.100}$$

As $t \Rightarrow \infty$, the exponential terms will approach zero and $A_e = A_e^\infty$

$$\therefore A_e^\infty = \frac{k_e FD}{K} = f_e FD \tag{5.101}$$

The substitution of Equation 5.101 in Equation 5.100 and solving for the amount remaining to be excreted yields

$$(A_e^\infty - A_e) = \frac{A_e^\infty}{k_a - K}(k_a e^{-Kt} - Ke^{-k_a t}) \tag{5.102}$$

Equation 5.102 also provides a curve similar to the Rate Plot. Under the condition that $k_a \gg K$, the extrapolated and residual lines provide the estimates of the overall elimination and absorption

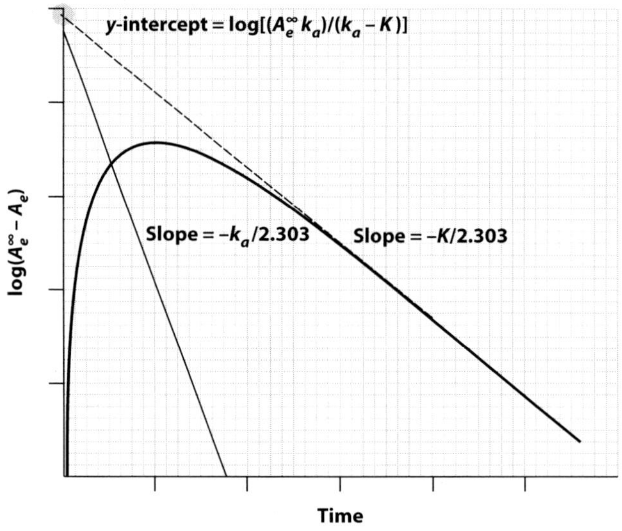

FIGURE 5.15 Theoretical plot of logarithm of amount remaining to be excreted unchanged in urine versus time for a single dose of xenobiotic administered through an extravascular route of administration.

rate constants, respectively (Figure 5.15). The total amount excreted unchanged, A_e^∞, can be estimated by multiplication of the y-intercept of the extrapolated line by $(k_a - K/k_a)$:

$$A_e^\infty = \frac{(y\text{-intercept}) \times (k_a - K)}{k_a} \tag{5.103}$$

The excretion rate constant, k_e, can then be calculated as

$$k_e = \frac{K \times A_e^\infty}{FD} \tag{5.104}$$

5.5.5 PK/TK Analysis of Urinary Excretion Data—Two-Compartment Model, Bolus Input

The diagram of the model is presented in Figure 5.16. It assumes the body as two distinct compartments of A_1 and A_2 with distribution between the two compartments and elimination only from A_1, the central compartment, which consists of systemic circulation and highly perfused organs/tissues, including the kidneys. The second compartment A_2 is the peripheral compartment and represents the less perfused organs/tissues (see also Chapter 10, Section 10.3).

The differential equation representing the excretion rate from the central compartment is

$$\frac{dA_e}{dt} = k_e A_1 \tag{5.105}$$

where A_1 is the amount in the central compartment at time t and is equal to

$$A_1 = \frac{Dose(\alpha - k_{21})}{\alpha - \beta} e^{-\alpha t} + \frac{Dose(k_{21} - \beta)}{\alpha - \beta} e^{-\beta t} \tag{5.106}$$

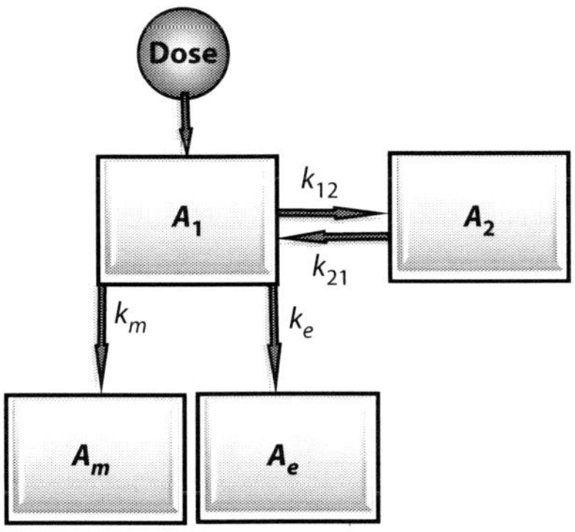

FIGURE 5.16 The diagram of a two-compartment model with instantaneous administration of a single dose of xenobiotic, first-order distribution between central and peripheral compartment, and first-order metabolism and excretion from the central compartment.

$$\therefore \frac{dA_e}{dt} = \underbrace{\boxed{\frac{k_e D(\alpha - k_{21})}{\alpha - \beta}}}_{\bar{A}} e^{-\alpha t} + \underbrace{\boxed{\frac{k_e D(k_{21} - \beta)}{\alpha - \beta}}}_{\bar{B}} e^{-\beta t} \tag{5.107}$$

$$\therefore \frac{dA_e}{dt} = \bar{A}e^{\alpha t} + \bar{B}e^{\beta t} \tag{5.108}$$

Adding the two coefficients of $\bar{A}$ and $\bar{B}$ yields the following relationship:

$$\bar{A} + \bar{B} = \frac{k_e D\alpha - k_e D k_{21} + k_e D k_{21} - k_e D\beta}{\alpha - \beta} = k_e D \tag{5.109}$$

Thus the excretion rate constant can be estimated as

$$\therefore k_e = \frac{\bar{A} + B}{D} \tag{5.110}$$

The urinary data is then fitted to the following equation of Rate Plot to determine $\bar{A}$ and $\bar{B}$:

$$\therefore \frac{\Delta A_e}{\Delta t} = \bar{A}e^{-\alpha t_{midpoint}} + \bar{B}e^{-\beta t_{midpoint}} \tag{5.111}$$

The extrapolated line of the linear terminal portion of the curve has a y-intercept of $\bar{B}$ and a slope of $-\beta/2.303$; the residual line has the y-intercept of $\bar{A}$ and slope of $-\alpha/2.303$ (Figure 5.17).

The integration of the following simultaneous differential equations of the model by the Laplace transform yields the integrated equation of the amount excreted unchanged in urine, which is then used to establish the equation of the ARE Plot:

$$\frac{dA_1}{dt} = k_{21}A_2 - k_{12}A_1 - k_{10}A_1 \tag{5.112}$$

$$\frac{dA_2}{dt} = k_{12}A_1 - k_{21}A_2 \tag{5.113}$$

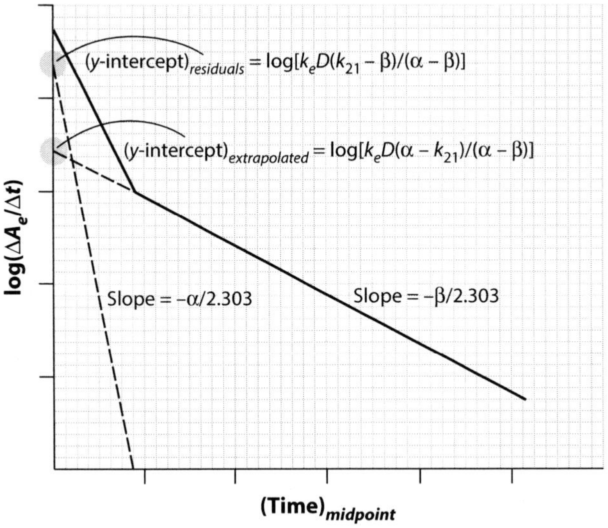

FIGURE 5.17 Rate Plot of urinary excretion of a xenobiotic that follows two-compartment model.

$$\frac{dA_e}{dt} = k_e A_1 \tag{5.114}$$

$$\frac{dA_m}{dt} = k_m A_1 \tag{5.115}$$

where $k_{10} = k_e + k_m + k_{nr}$ and k_{nr} is the first-order rate constant of non-renal/non-metabolic elimination from the body, if significant; k_{12} and k_{21} are the first-order distribution rate constants between the central and peripheral compartments.

The integrated equation for the amount of xenobiotic excreted unchanged is (Gibaldi and Perrier, 1982)

$$A_e = \frac{k_e D k_{21}}{\alpha\beta} + \frac{k_e D(k_{21} - \alpha)}{-\alpha(\beta - \alpha)} e^{-\alpha t} + \frac{k_e D(k_{21} - \beta)}{-\beta(\alpha - \beta)} e^{-\beta t} \tag{5.116}$$

Equation 5.116 can be further simplified to Equation 5.117

$$A_e = \frac{k_e D k_{21}}{\alpha\beta} - k_e D\left(\frac{\alpha\beta - \beta k_{21}}{\alpha\beta(\alpha - \beta)} e^{-\alpha t} + \frac{\alpha k_{21} - \alpha\beta}{\alpha\beta(\alpha - \beta)} e^{-\beta t}\right) \tag{5.117}$$

The hybrid first-order rate constants of α and β, the roots of quadratic equations developed during the integration (see Chapter 10, Section 10.3), are defined as

$$\alpha\beta = k_{21} k_{10} \tag{5.118}$$

$$\alpha + \beta = k_{12} + k_{21} + k_{10} \tag{5.119}$$

The substitution of Equation 5.118 in Equation 5.117, yields the following relationship that defines the amount of unchanged drug excreted in urine at time t:

$$A_e = \frac{k_e D}{k_{10}} - \frac{k_e D}{k_{10}}\left(\frac{k_{10}-\beta}{\alpha-\beta}e^{-\alpha t} + \frac{\alpha-k_{10}}{\alpha-\beta}e^{-\beta t}\right) \quad (5.120)$$

As $t \Rightarrow \infty$, the exponential terms of Equation 5.120 will approach zero and the total amount of xenobiotic that excrete unchanged in urine is

$$A_e^{\infty} = \frac{k_e D}{k_{10}} = f_e D \quad (5.121)$$

The substitution of Equation 5.121 in Equation 5.120 and solving for the amount remaining to be excreted yields the equation of the ARE Plot

$$A_e^{\infty} - A_e = A_e^{\infty}\left(\frac{k_{10}-\beta}{\alpha-\beta}e^{-\alpha t} + \frac{\alpha-k_{10}}{\alpha-\beta}e^{-\beta t}\right) \quad (5.122)$$

Or,

$$(A_e^{\infty} - A_e) = \ddot{A}e^{-\alpha t} - \ddot{B}e^{-\beta t} \quad (5.123)$$

The plot of $(A_e^{\infty} - A_e)$ versus time yields a biphasic curve with a linear terminal phase with a slope of $-\beta/2.303$ that can be extrapolated to provide the y-intercept of $\ddot{B}$. Subtraction of the extrapolated line from the biphasic curve generates a residual line with slope of $-\alpha/2.303$ and y-intercept of $\ddot{A}$ (Figure 5.18).

The addition of the two y-intercepts gives the calculated value of total amount eliminated unchanged in urine

$$\ddot{A} + \ddot{B} = A_e^{\infty}\left(\frac{k_{10}-\beta}{\alpha-\beta}\right) + A_e^{\infty}\left(\frac{\alpha-k_{10}}{\alpha-\beta}\right) = A_e^{\infty}\left(\frac{k_{10}-\beta+\alpha-k_{10}}{\alpha-\beta}\right) = A_e^{\infty} \quad (5.124)$$

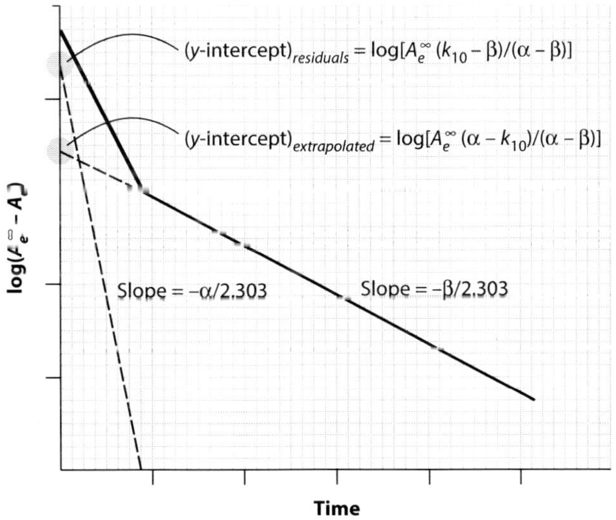

FIGURE 5.18 ARE Plot of urinary excretion of a xenobiotic that follows two-compartment model.

Therefore, the fraction of the dose excreted unchanged can be calculated as

$$f_e = \frac{\ddot{A} + \ddot{B}}{D} = \frac{A_e^\infty}{D} \tag{5.125}$$

The excretion rate constant can be estimated from the overall elimination rate constant as follows:

$$\ddot{B} = A_e^\infty \frac{\alpha - k_{10}}{\alpha - \beta} = (\ddot{A} + \ddot{B})\frac{\alpha - k_{10}}{\alpha - \beta} \tag{5.126}$$

$$\cancel{\ddot{B}\alpha} - \ddot{B}\beta = \ddot{A}\alpha + \cancel{\ddot{B}\alpha} - (\ddot{A} + \ddot{B})k_{10} \tag{5.127}$$

$$k_{10} = \frac{\ddot{A}\alpha + \ddot{B}\beta}{\ddot{A} + \ddot{B}} \tag{5.128}$$

$$\therefore k_e = f_e k_{10} \tag{5.129}$$

5.5.6 PK/TK Analysis of Urinary Excretion Data—Multi-Compartment Model (Oral Administration)

For the PK/TK software, capable of iteration of differential equations, the plasma concentration and urinary data are usually fitted simultaneously according to the following equations of multi-compartment models (Mitchel et al., 2001):

Oral administration:

$$Cp = \left(\sum_{i=1}^{n} Cp_i e^{-\lambda_i t}\right) + Cp^0 e^{-\lambda_n (t - t_0)} \tag{5.130}$$

$$\frac{dA_e}{dt} = \left(Cl_r \sum_{i=1}^{n} Cp_i e^{-\lambda_i t_{midpoint}}\right) + \left(\frac{dA_e}{dt}\right)_0 e^{-\lambda_n (t_{midpoint} - t_0)} \tag{5.131}$$

where (λ_i) represents the hybrid rate constants of the multi-compartmental model and λ_n is the hybrid rate constant of the terminal exponential term.

5.5.7 PK/TK Analysis of Urinary Excretion Data—Non-Compartmental Analysis (Moment Analysis)

In non-compartmental analysis, the process of renal excretion is quantified by mean excretion time (MET), which is the residual area divided by the total amount excreted unchanged in urine. The residual area is the difference between the area of the cumulative amount excreted unchanged up to time t and the area of the total amount excreted unchanged at $t = \infty$, that is, the area of the amount remaining to be excreted (ARE)

$$\text{Mean Excretion Time (MET)} = \frac{\text{Area under amount remaining to be excreted}}{\text{Total amount excreted unchanged}}$$

$$MET = \frac{\int_0^{\infty} [A_e^{\infty} - A_e^t]dt}{A_e^{\infty}} = \frac{\int_0^{\infty} ARE}{A_e^{\infty}} \tag{5.132}$$

The other parameter used for urinary excretion data is the mean transit time for urine outflow, which is based on the excretion rate as follows (Saito et al., 1991, Ito et al., 2000):

$$(MTT)_{urine} = \frac{\int_0^t t(dAe/dt)dt}{\int_0^t (dA_e/dt)dt} \tag{5.133}$$

The $(MTT)_{urine}$ is the combination of mean transit time for the amount or fraction filtered $(MTT)_{GF}$, the mean transit time of amount or fraction secreted $(MTT)_{secretion}$ and the mean transit time of amount of fraction reabsorbed $(MTT)_{reabsorption}$.

The mean residence time in renal epithelial cell $(MRT)_{epithelial\ cell}$ can then be estimated as the difference between the mean transit time of filtered and secreted fractions

$$(MRT)_{epithelial\ cells} = (MTT)_{GF} - (MTT)_{secretion} \tag{5.134}$$

5.6 RENAL METABOLISM

In addition to the liver, which is the major site of metabolism, the kidneys also contribute to the overall metabolism of xenobiotics (Goldberg and Anderson, 1985; Lock and Reed, 1998). All enzymes of Phase I metabolism are predominantly active in the proximal portion of tubules and responsible for oxidative, reductive, and hydrolytic reactions. The enzymes of Phase II metabolism, although not distribute uniformly in the kidneys, have also the highest activity in the proximal tubules forming glucuronide, glutathione, and sulfate conjugates. The metabolism of xenobiotics is discussed in Chapter 6.

5.7 ESTIMATION OF PK/TK PARAMETERS AND CONSTANTS OF ELIMINATION WHEN USING DIALYSIS

5.7.1 INTRODUCTION

Peritoneal dialysis and hemodialysis are two methods of enhancing the elimination of toxic substances in poisoning, or restoring the normal dynamic equilibrium between plasma and extracellular fluid in chronic kidney disease (CKD). The process of elimination in these patients during dialysis is a combination of metabolism, removal by dialysis, and the residual renal function plus other negligible routes of elimination. The fundamental principle in both dialyses is the removal of solutes and wastes from the systemic circulation via diffusion across a semi-permeable membrane, based on concentration gradient into a dialysis solution known as dialysate. The process is similar to glomerular filtration, but without the physiological selectivity of the kidney. Dialytic treatments remove xenobiotics from the systemic circulation and reduce their concentration in plasma. However, it is important to note that not all compounds are removed by dialysis. For example, phenytoin, beta blockers, benzodiazepines, and tricyclic antidepressants are not removed by dialysis, whereas ethylene glycol, barbiturates, lithium, aspirin, ethanol, and methanol are removed effectively by dialysis. Also, dialysis cannot replace the treatment with antidotes. For example, for acute poisoning by opioid and other narcotics, administration of naloxone and other antidotes are preferred. The procedure is also ineffective for treatment of fast acting cytotoxic poisons such as cyanide.

In CKD patients on medication, the plasma concentration of the therapeutic agent during dialysis is lower than interdialysis period. Thus, for drugs with narrow therapeutic range, a maintenance dose after dialysis may be required to uphold the plasma concentration at the therapeutic level before commencing dialysis.

Some of the biochemical criteria that are considered before initiating dialysis for CKD patients are

$$C_{cr}(GFR) \cong 5 \text{ mL/min}$$

$$BUN/S_{cr} \cong 3.5$$

$$S_{cr} \cong 10\text{–}20 \text{ mg/dL}$$

- Elevated serum uric acid
- Elevated serum inorganic phosphorous
- Presence of toxins such as guanidosuccinic acid, methyl guanidine, methyl histidine, and phenolic acids are in the blood

In general, factors influencing the removal of solutes by dialysis are molecular size of the compound, its partition coefficient, volume of distribution, free fraction in plasma, and plasma clearance. The characteristics of dialysis membrane, such as its pore size, surface area, and its geometry in the chamber, and the flow rates in and out of the system can also impact the removal process.

5.7.2 Hemodialysis

Hemodialysis is carried out with the use of dialyzers that provide the semi-permeable membrane for the removal of endogenous and exogenous compounds. They may fall into one of the following categories:

1. Disposable coil dialyzer
2. Disposable fiber dialyzer
3. Disposable flat plate dialyzer
4. Non-disposable flat plate dialyzer

The following is a brief description of the procedure to provide enough background for the kinetic discussion. For detailed and specialized information, related books and publications in nephrology should be consulted.

The procedure involves the aseptic insertion of two catheters into the artery and vein. Sometimes the catheters are inserted into the femoral vein approximately two inches apart. After the placement of catheters, a dialyzer, for example, coil dialyzer, is connected to the machine; dialysate is added to a container of water at 37°C. The dilution would provide a solution, which is expected to be similar to extracellular fluid. After priming the unit with heparinized saline, the inlet and outlet of the unit are connected to the catheters. The blood flow rate is usually maintained at a rate of 12 L/h. In long-term dialysis, the dialysate is changed every 2 to 3 hours and dialysis is usually carried out for 6–8 hours.

5.7.3 Peritoneal Dialysis

In this procedure the patient's peritoneum serves as the dialysis membrane. The procedure involves the aseptic insertion of a nylon catheter, approximately 3 mm in diameter and 25–30 cm in length with side perforations, that is attached to a metal stylet into the peritoneal cavity. The insertion of the catheter is through a small incision, about one third of the distance from the umbilicus to the pubic bone. The skin of the abdomen is prepared as for any surgery. There are times that the bladder should be emptied prior to inserting the catheter. The patient is placed in a supine position. Local anesthesia is administered around the insertion site. After the insertion, the tip of the stylet is withdrawn and the catheter is inserted as far as it can penetrate on the right side of the pelvis. The dialysate, approximately two liters for adults and 0.5–1.0 liter in children, is infused by gravity for 5–10 min. After approximately one hour and achieving equilibrium between blood and dialysate, the abdomen is drained. The procedure may be repeated depending on the severity of renal impairment or poisoning.

5.7.4 Composition of Dialysate

There are various commercially available dialysate formulations that generally contain electrolytes and hypertonic dextrose. The composition of the dialysate is formulated such that it meets the patients' metabolic needs, serum osmolality, potassium, sodium, and bicarbonate concentrations. One of the shortcomings of this type of formulation is that they are physiologic with respect to only few electrolytes. Therefore, all other dializable compounds that should not be removed by dialysis, such as amino acids, will also be removed. Obviously, such indiscriminate removal can become significant over time and may cause unwarranted side effects.

5.7.5 Dialysis Clearance

Dialysis clearance is determined as an indicator of dialysability of a drug; the meaning of clearance is as described in the previous chapters. It is the volume of plasma cleared from a compound per unit of time by a dialysis procedure, that is,

$$\text{Dialysis clearance} = (\text{Extraction ratio})_{dialysis} \times (\text{Blood flow})_{in} \quad (5.135)$$

Or,

$$Cl_{dialysis} = (Q)_{in}\left(\frac{C_{in} - C_{out}}{C_{in}}\right) \quad (5.136)$$

where $(Q)_{in}$ is the blood flow entering into the dialysis chamber with units of volume per time, C_{in} and C_{out} are the blood concentration entering and leaving the chamber, respectively. The dialysis clearance can also be estimated from the concentration of dialysate:

$$Cl_{dialysis} = \frac{[(Q_{dialysate})_{out} \times (C_{dialysate})_{out}] - [(Q_{dialysate})_{in} \times (C_{dialysate})_{in}]}{(C_{blood})_{in}} \quad (5.137)$$

A shorter version of Equation 5.137 is

$$Cl_{dialysis} = \frac{[(Q_{dialysate})_{out} \times (C_{dialysate})_{out}]}{(C_{blood})_{in}}$$

To define clearance in terms of plasma concentration, Equation 5.136 is modified to the following relationship (Frost and Kerr, 1977; Sargent and Gotch, 1980; Sprenger et al., 1983; Lee and Marburt, 1984; Robbins-Weilert, 2004):

$$Cl_{dialysis} = [(Q)_{in}(P_{coeff} \times Hct + (1 - Hct))] \times \frac{(Cp)_{in} - (Cp)_{out}}{(Cp)_{in}} \quad (5.138)$$

where $(Q)_{in}$ is the blood flow entering the chamber; P_{coeff} is the partition coefficient between blood cells and plasma; Hct is the hematocrit and $(Cp)_{in}$ and $(Cp)_{out}$ are the plasma concentration entering and exiting the chamber.

For compounds with high-protein binding with negligible uptake by blood cells, the Equation 5.138 is simplified to

$$Cl_{dialysis} = (Q)_{in} \times (1 - Hct) \times \frac{(Cp)_{in} - (Cp)_{out}}{(Cp)_{in}} \quad (5.139)$$

The extraction ratio of hemodialysis is also used to determine the efficiency of a hemodialysis unit:

$$ER = \frac{C_{in} - C_{out}}{C_{in}} = \frac{Q(C_{in} - C_{out})}{QC_{in}} \quad (5.140)$$

Or,

$$ER = \frac{(Cp)_{in} - (Cp)_{out}}{(Cp)_{in}} \quad (5.141)$$

where $Q(C_{in} - C_{out})$ is the rate of elimination by dialysis and QC_{in} is the rate of input in the unit.

As a general rule, compounds with large volume of distribution tend to stay longer in the body and only a small fraction is cleared per unit of time by dialysis (Kaloyanides, 1992). For example, the large distribution volume of tricyclic compounds indicates that the molecules are not easily removed by dialysis or other extracorporeal means (Haapanen, 1982; Heath et al., 1982; Henry, 1982; Bismuth, 1990; Pond, 1991). Since only unbound compounds can pass through the membrane, the ratio of percent unbound to volume of distribution is used as an index of the dialysability of a compound. When the ratio is greater than 80%, six hours of hemodialysis would be sufficient to remove up to 50% of the solute. A lower ratio, for example, under 20%, indicates that only a small amount of the compound is removed during six hours of dialysis.

Equations 5.136 through 5.139 are practical for hemodialysis. For peritoneal dialysis, the clearance is determined by the following equation:

$$Cl_{dialysis} = \frac{A_d}{t \times Cp_{midpoint}} \quad (5.142)$$

where A_d = total amount of drug removed by dialysis, t = the length of dialysis, and $Cp_{midpoint}$ is the plasma level at the midpoint of peritoneal dialysis.

5.7.6 Effects of Dialysis on Pharmacokinetic Parameters and Constants

The overall elimination of an exogenous compound during dialysis depends mainly on hepatic metabolism, dialysis and residual renal function (Parker and Parker, 1982). As the dialysis is based on the concentration gradient and simple diffusion through a semi-permeable membrane, it is considered a first-order process with a rate constant of k_d. Therefore, the overall elimination rate constant during dialysis can be expressed as

$$\dddot{K} = \bar{k}_e + k_m + k_d \quad \text{during dialysis} \tag{5.143}$$

$$\bar{K} = \bar{k}_e + k_m \quad \text{between dialysis} \tag{5.144}$$

where $\dddot{K}$ and $\bar{K}$ are the overall rate constants of elimination during dialysis and between dialysis, respectively; $\bar{k}_e$ is the excretion rate constant representing the residual kidney function and k_m is the metabolic rate constant. The assumption of Equations 5.143 and 5.144 is that metabolic function remains unperturbed. The effect of dialysis on the profile of plasma concentration–time curve is illustrated in Figures 5.19 (intravenous bolus) and 5.20 (first-order input). The amount of solute removed by dialysis depends on whether the dialysis procedure is initiated immediately after the administration of the compound or sometime later.

Thus, the fractions of dose excreted, metabolized, or removed by dialysis during dialysis can be estimated individually with the following equations:

$$\dddot{f}_e = \frac{\bar{k}_e}{\dddot{K}} \tag{5.145}$$

$$\dddot{f}_m = \frac{k_m}{\dddot{K}} \tag{5.146}$$

$$f_d = \frac{k_d}{\dddot{K}} \tag{5.147}$$

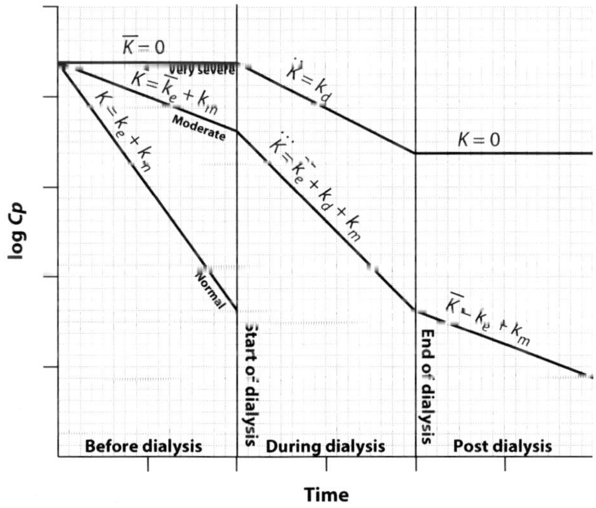

FIGURE 5.19 Changes in the profile of logarithm of plasma concentration versus time of a xenobiotic given intravenously before, during, and after dialysis when there is no elimination from the body ($\bar{K} = 0$) and when the compound is eliminated by metabolism and residual kidney function ($\bar{K} = \bar{k}_e + k_m$), in comparison to the normal elimination ($K = k_e + k_m$).

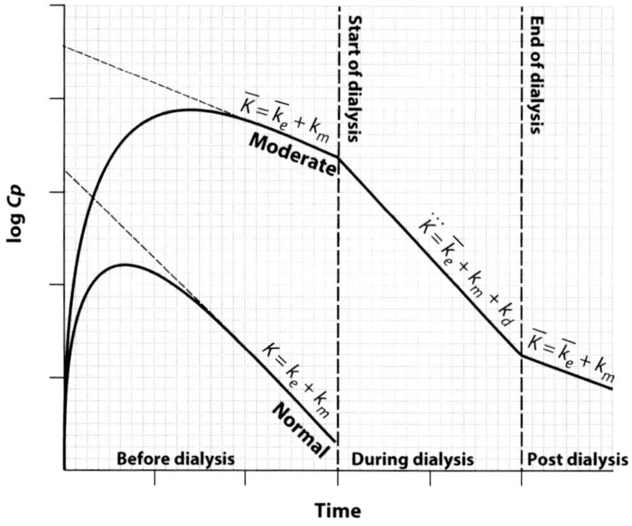

FIGURE 5.20 Changes in the profile of logarithm of plasma concentration of a xenobiotic following a single oral dose before, during, and after dialysis when the compound is eliminated by metabolism and residual kidney functions ($\bar{K} = \bar{k}_e + k_m$) in comparison to the normal elimination.

The fraction removed by dialysis is also determined by dividing the total amount of unchanged solute (drug) removed by dialysis to the administered dose

$$f_d = \frac{\text{Concentration of dialysate} \times \text{Volume of dialysate}}{\text{Dose}} \tag{5.148}$$

If the dose is injected during dialysis, the fraction of dose in the body and fraction of dose eliminated at time t can be expressed as

$$\dddot{f}_b = e^{-\ddot{K}t} \tag{5.149}$$

$$\dddot{f}_{el} = 1 - e^{-\ddot{K}t} \tag{5.150}$$

The following equation can be employed to estimate what fraction of dose eliminated at time t is removed by dialysis:

$$(f_{el})_{dialysis} = \dddot{f}_{el} \times f_d \tag{5.151}$$

The plasma concentration of the drug during dialysis is defined as

$$\dddot{Cp} = Cp^0 e^{-\ddot{K}t_{dialysis}} \tag{5.152}$$

where (Cp^0) is the initial plasma concentration during dialysis. If the dialysis were initiated sometime after the injection of the therapeutic agent, the plasma concentration at any time during dialysis is

$$\dddot{Cp} = (Cp^0 e^{-\bar{K}t})e^{-\ddot{K}t_{dialysis}} \tag{5.153}$$

where $Cp^0e^{-\bar{K}t}$ is the plasma concentration at the start of dialysis; (t) is the time between the injection and the start of dialysis; and $t_{dialysis}$ is the time of dialysis, or any time points during dialysis.

A similar approach can be employed to estimate the plasma concentration at any time after the end of dialysis, that is,

$$Cp = ((Cp^0e^{-\bar{K}t})e^{-\dddot{K}t_{dialysis}})e^{-\bar{K}t_p} \quad (5.154)$$

where (t_p) is anytime during post-dialysis period. Similar approaches can be used for more complex linear PK/TK models.

If the compound is administered orally, depending on the timing of the administration we may have the following cases:

1. The therapeutic agent is administered before dialysis and dialysis is initiated either during the absorptive phase or post-absorptive phase (Figure 5.20).
2. The agent is administered during the dialysis (Figure 5.21).

To calculate the plasma concentration during dialysis (Figure 5.21) we may use the following equation for prediction of plasma concentration during dialysis:

$$Cp_{during} = \frac{FDk_a}{Vd(k_a - \dddot{K})}(e^{-\dddot{K}t_{dialysis}} - e^{-k_a t_{dialysis}}) \quad (5.155)$$

where F is the fraction of dose absorbed; FD is the total amount absorbed; Vd is the apparent volume of distribution; and k_a and $\dddot{K}$ are the rate constants of absorption and elimination during dialysis, respectively. The assumption is that the absorption is unaltered during dialysis. If dialysis is discontinued in the post-absorptive phase of the curve, the plasma concentration in the post-dialysis phase can be estimated by the following relationship:

$$Cp_{post\text{-}dialysis} = \left(\frac{FDk_a}{Vd(ka - \dddot{K})}e^{-\dddot{K}t_{dialysis}}\right)e^{-\bar{K}t} \quad (5.156)$$

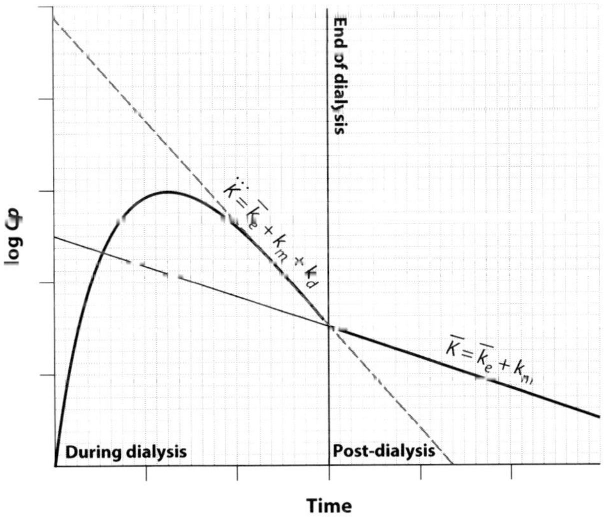

FIGURE 5.21 Changes in the profile of logarithm of plasma concentration of a xenobiotic administered as a single oral dose during dialysis and the change in decline of the compound in the post-dialysis period.

To predict plasma concentrations of compounds administered before the start of dialysis, the following equation can be used:

$$Cp_{during} = \frac{FDk_a}{Vd(ka - \bar{K})}(e^{-\bar{K}t} - e^{-k_a t})e^{-\ddot{K}t_{dialysis}} \tag{5.157}$$

The estimation of plasma concentration during or after dialysis is mostly to determine whether the concentration is within the therapeutic range, or in the case of toxins is within the safe levels. Also, by knowing the PK/TK parameters and constants of the compound the required length of dialysis can be estimated. The plasma levels can also be used to determine the maintenance dose for achieving optimum concentration after dialysis. The maintenance dose to be given at the end of dialysis to achieve, for instance, the initial plasma concentration would be

$$D_M = Vd(Cp^0 - Cp_{end\ of\ dialysis}) \tag{5.158}$$

An application of the approach presented in Equation 5.158 is in multiple-dosing kinetics. Figure 5.21 depicts the effect of dialysis on plasma concentration of a multiple-dosing regimen. The maintenance dose required to achieve the peak steady-state plasma concentration is

$$D_M = Vd[(Cp_{max})_{ss} - (Cp)_{end\ of\ dialysis}] \tag{5.159}$$

REFERENCES

Agarwal, R., Nicar, M. 1994. A comparative analysis of formulas used to predict creatinine clearance. *J Am Soc Nephrol* 5:386.

Bismuth, C. 1990. Biological valuation of extra-corporeal techniques in acute poisoning. *Acta Clin Belg Suppl* 13:20–8.

Burckhardt, G., Wolff, N. A. 2000. Structure of renal organic anion and cation trasporters. *Am J Physiol Renal Physiol* 278:F853–66.

Boroujerdi, M., Mattocks, A. M. 1979. Creatinine kinetics in rabbit. *J Pharmacokinet Biopharm* 7(3):291–301.

Boroujerdi, M. 1982. The comparability of pharmacokinetics of creatinine in rabbit and human: A mathematical approach. *J Theor Biol* 95(2):368–80.

Cha, S. H., Sekine, T., Fukushima, J. I., Kanai, Y., Kobayashi, Y., Goya, T., Endou, H. 2001. Identification and characterization of human organic anion transporter 3 expressing predominantly in the kidney. *Mol Pharmacol* 59(5):1277–86.

Cockcroft, D. W., Gault, M. H. 1976. Prediction of creatinine clearance from serum creatinine. *Nephrol* 16:31–41.

Cortney, M. A., Sawin, L. L., Weiss, D. D. 1970. Renal tubular protein absorption in the rat. *J Clin Invest* 49:1–4.

Felmlee, M. A., Dave, R. A., Morris, M. E. 2013. Mechanistic models describing active renal reabsorption and secretion: A simulation-based study. *AAPS J* 15(1):278–87.

Felmlee, M. A., Krzyzanski, W., Morse, B. L., Morris, M. E. 2011. Use of a local sensitivity analysis to inform study design based on a mechanistic toxicokinetic model for gamma-hydroxybutyric acid. *AAPS J* 13:240–54.

Felmlee, M. A., Wang, Q., Cui, D., Roiko, S. A., Morris, M. E. 2010. Mechanistic toxicokinetic model for gamma-hydroxybutyric acid: Inhibition of active renal reabsorption as a potential therapeutic strategy. *AAPS J* 12(3):407–6.

Finney, H., Newman, D. J., Price, C. P. 2000. Adult reference ranges for serum cystein C, creatinine and predicted creatinine clearance. *Acta Clin Biochem* 37:49–59.

Frost, T. H., Kerr, D. N. S. 1977. Kinetics of hemodialysis: A theoretical study of the removal of solutes in chronic renal failure compared to normal health. *Kidney Int* 12:41–50.

Gary, L., Myers, G. L., Miller, W. G., Coresh, J., Fleming, J., Greenberg, N., Greene, T., Hostetter, T., Levey, A. S., Panteghini, M., Welch, M., Eckfeldt, J. H., and the National Kidney Disease Education Program

Laboratory Working Group. 2006. Recommendations for improving serum creatinine measurement: A Report from the Laboratory Working Group of the National Kidney Disease Education Program Group, *Clin Chem* 52(1):5–18.

Gibaldi, M., Perrier, D. 1982. *Pharmacokinetics*, 2nd Edition. New York: Marcel Dekker.

Goldberg, J. P., Anderson, R. J. 1985. Renal metabolism and excretion of drugs. In: The *Kidney: Physiology and Pathophysiology*, eds. D. W. and Seldin, G. Giebisch, New York: Raven Press, 2097–110.

Grubb, A. 1992. Diagnostic value of analysis of cystatin C and protein HC in biological fluids. *Clin Nephrol* 38:S20–S27.

Grubb, A. 2010. Non-invasive estimation of glomerular filtration rate (GFR). The Lund model: Simultaneous use of cystatin C- and creatinine-based GFR-prediction equations, clinical data and an internal quality check. *Scand J Clin Lab Invest* 70:65–70.

Grubb, A., Nyman, U., Jonas Björk, J. 2012. Improved estimation of glomerular filtration rate (GFR) by comparison of $eGFR_{cystatin\ C}$ and $eGFR_{creatinine}$. *Scand J Clin Lab Invest* 72(1):73–77.

Grubb, A., Nyman, U., Björk, J., Lindström, V., Rippe, B., Sterner, G., Christensson, A. 2005. Simple cystatin C-based prediction equations for glomerular filtration rate compared with the modification of diet in renal disease prediction equation for adults and the Schwartz and the Counahan-Barratt prediction equations for children. *Clin Chem* 51:1420–31.

Haapanen, E. J. 1982. Hemoperfusion in acute intoxication. Clinical experience with 48 cases. *Acta Med Scand Suppl* 668:76–81.

Hall, S. L., Wasserman, M., Dall, L., Shubert, T. 1983. Acute renal failure secondary to myoglobinuria associated with Legionnaires' disease. *Chest* 84(5):633–35.

Hasegawa, M., Kusuhara, H., Endou, H., Sugiyama, Y. 2003. Contribution of organic anion transporters to the renal uptake of anionic compounds and nucleoside derivatives in rat. *J Pharmacol Exp Ther* 305:1087–97.

Heath, A., Wickstrom, I., Martensson, E. 1982. Treatment of antidepressant poisoning with resin hemoperfusion. *Hum Toxicol* 1:361–71.

Hediger, M. A., Kanai, Y., You, G., Nussburger, S. 1995. Mammalian ion-coupled solute transports. *J Physiol* 482:7S–17S.

Hellerstein, S., Berenbom M., Alon, U. S., Warady, B. A. 1998. Creatinine clearance following cimetidine for estimation of glomerular filtration rate. *Pediatr Nephrol* 12:49–54.

Henry, J. A. 1982. Can tricyclic antidepressants be removed by haemoperfusion? *Hum Toxicol* 1:359–60.

Hull, J. H., Hak, L. J., Koch, G. G., Wargin, W. A., Chi, S. L., Mattocks, A. M. 1981. Influence of range of renal function and liver disease on predictability of creatinine clearance. *Clin Pharmacol Ther* 29(4):516–21.

Ito, T., Yano, I., Hashimoto, Y., Inui, K.-I. 2000. Transepithelial transport of levofloxacin in the isolated perfused rat kidney. *Pharm Res* 17(2):236–41.

Jelliffe, R. W. 1973. Creatinine clearance: Bedside estimate. *Ann Intern Med* 79:604–5.

Kaloyanides, G. J. 1992. Drug-phospholipid interactions: Role in aminoglycoside nephrotoxicity. *Ren Fail* 14:351–7.

Koepsell, H., Lips, K., Volk, C. 2007. Polyspecific organic cation transporters: Structure, function, physiological roles, and biopharmaceutical implications. *Pharm Res* 24:1227–51.

Kountz, S. L., Yeh, S. H., Wood, J., Cohen, R., Kriss, J. P. 1967. Technetium-99 m(V)–Citrate complex for estimation of glomerular filtration rate. *Nature* 215:1397–99.

Launay-Vacher, V., Izzedine, H., Karie, S., Hulot, J. S., Baumelou, A., Deray, G. 2006. Renal tubular drug transporters. *Nephron Physiol* 103:97–106.

Lee, C. C., Marburt, T. C. 1984. Drug therapy in patients undergoing hemodialysis. *Clin Pharmacokinet* 9:42–66.

Leggett, R. W., Williams, L. R. 1991. Suggested reference values for regional blood volumes in humans. *Health Phys* 60(2):139–54.

Leggett, R. W., Williams, L. R. 1995. A proposed blood circulation model for reference man. *Health Phys* 69:187–201.

Lock, E. A., Reed, C. J. 1998. Xenobiotic metabolizing enzymes of the kidney. *Toxicol Pathol* 26:18–25.

Masereeuw, R., Terlouw, S. A., Van Aubel, R. A. M. H., Russel, F. G. M., Miller, D. S. 2000. Endothelin B receptor-mediated regulation of ATP-driven drug secretion in renal proximal tubule. *Mol Pharmacol* 57:159-67.

Mawer, G. E., Lucas, S. B., Knowles, B. R., Stirland, R. M. 1972. Computer-assisted prescribing of kanamycin for patients with renal insufficiency. *Lancet* 1(7740):12–15.

Müller, F., König, J., Glaeser, H., Schmidt, I., Zolk, O., Fromm, M. F., Maas, R. 2011. Molecular mechanism of renal tubular secretion of the antimalarial drug chloroquine. *Antimicrob Agents Chemother* 55(7):3091–98.

Mitchell, D. Y., Barr, W. H., Eusebio, R. A., Pallone Stevens, K. A., Duke, F. P., Russell, D. A., Nesbitt, J.D., Powell, J. H., Thompson, G. A. 2001. Risedronate pharmacokinetics and intra- and inter-subject variability upon single-dose intravenous and oral administration. *Pharm Res* 18(2):166–70.

National Kidney Disease Education Program (NKDEP) - Creatinine Standardization Recommendations. 2012. IDMS-traceable MDRD Study Equation. http://www.nkdep.nih.gov/lab-evaluation/gfr/creatinine-standardization/recommendations.shtml

Ohashi, R., Tamai, I., Yabuuchi, H., Nezu, J., Oku, A., Sai, Y., Shimane, M., Tsuji, A. 1999. Na1-dependent carnitine transport by organic cation transporter (OCTN2): Its pharmacological and toxicological relevance. *J Pharmacol Exp Ther* 291:778–84.

Parker, P. R., Parker, W. A. 1982. Pharmacokinetic consideration in hemodialysis of drugs. *J Clin Hosp Pharm* 7(2):87–99.

Pond, S. M. 1991. Extracorporeal techniques in the treatment of poisoned patients. *Med J Aust* 154:617–22.

Pritchard, J. B., Miller, D. S. 1993. Mechanisms mediating renal secretion of organic anions and cations. *Physiol Rev* 73:765–96.

Robbins-Weilbert, D. 2004. Design, conduct, and analysis of studies in patients with renal impairment. In: *Pharmacokinetics in Drug Development: Clinical Study Design and Analysis*, ed. P. L. Bonate, and D. R. Howard, 177–208. U.S.A.: AAPS Press.

Saito, Y., Tanigawara, Y., Okamura, N., Shimizu, H., Kamiya, A., Hori, R. 1991. Moment analysis of drug disposition in rat kidney: Transport systems for quinolones may exist on the brush-border Role of basolateral membrane transport in renal transepithelial membrane and play an important role in the renal secretion transport of p-aminohippurate. *J Pharm Pharmacol* 43:311–6.

Salazar, D. E., Corcoran G. B. 1988. Predicting creatinine clearance and renal drug clearance in obese patients from estimated fat-free body mass. *Am J Med* 84:1053–60.

Sargent J. A., Gotch, F. A. 1980. Mathematical modeling of dialysis therapy. *Kidney Int* 18:(Suppl 10)S-2.

Schwartz, G. J., Furth, S., Cole, S., Warady, B., Muñoz, A. 2006. Glomerular filtration rate via plasma iohexol disappearance: Pilot study for chronic kidney disease in children. *Kidney Int* 69:2070–77.

Schwartz, G. J., Haycock, G. B., Edelmann, C. M. Jr, Spitzer, A. 1976. A simple estimate of glomerular filtration rate in children derived from body length and plasma creatinine. *Pediatrics* 58:259–63.

Sekine, T., Cha, S. H., Endou, H. 2000. The multispecific organic anion transporter (OAT) family. *Pflugers Arch* 440:337–50.

Shitara, Y., Sato, H., Sugiyama, Y. 2005. Evaluation of drug–drug interaction in the hepatobiliary and renal transport of drugs. *Annu Rev Pharmacol Toxicol* 45:689–723.

Shull, B. C., Haughey, D., Koup, J. R., Baliah, T., Li, P. K. 1978. A useful method for predicting creatinine clearance in children. *Clin Chem* 24(7):1167–69.

Simonsen, O., Grubb, A., Thysell, H. 1985. The blood serum concentration of cystatin C as a measure of glomerular filtration rate. *Scand J Clin Lab Invest* 45:97–101.

Sprenger, K. B. G., Kratz, W., Lewis, A. E., Stadtmüller, U. 1983. Kinetic modeling pf hemodialysis, hemofiltration and hemodiafiltration. *Kidney Int* 24:143–151.

Tamai, I., Nakanishi, T., China, K., Kosugi, Y., Kobayashi, D., Nezu, J., Sai, Y., Tsuji, A. 2004. Involvement of OCTN1 (SLC22A4) in pH-dependent transport of organic cations. *Mol Pharm* 1:57–66.

Tanihara, Y., Masuda, S., Sato, T., Katsura, T., Ogawa, O., Inui, K-I. 2007. Subtrate specificity of MATE1 and MATE2-K, human multidrug and toxin extrusions/H+-organic cation antiporters. *Biochem Pharmacol* 74(2):359-71.

Toto, R. D., Kirk, K. A. 1995. Conventional measurement of renal function utilizing serum creatinine, creatinine clearance, inulin and para-aminohippuric acid clearance. *Curr Opin Nephrol Hypertens* 4:505–9.

Walton, K., Dorne, J. L. C. M., Renwick, A. G. 2004. Species-specific uncertainty factors for compounds eliminated principally by renal excretion in humans. *Food Chem Toxicol* 42:261–74.

6 Biotransformation of Xenobiotics

6.1 INTRODUCTION

Metabolism, the principal mechanism for biotransformation and elimination of xenobiotics, occurs at multiple sites in the body. Although the liver is the main site for metabolism, other organs such as the kidneys, GI tract, skin, and nasal mucosa are also capable of contributing quantitatively to the overall metabolism of xenobiotics.

6.2 LIVER

The liver weighs about one fiftieth of the body weight (≈between 1 and 1.5 kg). It has two lobes; the right lobe is approximately six times larger than the left one. A number of factors influence its weight in normal population. Its weight is higher in the young male group than female or aged population; also, for the same body mass index, there are differences among races. For example, the Asians have greater visceral adiposity than Caucasians. Using the following empirical equations, one can estimate the weight of liver:

$$\text{Estimated Standard Liver Weight (g)} = (bwt(\text{kg}) + 218) \times 12.3 + (\text{Gender Factor}) \times 51 \quad (6.1)$$

The gender factor for male is one and for female is zero, that is,

$$\text{Male:} \quad \text{Liver weight (g)} = (\text{Body weight (kg)} + 218) \times 12.3 + 51 \quad (6.2)$$

$$\text{Female:} \quad \text{Liver weight (g)} = (\text{Body weight (kg)} + 218) \times 12.3 \quad (6.3)$$

Another age and weight related measurement of the liver is its volume. The volume may shrink in disease states such as cirrhosis, or increase with a hepatic tumor. The imperical formulas for estimating the standard liver volume are

$$LV(\text{mL}) = 706.2 \times BSA(\text{m}^2) + 2.4 \quad (6.4)$$

$$LV(\text{mL}) = 2.223 \times bwt(\text{kg})^{0.426} \times height(\text{cm})^{0.682} \quad (6.5)$$

Physiologically, the liver is responsible for more than 500 various functions and its principal roles are

- Endocrine functions (including the IGF and HGF synthesis and catabolism of other hormones)
- Exocrine function (including formation of bile and urea)
- Regulation of homeostasis (including coagulation and fibrinolysis)
- Energetic and structural metabolism (including biosynthesis and metabolism of carbohydrates, lipids and proteins)
- Synthesis of albumin and proteins

- Immunomodulation and regulation of inflammatory response; storage (including glycogen, nutrients, vitamins and minerals)
- Biotransformation of endogenous and xenobiotics (including enzymatic reactions of Phase I and Phase II metabolism)
- Function of transporters in the liver, often identified as Phase III transfer (Nahmias et al., 2006; Arias et al., 2009; Mazzoleni and Steimberg, 2012); essentially the transporters transfer the conjugates into the bile for elimination

The vessels carrying blood to the liver are the hepatic portal vein and the hepatic artery. The former conveys venous blood from splenic, inferior and superior mesenteric veins, which originate from the small and large intestines, spleen, pancreas, stomach, the double layer of peritoneum (omentum) that is attached to stomach, and all organs of the abdomen including the gallbladder. Venous blood provides approximately 75% of blood volume, rich in nutrients and poor in oxygen, to the liver. The liver does not control the blood flow of the hepatic portal vein; the flow rate of the vein is the combination of blood flow out of all organs of the abdomen that forward the blood to the liver. Xenobiotics administered orally and absorbed via the gastrointestinal tract enter the capillaries and the portal blood, transfer to the liver, and depending on their physicochemical characteristics may metabolize by the enzymes of the liver. This process known as first-pass metabolism reduces the bioavailability of the administered dose (Galetin and Houston, 2006). The hepatic artery, which originates from the aorta by way of the celiac trunk, provides arterial blood rich in oxygen to the liver. Its flow rate depends on the flow rate of hepatic portal vein such that it maintains the balance of total blood flow through the liver and total blood volume of about 20%–25%. Based upon the oxygen supply, the liver is functionally divided into three zones. Zone 1, known as periportal, consists of parenchymal cells encircling the portal tracts where portal and arterial blood join and mix. Oxidative energy metabolism; glucose release; secretion of bile acids and bilirubin; fatty acid oxidation; amino acid utilization, degradation, and conversion to glucose; and urea formation are predominantly take place in zone 1. Zone 3 is the periphery region around the hepatic central vein and zone 2 is the region between zone 1 and 3. All outlets of the liver are located in zone 3. The major function in zone 3 includes glucose uptake, glycolysis, ketogenesis, synthesis of glycogen from glucose, synthesis of bile salts, lipogenesis, and bio transformation (Thurman and Kauffman, 1985; Jungermann, 1995). The vascular structure is such that the transfer through parenchymal cells is facilitated from zone 1 to zone 3, but not from zone 3 to zone 1. The basic structure of the liver lobule and its essential components are depicted in Figure 6.1. Approximately 78% of the cells are hepatocytes, 3% endothelial cells, 2% Kupffer cells and 1% Ito cells (fat storing cells), the remaining 16% is related to spaces, which includes the sinosuidal lumen, bile passageway, canaliculi, and the space between sinusoidal endothelium and hepatocytes, the space of Disse. Zone 3 has higher quantity of CYP450 subfamilies and the highest xenobiotic-related toxicity; for example, necrosis caused by acetaminophen, pyrrolizidine alkaloids, mushroom poisoning, hydrocarbons, halothane, carbon tetrachloride (CCl4), etc. Toxicity in zone 2 is very rare. Zone 1 has also toxicity manifestations like necrosis caused by allyl alcohol and phosphorous. The function of hepatocytes in the metabolism depends on what zone the cells are located (Haussinger, 1988; Bhatia et al., 1996). In addition to hepatocytes that comprise approximately 80% of the liver, there are other cells, at least a dozen, which play important functional roles in the normal and diseased liver (Malarkey et al., 2005).

6.3 BIOTRANSFORMATION (METABOLISM)

Essentially all xenobiotics entering the body are subject to some degree of metabolism. The outcome of this biotransformation is the formation of chemically modified molecules of xenobiotics known as metabolites, which are hydrophilic derivatives of the parent compound and are more easily eliminated through urinary or biliary elimination. This biotransformation occurs with the help of a number of enzyme systems that are utilized for metabolism of dietary constituents. Genes

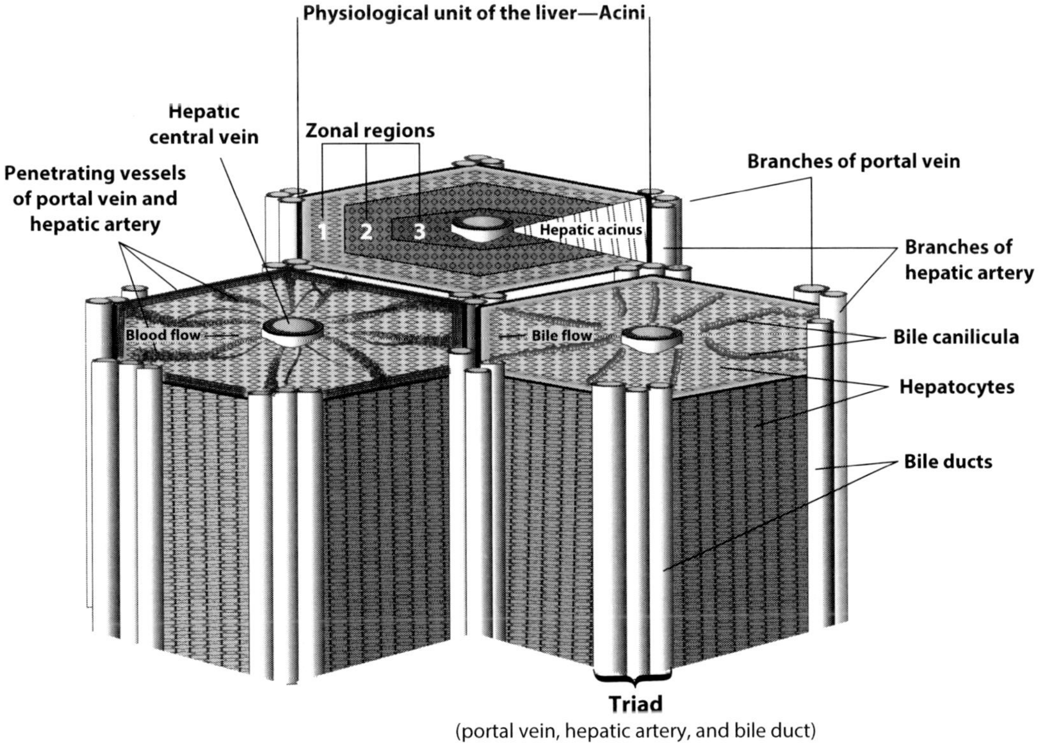

FIGURE 6.1 Schematic illustration of the liver lobule and its essential components.

encoding these enzyme systems have functioned in every eukaryotic cell. At a larger scale, each person has their own unique alleles coding for metabolic enzyme systems. Interindividual variations in metabolism of xenobiotics, mostly heritable, have given rise to the field of pharmacogenetics. Humans are continuously exposed to all sources of xenobiotics, from food constituents and additives to cosmetics to herbal medications to environmental pollutants and therapeutic agents. All go under biotransformation by the same enzyme systems and may set off a series of xenobiotic-xenobiotic interactions based on the inhibition or induction of the metabolizing enzyme systems (Sugimura and Sato, 1983).

The metabolites can be primary or secondary metabolites. There are compounds that their primary metabolites are more hydrophobic and reactive than the parent compound and require further metabolism for elimination. This is a paradoxical role of metabolism because elimination of xenobiotics from the body is useful for activation of inactive therapeutic agents known as prodrugs. The reactive or electrophilic metabolites may also interact with the cellular membrane or nucleophilic macromolecules, such as DNA, RNA, and protein, forming repairable or non-repairable adducts (Dipple, 1983). The interaction of the reactive metabolites with DNA may result in the mutation of genes and initiation of cancer.

The metabolic pathways catalyzed by the enzyme system occurs in two distinct phases known as Phase I and Phase II metabolism. Xenobiotics are subjected to one or multiple biotransformation pathways offered by the Phase I and II enzyme systems in succession. Depending upon chemical structure of a xenobiotic, enzymes of Phase I metabolism create functional groups such as NH_2, –OH, –SH, -*O*-,or, –COOH in the molecular structure, whereas the enzymes of Phase II metabolism add additional groups such as glucuronic acid, glutathione, sulfate, acetyl, or methyl group to inactivate and increase the water solubility of a xenobiotic and/or its metabolites. There are significant differences among species in Phase I and II metabolism of xenobiotics, attributable to the dietary and environmental conditions during the course of evolution. The differences in species metabolism can be qualitative or quantitative. The qualitative differences exist when the species use different pathways of metabolism to metabolize a compound. The quantitative differences are when the kinetic parameters of the same pathways are different for example, the rate of metabolism, metabolic clearance, etc.

6.3.1 Phase I Metabolism

Phase I reactions include the modification of the molecular structure of xenobiotics by creating new functional group(s) and/or modifying existing functional group(s). The products of Phase I metabolism, also known as primary metabolites, can be reactive and toxic or non-toxic and hydrophilic. In general, any enzymatic reaction that forms non-reactive and non-toxic metabolites from a toxic parent compound, or toxic and reactive primary metabolites, is known as detoxification pathway, which includes certain reactions of Phase I and all pathways of Phase II metabolism. The Phase I enzyme systems of metabolism are as follows.

6.3.1.1 Flavin-Containing Monooxygenases (FMO)

The flavin-containing monooxygenases catalyze the oxygenation of structurally diverse molecules containing nitrogen, sulphur, phosphorus, and selenium via the bound cofactor flavin. FMO is considered a complementary enzyme system to the CYP450 family of enzyme systems (Cashman, 2002). The metabolites of FMO systems are polar molecules, which eliminate easily from the body. These characteristics are considered advantages in designing therapeutic agents that are metabolized by FMO and not by CYP450 isozymes. However, sometimes FMO generate metabolites that are reactive materials and can cause toxicity (Cashman and Zhang, 2006). There are at least six FMO in the mammalian system. FMO1 mediates oxidative metabolism of lipophilic compounds such as imipramine and orphenadrine (Koukouritaki et al., 2002). It is FMO form 3 (FMO3) that is the prominent form in the adult human liver and is likely to be associated with the bulk of FMO-mediated metabolism (Larsen-Su and Williams, 1996). FMO1 and FMO3 S-oxygenate a number of nucleophilic sulfur-containing substrates and are sensitive to steric features of the substrate and aliphatic amines with linkages between the nitrogen atom and a large aromatic group such as a phenothiazine. Small amine molecules such as phenethylamines are efficiently N-oxygenated by human FMO3. For this reason, the cimetidine S-oxygenation or ranitidine N-oxidation is used as a functional probe of human FMO3. Trimethylamine is another amine molecule metabolized by FMO3 in the body to non-odorous trimethilamine N-oxide (TMA N-oxide). Mutation of the FMO3 gene reduces or eliminates TMA N-oxide formation, which results in trimethyaminuria, a disorder manifested by a body odor for affected individuals (Motika et al., 2009). In addition to being the prominent form, FMO3 is species- and tissue-specific, and contrary to human cytochrome P450, it is not easily inducible or inhibited, which minimizes drug-drug interaction and provides opportunities in design of molecules that metabolize by this enzyme system (Cashman and Zhang, 2006). There are significant inter-individual variations in FMO3-dependent metabolism of xenobiotics, an important issue in population PK/TK analysis (see also Chapter 17). Although selective substrates of FMO enzyme systems are known, selective inhibitors of the systems are yet to be determined (Harper and Brassil, 2008).

6.3.1.2 Flavin-Containing Amine Oxidoreductases

The members of this enzyme system are polyamine oxidase (PAO), L-amino acid oxidases (LAO), and various flavin containing monoamine oxidases (MAO).

Polyamine oxidase (PAO) with a molecular weight of 62 kDa is a FAD-dependent enzyme present in most tissues of vertebrates. Molecules with two positively charged amino groups, with one or both with alkyl substituent, separated by a short carbon chain are substrates for PAO. Spermine and the monoacetyl derivatives like N1-acetylspermine and N1-acetylspermidine are natural substrates of PAO (Seiler, 1995). PAO plays a major role in the intracellular regulatory system and maintenance of polyamine homeostasis in non-proliferating cells such as brain cells. The dynamics of polyamine homeostasis, identification of PAO, and spermine oxidase sequences in conjunction with metabolic and cellular responses have provided new opportunities to evaluate clinically relevant polyamine analogues and inhibitors (Vujcic et al., 2003).

L-amino acid oxidases (LAO) catalyze most amino acids of an L configuration. Methionine and other sulfur-containing molecules, for example, cystine, homocystine, and thialysine are the best substrates of this enzyme system (Cavallini et al., 1982). LAO have antibacterial, antifungal, antiprotozoal, antiviral, antiproliferative, and antitumor properties, with an unclear effect on platelet aggregation (Lukasheva et al., 2011).

MAO is a FAD-containing enzyme of the outer mitochondrial membrane. In humans, there are two isozymes, which are identified as MAO-A and MAO-B. They differ in their substrate specificity and inhibited by different inhibitors (Shih, 1991; Gargalidis-Moudanos et al., 1997). Their important role is in regulating the intracellular levels of amines via their oxidation; these include various neurotransmitters, neurotoxins, and trace amines. MAO-A is found in the liver, GI tract, placenta, neuron, and astroglia; MAO-B is found in blood platelets, neurons, and astroglia. MAO-A metabolizes melatonin, adrenaline, noradrenalin, and serotonin and MAO-B metabolizes phenethylamine and benzylamine. Both forms break down dopamine, tyramine, and tryptamine.

6.3.1.3 Epoxide Hydrolases (EHs)

Epoxide hydrolases, also known as epoxide hydratases, function in the detoxification of metabolites of xenobiotics. They catalyze the hydration of reactive endogenous and exogenous epoxides to dihydrodiols, which can then be conjugated and excreted from the body (Oesch and Arand, 1999; Arand and Oesch, 2002; Arand et al., 2003). Epoxide hydolase 1, identified as mEH, is localized in the endoplasmic reticulum and microsome, and epoxide hydrolase 2, a cytosolic enzyme known as soluble EH or sEH, plays a critical role in defending the body against highly reactive, toxic, and often carcinogenic epoxides. Epoxides are mostly produced as intermediate metabolites of lipophilic compounds that are considered carcinogenic. Many epoxides are strong electrophiles and react readily with the DNA forming repairable or non-repairable DNA adducts which cause mutation and initiate the cancer (Boroujerdi et al., 1981). EHs detoxify a broad range of structurally very different epoxides such as benzo (a) pyrene or styrene oxide. Members of EHs are used as the therapeutic targets for regulation of blood pressure, inflammation, cancer progression, and the onset of several other diseases (Morisseau et al., 2005).

6.3.1.4 Cytochrome P450s (CYP450)

CYP450 represents a superfamily of heme containing enzymes with the potential to catalyze dehydrogenation, hydroxylation, epoxidation, oxygenation, dealkylation, and ring opening (Porter et al., 1991; Guengerich, 1997). To represent the diversity of CYP450 family members, a nomenclature has been created to identify different families and subfamilies (Nebert et al., 1991). This nomenclature is based on the amino acid sequence of the protein and their assignment to different gene families. When 40% of amino acids of CYP450 are identical, they belong to a family, for example, CYP1, CYP2, CYP3, CYP4, etc. When more than 55% of amino acids are identical they are in a subfamily labeled with letters, for example, CYP1A or B or C, or CYP2A or B or C, etc. The final identification number is a cataloging number to identify the individual enzyme as a member of the subfamily, for example, CYP1A1, CYP2B6, etc. The CYP450 family members are present practically in all mammalian organs, but predominantly in the liver, GI tract, skin, nasal epithelia, lung, kidney, and other organs (Nebert et al., 1991; Nelson et al., 1993, 1996; Agundez, 2004). The following sections are a brief review of the role, characteristics, and related metabolic reactions of each member of CYP450 family.

6.3.1.4.1 CYP1A Subfamily

The members of CYP1A subfamily are predominantly involved in detoxification and bioactivation of environmental pollutant (Goldstone and Stegeman, 2006). In humans, the subfamily has two members, CYP1A1 and CYP1A2. CYP1A1 is expressed at low levels in the liver and variable in extrahepatic tissues, for example, lung, placenta, lymphocytes, and intestine. CYP1A2, on the other hand, is highly expressed in the liver (Roins and Ingelman-Sunderg, 1999). Both enzymes are inducible, but CYP1A1 is more sensitive to the inducers (Whitlock, 1999). The induction of CYP1A enzymes may also lead to catalytic activation of xenobiotics to their reactive metabolites, which may result in toxicity (Digiovanna et al., 1979; Gelboin, 1980). CYP1A1 and CYP1A2 catalyze the metabolic activation of two important classes of environmental carcinogens, polycyclic aromatic hydrocarbons and arylamines (Conney, 1982; Kamataki et al., 1983; Shimada, 2006). The CYP1A subfamily members are also expressed in experimental animals (mice, rats, dogs, and monkeys), and are involved in stereoselective metabolism of certain compounds in Caco-2 cells (Ishida et al., 2009).

6.3.1.4.2 CYP1B Subfamily

CYP1B, a member of this subfamily, is involved in the metabolism of anticancer agents used in the treatment of hormone-mediated cancer. This isozyme is present in human and experimental animals, mouse, rat, dog, and monkey. In humans, CYP1B1 catalyzes estrogens to active 4-hydroxylated derivatives that may cause breast cancer. CYP1B1 is expressed in most organs including the liver, heart, lung, kidney, placenta, prostate, and brain (Sutter et al., 1994; Willey et al., 1997; Rieder et al., 1998; Cheung et al., 1999; McFadyen et al., 1999a,b; Muskhelishvili et al., 2001; Spivak et al., 2001). The majority of the human tissues that are positive for CYP1B1 are those that are receptive to steroid hormones like prostate, testes, endometrium, and mammary tissue (Eltom et al., 1998; Larsen et al., 1998; Tang et al., 1999; Bofinger et al., 2001). Its expression is much higher in tumor cells (Murray et al., 1997; McFadyen et al., 1999a,b, 2001) and its induction in the body is considered a biomarker for determining a hormone-mediated risk of cancer and as a target for the development of anticancer agents (McFadyen et al., 1999a,b; Doostdar et al., 2000; Bofinger et al., 2001; Potter et al., 2002).

6.3.1.4.3 CYP2A Subfamily

The members of CYP2A have diverse metabolic role in different species. In humans, the two important functional members of this subfamily are CYP2A6 and CYP2A13 (Yamano et al., 1990; Su et al., 2000). CYP2A6 is expressed in the liver whereas CYP2A13 is expressed in extrahepatic tissues, mainly in respiratory tract (Koskela et al., 1999; Gu et al., 2000; Su et al., 2000). Two identified members of this subfamily in mice are *CYP2a4* and *CYP2a5*. The former hydroxylates steroid and the latter encodes coumarin 7-hydroxylase (Negishi et al., 1989). CYP2A6 of human functionally is similar to the CYP2A5 of mice (Yamano et al., 1990). CY2A6 and CYP2A13 participate in oxidation of nicotine, cyclophosphamide and aflatoxin B1, hydroxylation of coumarin, and metabolism of tobacco-specific nitrosamines derivatives (Yamano et al., 1990; Tiano et al., 1993; Nakajima et al., 2006; von Weymarn and Murphy, 2003; Bao et al., 2005). A number of xenobiotics such as phenobarbital, rifampicin, dexamethasone, and nicotine can induce both isoforms in humans (Honkakoski and Negishi, 1997). Other members of the same subfamily such as CYP1A2 have overlapping substrate specificity as CYP1A13 (Fukami et al., 2007). Other isozymes of this subfamily are CYP2A7 in human; CYP2A4, CYP2A5, CYP2A12 and CYP2A22 in mice; CYP2A1, CYP2A2, and CYP2A3 in rats; CYP2A13 and CYP2A25 in dogs; and CYP 2A23 and CYP2A24 in monkeys.

6.3.1.4.4 CYP2B Subfamily

The isozymes of CYP2B are present in the liver and other organs such as kidney, brain, intestine, and lung (Gervot et al., 1999). In rodents, they are in the intestine and two isozymes of CYP2B1 and CYP2B2 are expressed in rat brain (Rosenbrock et al., 2001). In humans, CYP2B6 is the major isozyme of the subfamily (Ekins et al., 1997, 1998). It is responsible for the metabolism of several therapeutic agents including propofol (Court et al., 2001); promazine (Wójcikowski et al., 2003); tamoxifen (Sridar et al., 2012); cyclophosphamide (Chang et al., 1993; Melanson et al., 2010); efavirenz (Lindfelt et al., 2010), testosterone and androstenedione (Domanski et al., 1999); and metabolism of carcinogens such as dibenzanthracin and aflatoxin B1 is also by human CYP2B isozymes. The metabolism of antidepressant/antismoking agent bupropion is exclusively by CYP2B6 and thus it is used as a marker for CYP2B6 (Faucette et al., 2000). The specific inhibitors of CYP2B6 include clopidorel (Richter et al., 2004) and thiotepa (Rae et al., 2002; Richter et al., 2004). Significant interindividual variability has been observed between men and women (♀ ≫ ♂) and between the females of different ethnicities (Shimada et al., 1994; Code et al., 1997; Lang et al., 2001; Lamba et al., 2003; Elens et al., 2010; Jamshidi et al., 2010) for CYP2B6.

Another isozyme of this subfamily in human is CYP2B7, which is present in normal and neoplastic lung tissues (Czerwinski et al., 1994).

Other isozymes of the subfamily in experimental animals are CYP2B9 and CYP2B10 in mice; CYP2B1, CYP2B2, and CYP2B3 in rats; CYP2B11 in dogs; and CYP2B17 in monkeys.

6.3.1.4.5 CYP2C Subfamily

The members of the CYP2C subfamily are involved in the metabolism of a significant number of xenobiotics (Goldstein and de Morais, 1994). The four members of the human CYP2C subfamily, namely CYP2C8, CYP2C9, CYP2C18, and CYP2C19, are expressed in different organs in the body: CYP2C18 is mostly in the skin (Zaphiropoulos, 1997) and CYP2C8, CYP2C9, and CYP2C19 are mainly in the liver. CYP2C8 is responsible for metabolism of paclitaxel, arachidonic acid, and benzo[a]pyrene (Rahman et al., 1994; Klose et al., 1999). CYP2C9 is the major isozyme and most abundant one involved in the metabolism of antidiabetic, anticonvulsant, anti-inflammatory, and antihypertensive therapeutic agents, such as tolbutamide, ibuprofen, phenytoin, warfarin, diclofenac, piroxicam, tenoxicam, mefenamic acid, losartan glipizide, and torasemide (Scott and Poffenbarger, 1979; Goldstein and de Morais, 1994; Kidd et al., 1999; McCrea et al., 1999; Miners et al., 2000). In addition to the liver, CYP2C19 is also present in the duodenum, and involved in the metabolism of many xenobiotics including antidepressants, proton pump inhibitors, and anxiolytic drugs, such as omeprazole, imipramin, diazepam, and many other compounds (Kupfer and Branch, 1985; Ward et al., 1989, 1991; Andersson et al., 1992; Sindrup et al., 1993). Because of the importance of this isozyme and its substrate specificity, humanized mouse models are available for the investigation of metabolism and disposition of CYP2C9 substrates (Scheer et al., 2012). Compounds like phenobarbital, rifampicin and dexamethasone induce CYP2C8, CYP2C9, and CYP2C19, whereas antifungal drugs fluconazole, voriconazole inhibit their activity.

The members of this subfamily are more complex and numerous in rodents and expressed mostly in the liver and extrahepatic tissues. In mice the members of the subfamily include CYP2C9, CYP2C37, CYP2C38, CYP2C39, CYP2C40, CYP2C44, CYP2C50, CYP2C54, and CYP2C55; in rats the isozymes are CYP2C6, CYP2C7, CYP2C11, CYP2C12, CYP2C13, CYP2C22, and CYP2C23; in dogs the isozymes are CYP2C21 and CYP2C41; and in monkeys they are CYP2C20 and CYP2C43. In dogs, CYP2C family members are neither well developed nor similar to humans. The CYP2C43 of monkeys is structurally similar to CYP2C9 of humans, but functionally is comparable to CYP2C19 (Matsunaga et al., 2002).

6.3.1.4.6 CYP2D Subfamily

The members of the CYP2D subfamily are responsible for monooxygenation of a number of xenobiotics; for example, methadone, propranolol, tamoxifen, dectromethorphan, sparteine, bufuralol, and desipramin in humans and experimental animals (Bogaards et al., 2000; Hiroi et al., 2002; Teh et al., 2012). In humans, there are three isozymes of CYP2D6, -2D7 and -2D8, but only CYP2D6 is the most active and studied isozyme of the subfamily in humans. It is present not only in the liver, but also in the kidney, lung, intestine, placenta, breast, and brain (Niznik et al., 1990; Huang et al., 1997).

Genetic polymorphism of this isozyme is the cause of variability in temoxifen metabolism, which attributes to the lack of therapeutic efficacy in certain patients and reoccurrence of breast cancer (Teh et al., 2012). This individual variation is the result of inherited mutant CYP2Ds alleles as an autosomal recessive trait in approximately 7%–10% of the Caucasian population (Mahgoub et al., 1977). Fluoxetine (Prozac) and its active metabolite, norfluoxetin inhibit CYP2D6 in the liver (Crewe et al., 1992; Stevens and Wrighton, 1993) and quinidine is its most potent inhibitor (Smith and Jones, 1992).

There are more members of this subfamily in experimental animals. In mice there are at least ten isozymes: CYP2D9, CYP2D10, CYP2D11, CYP2D12, CYP2D13, CYP2D22, CYP2D26, CYP2D34, and CYP2D40. The mouse CYP2D22 function is somewhat similar to the human CYP2D6 (Blume et al., 2000). In rats there are six isoforms: CYP2D1, CYP2D2, CYP2D3, CYP2D4, CYP2D5, and CYP2D18. The isozyme CYP2D1 is the rat orthologue of CYP2D6 of the human. The major isozyme in dogs is CYP2D15, which enzymatically is similar to the human CYP2D6 (Zuber et al., 2002) and is inhibited by quinidine and the HIV-I protease inhibitor. The CYP2D isozymes in monkeys are strain related. The CYP 2D42 of the Rhesus monkey, CYP2D30 of the Marmoset monkey and CYP2D17 of the Cynomolgus monkey are an orthologue of human CYP2D6.

6.3.1.4.7 CYP2E Subfamily

A member of the CYP2E subfamily, CYP2E1, is present in the liver, lung, and the tissues of the nose and pharynx. CYP2E1 has dual physiological roles of providing nutritional support and protective detoxification (Lieber, 1997), and under certain conditions may generate superoxide radical with pathlogical consequences of liver injury and interaction with cellular macromolecules (Wu and Cederbaum, 2003). It plays two major protective roles in the body. It detoxifies certain xenobiotics, particularly organic solvents such as ethanol (Wu and Cederbaum, 2003). It also plays an important role in the nutritional balance of the body during starvation, metabolizing lipids, and converting ketones to glucose (Lieber, 2004). It is also involved in the formation of reactive metabolites of carcinogens like acrylnitrile, styrene, isoniazid, nitrosamine, and benzene, which subsequently interacts with cellular DNA, forming carcinogen-DNA adducts (Siegers et al., 1983; Hetu et al., 1983; Dai et al., 1993; Wu and Cederbaum, 2003). To prevent the liver injury by CYP2E1-mediated hepatotoxicity induced by compounds like isoniazid, co-administration of compounds like kaempferol is recommended (Shih et al., 2013). CYP2E1 metabolizes xenobiotics such as chlorzoxazone, caffeine, and acetaminiohen. These compounds are also considered marker coe CYP2E1 activity (Lofgren et al., 2004).

Like humans, ethanol induces CYP2E1 in all species. Rodents are considered relevant animal models to study the CYP2E1 metabolism of xenobiotics in humans (Zuber et al., 2002). CYP2E1 of rats is one the few enzymes that functionally is similar to humans and extrapolation seems to be warranted. However, the comparative data of CYP2E1-dependent metabolism of dogs and monkeys with those of humans have reported to be inconsistent (Bogaards et al., 2000).

6.3.1.4.8 CYP3A Subfamily

The members of this subfamily are central to the metabolism of most xenobiotics. They are expressed in a number of organs and tissues including the liver, gastrointestinal tract, lung, and kidney. They metabolize a wide range of therapeutic agents such as benzodiazepines, lidocaine, quinidine, dextromethorphan, midazolam, triazolam, carbamazepine, nifedipine, erythromycin, terfenadine, tacrolimus, and many more (Nebert and Russell, 2002; Zuber et al., 2002). Furthermore, with their broad substrate specificity, they are also involved in oxidation of endogenous compounds like steroids (Hirita and Matsubara, 1993), retinoic acid, and bile acids, and oxidation of potent carcinogens notably aflatoxin B_1, aflatoxin G_1, and benzo (a) pyrene. The substrates of this subfamily or their metabolites are also substrates for a number of other subfamilies like CYP1A, CYP2A, CYP2E, and so on.

The CYP3A subfamily members in humans are CYP3A4, CYP3A5, -3A7, and CYP3A43. Among the members of the subfamily and because of their presence in the liver, stomach, lungs, kidneys, and duodenum, jejunum, and ileum, CYP3A4 and CYP3A5 are the most prominent isozymes. CYP3A4 is abundantly present in the human liver (Kolars et al., 1994; Dresser et al., 2000) and the tip of the enterocytes, and play a significant role in the intestinal and hepatic first-pass metabolism. Thus, they can reduce the bioavailability of xebobiotics significantly (Paine et al., 1997). Pgp is present in the same region of the small intestine and they often share the same substrate specificity. The presence of both proteins in the same locality of the G.I. tract prevents the absorption of xenobiotics. The efflux of xenobiotics by Pgp increases their concentration in the lumen and enhances their interaction with CYP3A4, suggesting a cooperative role between the two proteins for removal and metabolism of xenobiotics. CYP3A4 is present in the periphery of the endoplasmic reticulum of the enterocytes, and Pgp is present at the apical membrane of the enterocyte. The Pgp of the liver also effluxes the xenobiotics into the bile ducts, which enters into the small intestine and will be available for interaction with CYP3A4 (Van der Valk et al., 1990; Guengerich, 1992; McKinnon et al., 1995; Hall et al., 1999; Ueda et al., 1999; Benet et al., 2003; Liu and Pang, 2005).

The other members of the subfamily are: CYP3A5—more prominent in stomach and colon; CYP3A7—more expressed in fetal liver, and CYP3A43 with limited activity in the liver (Domanski et al., 2001; Gellner et al., 2001).

CYP3A isozymes in different species are diverse and dissimilar. In mice, there are six isozymes of CYP3A11, CYP3A13, CYP3A16, CYP3A25, CYP3A41, and CYP3A44 (Yanagimoto et al., 1992; Sakuma et al., 2000; Schellens et al., 2000). The aflatoxin B1 metabolism by CYP3A of mice is similar to humans (Gallagher et al., 1996). In rats the isoforms include CYP3A1, CYP3A2, CYP3A9, CYP3A18, CYP3A32, and CYP3A62 (Kirita et al., 1993; Strotkamp et al., 1995; Wang et al., 1996; Matsubara et al., 2004). Although CYP 3A1 is expressed in the liver, its induction or inhibition does not resemble the human CYP3A4 and in general, rats are not considered appropriate animal models for human CYP3A studies. In dogs, there are two members of the subfamily, CYP3A12 and CYP3A26, which are expressed in the liver and show similar characteristics as the CYP3A4 and CY3A5 of humans (Fraser et al., 1997). The CYP3A isofom identified in the Cynomolgus monkey is CYP3A8, which is to some extent similar to human's CYP3A4 and CYP3A5 (Komori et al., 1992; Ohmori et al., 1993).

The importance of the CYP3A subfamily in the metabolism of endogenous compounds and xenobiotics is well established and recognized. However, the species differences in the metabolism of xenobiotics are significant and the induction and/or inhibition of the isoforms are quite dissimilar. Thus, the extrapolation of metabolic data of the CYP3A subfamily among species is debatable.

6.3.1.4.9 CYP4A Subfamily

CYP4A subfamily members are fatty acid hydroxylases, present in most mammalian organs, and responsible for ω-hydroxylation and ω-1-hydroxylation of endogenous saturated and unsaturated fatty acids, prostaglandin, and arachidonic acid (Sharma et al., 1989; Aoyama et al., 1990; Su et al., 1998; Okita and Okita, 2001; Cowart et al., 2002; Savas et al., 2003). The isozymes facilitate the elimination of excess fatty acids to produce energy. Their role in xenobiotic metabolism is not quite clear. CYP4A11 and CYP4A22 are the isoforms of the subfamily in humans. CYP4A11 is in the human liver and kidney, and acts as fatty acid ω-hydroxylating enzyme to regulate cell and/or organ physiology by conversion of arachidonic acid to the vasoactive and natriuretic eicosanoid 20-hydroxyeicosatetraenoic acid. Three members of the subfamily in rats include CYP4A1, CYP4A2, CYP4A3, and CYP4A8; in rabbit they are CYP4A4, CYP4A5, and CYP4A7. There are functional and structural similarities between CYP4A11 of human, CYP4A1 and CYP4A3 of rats, and CYP4A6 of rabbits (Imaoka et al., 1993). CYP4A1 is the major fatty acid hydroxylase of the rat liver, regulated by physiological conditions including diabetes and fasting and induced by different compounds, such as fatty acids and peroxisome proliferators like hypolipidemic. These effects are mediated via the peroxisome proliferator-activated receptor alpha (PPAR) (Savas et al., 2003). Peroxisome proliferator-activated receptors (PPARs) are ligand-activated nuclear receptors, and exist as three different subtypes in mammals (PPARα, PPARδ and PPARγ [with isoforms γ1, γ2 and $\gamma 3$]). The nuclear receptors are involved in lipid homeostasis regulation of body weight and food intake, control of inflammation, wound healing, and induction of CYP4A isoforms.

6.3.1.4.10 CYP4B Subfamily

The members of this subfamily belong to the CYP4 family of the mammalian enzyme system, which includes CYP4A, CYP4F, CYP4V, CYP4X, and CYP4B1 subfamilies. The subfamilies have the capacity of participating in ω-hydroxylation of medium-chain fatty acids. CYP4B1, a member of CYP4B subfamily, is present in humans, rats, mice, dogs, pigs, chimpanzees, goats, rabbits, and cow. Furthermore, CYP4B2 is identified in goats and cattle, and CYP4B3 in wallabys. CYP4B1 also participate in the metabolism of some structurally unrelated xenobiotics that are protoxic such as aromatic amines, valproic acid, 4-ipomeanol, 3-methylindole, and 3-methoxy-4-aminoazobenzene. The metabolites of these compounds cause tissue- or organ-specific toxicity (Baer and Rettie, 2006). The pathway of hydroxylation arylamines not only is mediated by CYP4B1, but also by CYP1A1 and CYP1A2 (Windmill et al., 1997). Due to heterologous expression of CYP4B1, its metabolic capabilities are less clear in different human populations.

6.3.1.4.11 CYP4F Subfamily

The members of this subfamily are present in different locations in the body. For example, CYP4F2, CYP4F3, CYP4F11 and CYP4F12 are expressed in the liver and kidney (Christmas et al., 1999, 2001, 2003; Cui et al., 2000; Hirani et al., 2008), whereas CYP4F3A is found only in myeloid tissue, and CYP4F8 is expressed exclusively in seminal vesicles (Bylund et al., 2000). Similar to the enzymes of CYP4 family, CYP4F subfamily is involved in ω-hydroxylation of long and very long-chain fatty acids and the metabolism of vitamin E (Sontag and Parker, 2002; Sanders et al., 2006). The isozymes have the unique ability to oxidize or ω-hydroxylate the terminal methyl group present in saturated and unsaturated fatty acids of different chain. The isoform CYP4F2 participates in the metabolism of arachidonic acid into 20-hydroxyeicosatetraenoic acid (20-HETE), a strong vasoconstrictor, inhibitor of ion transport, and cellular proliferation agent.

6.3.1.5 Alcohol Dehydrogenase

Alcohol dehydrogenase (ADH), a member of the oxidoreductase family of enzymes, is present in high concentrations in the human liver and kidney. There are at least nine known isoforms of ADH in the liver and a form known as the sigma form in the G.I. tract (Farrés et al., 1994). In humans, the cytosolic fraction of the liver and other tissues contains several isozymes. The isozyme active in metabolism of ethanol is known as "class I ADH." The primary role of this subfamily is the metabolism and detoxificatrion of alcohols. The products of metabolism are often more toxic than the parent compounds, but quickly undergo detoxification through secondary metabolism; for example, ethanl and methanol metabolize to the more toxic acetaldehyde and formaldehyde, respectively (Hammes-Schiffer and Benkovic, 2006). The interaction is

$$\text{Ethanol} + \text{NAD} \rightarrow \text{Acetaldehyde} + \text{NADH}$$

The human "class I ADH" is a zinc metalloenzyme, consisted of three subunits of α, β, and γ either as homodimer or heterodimer isozymes encoded by the ADH1, ADH2, and ADH3 genes. All isozymes have two different monomers of type E for ethanol active and type S for steroid active and the possible combinations of monomers are EE, SS, and ES. Furthermore, there are five classes of ADH in humans, identified as Class I–V. Class I is the primary enzyme system in the liver. Each monomer has two subunits and each subunit has binding sites for NAD and two Zn^{2+}. The isoforms are inhibited by heavy metals and chelating agents.

In addition to ethanol and methanol, the isoforms of this enzyme metabolize other alcohols, retinol, steroids, and fatty acids (Hellgren et al., 2007; Duester, 2008). The catalysis is in the presence of NAD^+ or $NADP^+$ as the electron acceptor and form primary, secondary, or tertiary alcohols, aldehyde, and ketone. Their natural role is to protect the body against toxins. The metabolites of this enzyme system are substrates for other enzymatic reactions in the body.

6.3.1.6 Diamine Oxidase (Histaminase)

Diamine oxidase is present in high concentrations in the intestinal mucosa and plasma, and its role is the deamination of diamines, for example, cadaverine and putrescin. This enzyme is specific only for diamines and does not deaminate monoamines. The level of this enzyme in plasma correlates with its intestinal level. In disease states that cause mucosal damage, the level of the enzyme in plasma decreases parallel to enzyme level of intestinal mucosa. Thus, the plasma level is a marker to evaluate the integrity of intestinal mucosa (Luk et al., 1980). In addition, patients with medullary thyroid carcinoma tissue, spread into the lung, exhibit an elevated level of blood histaminases (Baylin et al., 1975). In acute viral hepatitis, the plasma level of the enzyme declines significantly (Gäng et al., 1976). In addition to its role of deamination of diamines, it generally plays the role of biomarker for different disease states.

6.3.1.7 Aldehyde Dehydrogenases (ALDH)

Different forms of this enzyme metabolize aldehydes to less reactive forms. Aldehydes, the by-products of metabolism of endogenous and exogenous compounds (e.g., metabolism of amino acids, certain xenobiotics, carbohydrate, retinoic acid, vitamins, steroids and lipid form different aldehydes), are substrates for the following enzymes of ALDH family:

- ALDH1A1: Involves in the formation of secondary metabolites of alcohols and helps transparency of cornea (Jester et al., 1999).
- ALDH1A2: Catalyzes the synthesis of retinoic acid, the derivative vitamin A from retinaldehyde (Ono et al., 1998).
- ALDH1A3, ALDH1B1, ALDH3B1, ALDH3B2, and ALDH7A1: Play major role in detoxification of aldehyde metabolites of alcohols and lipid peroxidation (Hsu and Chang, 1996; Hsu et al., 1997; Yoshida, 1993; Yoshida et al., 1998; Wang et al., 2009; Brocker et al., 2011).
- ALDHL1 and ALDHL2: Catalyze the metabolism of 10-formyltetrahydrofolate in the presence of NADP and water to tetrahydrofolate, NADPH, and CO^2 (Hong et al., 1999; Krupenko and Oleinik, 2002).
- ALDH2: Catalyzes the formation of acetic acid from acetaldehyde (Hempel et al., 1985; 1987; Agarwal and Goedde, 1987; Seitz and Meier, 2007).
- ALDH3A1: Metabolizes various aldehydes and participates in the metabolism of neurotransmitters, corticosteroids, and biogenic amines (Rekha et al., 1998; Yang et al., 2002).
- ALDH3A2: Involved in a neurocutaneous disease known as SLS or Sjögren–Larsson syndrome (Freedberg, 2003).
- ALDH4A1: Catalyzes the formation of glutamate from purroline-5-carboxylate, which is the second phase of proline catalysis (Goodman et al., 1974; Geraghty et al., 1998).
- ALDH5A1: Oxidizes aldehydes to carboxylic acids and is present in the liver and many other tissues (Crabb et al., 2004).
- ALDH6A1: Involved in the catabolic pathways of pyrimidine and valine (Kuiper et al., 2005).
- ALDH8A1: Involved in biosynthesis of 9-cis-retinoic acid biosynthesis (Lin and Napoli, 2000).
- ALDH9A1: Catalyzes the dehydrogenation of gamma-aminobutyraldehyde to GABA, gamma-amino butyric acid (McPherson et al., 1994; Kikonyogo and Pietruszko, 1996; Lin et al., 1996).
- ALDH16A1: Involved in the interaction with SPG21 protein ACP33/maspardin and spastic paraplegia (Hanna and Blackstone, 2009).
- ALDH18A1: Involved in biosynthesis of proline, ornithine, and arginine (Baumgartner et al., 2001, 2005).

All the isoforms of ALDH are detoxifying enzyme systems of Phase I metabolism and are in three different classes in the mammalian system: class 1 (cytosolic), class 2 (mitochondrial), and class 3 (expressed in tumors, stomach, and cornea).

6.3.1.8 Xanthine Oxidase (XOD)

Xanthine oxidase has a large and very complex structure with a molecular weight of about 270 kDa. It has two molybdenum atoms, two flavin nucleotide, and eight iron atoms, which are present as ferredoxin iron-sulfur groups and four redux centers that carry electrons to oxygen to yield superoxide ion (O_2^-) (Enroth et al., 2000, Hadizadeh et al., 2009). As a cytosolic enzyme, this metalloprotein catalyzes the metabolism of purine bases and hypoxanthine to xanthine and finally to uric acid. Due to its dual metabolic role, this enzyme system is often called xanthine oxidoreductase. In certain disease states such as gout, hypouricaemia, and chronic heart failure, the production of purine is increased, and for therapy, the major need is to reduce serum uric acid levels (Landmesser et al., 2002; Gibbings et al., 2011). Thus, this enzyme is the target for therapeutic agents like allopurinol to inhibit the enzyme in the related disease states (Pacher et al., 2006; Puntoni et al., 2013). The parent compound, allopurinol, not only is a substrate for xanthine oxidase, its metabolite oxypurinol (alloxanthine) is also an inhibitor. The complexity of interaction and the dual role of this enzyme in conjunction with the investigation of its inhibition in various disease states have encouraged continued research and inquiries by many researchers in different fields (Guthikonda et al., 2003; Becker et al., 2005).

6.3.1.9 Carboxylesterases (CES)

The members of this enzyme system catalyze the hydrolysis of a large number of structurally diverse endogenous and exogenous esters, amides, thioesters, and carbamates in humans (Satoh and Hosokawa, 1998, 2006). The diversity of the members among different species is significant. In humans, the members include CES1A1, CES1A2, CES1A3, CES2A1, CES3A1, CES4C1, and CES5C1. They are present in the endoplasmic reticulum of the liver, kidney, lung, testis and small intestine (Hosokawa et al., 1990, 1995, 2001). Environmental pollutants and other xenobiotics induce the members of this enzyme system. The metabolism of esters by this enzyme system is of interest in the development of ester prodrug. The ester prodrug is therapeutically inactive until it is hydrolyzed to its therapeutically active metabolite(s) by the enzyme system in the body. CES also catalyzes the hydrolysis of a number of therapeutic agents to its inactive metabolites, among them are opiate analgesics, anticancer drugs, angiotensin-converting enzyme inhibitors, and antiplatelet drugs; it also catalyzes the hydrolysis of illicit recreational drugs, cocaine and heroin (Redinbo and Potter, 2005). The transesterification of cocaine by this enzyme system in the presence of alcohol generate cocaethylene, a toxic metabolite. The impact of inhibition or induction of CES enzymes on metabolism and pharmacokinetics of xenobiotics are yet to be clarified.

6.3.1.10 Peptidase (Protease/Proteinase)

The enzymes of this family hydrolyze the peptide bonds of proteins and peptides. Based on the target link of amino acid in a protein or peptide, subfamilies of this enzyme system have been categorized as aspartate peptidases/proteases, cysteine peptidases/proteases, glutamic acid peptidases/proteases, metalloproteases/metallopeptidases, serine peptidases/proteases, and threonine peptidases/proteases. Each subfamily has its own members and each member has its own isoforms. For example, a member of serine peptidase is dipeptidyl peptidase-4 (DPP-IV) (Mentlein et al., 1993), DPP-7, DPP-8, DPP-9, etc. Dipeptidyl peptidase-4 (DPP4), also known as adenosine deaminase complexing protein 2 or CD26 (cluster of differentiation 26) is an antigenic enzyme expressed on the surface of most cell types and is associated with immune regulation, signal transduction and apoptosis. The substrates of CD26/DPPIV are proline (or alanine)-containing peptides and include growth factors, chemokines, neuropeptides, vasoactive peptides, and gastric inhibitory peptide/glucose-dependent insulinotropic peptide (GIP) with terminal Tyr-Ala, as well as glucagon-like peptide-1amide/insulinotropin [GLP-1(7–36)amide] and peptide histidine methionine (PHM) with terminal His-Ala hydrolyzed by dipeptidyl-peptidase IV. Proteases like thrombin and plasmin are present in blood; acid proteases such as pepsin are present in stomach; serine proteases like trypsin and chymotripsin are present in small intestine; and elastase and cathepsin G are proteases in white blood cells (Demuth, 1990; Hedstrom, 2002; Barrett et al., 2003). The peptidases/proteases hinder the delivery of therapeutic peptides or proteins. An example of this hindrance is the enzymatic barrier of nasal epithelium for nasal delivery of peptide drugs.

6.3.2 Phase II Metabolism: Conjugation

All xenobiotics, endobiotics, and their metabolites with appropriate functional groups can undergo metabolic conjugation in the body by interaction with charged ligands such as glucuronic acid, glycine, sulfate, and glutathione. The conjugated compounds are less reactive, more soluble than the parent compounds and eliminate easier from the body. The enzymes of Phase II metabolism, which catalyze the conjugation reactions, belong to a family of transferases with broad specificity. The conjugation reactions are the inactivation processes or detoxification pathways. However, conjugation may not always result in detoxification or full inactivation of active substrates.

The conjugation reactions, the related enzymes, and the appropriate functional groups of reactive compound are

1. Glucuronidation: Catalyzed by uridine 5′-diphospho-glucuronosyltransferase (UDP-glucuronosyltransferase) in endoplasmic reticulum. The required functional groups of substrates to interact with the glucuronic acid are –OH, $-NH_2$,–COOH, –SH, =NH, ≡CH.
2. Sulfation: Catalyzed by sulfotransferase in cytosol. The required functional groups are $-NH_2$, –OH.
3. Methylation: Catalyzed by methyltransferase in cytosol and endoplasmic reticulum. The required functional groups are –OH, $-NH_2$, –SH.
4. Acetylation: Catalyzed by acetyltransferase in cytosol. The required functional groups are $-NH_2$, –OH.
5. Gluthathion conjugation: Catalyzed by glutathione S-transferase in cytosol and endoplasmic reticulum. The required functional groups are epoxide and organic halide.
6. Amino acid conjugation (glycine, glutamine, arginine, or taurine conjugate): Catalyzed by acyl-CoA (amino acid N-acyltransferase) in mitochondria. The required functional group is –COOH.

6.3.2.1 Glucuronidation

Glucuronide conjugation is one of the most important and common detoxification pathways in the mammalian system (Kaivosaari et al., 2011). The body relies heavily on this pathway for the elimination of a significant number of exogenous and endogenous compounds and their metabolites (Kroemer and Klotz, 1992). Xenobiotics that eliminate by glucuronidation include nonsteroidal anti-infammatory compounds, sedatives, analgesics, and antidepressants (Baron and Sandler, 2000; Kuehl et al., 2005; Smith, 2009). Examples of endogenous substrates are thyroxin, steroids, bilirubin, catecholamines, vitamins, and bile acids.

Interspecies difference in hepatic glucuronidation is rather significant. The glucuronide conjugates are formed in two rather simultaneously occurring steps of activation followed by transfer reaction. The activation step is the formation of glucuronic acid by binding α-D-glucose 1-phosphate to UTP in the presence of pyrophosphorylase to produce uridin-5′-diphopho-α-D-glucose (UDPG). Through the process of dehydrogenation in the presence of UDPG-dehydrogenase, UDPG is then changed to uridine-5′-diphospho-α-D-glucuronic acid (UDPGA), which is the donor of glucuronic acid for interaction with reactive metabolites or compounds. The transfer reaction, catalyzed by UDP-glucuronyl transferase, forms glucuronide conjugate and UDP is released, as summarized below:

Activation:

$$\text{Glucose 1-Phosphate} + \text{UTP} \xrightarrow{\text{Pyrophodphorylase}} \text{UDPG} + \text{PP}$$

$$\text{UDPG} + 2\text{NAD}^+ + \text{H}_2\text{O} \xrightarrow{\text{UDP-glucuronyltranferase}} \text{UDPGA} + 2\text{NADH} + 2\text{H}^+$$

Transfer Reaction:

$$\text{UDPGA} + \text{Reactive Xenobiotic (RX)} \xrightarrow{\text{UDP-glucuronyltransferase}} \text{RX} - \text{Glucuronic acid} + \text{UDP}$$

Depending on the functional group of reactive xenobiotic or endobiotic, glucuronide conjugation can be divided into the following categories:

6.3.2.1.1 O-Glucuronide Conjugation

O-Glucuronides are formed when UDPGA interacts with primary, secondary, and tertiary alcohols, phenols, and carboxylic acid-containing molecules. Carboxylic acid-containing molecules form "ester," whereas alcohols and phenols form "ether." *O*-Glucuronides eliminate from the body through biliary excretion into the small intestine. Although it is generally accepted that glucuronide metabolites are pharmacologically inactive, for compounds like morphine the formation of glucuronide metabolites, namely morphine-6-glucuronide and morphine-3-glucuronide, have significant pharmacological activities with interactive ability to bind to opiod receptors (Milne et al., 1996).

O-Glucuronidation is an important elimination pathway for many xenobiotics, including morphine, naphthol, chloramphenicol, acetaminophen, codeine, hexobarbital, salicylic acid, bilirubin, diclofenac, naproxen, valproic acid, etc. The β-glucuronidase present in small intestine may hydrolyze the glucuronide conjugates transferred by the bile and release their primary xenobiotics or their metabolites. The released xenobiotic may then reabsorb and enter into the systemic circulation, or form a chemically modified compound in the alkaline environment of the intestine.

There are marked species regioselectivity differences in *O*-glucuronidation. This selectivity is due to the genetic differences between humans and experimental animals, namely human UDP-glucuronosyltransferase (UGT) genes and the related subfamilies.

6.3.2.1.2 N-Glucuronide Conjugation

The substrates of *N*-glucuronidation that interact with UDPGA are alkylamines, arylamines, primary aromatic amines, tertiary aliphatic amines, hydroxylamines, amides, and various aromatic heterocyclic compounds. In general, the substrates can be classified into compounds that form non-quaternary *N*-conjugates, such as sulfonamides, arylamines, and alicyclic, cyclic, and heterocyclic amines, and those that form the quaternary nitrogen conjugates, identified as N^+-glucuronides (Kaku et al., 2004). In humans, N^+-glucuronidation plays a significant role in the metabolism of many aliphatic tertiary amines such as tricyclic antipsychotics, antihistamines, and antidepressants. *N*-Glucuronides are generally safe, water-soluble metabolites, but may contribute to the carcinogenicity of primary arylamines (Green and Tephly, 1998).

The interspecies variability in *N*-glucuronidation is high, particularly for aliphatic tertiary amines and aromatic *N*-heterocyclic compounds. All common experimental animals (mouse, rat, guinea pig, rabbit, and dog) have *N*-glucuronidation pathways (Chiu and Huskey, 1998; Soars et al., 2001; Kaji and Kume, 2005; Xu et al., 2006; Shiratani et al., 2008). However, the variability among them stems from compound-dependency of their *N*-glucuronidation. For tertiary amines, most notably the tricyclic antidepressant and antihistamine drugs, N^+-glucuronidation is commonly observed in chimpanzees, marmoset monkeys, humans, and rabbits, but the metabolic pathway for the compounds is generally absent in rats, mice, guinea pigs, and cynomolgus monkeys. Another example is *N*-glucuronidation of molecules with the triazole ring, which occur at the same position of the ring in humans, rats, and dogs, but results in the generation of four different conjugates (Nakazawa et al., 2006). A better understanding of interspecies differences remains to be documented from the glucuronosyltransferase genes of animal species other than humans.

N-Glucuronidation represents a major elimination pathway for many xenobiotics, including aniline, *N*-hydroxyarylamines, benzidine, imipramine lamotrigine, olanzapine, carbamazepine, amitriptyline, retigabine, tioconazole, ketotifen, etc. (Zenser et al., 1998). Similar to *O*-glucuronides, *N*-glucuronides can breakdown by the intestinal β-glucuronidase and release their substrates in the intestinal environment.

6.3.2.1.3 *S-Glucuronide Conjugation*

Thiol-containing xenobiotics, such as diethyldithiocarbamate, thiophenol, 2-mercaptothiazole, etc., also interact with UDPGA to form *S*-glucuronides (Buchheit et al., 2011). The formation of this conjugate is less frequent than *O*- and *N*-glucuronide conjugates and the documented information on species differences is not as comprehensive. Similar to the other glucuronides, the elimination of *S*-glucuronides is through both renal and biliary elimination and can be subjected to enterohepatic recirculation.

6.3.2.1.4 *C-Glucuronide Conjugation*

This is a rare glucuronide conjugation pathway in humans. Xenobiotics with 1,3 dicarbonyl functional group, such as phenylbutazone and sulfinpyrazone, can interact with human UDPGA to form *C*-glucuronid conjugates (Kerdpin et al., 2006). Human liver and kidney microsomes form *C*-glucuronide of phenylbutazon. The *in vivo* study is also indicative of a significant elimination of phenylbutazone *C*-glucuronide in the urine. Because of the rarity of compounds, with dicarbonyl functional group at positions 1 and 3, only a few examples are known or documented in humans. In experimental animals, an acetylenic *C*-glucuronide of the sedative-hypnotic drug ethchlorvynol is identified as a major metabolite in rabbit's urine.

6.3.2.2 Sulfation

Sulfate conjugation, similar to glucuronide conjugation, is a critical metabolic pathway for elimination of xenobiotics and endobiotics. Sulfation is a major conjugation pathway for phenols, alcohols, arylamines, N-hydroxy compounds and, to a lesser extent, thiols; examples of endogenous substrates are steroids and carbohydrates. Sulfation occurs in vertebrates, invertebrates, fungi, and bacteria. Similar to glucuronide conjugations, the transfer of sulfate to a reactive substrate occurs in two steps of activation followed by transfer reaction. The activation involves the interaction of sulfate with ATP in the presence of ATP-sulfutransferase (or ATP sulfurylase) to form adenosine-5′-phosphosulfate (APS), which then interacts again with ATP in the presence of APS-phosphokinase to form 3′-phosphoadenosine-5′-phosphosulfate (PAPS) (Klaassen and Boles, 1997). The transfer reaction involves the transfer of sulfate from PAPS to the molecule of substrate in the presence of sulfokinase (or sulfotransferase) and the release of 3′-phosphoadenosine-5′-phosphate (PAP). The activation and transfer reactions are summarized below:

Activation:

$$SO_4^{--} + ATP \xrightarrow{\text{ATP-sulfutransferase}} \text{adenosine-5}'\text{-phosphosulfate (APS)} + PPi$$

$$APS + ATP \xrightarrow{\text{APS-Phosphokinase}} 3'\text{-phosphoadenosine-5}'\text{-phosphosulfate (PAPS)} + ADP$$

Transfer reaction:

$$PAPS + \text{Reactive endobiotic or xenobiotic (RX)} \xrightarrow{\text{Sulfotransferase}} \left[RX - SO_3^-\right] + PAP$$

The enzyme "sulfotransferase" catalyzes the transfer of sulfonate (SO_3^-) and not sulfate. The reaction is often called sulfonation rather than sulfation. Sulfation occurs less frequently than does glucuronidation mainly because of the lower concentration of PAPS at the cellular level (75 mM) as compared to UDPGA (350 mM), the availability of sulfate and the presence of the two enzymes of activation. Exogenous or endogenous compounds with functional groups of –OH (phenols and alcohols), $-NH_2$ (arylamines) or –NH – OH (N-hydroxy compounds) form sulfate conjugates. This *N*-sulfation or *O*-sulfation can occur for xenobiotics or endobiotics with an appropriate functional group. Sulfation of xenobiotics increases the hydrophilicity of molecules prior to biliary and urinary eliminations. Similar to the other Phase II metabolic pathways, sulfate conjugation is an inactivation or detoxification reaction of the parent compound and/or its metabolites. However, sulfation can also lead to the formation of biologically active compounds (Glatt et al., 1998; Funk et al., 2001), for example:

- Various environmental chemicals are metabolized to chemically reactive sulfuric acid esters, which may covalently bind to cellular macromolecules and induce mutations and tumors.
- Hepatic xenobiotic sulfation can lead to activation of prodrugs.
- Sulfate conjugates may inhibit salt efflux pump (e.g., troglitazone), or form DNA adducts (e.g., tamoxifen sulfate).

and so on.

It is worth noting that the formation of pharmacologically active conjugates, or any metabolites, with the same pharmacological or toxicological response as the parent compound should be taken into consideration for quantitation of pharmacodynamic/toxicodynamic parameters used in the estimation of PK/PD or TK/TD parameters.

6.3.2.3 Methylation

In comparison to glucuronide and sulfate conjugations, methylation of xenobiotics is considered an important but minor pathway, and most often generates compounds with less solubility and/or more pharmacological activities (e.g., formation of epinephrine from norepinephrine), but less toxicity. The pathway is an important biochemical reaction for endogenous compounds. The endogenous substrates of the pathway include biogenic amines, histamine, serotonine, dopamine, and catecholamins.

Methylation of the polar functional group of xenobiotics precludes the molecules from further conjugation by other pathways. The transfer of the methyl group to the appropriate functional group of acceptor molecules (e.g., phenols, catechols, amines, *N*-heterocyclic compounds, and sulfhydryl containing compounds) and metals (such as mercury and arsenic that form mono- and di-methylated conjugates, and inorganic selenium that forms trimethylated conjugate) take place in two steps of activation (formation of coenzyme), followed by transfer reaction:

Activation (formation of coenzyme):

$$ATP + L\text{-}methionine \xrightarrow[Mg^{2+}]{\text{L-methionine adenosyltransferase}} S - adenosylmethionine(SAM) + PPi$$

Transfer reaction:

$$\text{Acceptor molecule} + \text{SAM} \xrightarrow{\text{methyl transferase}} \text{Methylated molecule} + S\text{-adenosylhomocysteine}$$

There are different methyl transferases in the body, which facilitate the transfer of the methyl group from SAM to the acceptor molcules. Depending on the functional group(s) of the acceptor molecules, the related reactions are

- *O*-Methylation:
  - Catechol *O*-methyltransferase (COMT)
  - Phenol *O*-methyltransferase (POMT)
- *N*-Methylation:
  - Phenylethanolamine *N*-methyltransferase (PNMT)
  - Imidazole *N*-methyltransferase (histamine *N*-methyltransferase (HNMT) Nicotinamide *N*-methyltransferase
  - Non-specific *N*-methyltransferase
  - Guanidinoacetic acid *N*-methyltransferase
  - Phosphatadylaminoethanol *N*- methyltransferase
- *S*-Methylation:
  - Thiopurine *S*-methyltransferase (TPMT)
  - Thiol methyltransferase (TMT)

Catechol *O*-methyltransferase (COMT) is present in the liver, kidney, skin and nerve tissues; it is a cytosolic and microsomal enzyme, specific for catechols (e.g., dopamine, norepinephrine, epinephrine, dopamine, L-DOPA, catechol estrogens, etc.), but not phenols. The transfer of the methyl group from SAM by this enzyme requires the presence of divalent ions (Mg^{2+}, Co^{2+}, Fe^{2+}, Mn^{2+}, etc.) (Männistöl et al., 1999). Phenol *O*-methyltransferase (POMT) is a microsomal enzyme that catalyzes the transfer of the methyl group from SAM to phenols. Phenylethanolamine *N*-methyltransferase (PNMT) is present in adrenal tissue and is responsible for the methylation of the neurotransmitter norepinephrine, forming epinephrine. It also methylates norephedrine and normetanephedrine (Axelrod and Daly, 1968). Imidazole *N*-methyltransferase is located mainly in the liver, but it also present in other tissues, including the brain. This enzyme catalyzes the methylation of histamine and is often called histamine *N*-methyltransferase (HNMT) (Weinshilboum et al., 1999). Nicotinamide *N*-methyltransferase (NNMT) methylates serotonin and tryptophan, and pyridine containing compounds such as nicotinamide and nicotine (Aksoy et al., 1994). Non-specific *N*-methyltransferase methylate the demethylated products of Phase I metabolism (Saavedra et al., 1973). Guanidinoacetic acid *N*-methyltransferase and phosphatadylaminoethanol *N*-transferase are specific for methylation of endogenous compounds, for example, methylation of guanidinoacetic acid and formation of creatine (Caldeira Araújo et al., 2005). Thiopurine *S*-methyltransferase (TPMT) catalyzes the S-methylation of aromatic and heterocyclic sulfhydryl compounds; thiol methyltransferase (TMT) is an important pathway in the metabolism of xenobiotics with the sulfhydryl group (e.g., captopril) (Wang et al., 2005).

6.3.2.4 Acetylation (Acylation)

Xenobiotics containing an aromatic amine (R-NH2) or a hydrazine group (R-NH-NH2) converted to aromatic amides (R-NH-COCH3) and hydrazides (R-NH-NH-COCH3), carboxylic acid and alcohol (e.g., choline) undergo acetylation to form amide before elimination from the body. The endogenous substrate example is serotonin. The general sequence of acetylation is also in two steps: formation of acetyl-coenzymeA (acetyl-CoA) followed by transfer of acetyl moiety to the appropriate functional group of xenobiotics in the presence of *N*-acetyl transferases.

Formation of acetyl-CoA:

$$R - COOH + ATP \longrightarrow R - CO - AMP + PPi$$

$$R - CO - AMP + CoA - SH \xrightarrow{\text{CoA-S-acetyl transferase}} R - CO - S - CoA + AMP$$

Transfer reaction:

$$RCO - S - CoA + R'NH_2 \xrightarrow{\text{R-amine } N\text{-acetyl transferase}} RCONHR' + CoA - SH$$

N-Acetyl transferases are cytosolic enzymes present in the liver, mainly in Kupffer cells, and many other mammalian tissues such as spleen, lungs, and gut. In humans, similar to rats and hamsters, two distinct enzymes with different substrate specificity are present: *N*-acetyl transferase 1 and *N*-acetyl transferase 2 (NAT-1 and NAT-2) (Levy et al., 1998). Mice express three types of NAT and dogs are unable to form acetyl conjugate. The most common NAT in humans is NAT-1, which is present in several tissues and organs, whereas NAT-2 is present in the liver and gut. Although the primary aliphatic amines are rarely substrates for *N*-acetylation, cystein can form acetyl conjugates, which convert to mercapturic acids. Similar to methyl conjugation, *N*-acetylation forms less polar molecules with low water solubility than the parent compound. Acetyl conjugates do not undergo further conjugation by other Phase II pathways. Thus, the acetylated less-polar molecules remains longer in the body; it may reside in the kidney and cause renal toxicity (e.g., acetyl conjugate of sulfonamide) or increase the half-life of this conjugate in the body, thus potentiating the activity of the parent compound (e.g., acetylated procainmaide). Furthermore, there are significant interindividual genetic, gender, and age differences (Meisel, 2002). The polymorphism of NAT-2 gene is the reason for the interindividual variability as slow acetylators versus fast acetylators (Blum et al., 1991; Levy et al., 1998). The slow acetylators are less efficient in metabolism and elimination of xenobiotics containing primary aromatic amine or hydrazine groups such as caffeine, isoniazid, sulfamethazine, procainamide, and hydralazine, as well as carcinogens such as benzidine and 2-aminofluorene (Ilett et al., 1987; Pawlik et al., 2002).

6.3.2.5 Glutathione Conjugation

Glutathione (GSH) is a tripeptide formed by interaction of L-γ-glutamate (Glu) and cysteine (Cys) in the presence of enzyme gamma-glutamylcysteine synthetase forming L-γ glutamylcysteine dipeptide, which then interacts with glycine (Gly) in the presence of glutathione synthetase to form glutathione or L-γ-glutamyl-L-cysteinylglycine. This tripeptide is present mainly in intracellular environments with its two characteristic features of having a sulfhydryl (SH) functional group and a γ-glutamyl linkage. The thiol group (SH) of cysteine serves as a proton donor and is responsible for the biological activity of glutathione. It exists in both reduced (GSH) and oxidized (GSSG) states and the intracellular ratio of GSH:GSSH should be in favor of GSH (greater than 90%) in order to maintain the cellular integrity and is often used as an indicator of cellular to toxicity of oxidative stress. The endogenous substrates are the metabolites of arachidonic acid.

The significant physiological role of glutathione in the body is the formation of glutathione conjugates of reactive metabolites and harmful electrophilic compounds, for example, free radicals and oxygen radicals such as peroxides (H_2O_2), superoxides (O_2^-) and hydroxyl radicals. The free radicals and reactive oxygen species have the propensity to cause with consequent modification of cellular function. Glutathione conjugation protects the body and prevents damage to important cellular macromolecules. It is the major endogenous antioxidant. In addition to detoxifying electrophilic carcinogen and antitumor agents, the conjugation inactivates endogenous epoxides, aldehydes, and quinines. The GSA transferases have other biological roles in the body such as DNA repair process, amino acid transport, and biosynthesis of leukotrienes, progesterone, testosterone, prostaglandins, and protein biosynthesis (Hayes et al., 2005).

The formation of glutathione conjugates occurs in two steps of formation of co-substrate followed by transfer of glutathione. However, the glutathione S-conjugates undergo further metabolic reaction catalyzed by glutamyltranspeptidase (also known as γ-glutamyltransferase) to form Cys-Gly S-conjugates, which is a dipeptide, and free γ-glutamyl moiety. The newly formed dipeptide can undergo additional metabolism catalyzed by a number of dipepdiases and aminopeptidases to free glycine and form cysteine S-conjugates. *N*-acetylation of cysteine conjugates in the presence of *N*-acetyltransferase forms *N*-acetylcysteine, which is also known as mercapturic acid. The summary of the reactions is as follows:

Formation of q co-substrate:

$$L-\gamma \text{ glutamate} + \text{Cysteine} \xrightarrow{\text{glutamylcysteine synthetase}} L-\gamma \text{ glutamylcysteine}$$

$$L-\gamma \text{ glutamylcysteine dipeptide} + \text{ Glycine} \xrightarrow{\text{glutathione synthetase}} L-\gamma-\text{Glutamyl} - L - \text{Cysteinylglycine (glutathione)}$$

Transfer reaction:

$$\text{Glutathion} + \text{Electrophilic compound} \xrightarrow{\text{glutathione S-transferases}} \text{Glutathione conjugate}$$

Further metabolism:

1. Formation Cys-Glyc dipeptide

$$\text{Glutathione S} - \text{Conjugates} + H_2O \xrightarrow{\text{glutamyltranspeptidase}} \text{Cys} - \text{Gly S} - \text{Conjugates} + \text{Glutamate}$$

2. Formation of cystein conjugates

$$\text{Cys} - \text{Gly S} - \text{Conjugates} \xrightarrow{\text{dipepdiases or aminopeptidases}} \text{Glycine} + \text{Cys S} - \text{Conjugates}$$

3. Formation of *N*-acetylcysteine (mercapturic acid) conjugates

$$\text{Cys S} - \text{Conjugates} + \text{Acetyle coenzyme A} \xrightarrow{N-\text{acetyltransferase}} N - \text{Acetylcysteine (mercapturic acid)}$$

Approximately 90% of GSH is synthesized in the cytosol of all mammalian cells, 10% in the mitochondria, and a negligible amount in the endoplasmic reticulum. The conjugates of GSH forms in the cells and then transports out of the cells. In the liver, which is the major site of glutathione conjugation, the conjugates exit the hepatocytes and enter into the bile. The metabolism of GSH conjugates to cysteine and mercapturic acid conjugates begins mainly in the kidney, followed by biliary system, and small intestine. The GSH conjugates are mainly in the bile and mercapturic acid conjugates are predominantly in the urine.

Although the major role of glutathione conjugates is detoxification of harmful molecules, occasionally the conjugate itself plays an important role in the formation of cytotoxic, genotoxic, or mutagenic metabolites (Koob and Dekant, 1991). For example, the glutathione conjugate of *p*-aminophenol is nephrotoxic (Klos et al., 1992); the polyphenolic-glutathione (GSH) conjugates retain the electrophilic and redox properties of the parent polyphenols and contributes to their nephrotoxicity nephrocarcinogenicity, and neurotoxicity (Monks et al., 1998); or GSH conjugate of dibromoethane, is a mutagen (van Bladeren et al., 1980).

There are more than 20 different human glutathione S-transferases, which some are in the cytosolic fraction and some in the microsomal fraction. The GSH-transferases in the cytosolic fractions are more specific toward xenobiotics and those in the microsomal fraction are more involved in the metabolism of endobiotics. There are marked species differences in the activities of GSH transferases for a given compound. In humans there are seven subfamilies of cytosolic GSH-transferases:

- alpha (GSTA1 and 2) ⇒ $\alpha GSTA1$ and $\alpha GSTA1$
- mu (GSTM1 through 5) ⇒ $\mu GSTM1$, $\mu GSTM2$, $\mu GSTM3$, $\mu GSTM4$, $\mu GSTM5$
- omega (GSTO1) ⇒ $\omega GSTO1$
- pi (GSTP1) ⇒ $\pi GSTP1$

- sigma (GSTS1) ⇒ σ*GSTS*1
- theta (GSTT1 and 2) ⇒ θ*GSTT*1, θ*GSTT*2
- zeta (GSTZ1) ⇒ ζ*GSTZ*1

The population studies have shown the subfamilies are polymorphic (Long et al., 2006). The Null activity in the GSTT1 gene is associated with adverse side effects and toxicity in cancer chemotherapy with cytostatic drugs, and so on.

6.3.2.6 Amino Acid Conjugation

Amino acid conjugation is a relatively minor and unique pathway of Phase II metabolism. In humans, it involves the initial formation of a xenobiotic acyl-CoA thioester, which then conjugates with amino acids glycine, glutamine, glutamate, arginine, or taurine, glycine being the principle amino acids (Hutt and Caldwell, 1990; Shirley et al., 1994). In rats and rabbits, glutamine plays the principle role. Thus, the essential functional group of xenobiotics for reaction is the carboxylic acid moiety and the conjugation reaction is a form of *N*-acylation. Briefly, the carboxylic acid moiety interacts with ATP generating an acyl adenylate and pyrophosphate, which acyl adenylate interacts with coenzyme A forming xenobiotic-CoA thioester followed by catalytic transfer of the activated acyl group to the amino group of the acceptor amino acid in the presence acylCoA-amino acid-*N*-acyltransferase. The summary is as follows:

Activation reaction:

$$\text{R-COOH} + \text{ATP} \longrightarrow \text{R-CO-AMP} + \text{PPi}$$

$$\text{R-CO-AMP} + \text{CoASH} \xrightarrow{\text{ATP-dependent acid CoA synthetase}} \text{R-CO-S-CoA} + \text{AMP}$$

Transfer reaction:

$$\text{R-CO-S-CoA} + \text{R}'\text{-NH}_2 \xrightarrow{\text{acylCoA-amino acid- Nacyltransferase}} \text{R-CO-NH-R}' + \text{CoASH}$$

Amino acid conjugation, similar to the other pathways of Phase II metabolism, increases the solubility of xenobiotics (e.g., herbicides, benzoates, salicylic acid, valproic acid, etc.) and/or their metabolites and facilitates their biliary and urinary elimination. The endobiotics such as bile acids and branched chain fatty acids also form conjugates with amino acids.

6.3.3 *In Vitro* Systems for Xenobiotics Metabolism Study

The *in vitro* qualitative and quantitative screening methods of xenobiotics metabolism have broadened in recent years. The qualitative evaluations are to identify the metabolic profile of a compound, its major metabolic reactions, important enzymes, cofactors involved, and possibly the complete picture of its biotransformation (Houston, 1994; Houston and Carlile, 1997; Houston et al., 2012). The *in vitro* quantitative evaluations are mainly for the purpose of *in vivo* extrapolation/correlation in humans, which require proper experimental design using preferably human bio-tissues. In general, the *in vitro* metabolism data are essential in areas such as preclinical evaluation in drug discovery and development, environmental health sciences, cancer research, drug metabolism and toxicology studies, etc. The following is a short list of applications of *in vitro* metabolism studies:

- Evaluation of metabolic and kinetic profiles of xenobiotics
- Selection of the appropriate animal models to study metabolism and toxicity of a compound
- Determination of confidence limits for metabolic kinetic parameters
- Development of high throughput assays
- Determination of intrinsic clearance of xenobiotics

- Verification of reaction phenotyping
- Prediction of potential for drug-drug interactions
- Investigation of the influence of inducers or inhibitors on metabolism of xenobiotics
- Identification of metabolic activation and deactivation
- Formation of reactive and non-reactive metabolites
- Investigation of interspecies differences of xenobiotic metabolism

The *in vitro* systems can be classified as

Subcellular fractions (lung, liver, kidney, intestine, etc.):
  Induced subcellular fractions
Organ fractions:
  In Situ and isolated liver perfusion techniques
  Antibodies against CYP proteins
  bDNA probes

6.3.4 Subcellular Fractions

The subcellular fractions include (1) tissue or organ homogenate, (2) homogenate without mitochondria also known as the S9 fraction, (3) cytosolic fraction, and (4) micromal fraction. For routine *in vitro* metabolic studies, the subcellular fractions are from the liver. Subcellular fractions of other tissues/organs are also used to evaluate the *in vitro* extrahepatic metabolism and initial screening of xenobiotics (Dallner, 1978). Except the homogenate, which is used infrequently due to the bulkiness of the fraction and presence of debris and coarse fragments, other fractions are commonly used for *in vitro* metabolism studies. The S9 fraction contains the cytosolic and microsomal enzymes without the mitochondrial fraction. This fraction in the presence of UDPGA and PAPS is used to investigate Phase II metabolism. The cytosolic and microsomal fractions, under well-defined experimental design, are used in evaluation of metabolic stability, reaction phenotyping, species differences/similarities, and determination of the *in vitro* intrinsic clearance and/or extrahepatic metabolism of a compound. The cytosolic fraction is the soluble portion of the homogenate and contains glutathione S-transferases; sulfotransferases; alcohol dehydrogenases; diamine oxidases; xanthine oxidases; carboxylesterases; aldehyde oxidases; and soluble epoxide hydrolyses. The microsomal fraction is prepared by the ultracentrifugation of the S9 fraction consisting of vesicles that are formed from endoplasmic reticulum containing enzymes responsible for oxidation and glucuronidation of xenobiotics and their metabolites. It contains CYP 450, UDP-glucuronylotransferases, epoxide hydrolyses, flavin-monooxygenases and carboxylesterases, and other enzymes. The subcellular fractions, in particular microsomes, are widely used to evaluate species' metabolic differences/similarities. Microsomes of other tissues/organs are also valuable for the extrahepatic metabolism of a compound; for example, microsomes from the small intestine and liver may provide information on first-pass metabolism of orally administered xenobiotics. The schematic diagram of isolated subcellular fractions and preparation methods for S9, cytosol, and microsome are presented in Figure 6.2.

The subcellular fractions can also be prepared from pre-treated animals with specific enzyme inducers. The induced fractions are often used as the positive control for metabolism of compounds given on chronic basis to humans or experimental animals. Some of the commonly used inducers for CYP450 and UDPGA-glucuronidase are phenobarbital, rifampin, β-naphtoflavon, pyrazole, and other compounds (Emary et al., 1998).

6.3.5 Pure and Recombinant Enzymes

Relatively pure enzymes are used individually or in combination when specific information on metabolism of xenobiotics is investigated. The incubation of a xenobiotic with an enzyme provides

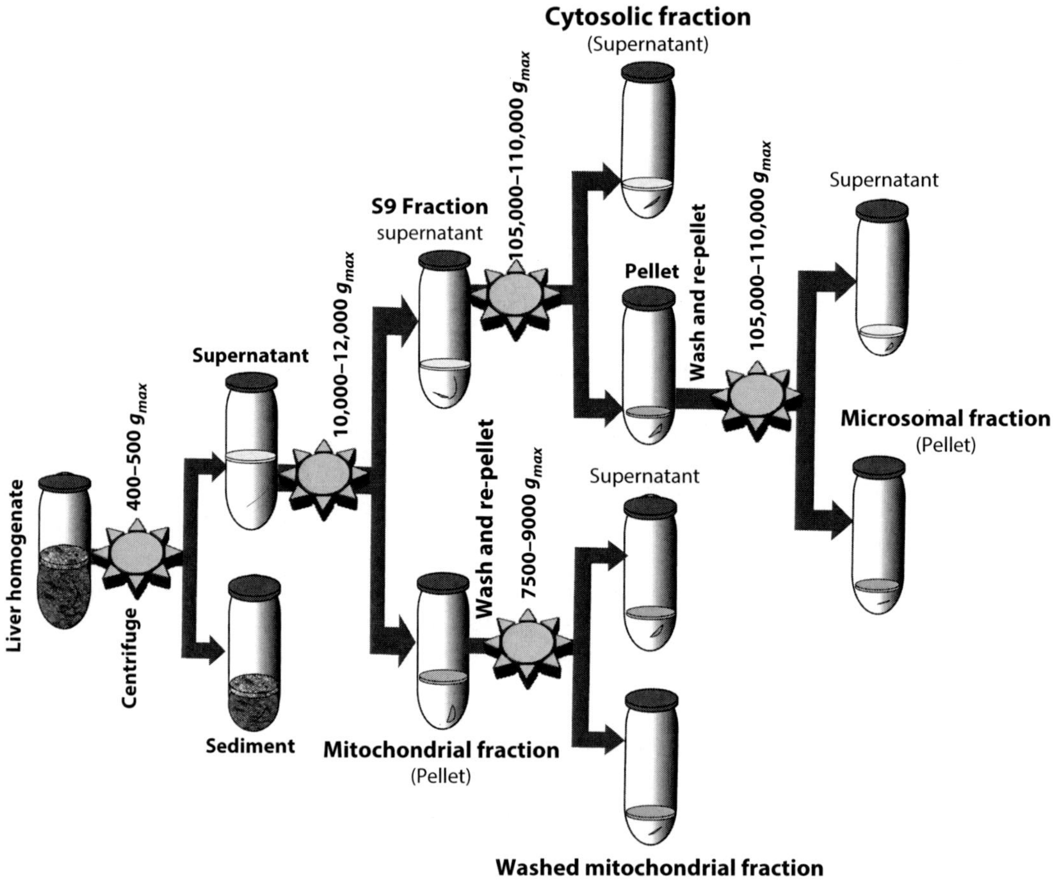

FIGURE 6.2 The schematic diagram of the isolation of subcellular fractions (S9, cytosol and microsome) and the preparation method.

a more accurate estimation of its kinetic parameters by the specific enzyme. Recombinant human enzymes, in particular, members of CYP450 families and subfamilies, are expressed in bacteria using *E. coli* and are commercially available as Bactosomes®.

6.3.6 Cellular Fractions—Hepatocytes

The parenchyma cells of the liver known as hepatocytes accounts for approximately 70%–80% of the liver cytoplasmic mass. They are involved in a number of significant physiological processes including biotransformation of xenobiotics and carbohydrates, protein synthesis and storage, phospholipid, cholesterol, and bile salts and formation of bile. The multiplicity and complexity of their role in the body is unmatched (Arber et al., 1988). Hepatocytes, isolated from the liver, is prepared by infusing the liver with a medium that lacks calcium to separate the cells from each other followed by infusion of a proteolytic enzyme such as collagenase to digest the extracellular proteins (Berry et al., 1991). The procedure for the separation and culture of hepatocytes requires special care to prepare the cells without bacterial contamination and free from damaged cells. Hepatocytes are effective tools for the Phase I and Phase II *in vitro* metabolic investigation of xenobiotics and represent the ideal *in vitro* system for studying xenobiotics (Worboys et al., 1996). They are whole cell models, expected to have the required cofactors and under experimental methodology that maintains their viability and *in vivo* functionality can generate metabolic data suitable for *in vitro/*

in vivo correlation and extrapolation. The hepatocytes of different species, including humans, can be isolated using fresh or as cryopreserved hepatocytes (Hengstler et al., 2000).

6.3.7 Organ Fractions (Precision-Cut Liver Slices)

The use of liver slices in drug metabolism study is about a century old. Initially it was prepared manually with razor blades, which resulted in thickness inconsistency of the slices. The developments of precision-cut tissue slicers and associated new incubation methodology have revived the application of liver slices for *in vitro* drug metabolism (Ekins, 1996). The advantages of liver slices over other methodology, for example, hepatocytes, are (1) maintaining the normal arrangement and structural design of hepatic cells, and (2) requiring no special treatment such as digestive enzymes that are used in preparation of hepatocytes. The shortcomings of slices are (1) insufficient diffusion of the medium during incubation/exposure period, (2) presence of splintered cells on the outer edges, and (3) a short period of viability (maximum 8 h). For a number of xenobiotics (e.g., tolbutamide, caffeine, 7-ethoxycoumarin, phenytoin, etc.) the estimated intrinsic clearance using precision-cut liver slices is consistent with the one calculated using hepatocytes. In addition, qualitative similarity in the production of metabolites of a few compounds has been shown between human liver slices and microsomes (Harris et al., 1994; Carlile et al., 1999). Cryopreservation of liver slices, due to the diversity of cell aggregates, is rather complicated and requires the optimization of concentration and composition of buffer and cryoprotectant in addition to the cooling and warming rates.

6.3.8 *In Situ* and Isolated Liver Perfusion Techniques

The isolated perfused liver is a methodology used frequently to estimate metabolic clearance and to determine metabolic profile and toxicity of xenobiotics. Its use for physiological and biochemical evaluation of endogenous compounds dates back to the beginning of the twentieth century. Other isolated organs are also employed for the evaluation of metabolic as well as physiological and toxicological studies. The term "perfused organ" is often referred to as an "*ex vivo*" method of study. The objective of using isolated perfused organ such as the liver is to mimic *in vivo* conditions in an *ex vivo* environment with all variables associated with the performance of the organ *in vivo*. There are a few requirements for the evaluation of an isolated organ:

- The function should be specific to the organ and not contingent on the performance of the other organs in the body.
- The study requires suitable circulatory fluid in an environment comparable to the *in vivo* condition.
- The viability of the organ during perfusion should be tested and maintained.

The *ex vivo* technique of isolated liver perfusion to study the metabolism of xenobiotics is different from *in situ* technique. The *ex vivo* technique requires complete vascular separation of the organ from the body. The organ is then maintained viable by an appropriate fluid mechanically circulated through its vascular network. It requires surgical removal of the liver and in most cases attachment to a special contraption to facilitate perfusion. The binding of the parent compound and/or its metabolites to the glassware, tubing, and perfusion apparatus should be prevented or corrected during the experiment. By means of the *in situ* technique, however, the organ is isolated but maintained within the *in vivo* environment. The change and influence of the *in vivo* environment on the function and performance of the liver should be explored and tested during the investigation to evaluate whether the *in situ* system represents the function of the liver in isolation without the influence of other organs. The main advantage of the *in situ* procedure over the *ex vivo* technique is minimizing the physical trauma to the organ during surgery and attachment to the apparatus.

The *ex vivo* techniques have several advantages over other *in vitro* methods:

- The role of the liver in metabolism of a compound can be established more definitely by the *ex vivo* technique.
- Unlike the other *in vitro* systems, the structural and functional integrity of the liver is sustained.
- Investigator controls the parameters of the perfusion, such as the concentration of xenobiotic or rate of perfusion, and thus influence of various parameters can be evaluated.
- The biliary elimination of a compound can also be evaluated.
- The hepatic clearance estimated by this method represents the role of metabolic enzymes and transport proteins. The enzymes and transport proteins of the intact liver include:
  - Enzymes of Phase I metabolism
    - CYP450 isoforms
    - Other Phase I metabolism (see Phase I metabolism)
  - Enzymes of Phase II metabolism
    - UDP-glucuronyltransferases
    - Sulfotransferases
    - Glutathione S-transferases
    - Etc. (see Phase II metabolism)
  - Transport Proteins
    - MDR1 (multidrug resistance protein 1)
    - MDR2 (multidrug resistance protein 2)
    - BCRP (breast cancer resistant protein)
    - BSEP (bile site export protein)
    - OATPs (organic anion transporting polypeptides)
    - OCTs (organic cation transporters)
    - MCT2 (monocarboxylic acid transporter 2)
    - Etc.

The main limitations of isolated perfused liver in a drug metabolism study are maintaining the organ viability during the course of study, which most often is rather short; setting up the system, which would require special apparatus, expertise in dealing with the rarity of human liver; and difficulty in duplicating *in vivo* conditions in an *ex vivo* setting.

The kinetic analysis of data obtained from isolated perfused liver is

$$(Rate)_{input} = Q_{perfusate} \times C_{in} \tag{6.6}$$

The rate of input into the liver is the input flow rate of perfusate ($Q_{perfusate}$) multiplied by the compound's concentration (C_{in}). The rate of output is the exit flow rate multiplied by the concentration of compound:

$$(Rate)_{output} = Q_{perfusate} \times C_{out} \tag{6.7}$$

The biliary excretion rate is

$$(Rate)_{biliary} = Q_{bile} \times C_{bile} \tag{6.8}$$

The average biliary rate is the total amount excreted in the bile divided by the time of collection. The biliary excretion ratio or biliary extraction ratio is estimated as

$$E_{biliary} = \frac{(Rate)_{biliary}}{(Rate)_{input}} \quad (6.9)$$

The estimated extraction ratio and the fraction that escapes metabolism are the same as Equations 4.9, 4.10, and 4.12, Chapter 4.

6.3.9 Antibodies against CYP Proteins

Among various approaches to determine the precise role of an isoform of CYP450 in the metabolism of a xenobiotic, monoclonal antibodies (MAbs) specific for individual isoform are used more confidently (Gelboin, 1993; Gelboin et al., 1996; Gelboin and Krausz, 2006). These antibodies are used to determine the protein content in subcellular fractions, the role of a specific isoform by its inhibition of substrate-specific metabolism, and the specific enzyme involved in metabolism of a compound. Although significant numbers of monoclonal antibodies have been identified toward human CYP450 isoforms and are commercially available, the list is far from completion. There are several issues in use of MAbs:

- The inhibition of the isoform must be complete. Any partial inhibition would defeat the purpose of using MAbs.
- Considering the closely resembled structure of different members of a CYP450 subfamily, for example, CYP3A4 and CYP3A5, the inhibition must be specific to an isoform only.
- The complex MAbs-CYP-substrate should not have any enzymatic activities.

It should be noted that understanding of the role of a CYP450 isoform in the metabolism of a xenobiotic by using MAbs is important in understanding of genetic polymorphism, drug-drug interaction, toxicity, and carcinogenicity.

6.3.10 bDNA Probes

The bDNA probes (branched DNA) are designed to identify mRNAs encoding CYP450 isoforms and a number of other xenobiotic-metabolizing enzyme systems (Hartley and Klaassen, 2000). The probs are mostly used in conjunction with hepatocytes, which provide different expressions of mRNA in a single 60 mm dish. Each probe, usually consisting of short oligonucleotides, is designed to various chemically inducible CYP450 mRNA transcripts and is capable of capturing specific mRNA. The commercially available probes give a chemiluminescent signal with intensity proportional to the level of mRNA, which are normalized with the levels of glyceraldehydes-3-phosphate dehydrogenase (GAPDH) mRNA. For example, the probes can measure the induction of CYP1A2 in omeprazole treated hepatocytes, which is normalized to GAPDH. The probes are often used as a valuable high-throughput assay for measuring enzyme induction of xenobiotics.

6.4 *IN VIVO* SAMPLES FOR DRUG METABOLISM STUDY

The *in vivo* metabolism study requires use of biological specimens such as serum, plasma, urine, bile, and tissues. In human urine, plasma and serum, and in experimental animals all samples are used routinely. Adherence to all relevant ethical and safety regulations and guidelines for the collection and use of biological samples in humans and experimental animals are essential in all *in vivo* studies. Experimental protocols for collection of invasive or non-invasive samples for metabolism studies from humans require advance approval by the Institutional Review Board (IRB), according to the federal guidelines. Furthermore, protocols for collection of same biological samples in experimental animals, including cannulation of bile ducts, jugular vein, portal vein, or placing probes such as microdialysis probes in tissues/organs, require approval of the Institutional Animal Care and Use Committee (IACUC) in accordance with the federal guidelines.

6.4.1 Serum and Plasma Samples

A serum sample is prepared by taking the blood sample and allowing it to clot naturally without adding any anticoagulants. The serum is then removed gently from the clot. Whereas a plasma sample is prepared by adding an anticoagulant followed by centrifugation, preferentially at a low temperature (4°C) to separate plasma from blood cells (red, white, and platelets). Serums require longer time to form, but plasma can be prepared after adding the anticoagulant and centrifugation. The centrifugation should be at an optimum speed that preserves the integrity of blood cells and prevents release of hemoglobin from red cells into plasma. Both samples should be prepared within 60 minutes after blood collection. The difference between serum and plasma is fibrinogen, which is removed by clotting in the serum and retained in plasma by an anticoagulant agent. There are, however, other minor differences between the two samples. For example, glucose concentrations are lower in plasma than in serum due to the effect of anticoagulants, which shift fluid from erythrocytes to plasma. To maintain consistency in a metabolism study, care should be taken to ensure that only one sample type is used throughout the investigation. A number of anticoagulants are used for preparing plasma samples, among them are: sodium citrate, potassium EDTA, or lithium heparin. For metabolism studies, lithium heparin is preferred because EDTA or citrate can interfere with metabolic profiling and analysis. It is highly recommended that after preparation of serum or plasma, aliquots of samples are frozen and stores at −80°C until analysis. This would stabilize to some extent the unstable metabolites and reduce the activity of enzymes present in the plasma or serum. Multiple freeze/thaw of the same aliquots should also be avoided. The concentration of primary metabolites, circulated in blood due to the dilution in systemic circulation is usually low and their detection would require sensitive detectors.

6.4.2 Urine Sample

The convenience of collection, lack of proteins, and a high concentration of eliminated metabolites make urine a key biological sample in PK/PD and TK/TD studies. The presence of an unchanged parent compound and metabolites in urine, their rate of urinary excretion and cumulative values over time of observation provide valuable information for mass balance study. From the urinary data a more accurate quantitative analysis of PK/TK parameters and constants, such as excretion or metabolic rate constant, can be obtained. The presence of conjugates in urine provide quantitative values of Phase II metabolism, and an enzymatic breakdown of the conjugates in urine samples can also provide information on Phase I metabolites, that is, primary metabolites, or the parent compound that was used as substrate for the conjugations.

6.4.3 Bile Sample

A major function of the liver is the production of bile with a rate of about 600–1200 mL/day. Bile is formed at canaliculi, which represent specialized extracellular spaces between hepatocytes. The canaliculi unite to form bile preductules that lead to bile ductules and finally hepatic duct. The hepatic duct is joined with the gallbladder duct to form the bile duct that connects through the sphincter of Oddi to the duodenum. Like urine, bile is a valuable biological sample for evaluation of metabolism and elimination of xenobiotics. Bile can be collected through bile duct cannulation, which is an invasive method and is used in humans only in certain disease states, patient condition permitted. Most bile cannulations for metabolism studies, however, are done on experimental animals such as different strains of rats or mice under anesthesia. Briefly, the procedure involves making an inch incision on the surgically scrubbed abdomen of the animal. The bile duct located under the lobes of the liver is then isolated by removing the surrounding connective tissues. Two pieces of silk suture are then passed under the bile duct; the distal suture is tied to prevent the bile flow to the intestine, followed by making a midline cut on the bile duct and inserting a polyethylene catheter (PE 10) toward the liver. The proximal silk suture then secures the catheter.

Most xenobiotics undergo Phase II metabolism before excretion into the bile. Thus, for most compounds the bile samples provide valuable data on conjugated metabolites. However, there are xenobiotics (e.g., anthracyclines) that the parent compound and metabolites of Phase I metabolism also excrete in the bile and enter in the duodenum, which, depending on their liophilicity may reabsorb and enter into systemic circulation. This occurrence is known as enterohepatic re-circulation. The enterohepatic re-circulation often times masks the biliary elimination of xenobiotics and may present itself as a broad peak in the plasma-concentration time curve.

The rate of biliary excretion depends to a large degree on the polarity of the molecules and their molecular weight. The polar and ionizable molecules can be conjugates of Phase II metabolism, parent compounds, and/or primary metabolites of Phase I metabolism. Thus, in the analysis of bile samples of compounds that undergo metabolism, one may observe the parent compound and products of Phase I and II metabolism. Protein bound molecules also undergo dissociation in the liver, followed by metabolism of unbound molecules and excretion in the bile. A number of biological factors such as sex, age, fasting, and pregnancy influence the biliary excretion of xenobiotics in humans. It is worth noting that although excreted parent compounds and/or their metabolites in the bile eventually leave the body through defecation, use of feces as a biological sample to estimate biliary elimination is neither accurate nor analytically practical.

6.4.4 Portal Vein Cannulation

The portal vein is one of the blood vessels that bring blood to the liver. It carries venous blood from the gut, spleen, pancreas, and gallbladder to the liver. By comparing the drug concentration of this vein before entering into the liver with blood levels of systemic circulation, we can obtain important information about the absorption of drug and the liver first-pass metabolism.

REFERENCES

Agarwal, D. P., Goedde, H. W. 1987. Human aldehyde dehydrogenase isozymes and alcohol sensitivity. *Isozymes-Curr T Biol* 16:21–48.

Agundez, J. A. 2004. Cytochrome P450 gene polymorphism and cancer. *Curr Drug Metab* 5:211–24.

Andersson, T., Regardh, C. G., Lou, Y. C., Zhang, Y., Dahl, M. L., Bertilsson, L. 1992. Polymorphic hydroxylation of S-mephenytoin and omeprazole metabolism in Caucasian and Chinese subjects. *Pharmacogenetics* 2:25–31.

Aksoy, S., Szumlanski, C. L., Weinshilboum, R. M. 1994. Human liver nicotinamide N-methyltransferase: cDNA cloning, expression, and biochemical characterization. *J Biol Chem* 269:14835–40.

Aoyama, T., Hardwick, J. P., Imaoka, S., Funae, Y., Gelboin, H. V., Gonzalez, F. J. 1990. Clofibrate inducible rat hepatic P450s IVA1and IVA3 catalyze the omega- and (omega-1)-hydroxylation of fatty acids and the omega-hydroxylation of prostaglandins E1 and F2 alpha. *J Lipid Res* 31:1477–82.

Arand, M., Oesch, F. 2002. Mammalian xenobiotic epoxide hydrolases. In: *Enzyme Systems that Metabolize Drugs and Other Xenobiotics*, ed. C. Ioannides, 459–83. New Jersey: John Wiley & Sons Ltd.

Arand, M., Plana, M. E. H., Hengstler, J. G., Lohmann, M., Cronin, A., Oesch, F. 2003. Detoxification strategy of epoxide hydrolase the basis for a threshold in chemical carcinogenesis. *Excli J* 2:22–30.

Arber, N., Zajicek, G., Ariel, I. 1988. The streaming liver. II The hepatocyte life history. *Liver* 8(2):80–7.

Arias, I., Wolkoff, A., Boyer, J., Shafritz, D., Fausto, N., Alter, H., Cohen, D. 2009. *The Liver: Biology and Pathobiology*, 5th Edition. Hoboken, USA: John Wiley & Sons, Inc.

Axelrod, J., Daly, J. 1968. Phenol-O-methyltransferase. *Biochim Biophys Acta* 159(3):472–8.

Baer, B. R., Rettie, A. E. 2006. CYP4B1: An enigmatic P450 at the interface between xenobiotic and endobiotic metabolism. *Drug Metab Rev* 38(3):451–76.

Bao, Z., He, X. Y., Ding, X., Prabhu, S., Hong, J. Y. 2005. Metabolism of nicotine and cotinine by human cytochrome P450 2A13. *Drug Metab Dispos* 33:258–61.

Barrett, A. J., Rawlings, N. D., Woessner, J. F. 2003. *The Handbook of Proteolytic Enzymes*, 2nd Edition. London: Academic Press.

Baron, J. A., Sandler, R. S. 2000. Nonsteroidal anti-inflammatory drugs and cancer prevention. *Annu Rev Med* 51:511–23.

Baumgartner, M. R., Hu, C. A., Almashanu, S. et al. 2001. Hyperammonemia with reduced ornithine, citrulline, arginine and proline: A new inborn error caused by a mutation in the gene encoding delta(1)-pyrroline-5-carboxylate synthase. *Hum Mol Genet* 9(19):2853–8.

Baumgartner, M. R., Rabier, D., Nassogne, M. C. et al. 2005. Delta1-pyrroline-5-carboxylate synthase deficiency: Neurodegeneration, cataracts and connective tissue manifestations combined with hyperammonaemia and reduced ornithine, citrulline, arginine and proline. *Eur J Pediatr* 164(1):31–6.

Baylin, S. B., Abeloff, M. D., Wieman, K. C., Tomford, J. W., Ettinger, D. S. 1975. Elevated histaminase (Diamine Oxidase) activity in small-cell carcinoma of the lung *N Engl J Med* 293:1286–90.

Becker, M. A., Schumacher Jr, H. R., Wortmann, R. L., MacDonald, P. A., Palo, W. A., Eustace, D., Vernillet, L., Joseph-Ridge, N. 2005. Febuxostat, a novel nonpurine selective inhibitor of xanthine oxidase: A twenty-eight–day, multicenter, phase II, randomized, double-blind, placebo-controlled, dose-response clinical trial examining safety and efficacy in patients with gout. *Arthritis Rheum* 52(3):916–23.

Benet, L. Z., Cummins, C. L., Wu, C. Y. 2003. Transporter-enzyme interactions: Implications for predicting drug-drug interactions from in vitro data. *Curr Drug Metab* 4:393–8.

Berry, M. N., Edwards, A. M., Barritt, G. J. 1991. Monolayer culture of hepatocytes. Isolated Hepatocytes: Preparation, properties and application. In: *Laboratory Techniques in Biochemistry and Molecular Biology*, eds. R. H. Burdon and P. H. Knippenberg, 1–460. Amsterdam: Elsvier.

Bhatia, S. N., Toner, M., Foy, B. D., Rotem, A., O'Neil, K. M., Tompkins, R. G.,Yarmush., M. L. 1996. Zonal liver cell heterogeneity: Effects of oxygen on metabolic functions of hepatocytes. *Cell Eng* 1:125–35.

Blum, M., Demierre, A., Grant, D. M., Markus, H., Meyer, U. 1991. Molecular mechanism of slow acetylation of drugs and carcinogens in humans. *Proc Nati Acad Sci USA* 88:5237–41.

Blume, N., Leonard, J., Xu, Z. J., Watanabe, O., Remotti, H., Fishman, J. 2000. Characterization of Cyp2d22, a novel cytochrome P450 expressed in mouse mammary cells. *Arch Biochem Biophys* 381:191–204.

Bofinger, D. P., Feng, L., Chi, L. H., Love, J., Stephen, F. D., Sutter, T. R., Osteen, K. G. et al. 2001. Effect of TCDD exposure on CYP1A1 and CYP1B1 expression in explant cultures of human endometrium. *Toxicol Sci* 62:299–314.

Bogaards, J. J., Bertrand, M., Jackson, P., Oudshoorn, M. J., Weaver, R. J., van Bladeren, P. J., Walther, B. 2000. Determining the best animal model for human cytochrome P450 activities: A comparison of mouse, rat, rabbit, dog, micropig, monkey and man. *Xenobiotica* 30:1131–52.

Boroujerdi, M., Wilson, A. G. E., Kung, H. C., Anderson, M. W. 1981. Metabolism and DNA binding of benzo(a)pyrene *in vivo* in the rat. *Cancer Res* 41:951–7.

Brocker, C., Cantore, M, Failli, P., Vasiliou, V. 2011. Aldehyde dehydrogenase 7A1 (ALDH7A1) attenuates reactive aldehyde and oxidative stress induced cytotoxicity. *Chem Biol Interact* 191(1–3):269–77.

Buchheit, D., Schmitt, E. I., Bischoff, D., Ebner, T., Bureik, M. 2011. S-Glucuronidation of 7-mercapto-4-methylcoumarin by human UDP glycosyltransferases in genetically engineered fission yeast cells. *Biol Chem* 392(12):1089–95.

Bylund, J., Hidestrand, M., Ingelman-Sundberg, M., Oliw, E. H. 2000. Identification of CYP4F8 in human seminal vesicles as a prominent 19-hydroxylase of prostaglandin endoperoxides. *J Biol Chem* 275(29):21844–9.

Caldeira Araújo, H., Smit, W., Verhoeven, N. M., Salomons, G. S., Silva, S., Vasconcelos, R., Tomás, H., Tavares de Almeida, I., Jakobs, C., Duran, M. 2005. Guanidinoacetate methyltransferase deficiency identified in adults and a child with mental retardation. *Am J Med Genet A* 133A(2):122–7.

Carlile, D. J., Hakooz, N., Houston, J. B. 1999. Kinetics ofdrug metabolism in rat liver slices: IV. Comparison of ethoxycoumarin clearance by liver slices, isolated hepatocytes, and hepatic microsomes from rats pretreated with known modifiers of cytochrome P450 activity. *Drug Metab Disp* 27(4):526–32.

Cashman, J. R. 2002. Human flavin-containing monooxygenase substrate specificity and role in drug metabolism. *Curr Drug Metab* 1(2):181–91.

Cashman, J. R., Zhang, J. 2006. Human flavin-containing monooxygenases. *Annu Rev Pharmacol Toxicol* 46:65–100.

Cavallini, D., Ricci, G., Federici, G, Costa, M., Pensa, B., Matarese, R. M., Achilli, M. 1982. The oxidation of sulfur-containing amino acids by L-amino acid oxidases. *Adv Exp Med Biol* 148:359–74.

Chang, T. K., Weber, G. F., Crespi, C. L., Waxman, D. J. 1993. Differential activation of cyclophosphamide and ifosphamide by cytochromes P-450 2B and 3A in human liver microsomes. *Cancer Res* 53:5629–37.

Cheung, Y. L., Kerr, A. C., McFadyen, M. C., Melvin, W. T., Murray, G. I. 1999. Differential expression of CYP1A1, CYP1A2, CYP1B1 in human kidney tumors. *Cancer Lett* 139:199–205.

Chiu, S. H., Huskey, S. W. 1998. Species differences in N-glucuronidation. *Drug Metab Dispos* 26:838–47.

Christmas, P., Carlesso, N., Shang, H., Cheng, S. M., Weber, B. M., Preffer, F. I., Scadden, D. T., Soberman, R. J. 2003. Myeloid expression of cytochrome P450 4F3 is determined by a lineage-specific alternative promoter. *J Biol Chem* 278(27):25133–42.

Christmas, P., Jones, J. P., Patten, C. J., Rock, D. A., Zheng, Y., Cheng, S. M., Weber, B. M. et al. 2001. Alternative splicing determines the function of CYP4F3 by switching substrate specificity. *J Biol Chem* 276(41):38166–72.

Christmas, P., Ursino, S. R., Fox, J. W., Soberman, R. J. 1999. Expression of the CYP4F3 gene. Tissue-specific splicing and alternative promoters generate high and low K(m) forms of leukotriene B(4) omega-hydroxylase. *J Biol Chem* 274(30):21191–9.

Code, E. L., Crespi, C. L., Penman, B. W., Gonzalez, F. J., Chang, T. K., Waxman, D. J. 1997. Human cytochrome P4502B6: Interindividual hepatic expression, substrate specificity, and role in procarcinogen activation. *Drug Metab Dispos* 25:985–93.

Conney, A. H. 1982. Induction of microsomal enzymes by foreign chemicals and carcinogenesis by polycyclic aromatic hydrocarbons: G.H.A. Clowes memorial lecture. *Cancer Res* 42:4875–917.

Court, M. H., Duan, S. X., Hesse, L. M., Venkatakrishnan, K., Greenblatt, D. J. 2001. Cytochrome P-450 2B6 is responsible for interindividual variability of propofol hydroxylation by human liver microsomes. *Anesthesiology* 94:110–19.

Cowart, L. A., Wei, S., Hsu, M. H., Johnson, E. F., Krishna, M. U., Falck, J. R., Capdevila, J. H. 2002. The CYP4A isoforms hydroxylate epoxyeicosatrienoic acids to form high affinity peroxisome proliferator-activated receptor ligands. *J Biol Chem* 277:35105–12.

Crabb, D. W., Matsumoto, M., Chang, D., You, M. 2004. Overview of the role of alcohol dehydrogenase and aldehyde dehydrogenase and their variants in the genesis of alcohol-related pathology. *Proc Nutrition Soc* 63(1):49–63.

Crewe, H. K., Lennard, M. S., Tucker, G. T., Woods, F. R., Haddock, R. E. 1992.The effect of selective serotonin re-uptake inhibitors on cytochrome P450D6 (CYP2D6) activity in human liver microsomes. *Br J Clin Pharmacol* 34(3), 262–5.

Cui, X., Nelson, D. R., Strobel, H. W. 2000. A novel human cytochrome P450 4F isoform (CYP4F11): cDNA cloning, expression, and genomic structural characterization. *Genomics* 68(2):161–6.

Czerwinski, M., McLemore, T. L., Gelboin, H. V., Gonzalez, F. J. 1994. Quantification of CYP2B7, CYP4B1, and CYPOR messenger RNAs in normal human lung and lung tumors. *Cancer Res* 54(4):1085–91.

Dai, Y., Rashba-Step, J., Cederbaum, A. I. 1993. Stable expression of human cytochrome P4502E1 in HepG2 cells: Characterization of catalytic activities and production of reactive oxygen intermediates. *Biochemistry* 32:6928–37.

Dallner, G. 1978. Isolation of microsomal subfractions by use of density gradient. *Methods Enzymol* 52:71–82.

Demuth, H-U. 1990. Recent developmentsin inhibiting cysteine and serine proteases. *J Enzym Inhib Med Ch* 3(4):249–78.

Digiovanna, J., Berry, D. L., Juchau, M. R., Slaga, T. J. 1979. 2,3,7,8-Tetrachlorodibenzo-p-dioxin: Potent anticarcinogenic activity in CD-1 mice. *Biochem Biophys Res Commun* 86:577–84.

Dipple, A. 1983. Formation, metabolism, and mechanism of action of polycyclic aromatic hydrocarbons. *Cancer Res* 43:2422–5.

Domanski, T. L., Schultz, K. M., Roussel, F., Stevens, J. C., Halpert, J. R. 1999. Structure function analysis of human cytochrome P-450 2B6 using a novel substrate, site-directed mutagenesis, and molecular modeling. *J Pharmacol Exp Ther* 290:1141–7

Domanski, T. L., Finta, C., Halpert, J. R., Zaphiropoulos, P. G. 2001. cDNA cloning and initial characterization of CYP3A43, a novel human cytochrome P450. *Mol Pharmacol* 59:386–92.

Doostdar, H., Burke, M. D., Mayer, R. T. 2000. Bioflavonoids: Selective substrates and inhibitors for cytochrome P450 CYP1A and CYP1B1. *Toxicology* 144:31–8.

Dresser, G. K., Spence, J. D., Bailey, D. G. 2000. Pharmacokinetic–pharmacodynamic consequences and clinical relevance of cytochrome P450 3A4 inhibition. *Clin Pharmacokinet* 38:41–57.

Duester, G. 2008. Retinoic acid synthesis and signaling during early organogenesis. *Cell* 134(6):921–31.

Elens, L., Vandercam, B., Yombi, J.-C., Lison, D., Wallemacq, P., Haufroid, V. 2010. Influence of host genetic factors on efavirenz plasma and intracellular pharmacokinetics in HIV-1-infected patients. *Pharmacogenomics* 11:1223–34.

Ekins, S. 1996. Past, present, and future applications of precision-cut liver slices for *in vitro* xenobiotic metabolism. *Drug Metab Rev* 28(4):591–623.

Ekins, S., Vandenbranden, M., Ring, B. J., Gillespie, J. S., Yang, T. J., Gelboin, H. V., Wrighton, S. A. 1998. Further characterization of the expression in liver and catalytic activity of CYP2B6. *J Pharmacol Exp Ther* 286:1253–9.

Ekins, S., Vanden Branden, M., Ring, B. J., Wrighton, S. A. 1997. Examination of purported probes of human CYP2B6. *Pharmacogenetics* 7:165–79.

Eltom, S. E., Larsen, M. C., Jefcoate, C. R. 1998. Expression of CYP1B1 but not CYP1A1 by primary cultured human mammary stromal fibroblasts constitutively and in response to dioxin exposure: Role of the Ah receptor. *Carcinogenesis* 19:1437–44.

Emary, W. B., Toren, P. C., Mathews, B., Huh, K. 1998. Disposition and metabolism of rifapentine, a rifamycin antibiotic, in mice, rats and monkeys. *Drug Metab Dispos* 26:725–31.

Enroth, C., Eger, B. T., Okamoto, K., Tomoko Nishino, T., Nishino, T., Pai, E. F. 2000. Crystal structures of bovine milk xanthine dehydrogenase and xanthine oxidase: Structure-based mechanism of conversion. *Proc Natl Acad Sci USA* 97(20):10723–8.

Farrés, J., Moreno, A., Crosas, B., Peralba, J. M., Allali-Hassani, A., Hjelmqvist, L., Jörnvall, H., Parés, X. 1994. Alcohol dehydrogenase of class IV (σσ-ADH) from human stomach—cDNA sequence and structure/function relationships. *Eur J Biochem* 224(2):549–57.

Faucette, S. R., Hawke, R. L., Lecluyse, E. L., Shord, S. S., Yan, B., Laethem, R. M., Lindley, C. M. 2000. Validation of bupropion hydroxylation as a selective marker of human cytochrome P4502B6 catalytic activity. *Drug Metab Dispos* 28:1222–30.

Fraser, D. J. Feyereisen, R. Harlow, G. R. Halpert, J. R. 1997. Isolation, heterologous expression and functional characterization of a novel cytochrome P450 3A enzyme from a canine liver cDNA library. *J Pharmacol Exp Ther* 283:1425–32.

Freedberg, I. M. 2003. *Fitzpatrick's Dermatology in General Medicine*, 6th Edition. New York: McGraw-Hill.

Fukami, T., Nakajima, M., Sakai, H., Katoh, M.,Yokoi, T. 2007. CYP2A13 Metabolizes the substrates of human CYP1A2, phenacetin, and theophylline. *Drug Metab Dispos* 35(3):335–9.

Funk, C., Ponelle, C., Scheuermann, G., Pantze, M. 2001. Cholestatic potential of troglitazoneas a possible factor contributing to troglitazone-induced hepatotoxicity: *In vivo* and *in vitro* interaction at the canalicular bile salt export pump in the rat. *Mol Pharmacol* 59:627–35.

Galetin, A., Houston, J. B. 2006. Intestinal and hepatic metabolic activity of five cytochrome P450 enzymes: Impact on prediction of first-pass metabolism. *J Pharmacol Exp Ther* 318:1220–9.

Gallagher, E. P., Kunze, K. L., Stapleton, P. L., Eaton, D. L. 1996. The kinetics of aflatoxin B1 oxidation by human cDNA-expressed and human liver microsomal cytochromes P450 1A2 and 3A4. *Toxicol Appl Pharmacol* 141:595–606.

Gäng, V., Baldus, M., Kadereit, M. 1976. Serum level changes of endogenous and postheparin diamine oxidase (histaminase) in clinical and experimental hepatitis. *Acta Hepato-Gastro* 23(2):104–9.

Gargalidis-Moudanos, C., Remaury, A., Parini, A. 1997. Predominant expression of monoamine oxidase B isoform in rabbit renal proximal tube: Regulation by I2 imidazoline ligands in intact cells. *Mol Pharmacol* 51:637–43.

Gelboin, H. V. 1980. Benzo[a]pyrene metabolism, activation and carcinogenesis: Role of mixed function oxidases and related enzymes. *Pharmacol Rev* 60:1107–66.

Gelboin, H. V. 1993. Cytochrome P450 and monoclonal antibodies. *Pharmacol Rev* 45:413–53.

Gelboin, H. V., Goldfarb, I., Krausz, K. W., Grogan, J., Korzekwa, K. R., Gonzalez, F. J., Shou, M. 1996. Inhibitory and noninhibitory monoclonal antibodies to human cytochrome P450 2E1. *Chem Res Toxicol* 9:1023–30.

Gelboin, H. V., Krausz, K. 2006. Monoclonal antibodies and multifunctional cytochrome P450: Drug metabolism as paradigm. *J Clin Pharmacol* 46:353–72.

Gellner, K., Eiselt, R., Hustert, E., Arnold, H., Koch, I., Haberl, M., Deglmann, C. J. et al. 2001. Genomic organization of the human CYP3A locus: Identification of a new, inducible CYP3A gene. *Pharmacogenetics* 11:111–21.

Geraghty, M. T., Vaughn, D., Nicholson, A. J., Lin, W. W., Jimenez-Sanchez, G., Obie, C., Flynn, M. P., Valle, D., Hu, C. A.1998. Mutations in the Delta1-pyrroline 5-carboxylate dehydrogenase gene cause type II hyperprolinemia. *Hum Mol Genet* 7(9):1411–5.

Gervot, L., Rochat, B., Gautier, J. C., Bohnenstengel, F., Kroemer, H., de Berardinis, V., Martin, H., Beaune, P., de Waziers, I. 1999. Human CYP2B6: Expression, inducibility and catalytic activities. *Pharmacogenetics* 9:295–306.

Gibbings, S., Elkins, N. D., Fitzgerald, H., Janice Tiao, J., Weyman, M. E., Shibao, G., Fini, M. A., Wright, R. M. 2011. Xanthine oxidoreductase promotes the inflammatory state of mononuclear phagocytes through effects on chemokine expression, peroxisome proliferator-activated receptor-γ sumoylation, and HIF-1α. *J Biol Chem* 286:961–75.

Glatt, H., Davis, W., Meinl, W., Hermersdorfer, H., Venitt, S., Phillips, D. H. 1998. Rat, but not human, sulfotransferase activates a tamoxifen metabolite to produce DNA adducts and gene mutations in bacteria and mammalian cells in culture. *Carcinogenesis* 19:1709–13.

Goldstein, J. A., de Morais, S. M. 1994. Biochemistry and molecular biology of the human CYP2C subfamily. *Pharmacogenetics* 4:285–99.

Goldstone, H. M., Stegeman, J. J. 2006. A revised evolutionary history of the CYP1A subfamily: Gene duplication, gene conversion, and positive selection. *J Mol Eval* 62(6):708–17.

Goodman, S. I., Mace, J. W., Miles, B. S., Teng, C. C., Brown, S. B. 1974. Defective hydroxyproline metabolism in type II hyperprolinemia. *Biochem Med* 10(4):329–36.

Green, M. D., Tephly, T. R. 1998. Glucuronidation of amine substrates by purified and expressed UDP glucuronosyltransferase proteins. *Drug Metab Dispos* 26:860–7.

Gu, J., Su, Y., Chen, Q. Y., Zhang, X., Ding, X. 2000. Expression and biotransformation enzymes in human fetal olfactory mucosa: Potential roles in developmental toxicity. *Toxicol Appl Pharmacol* 165:158–62.

Guengerich, F. P. 1992. Characterization of human cytochrome P450 enzymes. *FASEB J* 6:745–8.

Guengerich, F. P. 1997. Comparisons of catalytic selectivity of cytochrome P450 subfamily enzymes from different species. *Chem Biol Interact* 106:161–82.

Guthikonda, S. Sinkey, C., Barenz, T., Haynes, W. G. 2003. Xanthine oxidase inhibition reverses endothelial dysfunction in heavy smokers. *Circulation* 107:416–21.

Hadizadeh, M., Keyhani, E., Keyhani, J., Khodadadi, C. 2009. Functional and structural alterations induced by copper in xanthine oxidase. *Acta Biochim Biophys Sin* 41(7):603–17.

Hall, S. D., Thummel, K. E., Watkins, P. B., Lown, K. S., Benet, L. Z., Paine, M. F., Mayo, R. R. et al. 1999. Molecular and physical mechanisms of first pass extraction. *Drug Metab Dispos* 27:161–6.

Hammes-Schiffer, S., Benkovic, S. J. 2006. Relating protein motion to catalysis. *Annu Rev Biochem* 75:519–41.

Hanna, M. C., Blackstone, C. 2009. Interaction of the SPG21 protein ACP33/maspardin with the aldehyde dehydrogenase ALDH16A1. *Neurogenetics* 10(3):217–28.

Harper, T. W., Brassil, P. J. 2008. Reaction phenotyping: Current industry efforts to identify enzymes responsible for metabolizing drug candidates. *AAPS J* 10(1):200–7.

Harris, J. W., Rahman, A., Kim, B-R, Guengerich, F. P., Collins, J. M. 1994. Metabolism of taxol by human hepatic microsomes and liver slices: Participation of cytochrome P450 3A4 and an unknown P450 enzyme. *Cancer Res* 54(15):4026–35.

Hartley, D. P., Klaassen, C. D. 2000. Detection of chemical-induced differential expression of rat hepatic cytochrome P450 mRNA transcripts using branched DNA signal amplification technology. *Drug Metab Dispos* 28(5):608–16.

Haussinger, D. 1988. Hepatic sublobular synthesis of urea: Control by oxygen or by pH gradient? *Hepatol* 8:184–6.

Hayes, J. D., Flanagan, J. U., Jowsey, I. R. 2005. Glutathione transferases. *Annu Rev Pharmacol Toxicol* 45:51–88.

Hedstrom, L. 2002. Serine protease mechanism and specificity. *Chem Rev* 102:4501–23.

Hellgren, M., Strömberg, P., Gallego, O., Martras, S., Farrés, J., Persson, B., Parés, X., Höög, J. O. 2007. Alcohol dehydrogenase 2 is a major hepatic enzyme for human retinol metabolism. *Cell Mol Life SCi* 64(4):498–505.

Hempel, J., Höög, J. O., Jörnvall, H. 1987. Mitochondrial aldehyde dehydrogenase. Homology of putative targeting sequence to that of carbamyl phosphate synthetase I revealed by correlation of cDNA and protein data. *FEBS Lett* 222(1):95–8.

Hempel, J., Kaiser, R., Jörnvall, H. 1985. Mitochondrial aldehyde dehydrogenase from human liver. Primary structure, differences in relation to the cytosolic enzyme, and functional correlations. *Eur J Biochem* 153(1):13–28.

Hengstler, J. G., Ringel, M., Biefang, K., Diener, B., Platt, K., Böttger, T., Steinberg, P. Oesch, F. 2000. Cultures with cryopreserved hepatocytes applicability for studies of enzyme induction. *Chem Biol Interact* 125:51–73.

Hetu, C., Dumont, A., Joly, J. G. 1983. Effect of chronic ethanol administration on bromobenzene liver toxicity in the rat. *Toxicol Appl Pharmacol* 67:166–77.

Hirani, V., Yarovoy, A., Kozeska, A., Magnusson, R. P., Lasker, J. M. 2008. NIHMSID: NIHMS71438 expression of CYP4F2 in human liver and kidney: Assessment using targeted peptide antibodies. *Arch Biochem Biophys* 478(1):59–68.

Hirita, S., Matsubara, T. 1993. cDNA cloning and characterization of a novel member of steroid induced cytochrome P450 3A in rats. *Arch Biochem Biophys* 307:253–8.

Hiroi, T., Chow, T., Imaoka, S., Funae, Y. 2002. Catalytic specificity of CYP2D isoforms in rat and human. *Drug Metab Dispos* 30:970–6.

Hong, M., Lee, Y., Kim, J. W., Lim, J. S., Chang, S. Y., Lee, K. S., Paik, S. G., Choe, I. S. 1999. Isolation and characterization of cDNA clone for human liver 10-formyltetrahydrofolate dehydrogenase. *Biochem Mol Biol Int* 47(3):407–15.

Honkakoski, P., Negishi, M. 1997. The structure, function, and regulation of cytochrome P450 2A enzymes. *Drug Metab Rev* 29:977–96.

Hosokawa, M., Endo, Y., Fujisawa, M., Hara, S., Iwata, N., Sato, Y., Satoh. T. 1995. Interindividual variation in carboxylesterase levels in human liver microsomes. *Drug Metab Dispos* 23:1022–7.

Hosokawa, M., Maki, T., Satoh, T. 1990. Characterization of molecular species of liver microsomal carboxylesterases of several animal species and humans. *Arch Biochem Biophys* 277:219–27.

Houston, J. B. 1994. Utility of *in vitro* drug metabolism data in predicting *in vivo* metabolic clearance. *Biochem Pharmacol* 47(9):1469–79.

Houston, J. B., Carlile, D. J. 1997. Prediction of hepatic clearance from microsomes, hepatocytes, and liver slices. *Drug Metab Rev* 29(4):891–922.

Houston, J. B., Rowland-Yeo, K., Zanelli, U. 2012. Evaluation of the novel *in vitro* systems for hepatic drug clearance and assessment of their predictive utility. *Toxicol in vitro* 26:1265–71.

Hsu, L. C., Chang, W. C. 1996. Sequencing and expression of the human ALDH8 encoding a new member of the aldehyde dehydrogenase family. *Gene* 174(2):319–22.

Hsu, L. C., Chang, W. C., Yoshida, A. 1997. Human aldehyde dehydrogenase genes, ALDH7 and ALDH8: Genomic organization and gene structure comparison. *Gene* 189(1):89–94.

Huang, Z., Fasco, M. J., Kaminsky, L. S. 1997. Alternative splicing of CYP2D mRNA in human breast tissue. *Arch Biochem Biophys* 343:101–8.

Hutt, A. J., Caldwell, J. 1990. Amino acid conjugation. In *Conjugation Reactions in Drug Metabolism*, ed. G. J. Mulder, 273–305. London: Taylor & Francis.

Hosokawa, M., Suzuki, K., Takahashi, D., Mori, M., Satoh, T., Chiba, K. 2001. Purification, molecular cloning, and functional expression of dog liver microsomal acyl-CoA hydrolase: A member of the carboxylesterase multigene family. *Arch Biochem Biophys* 389:245–53.

Ilett, K. F., David, B. M., Detchon, P., Castleden, W. M., Rose Kawa, R. 1987. Acetylation phenotype in colorectal carcinoma. *Cancer Res* 47:1466–9.

Imaoka, S., Ogawa, H., Kimura, S., Gonzalez, F. J. 1993. Complete cDNA sequence and cDNA-directed expression of CYP4A11, a fatty acid omega-hydroxylase expressed in human kidney. *DNA Cell Biol* 12(10):893–9.

Ishida, K., Taguchi, M., Akao, T., Hashimoto, Y. 2009. Involvement of the CYP1A subfamily in stereoselective metabolism of carvedilol in beta-naphthoflavone-treated Caco-2 cells. *Biol Pharm Bull* 32(3):513–6.

Jamshidi, Y., Moreton, M., McKeown, D. A., Anfrews, S., Nithyananthan, T., Tinworth, L., Holt, D. W., Sadiq, S. T. 2010. Tribal ethnicity and CYP2B6 genetics in Ugandan and Zimbabwean populations in the UK: Implications for efavirenz dosing in HIV infection. *J Antimicrob Chemother* 65(12):2614–9.

Jester, J. V., Moller-Pedersen, T., Huang, J., Sax, C. M., Kays, W. T., Cavangh, H. D., Petroll, W. M., Piatigorsky, J. 1999. The cellular basis of corneal transparency: Evidence for corneal crystallins. *J Cell Sci* 112(5):613–22.

Jungermann, K. 1995. Zonation of metabolism and gene expression in liver. *Histochemistry* 103:81–91.

Kaivosaari, S., Finel, M., Koskinen, M. 2011. N-glucuronidation of drugs and other xenobiotics by human and animal UDP-glucuronosyltransferases. *Xenobiotica* 41(8):652–69.

Kaji, H., Kume, T. 2005. Regioselective glucuronidation of denopamine: Marked species differences and identification of human UDP-glucuronosyltransferase isoform. *Drug Metab Dispos* 33:403–12.

Kaku, T., Ogura, K., Nishiyama, T., Ohnuma, T., Muro, K., Hiratsuka, A. 2004. Quaternary ammonium-linked glucuronidation of tamoxifen by human liver microsomes and UDP-glucuronosyltransferase 1A4. *Biochem Pharmacol* 67:2093–102.

Kamataki, T., Maeda, K., Yamazoe, Y., Matsuda, N., Ishii, K., Kato, R. 1983. A high-spin form of cytochrome P-450 highly purified from polychlorinated biphenyl-treated rats. Catalytic characterization and immunochemical quantitation in liver microsomes. *Mol Pharmacol* 23:146–55.

Kerdpin, O., Elliot, D. J., Mackenzie, P. I., Miners, J. M. 2006. PeterI.Mackenzie. Sulfinpyrazone C-glucuronidation is catalyzed selectively by human UDP-glucuronosyltransferase 1A9. *Drug Metab Dispos* 34(12):1950–3.

Kidd, R. S. Straughn, A. B., Meyer, M. C., Blaisdell, J., Goldstein, J. A., Dalton, J. T. 1999. Pharmacokinetics of chlorpheniramine, phenytoin, glipizide and nifedipine in an individual homozygous for the CYP2C9*3 allele. *Pharmacogenetics* 9:71–80.

Kikonyogo, A., Pietruszko, R. 1996. Aldehyde dehydrogenase from adult human brain that dehydrogenates gamma-aminobutyraldehyde: Purification, characterization, cloning and distribution. *Biochem J* 316(1):317–24.

Kirita, S., Matsubara, T. 1993. cDNA cloning and characterization of a novel member of steroid-induced cytochrome P450 3A in rats. *Arch Biochem Biophys* 307:253–8.

Klaassen, C. D., Boles, J. W. 1997. Sulfation and sulfotransferases 5: The importance of 3′-phosphoadenosine 5′-phosphosulfate (PAPS) in the regulation of sulfation. *FASEB J* 11(6):404–18.

Klos, C., Koob, M., Kramer, C., Wolfgang Dekant, W. 1992. p-Aminophenol nephrotoxicity: Biosynthesis of toxic glutathione conjugates. *Toxicol Appl Pharmacol* 115(1):98–106.

Klose, T. S., Blaisdell, J. A., Goldstein, J. A. 1999. Gene structure of CYP2C8 and extrahepatic distribution of the human CYP2Cs. *J Biochem Mol Toxicol* 13:289–95.

Kolars, J. C., Lown, K. S., Schmiedlin-Ren, P., Ghosh, M., Fang, C., Wrighton, S. A., Merion, R. M., Watkins, P. B. 1994. CYP3A gene expression in human gut epithelium. *Pharmacogenetics* 4:247–59.

Komori, M., Kikuchi, O., Sakuma, T., Funaki, J., Kitada, M., Kamataki, T. 1992. Molecular cloning of monkey liver cytochrome P450 cDNAs: Similarity of the primary sequences to human cytochromesP450. *Biochim Biophys Acta* 1171:141–6.

Koob, M., Dekant, W. 1991. Bioactivation of xenobiotics by formation of toxic glutathione conjugates. *Chem Biol Interact* 77(2):107–36.

Koskela, S., Hakkola, J., Hukkanen, J., Pelkonen, O., Sorri, M., Saranen, A., Anttila, S., Fernandez-Salguero, P., Gonzalez, F., Raunio, H. 1999. Expression of CYP2Agenes in human liver and extrahepatic tissues. *Biochem Pharmacol* 57:1407–13.

Koukouritaki, O. S. B., Simpson, P., Yeung, C. K., Rettie, A. E., Hines, R. N. 2002. Human hepatic flavin-containing monooxygenases 1(FMO1) and 3 (FMO3) developmental expression. *Pediatr Res* 51:236–43.

Kroemer, H. K., Klotz, U. 1992. Glucuronidation of drugs—A reevaluation of the pharmacological significance of the conjugates and modulating factors. *Clin Pharmacokint* 23(4):292–310.

Krupenko, S. A., Oleinik, N. V. 2002. 10-formyltetrahydrofolate dehydrogenase, one of the major folate enzymes, is down-regulated in tumor tissues and possesses suppressor effects on cancer cells. *Cell Growth Differ* 13(5):227–36.

Kuehl, G. E., Lampe, J. W., Potter, J. D., Bigler, J. 2005. Glucuronidation of nonsteroidal anti-inflammatory drugs. Identifying the enzymes responsible in human liver microsome. *Drug Metab Dispos* 33(7):1027–35.

Kuiper, H., Spötter, A., Williams, J. L. 2005. Physical mapping of CHX10, ALDH6A1, and ABCD4 on bovine chromosome 10q34. *Cytogenet Genome Res* 109(4):533.

Kupfer, A., Branch, R. A. 1985. Stereoselective mephobarbital hydroxylation cosegregates with mephenytoin hydroxylation. *Clin Pharmacol Ther* 38:414–18.

Lamba, V., Lamba, J., Yasuda, K., Strom, S., Davila, J., Hancock, M. L., Fackenthal, J. D. et al. Hepatic CYP2B6 expression: Gender and ethnic differences and relationship to CYP2B6 genotype and CAR (constitutive androstane receptor) expression. *J Pharmaco Exp Ther* 307:906–22.

Landmesser, U., Spiekermann, S., Dikalov, S., Tatge, H., Ragna Wilk, R., Kohler, C., Harrison, D. G., Hornig, B., Helmut Drexler, H. 2002. Vascular xidative stress and endothelial dysfunction in patients with chronic heart failure role of xanthine-oxidase and extracellular superoxide dismutase. *Circulation* 106:3073–8.

Lang, T., Klein, K., Fischer, J., Nussler, A. K., Neuhaus, P., Hofmann, U., Eichelbaum, M., Schwab, M., Zanger, U. M. 2001. Extensive genetic polymorphism in the human CYP2B6 gene with impact on expression and function in human liver. *Pharmacogenetics* 11:399–415.

Larsen, M. C., Angus, W. G., Brake, P. B., Eltom, S. E., Sukow, K. A., Jefcoate, C. R. 1998. Characterization of CYP1B1 and CYP1A1 expression in human mammary epithelial cells: Role of the aryl hydrocarbon receptor in polycyclic aromatic hydrocarbon metabolism. *Cancer Res* 58:2366–74.

Larsen-Su, S., Williams, D. E. 1996. Dietary indole-3-carbinol inhibits FMO activity and the expression of flavin-containing monooxygenase form 1 in rat liver and intestine. *Drug Metab Dispos* 24:927–31.

Levy, M., Caraco, Y., Geisslinger, G. 1998. Drug acetylation in liver disease. *Clin Pharmacokinet* 34(3):219–26.

Lieber, C. S. 2004. The discovery of the microsomal ethanol oxidizing system and its physiologic and pathologic role. *Drug Metab Rev* 36:511–29.

Lieber, C. S. 1997. Cytochrome P-4502E1: Its physiological and pathological role. *Physiol Rev* 77:517–44.

Lin, M., Napoli, J. L. 2000. cDNA cloning and expression of a human aldehyde dehydrogenase (ALDH) active with 9-cis-retinal and identification of a rat ortholog, ALDH12. *J Biol Chem* 275(51):40106–12.

Lin, S. W., Chen, J. C., Hsu, L. C., Hsieh, C. L., Yoshida, A. 1996. Human gamma-aminobutyraldehyde dehydrogenase (ALDH9): cDNA sequence, genomic organization, polymorphism, chromosomal localization, and tissue expression. *Genomics* 34(3):376–80.

Lindfelt, T., O'Brien, J., Song, J. C., Patel, R., Winslow, D. L. 2010. Efavirenz plasma concentrations and cytochrome 2B6 polymorphisms. *Ann Pharmacother* 44(10):1572–8.

Liu, L., Pang, K. S. 2005. The roles of transporters and enzymes in hepatic drug processing. *Drug Metab Dispos* 33:1–9.

Lofgren, S., Hagbjork, A. L., Ekman, S., Fransson-Steen, R., Terelius, Y. 2004. Metabolism of human cytochrome P450 marker substrates in mouse: A strain and gender comparison. *Xenobiotica* 34:811–34.

Long, X. D., Ma, Y., Wei, Y. P., Deng, Z. L. 2006. The polymorphisms of GSTM1, GSTT1, HYL1*2, and XRCC1, and aflatoxin B1-related hepatocellular carcinoma in Guangxi population, China. *Hepatol Res* 36:48–55.

Luk, G. D., Bayless, T. M., SBBayli, S. B. 1980. Diamine oxidase (histaminase). A circulating marker for rat intestinal mucosal maturation and integrity. *J Clin Invest* 66(1):66–70.

Lukasheva, E. V., Efremova, A. A., Treshalina, E. M., Arinbasarova, A. Yu., Medentzev, A. G., Berezova, T. T. 2011. L-Amino acid oxidases: Properties and molecular mechanisms of action. *Biochem (Moscow—Suppl. Series B—Biomedical Chemistry)* 5(4):337–45.

Mahgoub, A., Idle, J. R., Dring, L. G., Lancaster, R., Smith, R. L. 1977. Polymorphic hydroxylation of debrisoquine in man. *Lancet* 2:584–6.

Malarkey, D. E., Johnson, K., Ryan, L., Boorman, G., Maronpot, R. R. 2005. New insights into functional aspects of liver morphology. *Toxicol Pathol* 33(1):27–34.

Männistöl, P. T., Seppo Kaakkola, S. 1999. Catechol-O-methyltransferase (COMT): Biochemistry, molecular biology, pharmacology, and clinical efficacy of the new selective COMT inhibitors. *Pharmacol Rev* 51(4):593–628.

Matsubara, T., Kim, H. J., Miyata, M., Shimada, M., Nagata, K., Yamazoe, Y. 2004. Isolation and characterization of a new major intestinal CYP3A form, CYP3A62, in the rat. *J Pharmacol Exp Ther* 309:1282–90.

Matsunaga, T., Ohmori, S., Ishida, M., Sakamoto, Y., Nakasa, H., Kitada, M. 2002. Molecular cloning of monkey CYP2C43 cDNA and expression in yeast. *Drug Metab Pharmacokinet* 17:117–24.

Mazzoleni, G., Steimberg, N. 2012. New models for the *in vitro* study of liver toxicity: 3D culture systems and the role of bioreactors. In: *The Continuum of Health Risk Assessments*, ed. M. G. Tyshenko, 162–94. Croatia: Tech Europe University Campus STeP Ri.

McCrea, J. B., Cribb, A., Rushmore, T., Osborne, B., Gillen, L., Lo, M. W., Waldman, S., Bjornsson, T., Spielberg, S., Goldberg, M. R. 1999. Phenotypic and genotypic investigations of a healthy volunteer deficient in the conversion of losartan to its active metabolite E-3174. *Clin Pharmacol Ther* 65:348–52.

McFadyen, M. C., Breeman, S., Payne, S., Stirk, C., Miller, I. D., Melvin, W. T., Murray, G. I. 1999a. Immunohistochemical localization of cytochrome P450 CYP1B1 in breast cancer with monoclonal antibodies specific for CYP1B1. *J Histochem Cytochem* 47:1457–64.

McFadyen, M. C., Cruickshank, M. E., Miller, I. D., McLeod, H. L., Melvin, W. T., Haites, N. E., Parkin, D., Murray, G. I. 2001. Cytochrome P450 CYP1B1 over-expression in primary and metastatic ovarian cancer. *Br J Cancer* 85:242–6.

McFadyen, M. C., Murray, G. I., Melvin, W. T. 1999b. Cytochrome P450 CYP 1B1 mRNA in normal human brain. *Mol Pathol* 52:164.

McKinnon, R. A., Burgess, W. M., Hall, P. M., Roberts-Thomson, S. J., Gonzalez, F. J., McManus, M. E. 1995. Characterisation of CYP3A gene subfamily expression in human gastrointestinal tissues. *Gut* 36:259–67.

McPherson, J. D., Wasmuth, J. J., Kurys, G., Pietruszko, R. 1994 Human aldehyde dehydrogenase: Chromosomal assignment of the gene for the isozyme that metabolizes gamma-aminobutyraldehyde. *Hum Genet* 93(2):211–2.

Meisel, P. 2002. Arylamine N-acetyltransferases and drug response. *Pharmacogenomics* 3:349–66.

Melanson, S. E. F., Stevenson, K., Kim, H., Antin, J. H., Court, M. H., Ho, V., Ritz, J. et al. 2010. Allelic variations in CYP2B6 and CYP2C19 and survival of patients receiving cyclophosphamide prior to myeloablative hematopoietic stem cell transplantation. *Am J Hematol* 85(12):967–71.
Mentlein, R., Gallwitz, B., Schmidt, W. F. 1993. Dipeptidyl-peptidase IV hydrolyses gastric inhibitory polypeptide, glucagon-like peptide-1(7–36) amide, peptide histidine methionine and is responsible for their degradation in human serum. *Eur J Biochem* 214(3):829–35.
Milne, R. W., Nation, R. L., Somogyi, A. A. 1996. The disposition of morphine and its 3- and 6-glucuronide metabolites in humans and animals, and the importance of the metabolites to the pharmacological effects of morphine. *Drug Metab Rev* 28(3):345–72.
Miners, J. O., Coulter, S., Birkett, D. J., Goldstein, A. 2000. Torsemide metabolism by CYP2C9 variants and other human CYP2C subfamily enzymes. *Pharmacogenetics* 10:267–70.
Miners, J. O., Coulter, S., Birkett, D. J., Goldstein, J. A. 2000. Torsemide metabolism by CYP2C9 variants and other human CYP2C subfamily enzymes. *Pharmacogenetics* 10:267–70.
Monks, T. J., Serrine, S., Lau, S. S. 1998. The pharmacology and toxicology of polyphenolic-glutathione conjugates. *Annu Rev Pharmacol Toxicol* 38:229–55.
Morisseau, C., Bruce, D. Hammock, B. D. 2005. Epoxide hydrolases: Mechanisms, inhibitor designs, and biological roles. *Ann Rev Pharmacol Toxicol* 45:311–33.
Motika, M. S., Zhang, J., Zheng, X., Kiersten Riedler, K., John, R., Cashman, J. R. 2009. Novel variants of the human flavin-containing monooxygenase 3 (FMO3) gene associated with trimethylaminuria. *Mol Genet Metab* 97(2):128–135.
Murray, G. I., Taylor, M. C., McFadyen, M. C., McKay, J. A., Greenlee, W. F., Burke, M. D., Melvin, W. T. 1997. Tumor-specific expression of cytochrome P450 CYP1B1. *Cancer Res* 57:3026–31.
Muskhelishvili, L., Thompson, P. A., Kusewitt, D. F., Wang, C., Kadlubar, F. F. 2001. *In situ* hybridization and immunohistochemical analysis of cytochrome P450 1B1 expression in human normal tissues. *J Histochem Cytochem* 49:229–36.
Nahmias, Y., Berthiaume, F., Yarmush, M. L. 2006. Integration of technologies for hepatic tissue engineering. In: *Advances in Biochemical Engineering/Biotechnology*, ed. T. Scheper, Vol. 103, 309–29. Berlin Heidelberg: Springer-Verlag.
Nakajima, M., Itoh, M., Sakai, H., Fukami, T., Katoh, M., Yamazaki, H., Kadlubar, F. F., Imaoka, S., Funae, Y., Yokoi, T. 2006. CYP2A13 expressed in human bladder metabolically activates 4-aminobiphenyl. *Int J Cancer* 119:2520–6.
Nakazawa, T., Miyata, K., Omura, K., Iwanaga, T., Nagata, O. 2006. Metabolic profile of FYX-051 (4-(5-pyridin-4-yl-1*H*-[1,2,4]triazol-3-yl) pyridine-2-carbonitrile) in the rat, dog, monkey, and human: Identification of *N*-glucuronides and *N*-glucosides. *Drug Metab Dispos* 34:1880–6.
Nebert, D. W., Nelson, D. R., Conn, M. J., Estrabrook, R. W., Feyerseisen, R., Fujuii-Kuriyama, Y., Gonzalez, F. J. et al. 1991. The P450 superfamily: Update on new sequences gene mapping, and recommended nomenclature. *DNA Cell Biol* 10:1–14.
Nebert, D. W., Russell, D. W. 2002. Clinical importance of the cytochromes P450. *Lancet* 360:1155–62.
Negishi, M., Lindberg, R., Burkhart, B., Ichikawa, T, Honkakoski, P., Lang, M. 1989. Mouse steroid 15.alpha.-hydroxylase gene family: Identification of type II P-45015.alpha. as coumarin 7-hydroxylase. *Biochemistry* 28(10):4169–72.
Nelson, D. R., Kamataki, T., Waxman, D. J., Guengerich, F. P., Estabrook, R. W., Feyereisen, R., Gonzalez, F. J., Coon, M. J., Gunsalus, I. C., Gotoh, O. et al. 1993. The P450 superfamily: Update on new sequences, gene mapping, accession numbers, early trivial names of enzymes, and nomenclature. *DNA Cell Biol* 12(1):1–51.
Nelson, D. R., Koymans, L., Kamataki, T., Stegeman, J. J., Feyereisen, R., Waxman, D. J., Waterman, M. R. et al. 1996. P450 superfamily: Update on new sequences, gene mapping, accession numbers and nomenclature. *Pharmacogenetics* 6(1):1–42.
Niznik, H. B., Tyndale, R. F., Sallee, F. R., Gonzalez, F. J., Hardwick, J. P., Inaba, T., Kalow, W. 1990. The dopamine transporter and cytochrome P45OIID1 (debrisoquine 4-hydroxylase) in brain: Resolution and identification of two distinct [3H]GBR-12935 binding proteins. *Arch Biochem Biophys* 276:424–32.
Oesch, F., Arand, M. 1999. Xenobiotic metabolism. In: *Toxicology*, eds. H. Marquardt, S. Schäfer, D. McLellan and C. Welsch, 84–110. San Diego: Academic Press.
Ohmori, S., Horie, T., Guengerich, F. P., Kiuchi, M., Kitada, M. 1993. Purification and characterization of two forms of hepatic microsomal cytochrome P450 from untreated cynomolgus monkeys. *Arch Biochem Biophys* 305:405–13.
Okita, R. T., Okita, J. R. 2001. Cytochrome P450 4A fatty acid omega hydroxylases. *Curr Drug Metab* 2:265–81.

Ono, Y., Fukuhara, N., Yoshie, O. 1998. TAL1 and LIM-Only proteins synergistically induce retinaldehyde dehydrogenase 2 expression in T-cell acute lymphoblastic leukemia by acting as cofactors for GATA3. *Mol Cell Biol* 18(12):6939–50.

Pacher, P., Nivorozhkin, N., Szabo, C. 2006. Therapeutic effects of xanthine oxidase inhibitors: Renaissance half a century after the discovery of allopurinol. *Pharmacol Rev* 58:187–114.

Paine, M. F., Khalighi, M., Fisher, J. M., Shen, D. D., Kunze, K. L., Marsh, C. L., Perkins, J. D., Thummel, K. E. 1997. Characterization of interintestinal and intraintestinal variations in human CYP3A-dependent metabolism. *J Pharmacol Exp Ther* 283:1552–62.

Pawlik, A., Ostanek, L., Brzosko, I., Gawroska-Szklarz, B., Marek Brzosko, M., Dabrowska-Zamojcin, E. 2002. Increased genotype frequency of N-acetyltransferase 2 slow acetylation in patients with rheumatoid arthritis. *Clin Pharmacol Ther* 72:319–25.

Potter, G. A., Patterson, L. H., Wanogho, E., Perry, P. J., Butler, P. C., Ijaz, T., Ruparelia, K. C. et al. 2002. The cancer preventative agent resveratrol is converted to the anticancer agent piceatannol by the cytochrome P450 enzyme CYP1B1. *Br J Cancer* 86:774–8.

Porter, T. D., Coon, M. J. Cytochrome P-450. 1991. Multiplicity of isoforms, substrates, catalytic and regulatory mechanisms. *J Biol Chem* 266(21):13469–72.

Puntoni, M., Branchi, D., Argusti, A., Zanardi, S., Crosta, C., Meroni, E., Munizzi, F. et al. 2013. A randomized, placebo-controlled, preoperative trial of allopurinol in subjects with colorectal adenoma. *Cancer Prev Res* 6:74–81.

Rae, J. M., Soukhova, N. V., Flockhart, D. A., Desta, Z. 2002. Triethylenethiophosphoramide is a specific inhibitor of cytochrome P450 2B6: Implications for cyclophosphamide metabolism. *Drug Metab Dispos* 30:525–30.

Rahman, A., Korzekwa, K. R., Grogan, J., Gonzalez, F. J., Harris, J. W. 1994. Selective biotransformation of taxol to 6 alpha-hydroxytaxol by human cytochrome P450 2C8. *Cancer Res* 54:5543–6.

Redinbo, M. R., Potter, P. N. 2005. Mammalian carboxylesterases: From drug targets to protein therapeutics. *Drug Discovery Today* 10:313–20.

Rekha, G. K., Devaraj, V. R., Sreerama, L., Lee, M. J., Nagasawa, H. T., Sladek, N. E. 1998. Inhibition of human class 3 aldehyde dehydrogenase, and sensitization of tumor cells that express significant amounts of this enzyme to oxazaphosphorines, by chlorpropamide analogues. *Biochem Pharmacol* 55(4):465–74.

Richter, T., Murdter, T. E., Heinkele, G., Pleiss, J., Tatzel, S., Schwab, M., Eichelbaum, M., Zanger, U. M. 2004. Potent mechanism-based inhibition of human CYP2B6 by clopidogrel and ticlopidine. *J Pharmacol Exp Ther* 308:189–97.

Rieder, C. R., Ramsden, D. B., Williams, A. C. 1998. Cytochrome P450 1B1 mRNA in the human central nervous system. *Mol Pathol* 51:138–42.

Roins, M. J. J., Ingelman-Sunderg, M. 1999. Induction of human drug-metabolizing enzymes: Mechanisms and implications. In: *Handbook of Drug Metabolism*, ed. T. F. Woolf, 239–62. New York: Marcel Dekker.

Rosenbrock, H., Hagemeyer, C. E., Ditter, M., Knoth, R., Volk, B. 2001. Expression and localization of the CYP2B subfamily predominantly in neurones of rat brain. *J Neurochem* 76(2):332–40.

Saavedra, J. M., Coyle, J. T., Axelrod, J. 1973. The distribution and properties of the nonspecific N-methyltransferase in brain. *J Neurochem* 20(3):743–52.

Sakuma, T., Takai, M., Endo, Y., Kuroiwa, M., Ohara, A., Jarukamjorn, K., Honma, R., Nemoto, N. 2000. A novel female-specific member of the CYP3A gene subfamily in the mouse liver. *Arch Biochem Biophys* 377:153–62.

Sanders, R. J., Ofman, R., Duran, M., Kemp, S., Wanders, R. J. 2006. Omega-oxidation of very long-chain fatty acids in human liver microsomes. Implications for X-linked adrenoleukodystrophy. *J Biol Chem* 281(19):13180–7.

Satoh, T., Hosokawa, M. 1998. The mammalian carboxylesterases: From molecules to functions. *Ann Rev Pharmacol Toxicol* 38:257–88.

Satoh, T., Hosokawa, M. 2006. Structure, function and regulation of carboxylesterases. *Chem Biol Interact* 162:195–211.

Savas, Ü., Hsu, M. H., Johnson, E. F. 2003. Differential regulation of human CYP4A genes by peroxisome proliferators and dexamethasone. *Arch Biochem Biophy* 409(1):212–20.

Scheer, N., Kapelyukh, Y., Chatham, L., Rode, L., Buechel, S., Wolf, C. R. 2012. Generation and characterization of novel cytochrome P450 Cyp2c gene cluster knockout and CYP2C9 humanized mouse lines. *Mol Pharmacol* 82(6):1022–9.

Schellens, J. H., Malingré, M. M., Kruijtzer, C. M., Bardelmeijer, H. A., van Tellingen, O., Schinkel, A. H., Beijnen, J. H. 2000. Modulation of oral bioavailability of anticancer drugs: From mouse to man. *Eur J Pharm Sci* 12:103–110.

Scott, J., Poffenbarger, P. L. 1979. Pharmacogenetics of tolbutamide metabolism in humans. *Diabetes* 28:41–51.

Seiler, N. 1995. Polyamine oxidase, properties and functions. *Prog Brain Res* 106:333–44.

Seitz, H. K., Meier, P. 2007. The role of acetaldehyde in upper digestive tract cancer in alcoholics. *J Lab Clin Med* 149(6):293–7.

Sharma, R. K., Doig, M. V., Lewis, D. F. V., Gibson, G. G. 1989. Role of hepatic and renal cytochrome P-450 IVA1 in the metabolism of lipid substrates. *Biochem Pharmacol* 38:3621–9

Shih, J. C. 1991. Molecular basis of human MAO A and B. *Neuropsychopharm* 4:1–7.

Shih, T.-Y., Young, T-H., Lee, H.-S., Hsieh, C.-B., Hu, O. Y.-P. 2013. Protective effects of kaempferol on isoniazid- and rifampicin-induced hepatotoxicity. *AAPS J* 15(3):753–62.

Shimada, T. 2006. Xenobiotic-metabolizing enzymes involved in activation and detoxification of carcinogenic polycyclic aromatic hydrocarbons. *Drug Metab Pharmacok* 21:257–76.

Shimada, T., Yamazaki, H., Mimura, M., Inui, Y., Guengerich, F. P. 1994. Interindividual variations in human liver cytochrome P-450 enzymes involved in the oxidation of drugs, carcinogens and toxic chemicals: Studies with liver microsomes of 30 Japanese and 30 Caucasians. *J Pharmacol Exp Ther* 270:414–23.

Shiratani, H., Katoh, M., Nakajima, M., Yokoi, T. 2008. Species differences in UDPglucuronosyltransferase activities in mice and rats. *Drug Metab Dispos* 36:1745–52.

Shirley, M. A., Guan, X., Kaiser, D. G., Halstead, G. W., Baillie, T. A. 1994. Taurine conjugation of ibuprofen in humans and in rat liver in vitro. Relationship to metabolic chiral inversion. *J Pharmacol Exp Ther* 269:1166–75.

Siegers, C. P., Heidbuchel, K., Younes, M. 1983. Influence of alcohol, dithiocarb and (+)-catechin on the hepatotoxicity and metabolism of vinylidene chloride in rats. *J Appl Toxicol* 3:90–5.

Sindrup, S. H., Brosen, K., Hansen, M. G. J., Aaes-Jorgensen, T., Overo, K. F., Gram, L. F. 1993. Pharmacokinetics of citalopram in relation to the sparteine and the mephenytoin oxidation polymorphisms. *Ther Drug Monit* 15:11–17.

Smith, D. A., Jones, B. C. 1992. Speculations on the substrate structure–activity relationship (SSAR) of cytochrome P450 enzymes. *Biochem Pharmacol* 44.2089–98.

Smith, H. S. 2009. Opioid Metabolism. *Mayo Clin Proc* 84(7):613–24.

Soars, M. G., Riley, R. J., Findlay, K. A., Coffey, M. J., Burchell, B. 2001. Evidence for significant differences in microsomal drug glucuronidation by canine and human liver and kidney. *Drug Metab Dispos* 29:121–6.

Sontag, T. J., Parker, R. S. 2002. Cytochrome P450 omega-hydroxylase pathway of tocopherol catabolism. Novel mechanism of regulation of vitamin E status. *J Biol Chem* 277(28):25290–6.

Spivak, S. D., Hurteau, G. J., Reilly, A. A., Aldous, K. M., Ding, X., Kaminsky, L. S. 2001. CYP1B1 expression in human lung. *Drug Metab Dispos* 29:916–22.

Sridar, C., D'Agostino, J., Hollenberg, P. F. 2012. Bioactivation of the cancer chemopreventive agent tamoxifen to quinone methides by cytochrome P4502B6 and identification of the modified residue on the apoprotein. *Drug Metab Dispos* 40(12):2280–88.

Stevens, J. C., Wrighton, S. A. 1993. Interaction of the enantiomers of fluoxetine and norfluoxetine with human liver cytochrome P450. *J Pharmacol Exp Ther* 266:964–71.

Strotkamp, D., Roos, P. H., Hanstein, W. G. 1995. A novel CYP3 gene from female rats. *Biochim Biophys Acta* 1260:341–4.

Su, P., Kaushal, K. M., Kroetz, D. L. 1998. Inhibition of renal arachidonic acid omega-hysroxylase activity with ABT reduces blood pressure in the SHR. *Am J Physiol* 275: R426–38.

Su, T., Bao, Z., Zhang, Q. Y., Smith, T. J., Hong, J. Y., Ding, X. 2000. Human cytochrome P450 CYP2A13: Predominant expression in the respiratory tract and its high efficiency metabolic activation of a tobacco-specific carcinogen, 4-(methylnitrosamino)-1-(3-pyridyl)-1-butanone. *Cancer Res* 60:5074–9.

Sugimura, T., Sato, S. 1983. Mutagens-carcinogens in foods. *Cancer Res* 43:2415–21.

Sutter, T. R., Tang, Y. M., Hayes, C. L., Wo, Y. Y., Jabs, E. W., Li, X., Yin, H., Cody, C. W., Greenlee, W. F. 1994. Complete cDNA sequence of a human dioxin-inducible mRNA identifies a new gene subfamily of cytochrome P450 that maps to chromosome 2. *J Biol Chem* 269:13092–99.

Tang, Y. M., Chen, G. F., Thompson, P. A., Lin, D. X., Lang, N. P., Kadlubar, F. F. 1999. Development of an antipeptide antibody that binds to the C-terminal region of human CYP1B1. *Drug Metab Dispos* 27:274–80.

Teh, L. K., Mohamed, N. I., Salleh, M. Z., Rohaizak, M., Shahrun, N. S, Saladina, J. J., Shia, J. K. et al. 2012. The risk of recurrence in breast cancer patients treated with tamoxifen: Polymorphisms of CYP2D6 and ABCB1. *AAPS J* 14(1):52–9.

Thurman, R. G., Kauffman, F. C. 1985. Subcellular compartmentation of pharmacologic events (SCOPE): Metabolic fluxes in peripheral and pericentral region of the liver lobule. *Hepatology* 5:144–51.

Tiano, H. F., Hosokawa, M., Chulada, P. C., Smith, P. B., Wang, R. L., Gonzalez, F. J., Crespi, C. L., Langenbach, R. 1993. Retroviral mediated expression of human cytochrome P450 2A6 in C3H/10T1/2 cells confers transformability by 4-(methylnitrosamino)-1-(3-pyridyl)-1-butanone (NNK). *Carcinogenesis* 14:1421–7.

Ueda, K., Yoshida, A., Amachi, T. 1999. Recent progress in P-glycoprotein research. *Anticancer Drug Des* 14:115–21.

van Bladeren, P. J.,Breimer, D. D., Rotteveel-Smijs, G. M. T., De Jong, R. A. W., Buijs, W., van der Gen, A., Mohn, G. R. 1980. The role of glutathione conjugation in the mutagenicity of 1,2-dibromoethane. *Biochem Pharmacol* 29(21):2975–82.

Van der Valk, P., van Kalken, C. K., Ketelaars, H., Broxterman, H. J., Scheffer. G., Kuiper, C. M., Tsuruo, T. et al. 1990. Distribution of multidrug resistance-associated P-glycoprotein in normal and neoplastic human tissues. *Ann Oncol* 1:56–64.

von Weymarn, L. B., Murphy, S. E. 2003. CYP2A13-catalysed coumarin metabolism: Comparison with CYP2A5 and CYP2A6. *Xenobiotica* 33:73–81.

Vujcic, S., Liang, P., Diegelman, P., Kramer, D. L., Porter, C. W. 2003. Genomic identification and biochemical characterization of the mammalian polyamine oxidase involved in polyamine back-conversion. *Biochem J* 370:19–28.

Wang, H., Kawashima, H., Strobel, H. W. 1996. cDNA cloning of a novel CYP3A from rat brain. *Biochem Biophys Res Commun* 221:157–62.

Wang, L., Nguyen, T. V., McLaughlin, R. W., Sikkink, L. A., Ramerez Alvarado, M., Weinshilboum, R. M. 2005. Human thiopurine S-methyltransferase pharmacogenetics: Variant allozyme misfolding and aggresome formation. *Pr Natl Acad Sci USA* 102(26):9394–9.

Wang, Y, Hu, Y, Fang, Y., Zhang, K., Yang, H., Ma, J., Xu, Q., Shen, Y. 2009. Evidence of epistasis between the catechol-O-methyltransferase and aldehyde dehydrogenase 3B1 genes in paranoid schizophrenia. *Biol Psychiatry* 65(12):1048–54.

Ward, S. A., Helsby, N. A., Skjelbo, E., Brøsen, K., Gram, L. F., Breckenridge, A. M. 1991. The activation of the biguanide antimalarial proguanil co-segregates with the mephenytoin oxidation polymorphism—A panel study. *Br J Clin Pharmacol* 31:689–92.

Ward, S. A., Walle, T., Walle, U. K., Wilkinson, G. R., Branch, R. A. 1989. Propranolol metabolism is determined by both mephenytoin and debrisoquin hydroxylase-activities. *Clin Pharmacol Ther* 45:72–9.

Weinshilboum, R. M. Otterness, D. M., Carol, L.. Szumlanski, C. L. 1999. Methylation pharmacogenetics. Catechol O-Methyltransferase, Thiopurine Methyltransferase, and Histamine N-Methyltransferase. *Ann Rev Pharmacol Toxicol* 39:19–52.

Whitlock, J. P., Jr. 1999. Induction of cytochrome P4501A1. *Annu Rev Pharmacol Toxicol* 39:103–25.

Willey, J. C., Coy, E. L., Frampton, M. W., Torres, A., Apostolakos, M. J., Hoehn, G., Schuermann, W. H. et al. 1997. Quantitative RT-PCR measurement of cytochromes p450 1A1, 1B1, and 2B7, microsomal epoxide hydrolase, and NADPH oxidoreductase expression in lung cells of smokers and nonsmokers. *Am J Respir Cell Mol Biol* 17:114–24.

Windmill, K. F., McKinnon, R. A., Zhu, X, Gaedigk, A, Grant, D. M., McManus, M. E. 1997. The Role of xenobiotic metabolizing enzymes in arylamin toxicity and carcinogenesis: Functional and localization studies. *Muta Res* 376(1–2):153–60.

Wójcikowski, J., Pichard-Garcia, L., Maurel, P., Daniel, W. A. 2003. Contribution of human cytochrome p-450 isoforms to the metabolism of the simplest phenothiazine neuroleptic promazine. *Br J Pharmacol* 138:1465–74.

Worboys, P. D., Bradbury, A., Houston, J. B. 1996. Kinetics of drug metabolism in rat liver slices. II. Comparison of clearance by liver slices and freshly isolated hepatocytes. *Drug Metab Dispos* 24:676–81.

Wu, D., Cederbaum, A. I. 2003. Alcohol, oxidative stress, and free radical damage. *Alcohol Res Health* 27:277–84.

Xu, L., Krenitsky, D. M., Seacat, A. M., Butenhoff, J. L., Tephly, T. R., Anders, M. W. 2006. N-glucuronidation of perfluorooctanesulfonamide by human, rat, dog, and monkey liver microsomes and by expressed rat and human UDP-glucuronosyltransferases. *Drug Metab Dispos* 34(8):1406–10.

Yamano, S., Tatsuno, J., Gonzalez, F. J. 1990. The CYP2A3 gene product catalyzes coumarin 7-hydroxylation in human liver microsomes. *Biochemistry* 29:1322–9.

Yanagimoto, T., Itoh, S., Muller-Enoch, D., Kamataki, T. 1992. Mouse liver cytochrome P450 (P450IIIAM1): Its cDNA cloning and inducibility by dexamethasone. *Biochim Biophys Acta* 1130:329–32.

Yang, M, Coles BF, Delongchamp R, Lang, N. P., Kadlubar, F. F. 2002. Effects of the ADH3, CYP2E1, and GSTP1 genetic polymorphisms on their expressions in Caucasian lung tissue. *Lung Cancer* 38(1):15–21.

Yoshida, A. 1993. Molecular genetics of human aldehyde dehydrogenase. *Pharmacogenetics* 2(4):139–47.

Yoshida, A., Rzhetsky, A., Hsu, L. C., Chang, C. 1998. Human aldehyde dehydrogenase gene family. *Eur J Biochem* 251(3):549–57.

Zaphiropoulos, P. G. 1997. Exon skipping and circular RNA formation in transcripts of the human cytochrome P-450 2C18 gene in epidermis and of the rat androgen binding protein gene in testis. *Mol Cell Biol* 17(6):2985–93.

Zenser, T. V., Lakshmi, V. M., Davis, B. B. 1998. N-glucuronidation of benzidine and its metabolites. Role in bladder cancer. *Drug Metab Dispos* 26:856–9.

Zuber, R., Anzenbacherova, E., Anzenbacher, P. 2002. Cytochromes P450 and experimental models of drug metabolism. *J Cell Mol Med* 6:189–98.

7 Kinetics of *In Vitro* Metabolism

Assimilation of Intrinsic Clearance in Hepatic Clearance

7.1 INTRODUCTION

At the molecular level, the basic interaction of a xenobiotic with an enzyme system is characterized by a two-step chemical catalysis (Equation 7.1). The first step is the reversible formation of xenobiotic-enzyme complex, which is basically a fast, relatively weak hydrogen bonding, ionic or hydrophobic associations, and induce conformational change in the enzyme. The second step is the slower step of irreversible formation of metabolite from enzyme-xenobiotic complex. The basics of kinetics of this catalysis, its relevance to estimation of intrinsic clearance and hepatic clearance are discussed in this chapter.

7.2 MICHAELIS–MENTEN KINETICS

Equation 7.1 is the interaction of one compound with one enzyme and does not represent the interaction of two compounds or more with one enzyme. Under controlled and optimized conditions, the *in vitro* data from subcellular fractions follow Equation 7.1.

$$\text{Enzyme }[E] + \text{Xenobiotic}[D] \underset{k_2}{\overset{k_1}{\rightleftarrows}} \text{Enzyme-xenobiotic complex}[E.D]$$
$$\xrightarrow{k_3} \text{Metabolite}[M] + \text{Enzyme }[E] \tag{7.1}$$

where k_1 and k_2 are the rate constants of reversible reaction and k_3 is the rate constant of irreversible formation of metabolite, which is also known as the catalytic rate constant. The following differential equations define the kinetics of Equation 7.1:

$$\frac{d[D]}{dt} = -k_1[E][D] + k_2[E.D] \tag{7.2}$$

$$\frac{d[E]}{dt} = -k_1[E][D] + (k_2 + k_3)[E.D] \tag{7.3}$$

$$\frac{d[E.D]}{dt} = k_1[E][D] - (k_2 + k_3)[E.D] \tag{7.4}$$

$$\frac{d[M]}{dt} = k_3[E.D] = v \tag{7.5}$$

The rate of metabolism or the turnover rate of enzyme-xenobiotic complex to metabolite as defined by Equation 7.5 can be considered under either the assumption equilibrium condition or a pseudo steady-state condition.

Under the assumption of equilibrium

$$K_{eq} = \frac{[E.D]}{[E][D]} = \frac{k_1}{k_2} \tag{7.6}$$

where K_{eq} is the equilibrium constant, $[E.D]$ concentration of enzyme-xenobiotic complex, and $[E]$ and $[D]$ are concentration of enzyme and xenobiotic, respectively.

Therefore,

$$[E.D] = K_{eq}[E][D] \tag{7.7}$$

The enzyme is a catalyst and its total concentration at any time during the experiment remains constant and is equal to the concentration of free and bound enzyme, that is, $[E_0] = [E] + [E.D]$. Thus, the free enzyme can be estimated as

$$[E] = [E_0] - [E.D] \tag{7.8}$$

Substituting Equation 7.8 in Equation 7.7 yields

$$[E.D] = K_{eq}[D]([E_0] - [E.D]) \tag{7.9}$$

$$\therefore [E.D]\left(\frac{1}{K_{eq}} + [D]\right) = [D][E_0] \tag{7.10}$$

and

$$[E.D] = \frac{[E_0][D]}{(1/K_{eq}) + [D]} \tag{7.11}$$

According to Equation 7.1, the rate of the formation of metabolite, that is, the rate of metabolism is

$$v = k_3[E.D] \tag{7.12}$$

Substituting Equation 7.11 in Equation 7.12 yields the rate of metabolism as

$$v = \frac{k_3[E_0][D]}{(1/K_{eq}) + [D]} \tag{7.13}$$

Equation 7.13 leads to the traditional Michaelis–Menten equation, which is named after Leonor Michaelis and Maud Leonora Menten. The equilibrium assumption offers $k_3[E_0]$ as the maximum rate of metabolism known as $V_{\max}$ (i.e., $V_{\max} = k_3[E_0]$) and $1/K_{eq}$ is branded as the Michaelis–Menten constant, K_M (i.e., $K_M = 1/K_{eq}$). The final form of the equation after substituting K_M and V_{max} and setting the concentration of xenobiotic as C_D is

$$v = \frac{V_{max} \times C_D}{K_M + C_D} \tag{7.14}$$

Under the steady-state condition, the rate of change of an enzyme-xenobiotic complex with respect to time is defined as

$$\frac{d[E.D]}{dt} = k_1[E][D] - k_2[E.D] - k_3[E.D] \tag{7.15}$$

where $d[E.D]/dt$ is the rate of change of the complex with respect to time, $k_1[E][D]$ is the rate of formation of complex, $k_2[E.D]$ is the rate of breakdown of the complex to the enzyme $[E]$ and xenobiotic $[D]$, and $k_3[E.D]$ is the rate of conversion of the complex to metabolite and $[E]$. Under the steady-state condition the rate of formation of the complex is equal to the rate of its breakdown, plus its rate of conversion to metabolite, that is, $k_1[E][D] = k_2[E.D] + k_3[E.D]$. Therefore, as in any steady-state condition, the rate of the change of complex with respect to time (Equation 7.15) is equal to zero

$$\frac{d[E.D]}{dt} = 0 \tag{7.16}$$

$$\therefore k_1[E][D] - k_2[E.D] - k_3[E.D] = 0 \tag{7.17}$$

Solving for $[E.D]$ yields

$$[E.D] = \frac{k_1[E][D]}{k_2 + k_3} \tag{7.18}$$

Substituting Equation 7.8 in Equation 7.17 and setting $(k_2 + k_3)/k_1 = K_M$ yields

$$[E.D] = \frac{[E_0][D] - [E.D][D]}{K_M} \tag{7.19}$$

$$\therefore [E.D] \times K_M = [E_0] \times [D] - [E.D][D] \tag{7.20}$$

$$[E.D](K_M + [D]) = [E_0][D] \tag{7.21}$$

$$[ED] = \frac{[E_0][D]}{K_M + [D]} \tag{7.22}$$

According to Equation 7.1, the rate of metabolism is k_{13} $[ED]$, that is, k_{13} multiplied by Equation 7.22:

$$k_{13}\left[ED\right] = \frac{k_{13}\left[E_0\right]\left[D\right]}{K_M + \left[D\right]} \tag{7.23}$$

where k_{13} $[ED] = v =$ the rate of metabolism and k_{13} $[E_0] = V_{max} =$ the maximum rate of metabolism.

$$\therefore\ v = \frac{V_{max}\left[D\right]}{K_M + \left[D\right]} = \frac{V_{max} \times C_D}{K_M + C_D} \tag{7.24}$$

Thus, the equation developed under the steady-state condition is the same Michaelis–Menten equation that was developed under the equilibrium assumption. The two approaches differ in the condition of the enzymatic reaction and the mathematical definition of Michaelis–Menten constant. K_M is not a rate constant, nor a dissociation constant, nor an affinity constant. It is merely a constant of convenience and represents the concentration of drug at the half-maximum rate of metabolism, and has the units of concentration:

$$\begin{aligned} \frac{V_{max}}{2} &= \frac{V_{max} \times C_D}{K_M + C_D} \\ \therefore\ 2C_D &= K_M + C_D \\ \therefore\ K_M &= C_D \end{aligned} \tag{7.25}$$

The Michaelis–Menten constant, (K_M) reflects the affinity of an enzyme for its substrate. The larger the value of K_M, the lower is the affinity. In other words, a small K_M corresponds to tight binding between an enzyme and its substrate, and a large K_M represents weak binding. The maximum rate of metabolism, (V_{max}), has units of mass/time and is constant for the interaction of an enzyme with its substrate. It represents the rate when the binding sites of all molecules of an enzyme, that is, binding sites of $[E_0]$, are occupied by substrate molecules and are present as a enzyme-substrate complex, $[ED]$.

The Michaelis–Menten equation defines the relationship between the rate of metabolism and the concentration of xenobiotic. The equation predicts a rectangular hyperbolic curve that represents the capacity-limited nature of an enzymatic metabolism (Figure 7.1). The rate of metabolism asymptotically approaches V_{max} as the xenobiotic concentration increases.

The parameters K_M and V_{max} provide the estimation of intrinsic clearance, quantitative comparison of different compounds, and comparison of different isozymes from different tissues or different species. When the concentration of a compound, C_D, is much smaller than K_M, the rate of metabolism follows first-order kinetics, that is,

$$\begin{aligned} K_M &\gg C_D \\ v &= \frac{V_{max}}{K_M} C_D = Cl_{int} \times C_D \end{aligned} \tag{7.26}$$

The ratio V_{max}/K_M in Equation 7.26 is the intrinsic metabolic clearance with units of volume per time, that is,

$$\frac{V_{max}(\text{Mass/Time})}{K_M(\text{Mass/Volume})} = Cl_{int}(\text{Volume/Time}) \tag{7.27}$$

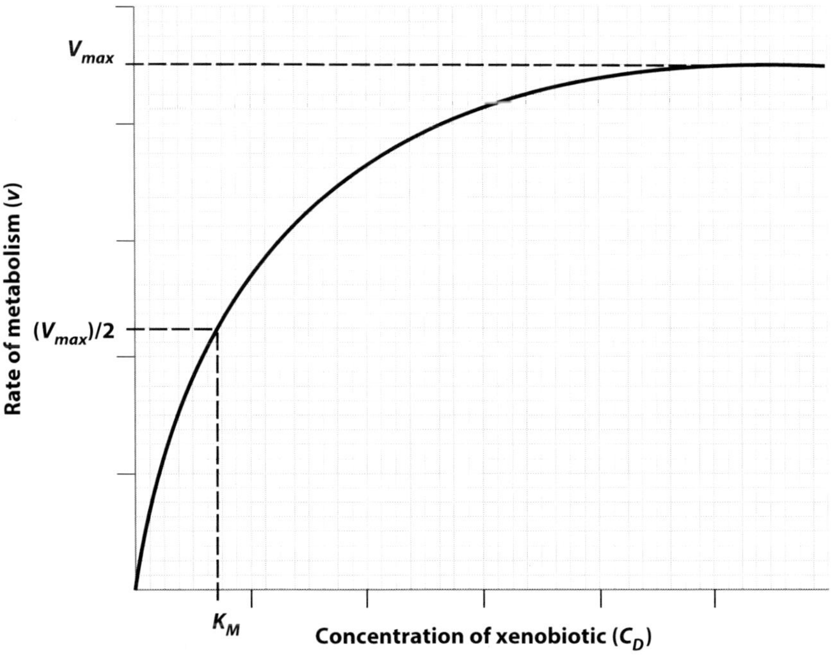

FIGURE 7.1 Rectangular hyperbolic curve of the Michaelis–Menten equation; the plateau rate corresponds to the maximum rate of metabolism (V_{max}) and the concentration at half-maximum rate corresponds to the Michaelis–Menten constant (K_M).

When C_D is much greater than K_M, the rate follows zero-order kinetics, that is,

$$\begin{aligned} K_M &\ll C_D \\ v &= V_{max} \end{aligned} \tag{7.28}$$

The parameters K_M and V_{max} are estimated by measuring the rate of metabolism at different concentrations of a xenobiotic.

7.2.1 Intrinsic Metabolic Clearance

The intrinsic clearance (Equation 7.27) is an *in vitro* measure of catalytic activity of an enzyme toward a compound. It is independent of all physiological factors that may influence the estimation of clearance *in vivo*, such as blood flow, binding to blood cells and/or proteins, etc.

The intrinsic clearance (Cl_{int}) is also a constant of proportionality between the rate of metabolism and concentration of the free xenobiotic, and remains constant as long as the metabolism of a compound follows first-order kinetics (Equation 7.26). The free concentration of xenobiotic (C_D) is a variable and the rate of metabolism (v) is a dependent variable of the free concentration. The ratio of the two interrelated variables yields the intrinsic clearance, which is a constant of proportionality with units of volume/time. For example, if it is calculated from the incubation of subcellular fractions, the units can be reported as volume/time/mg of protein in the fraction (e.g., μL/min/mg of protein), or in the case of hepatocyte it would be reported as volume/time/million cells (e.g. μL/min/# of cells).

7.2.2 The Catalytic Efficiency and Turnover Number

The ratio of the metabolic rate constant (k_3) to (K_M) at low concentrations of a compound (i.e., k_3/K_M) represents the catalytic strength or efficiency of an enzyme. The ratio has units of 1/(time × Concentration) or ($\text{time}^{-1}\text{M}^{-1}$) which corresponds to the units of a rate constant of a pseudo-second-order reaction. The upper limit of the efficiency can be range of 10^8 or 10^9 s^{-1} M^{-1}.

The turnover number is the ratio of the maximum rate of metabolism to the total amount of enzyme

$$\text{Turnover number} = \frac{V_{max}}{[E_0]} \tag{7.29}$$

The turnover number represents the number of molecules of parent compound converted to metabolite per units of time.

7.2.3 Estimation of Parameters and Constants of the Michaelis–Menten Equation

There are various approaches for determining K_M V_{max}, Cl_{int}, etc., and most are based on linearization of the Michaelis–Menten equation or any other definable relationship.

7.2.3.1 Lineweaver–Burk Plot or Double Reciprocal Plot

Taking the reciprocal of the Michaelis–Menten equation yields the following linear relationship, where a dependent variable (y) is the reciprocal of the metabolic rate and an independent variable (x) is the reciprocal of substrate concentration (Lineweaver and Burk, 1934):

$$\frac{1}{v} = \frac{K_M}{V_{max}} \cdot \frac{1}{C_D} + \frac{1}{V_{max}} \tag{7.30}$$

Thus, a plot of $1/v$ versus $1/C_D$ should yield a straight line as presented in Figure 7.2 with the slope of K_M/V_{max} and y-intercept of $1/V_{max}$.

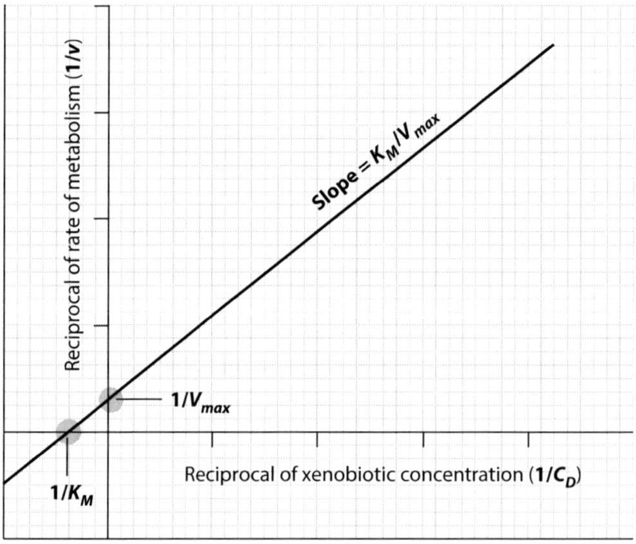

FIGURE 7.2 Plot of reciprocal of metabolism rate versus reciprocal of xenobiotic concentration (Lineweaver–Burk or double reciprocal plot).

Although this method provides the best straight line, it suffers from the distortion of the errors associated with the measurements. The reciprocal of small errors appear as large deviations and the reciprocal of large errors appear as small deviations. Therefore, the estimation of K_M and V_{max} from slope and y-intercept of the unweighted data is likely to contain significant errors. This method is used often to compare inhibition mechanisms.

7.2.3.2 Hanes–Woolfe Plot

Multiplying the Lineweaver–Burk equation by xenobiotic concentration, C_D, provides the following relationship, as presented in Figure 7.3, with C_D/v as the dependent variable (y) and C_D as the independent variable (x):

$$\frac{C_D}{v} = \frac{K_M}{V_{max}} + \frac{C_D}{V_{max}} \tag{7.31}$$

The plot is known as the Hanes plot (Hanes, 1932), or the Hanes–Woolfe plot (Haldane, 1957) with the slope = $1/V_{max}$ and y-intercept = K_M/V_{max}.

The V_{max} estimated from the slope of the line is more precise than the Lineweaver–Burk plot. However, the errors in measurements of metabolic rate (v) contribute to the uncertainty about the precision of y-intercept (K_M/V_{max}).

7.2.3.3 Eadie–Hofstee Plot

Multiplying the Lineweaver–Burk equation by ($V_{max} \times v$) provides a different rearrangement of the Michaelis–Menten equation known as the Eadie–Hofstee plot (Eadie, 1942; Hofstee, 1959)

$$v = V_{max} - K_M \frac{v}{C_D} \tag{7.32}$$

The dependent variable is the rate of metabolism, (v), and the independent variable is the normalized rate of metabolism with respect to the concentration of substrate, (v/C_D). The linear relationship provides a slope of $-K_M$ and y-intercept is V_{max} (Figure 7.4).

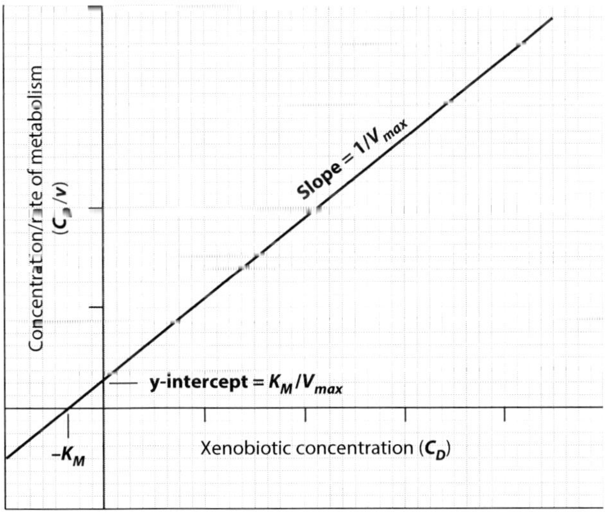

FIGURE 7.3 Plot of xenobiotic concentration normalized with respect to the rate of metabolism versus the concentration (Hanes plot).

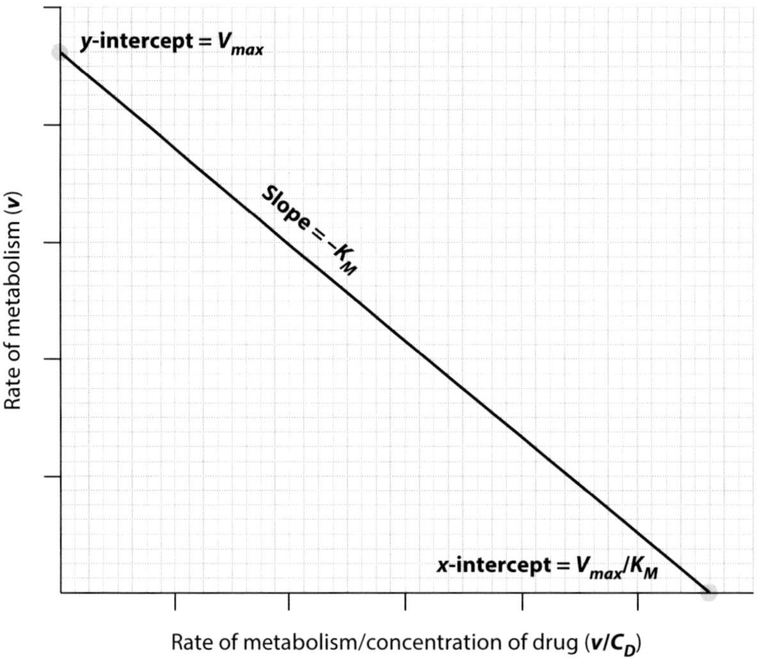

FIGURE 7.4 Plot of rate of metabolism versus rate of metabolism normalized with respect to the concentration of xenobiotic (Eadie–Hofstee plot).

The estimation of K_M and V_{max} by this method is straightforward and are obtained directly from the graph or the regression equation. However, the measurement of metabolic rate, v, a quantity that is subject to more random errors than the xenobiotic concentration, can influence the values of slope and y-intercept.

7.2.3.4 Direct Linear Plot

In this method the metabolic rate (v) is plotted on the y-axis and the corresponding concentration of xenobiotic, (C_D), is multiplied by –1 and plotted on the x-axis. The two points (i.e., $+v$ and $-C_D$) are connected together and extrapolated into a positive quadrant, that is, ($+v$) and ($+C_D$) (Henderson, 1992). Therefore, for each pair of data a line will be generated and extrapolated. The point of intersection of the lines in the positive quadrant will have the coordinates of K_M and V_{max} (Figure 7.5). If the lines form more than one point of intersection, the median of coordinates of intersections is selected for the estimation of the parameters of the Michaelis–Menten equation (Figure 7.6).

In a direct plot no transformation of v or C_D is needed, and the estimation of K_M and V_{max} does not require regression analysis. However, the disadvantage of this method is that if the relationship between the rate of metabolism and the xenobiotic concentration depart from the simple Michaelis–Menten hyperbolic relationship, it would not be apparent in the direct plot.

In general, the direct plot followed by the Hanes plot are preferred for estimation of V_{max} and K_M.

7.2.3.5 Hill Plot

Often the relationship between the rate of metabolism (v) and the concentration of xenobiotic (C_D) is sigmoidal and does not exhibit the classical rectangular hyperbolic curve of the Michaelis–Menten kinetics. This may indicate that the enzyme responsible for the metabolism has more than one binding site for the compound and the binding of one molecule may facilitate (positive co-operativity) or delay (negative co-operativity) the binding of other molecules. Under this condition,

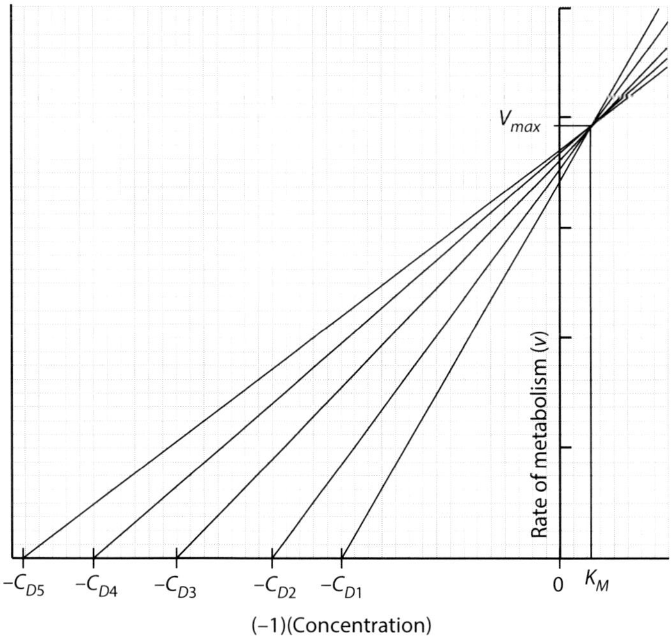

FIGURE 7.5 Direct plot with one point of intersection.

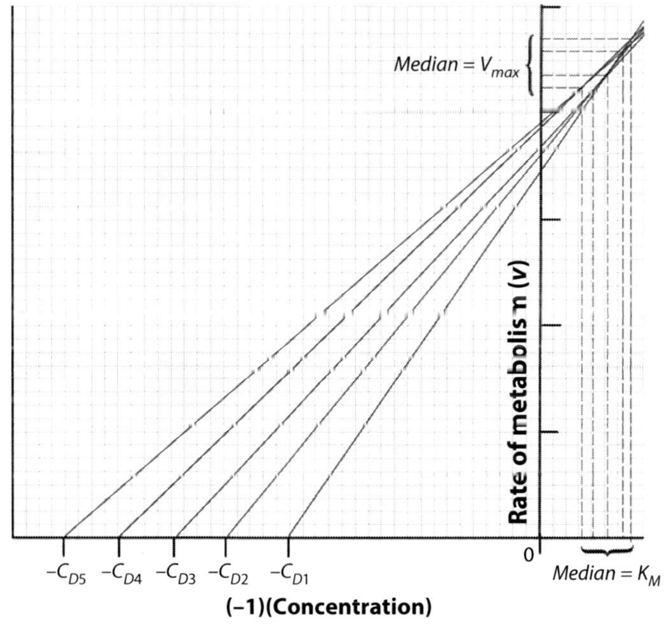

FIGURE 7.6 Direct plot with more than one point of intersection.

the rate of metabolism is governed by the following relationship known as the Hill equation (Hill, 1910):

$$v = \frac{V_{max} \times [C_D]^n}{K + [C_D]^n} \tag{7.33}$$

where K and n are known as the Hill coefficients.

Cross multiplication of Equation 7.33, solving for K and substituting back into the equation, and taking logarithm of both sides yields the linear relationship of Equation 7.36

$$V_{max}[C_D]^n = vK + v[C_D]^n \tag{7.34}$$

$$K = \frac{[C_D]^n (V_{max} - v)}{v} \tag{7.35}$$

$$\log \frac{v}{V_{max} - v} = n \log[C_D] - \log K \tag{7.36}$$

Thus, a plot of log $v/(V_{max} - v)$ versus log $[C_D]$ yields a straight line with slope = n and the y-intercept = log K. Determination of K and n requires a known value for V_{max}. Setting $v = 1/2V_{max}$ would make $v/(V_{max} - v)$ equal to one and log $v/(V_{max} - v)$ equal to zero. Therefore, the zero of the y-axis corresponds to the logarithm of the xenobiotic concentration log $[C_D]$ on the x-axis, which yields a rate of metabolism equal to $1/2V_{max}$ (Figure 7.7).

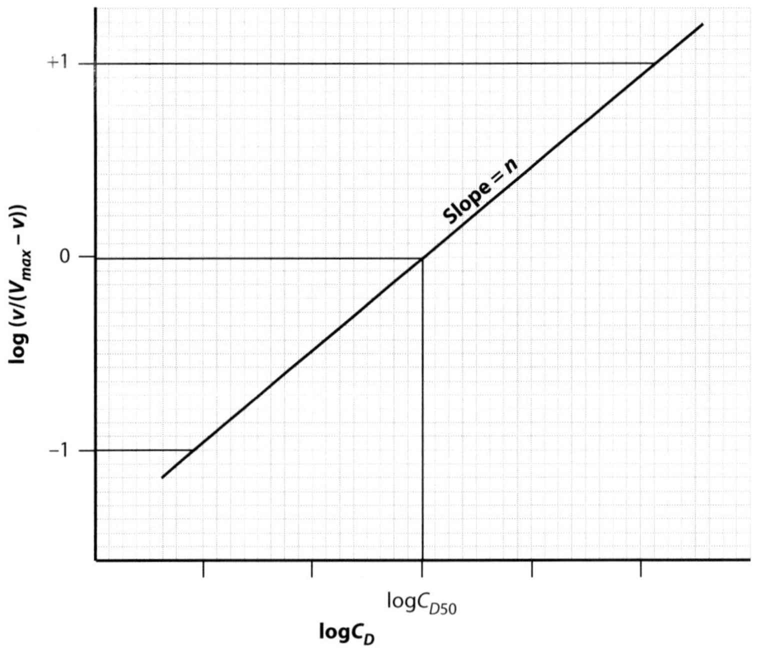

FIGURE 7.7 Hill plot.

$$\log \frac{v}{V_{max} - v} = 0$$
$$n \log C_{D50} = \log K \tag{7.37}$$
$$\therefore K = C_{D50}^{n}$$

7.3 ASSIMILATION OF INTRINSIC CLEARANCE IN HEPATIC CLEARANCE USING LIVER MODELS

There are theoretical models that can integrate an intrinsic clearance in the calculation of the hepatic clearance (i.e., organ clearance). These models were discussed in Chapter 4, Section 4.4.1. Briefly, they are

- Well-stirred model, also known as venous equilibration model
- Parallel-tube model, also known as undistributed sinusoidal model
- Dispersion model
- Zonal liver model

The assumptions associated with the development and use of the models are

- The distribution of xenobiotics in the liver is perfusion-limited distribution (Chapter 3, Section 3.2.2.1).
- The diffusion of the compounds into the hepatocytes is instantaneous with no diffusional impediment.
- The enzyme interacts only with the free and unbound compound.
- The interaction of the enzyme with the compound is also instantaneous with no threshold.

7.3.1 The Well-Stirred Model (or Venous Equilibration Model)

The model is discussed in Chapter 4 Section 4.4.1.1 in more detail (Gillette, 1971; Rowland et al., 1973; Wilkinson Shand, 1975; Pang and Rowland, 1977; Benet, 2010). The relationship relevant to this section is

$$Cl_H = \frac{Q_H \times Cl_{int} \times f_u}{Q_H + Cl_{int} \times f_u} \tag{7.38}$$

where (Cl_H) is the hepatic clearance, Q_H is hepatic blood flow, Cl_{int} is the intrinsic metabolic clearance and f_u is the free fraction of compound in the blood. As the names of the model imply, the assumptions are (1) the liver acts as a single homogeneous and well-stirred organ, and (2) the concentration of the unbound compound in venous blood leaving the organ is in equilibrium with unbound concentration in the hepatic tissue. Equation 7.38 integrates the intrinsic clearance, estimated by the *in vitro* methodologies, into the estimation of organ clearance using physiological parameters of blood flow and protein binding. The intrinsic clearance can also be used to estimate the extraction ratio of a compound by the liver:

$$ER = \frac{f_u \times Cl_{int}}{Q + f_u \times Cl_{int}} \tag{7.39}$$

7.3.2 The Parallel-Tube Model (or Undistributed Sinusoidal Model)

The model considers the liver a series of parallel tubes or sinusoids, with a concentration gradient present along the theoretical tubes with the irreversible removal of compound by the metabolic process. (Bass et al., 1976; Bass and Bracken, 1977, 1978; Pang and Rowland, 1977; Pang and Gillette,

1978; Bass, 1979; Bracken and Bass, 1979; Lau et al., 2002; Ito and Houston, 2004). The metabolic process occurs in the hepatocytes that encircle each tube, through which blood containing xenobiotic flows at a constant rate. The concentration of the compound along the tubes declines exponentially. The equation of the model that incorporates the intrinsic clearance (Chapter 4, Equation 4.27) is

$$Cl_H = Q_H\left(1 - e^{-\left[\frac{f_u \times Cl_{int}}{Q_H}\right]}\right) \tag{7.40}$$

Comparison of Equation 7.40 with Equation 7.41 identifies the hepatic extraction ratio in terms of intrinsic clearance (Equation 7.42):

$$Cl_H = \frac{Q(C_{in} - C_{out})}{C_{in}} = Q(ER) \tag{7.41}$$

$$ER = \left(1 - e^{-\left[\frac{f_u \times Cl_{int}}{Q_H}\right]}\right) \tag{7.42}$$

7.3.3 The Dispersion Model

As discussed in Chapter 4 (Section 4.4.1.3), the dispersion model is a more complex model for defining the liver and integrating the intrinsic clearance with the hepatic clearance. Its basic principles are from the chemical engineering nonideal flow longitudinal dispersion model. There are two important parameters in this model: the efficiency number and dispersion numbers (Wehner and Wilhelm, 1956; Roberts and Rowland, 1986).

The efficiency number (R_n) reflects the irreversible removal of xenobiotic under first-order condition from the blood by the liver

$$R_n = \frac{f_u Cl_{int}}{Q_H} \tag{7.43}$$

If the number of compounds present in the system is more than one, for example the parent compound plus two primary metabolites, the equation changes to

$$R_n = \frac{(f_u)_i (Cl_{int})_i \rho_i}{Q_H} \tag{7.44}$$

where (i) represents the ith compound and ρ_i is the transmembrane clearance of the ith compound defined in Chapter 4, Equation 4.31, that is, $\rho = P_{coeff}/(P_{coef} + Cl_{int})$.

The axial dispersion number, D_n, is a flow independent parameter representing the stochastic measure of dispersion for a xenobiotic during the time it passes through the liver. According to this model, the dispersion in the liver is due to the nonideal blood flow through a maze of interconnected sinusoids, which causes variations in transit time of the drug through the liver. A common value of 0.17 has been recommended for this parameter.

The formula for hepatic clearance according to the dispersion model is

$$Cl_H = Q_H\left[1 - \frac{4a}{(1+a)^2 e^{[(a-1)/2D_n]} - (1-a)^2 e^{[-(a-1)/2D_n]}}\right] \tag{7.45}$$

where

$$a = (1 + 4R_n D_n)^{1/2} \tag{7.46}$$

When the axial dispersion number, D_n, approaches zero, Equation 7.45 changes to Equation 7.47, which is the same as Equation 7.40:

$$Cl_H = Q_H(1 - e^{-R_n}) \tag{7.47}$$

On the other hand, if the number approaches infinity (i.e., $D_n \rightarrow \infty$), Equation 7.45 reduces to Equation 7.48 after substituting Equation 7.44 for R_n. Equation 7.48 defines the hepatic clearance as

$$Cl_H = \frac{R_n}{1 + R_n} \tag{7.48}$$

It should be noted that no mathematical model could comprehensively delineate the physiological complexity of the liver and its numerous roles in the body. The above theoretical models may seem abstract, but they can approximate the hepatic clearance based on the accessible data. A modified version of the dispersion model known as convection-dispersion model has also been introduced that describes the disposition of the extracted solute, hepatic clearance and bioavailability in both experimental animals and clinical situations (Roberts and Anissimov, 1999).

7.3.4 Physiologically Based Pharmacokinetic Model for the Liver

Based on the principles of physiological modeling, which include blood flow, binding to plasma proteins and blood cells, and other important elements of the liver such as bile flow and functions of enzymes and efflux proteins, physiological organ models are proposed for refinement of interrelationships between organ clearance and intrinsic clearance (Figure 7.8). Models similar to the liver model (Pang and Rowland, 1977; Sun et al., 2006; Pang et al., 2007) are also proposed for other organs such as the kidney (de Lannoy et al., 1990).

The model depicted in Figure 7.8 is an example of an organ-related, physiologically based model for the liver. The following relationships are developed based on the models that demonstrate the relationship between the organ clearance and intrinsic clearance of the liver (Pang et al., 2007):

$$(Cl_m)_{hepatic} = \frac{Q_{liver} \times f_u \times Cl_{influx} \times Cl_{int}}{Q_{liver}(Cl_{efflux} + Cl_{int} + Cl_{sec}) + f_u \times Cl_{influx}(Cl_{in} + Cl_{sec})} \tag{7.49}$$

$$(Cl)_{biliary} = \frac{Q_{liver} \times f_u \times Cl_{influx} \times Cl_{sec}}{Q_{liver}(Cl_{efflux} + Cl_{int} + Cl_{sec}) + f_u \times Cl_{influx}(Cl_{in} + Cl_{sec})} \tag{7.50}$$

$$(Cl_{organ})_{liver} = \frac{Q_{liver} \times f_u \times Cl_{influx}(Cl_{int} + Cl_{sec})}{Q_{liver}(Cl_{efflux} + Cl_{int} + Cl_{sec}) + f_u \times Cl_{influx}(Cl_{int} + Cl_{sec})} \tag{7.51}$$

where Q_{liver} is the blood flow through the liver with the unbound fraction of xenobiotic f_u; Cl_{influx} and Cl_{int} are influx and intrinsic metabolic clearances, respectively; Cl_{efflux} and Cl_{sec} are the efflux and intrinsic secretion clearances, respectively. Equation 7.51 also demonstrates that the total hepatic

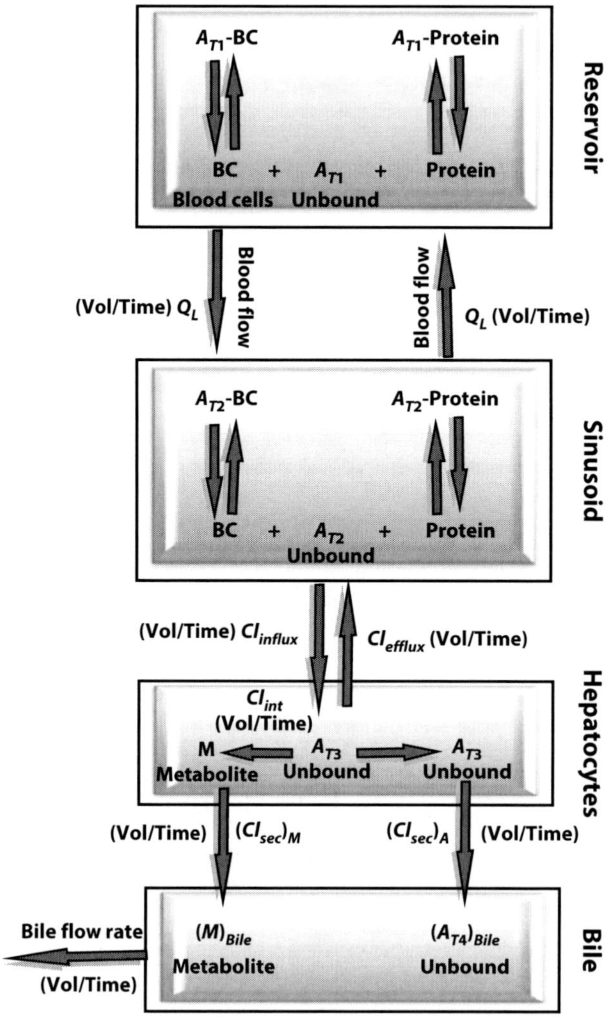

FIGURE 7.8 Schematic diagram of a physiologically based organ model for conceptualizing the interrelationship between hepatic clearance and intrinsic clearance.

clearance that is, $(Cl_{organ})_{liver}$ is the summation of biliary elimination $(Cl)_{biliary}$ and the hepatic metabolic clearance $(Cl_m)_{hepatic}$ with intrinsic metabolic clearance incorporated in all clearance terms.

When using hepatocytes incubation, the extrapolations of intrinsic metabolic clearance to hepatic clearance can be achieved through the use of Equation 7.52

$$Cl_{medium} = Cl_{\text{int}} \frac{Cl_{passive} + Cl_{active}}{Cl_{passive} + Cl_{efflux} + Cl_{\text{int}}} \tag{7.52}$$

In addition, substituting Equation 7.49 into Equation 7.38 also yields the equation of hepatic clearance based on the well-stirred model (Equation 7.50)

$$Cl_H = \frac{Q_H \times f_u \times Cl_{int}(Cl_{passive} + Cl_{active}/Cl_{passive} + Cl_{efflux} + Cl_{int})}{Q_H + f_u \times Cl_{int}(Cl_{passive} + Cl_{active}/Cl_{passive} + Cl_{efflux} + Cl_{int})} \tag{7.53}$$

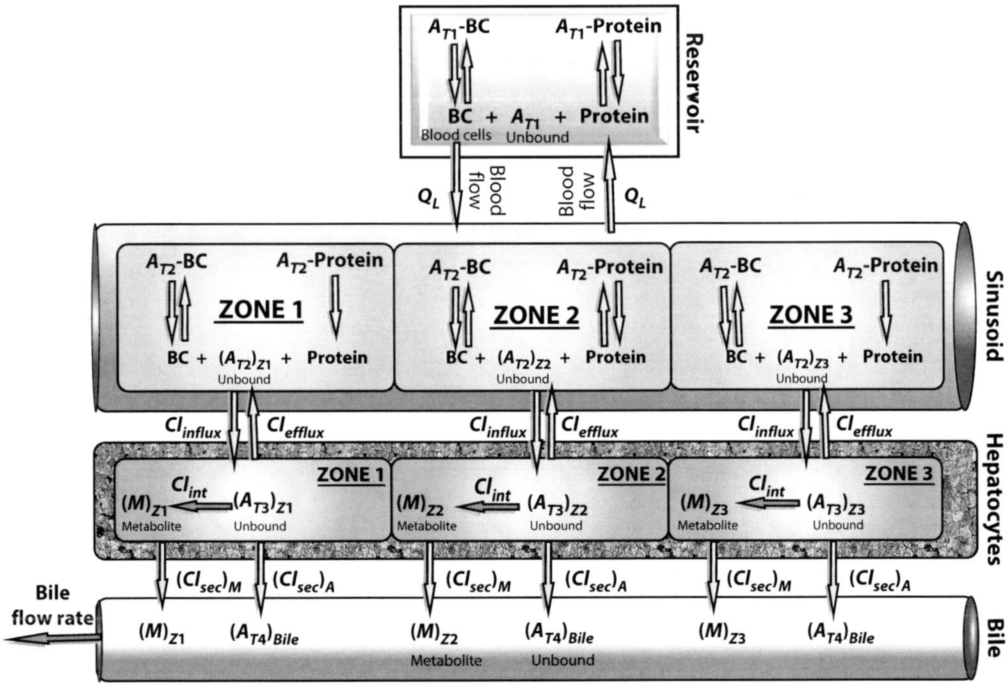

FIGURE 7.9 Zonal liver model with sub-compartments and differentiation of the intrinsic metabolism by different hepatic zones.

7.3.5 Zonal Liver Model

As noted in Chapter 6, Section 6.2, the liver has the most complex histochemical heterogeneity and physiological functions in the body. Its complex roles combined with the diversity of its transport proteins, efflux proteins, receptors, and enzyme systems provide challenges in developing relevant theoretical and applicable models for hepatic metabolism of xenobiotics. An approach for refinement of the liver models is based on inclusion of metabolic zonation of the liver in the model (Knapp et al., 1988; Homma et al., 1997; Oinonen and Lindros, 1998; Tirona and Pang, 1999; Abu-Zahra et al., 2000; Pang et al., 2007). Depending upon the involvement of the periportal, midzonal, and prevenous zones (Chapter 6, Figure 6.1) in the metabolism of a compound, each zone is included as sub-compartments of the liver model with their specific intrinsic hepatic clearances (Figure 7.9). Contrary to well-stirred, parallel, and dispersion models, the zonal liver model is compound-specific and is developed based on the compound's intrinsic metabolism by different hepatic zones. Furthermore, the model takes into consideration the intrinsic influx and efflux of the parent compounds and their metabolites by the transport and efflux proteins of hepatocytes and sinusoid. The specific information related to the chemical structure of a xenobiotic—specific function of the relevant enzyme systems, transporters and efflux proteins acting on the compound in a specific zone—are considered in this model when it is developed for a given compound.

7.4 INHIBITION OF XENOBIOTIC METABOLISM

Metabolic induction or inhibition of a compound by another is one of the main reasons for potentiating or inhibiting pharmacological or toxicological response. Other reasons are influence of one compound on the protein binding of the other, and/or modulation/inhibition of transport and efflux proteins, such as Pgp, OAT, OCT, etc. These interactions can modify the *in vitro* parameters of metabolism, the *in vivo* pharmacokinetic profile, and the therapeutic or toxic response. For therapeutic agents the interactions occurs mainly in the course of chronic combination therapy with different agents, drug–drug interactions.

In the case of chronic exposure to environmental/industrial chemicals, the interactions can potentiate or inhibit toxic response, promote carcinogenicity, or cause death.

7.4.1 Inhibition Classifications of Xenobiotic Metabolism

The various scenarios for inhibition of metabolic reactions are competitive, simple non-competitive, uncompetitive, mixed non-competitive and suicide inhibitions, which are discussed in the following sections. It should be noted that the inhibition of efflux transporters such as Pgp follow the same general rules of inhibition. The inhibition of efflux proteins and/or metabolic enzymes can influence drug disposition, efficacy, and safety. The inhibition can be beneficial by increasing the extent of absorption, or to reverse multidrug resistance, and it can be harmful by elevating the concentration of free compound to an unexpected toxic levels.

7.4.1.1 Competitive Inhibition

In this scenario, the inhibitor (i.e., the second compound, which closely resembles to the chemical structure and molecular geometry of the first compound) has a strong affinity for the same binding site of the first compound, and if it were present in a concentration comparable to the first compound, it would compete for binding to the same site. It is a competitive process between the first and second compound. The enzyme can interact with only one compound, but not both at the same time, and none of the compounds can occupy all binding sites completely. Thus, during the metabolism there will be enzyme-compound and enzyme-inhibitor complexes. The related reaction is

$$\text{Enzyme } [E] + \text{Compound } [C_D] \underset{k_2}{\overset{k_1}{\rightleftarrows}} \text{Enzyme-compound} [E.C_D] \xrightarrow{k_3} \text{Metabolite} [M]$$
$$+ \text{ Enzyme } [E] + \text{ Inhibitor } [I] \overset{K_i}{\rightleftarrows} \text{Enzyme-inhibitor } [E.I] \tag{7.54}$$

According to the above reaction, the inhibitor binds only to an enzyme, not to the $[E.C_D]$ complex. Therefore, the total free enzyme at a given time is

$$[E] = [E_0] - ([E.C_D] + [E.I]) \tag{7.55}$$

Applying the stepwise approach in development of the Michaelis–Menten relationship (Equations 7.2 through 7.14), the rate of metabolism in the presence of inhibitor is

$$v = \frac{V_{max} \times C_D}{K_M(1 + I/K_i) + C_D} \tag{7.56}$$

The double reciprocal (Lineweaver–Burk plot) of Equation 7.56 is

$$\frac{1}{v} = \frac{K_M}{V_{max}}\left(1 + \frac{I}{K_i}\right)\frac{1}{C_D} + \frac{1}{V_{max}} \tag{7.57}$$

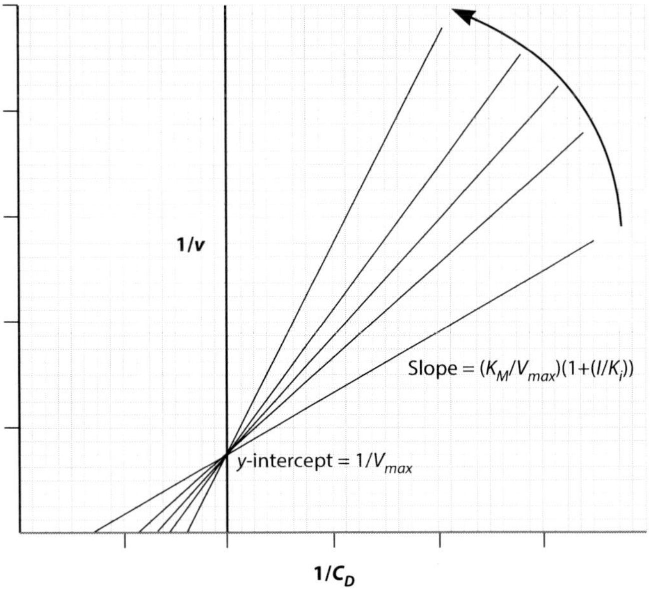

FIGURE 7.10 Double reciprocal plot of metabolism rate in the presence of a competitive inhibitor.

According to Equation 7.57, a plot of $1/v$ versus $1/C_D$ yields a straight line with the following parameters (Figure 7.10):

$$y\text{-intercept} = \frac{1}{V_{max}} \tag{7.58}$$

$$x\text{-intercept} = \frac{-1}{K_M(1 + (I/K_i))} \tag{7.59}$$

$$\text{Slope} = \left(\frac{K_M}{V_{max}}\right)\left(1 + \frac{I}{K_i}\right) \tag{7.60}$$

where (I) is the concentration of the inhibitor and K_i is the dissociation constant of the enzyme-inhibitor complex and is estimated as

$$K_i - \frac{[E][I]}{[E.I]} \tag{7.61}$$

In competitive inhibition, V_{max} and y-intercept are constant for an enzyme, but the x-intercept is a variable and changes with the concentration of inhibitor.

7.4.1.2 Noncompetitive Inhibition

Noncompetitive inhibition occurs when the second compound, the inhibitor, has an affinity for interaction with a different binding site on the same enzyme. Thus, there is no competition between the two compounds for interaction with the same binding site, and the inhibitor can interact not only with the enzyme [E] but also with enzyme-drug complex [$E.C_D$], that is,

$$
\begin{array}{lll}
[E]+[C_D] \underset{k_2}{\overset{k_1}{\rightleftharpoons}} [E.C_D] \xrightarrow{k_3} [M]+[E] \\
+ \qquad\qquad\quad + \\
[I] \qquad\qquad\quad [I] \\
\uparrow\downarrow K_i \qquad\quad\; \uparrow\downarrow K_i \\
[E.I]+[C_D] \rightleftharpoons [E.C_D.I]
\end{array}
\tag{7.62}
$$

The concentration of the enzyme-substrate-inhibitor complex is [$E.C_D.I$]. The interaction of the inhibitor with the enzyme results in energy requiring a conformational change of the protein, which may influence the binding of the compound to its binding site. The assumption here is that the affinity of the inhibitor to interact with an enzyme and enzyme-substrate complex is the same and the extent of inhibition is independent of xenobiotic concentration. The rate of metabolism with a noncompetitive inhibition is

$$v = \frac{\left[V_{max} \div (1 + (I/K_i))\right] \times C_D}{K_M + C_D} \tag{7.63}$$

The linear form of Equation 7.63 according to the Lineweaver–Burk plot is

$$\frac{1}{v} = \frac{K_M}{V_{max}}\left(1 + \frac{I}{K_i}\right)\frac{1}{C_D} + \frac{1}{V_{max}}\left(1 + \frac{I}{K_i}\right) \tag{7.64}$$

The plot of $1/v$ versus $1/C_D$ yields a straight line with the following parameters (Figure 7.11):

$$\text{Slope} = \frac{K_M}{V_{max}}\left(1 + \frac{I}{K_i}\right) \tag{7.65}$$

$$y\text{-intercept} = \frac{1}{V_{max}}\left(1 + \frac{I}{K_i}\right) \tag{7.66}$$

$$x\text{-intercept} = -\frac{1}{K_M} \tag{7.67}$$

In noncompetitive inhibition, the value of K_M remains constant; the inhibitor influences k_3 and slows down the metabolism and V_{max} changes to

$$(V_{max})_{noncompetitive} = V_{max}/(1 + I/K_i) \tag{7.68}$$

7.4.1.3 Uncompetitive Inhibition

The uncompetitive inhibition occurs when the inhibitor does not bind to the free enzyme directly, but interacts with the enzyme after the formation of an enzyme-xenobiotic complex. In other words, due to the interaction of the compound with the enzyme and resulting conformational change of the

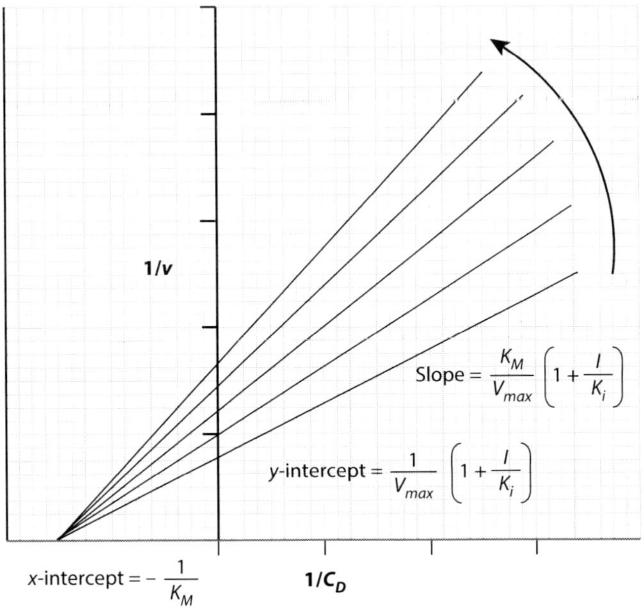

FIGURE 7.11 Double reciprocal plot of metabolism rate in the presence of a noncompetitive inhibitor.

protein, the binding site(s) for the inhibitor will be exposed and available for the binding of inhibitor. The binding of the inhibitor yields an inactive $[E.C_D.I]$ complex and prevents the formation of a metabolite and further binding of substrate to the enzyme

$$\begin{array}{c} [E]+[C_D]\underset{k_2}{\overset{k_1}{\rightleftarrows}}[E.C_D]\xrightarrow{k_3}[M]+[E] \\ + \\ [I] \\ \uparrow\downarrow K_i \\ [E.C_D.I] \end{array} \tag{7.69}$$

The rate of metabolism and its double-reciprocal linear relationship are

$$v = \frac{[V_{max} \div (1+(I/K_i))] \times C_D}{K_M \div (1+(I/K_i)) + C_D} \tag{7.70}$$

$$\frac{1}{v} = \frac{K_M}{V_{max}}\frac{1}{C_D} + \frac{1}{V_{max}}\left(1+\frac{I}{K_i}\right) \tag{7.71}$$

The parameters of the linear plot are

$$\text{Slope} = \frac{K_M}{V_{max}} \tag{7.72}$$

$$y\text{-intercept} = \frac{1}{V_{max}}\left(1+\frac{I}{K_i}\right) \tag{7.73}$$

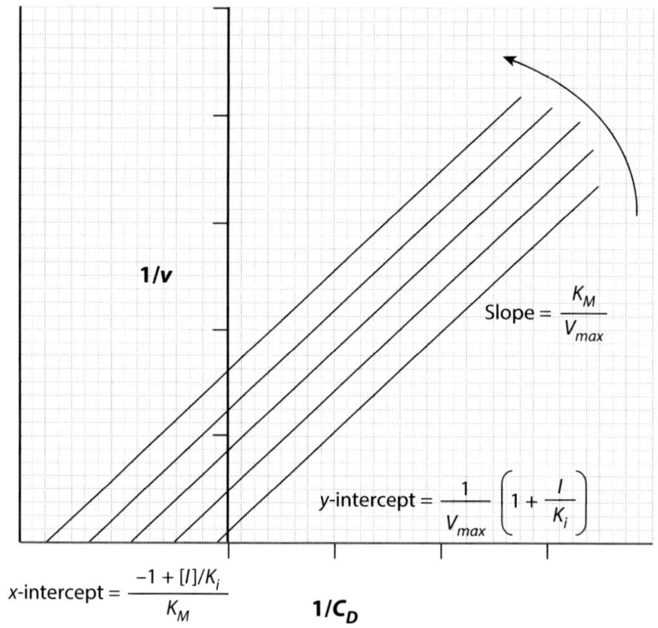

FIGURE 7.12 Double reciprocal plot of metabolism rate in the presence of a noncompetitive inhibitor.

$$x\text{-intercept} = \frac{-1+[I]/K_i}{K_M} \tag{7.74}$$

In this type of inhibition, both K_M and V_{max} are changed (Figure 7.12)

$$(K_M)_{uncompetitive} = K_M/(1+[I]/K_i) \tag{7.75}$$

$$(V_{max})_{uncompetitive} = V_{max}/(1+[I]/K_i) \tag{7.76}$$

7.4.1.4 Mixed Noncompetitive Inhibition

The mixed type inhibition is similar to noncompetitive inhibition. The main difference is that the binding of inhibitor to the enzyme reduces the binding of the compound and vice versa. Thus, the binding of the compound decreases when the inhibitor is present. The related scheme is

$$\begin{array}{lll} [E]+[C_D] \xrightleftharpoons{K_D} [E.C_D] \xrightarrow{k_3} [M]+[E] \\ + \qquad\qquad\qquad + \\ [I] \qquad\qquad\quad [I] \\ \uparrow\downarrow K_i \qquad\quad \uparrow\downarrow \bar{K}_i \\ [E.I]+[C_D] \xrightleftharpoons{\bar{K}_D} [E.C_D.I] \end{array} \tag{7.77}$$

The rate of metabolism for this type of inhibition is

$$v = \frac{V_{max} \times C_D}{K_D(1+(I/K_i)) + C_D(1+(I/\bar{K}_i))} \tag{7.78}$$

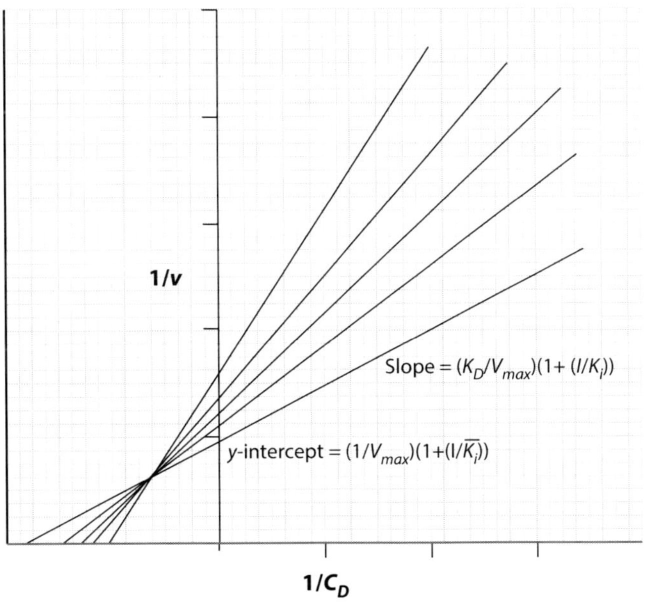

FIGURE 7.13 Double reciprocal plot of metabolism rate in the presence of a mixed noncompetitive inhibitor.

The linear conversion of Equation 7.78 is

$$\frac{1}{v} = \underbrace{\frac{K_D}{V_{max}}(1 + (I/K_i))}_{\text{Slope}}\frac{1}{C_D} + \underbrace{\frac{1}{V_{max}}(1 + (I/\bar{K}_i))}_{y\text{-intercept}} \tag{7.79}$$

When $K_i \neq \bar{K}_i$, K_M and V_{max} are

$$K_M = \left(K_D + \frac{K_D \times I}{K_i}\right) \div \left(1 + \frac{I}{\bar{K}_i}\right) \tag{7.80}$$

$$V_{max} = \frac{V_{max}}{1 + (I/\bar{K}i)} \tag{7.81}$$

The quandary here is how to estimate the parameters explicitly from the double reciprocal plot (Figure 7.13). To overcome this dilemma, most often the slope and y-intercept of the line is determined at different concentrations of the inhibitor and the values of slope and intercept are then plotted against inhibitor concentrations. The x-intercept of plot of slope values versus the concentrations of the inhibitor is K_i and the plot x-intercept of y-intercept values versus the concentration of the inhibitor is $\bar{K}_i$.

7.4.1.5 Suicide Inhibition

When an inhibitor with a structure similar to the substrate interacts with a functional group of the binding site covalently, that is, irreversibly, it is called suicide inhibition. This type of interaction disables the enzyme from binding to the compound. Thus, the metabolism is reduced merely because of the reduction of free enzyme. The rate and double reciprocal equations and plots resemble the noncompetitive inhibition.

$$\begin{array}{l} [E]+[C_D] \underset{k2}{\overset{k1}{\rightleftarrows}} [E.C_D] \xrightarrow{k3} [M]+[E] \\ + \\ SI \xrightarrow{irreversible} \boxed{E.SI} \end{array} \quad (7.82)$$

where SI is the suicide inhibitor and $E.SI$ is the inactive enzyme.

7.4.1.6 Bimolecular Enzymes

There are enzymes that can interact with two different compounds and form two different metabolites. The interaction can occur in an orderly manner or randomly mode. The orderly interaction is when the enzyme interacts with both compounds according to the following formula:

$$[E]+[C_{DA}]+[C_{DB}] \underset{k_2}{\overset{k_1}{\rightleftarrows}} [E.C_{DA}.C_{DB}] \underset{k_4}{\overset{k_3}{\rightleftarrows}} [E.M_A.M_B] \underset{k_6}{\overset{k_5}{\rightleftarrows}} E + M_A + M_B \quad (7.83)$$

where C_{DA} and M_A are compound A and its metabolite, and C_{DB} and M_B are compound B and its metabolite, respectively. All steps of the interactions are reversible and sequential. This interaction is also known as single displacement mechanism, both compounds should interact with the enzyme before any metabolites are released.

The interaction can also occur as a double displacement reaction known as the ping-pong mechanism in which the interaction of the first compound modifies the enzyme structurally; the modified enzyme will then interact with the second compound

$$\begin{array}{l} [E]+[C_{DA}] \underset{k_2}{\overset{k_1}{\rightleftarrows}} [E.C_{DA}] \underset{k_4}{\overset{k_3}{\rightleftarrows}} [E^*.M_A] \underset{k_6}{\overset{k_5}{\rightleftarrows}} \boxed{E^*}+[M_A] \\ \boxed{E^*}+[C_{DB}] \underset{k_8}{\overset{k_7}{\rightleftarrows}} [E^*.C_{DB}] \underset{k_{10}}{\overset{k_9}{\rightleftarrows}} [E.M_B] \underset{k_{12}}{\overset{k_{11}}{\rightleftarrows}} [E]+[M_B] \end{array} \quad (7.84)$$

When the interaction is uncompetitive, the rate equation is

$$v = \frac{V_{max}[C_{DA}][C_{DB}]}{K_A[C_{DB}]+K_B[C_{DA}]+[C_{DA}][C_{DB}]} \quad (7.85)$$

When the interaction is a mixed one, the rate is

$$v = \frac{V_{max}[C_{DA}][C_{DB}]}{[C_{DA}][C_{DB}]+K_A[C_{DB}]+K_B[C_{DA}]+K_AK_B} \quad (7.86)$$

The single displacement kinetics can be differentiated from ping-pong by measuring the rate of the reaction at a different concentration of one compound while keeping the concentration of the second compound constant. Then, repeating the same experiment by changing the concentration of the second compound while keeping the concentration of the first compound constant. For the

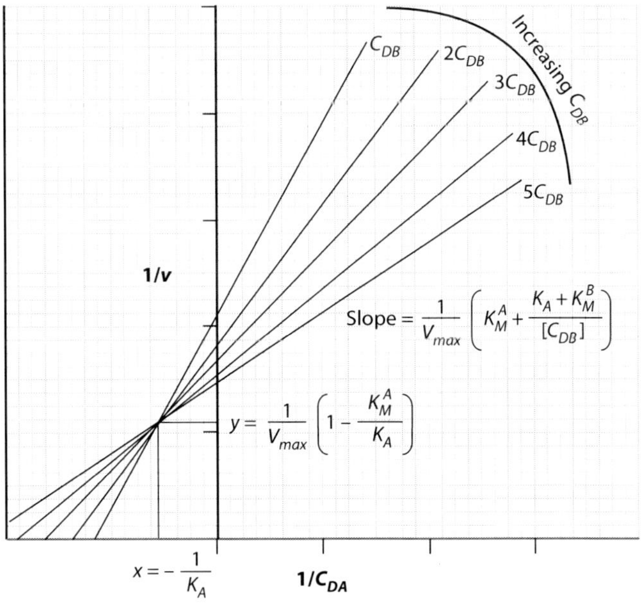

FIGURE 7.14 Double reciprocal plot of metabolism rate for bimolecular reactions.

single displacement (i.e., sequential interaction), the Lineweaver–Burk plot of the data generates a series of straight lines that interconnect to the left of y-axis (Figure 7.14).

The Lineweaver–Burk equation for the plot of $1/v$ versus $1/C_{DA}$ (i.e., when $[C_{DA}]$ is kept constant and $[C_{DB}]$ is changed) is as follows:

$$\frac{1}{v} = \frac{1}{V_{max}}\left(K_M^A + \frac{K_A + K_M^B}{[C_{DB}]}\right)\left(\frac{1}{[C_{DA}]} + \frac{1}{V_{max}}\left(1 + \frac{K_M^B}{[C_{DB}]}\right)\right) \quad (7.87)$$

where

$$\text{Slope} = \frac{1}{V_{max}}\left(K_M^A + \frac{K_A + K_M^B}{[C_{DB}]}\right) \quad (7.88)$$

The coordinates of the point of interconnection of the lines on the left side are

$$x = -\frac{1}{K_A} \text{ and } y = \frac{1}{V_{max}}\left(1 - \frac{K_M^A}{K_A}\right) \quad (7.89)$$

When the binding of $[C_{DB}]$ is independent of $[C_{DA}]$, the double reciprocal lines of $1/v$ versus $1/C_{DA}$ meet on the x-axis with the x-intercept of $-(1/K_M^A)$.

In random interaction, that is, random sequential, there is no preference for binding of either compound to the enzyme and the interaction follows the scheme presented in Figure 7.15, as compared to ordered sequential.

The rate and double reciprocal equations for the random sequential are similar to the ordered sequential and it is difficult to differentiate between the random and ordered sequential by using the linear plot.

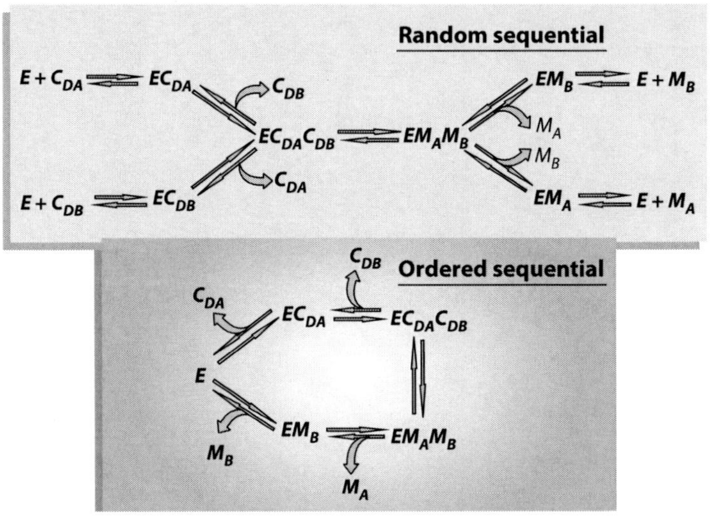

FIGURE 7.15 Comparison of random-sequential and ordered-sequential for a bimolecular interaction.

7.4.2 Product Inhibition

The product inhibition occurs when the metabolite has a chemical structure similar to the parent compound and interacts with the enzyme at the binding site of the parent compound. Thus, the metabolite acts as an inhibitor and competes with the parent compound for binding to the free enzyme.

$$\begin{array}{l} [E]+[C_D] \rightleftharpoons [E.C_D] \longrightarrow [M]+[E] \\ + \\ [M] \rightleftharpoons [E.M] \end{array} \tag{7.90}$$

At low concentrations of substrate, the interaction resembles the competitive inhibition. However, as the concentration of the parent compound increases, more metabolite is formed and the rate of metabolism declines and the Lineweaver–Burk plot loses its linear characteristics.

7.5 INDUCTION OF XENOBIOTIC METABOLISM

The induction of enzyme systems by xenobiotics can occur through two general mechanisms: stabilization of the enzyme or mRNA, and/or augmented gene transcription. The induction increases the amount of enzyme, elevates its metabolic activity, amplifies its intrinsic metabolic clearance and ultimately enhances the total body clearance of its substrates. The induction can have a profound effect on the PK/TK parameters and constants of xenobiotics both *in vitro* and *in vivo*. The *in vitro* parameters including the amount of enzyme available for interaction with a substrate, the metabolic parameters of V_{max} and K_M, and the intrinsic clearance may increase due to the induction. The *in vivo* PK/TK constants, such as the metabolic rate constant, the overall elimination rate constant, the total body clearance, and half-life can increase significantly. Whereas, in PK/TK parameters, such as the area under plasma concentration–time curve (AUC), maximum plasma concentration, and bioavailability decrease accordingly. Consequently, the induction of the enzyme responsible for metabolism of a compound lowers its optimum plasma concentration, reduces its duration of action, and thus, influences negatively on the pharmacological response and its therapeutic outcome. An example of influence of enzyme induction on PK/TK parameters is the co-administration of St. John's Wort, a known inducer of CYP3A4, with midazolam, a substrate of the enzyme. The co-administration decreases the maximum plasma concentration by 65% and AUC by 79% (Mueller et al., 2006). This scenario is an example of having two compounds, one being the substrate and the other inducer. There are compounds, for example, carbamazepine, which induce the enzyme system responsible for their metabolism and exhibit autoinduction (Bertilsson et al., 1980; Simonsson et al., 2003).

The induction and inhibition of enzymes, protein transporters, and efflux proteins are the main reasons for drug–drug interaction. In drug discovery and development, The assessment of lead compound metabolism by the following members of the CYP450 subfamilies are recommended: all members of the CYP3A subfamily and CYP1A2, CYP2B6, CYP2C8, CYP2C9, CYP2C19, and CYP2D6 (Zhang et al., 2009).

REFERENCES

Abu-Zahra, T. N., Wolkoff, A. W., Kim, R. B., Pang, K. S. 2000. Uptake of enalapril and expression of organic anion transporting polypeptide 1 in zonal, isolated rat hepatocytes. *Drug Metab Dispos* 28:801–6.

Bass, L. 1979. Current models of hepatic elimination. *Gastroenteroloa* 76:1504–5.

Bass, L., Bracken, A. J. 1977. Time-dependent elimination of substrates flowing through the liver or kidney. *J Theor Biol* 67:637–52.

Bass, L., Keiding, E. S., Winkler, K., Tygstrup, N. 1976. Enzymatic elimination of substrates flowing through the intact liver. *J Theor Biol* 61:393–409.

Bass, L., Robinson, P., Bracken, A. J. 1978. Hepatic elimination of flowing substrates: The distributed model. *J Theor Biol* 72:161–84.

Benet, L. Z. 2010. Clearance (née Rowland) concepts: A downdate and an update. *J. Pharmacokinet Phar* 37(6):529–39.

Benet, L. Z., Cummins, C. L., Wu, C. Y. 2003. Transporter-enzyme interactions: Implications for predicting drug-drug interactions from *in vitro* data. *Curr Drug Metab* 4:393–8.

Bertilsson, L., Hojer, B., Tybring, G., Osterloh, J., Rane. A. 1980. Autoinduction of carbamazepine metabolism in children examined by a stable isotope technique. *Clin Pharmacol Ther* 27(1):83–8.

Bracken, A. J., Bass, L. 1979. Statistical mechanics of hepatic elimination. *Mathl Bi Sci* 44:97–120.

de Lannoy I. A. M., Hirayama, H., Pang, K. S. 1990. A physiological model for renal drug metabolism: Enalapril esterolysis to enalaprilat in the isolated perfused rat kidney. *J Pharmacokinet Biopharm* 18:561–87.

Eadie, G. S. 1942. The inhibition of cholinesterase by physostigmine and prostigmine. *J Biol Chem* 146:85–93.

Gillette, J. R. 1971. Factors affecting drug metabolism. *Ann NY Acad Sci* 179:43–66.

Haldane, J. B. S. 1957. Graphical methods in enzyme chemistry. *Nature* 179:832–5.

Hanes, C. S. 1932. Studies on plant amylases: The effect of starch concentration upon the velocity of hydrolysis by the amylase of germinated barley. *Biochem J* 26(5):1406–21.

Henderson, P. J. F. 1992. Statistical analysis of enzyme kinetic data. In *Enzyme Assays – A Practical Approach*, eds. R. Eisenthal and M. J. Danson, 284–9. Oxford: IRL Press, Oxford University Press.

Hill, A. V. 1910. The possible effects of the aggregation of molecules of hemoglobin on its dissociation curve. *J Physiol* 40(Suppl): iv–vii.

Hofstee, B. H. J. 1959. Non-Inverted versus inverted plots in enzyme kinetics. *Nature* 184(4695):1296–8.

Homma, H., Tada, M., Nakamura, T., Yamagata, S., Matsui, M. 1997. Heterogeneous zonal distributon of sulfotransferase isoenzymes in rat liver. *Arch Biochem Biophys* 339:235–41.

Ito, K., Houston, J. B. 2004. Comparison of the use of liver models for predicting drug clearance using *in vitro* kinetic data from hepatic microsomes and isolated hepatocytes. *Pharm Res* 21(5):2004.

Knapp, S. A., Green, M. D., Tephly, T. R., Baron J. 1988. Immunohistochemical demonstration of isozyme- and strain-specific differences in the intralobular localizations and distributions of UDP-glucuronosyltransferases in livers of untreated rats. *Mol Pharmacol* 33:14–21.

Lau, Y. Y., Krishna, G., Yumibe, N. P., Grotz, D. E., Sapidou, E., Norton, L., Chu, I., Chen, C., Soares, A. D., Lin. C. C. 2002. The use of *in vitro* metabolic stability for rapid selection of compounds in early discovery based on their expected hepatic extraction ratios. *Pharm Res* 19(11):1606–10.

Lineweaver, H., Burk, D. 1934. The determination of enzyme dissociation constants. *J Am Chem Soc* 56(3):658–66.

Mueller, S. C., Majcher-Peszynska, J., Uehleke, B., Klammt, S., Mundkowski, R. G., Miekisch, W., Sievers, H. et al. 2006. The extent of induction of CYP3A by St. John's wort varies among products and is linked to hyperforin dose. *Eur J Clin Pharmacol* 62(1):29–36.

Oinonen, T., Lindros, K. O. 1998. Zonation of hepatic cytochrome P-450 expression and regulation. *Biochem J* 55:413–21.

Pang, K. S., Gillette, J. R. 1978. Complications in the estimation of hepatic blood flow *in vivo* by pharmacokinetic parameters. *Drug Metab Disp* 6:567–76.

Pang, K. S., Rowland, M. 1977. Hepatic clearance of drugs. I. Theoretical considerations of a "well stirred" model and a "parallel tube" model. Influence of hepatic blood flow, plasma and blood cells binding, and the hepatocellular enzymatic activity on hepatic drug clearance. *J Pharmacokinet Biopharm* 5:625–53.

Pang, K. S., Weiss, M., Macheras, P. 2007. Advanced pharmacokinetic models based on organ clearance, circulatory, and fractal concepts. *AAPS J* 9(2):E268–83.

Roberts, M. S., Anissimov, Y. G. 1999. Modeling of hepatic elimination and organ distribution kinetics with the extended convection-dispersion model. *J Pharmacokinet Biopharm* 27:343–82.

Roberts, M. S., Rowland, M. 1986. A dispersion model of hepatic elimination: 1. Formulation of the model and bolus consideration. *J Pharmacokin Biopharm* 14:227–60.

Rowland, M., Benet, L. Z., Graham, G. G. 1973. Clearance concepts in pharmacokinetics. *J Pharmacokinet Biopharm* 1:123–36.

Simonsson, U. S., Jansson, B., Hai, T. N., Huong, D. X.,Tybring, G., Ashton, M. 2003. Artemisinin autoinduction is caused by involvement of cytochrome P450 2B6 but not 2C9. *Clin Pharmacol Ther* 74(1):32–43.

Sun, H., Liu, L., Pang, K. S. 2006. Increased estrogen sulfation of estradiol 17 β-D glucuronide in rat metastasis tumor livers. *J Pharmacol Exp Ther* 319:818–31.

Tirona, R. G., Pang, K. S. 1999. Bimolecular glutathione conjugation of ethacrynic acid and efflux of the glutathione adduct by periportal and perivenous rat hepatocytes. *J Pharmacol Exp Ther* 290:1230–41.

Wehner, J. F., Wilhelm, R. H. 1956. Boundary conditions of flow reactor. *Chem Eng Sci* 6:89–93.

Wilkinson, G. R., Shand, D. G. 1975. Commentary: A physiological approach to hepatic drug clearance. *Clin Pharmacol Ther* 18:377–390.

Zhang, L., Yuanchao (Derek) Zhang, Y. D., Zhao, P., Huang, S-M. 2009. Predicting drug–drug interactions: An FDA perspective. *AAPS J* 11(2):300–6.

8 Pharmaceutical, Biopharmaceutical, and Physiological Factors Influencing Oral Absorption

Mechanistic Absorption Models

8.1 INTRODUCTION

The absorption from the gastrointestinal (GI) tract is a complex process, which includes dissolution, permeation, degradation and intestinal and/or hepatic metabolism. These processes are often influenced by physiological mechanisms of gastric emptying, intestinal motility and the action of influx and efflux proteins. Furthermore, the rate and extent of absorption from the GI tract depends on the physicochemical properties of the administered compound and its functioning in different environments of the GI tract. The absorption into the systemic circulation entails the permeation of the compound from the lumen to gut wall and subsequent diffusion through capillary wall into the blood stream. A summary of the factors influencing the absorption from the GI tract is presented in Figure 8.1.

8.2 CLASSIFICATION OF FACTORS INFLUENCING GI TRACT ABSORPTION

The factors influencing the gastrointestinal absorption of xenobiotics that can be categorized are: (1) physiologic and dynamic attributes of the GI tract, (2) physicochemical characteristics of the administered xenobiotic, and (3) other factors that can influence the absorption process directly or indirectly, such as food or disease state. The factors associated with each category are

1. Physiologic and dynamic attributes of the GI tract:
    - Regional pH of GI tract and pKa of xenobiotics
    - Surface area for absorption
    - Gastric emptying rate
    - Intestinal motility, small intestinal transit time (SITT)
    - Effect of bile salts
    - Hepatic first-pass metabolism (presystemic hepatic extraction)
    - Intestinal metabolism—Role of CYP3A4 (presystemic intestinal metabolism)
    - Influence of efflux proteins
    - Role of gastrointestinal microbiotas
2. Physicochemical factors:
    - Polymorphism and the chemical nature of the compound
    - Partition coefficient and molecular properties
    - Particle size, porosity, and wettability

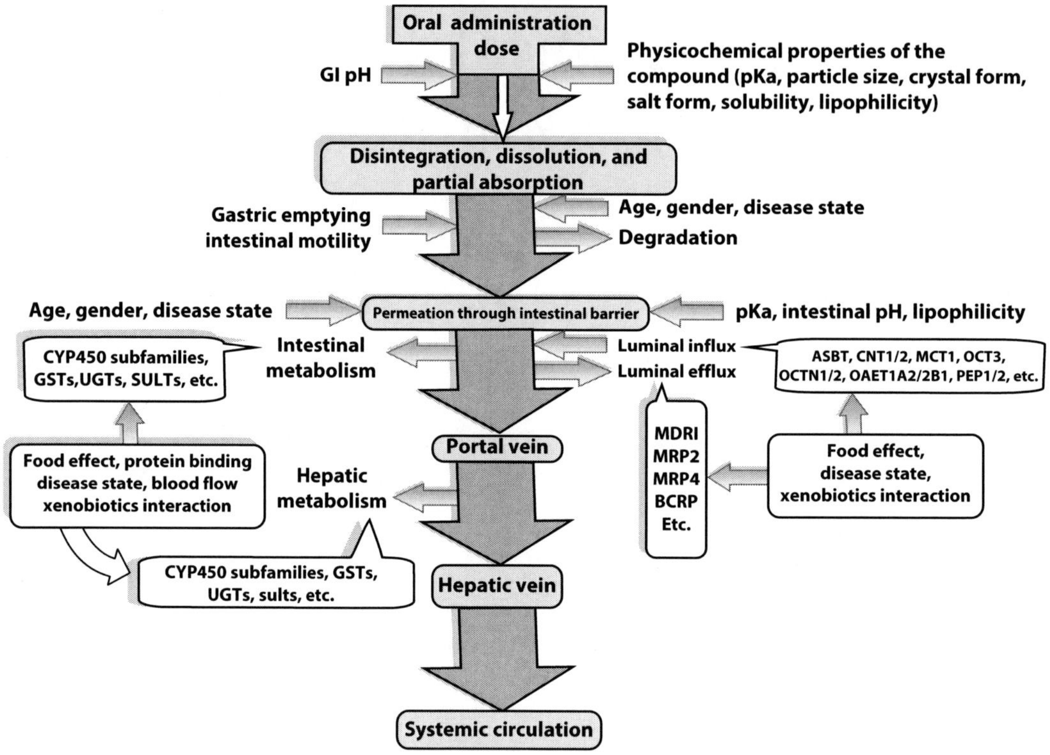

FIGURE 8.1 Schematic diagram of significant factors influencing the absorption of xenobiotics from the GI tract.

- Dosage factors
- Disintegration and dissolution
- Chirality and enantiomers

3. Other factors:
   - Type of food and drink
   - Disease states
   - Genetic polymorphism

8.2.1 Physiologic and Dynamic Attributes of the GI Tract

8.2.1.1 Regional pH of GI Tract and pH-Partition Theory

Orally ingested compounds journey through two different and distinct environments: acidic in the stomach and alkaline in the small intestine. Aside from the anatomical differences of the stomach and intestine, gradual changes in the pH throughout the tract influence the absorption of weakly acidic and weakly basic compounds. Gastric pH is highly acidic during the fasting state and is comparable to the pH of 0.15 M HCl containing pepsin. Age, disease states, and foods influence the pH of the GI tract (Russell et al., 1993; Zimmermann et al., 1994; Charman et al., 1997; Martinez and Amidon, 2002). The presence of food alters the pH, content viscosity, and transit rates. It also influences the disintegration and dissolution processes (Gai et al., 1997). Because of the daily intake of food and beverages, the gastric pH fluctuates during the day. The alkaline environment of the small intestine, on the other hand, remains essentially consistent during the day and throughout life. The regional pH of the small intestine starts from 5.5 to 6.0 in the duodenum and gradually increases to 7.0–9.0 in the jejunum and about 9–11 in the ileum. The chemical stability, dissolution, and absorption of xenobiotics may also be affected by the pH along the GI tract (Charman et al., 1997; Ward, 2010).

Most xenobiotics are either weakly acidic or weakly basic compounds, and according to the pH-partition theory only the unionized form of molecules are able to partition through biological membrane, that is, the low pH of the stomach unionize weakly acidic compounds and the high pH of the intestines unionize weakly basic compounds. The solubility of a compound also depends on its pKa and pH of the GI tract (Dressman et al., 2007). The classical Henderson-Hasselbalch equation links the pKa of a compound to the pH of the GI tract to estimate the degree of ionization and the ratio of unionized to ionized forms of the molecules at a given pH

$$\left(\frac{[unionized]}{[ionized]}\right)_{acidic} = 10^{(pKa-pH)} \tag{8.1}$$

$$\left(\frac{[unionized]}{[ionized]}\right)_{basic} = 10^{(pH-pKa)} \tag{8.2}$$

For example, consider compound A with a pKa of 3 and compound B with a pKa of 8 that are placed in pH 2 and 7, representing the pH of stomach and jejunum, respectively

$$\left(\frac{[unionized]}{[ionized]}\right)_{drugA} = 10^{(3-2)} = 10$$

$$\left(\frac{[unionized]}{[ionized]}\right)_{drugB} = 10^{(2-8)} = 10^{-6}$$

According to the calculation, at pH 2 for every 10 unionized molecules of A with a pKa of 3, only one molecule would be ionized. The ratio changes significantly for B with a pKa of 8, where for every one unionized molecule 10^6 molecules would be ionized. At pH 7 the ratios for A and B are 10^{-4} and 10^{-1}, respectively. Thus at a higher pH, the weakly basic compounds are more unionized and acidic ones are more ionized, and the opposite is true at low pH.

It should be noted, however, that although weakly acidic compounds are more unionized in the stomach, they are mainly absorbed from the intestine. The reasons are the limited surface area of the stomach, impermeability of the absorption barrier and the short residence time in the stomach. Thus, the unionized form of the weakly acidic compounds at a high pH of intestine, although in low concentration compared to the ionized molecules, contributes more significantly to the overall absorption of weakly acidic xenobiotics. This is mainly due to the large surface area of the intestine and the long intestinal transit time. The longer residence time in the small intestine is considered a favorable condition for the absorption of weakly acidic compounds (Löbenberg and Amidon, 2000a,b).

The absorption of neutral, acidic, basic, and amphoteric compounds has been studied in various models. The simultaneous chemical equilibria and mass transfer were studies in physical two- and multi-compartment unstirred diffusional models (Suzuki et al., 1970a,b). The multidiffusional model consists of an aqueous compartment with a diffusion layer followed by n compartments and a perfect sink condition. The significance of the unstirred diffusional model, the role of pH and pKa, and the degree of ionization of a compound is discussed here based on the following ionic equilibria of an amphoteric compound in water:

$$R_w^+ \underset{}{\overset{K_1}{\rightleftarrows}} R_w^0 + H^+ \underset{}{\overset{K_2}{\rightleftarrows}} R_w^- + 2H^+ \tag{8.3}$$

where R_w^+, R_w^-, and R_w^0 are the concentrations of the cationic, anionic, and unionized of the compound R, respectively. The subscript "w" represents the aqueous phase; K_1 and K_2 are the dissociation constants and H^+ is the hydrogen ion concentration. For a basic compound, K_1 is the dissociation constant with K_2 = zero. For an acidic compound, K_2 is its dissociation constant with $K_1 = \infty$. For neutral compound, $K_1 = \infty$ and K_2 = zero (Suzuki et al., 1970a,b). According to the Fick's first law (Chapter 2, Section 2.2.1.4), the total flux of different species of the compound is

$$J = -(D_{coeff})_w^+ \frac{d[R_w^+]}{dx} - (D_{coeff})_w^- \frac{d[R_w^-]}{dx} - (D_{coeff})_w^0 \frac{d[R_w^0]}{dx} \tag{8.4}$$

where (J) is the total flux and (D_{coeff}) is the diffusion coefficient.

A xenobiotic must be in solution form to be absorbed. The solubility of the compound at the site of absorption depends on the state of ionization of the molecules. The solubility of weak acids or bases is fully influenced by the pH of the environment. Weak acids and bases are more soluble in their ionized (conjugate base or conjugate acid) forms than their unionized forms, so at a high pH, where the weakly acidic compounds are in ionized state, the solubility of the compounds increases, whereas the solubility of weakly basic drugs decreases. The reverse is also true for weakly basic compounds, when their solubility enhances at low pH.

When Equation 8.3 is at equilibrium, K_2 represents the equilibrium constant equal to

$$K_2 = \frac{\cancel{2}[H^+]\times[R_w^-]}{[R_w^0][\cancel{H}^+]} = \frac{[H^+]\times[R_w^-]}{[R_w^0]} \tag{8.5}$$

Solving for $[R_w^-]$ yields

$$[R_w^-] = \frac{K_2[R_w^0]}{[H^+]} \tag{8.6}$$

If the total solubility (S) consists of the concentrations of ionized and unionized forms of a compound, that is,

$$S = [R_w^-] + [R_w^0] \tag{8.7}$$

The substitution of Equation 8.6 in Equation 8.7 yields

$$S = \frac{K_2 \times [R_w^0]}{[H^+]} + [R_w^0] = [R_w^0]\left(1 + \frac{K_2}{[H^+]}\right) \tag{8.8}$$

Setting $[R_w^0]$ as the pH independent solubility, that is, the solubility of unionized forms, Equation 8.8 can be written as

$$S = S_0\left(1 + \frac{K_2}{[H^+]}\right) \tag{8.9}$$

where (S) is the pH dependent total solubility of a compound and S_0 is the pH independent solubility of the compound.

The solubility of ionized form is then equal to

$$(S - S_0) = K_2 \frac{S_0}{[H^+]} \tag{8.10}$$

Equation 8.10 can be used to estimate pH below which the compound remains in unionized form by taking the logarithm of the both sides

$$\log(S - S_0) = \log K_2 + \log S_0 - \log[H^+] \tag{8.11}$$

Setting $K_2 = K_a$, where K_a is the association constant, and solving for pH = –log $[H^+]$ yields

$$\therefore \text{pH} = pK_a + \log \frac{S - S_0}{S_0} \tag{8.12}$$

Several conclusions can be made from the above Equation 8.12. For example, because food increases the pH of the stomach, giving weakly basic therapeutic agents (e.g., dipyridamole, ketoconazole, etc.) after the meal may not help their solubility; or co-administration of weakly basic drugs with proton pump inhibitors has the same outcome; or solubility of poorly soluble weakly acidic drugs (e.g., indomethacin, furosemide) occur mainly in small intestine. Compounds in the free acid form are soluble at any pH of the GI tract.

The application of the pH-partition theory for defining the passive absorption of unionized molecules at the GI tract has been debated and questioned in the literature. For example, the relationship between the solubility, partition coefficient and pH has been investigated with the suggestion that the product of intrinsic solubility and intrinsic octanol-water partition coefficient is equal to the product of total solubility of a partially ionized solute and its octanol-water distribution coefficient at any pH.

(Solubility of a partially ionized solute) × (Octanol/water distribution coefficient)

Thus, it may not be necessary to consider either pH of the GI tract or pKa of a compound necessary to predict or evaluate the passive absorption of orally administered compounds (Ni et al., 2002).

It should also be noted that not all xenobiotics are either weak acids or weak bases. The behavior of classes of compounds such as ampholytes and zwitterions are distinctively different. The ordinary ampholytes ($pK_a^{acidic} > pK_a^{basic}$) are both weak acids and weak bases. Thus, they remain unchanged at neutral pH and their lipophilicity is similar to that of acids and bases. On the other hand, zwitterions at neutral pH have a negative and a positive charge and their lipophilicity is constant in a pH range defined by their two pKa (Pagliara et al., 1997; Bouchard et al., 2002). A more complex picture emerges when a compound is considered zwitterionic ampholytes. The discussion on the ionization equilibria of this group of compounds is beyond the scope of this chapter.

8.2.1.2 Absorptive Surface Area

An orally administered compound travels first through the esophagus, which is a tubular muscular structure 25–50 cm between the distal pharynx and gastroesophageal junction below diaphragm. Because of its peristaltic contraction waves of the upper and lower sphincters and the related propulsive force with the help of gravity, make this part of the GI tract unimportant for any kind of absorption.

The next stop for the compound is the stomach with a normal empty volume of 45–50 mL, which lies on the left side of abdominal cavity between two smooth muscle sphincters: esophageal and pyloric. It has four distinct regions of cardia (i.e., gastro-esophageal junction), fundus (the upper curvature), corpus (the central region) and pylorus (the lower region). Its walls are composed of a mucous layer, sub-mucosa layer, and muscular layer consisting of longitudinal, circular and oblique muscle fiber, and serosa or connective tissues that fasten the stomach to peritoneum. The surface of the gastric mucosa is composed of a layer of columnar cells and secretory cells responsible for secretion of about two liters of gastric fluid per day containing hydrochloric acid, gastrin, pepsin, and mucus (Löbenberg and Amidon, 2000a,b). Most weakly acidic compounds are expected to absorb in the stomach. Compounds like aspirin and alcohol absorb rather rapidly in the stomach. The empty stomach has a number of inner folds called rugae, which can expand significantly to accommodate the volume of food. In individuals with normal body weight the surface area of acid-producing mucosa of the stomach is between 383 and 784 cm^2 with an average of 583.48 cm^2.

After the stomach, the small intestine has the largest surface area for absorption of nutrients and xenobiotics; approximately 90% of all absorptive processes occur in the small intestine (Lennernäs, 1998; Fangerholm et al., 1996). Practically all mechanisms of absorption discussed in Chapter 2 occur in the small intestine. Its walls are more permeable than the stomach, and because of the gradual variability of the pH, weakly basic and weakly acidic drugs permeate through the walls more readily than the stomach. The walls are folded into long-wound crinkles that increase the surface area by a factor of 3; its epithelial layer is folded into villi, which can increase the surface area by a factor of 10; and the microvilli foldings of the villi enhances the surface area by a factor of 20. The epithelial cells that cover the surface of the villi are comprised of 90% enterocytes (Hillgren et al., 1995). The enterocytes have an apical side facing the lumen of the intestine, and a basolateral side or serosal side. The apical region of enterocytes has a brushed structure that forms microvilli. The enterocytes are connected together by a tight junction. The main biological barrier for absorption of xenobiotics in the small intestine is the combination of membrane and tight junctions of entrocytes (Salana et al., 2006; Fangerholm et al., 2007). The entire surface of the intestine is covered by a thin layer of 0.1 mm thick glycoprotein known as glycocalix layer. The length of the small intestine for a 70 kg man is about 6–7 m and its diameter is 2.5–3 cm; approximately 5% of the total length is the duodenum (the initial segment), 50% is the jejunum (the mid section of intestine), and 45% is the proximal and distal sections of the ileum (the terminal portion). Assuming the small intestine as a cylinder with a radius of 1.5 cm and length of 7 m, without its complex anatomical structure, its estimated surface area would be

$$SA = 2\pi r(r + h) = 2 \times 3.14 \times 1.5(1.5 + 700) = 6608\,\text{cm}^2 = 0.66\,\text{m}^2$$

Adding the surface of its complex anatomical structure (i.e., all the folding, villi and microvilli), its theoretical surface area of small intestine would be

$$(SA)_{SI} = 0.66\,\text{m}^2 \times 3 \times 10 \times 20 = 396\,\text{m}^2 = 4,262.50\,\text{ft}^2$$

where $(SA)_{SI}$ is surface area of small intestine. According to these theoretical calculations, the total surface area of small intestine approximately equals an NBA basketball court with dimensions of 50×94 ft. Because of this enormous surface area and because of being the hub of all enzymatic metabolism and transport proteins in the GI tract, the small intestine is the final stage and sole site for the absorption of nutrients and orally administered xenobiotics.

The large intestine has a length of approximately 1.5 m (5 feet) and a diameter larger than the small intestine. It starts from the cecum (the pouch connected to the distal portion of ileum) followed by the colon, which has four portions: ascending colon, traverse colon, descending colon, and the sigmoid colon, which is the final *S*-shape portion connected to the anal canal. The ileocecal valve located between the distal portion of ileum and cecum section serves as a barrier for guarding ileum against colonic content and microorganisms of the large intestine. The appendix, a closed tubular structure, is also in the cecum. The mucosal cells known as colonocytes have no microvilli and are covered with a thick layer (200–500 mm) of cucos and water, which prevents the absorption of xenobiotics. The major roles of the large intestine are absorption of residual water and electrolytes, providing an environment for microbiotas' activities, and also acts as a transport canal to get rid of waste products (Kararli, 1995; Wilson, 2000). In recent years, the colonic delivery systems of proteins and peptides have been developed for the localized and targeted treatment of disease (Patel et al., 2007).

The rectum is a highly vascular region of the GI tract with a limited surface area, which begins at the rectosigmoid junction and ends in an anorectal junction. Its average length is about 13 cm (5 in) and because of its vascular walls is used for rectal administration of therapeutic agents (see Chapter 1, Section 1.5.1.7). The contribution of this terminal section of the GI tract to the absorption of an orally administered compound is insignificant.

8.2.1.3 Gastric Emptying

The gastric emptying process, regulated by neural and hormonal controls, is a carefully synchronized contractile movement between stomach and small intestine. Following mastication (chewing) and deglutition (swallowing), the esophagus through the coordinated contraction of its muscle propels materials toward the stomach and sphincters act in a coordinated manner with swallowing to prevent the materials from returning back to the esophagus (Johnson, 2007). A sustained contraction is initiated and the content of the upper part of stomach is pressed toward its lower part. The proximal stomach is responsible for the transfer of gastric content known as chyme from the stomach to the upper part of small intestine (duodenum). The lower part of the stomach triturates the content, prevents the duodenal reflux, and allows only liquids and small particles to pass into the duodenum. Peristaltic contractions of both the proximal and distal stomach are under neural control of vagus nerves and its autoregulation of acetylcholine release. The acetylcholine interacts with the appropriate receptor of the stomach muscle to stimulate the contraction and relaxes during the swallowing. Furthermore, several hormones participate in the stimulation or inhibition of the contraction. For example, cholecystokinin inhibits the contraction of the proximal stomach whereas it stimulates the contraction of the distal stomach; secretin and somatostatin both inhibit the contractions of the proximal and distal stomach.

The gastric emptying rate is the rate at which the stomach empties its content into the duodenum. The delay in the gastric emptying rate is responsible for a delay in the onset of action of certain therapeutic xenobiotics and dosage forms. For example, weakly basic drugs are in an ionized form in the stomach and therefore, a slow gastric emptying rate delays their onset of action. Furthermore, the gastric emptying rate often contributes to inconsistency in the absorption of an orally administered drug because of particle size, particle density, and viscosity of the content. The intensity of the gastric contraction and pyloric resistance changes during the fed state with a pattern of consistent low-medium amplitude contractions, which assist the small particles (2–7 mm) to be transferred to the small intestine. The influence of particle size cut off on gastric emptying is important, yet controversial. There are contradictory published data with results indicating that tablets and particles as large as 11 mm can leave the stomach during the fed state (Coupe et al., 1991a,b).

In fasting, the gastric emptying is a function of its cyclical contractions, which is characterized by the following phases:

Phase I: Calm phase (quiescence)—duration: 30–60 minutes
Phase II: Irregular phase with random medium-amplitude contractions—duration: 20–40 minutes

Phase III: Intense contraction phase with high amplitude contractions—duration: 5–15 minutes
Phase IV: decelerate phase with graduate reduction of contractions leading to Phase I

The cycle of fast-state contractions (Phase I–IV) are also known as "interdigestive migrating myoelectric complex" (IMMC). This cycle of contractions initiate from the part of the stomach adjoining the pylorus, known as pyloric antrum. Larger indigestible particles empty into the duodenum during the Phase III of fast-state gastric emptying, when the gastric contractions are stronger with high amplitude. The IMMC continues during the fast state until the food is introduced into the stomach, which changes the IMMC to frequent contractions with medium-low amplitudes (Burks et al., 1985; Oberle et al., 1990).

The kinetics of gastric emptying of non-nutrient liquids such as polyethylene glycol (PEG 4000) or phenol red has been shown to follow a first-order process after a lag time, attributed to a rapid emptying of the content at the beginning of fed-state contractions when the content volume is high. The literature data are not unanimous on the kinetics of the gastric emptying process.

The following are important factors that influence the gastric emptying rate:

1. Compounds that block gastric muscle receptors for acetylcholine delay the emptying rate (e.g., propanthelin).
2. The high acidity of gastric chyme also delays the emptying rate.
3. The chemical composition of chyme within the stomach determines the gastric emptying rate. In humans, liquids are emptied with a half-life of approximately 12 minutes and digestible solids at a half-life of about 2 hours depending upon the chemical composition of the chyme. Carbohydrates are emptied faster than proteins and proteins are faster than fats.
4. The gastric emptying rate is controlled by the stomach caloric content such that the number of calories transferred to the small intestine remains constant for different nutrients (carbohydrates, proteins or fats) over time and the emptying process is slower when the meal is rich in calories.
5. The gastric emptying rate can be modulated by the meal size. For example changing the solid meal size from 300 to 1692 g increases the half-life from 77 to 277 minutes; however, the size of the meal produces a pressure that stimulates the gastric emptying (Christian et al., 1980).
6. Stimulation of the small intestinal receptors (i.e., duodenojejunal receptors sensitive to osmotic pressure) by hypertonic or hypotonic solutions slows down the gastric emptying rate.
7. The temperature of the solid or fluid intake affect the gastric emptying rate. The temperature above or below the physiological 37°C proportionately reduces the gastric emptying rate.
8. Other factors such as anger or agitation may increase the rate whereas depression, trauma, or injury has been suggested to reduce the gastric emptying rate. Position of the body such as standing or lying on the right side may facilitate the transfer of the content to the small intestine by increasing the pressure in the proximal part of the stomach.

Delayed gastric emptying, using gastro-retentive drug delivery systems, enhances the gastric absorption of therapeutic agents by increasing gastric retention time and reducing the gastric emptying rate (Lee et al., 1999; Streubel et al., 2003; Mundada et al., 2008). A few examples of these systems are

- Low-density floating systems
- Gel-forming polymeric systems
- Gas-generating systems
- Expandable systems

8.2.1.4 Intestinal Motility (Small Intestinal Transit Time)

The major site of absorption in the GI tract is the small intestine. Stomach and colon have a small absorptive area and the colon has a lumen full of bacteria. The small intestine has a length of approximately 300–400 cm with an alkaline environment that starts from the pyloric sphincter and continues with the duodenum, jejunum, and proximal and distal portions of the ileum to the ileocecal valve of the large intestine. The duodenum is about 25 cm (10 inches) long, receives chyme from the stomach through the pyloric sphincter. Bile and pancreatic ducts deliver pancreatic juice (which contains a variety of enzymes, including trypsinogen, chymotrypsinogen, elastase, carboxypeptidase, pancreatic lipase, nucleases and amylase) and bile (which contains bile acids and waste products such as bilirubin) from the pancreas and liver, respectively. The jejunum is about 2.5 m (8 feet) long; it is the middle section of the small intestine. The ileum is about 3.6 m (12 feet) long and is the last section of the small intestine. It ends with the ileocecal valve (sphincter), which regulates the movement of chyme into the large intestine and prevents backward movement of material from the large intestine. As it was noted earlier (Section 8.2.1.1), the average pH of the duodenum is approximately 6 and increases gradually throughout the intestine. The average absorptive surface area of the small intestine is approximately 250 square meters, or about 2691 square feet, which is less than the theoretical value discussed in Section 8.2.1.2. The three features of its anatomy contribute to its absorptive surface area: (1) mucosal folds, (2) villi, and (3) microvilli (see Figure 2.1, Chapter 2 and Section 8.2.1.2). The microvillus border of intestinal epithelial cells is referred to as the "brush border."

The contents of the small intestine are highly rich in digestive enzymes such as lipase, protease, amylase, esterase, and nucleases. In addition, bile that is rich in micelles of bile salts is added to the content of the small intestine. The small intestinal motility is mostly segmental contractions that consist of mixing-contraction and propulsive-contraction.

The intestinal motility has two different patterns of digestive and interdigestive activities. The interdigestive activity cycle starts with about 30–40 minutes of no activity in both the stomach and intestine, followed by increasing contractions that are circular and migrate along the small intestine (peristaltic). This pattern of contraction removes everything that is left in the small intestine. Finally the contraction diminishes in occurrence and intensity and the cycle starts again by the rest period followed by the contraction. One entire cycle may take between 90 and 120 minutes in healthy and normal subjects. However, it has been shown that intra-individual differences can be significant (Davis et al., 1986; Coupe et al., 1991a,b). As it was noted in the gastric emptying process, the ingestion of food suspends the interdigestive cycles and establishes the intestinal motility pattern for digestive cycle (van der Ohe and Camillari, 1992). The digestive cycle constitutes the contractions that are mainly mixing contractions with few propulsive contractions. The small intestinal transit time (SITT) during the digestive cycle is rather difficult to determine considering the variability among the individuals and the type of food; it may not be as accurate as one would expect. The SITT of healthy individuals ranges from 3 to 5 hours. The esophago-duodenal-jejunal-ileal transit time, the gastric emptying time, in normal subjects receiving mashed potatoes with labeled sulfur has been reported as 378 ± 90 minutes. Various factors, including pharmaceutical excipients such as mannitol can reduce the SITT (Wilson, 2000). The transit time between the stomach and large intestine is one of the significant factors in bioavailability of xenobiotics.

Most of the factors that affect the gastric emptying time, such as the presence of fats or the volume of food, also influences the intestinal motility. For example, a large meal rich in fats requires longer and stronger intestinal contractions.

8.2.1.5 Role of Bile Salts

As discussed in Chapter 6, hepatocytes in the liver secrete bile into canaliculi, which flows into bile ducts and ultimately enters into the duodenum. Its composition consists of mainly water, bile acids/salts, inorganic salts, electrolytes, cholesterol, phospholipids, bilirubin, xenobiotics, and their metabolites. It is an important biological sample for xenobiotic *in vivo* metabolism studies, PK/TK analysis, and the mass balance studies of xenobiotics. Biliary components are critical for the digestion and absorption of lipid, lipophillic compounds including fat-soluble vitamins in intestinal milieu in the form bile salt phospholipid mixed with micelles and vesicular colloids (Hofmann and Borgstrom, 1962; Nordskog et al., 2001). Bile salts are surface-active agents that aggregate in an aqueous solution forming micelles and mixed micelles with phosphotydylcholine, glyceraldehydes, and fatty acids in the small intestine (Hoffman, 1962; Cabral and Small, 1989). It has been shown that biliary-derived endogenous fatty acids have a higher tendency to transport through the lymphatic network (Trevaskis et al., 2005). Bile secretion into the duodenum is due to the chyme stimulation and also secretion of secretin and cholecystokinin. Its total production in humans is about 400–800 mL per day. During fasting, the bile flow declines and most of that is diverted into the gallbladder. The bile stored in the gallbladder because of re-absorption of water and electrolytes gets concentrated gradually and used when it is needed.

Bile salts are the sodium salt of bile acids and polar derivatives of breakdown products of cholesterol in the liver by the cytochrome P450-mediated oxidation. The important intermediates of the reaction are trihydroxycoprostanoate and cholyl CoA. The activated carboxyl carbon of cholyl CoA interacts with the amino group of glycine or taurine to form glycocholate or taurocholate, respectively. In humans, the most important bile acids are cholic acid, deoxycholic acid, and chenodeoxycholic. Bile salts are amphipathic compounds and have both hydrophilic (polar hydroxyl group) and hydrophobic (nonpolar methyl group) regions, and are considered effective detergents in emulsifying lipids. Thus, bile acids/salts play a number of important roles in the body including elimination of cholesterol, removal of byproducts of catabolism, emulsifying lipids and fat-soluble endogenous and exogenous compounds, reduction of gut flora, and removal of xenobiotics and their metabolites through fecal elimination. Bile acids/salts are mostly reabsorbed into the body and similar to compounds that go through entrohepatic recirculation (Chapter 6) return to the liver through the hepatoportal vein to be reused. Approximately less than 10% of bile entering into the duodenum leaves from the body via fecal elimination.

8.2.1.6 Hepatic First-Pass Metabolism (Presystemic Hepatic Extraction)

Numerous aspects of oral absorption including poor intestinal absorption, poor solubility, poor permeability, intestinal metabolism, removal by efflux proteins, degradation, digestive proteolytic and microfloraenzymatic activity, and presystemic hepatic extraction can cause incomplete or poor bioavailability. The presystemic metabolism is a major factor in reducing the bioavailability of an administered dose. The intestinal metabolism and hepatic first pass metabolism are the main first-pass extractions for an orally administered compound and are often difficult to differentiate (Paine et al., 1996; Lee et al., 2001). The intestinal metabolism is discussed in the next section (Section 8.2.1.7). The hepatic extraction occurs when xenobiotics and nutrients absorb into the submucosal capillaries, transfer into the intestinal vein, which joins with the veins from the spleen and pancreas, and finally enter into the hepatoportal vein moving toward the liver. In the liver, the nutrients, xenobiotics, small particles, and droplets from the hepatoportal vein are absorbed and hepatocytes metabolize free nutrients and xenobiotics before reaching the systemic circulation.

The metabolic enzyme systems and pathways of the liver are discussed in Chapter 6. The amount of xenobiotics eliminated by the hepatic first-pass metabolism is usually expressed in terms of fraction or percentage removal, also known as extraction ratio (see also Chapters 6 and 7), which is the ratio of the rate of metabolism to the rate of input into the liver:

$$\text{Rate of input} - \text{Rate of output} = \text{Rate of extraction} \tag{8.13}$$

$$\text{Rate of extraction/Rate of input} = \text{Extraction ratio} \tag{8.14}$$

$$(Q_{liver} \times C_{liver})_{input} - (Q_{liver} \times C_{liver})_{output} = Q_{liver}(C_{liver_{input}} - C_{liver_{output}}) \quad (8.15)$$

$$\frac{Q_{liver}(C_{liver_{input}} \quad C_{liver_{output}})}{Q_{liver}(C_{liver})_{input}} = \frac{C_{liver_{input}} \quad C_{liver_{output}}}{C_{liver_{input}}} = ER \quad (8.16)$$

where Q is the blood flow, C is concentration and ER is the extraction ratio. The amount that will be available for the systemic effect can therefore be estimated in terms of the fraction that escapes the liver first-pass metabolism. That is,

$$F_{available} = 1 - ER \quad (8.17)$$

Under linear steady state condition, the rate of extraction by the liver is

Rate of extraction = (Hepatic blood clearance) (Blood concentration of the xenobiotic) (8.18)

$$\therefore Q_{liver}(C_{liver_{input}} - C_{liver_{output}}) = Cl_H \times C_{blood} \quad (8.19)$$

$$ER = \frac{Cl_H \times C_{blood}}{Q_{liver} \times C_{blood}} = \frac{Cl_H}{Q_{liver}} \quad (8.20)$$

$$\therefore F_{available} = 1 - \frac{Cl_H}{Q_{liver}} \quad (8.21)$$

where Cl_H is hepatic blood clearance.

From Equation 8.21, it can be concluded that the bioavailability of a compound after the hepatic first-pass metabolism depends on the hepatic blood clearance. A large Cl_H is indicative of a smaller fraction that escapes first pass hepatic metabolism for distribution in the body and systemic effect. To expand on the definition of hepatic blood clearance, one may apply the principles of mass balance to the amount of drug associated with different components of blood, namely plasma and blood cells:

Total amount in blood = Amount in plasma + Amount associated with blood cells (8.22)

$$\therefore (V_{blood} \times C_{blood}) = (V_p \times C_p) + (V_{bc} \times C_{bc}) \quad (8.23)$$

where V_p and V_{bc} are the volume of plasma and blood cells, respectively, C_p and C_{bc} are the related concentrations.

Dividing both sides of Equation 8.23 by V_{blood} and C_p yields

$$\frac{C_{blood}}{C_p} = \frac{V_p}{V_{blood}} + \frac{V_{bc} \times C_{bc}}{V_{blood} \times C_p} \quad (8.24)$$

The volume of blood cells divided by the total volume of blood is equal to the hematocrite (Hct) or the percentage of blood cells. Thus, the volume of plasma divided by the total volume of blood is equal to 1 – hematocrite (1 – Hct), that is,

$$\frac{C_{blood}}{C_p} = (1 - Hct) + Hct \times \frac{C_{bc}}{C_p} \quad (8.25)$$

Since the rate of elimination from the blood depends solely on the elimination from plasma, that is,

$$Cl_{blood} \times C_{blood} = Cl_p \times C_p \tag{8.26}$$

The total blood clearance would be equal to plasma clearance divided by Equation 8.25:

$$Cl_{blood} = Cl_p \times \frac{C_p}{C_{blood}} = \frac{Cl_p}{\left(\dfrac{C_{blood}}{C_p}\right)} = \frac{Cl_p}{(1 - Hct) + Hct(C_{bc}/C_p)} \tag{8.27}$$

Hepatic blood clearance is then equal to the fraction of the dose eliminated as metabolites (f_m) multiplied by the blood clearance (Cl_{blood})

$$Cl_H = f_m \times Cl_{blood} \tag{8.28}$$

Substituting Equation 8.27 into Equation 8.28 yields

$$Cl_H = \frac{f_m \times Cl_p}{(C_{blood}/C_p)} \tag{8.29}$$

Replacing Cl_H in Equation 8.21 with Equation 8.29 gives the bioavailability equation for a compound after the hepatic first-pass metabolism

$$F_{available} = 1 - \frac{f_m \times Cl_p}{Q_{liver} \times (C_{blood}/C_p)} \tag{8.30}$$

The fraction of the dose eliminated as metabolite, f_m, can also be estimated from the fraction of the dose excreted unchanged in urine, f_e, that is,

$$f_m = 1 - f_e \tag{8.31}$$

The above equations are based on the assumption that the metabolism follows first-order kinetics, the fraction of dose metabolized and excreted unchanged is constant for a given compound in a given population. However, when the metabolism is capacity-limited and nonlinear, the extraction ratio and other metabolic constants will be dose-dependent variables. Therefore, the fraction available for systemic effect (i.e., bioavailability) increases significantly when increasing the dose. Under the nonlinearity condition, hepatic first-pass metabolism follows Michaelis–Menten equation kinetics. To evaluate the first-pass metabolism, linear and nonlinear, within the framework of non-compartmental analysis, the following relationships have been proposed (Ueda et al., 2002):

$$\frac{d(A_t)_{portal}}{dt} = (Q_{liver})_{portal}(C_t^{portal} - C_t^{systemic}) \tag{8.32}$$

where $d(A_t)_{portal}/dt$ is the rate of absorption from the intestinal vein into the portal vein, C_t^{portal} and $C_t^{systemic}$ are concentrations of xenobiotic in the portal and systemic vein, respectively, and $(Q_{liver})_{portal}$ is the portal vein blood flow rate. The local moment curve is then defined as

$$F_{portal} = \int_0^{\infty}\left(\frac{d(A_t)_{portal}}{dt}\right)dt \Big/ D = (Q_{liver})_{portal}\frac{(AUC)_{portal} - (AUC)_{systemic}}{D} \tag{8.33}$$

where (F_{portal}) is the organ absorption ratio from the intestinal vein into the portal system. The mean absorption time from the GI tract into the portal system is

$$\bar{t}_{absorption} = \frac{\int_0^{\infty} t\frac{d(A_t)_{portal}}{dt}}{\int_0^{\infty}\frac{d(A_t)_{portal}}{dt}} = \frac{\frac{(MRT)_{portal}}{(AUC)_{portal}} - \frac{(MRT)_{systemic}}{(AUC)_{systemic}}}{(AUC)_{portal} - (AUC)_{systemic}} \tag{8.34}$$

Under the assumption that the hepatic pre-systemic extraction is linear and the total body clearance is a constant, the hepatic recovery ratio, that is, the escaped fraction from the first-pass metabolism, and the related mean absorption time (*MAT*) are

$$F_H = \frac{AUC_{oral}^{systemic} / AUC_{intra\text{-}arterial}^{systemic}}{F_{portal}} = \frac{F}{F_{portal}} \tag{8.35}$$

$$MAT = MRT_{oral}^{systemic} - MRT_{intra\text{-}arterial}^{systemic} \tag{8.36}$$

where $MRT_{oral}^{systemic}$ and $MRT_{intra\text{-}arterial}^{systemic}$ are the mean residence times of the compound given orally and intra-arterially, respectively.

Under nonlinear conditions and a capacity-limited metabolism, the disposition rate is defined as

$$Vd\frac{dC}{dt} = Q_{liver}(C_{in} - C_{out}) - \frac{V_{max} \times C_{out}}{K_M + C_{out}} \tag{8.37}$$

The volume of distribution is *Vd*, the blood flow rate of the hepatic vein is Q_{liver}, the concentration of the compound in and out of the liver is C_{in} and C_{out}, respectively. The Michaelis–Menten parameters are the maximum rate of metabolism, V_{max}, and Michaelis–Menten constant, K_M. Setting Equation 8.37 equal to zero and solving for the ratio of concentrations in and out of the liver yields

$$F_H = \frac{C_{out}}{C_{in}} = \frac{\left(C_{in} - K_M - \frac{V_{max}}{Q_{liver}}\right) + \sqrt{\left(C_{in} - K_M - \frac{V_{max}}{Q_{liver}}\right)^2 + 4C_{in} \times K_M}}{2C_{in}} \tag{8.38}$$

where (C_{out}/C_{in}) is the hepatic recovery ratio, F_H. Thus, the rate of drug reaching the systemic circulation is

$$\frac{d(A_t)_{systemic}}{dt} = (Q_{liver})_{portal}\left[C_t^{portal} - C_t^{systemic}\right]\frac{C_t^{out}}{C_t^{in}} \tag{8.39}$$

The integrated value is the total amount available in systemic circulation (Ueda et al., 2002), that is,

$$A = FD = \int_0^{\infty}(Q_{liver})_{portal}\left[C_t^{portal} - C_t^{systemic}\right]\frac{C_t^{out}}{C_t^{in}}\,dt \tag{8.40}$$

The mean absorption time is then estimated as

$$MAT = \frac{\int_0^\infty t \frac{d(A_t)_{systemic}}{dt}}{\int_0^\infty \frac{d(A_t)_{systemic}}{dt}} \tag{8.41}$$

8.2.1.7 Gastrointestinal First-Pass Metabolism—Role of CYP450 Isozymes

The small intestine, in addition to being the major site of absorption of orally administered xenobiotics, has the capability of metabolizing both beneficial and harmful compounds administered orally (Paine et al., 2006). Thus, similar to the hepatic first-pass metabolism, the intestinal presystemic metabolism reduces the bioavailability of orally administered doses and influences the PK/TK analysis (Mizuma, 2002). A large body of research both *in vitro* and *in vivo* has established the intestinal metabolism of therapeutic agents by CYP3A4 as a major determinant of the systemic bioavailability (Paine et al., 1996; Zhang et al., 1999; Kaminsky and Zhang, 2003). In the epithelial barrier of the small intestine the amount of CYP3A4 is about 60%–70% of the total CYP450 present in the small intestine. The presence of inhibitors (e.g., ketoconazole, erythromycin, itoconazole, and grapefruit juice) or inducers (e.g., rifampicin) of CYP3A4 can increase or decrease the bioavailability of an orally administered compound that is a substrate for the isozyme. While CYP3A4 is the most abundant isozyme present in the duodenum for oxidative metabolism of xenobiotics, other members of CYP450 subfamilies are also present, but to a lesser extent (Shimada et al., 1994; Zhang et al., 1999). The presence of other members of the CYP3A subfamily such as CYP3A5 (Lin et al., 2002) and CYP3A7 has been advocated and is yet to be confirmed in adults (Kivistö et al., 1996). Another subfamily of CYP450 present in the small intestine is CYP2C, which is reported as the second most abundant CYP450 subfamily (Klose et al., 1999; Obach et al., 2001; Lapple et al., 2003). Two members of this subfamily, CYP2C9 and CYP2C19, have been expressed in the small intestine of humans. Other CYP450s reportedly expressed in the human small intestine are: CYP2D6 (de Waziers et al., 1990; Madani et al., 1999), CYP1A1 (Prueksaritanont et al., 1996; Paine et al., 1999), CYP2S1, CYP4F12, and CYP2J2 (Matsumoto et al., 2002). Enzymes of conjugation, namely glutathione S-transferase, UDP-glucuronosyl transferase and *N*-acetyltransferate are also present in the small intestine (Gibbs et al., 1998; Fisher et al., 2000, 2001; Coles et al., 2002). There are polymorphisms in the coding and noncoding regions of the genes for expression of metabolic enzymes in humans, which explains the interindividual variability and inconsistencies for the presence or absence of some enzymes in the small intestine, for example, CYP3A5 or CYP3A7.

In addition, the majority of recent therapeutic agents are peptides and proteins, which cannot be administered through the oral route of administration mostly because of a poor membrane uptake and extensive presystemic metabolism. This presystemic metabolism in the gastrointestinal tract is mainly based on luminally secreted enzymes (pepsins, trypsin, chymotrypsin, elastase and carboxypeptidase A/B), on brush border membrane bound enzymes (carboxypeptidases and aminopeptidases) and on cytosolic enzymes.

When the PK/TK of orally administered xenobiotics are linear and the compounds are subjected to hepatic and intestinal first pass metabolism, without the enterohepatic recirculation, the systemic availability may be estimated by the following relationship:

$$F_{available} = \frac{(Q_{liver})_{arterial} \times (Q_{liver})_{portal}}{((Q_{liver})_{arterial} + (Cl_{int})_H)((Q_{liver})_{portal} + (Cl_{int})_{GI})} = \frac{AUC_{oral}}{AUC_{iv}} \tag{8.42}$$

where $(Q_{liver})_{arterial}$ and $(Q_{liver})_{portal}$ are hepatic arterial and portal vein blood flows, respectively.

$(Cl_{int})_H$ and $(Cl_{int})_{GI}$ are intrinsic clearances of the liver and gastrointestinal, respectively. AUC_{oral} and AUC_{iv} are the area under plasma concentration–time curve of the compound given orally and intravenously.

Without the pre-systemic intestinal metabolism, Equation 8.42 is simplified to Equation 8.21. In the presence of hepatic and intestinal presystemic metabolism, without taking into consideration the enterohepatic re-circulation, the extent of absorption, that is, the area under the plasma concentration curve, can be estimated by the following relationship:

$$AUC_{oral} = \frac{Dose \times (Q_{liver})_{arterial} \times (Q_{liver})_{portal}}{Q_{liver_{HV}}\left[((Cl_{int})_H + (Cl_{int})_B)((Q_{liver})_{portal} + (Cl_{int})_{GI}) + (Q_{liver})_{portal} \times (Cl_{int})_{GI}\right]} \quad (8.43)$$

where $(Cl_{int})_B$ is the intrinsic biliary clearance.

$$AUC_{iv} = \frac{Dose\left[((Q_{liver})_{arterial} + (Cl_{int})_H + (Cl_{int})_B)((Q_{liver})_{portal} + (Cl_{int})_{GI})\right]}{(Q_{liver})_{arterial}\left[((Cl_{int})_H + (Cl_{int})_B)((Q_{liver})_{portal} + (Cl_{int})_{GI}) + (Q_{liver})_{portal}(Cl_{int})_{GI}\right]} \quad (8.44)$$

The AUC of oral and intravenous dose in the presence of combined hepatic metabolism, intestinal pre-systemic metabolism, and enterohepatic recirculation can be estimated as

$$AUC_{oral} = \frac{Dose \times (Q_{liver})_{arterial} \times (Q_{liver})_{portal}}{(Q_{liver})_{arterial}\left[(Cl_{int})_H((Q_{liver})_{portal} + (Cl_{int})_{GI}) + (Cl_{int})_{GI}((Q_{liver})_{portal} + (Cl_{int})_B)\right]} \quad (8.45)$$

$$AUC_{iv} = \frac{Dose\left[((Q_{liver})_{arterial} + (Cl_{int})_H)((Q_{liver})_{portal} + (Cl_{int})_{GI}) + (Cl_{int})_B(Cl_{int})_{GI}\right]}{(Q_{liver})_{arterial}\left[(Cl_{int})_H((Q_{liver})_{portal} + (Cl_{int})_{GI}) + (Cl_{int})_{GI}((Q_{liver})_{portal} + (Cl_{int})_B)\right]} \quad (8.46)$$

The following equation defines the intestinal clearance when only the intestinal first-pass is present:

$$Cl_{intestinal} = \frac{Q_G \times f_u \times (Cl_{int})_{GI}}{Q_G + f_u \times (Cl_{int})_{GI}} \quad (8.47)$$

where $Cl_{intestinal}$ and Q_G are the intestinal clearance and blood flow to the intestine, respectively, f_u is the free fraction of compound in plasma and $(Cl_{int})_{GI}$ is the intrinsic intestinal clearance.

The fraction that escapes the intestinal metabolism is

$$\therefore F = \frac{Q_G}{Q_G + f_u \times (Cl_{int})_{GI}} \quad (8.48)$$

8.2.1.8 Role of Influx and Efflux Proteins

Intestinal enterocytes, in addition to forming a rigid epithelial monolayer preventing the paracellular diffusion of xenobiotics and expressing CYP450 subfamilies and other drug metabolizing enzymes, they also express membrane-associated efflux/influx transport proteins. The efflux proteins collectively establish a remarkable biochemical barrier to xenobiotics absorption through the GI tract wall (Bruyere et al., 2010). The adenosine triphosphate (ATP)-binding cassette (ABC) transporters, that is, efflux proteins (Chapter 2, Section 2.3) such as permeability glycoprotein multidrug resistance proteins (Pgp/MDR1), multidrug resistance-associated proteins (MRPs), breast cancer resistance proteins (BCRP), in conjunction with the inhibition or induction of influx proteins can facilitate or limit the absorption (Dahan and Amidon, 2009a,b). The influx transport proteins located on the apical side of the membrane or brush border side of enterocytes are ASBT (apical sodium-dependent bile acid transporter), CNT1/2 (concentrative nucleoside transporter), MCT1 (monocarboxylate transporter), PEP1/2 (peptide transporter), OATP1A2 and OATP2B1 (organic anion transporting polypeptides), OCT3 and OCTN1/2 (organic cation transporters). The efflux proteins of the epical side are BCRP, MDR1, MRP2, and MRP4. These influx and efflux proteins are in direct contact with the content of the small intestine that contains xenobiotics. On the basolateral side, the influx proteins are OATP3A1, OATP4A1, and OCT1/2, and efflux proteins are MRP1, MRP3, MRP4, MRP5, MCT1, and ENT1/2 (Equilibrative Nucleoside Transporter), which are in contact with blood in capillaries. The apical and basolateral proteins can be induced, inhibited, or modulated in the presence of xenobiotics (Suzuki and Sugiytama, 2000; Chan et al., 2004; Kis et al., 2010). Among these transporters, permeability glycoprotein (Pgp/MDR1) has been studied in detail for its influence on the bioavailability of single dose, multiple dose and concomitantly administered xenobiotics and its significant role in chemoresistance and functional interaction with intestinal CYP3A subfamily (Lown et al., 1997a,b; Ueda et al., 1999; Dietrich et al., 2003; Shugarts and Benet, 2009; Dahan et al., 2009).

The high concentration of Pgp and CYP3A isozymes are present in the villus tip of enterocytes and share significant similarities in substrate specificity. It has been well documented that extrusion of xenobiotics by intestinal Pgp allows CYP3A isozymes to have prolonged access to their substrates. Thus, Pgp reduces absorption of orally administered xenobiotics and can also modulate the influence of inhibitors and inducers of CYP3A isozymes-mediated metabolism. Contrary to CYP3A4, which is present in a limited region of the small intestine, the expression of Pgp is present along the length of intestine. The function of Pgp and CYP3A4 are mutually exclusive, yet synergistic, and their functions as the intestinal barrier are of equal value and complementary (Lown et al., 1997a,b; Ito et al., 1999; Perloff et al., 2000; Holmstock et al., 2012). Figure 8.2 depicts the localization of transport proteins and enzymes of enterocytes.

In addition to xenobiotics, food and excipients of orally administered dosage forms may also interact with Pgp and increase the bioavailability of orally administered drugs. The inhibitory effect of grapefruit and other citrus fruit juices on CYP3A4 and Pgp, which has been attributed to its flavonoids and furanocoumarins, and the inhibitory effects of excipients such as cremophor EL and cremophore RH 40 are a few examples (Lown et al., 1997a,b; Wang et al., 2001; Xu et al., 2003; Dahan et al., 2009).

MDR1 and other highly expressed efflux transporters, for example, BCRP, MRP2, and MRP4, are present in the human jejunum and colon, and extrude xenobiotics back into the lumen. The influx proteins facilitate the transport of xenobiotics into the enterocytes and systemic circulation. Both efflux and influx transporters require ATP and function as active transport proteins. In the evaluation of the bioavailability of xenobiotics that are substrate for these proteins, the function of both groups of transporters should be taken into consideration.

The kinetics of intestinal absorption of xenobiotics, taking into consideration the presence of efflux and influx proteins and CYP3A4, based on the model depicted in Figure 8.3 are theoretically developed (Ito et al., 1999) as follows:

The equations of the model defining the concentration of drug in the lumen (C_{lum}) and in epithelial cells (C_{epi}) are described by the following partial differential equations, which represent the mass balance equation of the model (Ito et al., 1999):

$$\frac{\partial C_{lum(x,t)}}{\partial t} = -\frac{Q_{lum}}{Area_{lum}}\frac{\partial C_{lum(x,t)}}{\partial x} - \frac{Cl_{influx}}{V_{lum}}C_{lum(x,t)} + Cl_{efflux} \times \frac{V_{epi}}{V_{lum}} \times C_{epi(x,0,t)} \tag{8.49}$$

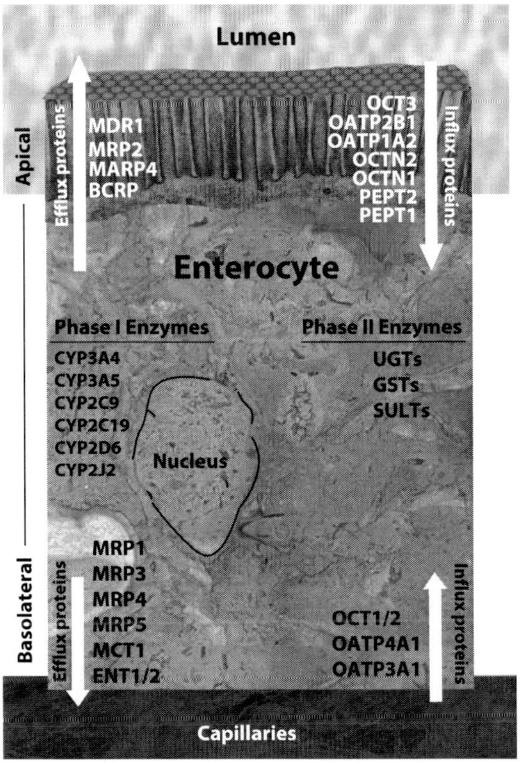

FIGURE 8.2 Depiction of localized transport proteins and metabolic enzymes of enterocytes.

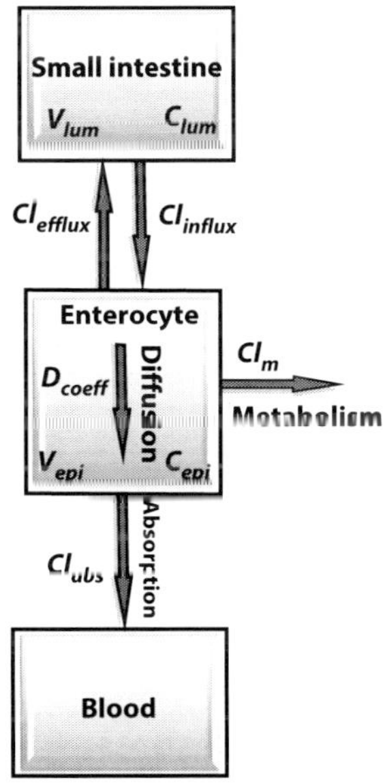

FIGURE 8.3 Theoretical model for defining the kinetics of intestinal absorption of xenobiotics in the presence of efflux and influx proteins and CYP3A4.

$$\frac{\partial C_{epi(x,y,t)}}{\partial t} = D_{coeff} \frac{\partial^2 C_{epi(x,y,t)}}{\partial y^2} - Cl_m \times C_{epi(x,y,t)} \tag{8.50}$$

where $(\partial C_{lum(x,t)}/\partial t)$ is the partial change of drug concentration in the gut lumen with respect to time, x represents the distance in the direction of luminal flow. (Q_{lum}) is the luminal flow rate and $(Area_{lum})$ is the sectional area of lumen. (Cl_{influx}) is the apparent influx clearance from the lumen into epithelial cells, (V_{lum}) is the volume of lumen and (Cl_{efflux}) is the apparent efflux clearance from cells to the lumen. (V_{epi}) is the volume of epithelial cells, $(C_{epi(x,0,t)})$ is drug concentration in epithelial cells and (D_{coeff}) is the diffusion coefficient in epithelial cells. $(\partial C_{epi(x,y,t)}/\partial t)$ is the partial change of drug concentration in epithelial cells with respect to time, $(\partial^2 C_{epi(x,y,t)}/\partial y^2)$ is the second partial derivative of the concentration of a drug in the epithelial cells with respect to the distance in the direction of drug absorption and (Cl_m) is the metabolic clearance in the epithelial cells. The drug concentration, $(C_{epi(x,y,t)})$, is considered with respect to three variables: the distance of luminal flow, x, the distance in the direction of drug absorption, y, and time, t.

Integrating the above partial differential equations by Laplace transformation and setting the following initial conditions at $t = 0$:

$$C_{lum(x,0)} = 0$$
$$C_{epi(x,y,0)} = 0$$

The boundary conditions at the apical and basal membranes of epithelial cells are

$$Cl_{influx} \times C_{lum(x,t)} = C_{efflux} \times V_{epi} \times C_{epi(x,0,t)} - D_{coeff} \times Area_{lum} \frac{\partial C_{epi(x,0,t)}}{\partial y}$$

$$Cl_{abs} \times V_{epi} \times C_{epi(x,l_{epi},t)} = -D_{coeff} \times Area_{lum} \times \frac{\partial C_{epi(x,l_{epi},t)}}{\partial y}$$

where (Cl_{abs}) is the absorption clearance from cells to the blood, and l_{epi} is the length of each epithelial cell in the direction of drug absorption.

The integrated equations under the assumption of steady-state drug concentration provide the fraction of a drug available in the lumen and the fraction absorbed as follows:

$$F_{lum} = \exp\left\{\frac{\phi\sqrt{D_{coeff} \times Cl_m}(1+\mathbb{R})}{Cl_{efflux} \times l_{epi} \times (1-\mathbb{R}) - \sqrt{D_{coeff} \times Cl_m}(1+\mathbb{R})}\right\} \tag{8.51}$$

where (F_{lum}) is the fraction of drug in the lumen, and ϕ and $\mathbb{R}$ are

$$\phi = \frac{Cl_{influx}}{Q_{lum}}$$

$$\mathbb{R} = \frac{l_{epi} \times Cl_{abs} + \sqrt{D_{coeff} \times Cl_m}}{l_{epi} \times Cl_{abs} - \sqrt{D_{coeff} \times Cl_m}} \exp\left(2 l_{epi}\sqrt{\frac{Cl_m}{D_{coeff}}}\right)$$

The fraction absorbed into the systemic circulation at a steady state, ($F_{available}$), is

$$F_{available} = (1 - F_{lum}) \times \left(\frac{Rate_{abs}}{Rate_{abs} + Rate_{metab}} \right) \tag{8.52}$$

$$F_{available} = \frac{2l_{epi} \times Cl_{abs}(1 - F_{lum})}{\left\{ (l_{epi} \times Cl_{abs} - \sqrt{D_{coeff} \times Cl_m})\exp\left(-l_{lum}\sqrt{\frac{Cl_m}{D_{coefrf}}} \right) + (l_{epi} \times Cl_{abs} + \sqrt{D_{coeff} \times Cl_m})\exp\left(l_m\sqrt{\frac{Cl_m}{D_{coeff}}} \right) \right\}} \tag{8.53}$$

When ($D_{coeff} \gg Cl_m$) and under the assumption that epithelial cells act as a well-stirred compartment, Equation 8.53 changes to

$$F_{available} = \left(\frac{Cl_{abs}}{Cl_{abs} + Cl_m} \right) \times \left\{ 1 - \exp\left(-\phi \times \frac{Cl_{abs} + Cl_m}{Cl_{efflux} + Cl_{abs} + Cl_m} \right) \right\} \tag{8.54}$$

8.2.1.9 Role of Gastrointestinal Microbiotas

The gastrointestinal microorganisms (microbiota, also known as microflora), are prokaryotic (90%) and eukaryotic (10%) microorganisms of different kinds that become inhabitants of the gastrointestinal environment soon after birth. They colonize not only in the large intestine but also in the distal portion of the ileum. All mammals, including humans, have a symbiotic life with microbial cells, which are implicated in immunology and nutrition of the host (Macfarlene and Macfarlane, 1997). Using ribosomal RNA gene sequencing methodology, eight prevailing kinds (phyla) have been identified that may have a common evolutionary ancestors; they are Firmicutes, Bacteroidetes, Proteobacteria, Fusobacteria, Verrucomicrobia, Cyanobacteria, Spirochaeates, and Actinobacteria. The intestinal microbiotas are inhabited in the distal portion of ileum cecum, colon, rectum, and anal canal and interact extensively with the host through metabolic exchanges and the metabolism of mostly orally administered xenobiotics and their metabolites (Wang et al., 2005; Eckburg et al., 2005; Egert et al., 2006). The microbial content in the human GI tract includes lactobacilli, streptococci, and yeasts in the stomach and duodenum; lactobacilli, streptococci, enterobacteriacaea, bifidobacteria, bacteriodes, and fusobacteria in jejunum and ileum; and lactobacilli, streptococci, enterobacteriacaea, bifidobacteria, bacteriodes, viellonella, yeasts, proteus, staphylococci, pseudomonas, and fusobacteria in the colon (Holzapfel et al., 1998; Iannitti and Palmieri, 2010).

Microbiotas, with their enormous metabolic capacity, can influence the disposition and toxicity of xenobiotics (Sousaa et al., 2008). The variability in population of microbiota, the inconsistency in their quantitative and qualitative metabolic activity, and their influence on absorption and bioavailability of xenobiotics can be linked to many cases of inter-individual PK/TK differences. The microbiota carries out biochemical reactions that are mostly hydrolysis and since the GI tract is practically an oxygen-free environment, no oxidation or reduction is expected. They are also capable of breaking down the conjugated metabolites such as glucuronides, sulfates, glutathiones, and other forms of conjugates. Other influences are reducing the intestinal transit time, hydrolysis of macromolecules, and dietary polymers in the large intestine. Contrary to humans the stomach, duodenum, jejunum, and proximal portion of ileum are sterile and free from microbiota; experimental animals such as rats, mice, guinea pigs, and rabbits have a stomach and small intestine heavily colonized with bacteria. Because of the animals' eating habit, type of food, and kind of bacteria, caution should be exercised in quantitative extrapolation from animal to human. For PK/TK studies, the plasma concentration–time profile of a compound is compared between a pseudo-germ-free animal model treated with antibiotics and a non-antibiotic-exposed animals administered with the same compound. An approach in estimating the extent of bacterial metabolism in humans is by comparing metabolic profile of a compound given orally with those estimated after the intravenous administration. A shortcoming of this approach is overlooking the biliary elimination of an intravenously administered compound and/or its metabolites, which may also undergo secondary or tertiary metabolism by bacteria.

8.2.2 Physicochemical Factors

8.2.2.1 Polymorphism and Chemical Nature of the Compound

The ability of a compound to crystallize more than one crystalline form is known as polymorphism. The different crystalline forms or polymorphs occur due to the different space lattice arrangements of regularly packed atoms or molecules or ions with three-dimensional repeating patterns. The number of polymorphs for a compound poses the question of which form would provide better solubility, dissolution, absorption, and thus better bioavailability? For example sulfonamids have 30 and steroid hormones have 42 polymorphs. The study of the polymorphic behavior of a compound is an important part of preformulating therapeutic oral dosage forms, and it includes both the active ingredients and excipients (Curatolo, 1998; Curatolo and Singhal, 2004). A challenging part of the study is the phase transition, which is the process of transition from one polymorph into another during storage or manufacturing processes (Aguiar et al., 1967; Aguiar and Zelmer, 1969; Singhal et al., 2004). There are a few terminologies that are related to polymorphism and deal with enhancing the solubility of polymorphs

- If solvent molecules are incorporated in the lattice of a polymorph, it is known as *solvates*, the solvates also form their own polymorphs, which are known as pseudopolymorphs.
- If the solvent in the lattice is water, the polymorphs are known as *hydrates.*
- If the regularly packed atoms and molecules of crystals are changed into randomly arranged molecules, for example by lyophilizing, it changes to *amorphous* state.

In general, the amorphous compounds are more soluble than their corresponding crystalline form; the hydrates have a lower dissolution rate than the anhydrous crystalline form. The polymorphs or pseudopolymorphs exist in two forms, the ones with the lowest energy, lowest solubility, and highest melting point are considered *stable* polymorphs, and those with a lower melting point, higher solubility, and higher energy are known as *metastable.* During the course of the dosage form development, the lowest energy polymorph, which is soluble and stable, is selected for the drug dosage form development. The selection of stable polymorph is usually carefully investigated to rule out the metastable polymorphs before the final formulation is decided. The solubility and dissolution behavior of polymorphs is an important factor in the absorption and bioavailability of an orally administered drug. The absorption of drugs with low solubility rate is considered solubility-limited absorption. It is common to choose a stable amorphous form of a compound to achieve the required solubility/dissolution and bioavailability. However, it should be noted that due to the greater molecular mobility, the stability of amorphous is generally less than polymorphs.

A theoretical methodology to determine the maximum absorbable amount based on the solubility of the compound is

$$Ab_{max} = S \times k_a \times SIWV \times SITT \tag{8.55}$$

where Ab_{max} is maximum absorbable amount if the small intestine could be saturated with a drug for 270 min (the average intestinal transit time), S is the solubility of the compound at pH 6.5 in mg/mL, k_a is the first-order absorption rate constant in min^{-1}, $SIWV$ is the small intestinal water volume in mL (≈250 mL) and $SITT$ is the small intestinal transit time in min. Equation 8.55 does not take into consideration the presystemic intestinal and hepatic metabolism.

8.2.2.2 Partition Coefficient and Molecular Properties

The octanol/water partition coefficient, CLOG, MLOG and distribution coefficient (Todeschini and Consonni, 2000; Moriguchi et al., 1992, 1994) are discussed in Chapter 2 (Sections 2.2.1.1 and 2.2.1.2). They are quantitative measurements of lipophilicity or hydrophobicity of a compound. The $MLOGP_{coeff}$ and $CLOGP_{coeff}$ have been used in various theoretical approaches to predict the solubility and permeability of a compound. One of these approaches is known as Lipinski's Rule of Five (Lipinski et al., 1997, 2001; Veber et al., 2002).

The Rule of Five is established based on the evaluation of known physicochemical characteristics of drugs in USAN (United States Adopted Names) entries, and relies essentially on four parameters of the drugs, namely partition coefficient, molecular weight, the number of H-bond donors, and the number of H-bond acceptors. It is called the Rule of Five because the cutoff values for each of the four parameters are five or a multiple of five. According to this rule, a drug candidate has poor absorption or permeation when:

- The calculated logarithm of partition coefficient exceeds 5, that is, $CLOGP_{coeff} > 5$, (the cutoff for $MLOGP_{coeff}$ is >4.15).
- The molecular weight of a drug is over 500.
- There are more than five NH bonds and OH bonds in the molecule. These bonds represent the hydrogen bond donating ability of the molecule. Therefore, if a drug molecule has less than five H-bond donors of the type OH and NH collectively, the permeability of the drug would be enhanced and if the H-bond donors are more than five the permeability would be reduced.
- The number of hydrogen bond acceptors of the type Os (oxygen) and Ns (nitrogen) exceeds ten.

The Rule of Five identifies a number of molecular properties that influence the absorption of drugs and help to predict the bioavailability of a compound. However, the Rule has some limitations and cannot be used as a blanket rule for all xenobiotics. For example, it is not applicable to the compounds that are considered substrates or modulators of protein transporters, or in certain cases the molecular weight cutoff at 500 does not significantly separate compounds with poor oral bioavailability from those with acceptable one. It is a qualitative rule, which identifies the compounds either to have good or poor absorption (Andrews et al., 2000).

In addition to the criteria defined in the Rule of Five, a number of other molecular properties have also been identified as important factors influencing the absorption. The examples are

- Water complexation with amide bonds negatively influences the bioavailability of a compound.
- Molecular flexibility helps the absorption and permeation of drugs through biological barriers.
- The polar surface area prevents the absorption of xenobiotics in the small intestine.
- Rotational bonds of 10 or fewer with polar surface area equal to or less than 140A° will have a high probability of acceptable bioavailability.

8.2.2.3 Particle Size, Porosity, and Wettability

The influence of particle size on absorption of xenobiotics may be considered from two different perspectives, (1) the absorption of particles, and (2) the influence of the particle size on the dissolution of particles at the site of absorption.

Absorption of particles: Particles of appropriate size (<100 nm) can cross the intestinal barrier through two pathways:

- Absorption via M-cells in the Peyer's patches and the isolated follicles of the gut-associated lymphoid tissue (GALT)
- Absorption through the intestinal enterocytes

Particles of a larger size (>1 μm) may get absorbed through GALT, but remains trapped in the Preyer's patchs. Physicochemical factors such as surface charge and hydrophobicity can also influence absorption of the particles. The study of particulate absorption is of interest in areas such as traditional solid dosage forms, nanoparticles, polymeric delivery systems, and biopharmaceuticals (Jani et al., 1990; Florence, 2005; Chen et al., 2011). The main dispute, however, is whether relying solely on absorption of particles would be sufficient to provide the intended optimum therapeutic outcomes.

Influence of the particle size on the dissolution of particles at the site of absorption: The size of particles usually is interpreted as the surface area in contact with the environment they are in contact with. For a known amount of a solid compound, decreasing the particle size is synonymous with increasing their surface for the contact with the dissolving fluid, thus reducing the time required for a compound to dissolve (i.e., a faster dissolution rate), and facilitates the absorption of the compound. In manufacturing of oral solid dosage forms, milling or micronization for particle size reduction are commonly performed as approaches to improve solubility, on the basis of the increase in surface area. The conventional methods of particle size reduction have long been employed to enhance the bioavailability of drugs. The size reduction limit of the conventional methods is approximately 2–5 μm, which for some compounds may not be enough to improve considerably the drug solubility in the small intestine. Furthermore, solid powders with very small particle size have low flow properties and high adhesion characteristics and often are difficult to handle. Small particles of hydrophobic compounds have a tendency to establish static surface charges, which often causes agglomeration in aqueous environments. For this reason solid dispersion methodology is used, which essentially is the molecular mixture of compounds with poor solubility in hydrophilic carries and solidifying the mixture by cooling or evaporation (Higuchi, 1963; Ozeki et al., 2005; Hou et al., 2013; Patel and Dave, 2013). As a result of solid dispersion, the dissolution rates increase and the absorption in the GI tract is facilitated. The solid dispersion is also used to decrease the solubility and release of a compound to provide sustained release dosage forms. It should be noted that reduction of particle size is not always desirable, for example

- Irritant compounds in fine particle size may cause more gastrointestinal irritation.
- Compounds with low therapeutic index with fast dissolution of their fine particle size may raise their plasma concentration rapidly and bring on unexpected side effects.
- Reductions of particle size of compounds that are unstable in the GI tract accelerate the instability of the compounds in the GI tract.

To enhance the solubility of a compound at the site of absorption, in addition to the reduction of its particle size, the porosity of the particles with larger sizes can also increase the surface area and enhance the solubility. Particles in solid dispersions have been found to have a higher degree of porosity.

Another factor that plays an important role in the solubility of hydrophobic particles at the site of absorption is the wettability of the particles. Additions of any agents that reduce the hydrophobic layer of the particles or carriers enhance their wettability and facilitate their solubility. Surfactants are often used to reduce the interfacial tension of particles and promote wetting of the particles.

8.2.2.4 Dosage Factors

The oral route of administration is the most convenient route for administration of therapeutic agents and it provides versatility for formulation of various dosage forms:

- Solutions
- Syrups
- Suspensions
- Emulsions

- Soft and hard gelatin capsules
- Compressed tablets (coated & uncoated)

The formulation of dosage forms is beyond the scope of this book. However, their important characteristics, which influence the absorption of the compounds from the GI tract, are discussed briefly in this section.

8.2.2.4.1 *Solutions and Syrups*

The active ingredient in these dosage forms is already in soluble form and its absorption is expected to be fast and complete. However, the chemical nature of compounds and their chemical interaction with the acidic pH of the stomach may hinder their absorption. Examples are salts of acidic drugs soluble in neutral pH, but insoluble as free acid in the stomach. Often the insoluble form aggregates and forms large particles, which further reduce the solubility of the compound.

8.2.2.4.2 *Suspensions*

Unlike the solutions and syrups, the active ingredient of suspensions is not in soluble form and the dosage form is prepared by suspending the drug particles in a liquid medium. The absorption of the particles then requires the dissolution of particles in the GI tract. The advantage of this dosage form is circumventing the disintegration process, which is required for solid dosage forms like tablets, and having the suspended small particles ready to be dissolved at the site of absorption. However, if the dissolution were slow, the absorption would become dissolution-rate limited process. Most suspensions have surfactants in their formulation, which helps the wettability of the particles and their dissolution. Another formulation factor that may influence the absorption of drugs is the viscosity of the suspension due to the addition of hydrophilic polymers such as sodium carboxymethyl cellulose, alginates, or gelatin, which may retard the passive diffusion of the drug and gastric emptying rate.

8.2.2.4.3 *Emulsions*

Emulsions are prepared by dissolving hydrophobic compounds in oil and dispersing it in water to form O/W emulsion, or dissolving hydrophilic compounds in water and dispersing it in oil to form W/O emulsion. There are also emulsions prepared by dispersing oil in water and then in oil O/W/O, or dispersing water in oil and then in water W/O/W emulsions. The oils used in emulsions are edible olive, corn, and peanut oil. Other compounds such as beeswax, long chain acids, and alcohol are also included. Surfactants and emulsifying agents such as gelatin, acacia, or tragacanth are also added. The bioavailability of drugs given in emulsion is comparable or better than suspension. The only concern is the entrapment of drugs within the micelles, which may delay the absorption of drugs given orally using this dosage form. In addition to the passive diffusion of compounds in emulsion form, the small droplets of emulsions or microscopic droplet of microemulsions can cross the barrier by pinocytosis.

8.2.2.4.4 *Soft and Hard Gelatin Capsules*

The basic formulation of soft gelatin is encapsulation of compounds dissolved in an appropriate vehicle, for example, polyethylene glycol or edible oils, in soft gelatin capsules. The soft shell dissolves or disintegrates in the stomach and its content is released into the GI environment. If the content is formulated as a suspension, the particles must be dissolved and then absorbed.

The hard gelatin capsules often contain powder, which after dissolution and/or breakdown of the hard gelatin shell in GI tract undergo dissolution at the site of absorption. Because of the fast disintegration of the shell and quick availability of the loose powder for dissolution, the hard gelatin capsules usually have a better bioavailability than the compressed tablets. Depending on its chemical nature, the compound released from the soft or hard gelatin capsules will be absorbed in stomach or small intestine by one of the absorption mechanisms discussed in Chapter 2.

8.2.2.4.5 Compressed Tablets (Uncoated & Coated)

The uncoated compressed tablet consists of active ingredient and inert additives that are compressed together under pressure. Depending on the compound and formulation, the compressed tablet can be taken orally, which requires disintegration and dissolution in the GI tract before the absorption, or if it designed to be dissolved in water before the administration, such as effervescent tablets. The dissolved molecules of active ingredient are then ready for absorption upon administration. In designing the uncoated tablets, the physicochemical properties of compound such as pKa, crystalline form, and amorphous form influence the solubility and absorption of compounds and must be taken into consideration. The amount of additives in a tablet (i.e., binders such as methyl hydroxyl ethyl cellulose, diluents like starch or avicel, disintegrants such as primogel and lubricant like magnesium stearate) and the applied compression of a tablet machine must be carefully optimized to prevent ill-timed disintegration and dissolution of the tablet in the GI tract and avoid turning the disintegration and dissolution into rate limiting steps for the absorption.

Coated tablets are compressed tablets coated with appropriate water-soluble polymers such as cellulose acetate phthalate polymer or neutral methyl cellulose, hydroxyl propyl methyl cellulose, or poly vinyl pyrrolidone followed by application of beeswax to polish the tablet. The water-soluble polymers dissolve in the GI tract and compressed tablets will undergo disintegration and dissolution. The enteric-coated tablets are coated with polymers that are insoluble in the acidic pH of the stomach.

8.2.2.5 Disintegration and Dissolution

The conventional solid dosage forms (compressed tablets and capsules) are the most economical and convenient-to-administer dosage forms for most therapeutic agents. The feasibility of developing a solid dosage form is considered initially for a new drug entity. The release of a drug from a solid dosage form, whether it is fast release or controlled release, is influenced and characterized by parameters related to the physicochemical properties of the drug and disintegration and dissolution characteristics of its dosage (Hörter and Dressman, 2001). The *in vitro* disintegration/dissolution testing does not guarantee therapeutic efficacy. However, dissolution testing in particular has been accepted as a reliable *in vitro* methodology, revealing qualitatively the behavior of solid dosage form in the GI tract and as an *in vitro* predictive marker for bioavailability and bioequivalence of the dosage form.

The disintegration apparatus recommended by the United States Pharmacopeia (USP) is known as basket–rack assembly, which determines whether a solid dosage form disintegrates in a given time under experimental conditions and in a liquid medium, which are described in the Pharmacopeia. Thus, the disintegration test neither implies nor tests for the dissolution of the drug or the dosage form. For dissolution testing, there are four USP recommended and standardized dissolution apparatuses (Basket, Paddle, Reciprocating Cylinder, and Flow-through Cell) with performance verification tests (PVT), which are used to determine the concentration of the active ingredient of the dosage form in a given medium at a specified time. The most widely used one is the Paddle apparatus.

Dissolution testing is usually performed to measure the release from the solid dosage form as a test for quality assurance and compliance with the dissolution requirement in the individual monograph. In certain cases, *in vitro–in vivo* correlation can be established between the release of a therapeutic agent from the dosage form and absorption of the active ingredient (Dressman et al., 1998). The U.S. Food and Drug Administration (FDA) requires that all drugs listed in the USP conform to the standards laid out in their respective monographs. Conforming to the standards ensures the reproducibility of the dosage forms and their bioavailability.

8.2.2.6 Dissolution Models

Various mathematical relationships have been proposed to describe the dissolution rates of solid dosage forms and factors that can influence the dissolution rate of a solid dosage.

8.2.2.6.1 Noyes–Whitney Model

This is the oldest and a classic model for defining the dissolution rate of solid particles

$$\frac{dC}{dt} = K_{Dis} \times S_{Area}(C_{sat} - C_t) \tag{8.56}$$

where the rate of dissolution is dC/dt; the dissolution constant is K_{Dis}; the surface area of dissolving particles is S_{Area}; the saturation concentration of a drug in the diffusion layer, which is the stagnant layer around the dissolving particles is C_{sat}; and the concentration in dissolution media at time t is C_t.

$$K_{Dis} = \frac{D_{coeff}}{x} \tag{8.57}$$

(D_{coeff}) is the diffusion coefficient and x is the thickness of diffusion layer.

When the concentration of drugs in the dissolution media is negligible compared to the saturation concentration, that is, $C_t \ll C_{sat}$, it implies the sink condition, which is normally achieved by keeping the volume of media large, under the sink condition Equation 8.56 changes to

$$\frac{dC}{dt} = K_{Dis} \times S_{Area} \times C_{sat} \tag{8.58}$$

In calculating the rate of dissolution, the Noyes–Whitney relationship assumes S_{area} a constant.

8.2.2.6.2 Hixson and Crowell's "Cube Root" Model

This model is based on the concept of a changing surface area with respect to time during dissolution without including it as a variable directly into the model

$$W_0^{1/3} - (W_0 - W_t)^{1/3} = Kt \tag{8.59}$$

where W_0 is the initial weight or mass, W_t is the mass remained at time t, K is a complex constant proportional to diffusion coefficient, solubility, the cube root of particles number, the particle size, and the thickness of the diffusion layer. This model is applied to the dissolution of monodispersed powder consisted of particles with uniform size. The plot $W_0^{1/3} - (W_0 - W_t)^{1/3}$ versus time will be linear with a slope of K.

The variable $(W_0 - W_t)^{1/3}$ is the cube root of amount dissolved at time t. ∴ $W_0^{1/3} - (W_0 - W_t)^{1/3} = W_0^{1/3} - W_t^{1/3}$ is the cube root of mass remained at time t and Equation 8.59 can be written as

$$W_0^{1/3} - W_t^{1/3} = Kt \tag{8.60}$$

Dividing both sides of Equation 8.60 by $W_0^{1/3}$ yields the following relationship, which describes the dissolution rate in terms of the fraction of mass at time zero:

$$1 - \frac{W_t^{1/3}}{W_0^{1/3}} = \frac{Kt}{W_0^{1/3}}$$

$$\therefore (1 - f_t)^{1/3} = K_N t \tag{8.61}$$

where the fraction of dose dissolved is f_t and the fraction of the initial mass remaining to be dissolved is $1 - f_t$. (K_N) is the normalized value of K with respect to the cube root of the initial mass. Thus, under sink conditions a plot of the fraction of initial mass remaining to be dissolved versus time is linear with a slope of K_N.

8.2.2.6.3 Wagner's First-Order Kinetic Model

There are two approaches in this model and both are based on first-order kinetics:

The first approach is when under the sink condition the surface area changes with time and decreases exponentially, that is,

$$(S_{Area}) = (S_{Area})_0 e^{-k_s(t-t_0)} \tag{8.62}$$

where the initial surface area at time zero is $(S_{Area})_0$ and k_s is the first-order rate constant for the reduction of surface area.

Under the sink condition, the rate of dissolution is equal to

$$\frac{dA}{dt} = K \times S_{Area} \times C_s \tag{8.63}$$

where the dissolution rate with dimension of mass/time is dA/dt, K is a constant with units of length/time, S_{Area} has units of length^2 and C_s is the aqueous solubility of drug with units of mass/volume.

Substituting Equation 8.62 in Equation 8.63 for $t \geq t_0$ yields

$$\frac{dA}{dt} = KC_s(S_{Area})_0 e^{-k_s(t-t_0)} \tag{8.64}$$

Integration of Equation 8.64 for $t \geq t_0$ yields

$$A_t = A_{t_0} + \frac{K}{k_s} \times C_s \times (S_{Area})_0[1 - e^{-k_s(t-t_0)}] \tag{8.65}$$

Setting $M = (K/k_s) \times C_s \times (S_{Area})_0$ simplifies Equation 8.65 to 8.66

$$A_t = A_{t_0} + M[1 - e^{-k_s(t-t_0)}] \tag{8.66}$$

where the amount dissolved at time t is A_t, at time zero is A_{t_0}, and M has units of mass. As $t \Rightarrow \infty$, when the dosage form is dissolved completely, $(A^\infty = A_{t_0} + M)$, therefore

$$A^\infty - A_t = Me^{-k_s(t-t_0)}, \text{or} \tag{8.67}$$

$$\log(A^\infty - A_t) = \log M - \frac{k_s}{2.303}(t - t_0) \tag{8.68}$$

Equations 8.67 and 8.68 are linear first-order equations in terms of the amount remaining to be dissolved at time t.

The second approach is based on the assumption that the surface area is a variable proportional to the amount remaining to be dissolved

$$S_{Area} = k_p(A^\infty - A_t) \tag{8.69}$$

where (k_p) is proportionality constant with units of area/mass.

Substituting Equation 8.69 in Equation 8.63 yields

$$\frac{dA}{dt} = KC_s k_p (A^\infty - A_t) \tag{8.70}$$

Setting $k = KC_s k_p$ changes Equation 8.70 to 8.71

$$\frac{dA}{dt} = k(A^\infty - A_t) \tag{8.71}$$

The constant (k) in Equation 8.71 have units of time^{-1}

$$\frac{\text{Length}}{\text{Time}} \times \frac{\text{Mass}}{\text{Volume}} \times \frac{\text{Length}^2}{\text{Mass}} = \text{Time}^{-1}$$

Integration of Equation 8.71 yields

$$\log(A^\infty - A_t) = \log A^\infty - \frac{kt}{2.303} \tag{8.72}$$

Equation 8.72 is another first-order linear equation in terms of the amount remaining to be dissolved. Both approaches assume first-order kinetics under sink conditions.

8.2.2.6.4 Kitazawa's Model

This model is similar to the Wagner model and is defined as

$$\ln A^\infty/(A^\infty - A_t) = K_D \times t \tag{8.73}$$

The variables have the same definitions as described before. In terms of fraction dissolved, the equation changes to

$$\ln 1/(1 - f_t) = Kt \tag{8.74}$$

Equations 8.73 and 8.74 are a linear relationship under sink conditions and the plot of $\ln A^\infty/(A^\infty - A_t)$ or $\ln 1/(1 - f_t)$ versus t should yield a straight line with a slope equal to the dissolution rate constant. The model may exhibit a biphasic straight line, where the first phase is attributed to the disintegration and the second phase to the dissolution of the dosage form.

8.2.2.6.5 Higuchi's "Square Root of Time Plot" Model

This model is proposed for the diffusion-controlled drug release from a matrix under a perfect sink condition. It assumes that the release of the drug is through diffusion from a planar surface of a constant area. The equation for a homogeneous matrix is

$$A_s = \sqrt{D_{coeff}(2A_{total} - C_s)C_s t} \tag{8.75}$$

where (A_s) is the amount of drug released per unit of surface area, (D_{coeff}) is diffusion coefficient, (A_{total}) is the total amount of drug in the homogeneous matrix, (C_s) is drug solubility in the matrix and t is the time.

Defining Equation 8.75 in terms of fraction of dose released from the matrix, that is,

$$f_t = K_H t^{1/2} \tag{8.76}$$

where K_H is Higuchi's release rate constant defined as

$$K_H = \frac{\sqrt{D_{coeff}(2A_{total} - C_s)C_s}}{A_{total}} \tag{8.77}$$

Under the assumptions of this model, a plot f_t versus square root of time yields a straight line with a slope of K_H.

The Higuchi's model for solid dosage forms with a heterogeneous matrix is

$$A_s = \sqrt{\frac{D_{coeff}E}{\tau}(2A_{total} - EC_s)C_s t} \tag{8.78}$$

where E and τ are the porosity and tortuosity of the matrix, respectively.

For a homogeneous or heterogeneous matrix, a plot of the fraction of the drug released from the matrix versus the square root of time is linear.

8.2.2.6.6 Weibull's Model

This model is also known as Langenbucher's model and it is based on the following equation, which defines the cumulative fraction of dose dissolved per units of time:

$$f_t = 1 - e^{-\frac{(t-T_{lag})^b}{a}} \tag{8.79}$$

In the above equation f_t is the ratio of A_t/A^∞, T_{lag} is the lag time of dissolution, "a" is a time scale constant and "b" is a constant of the curvature, for exponential curves ($b = 1$), for S-shaped curves ($b > 1$), and for parabolic curve ($b < 1$). The linear version of the model is

$$\log[-\ln(1 - f_t)] = b\log(t - T_{lag}) - \log a \tag{8.80}$$

For most solid dosage forms T_{lag} is equal to zero. The time scale "a" is usually defined as $a = (T_d)^b$, where T_d is the dissolution time or the time point that correspond to $-\ln(-1 - f_t) = 1$, that is, $f_t = 0.693$. Therefore, the dissolution time (T_d) is the time required for 63.2% of the solid dosage to dissolve in dissolution media.

8.2.2.6.7 Korsmeyer–Peppas Model

This model is known as the Power Law and relates the fraction of the dose released from a polymeric dosage form to the time and mechanism of release:

$$f_t = \frac{A_t}{A^\infty} = at^n \tag{8.81}$$

where "a" is a combined constant of structural and geometrical characteristics of the dosage form, the parameter "n" is the release exponent refers to the release mechanism of drug from the dosage forms. When the release is Fickian diffusion, "n" is equal to 0.5. When the release mechanism deviates from the Fick's law of diffusion, it is identified as a non-Fickian release and "n" is between 0.5

and 1.0 (i.e., $0.5 < n < 1$). Adjusting Equation 8.81 for the lag time of dissolution yields Equation 8.82, and Equation 8.83 in its linear logarithmic form

$$\left(\frac{A_{t-T_{lag}}}{A^{\infty}}\right) = a(t - T_{lag})^{n} \tag{8.82}$$

$$\log\left(\frac{A_{t-T_{lag}}}{A^{\infty}}\right) = \log a + n\log(t - T_{lag}) \tag{8.83}$$

8.2.2.6.8 Nernst and Brunner Model

This model is a modified version of the original Noyes–Whitney equation and defines the dissolution as a diffusion limited two-step process

$$\frac{dC}{dt} = \frac{D_{coeff} \times S_{area}}{V \times x}(S - C_i) \tag{8.84}$$

where dC/dt is the rate of dissolution, D_{coeff} and S_{area} are the diffusion coefficient and surface area respectively, V and x are the volume of dissolution media and thickness of diffusion layer, S is the solubility, and C_i is the concentration of the dissolved drug.

8.2.2.7 *In Vitro–In Vivo* Correlation (IVIVC) of Dissolution Data

The dissolution data are the appropriate *in vitro* properties of a solid dosage form that can be correlated with an appropriate pharmacokinetic parameter of an orally administered drug; parameters such as maximum plasma concentration or area under the plasma concentration–time curve. The definition of IVIVC recommended by the United States Pharmacopeia (USP, 2004) is "the establishment of a relationship between a biological property, or a parameter derived from a biological property produced from a dosage form, and a physicochemical property of the same dosage form." The significance of establishing such a correlation is to use the dissolution test as a surrogate for bioavailability studies and to reduce the requirement for frequent bioequivalence studies in humans, in particular at the initial stages of the drug approval process (Dressman et al., 1998; Sirisuth and Eddington, 2002; Emami, 2006).

The FDA definition of IVIVC is "a predictive mathematical model describing a relationship between an *in vitro* property of a dosage form and an *in vivo* response." The FDA guidance classifies levels A, B, C, and multiple C of correlation for IVIVC (FDA, CDER, 1995, 1997) as follows:

8.2.2.7.1 Level A

The level A correlation is a point-to-point relationship between the *in vitro* dissolution rate and the *in vivo* input rate. The relationship can be linear or nonlinear and is considered the most useful approach in defining a direct relationship between the *in vitro* dissolution rate and a pharmacokinetic parameter, such as the fraction of the dose absorbed at different time intervals. The A level correlation indicates that the *in vitro* dissolution rate is indeed a surrogate for an *in vivo* response.

8.2.2.7.2 Level B

Level B correlation is based on the application of the statistical moment theory. The correlation is the comparison of the mean *in vitro* dissolution time $(MDT)_{invitro}$ with *in vivo* mean residence time (MRT) or mean *in vivo* dissolution time $(MDT)_{invivo}$.

$$(MDT)_{invitro} = \int_0^{\infty} \frac{(A^{\infty} - A_t)dt}{A^{\infty}} \tag{8.85}$$

$$MRT = \frac{AUMC}{AUC} \tag{8.86}$$

$$(MDT)_{invivo} = (MRT)_{solid} - (MRT)_{solution} \tag{8.87}$$

In Equations 8.85 through 8.87, A^{∞} and A_t have the same definitions as before; $AUMC$ and AUC are areas under the moment curve and areas under plasma concentration–time curve, respectively; $(MRT)_{solid}$ and $(MRT)_{solution}$ are the mean residence time of the solid dosage form and solution of the same drug. Level B correlation relies on the comparison of calculated parameters; the correlation is not as analytical as level A.

8.2.2.7.3 Level C

Level C correlation is based on the comparison of a dissolution time, for example $t_{50\%}$, $t_{63\%}$, or $t_{90\%}$, which represents the time required for certain fraction of solid dosage form to be dissolved *in vitro* with the average of appropriate pharmacokinetic parameters like AUC or Cp_{max}, or T_{max}. This level of correlation establishes a single point relationship between a dissolution parameter and a PK parameter.

8.2.2.7.4 Multiple-Level C

The multiple-level C correlation relates the amount of compound dissolved at various time points to one or several appropriate pharmacokinetic parameters. This level of correlation establishes a relationship between several dissolution and pharmacokinetic parameters and is considered more realistic and dependable.

8.2.2.8 Biopharmaceutics Classification System (BCS)

BCS provides a scientific basis for the correlation of the *in vitro* dissolution of immediate release solid dosage form and *in vivo* bioavailability/bioequivalence studies. It helps to predict the *in vivo* performance of drug products from *in vitro* measurements of permeability and solubility. The system recognizes fundamental parameters of dissolution along with solubility and permeability characteristics of a drug. According to this system, a drug has rapid dissolution when more than 85% of the dose dissolves within 30 minutes in a volume of less than 900 mL using one of the USP dissolution apparatuses. It is considered highly soluble when the highest dose strength is soluble in 250 mL of water or less over a pH range of 1–7.5 at 37°C. It is considered highly permeable when the extent of absorption in the human is greater or equal to 90% of the administered dose, or in comparison with an intravenous reference dose.

The BCS deals with the parameters of dissolution and absorption through a set of dimensionless numbers as follows.

8.2.2.8.1 Absorption Number

Absorption number (A_n) is defined as the ratio of the residence time in the GI tract or GI transit time (T_R) to time required for complete absorption (T_{abs}) (Löbenberg and Amidon, 2000a,b; Martinez and Amidon, 2002)

$$A_n = \frac{T_R}{T_{abs}} = \frac{P_{eff}}{R}(T_R) \tag{8.88}$$

where P_{eff} is the effective permeability and R is the radius of GI tract.

8.2.2.8.2 Dissolution Number

Dissolution number (D_n) is the ratio of residence time in the GI tract to dissolution time

$$D_n = \frac{T_R}{T_{dis}} = \left(\frac{3D_{diff}}{r^2}\right)\left(\frac{S}{\rho}\right)(T_R) \tag{8.89}$$

where D_{diff} is the diffusivity, r is the particle radius, S is the solubility (mg/mL) and ρ is the density (Löbenberg and Amidon, 2000a,b; Martinez and Amidon, 2002).

8.2.2.8.3 Dose Number

(N_{dose}) is the ratio of drug in solution to its solubility, that is, defining the dose as a function of its solubility:

$$N_{dose} = \frac{(D_H/V_{water})}{S} \tag{8.90}$$

(D_H) is the highest dose and V_{water} is the volume of water taken with the dose (~250 mL).

According to the BCS, drug substances are categorized into four classes as follows (Amidon et al., 1995; Dahan et al., 2009).

8.2.2.8.4 Class I: Compounds with High Permeability and High Solubility

The compounds in this category are expected to dissolve rapidly and absorb rather completely. It is assumed that the high permeability and high solubility allow the drugs in this class to be present at high concentrations at the absorption site. Thus, they can saturate efflux, absorptive and metabolic proteins, and reduce the clinical significance of interaction with these proteins. The drugs in this class exhibit a high absorption number and high dissolution numbers. Their IVIVC is expected when the dissolution rate is slower than the gastric emptying rate and is considered the rate-limiting step for absorption. When the rate-limiting step for absorption is the gastric emptying rate, no correlation is required. Examples of drugs in Class I are propranolol, metoprolol, verapamil, midazolan, and diltiazem, of which most are considered Pgp substrate/modulators.

8.2.2.8.5 Class II: Compounds with High Permeability and Low Solubility

The Class II compounds have a high absorption number, but a low dissolution number. The rate-limiting step for absorption is the compound's dissolution. As a result, they have slower absorption than compounds in Class I; and because their low bioavailability is related to the low dissolution, IVIVC is expected for compounds in Class II, unless the dose is very high. The high permeability of compounds in this category indicates that they are lipophilic compounds and thus permeate rapidly into the enterocytes by passive diffusion. However, due to their low solubility, the amount present at the site of absorption is not at a level to saturate the intestinal efflux and metabolic proteins; hence, the proteins influence their bioavailability (Löbenberg and Amidon, 2000a,b). Examples of compounds in Class II are nifedinpine, ketoconazole, carbamezapine, phenytoin, ketoprofen, mefenamic acid, naproxen, danazol, and glibenclamide. Many of these compounds are substrates for Pgp and CYP3A4.

8.2.2.8.6 Class III: Compounds with Low Permeability and High Solubility

The low permeability of compounds in this class is the rate-limiting step for their absorption. Thus, any variability in the absorption of compounds would be due to the physiological factors governing the permeability of compounds in the GI tract and the role of intestinal apical transporters. Limited or no correlation is expected. Examples of compounds in Class III are acyclovir, cimetidine, ranitidine, neomycin B, atenolol, and captopril.

8.2.2.8.7 Class IV: Compounds with Low Permeability and Low Solubility

Class IV examples include furosemide, taxol, and hydrochlorothiazide. The compounds in this class have poor bioavailability, but may achieve sufficient solubility in the presence of biological surfactants and act similar to the compounds of Class III. Limited or no correlation is expected.

8.2.2.9 Biowaivers

The term "biowaivers" refers to the exceptions to the requirement to perform bioequivalence studies. The exceptions, determined by regulatory authorities, are based on reasonable IVIVC and application of BCS. Biowaivers reduce the exhaustive study on humans, thus making the dug development process more economical and fast. Between the four levels of IVIVC, level A or multiple-level C correlation can serve as a surrogate for *in vivo* investigations. The guidance for the waiver of bioavailability and bioequivalence studies has been adopted by FDA under the title of Biopharmaceutics Classification System (BCS).

As it was mentioned earlier, the application of IVIVC is mostly to establish dissolution specifications and criteria for biowaivers. The FDA guidance identifies five categories of biowaivers (CDER/FDA, 2000; Yu et al., 2002):

- Biowaivers without an IVIVC
- Biowaivers using an IVIVC—non-narrow therapeutic index compounds
- Biowaivers using an IVIVC—narrow therapeutic index compounds
- Biowaivers when *in vitro* dissolution is independent of dissolution test conditions
- Situations where an IVIVC is not recommended for biowaivers

8.2.2.10 Biopharmaceutical Drug Disposition Classification System (BDDCS)

As it was discussed earlier, the BCS (Biopharmaceutics Classification System) relies mainly on dissolution/solubility and permeability of drug products. Thus, the confidence in granting biowaivers is based on IVIVC and BCS predictions of the *in vivo* performance of the products. A modified version of BCS classification has been suggested, which takes into consideration the overall disposition of drugs in the body (Wu et al., 2005; Chen et al., 2011). The classification is known as Biopharmaceutics Drug Disposition Classification System (BDDCS). The disposition includes routes of drug elimination with emphasis on metabolism, influence of influx and efflux transport proteins, food effects, and transporter's influence on post-absorptive plasma levels following oral and intravenous dosing.

Based on evaluated PK parameters, it is suggested that (Wu et al., 2005)

- Class I and II are eliminated from the body predominantly through metabolism
- Class III and IV are eliminated unchanged through bile and urine
- The effect of transporter is the lowest for compounds in Class I
- The effect of efflux proteins for Class II is predominant, with transporter-enzyme interplay, mainly through CYP3A4 and Phase II metabolism
- Absorptive transporter effect is predominant for Class III
- Compounds in Class IV are substrates for absorptive and efflux proteins
- The effect of high-fat food on the extent of absorption for Class I compounds is not significant, but it increases the extent of absorption for Class II compounds, decreases for Class III, and difficult to predict for Class IV

It is further suggested that BDDCS expands the compounds in Class I and makes them eligible for a biowaiver (Wu et al., 2005; Chen et al., 2011).

8.2.2.11 Chirality and Enantiomers

A molecule or a system is considered chiral when it cannot be superimposed to its mirror image. Chirality can be considered a property of many molecules in nature and biological systems and arises from a chiral atom; most often is a carbon atom bonded to four different groups. The molecule and its mirror image are known as enantiomers. A mixture of equal amounts of enantiomers of a chiral xenobiotic is called racemic mixture. Enantiomers with different 3-dimensional shapes behave differently in their interaction with proteins and biological systems. They have different rates and extent of absorption, and have different pharmacological responses. On the basis of their optical rotation, the enantiomers are identified as R (from the Latin word "rectus") for the clockwise rotation and S (from the Latin word "sinister") for counterclockwise rotation. There is evidence that chiral molecules such as nonsteroidal anti-inflammatory drugs (e.g., ibuprofen, fenoprofen, 2 aryl-propionic and benoxaprofen) may undergo enantiomer interconversion in the GI tract (Foster and Jamali, 1988; Simmonds et al., 1980; Hutt and Caldwell, 1984; Caldwell et al., 1988; Evans, 1992; Jamali, 1994; Geisslinger et al., 1995), and a racemic mixture of drugs such as L-dopa, methotrexate, terbutaline, leucovorin, and cephalaxin exhibit selectivity in the absorption of one enantiomer. Often a racemic mixture may exhibit complete absorption, but has a low bioavailability due to the extensive first pass metabolism of one of the enantiomers; examples are chiral molecules of beta-blockers and calcium channel blockers (Wade et al., 1973; Hendel and Brodthagen, 1984; Echizen et al., 1985; Schilsky et al., 1989; Borgstrom et al., 1989). Thus, two enantiomers of a chiral xenobiotic may have (1) different rates and extents of absorption, (2) inconsistent interaction with metabolic enzymes, transporters and receptors, and (3) different potency and effectiveness; and significantly variable bioavailability, PK/PD or TK/TD (Tamai et al., 1988).

8.2.3 Other Factors

8.2.3.1 Effect of Food and Drink on Absorption

Food and drink can influence the absorption of xenobiotics and complicate the PK/TK analysis (Custodio et al., 2008). The influence is mostly on the rate and extent of absorption and thus the bioavailability of the administered dose (Fleisher et al., 1999; FDA Guidance for Industry, 2002). Depending on the type of compound, food and drink may delay its rate and extent of absorption, or accelerate the absorption, or remain unchanged. There are certain physiological changes associated with consumption of food that have impact on the absorption of xenobiotics:

- Food may increase splanchnic blood flow and thus increase the absorption, bioavalability, and first-pass effect (Liedholmand and Melander, 1986)
- Food may interact with specific transport proteins (Patil et al., 1998)
- Certain components of food may bind the compound and reduce bioavailability (Brownand and Juhl, 1976; Leyden, 1985)
- Food intake causes the fluctuation of gastric pH; it rises initially and then declines. (Dressman, 1986; Kararli, 1995)
- Food stimulates bile secretion, which may enhance the absorption of lipophilic compounds (Kararli, 1995)
- Food prevents diffusion of poorly permeable compounds (Fleisher et al., 1999)
- Food reduces gastric emptying rate (Singh, 1999; Schmidtand and Dalhoff, 2002)

Food has no effect on BCS Class I compounds, may increase absorption of Class II compounds, decrease the absorption of Class III and has mixed effect on Class IV (Fleisher et al., 1999; Radwan et al., 2012; Heimbach et al., 2013). The therapeutic agents with reduced absorption and bioavailability in the presence of food include the compounds that are unstable in gastric fluid and because of presence of food stay longer in the stomach. As well, the compounds that interact irreversibly with food components and are not absorbed, or absorbed later in their passage through the small intestine when they have already passed their site of absorption. On the other hand, the absorption of drugs that are poorly soluble in GI fluids may increase in the presence of food because of the secretion of bile salts. The secretion of bile salts is the direct result of food intake, which releases the contents of the gallbladder into the small intestine. The classical examples of these types of compounds are griseofulvin and cyclosporine, in which their extent of absorption increases in the presence of bile salts. Food with a high fat content is expected to influence the absorption and bioavailability of drugs that are a substrate for Pgp and CYP3A4; for example, Class II and III compounds of BCS could be included in this category. However, the effect of a high fat diet on the absorption of Class I compounds is expected to be insignificant. Ordinarily high fat meals contain significant amount of triglycerides, which are hydrolyzed in the GI tract to monoglycerides, known to inhibit Pgp. The type of food influences the absorption of drugs like theophylline, albuterol, and epinephrine. The high fat meal may increase their bioavailability, whereas carbohydrate meals may reduce the extent of absorption. The main reason is the influence of food constituents on the activity of proteins. The physical nature of food can also influence the absorption of therapeutic agents in solid dosage forms. For example, meals that enhance the viscosity of the GI tract fluid delay the disintegration and dissolution of solid dosage forms.

Fruit juices (grapefruit, orange, and apple) are well known to influence the bioavailability of drugs. Grapefruit juice, as it was discussed in previous chapters, inhibits intestinal CYP3A4, and Pgp, and thus, increases the bioavailability of these protein substrates. The herbal extracts containing compounds such as flavonoids, antioxidants, curcumin (the principal curcuminoid of the popular Indian spice turmeric), etc., could also potentially influence the absorption of drugs by inhibiting efflux proteins.

8.2.3.2 Effect of Disease States

Discussion of the effect pathophysiological factors of disease states and their influences on the process of absorption is beyond the scope of this chapter. It is, however, important to indicate that certain disease states such as ulcerative colitis, thyrotoxicosis (a toxic condition resulting from excessive amounts of thyroid hormones, hyperthyroidism), achlohydria (absence or reduction of hydrochloric acid in stomach), cirrhosis of the liver, biliary obstruction, and Crohn's disease (inflammatory bowel diseases) can affect the absorption and bioavailability of drugs significantly. Furthermore, it has been reported that gastric emptying can be delayed for patients with diabetes mellitus. Therefore, depending upon the type of disease, different changes in the process of drug absorption is expected.

8.2.3.3 Genetic Polymorphism

In addition to the various factors discussed in this chapter that can influence the absorption of xenobiotics from the GI tract, genetic variations in humans and experimental animals has also been recognized as an important determinant of individual variability of absorption. The influence of genetic polymorphism is through the control of metabolic enzymes and protein transporters. For example, ABCB1 gene encodes the efflux protein, Pgp, which is highly polymorphic in different human populations; or more than 20 variants of CYP3A4 are known across ethnic groups, which have altered enzymatic activities, ranging from very low to high catalytic efficiency. The mutation of the breast cancer resistance protein, BCRP, which is an ABC transporter, ABCG2, involved in the intestinal absorption and biliary excretion of xenobiotics, is widely present in ethnic groups (Gradhand and Kim, 2008; Zhu et al., 2008).

8.2.3.4 Other Factors Influencing Oral Absorption of Drugs

- Formulation of different dosage forms of a drug may result in different rates and extents of absorption. For example, a controlled-release oral dosage form has a different profile of absorption than the conventional solid dosage forms prepared by direct compression.
- The timing of drug administration relative to food intake is also an important factor that may influence the absorption of drugs. For example, taking penicillin G or erythromycin 1 hour before or 2 hours after a meal improves the rate and extent of absorption of these drugs. In this particular case, the reason is a reduction in gastric emptying time and length

of exposure to the acidic environment of the stomach. As it was discussed earlier, the rate and extent of absorption for some drugs, such as halofantrine, are improved by taking the drug with fatty foods.

- The presence of other substances, such as environmental chemicals, other drugs, dietary constituents, tobacco smoking, and alcohol use are all known to influence the absorption of drugs. Their influences are mainly by induction or inhibition of drug-metabolizing enzymes, such as CYP3A4, and modulation of drug transporters, such as Pgp, which alter drug efficacy, induce drug-drug and drug-chemical interactions and drug side effects.
- Factors such as age, gender, pregnancy, exercise, starvation, and circadian rhythm can also contribute to individual variation of the absorption and in general PK/PD of administered drugs.

8.3 MECHANISTIC ABSORPTION MODELS

A number of mechanistic/predictive models are reported in the literature with the intention of establishing strategies for the selection and formulation of a lead candidate and ultimately forecasting the rate and extent of oral absorption. These are predictive models used in drug discovery and development. They include:

- Absorption potential models
- Dispersion models
- Compartmental absorption and transit (CAT) model
- Gastrointestinal transit absorption (GITA) model
- Advanced compartmental absorption and transit (ACAT) model
- Advanced dissolution, absorption and transit (ADAM) model
- Grass model

8.3.1 Absorption Potential Models

The absorption potential is a unitless number calculated by the following equation to predict the fraction of the administered dose absorbed by passive diffusion (Dressman et al., 1985):

$$(Abs)_{potential} = \log \frac{P_{coeff} \times S_{int} \times f_u \times V_{lumen}}{Dose} \tag{8.91}$$

where P_{coeff} is the partition coefficient of the compound; S_{int} is the intrinsic solubility of the compound in the water, f_u is the unionized fraction at pH 6.5; and V_{lumen} is the volume of lumen in liters (≈250 mL = 0.25 L). The absorption potential estimated by Equation 8.91 correlated well with the fraction of the dose absorbed in selected chemicals.

Another suggested relationship for estimating absorption potential is (Balon et al., 1999).

$$(Abs)_{potential} = \log\left[\frac{[(Dis)_{coeff}]^{6.8} \times (S)^{6.8} \times V_{lumen}}{Dose}\right] \tag{8.92}$$

where $[(Dis)_{coeff}]^{6.8}$ is the distribution coefficient (Chapter 2, Section 2.2.1.2) at pH 6.8 and $(S)^{6.8}$ is the solubility at pH 6.8.

The next proposed relationship is based on the pH independence of the product of total solubility and the distribution coefficient (Sanghvi et al., 2001). In other words, the product of distribution coefficient and solubility at any pH is equal to the intrinsic values of partition coefficient and intrinsic solubility. Thus, the relationship for absorption potential, identified as a modified absorption potential, $(MAbs)_{potential}$ is

$$(MAb)_{potential} = \log\left[\frac{P_{coeff} \times S_{int} \times V_{lumen}}{Dose}\right] \tag{8.93}$$

For the lumen volume of 250 mL, the relationship can be simplified to

$$(MAb)_{potential} = \log\left[\frac{P_{coeff} \times S_{int}}{4Dose}\right] \tag{8.94}$$

8.3.2 Dispersion Models

The main assumption of the dispersion model is considering the small intestine as a uniform tube with constant dispersion behavior, constant concentration, and axial velocity. The model postulates that the dynamics of drug absorption is based on the following convection-dispersion equation (Ho et al., 1983; Yu et al., 1996a,b; Huang et al., 2009).

$$\frac{\partial C}{\partial t} = \alpha \frac{\partial^2 C}{\partial z^2} - v\frac{\partial C}{\partial z} - \gamma C \tag{8.95}$$

where C is the concentration of the drug in the GI tract; α is the hybrid dispersion coefficient for mixing, molecular diffusion, and physiological factors; z is the axial distance from the stomach; v is the velocity in the axial direction; and γ is the absorption rate constant

A different approach to the dispersion model is the inclusion of solubility and permeability effects on the absorption, and the assumption that the intestine is a tube with a variable diameter and dimension, with species-dependent physiological parameters (Willmann et al., 2003, 2004, 2007). The modified model allows the estimation of the fraction of an administered dose in a certain segment of the tube at time t. In rats the length of the duodenum is assumed at 10 cm and its initial pyloric diameter as 1.5 mm and proximal portion of the jejunum as 2.0 mm. The lengths of the jejunum and ileum are assumed at 87 cm and 3 cm, respectively, with a diameter of 2.0 mm for the jejunum and the proximal portion of the ileum and 1 mm for the distal portion of the ileum at the ileocecal connection. The administered dose amount that is absorbed passively into the portal vein in the region of $[z, z + dz]$ in a time interval of $[t, t + dt]$ is defined as

$$\frac{d^2 M_{PV}(z,t)}{dzdt} = P_{intestinal} C_{lumen}(z,t)\frac{dA_{eff}(z)}{dz} \tag{8.96}$$

where d^2M is the second derivative of administered dose per body weight; $P_{intestinal}$ is the apparent intestinal permeability coefficient of the drug, $A_{eff}(z)$ is the effective surface at the intestinal position, and C_{lumen} is the luminal concentration which can be treated as the compound's intestinal solubility, that is,

$$C_{lumen} = \begin{cases} C_{lumen} & if \quad C_{lumen} \le S_{intestinal} \\ S_{intestinal} & if \quad C_{lumen} > S_{intestinal} \end{cases} \tag{8.97}$$

The dispersion models have been developed within the framework of a physiology-based model for rats, monkeys, and humans. The related mathematical relationships have provided realistic predicted values for intestinal permeability, absorbable amount, luminal concentration, and other pertinent parameters when the presystemic metabolism is negligible (Willman et al., 2003, 2004, 2007; Huang et al., 2009).

8.3.3 Compartmental Absorption and Transit (CAT) Model

The CAT model assumes the intestine as a physiological compartment in a series with different volume, different flow rate, and linear transfer kinetics. The small intestinal transit time (SITT) (the residence time) is assumed to be the same in all compartments (Yu et al., 1996a,b; Huang et al., 2009), that is,

$$\frac{V_1}{Q_1} = \frac{V_2}{Q_2} = \dots\dots \frac{V_n}{Q_n} = \frac{V_N}{Q_N} = \frac{1}{K_t} = \frac{\langle T_{SITT} \rangle}{N} \tag{8.98}$$

where V terms represent the volumes; Q terms are the flow rates; K_t is the transit rate constant; N is the number of compartments; and $\langle T_{SITT} \rangle$ is the mean SITT.

The rate equation of the model is (Yu et al., 1996b)

$$\frac{dY_n}{dt} = K_t Y_{n-1} - K_t Y_n - K_a Y_n, \quad n - 1,2,3\dots,7 \tag{8.99}$$

where n is the number of compartments; Y_n is the percent of dose at the nth compartment; K_t and K_a are the rate constants of transit and absorption, respectively.

The model was first developed based on the assumption that the drug is neither absorbed nor degraded at the G.I. tract, and the percent rate of the dose entering the colon (dY_c/dt) that is, the transit rate is

$$\frac{dY_c}{dt} = K_t Y_n \tag{8.100}$$

where (Y_c), the percent dose leaving the small intestine, is

$$Y_c = 1 - \sum_{n-1}^{N} \frac{(K_t^* t)^{n-1}}{(n-1)!} e^{-K_t^* t} \tag{8.101}$$

(N) is the number of total compartment, ($K_t^* t$) is the convolution of (K_t) and (t), and K_t is estimated as

$$K_t = \frac{N}{\langle T_{SITT} \rangle} \tag{8.102}$$

The small intestinal absorption rate is then defined as

$$\frac{dY_a}{dt} = Ka \sum_{n=1}^{N=7} Y_n \tag{8.103}$$

The fraction of the dose absorbed, F_a, can be calculated as

$$F_a = K_a \int_0^{\%} \sum_{n=1}^{7} Y_n dt = 1 - \frac{K_t^7}{(K_t + K_a)^7} \tag{8.104}$$

The CAT model can easily be incorporated into the compartmental model, for example, for a multicompartmental model with an absorption rate into the central compartment and elimination from the central compartment, the differential equations are

$$\frac{dC_1}{dt} = \frac{D}{V}\frac{dY_a}{dt} - (k_{12} + k_{13} + k_e)C_1 + k_{21}C_2 + k_{31}C_3 \tag{8.105}$$

8.3.4 Gastrointestinal Transit Absorption (GITA) Model

The model was first developed for the analysis and prediction of plasma concentration after oral administration of drugs given in a solution to rats (Sawamoto et al., 1997). The model was then used for the solid dosage form (Fujioka et al., 2007) and also applied to human data to predict the plasma concentration (Kimura and Higaki, 2002). In the GITA model, the gastrointestinal tract is divided into eight sections and the absorption parameters of each section are incorporated into a linear first-order GITA kinetic model (Figure 8.4). The stomach is assumed as one section and the small intestine is divided into six segments of duodenum, upper jejunum, lower jejunum, upper ileum, lower ileum, and cecum; the large intestine is assumed as a separate section. The transit of drugs from the section that is not the absorption site of the drug to the next is assumed to follow first-order kinetics. The absorption at the absorption site(s) is also assumed to follow first-order kinetics (Figure 8.4).

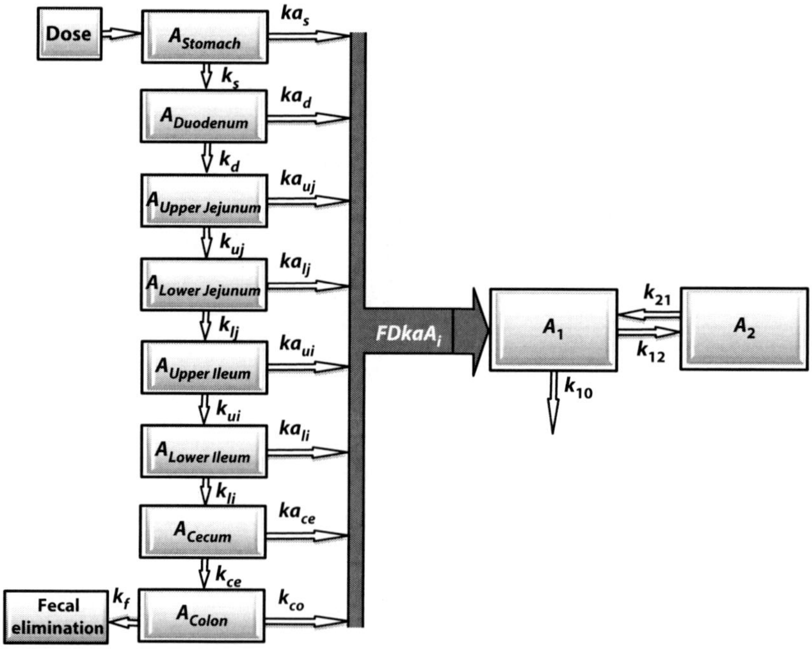

FIGURE 8.4 Linear first-order gastrointestinal transit absorption (GITA) model with eight physiological sections of GI tract and the absorption parameters of each section.

The gastric emptying rate and intestinal transit rate for each section is defined as

$$\left(\frac{dA}{dt}\right)_{stomach} = -(k_{stomach} + (k_a)_{stomach})A_{stomach} \tag{8.106}$$

where $(dA/dt)_{stomach}$ is the gastric emptying rate; $(A_{stomach})_{t=0} = Dose$; $k_{stomach}$ is the emptying rate constant; $(k_a)_{stomach}$ is the rate constant of absorption from stomach.

The intestinal transit rate is

$$\frac{dA_{i+1}}{dt} = A_i k_i - (k_{i+1} + k_{a_{i+1}})A_{i+1} \tag{8.107}$$

The subscript i stands for intestinal sections.

The following three mathematical steps are used to predict and estimate plasma concentration:

- Using the convolution method the time course of amount in each section can be estimated. The Laplace transform of the amount of drug in segment $i + 1$ is

$$\tilde{A}_{i+1}(s) = \frac{k_i \tilde{A}_i(s)}{s + k_{i+1} + k_{a_{i+1}}} \tag{8.108}$$

- The Laplace transform of absorbable fraction in each section is

$$\tilde{f}_{i+1}(s) = \frac{k_i \tilde{f}_i(s)}{s + k_{i+1} + k_{a_{i+1}}} \tag{8.109}$$

- The Laplace transform of the fraction of the dose absorbed in each section is

$$\tilde{f}_{a_i}(s) = k_{a_i} \tilde{f}_i(s) \tag{8.110}$$

The prediction of plasma concentration following oral administration is accomplished by means of the convolution method, assuming the total absorption rate-time data corresponds to input function and PK parameters after intravenous administration corresponds to transfer function. The Laplace transform of the plasma concentration is defined as

$$\tilde{C}p_{oral}(s) = \left(\frac{Dose_{oral}}{Dose_{iv}}\right)\left(\tilde{f}_{a_s}(s) + \sum_{i=d}^{co} \tilde{f}_{a_i}(s)\right)\tilde{C}p_{iv}(s) \tag{8.111}$$

When the orally administered drug is in a solid dosage form, the dissolution and solubility of the dosage form can become the rate-limiting step in the absorption process. The modified GITA model that takes into the consideration the dissolution of the solid dosage form (Fujioka et al., 2007) is depicted in Figure 8.5.

The Laplace transform of the amount of drug dissolved in the stomach and small intestine (Fujioka et al., 2007) are

$$(\tilde{A}_{sto})_{soln}(s) = \frac{(k_{sto})_{dis}}{s + (k_{sto})_{soln}} \times (\tilde{A}_{sto})_{solid}(s) \tag{8.112}$$

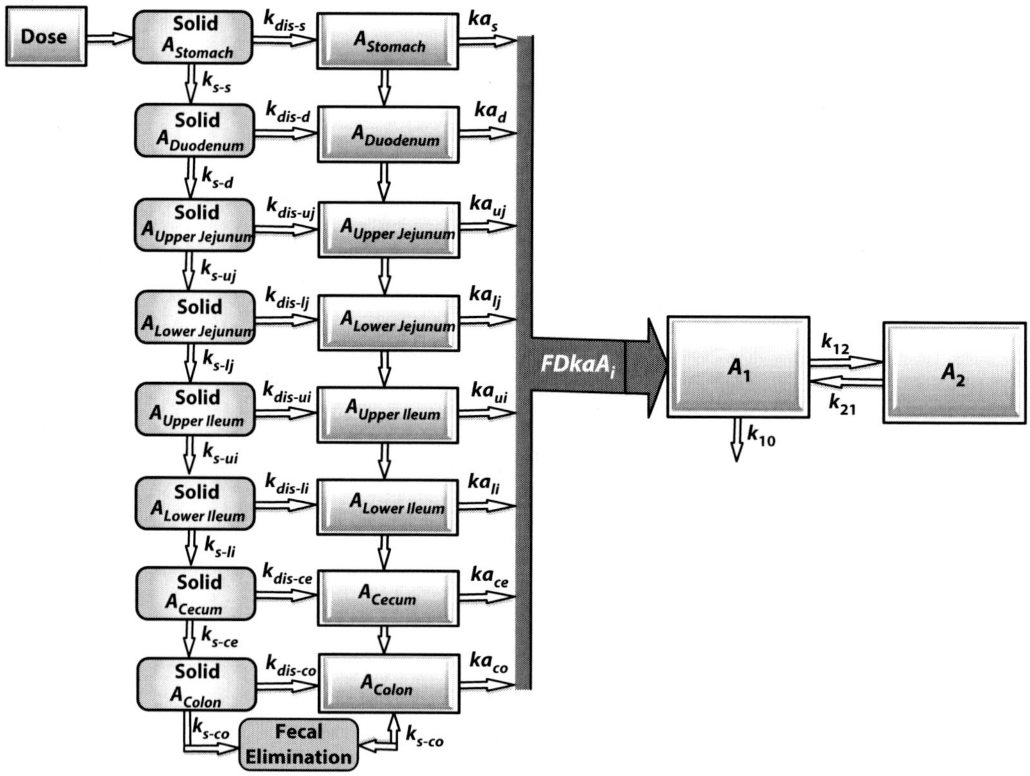

FIGURE 8.5 The modified GITA model that takes into consideration the dissolution of the solid dosage forms.

$$(\tilde{A}_{i+1})_{soln}(s) = \frac{(k_i)_{soln}}{s + (k_{i+1})_{soln} + k_{a_{i+1}}} \times (\tilde{A}_i)_{soln}(s) + \frac{(k_{i+1})_{dis}}{s + (k_{i+1})_{soln} + k_{a_{i+1}}} \times (\tilde{A}_{i+1})_{solid}(s) \quad (8.113)$$

where $(\tilde{A}_{sto})_{solid}(s)$ and $(\tilde{A}_{i+1})_{solid}(s)$ are the Laplace transform of the solid dosage form in the stomach (*sto*) and sections of intestine ($i + 1$), respectively; $(\tilde{A}_{sto})_{soln}(s)$ and $(\tilde{A}_{i+1})_{soln}(s)$ are the Laplace transform of the amount of the solid dosage form dissolved in the stomach and sections of intestine, respectively; $(k_{sto})_{dis}$ and $(k_{i+1})_{dis}$ are the dissolution rate constants; $(k_{sto})_{soln}$ and $(k_{i+1})_{soln}$ are the transit rate constants; and $k_{a_{i+1}}$ is the absorption rate constant in the intestine. Absorption from the stomach is considered negligible.

Based on Equation 8.113, the following relationship is developed to describe the Laplace transform of the fraction of the dose in solution form available for absorption at the site of absorption (Fujioka et al., 2007):

$$\tilde{f}_{i+1}(s) = \frac{(k_i)_{soln}}{s + (k_{i+1})_{soln} + k_{a_{i+1}}} \times (\tilde{f}_i)_{soln}(s) + \frac{(k_{i+1})_{dis}}{s + (k_{i+1})_{soln} + k_{a_{i+1}}} \times (\tilde{f}_{i+1})_{solid}(s) \quad (8.114)$$

where $\tilde{f}_{i+1}(s)$ is the fraction of dose available for absorption at the site of absorption; $(\tilde{f}_i)_{soln}(s)$ and $(\tilde{f}_{i+1})_{solid}(s)$ are the fraction of the dose dissolved in section i and the fraction remaining to be dissolved, respectively.

The Laplace transforms of the rate of absorption in section i and the predicted plasma concentration are

$$f_{a_i}(s) = k_{a_i} \times \tilde{f}_i(s) \tag{8.115}$$

$$\tilde{Cp}_{oral}(s) = \frac{D_{oral}}{D_{iv}} \times \sum_{i=d}^{cc} \tilde{f}_{a_i}(s) \times \tilde{Cp}_{iv}(s) \tag{8.116}$$

where $\tilde{Cp}_{oral}(s)$ and $\tilde{Cp}_{iv}(s)$ are the Laplace transforms of plasma concentration after oral and intravenous administration; D_{oral} and D_{iv} are the oral and intravenous doses, respectively. A convolution program can then estimate the inverse Laplace transforms and (Cp – *time*) profile following oral administration.

8.3.5 Advanced Compartmental Absorption and Transit (ACAT) Model

The principles and assumptions of the ACAT model are similar to those of the CAT and GITA models. The differences are the inclusion of first-pass metabolism and colon absorption (Yu et al., 1996a,b; Agoram et al., 2001; Bolger et al., 2009; www.simulations-plus.com; Parrott et al., 2009). The model considers the GI tract as nine linked compartments. The first and the last compartments are the stomach and the ascending colon, respectively. The seven other compartments, similar to GITA, represent seven regions of the small intestine (Figure 8.6). The transfer kinetics between the compartments is assumed linear; metabolism and transport are assumed nonlinear and the model considers six states for a solid dosage form in the GI tract. The six states are: unreleased, undissolved, dissolved, degraded, metabolized, and absorbed (Huang et al., 2009). The physicochemical

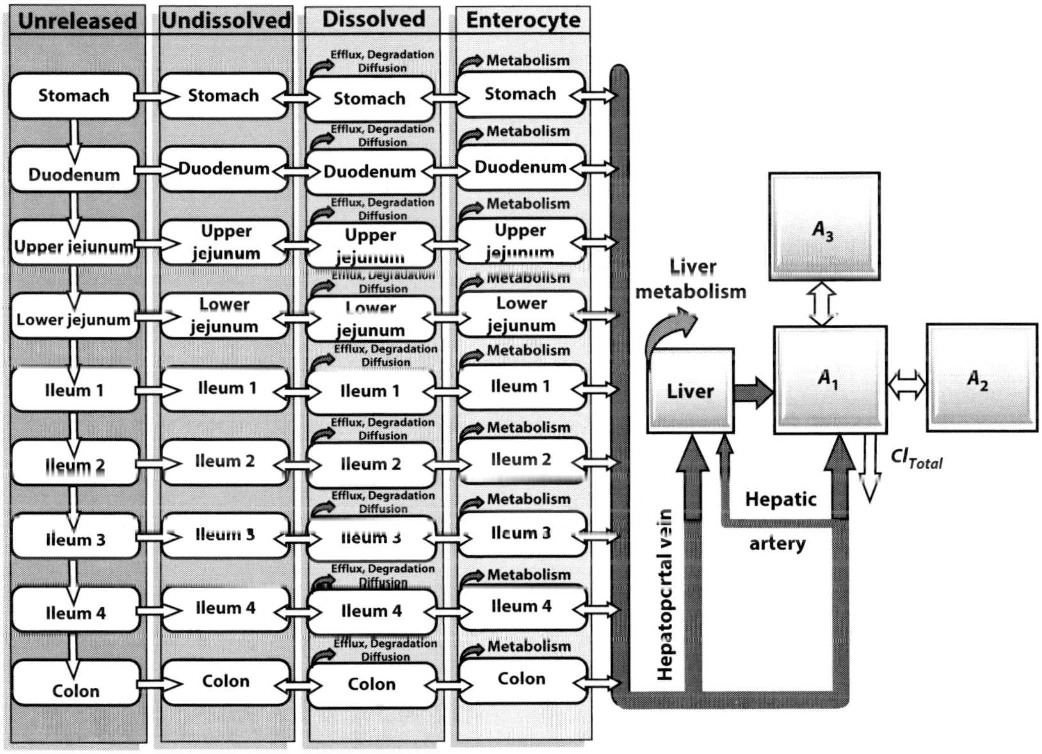

FIGURE 8.6 Advanced compartmental absorption and transit (ACAT) model with nine anatomical regions of GI tract as linked compartments with first-order transfer kinetics and six states of unreleased, undissolved, dissolved, degraded, metabolized, and absorbed.

parameters of the drug (e.g., pKa, solubility, particle size, particle density and permeability) and physiological parameters for each section/compartment (including pH, transit time, length, gastric emptying, luminal transport and metabolism) are set according to the published data (Parrott et al., 2009; Huang et al., 2009). The linear and nonlinear features of the ACAT model with the relevant parameters are included in the commercially available simulation software GastroPlus®.

8.3.6 Advanced Dissolution, Absorption, and Transit (ADAM) Model

ADAM is a similar model to GITA, CAT, and in particular to ACAT. It considers the GI tract as nine compartments that are anatomically identifiable (Figure 8.7). The absorption process is described in terms of formulation, dissolution, precipitation, degradation, permeability, metabolism, transport, and transit through the sections of the GI tract. Similar to previous models, the transit is governed by first-order kinetics. The commercially available software (Simcyp®—http://www.simcyp.com) is the incorporation of ADAM with population-based pharmacokinetics. Two processes of absorption from the stomach and metabolism at the colon are considered negligible and the rates of other processes are defined by a set of ordinary differential equations (Jamei et al., 2009)

$$\frac{dA_{S,n}}{dt} = -\frac{dA_{dis,n}}{dt} - k_{t,n}A_{S,n} + k_{t,n-1}A_{S,n-1} + \frac{dA_{F,n}}{dt} \tag{8.117}$$

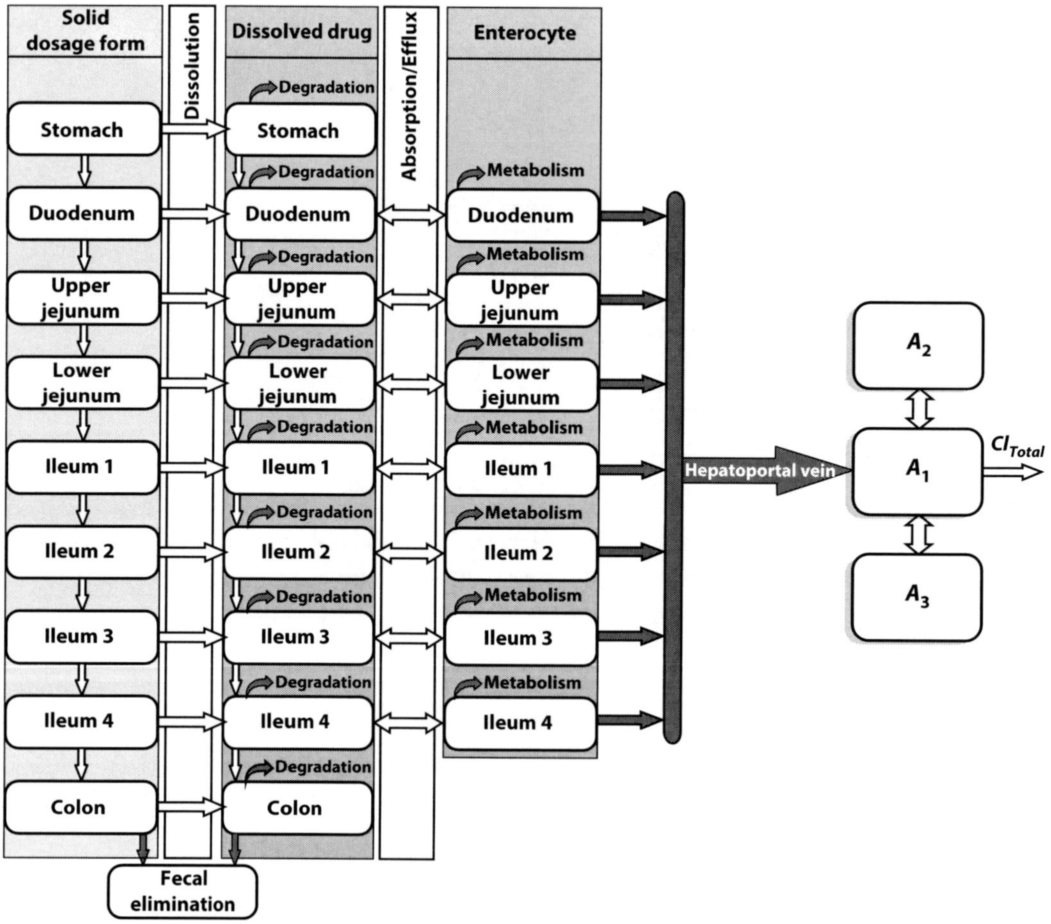

FIGURE 8.7 The advanced dissolution, absorption, and transit (ADAM) model.

$$\frac{dA_{D,n}}{dt} = \frac{dA_{dis,n}}{dt} - (k_{deg,n} + k_{a_n} + k_{t,n})A_{D,n} + k_{t,n-1}A_{D,n-1} + \gamma_n CL_{u_{int}-T,n} - f_{u_{gut}} C_{ent,n} \quad (8.118)$$

$$\frac{dC_{ent,n}}{dt} = \frac{1}{V_{ent,n}}(A_{dis,n}k_{a_n} - C_{ent,n}Q_{ent,n} - f_{u_{gut}} C_{ent,n}[Clu_{int-G,n} + Cl_{u_{int}-T,n}]) \quad (8.119)$$

where $A_{S,n}$ is the amount of solid dose available for dissolution; $dA_{dis,n}/dt$ is the dissolution rate; $dA_{F,n}/dt$ is the release rate of solid dosage form due to disintegration and deaggregation; $k_{\text{deg},n}$ and k_{a_n} are the degradation and absorption, respectively; γ_n is an adjustment factor for enterocyte uptake; $CL_{u_{int}-T,n}$ and $Cl_{u_{int}-G,n}$ are the intrinsic clearances due to enterocyte's efflux and metabolism, respectively; $f_{u_{gut}}$ is the unbound fraction in the enterocyte; $V_{ent,n}$ is the volume of enterocyte; and $Q_{ent,n}$ and $C_{ent,n}$ are the blood flow to the section and enterocyte concentration, respectively.

8.3.7 Grass Model

The Grass model is another multiple section model (stomach, duodenum, jejunum, ileum, and colon) for absorption of orally administered dosage forms. It takes into the consideration three parameters of solubility, permeability, and tissue surface area. The software for the model is used in conjunction with STELLA® (Structural Thinking Experimental Learning Laboratory with Animation) software (Grass, 1997).

REFERENCES

Agoram, B., Woltosz, W. S., Bolger, M. B. 2001. Predicting the impact of physiological and biochemical processes on oral drug bioavailability. *Adv Drug Deliv Rev* 50(1):S41–67.

Aguiar, A. J., Krc, J., Kinkel, A. W., Samyn, J. C. 1967. Effect of polymorphism on the absorption of chloramphenicol from chloramphenicol palmitate. *J Pharm Sci* 56:847–53.

Aguiar, A. J., Zelmer, J. E. 1969. Dissolution behavior of polymorphs of chloramphenicol palmitate and mefanamic acid. *J Pharm Sci* 58:983–7.

Amidon, G. L., Lennernäs, H., Shah, V. P., Crison, J. R. 1995. A theoretical basis for a biopharmaceutic drug classification: The correlation of *in vitro* drug product dissolution and *in vivo* bioavailability. *Pharm Res* 12(3):413–20.

Andrews, C. W., Bennett, L., Yu, L. W. 2000. Predicting human oral bioavailability of a compound: Development of a novel quantitative structure-bioavailability relationship. *Pharm Res* 17(6):639–44.

Balon, K., Riebesehl, B. U., Muller, B. W. 1999. Drug liposome partitioning as a tool for the prediction of human passive intestinal absorption. *Pharm Res* 16:882–8.

Bolger, M. B., Lukacova, V., Woltosz, W. S. 2009. Simulations of the nonlinear dose dependence for substrates of influx and efflux transportes in the human intestine. *AAPS J* 11(2):353–63.

Borgstrom, L., Nyberg, L., Jonsson, S., Lindberg, C., Paulson, J. 1989. Pharmacokinetic evaluation in man of terbutaline given as separate enantiomers and as the racemate. *Br J Clin Pharmacol* 27(1):49–56.

Bouchard, G., Pagliara, A., Pierre-Alain Carrupt, P. A., Bernard Testa, B., Gobry, V., Hubert, H., Girault, H. H. 2002. Theoretical and experimental exploration of the lipophilicity of zwitterionic drugs in the 1, 2-dichloroethane/water system. *Pharm Res* 19(8):1150–9.

Brownand, D., Juhl, R. 1976. Decreased bioavailability of digoxacin due to antacids and kaolin-pectin. *N Engl J Med* 19:1034–7.

Bruyere, A., Decleves, X., Bouzom, F., Ball, K., Marques, C., Treton, X., Pocard, M. et al. 2010. Effect of variations in the amounts of P-glycoprotein (ABCB1), BCRP (ABCG2) and CYP3A4 along the human small intestine on PBPK models for predicting intestinal first pass. *Mol Pharm* 7(5):1596–607.

Burks, T. F., Galligan, J. J., Porreca, F., Barker, W. D. 1985. Regulation of gastric emptying. *Fed Proc* 44:2897–901.

Cabral, D. J., Small, D. M. 1989. Physical chemistry of bile. In *Handbook of Physiology—The Gastrointestinal System III*, eds. S. G. Forte and B. B. Tauner, 621–661. New York: Waverly Press.

Caldwell, J., Hutt, A. J., Fournel-Gigleux, S. 1988. The metabolic chiral inversion and dispositional enantioselectivity of the 2-arylpropionic acids and their biological consequences. *Biochem Pharmacol* 37:105–14.

Chan, L. M., Lowes, S., Hirst, B. H. 2004. The ABCs of drug transport in intestine and liver: Efflux proteins limiting drug absorption and bioavailability. *Eur J Pharm Sci* 21(1):25–51.

Charman, M. N., Porter, C. J. H., Mithani, S., Dressman, J. 1997. Physicochemical and physiological mechanisms for the effects of food on drug absorption: The role of lipids and pH. *J Pharm Sci* 86(3):269–82.

Chen, H., Zheng, Y., Tian, G., Tian, Y., Zeng, X., Liu, G., Liu, K. et al. 2011. Oral delivery of DMAB-modified docetaxel-loaded PLGA-TPGS nanoparticles for cancer chemotherapy. *Nanoscale Res Lett* 6(1):4.

Chen, M-L., Amidon, G. L., Benet, L. Z., Lennernas, H., Yu, L. X. 2011. The BCS, BDDCS, and regulatory guidances. *Pharm Res* 28(7):1774–8.

Christian, P. E., More, J. G., Sorenson, J. A., Coleman, R. E., Welch, D. M. 1980. Effect of meal size and correction technique on gastric emptying time: Studies with two tracers and opposed detectors. *J Nucl Med* 21(9):883–5.

Coles, B. F., Chen, G., Kadlubar, F. F., Radominska-Pandya, A. 2002. Interindividual variation and organ-specific patterns of glutathione S-transferase alpha, mu and pi expression in gastrointestinal tract mucosa of normal individuals. *Arch Biochem Biophys* 403:270–6.

Coupe, A. J., Davis, S. S., Evans, D. F., Wilding, I. R. 1991a. Correlation of the gastric emptying of nondisintegrating tablets with gastrointestinal motility. *Pharm Res* 8(10):1281–5.

Coupe, A. J., Davis, S. S., Evans, D. F., Wilding, I. R. 1991b.Variation in gastrointestinal transit of pharmaceutical dosage forms in healthy subjects. *Pharm Res* 8:360–4,

Curatolo, W. 1998. Physical chemical properties of oral drug candidates in the discovery and exploratory development settings. *Pharm Sci Technol Today* 1:387–93.

Curatolo, W., Singhal, D. 2004. Drug polymorphism and dosage form design: A practical perspective. *Adv Drug Del Rev* 56:335–47.

Custodio, J. M., Wu, C-Y., Benet, L. Z. 2008. Predicying drug disposition, absorption/elimination/transporter interplay and the role of of food on drug absorption. *Adv Drug Deliv Rev* 60(6):717–33.

Dahan, A., Amidon, G. L. 2009a. Small intestinal efflux mediated by MRP2 and BCRP shifts sulfasalazine intestinal permeability from high to low, enabling its colonic targeting. *A J Physiol Gastrointest Liver Physiol* 297:371–7.

Dahan, A., Amidon, G. L. 2009b. Grapefruit juice and its constituents augment colchicine intestinal absorption: Potential hazardous interaction and the role of P-glycoprotein. *Pharm Res* 26(4):883–92.

Dahan, A., Miller, J. M., Amidon, G. L. 2009. Prediction of solubility and permeability class membership: Provisional BCS classification of the world's top oral drugs. *AAPS J* 11(4):740–6.

Davis, S. S., Hardy, J. G., Fara, J. W. 1986. Transit of pharmaceutical dosage forms through the small intestine. *Gut* 27:886–92.

de Waziers, I., Cugnenc, P. H., Yang, C. S., Leroux, J. P., Beaune, P. H. 1990. Cytochrome P 450 isoenzymes, epoxide hydrolase and glutathione transferases in rat and human hepatic and extrahepatic tissues. *J Pharmacol Exp Ther* 253:387–94.

Dietrich, C. G., Geier, A., Oude Elferink, R. P. 2003. ABC of oral bioavailability: Transporters as gatekeepers in the gut. *Gut* 52:1788–95.

Dressman, J. B. 1986. Comparison of canine and human gastrointestinal physiology. *Pharm Res* 3(3):123–31.

Dressman, J. B., Amidon, G. L., Fleisher, D. 1985. Absorption potential: Estimating the fraction absorbed for orally administered compounds. *J Pharm Sci* 74:588–9.

Dressman, J. B., Amidon, G. L., Reppas, C., Shah, V. P. 1998. Dissolution testing as a prognostic tool for oral drug absorption: Immediate release dosage forms. *Pharm Res* 15(1):11–22.

Dressman, J. B., Vertzoni, M., Goumas, K. 2007. Estimating drug solubility in the gastrointestinal tract. *Adv Drug Deliv Rev* 59:591–602.

Echizen, H., Vogelgesang, B., Eichelbaum, M. 1985. Effects of d, l-verapamil on atrioventricular conduction in relation to its stereoselective first-pass metabolism. *Clin Pharmacol Ther* 38:71–6.

Eckburg, P. B., Bik, E. M., Bernstein, C. N., Purdom, E., Dethlefsen, L., Sargent, M., Gill, S. R., Nelson, K. E., Relman, D. A. 2005. Diversity of the human intestinal microbial flora. *Science* 308:1635–8.

Egert, M., de Graaf, A. A., Smidt, H., de Vos, W. M., Venema, K. 2006. Beyond diversity: Functional microbiomics of the human colon. *Trends Microbiol* 14:86–91.

Emami, J. 2006. *In vitro–in vivo* correlation: From theory to applications. *J Pharm Pharm Sci* 9(2):169–89.

Evans, A. M. 1992. Enantioselectivity pharmacodynamics and pharmacokinetics of chiral nonsteroidal antiinflammatory drugs. *Eur J Clin Pharmacol* 42:237–56.

Fangerholm, U., Johansson, M., Lennernäs, H. 1996. Comparison between permeability coefficients in rat and human jejunum. *Pharm Res* 13:1336–42.

Fangerholm, U., Lindahl, A., Lennernäs, H. 2007. Regional intestinal permeability in rats of compounds with different physicochemical properties and transport mechanisms. *J Pharm Pharmacol* 49:687–90.

Fisher, M. B., Paine, M. F., Strelevitz, T. J., Wrighton, S. A. 2001. The role of hepatic and extrahepatic UDP-glucuronosyltransferases in human drug metabolism. *Drug Metab Rev* 33:273–97.

Fisher, M. B., VandenBranden, M., Findlay, K., Burchell, B., Thummel, K. E., Hall, S. D., Wrighton, S. A. 2000. Tissue distribution and interindividual variation in human UDP-glucuronosyltransferase activity: Relationship between UGT1A1 promoter genotype and variability in a liver bank. *Pharmacogenesis* 10:727–39.

Fleisher, D., Li, C., Zhou, Y., Pao, L-H., Karim, A. 1999. Drug, meal and formulation interactions influencing drug absorption after oral administration: Clinical implications. *Clin Pharmacokinet* 36(3):233–54.

Florence, A. T. 2005. Nanoparticle uptake by the oral route: Fulfilling its potential? *Drug Disc Today Tech* 2(1):75–81.

Food and Drug Administration. 2000. *Guidance for Industry, Waiver of in vivo Bioavailability and Bioequivalence Studies for Immediate Release Dosage Forms Based on a Biopharmaceutics Classification System*. Rockville, MD: Food and Drug Administration. Available at: http://www.fda.gov/cder/guidance/index.htm

Food and Drug Administration. 1997. Guidance for industry, extended release oral dosage forms: Development, evaluation and application of an *in vitro/in vivo* correlation. FDA,CDER. Available at: http://www.fda.gov/downloads/Drugs/GuidanceComplianceRegulatoryInformation/Guidances/ucm070239.pdf

Food and Drug Administration, 1995. Guidance for industry, immediate release solid oral dosage forms: Scale-up and post approval changes: Chemistry, manufacturing, and controls, *in vitro* dissolution testing, and *in vivo* bioequivalance documentation. FDA, CDER. Available at: http://www.fda.gov/downloads/Drugs/Guidances/UCM070636.pdf

Food and Drug Administration. Guidance for Industry. 2002. Food-effect bioavailability and fed bioequivalence studies. Food and Drug Administration; Rockville, MD. Available at: http://www.fda.gov/cder/guidance/index.htm.

Foster, R. T., Jamali, F. 1988. Stereoselective pharmacokinetics of ketoprofen in the rat. Influence of route of administration. *Drug Metab Dispos* 16:623–26.

Fujioka, Y., Kadono, Y., Fujie, Y., Metsugi, Y., Ogawara, K-I., Higaki, K., Kimura, T. 2007. Prediction of oral absorption of griseofulvin, a BCS class II drug, based on GITA model: Utilization of a more suitable medium for *in-vitro* dissolution study. *J Control Release* 119(2, 1):222–8.

Gai, M. N., Isla, A., Andonaeugui, M. T., Thielemann, A. M., Seitz, C. 1997. Evaluation of the effect of three different diets on the bioavailability of two sustained release theophylline matrix tablets. *Int J Clin Pharm Ther* 35:565–71.

Geisslinger, G., Menzel, S., Wissel, K., Brune, K. 1995. Pharmacokinetics of ketoprofen enantiomers after different doses of racemate. *Br J Clin Pharmacol* 40:73–75.

Gibbs, J. P., Yang, J. S., Slattery, J. T. 1998. Comparison of human liver and small intestinal glutathione S-transferase-catalyzed busulfan conjugation *in vitro*. *Drug Metab Dispos* 26:52–5.

Gradhand, U., Kim, R. B. 2008. Pharmacogenomics of MRP transporters (ABCC1–5) and BCRP (ABCG2). *Drug Metab Rev* 40:317–54.

Grass, G. M. 1997. Simulation models to predict oral drug absorption from *in vitro* data. *Adv Drug Del Rev* 23:1999–219.

Heimbach, T., Xia, B., Tsu-han Lin, T-H., He, H. 2013. Case studies for practical food effect assessments across BCS/BDDCS class compounds using in silico, *in vitro*, and preclinical *in vivo* data. *AAPS J* 15(1):143–58.

Hendel, J., Brodthagen, H. 1984. Enterohepatic cycling of methotrexate estimated by use of the D-isomer as a reference marker. *Eur J Clin Pharmacol* 26:103–7.

Higuchi, T. 1963. Theoretical analysis of rate of release of solid drugs dispersed in solid matrices. *J Pharm Sci* 52:1145–9.

Hillgren, K. M., Kato, A., Borchardt, R. T. 1995. *In vitro* systems for studying intestinal drug absorption. *Med Res Rev* 15(2):83–109.

Ho, N. F. H., Park, J. Y., Ni, P. F., Higuchi, W. I. 1983. Advanced quantitative and mechanistic approaches in interfacing gastrointestinal drug absorption studies in animals and humans. In *Animal Models for Oral Drug Delivery in Man: In situ and in vitro Approaches*, eds. W. Crouthanel and A. C. Sarapu, 27–106. Washington, DC: American Pharmaceutical Association.

Hoffman, A. F. 1962. The function of bile salts in fat absorption. *Biochem J* 89:57–68.

Hofmann, A. F., Borgstrom, B. 1962. Physico-chemical state of lipids in intestinal content during their digestion and absorption. *Fed Proc* 21:43–50.

Holmstock, N. F., Annaert, P. P., Augustijns, P. 2012. Boosting of HIV protease inhibitors by ritonvir: The relative role of CYP and P-gp inhibition based on Caco-2 monolayers versus *in situ* intestinal inhibitors by ritonavir in the intestine perfusion in mice. *Drug Metab Dispos* 40(8):1473–7.

Holzapfel, W. H., Haberer, P., Snel, J., Schillinger, U., Huisin't Veld, J. H. J. 1998. Overview of gut flora and probiotics. *Int J Food Microbiol* 41(2):85–101.

Hörter, D., Dressman, J. B. 2001. Influence of physicochemical properties on dissolution of drugs in the gastrointestinal tract. *Ad Drug Del Rev* 46(1–3):75–87.

Hou, P., Jian Ni, J., Cao, S., Lei, H., Zhengjun Cai, Z., Zhang, T., Yu, F., Tan, Q. 2013. Preparation and evaluation of solid dispersions of a new antitumor compound based on early-stage preparation discovery concept. *AAPS Pharm Sci Rech* 14(2):629–38.

Huang, W., Lee, S. L., Yu, L. X. 2009. Mechanistic approaches to predicting oral drug absorption. *AAPS J* 11(2):217–24.

Hutt, A. J., Caldwell, J. 1984. The importance of stereochemistry in the clinical pharmacokinetics of the 2-arylpropionic acid non-steroidal anti-inflammatory drugs. *Clin Pharmacokinet* (9):371–3.

Iannitti, T., Palmieri, B. 2010. Therapeutical use of probiotic formulations in clinical practice. *Clin Nutr* 29(6):701–25.

Ito, K., Kusuhara, H., Sugiyama, Y. 1999. Effect of intestinal CYP3A4 and P-glycoprotein on oral absorption-theoretical approach. *Pharm Res* 16(2):225–31.

Jamali, F. 1994. Pharmacokinetics of enantiomers of chiral non-steroidal antiinflammatory drugs. *Eur J Drug Metab Pharmacokinet* 26:259–74.

Jamei, M., Turner, D., Yang, J., Neuhoff, S., Polak, S., Rostami-Hodjegan, A. Tucker, G. G. 2009. Population-based mechanistic prediction of oral drug absorption. *AAPS J* 11(2):225–37.

Jani, P., Halbert, G. W., Langridge, J., Florence, A. T. 1990. Nanoparticle uptake by the rat gastro-intestinal mucosa: Quantitation and particle size dependency. *J Pharm Pharmacol* 42:821–6.

Johnson, L. R. 2007. *Gastrointestinal Physiology*. Philadelphia: Mosby, Elsevier.

Kaminsky, L. S., Zhang, Q-Y. 2003. The small intestine as a xenobiotic-metabolizing organ. *Drug Metab Disp* 31(12):1520–5.

Kararli, T. T. 1995. Comparison of the gastrointestinal anatomy, physiology, and biochemistry of humans and commonly used laboratory animals. *Biopharm Drug Dispos* 16:351–80.

Kimura, T., Higaki, K. 2002. Gastrointestinal transit and drug absorption. *Biol Pharm Bull* 25:149–64.

Kis, O., Robillard, K., Chan, G. N., Bendayan, R. 2010. The complexities of antiretroviral drug-drug interactions: Role of ABC and SLC transporters. *Trends Pharmacol Sci* 31(1):22–35.

Kivistö, K. T., Bookjans, G., Fromm, M. F., Griese E-U., Munzel, P., Kroemer, H. K. 1996. Expression of CYP3A4, CYP3A5 and CYP3A7 in human duodenal tissue. *Br J Clin Pharmacol* 42:387–9.

Klose, T. S., Blaisdell, J. A., Goldstein, J. A. 1999. Gene structure of CYP2C8 and extrahepatic distribution of human CYP2Cs. *J Biochem Mol Toxicol* 13:789–95.

Lapple, F., von Richter, O., Fromm, M. F., Richter, T., Thon, K. P., Wisser, H., Griese, E. U., Eichelbaum, M., Kivisto, K. T. 2003. Differential expression and function of CYP2C isoforms in human intestine and liver. *Pharmacogenetics* 13:565–75.

Lee, J. H., Park, T. G., Choi, H. K. 1999. Development of oral drug delivery system using floating microspheres. *J Microencapsul* 16:715–29.

Lee, Y-H., Perry, B. A., Lee, H-S. Kunta, J. R., Sutyak, J. P., Sinko, P. J. 2001. Differentiation of gut and hepatic first-pass effect of drugs: 1. Studies of verapamil in ported dogs. *Pharm Res* 18(12):1721–8.

Lennernäs, H. 1998. Human intestinal permeability. *J Pharm Sci* 87(4):403–10.

Leyden, J. J. 1985. Absorption of minocycline and tetracycline: Effect of food, milk and iron. *J Am Acad Dermatol* 12(2):308–12.

Liedholmand, H., Melander, A. 1986. Concomitant food intake can increase the bioavailability of propranolol by transient inhibition of its presystemic primary conjugation. *Clin Pharmacol Ther* 40:29–36.

Lin, Y. S., Dowling, A. L., Quigley, S. D., Farin, F. M., Zhang, J., Lamba, J., Schuetz, E. G., Thummel, K. E. 2002. Co-regulation of CYP3A4 and CYP3A5 and contribution to hepatic and intestinal midazolam metabolism. *Mol Pharmacol* 62:162–72.

Lipinski, C. A., Lombardo, F., Dominy, B. W., Feeney, P. J. 1997. Experimental and computational approaches to estimate solubility and permeability in drug discovery and development settings. *Adv Drug Del Rev* 23:3–25.

Lipinski, C. A., Lombardo, F., Dominy, B. W., Feeney, P. J. 2001. Experimental and computational approaches to estimate solubility and permeability in drug discovery and development setting. *Adv Drug Deliv Rev* 46(1–3):3–26.

Löbenberg, R., Amidon, G. I. 2000a. Modern bioavailability, bioequivalence and biopharmaceutics classification system. New scientific approaches to international regulatory standards. *Eur J Pharm Biopharm* 50(1):3–12.

Löbenberg, R., Amidon G. L. 2000b. Solubility as a limiting factor to drug absorption. In *Oral Drug Absorption, Prediction and Assessment, Drugs and the Pharmaceutical Sciences* ed. J. B. Dressman and Lennernas (136): 137–153, New York: Marcel Dekker, Inc.

Lown, K. S., Baily, D. G., Fontana, R. J., Janardan, S. K., Adair, C. H., Fortlage, L. A., Brown, M. B., Guo, W., Watkins, P. B. 1997a. Grapefruit juice increases felodipine oral availability in humans by decreasing intestinal CYP3A4 protein expression. *J Clin Invest* 99:2545–53.

Lown, K. S., Mayo, R. R., Leichtman, A. B., Hsiao, H-L., Turgeon, D. K., Schmiedlin-Ren, P., Brown, M. B. et al. 1997b. Role of intestinal P-glycoprotein (mdr1) in interpatient variation in the oral bioavailability of cyclosporin A. *Clin Pharmacol Ther* 62:248–60.

Macfarlane, G.T., Macfarlane, S. 1997. Human colonic microbiota: Ecology, physiology and metabolic potential of intestinal bacteria. *Scand J Gastroenterol* 222:3–9.

Madani, S., Paine, M. F., Lewis, L., Thummel, K. E., Shen, D. D. 1999. Comparison of CYP2D6 content and metoprolol oxidation between microsomes isolated from human livers and small intestines. *Pharm Res* 16:1199–205.

Martinez, M. N., Amidon, G. L. 2002. A mechanistic approach to understanding the factors affecting drug absorption: A review of fundamentals. *J Clin Pharmacol* 42:620–43.

Matsumoto, S., Hirama, T., Matsubara, T., Nagata, K., Yamazoe, Y. 2002. Involvement ofCYP2J2 on the intestinal first-pass metabolism of antihistamine drug, astemizole. *Drug Metab Dispos* 30:1240–5.

Mizuma, T. 2002. Kinetic impact of presystemic intestinal metabolism on drug absorption: Experiment and data analysis for the prediction of *in vivo* absorption from *in vitro* data. *Drug Metab Pharmacokin* 17(6):496–506.

Moriguchi, I., Hirono, S., Liu, Q., Nakagome, I., Matsushita, Y. 1992. Simple method of calculating octanol/water partition coefficient. *Chem Pharm Bull* 40(1):127–30.

Moriguchi, I., Hirono, S., Nakagome, I., Hirano, H. 1994. Comparison of log P values for drugs calculated by several methods. *Chem Pharm Bull* 42:976–8.

Mundada, A. S., Bhola, N. V., Avari, J. G. 2008. Gastro-retentive drug delivery: A technical Note. *Drug Deliv Technol* 8(7):44–7.

Ni, N., Sanghvi, T., Yalkowsky, S. H. 2002. Independence of the product of solubility and distribution coefficient of pH. *Pharm Res* 19(12):1862–6.

Nordskog, B. K., Phan, C. T., Nutting, D. F., Tso. P. 2001. An examination of the factors affecting intestinal lymphatic transport of dietary lipids. *Adv Drug Deliv Rev* 50:21–44.

Obach, R. S., Zhang, Q. Y., Dunbar, D., Kaminsky, L. S. 2001. Metabolic characterization of the major human small intestinal cytochrome P450s. *Drug Metab Dispos* 29:347–52.

Oberle, R. L., Chen, T. S., Lloyd, C., Barnett, J. L., Owyang, C., Meyer, J., Amidon, G. L. 1990. The influence of the interdigestive migrating myoelectric complex on the gastric emptying of liquids. *Gastroenterology* 99:1275–82.

Ozeki, T., Yuasa, H., Okada, H. 2005. Controlled release of drug via methylcellulose–caboxyvinylpolymer interpolymer complex solid dispersion. *AAPS Pharm Sci Tech* 6(2):231–6.

Pagliara, A., Carrupt, P. A., Caron, G., Gaillard, P., Testa, B. 1997. Lipophilicity profiles of ampholytes. *Chem Rev* 97:3385–400.

Paine, M. F., Hart, H. L., Ludington, S. S., Haining, R. L. Rettie, A. E., Zeldin, D. C. 2006. The human intestinal cytochrome P450 "PIE". *Drug Metab Disp* 34(5):880–6.

Paine, M. F., Schmiedlin-Ren, P., Watkins, P. B. 1999. Cytochrome P-450 1A1 expression in human small bowel: Interindividual variation and inhibition by ketoconazole. *Drug Metab Dispos* 27:360–4.

Paine, M. F., Shen, D. D., Kunze, K. L., Perkins, J. D., Marsh, C. L., McVicar, J. P., Barr, D. M., Gillies, B. S., Thummel, K. E. 1996. First-pass metabolism of midazolam by the human intestine. *Clin Pharmacol Ther* 60:14–24.

Parrott, N., Lukacova, V., Fraczkiewicz, G., Bolger, M. B. 2009. Predicting pharmacokinetics of drugs using physiologically based modeling—Application to food effects. *AAPS J* 11(1):45–53.

Patel, M., Shah, T., Amin, A. 2007. Therapeutic opportunities in colon-specific drug-delivery systems. *Crit Rev Ther Drug Carr Syst* 24(2):147–202.

Patel, V. I., Dave, R. H. 2013. Evaluation of colloidal solid dispersion: Physiochemical considerations and *in vitro* release profile. *AAPS Pharm Tech* 14(2):620–8.

Patil, S. D., Ngo, L. Y., Glue, P., Unadkat, J. D. 1998. Intestinal absorption of ribavirin is preferentially mediated by the Na^+-nucleoside purine (N1) transporter. *Pharm Res* 15:950–2.

Perloff, M. D., Von Moltke, L. L., Fahey, J. M., Daily, J. P., Greenblatt, D. J. 2000. Induction of P-glycoprotein expression by HIV protease inhibitors in cell culture. *AIDS* 14(9):1287–9.

Prueksaritanont, T., Gorham, L. M., Hochman, J. H., Tran, L. O., Vyas, K. P. 1996. Comparative studies of drug-metabolizing enzymes in dog, monkey, and human small intestines and in Caco-2 cells. *Drug Metab Dispos* 24:634–42.

Radwan, A., Amidon, G. L., Langguth, P. 2012. Mechanistic investigation of food effect on disintegration and dissolution of BCS class III compound solid formulations: The importance of viscosity. *Biopharm Drug Disp* 33(7):403–16

Russell, T. L., Berardi, R. R., Barnett, J. L., Dermentzoglou, L. C., Jarvenpaa, K. M., Schmaltz, S. P., Dressman, J. B. 1993. Upper gastrointestinal pH in seventy-nine healthy, elderly, North American men and women. *Pharm Res* 10(2):187–96.

Salana, N. N., Eddington, N. D., Fasano, A. 2006. Tight junction modulation and its relationship to drug delivery. *Adv Drug Del Rev* 58:15–28.

Sanghvi, T., Ni, N., Yalkowsky, S. H. 2001. A simple modified absorption potential. *Pharm Res* 18(12):1794–6.

Sawamoto, T., Haruta, S., Kurosaki, Y., Higaki, K., Kimura, T. 1997. Prediction of the plasma concentration profiles of orally administered drugs in rats on the basis of gastrointestinal transit kinetics and absorbability. *J Pharm Pharmacol* 49:450–7.

Schilsky, R. L., Choi, K. E., Vokes, E. E., Guaspari, A., Guarnieri, C., Whaling, S., Liebnet, M. A. 1989. Clinical pharmacology of the stereoisomers of leucovorin during repeated oral dosing. *Cancer* 63(suppl 6):1018–21.

Schmidtand, L., Dalhoff, K. 2002. Food-drug interactions. *Drugs* 62:1481–502.

Shimada, T., Yamazaki, H., Mimura, M., Inui, Y., Guengerich, F. P. 1994. Interindividual variations in human liver cytochrome P-450 enzymes involved in the oxidation of drugs, carcinogens and toxic chemicals: Studies with liver microsomes of 30 Japanese and 30 Caucasians. *J Pharmacol Exp Ther* 270:414–23.

Shugarts, S., Benet, L. Z. 2009. The role of transporters in the pharmacokinetics of orally administered drugs. *Pharm Res* 26(9):2039–54.

Simmonds, R. G., Woodage, T. J., Duff, S. M., Green, J. N. 1980. Stereospecific inversion of R-benoxaprofen in rat and man. *Eur J Metab Pharmacokinet* 5:169–72.

Singh, B. 1999. Effects of food on clinical pharmacokinetics. *Clin Pharmacokinet* 37:213–55.

Singhal, D., Curatolo, W. 2004. Drug polymorphism and dosage form design—A practical perspective. *Adv Drug Deliv Rev* 56:335–47.

Sirisuth, N., Eddington, N. D. 2002 *In vitro–in vivo* correlation, definitions and regulatory guidance. *Int J Generic Drugs* Part 2:1–11.

Sousaa, T., Patersonb, R., Mooreb, V., Carlssonc, A., Abrahamssond, B., Basita, A. W. 2008. The gastrointestinal microbiota as a site for the biotransformation of drugs. *Int J Pharm* 363:1–25.

Streubel, A., Siepmann, J., Bodmeier, R. 2003. Floating matrix tablets based on low density foam powder: Effect of formulation and processing parameters on drug release. *Eur J Pharm Sci* 18:37–45.

Suzuki, A., Higuchi, W. I., Ho, N. F. H. 1970a. Theoretical model studies of drug absorption and transport in the gastrointrointestinal tract—I. *J Pharm Sci* 59(5):644–51.

Suzuki, A., Higuchi, W. I., Ho, N. F. H. 1970b. Theoretical model studies of drug absorption and transport in the gastrointrointestinal tract—II. *J Pharm Sci* 59(5):651–9.

Suzuki, H., Sugiytama, Y. 2000. Role of metabolic enzymes and efflux transporters in the absorption of drugs from the small intestine. *Eur J Pharm Sci* 12:3–12.

Tamai, I., Ling, H-Y., Timbul, S-M., Nishikido, J., Tsuji, A. 1988. Stereo-specific absorption and degradation of cephalexin. *J Pharm Pharmacol* 40:320–4.

Todeschini, R., Consonni, V. 2000. Handbook of molecular descriptors. In *Methods and Principles in Medicinal Chemistry*, Vol. 11, eds. R. Mannold, H. Kubinyi and H, Timmerman, 276pp. West Sussex, United Kingdom: Wiley-VCH Verlag GmbH.

Trevaskis, N. L., Porter, C. J. H., Charman, W. N. 2005. Bile increases intestinal lymphatic drug transport in the fasted rat. *Pharm Res* 22(11):1863–70.

Ueda, K., Yoshida, A., Amachi, T. 1999. Recent progress in P-glycoprotein research. *Anticancer Drug Des* 14:115–21.

Ueda, S., Yamaoka, K., Yui, J., Shigematsu, A., Terumichi Nakagawa, T. 2002. Evaluation of capacity-limited first-pass effect through liver by three-points sampling in portal and hepatic veins and systemic artery. *Pharm Res* 19(6):852–7.

United States Pharmacopoeia. 2004. *In vitro and in vivo Evaluations of Dosage Forms*, 27th Edition. Easton, PA: Mack Publishing Co.

Veber, D. F., Johnson, S. R., Cheng, H-Y., Smith, B. R., Ward, K. W., Kopple, K. D. 2002. Molecular properties that influence the oral bioavailability of drug candidates. *J Med Chem* 45:2615–23.

van der Ohe, M., Camillari, M. 1992. Measurements of small bowel and colonic transit: Indications and methods. *Mayo Clin Proc* 67:1169–79.

Wade, D. N., Mearrick, P. T., Morris, J. L. 1973. Active transport of L-dopa in the intestine. *Nature* 242:463–5.

Wang, M., Ahrne, S., Jeppsson, B., Molin, G. 2005. Comparison of bacterial diversity along the human intestinal tract by direct cloning and sequencing of 16S rRNA genes. *FEMS Microbiol Ecol* 54:219–31.

Wang, E. J., Casciano, C. N., Clement, R. P., Johnson, W. W. 2001. Inhibition of P-glycoprotein transport function by grapefruit juice psoralen. *Pharm Res* 18(4):432–8.

Ward, N. 2010. The impact of intestinal failure on oral drug absorption: A review. *J Gastrointest Surg* 14:1045–51.

Willmann, S., Edginton, A. N., Dressman, J. B. 2007. Development and validation of a physiology-based model for the prediction of oral absorption in monkeys. *Pharm Res* 24:1275–82.

Willmann, S., Schmitt, W., Keldenich, J., Dressman, J. B. 2003. A physiologic model for simulating gastrointestinal flow and drug absorption in rats. *Pharm Res* 20:1766–71.

Willmann, S., Schmitt, W., Keldenich, J., Lippert, J., Dressman, J. B. 2004. A physiological model for the estimation of the fraction dose absorbed in humans. *J Med Chem* 47:4022–31.

Wilson, C. G. 2000. Gastrointestinal transit and drug absorption. In *Oral Drug Absorption, Prediction and Assessment*, eds. J. B. Dressman and H. Lennernäs, 1–10. New York: Dekker.

Wu, C-Y., Leslie, Z. Benet, L. Z. 2005. Predicting drug disposition via application of BCS: Transport/absorption/elimination interplay and development of a biopharmaceutics drug disposition classification system. *Pharm Res* 22(1):11–23.

Xu, J., Go, M. L., Lim, L-Y. 2003. Modulation of digoxin transport across caco-2 cell monolayers by citrus fruit juices: Lime, lemon, grapefruit, and pummelo. *Pharm Res* 20(2):169–76.

Yu, L. X., Amidon, G. L., Polli, J. E., Zhao, H., Mehta, M. U. M., Dale, P., Conner, D. P. et al. 2002. Biopharmaceutics classification system: The scientific basis for biowaiver extensions. *Pharm Res* 19(7):921–5.

Yu, L. X., Crison, J. R., Amidon, G. L. 1996a. Compartmental transit and dispersion model analysis of small intestinal flow in humans. *Int J Pharm* 140:111–18.

Yu, L. X., Lipka, E., Crison, J. R., Gordon L. Amidon, G. L. 1996b. Transport approaches to the biopharmaceutical design of oral drug delivery systems: Prediction of intestinal absorption. *Adv Drug Del Rev* 19:359–76.

Zhang, Q. Y., Dunbar, D., Ostrowska, A., Zeisloft, S., Yang, J., Kaminsky, L. S. 1999. Characterization of human small intestinal cytochromes P-450. *Drug Metab Dispos* 27:804–9.

Zhou, S. F., Di, Y. M., Chan, E., Du, Y. M., Chow, V. D., Xue, C. C., Lai, X., Wang, J. C., Li, C. G., Tian, M., Duan, W. 2008. Clinical pharmacogenetics and potential application in personalized medicine. *Curr Drug Metab* 9:738–84.

Zimmermann, T., Yeates, R. A., Laufen, H., Pfaff, G. Wildfeuer, A. 1994. Influence of concomitant food intake on the oral absorption of two triazole antifungal agents, itraconazole and fluconazole. *Eur J Clin Pharmacol* 46:147–50.

9 Mathematical Modeling in Pharmacokinetics and Toxicokinetics

9.1 INTRODUCTION

Mathematical modeling has been used to study the interconnections between singular or pooled physiological compartments and elucidate the functional mechanisms that comprise the biological system. The system can be as simple as the cell culture or as complex as the *in vivo* studies and human risk assessment. The PK/TK modeling entails the appraisal of kinetic parameters and constants of absorption, distribution, metabolism, and the elimination of xenobiotics and their corresponding pharmacodynamic or toxicodynamic parameters. The goal in PK/TK modeling is to develop a scientific, practical, and meaningful mathematical model that represents the biological system faithfully. The achievement of this goal reveals how the components and compartments of the system interact and cooperate to yield the functional properties of the system and how the system ultimately handles the exposure to xenobiotics, whether they are therapeutic agents or environmental pollutants. The development of the model has immense practical implications for diagnosis and treatment of disease, and quantifying human response. It is the precise and abstract nature of the mathematical terms that enable us to explore the interconnection and influence of physiological parameters and constants.

Within the past four decades, there has been significant progress in PK/TK modeling and it will become progressively more complex and sophisticated. However, a link between experimental sciences and predictive modeling must exist in order to maintain a firm base of reality. Lord Bertrand Russell, the 20th century philosopher and mathematician (1872–1970), in his *A History of Western Philosophy*, wrote:

> Mathematics, rightly viewed, possesses not only truth, but supreme beauty—a beauty cold and austere, like that of sculpture, without appeal to any part of our weaker nature, without the gorgeous trappings of painting or music, yet sublimely pure, and capable of a stern perfection such as only the greatest art can show.

One traditional approach in PK/TK modeling, known as compartmental analysis, is to develop the model based on concentration–time curves of xenobiotics and/or their metabolite(s) or other relevant biological variables in the body. Thus, the data define the model, and the model is then tested to see how closely it agrees with the observed data, and hence the model describes the data. This type of modeling is used to summarize the data, and to predict the behavior of a compound in the system as a function of biological processes. A different conventional approach in PK/TK modeling is to rely on the anatomical and physiological regions of the body comprising of all relevant tissues and organs of disposition connected via the vascular system. There are other approaches in PK/TK modeling such as stochastic or neuronal network models that are based on different fundamentals than the traditional modeling. The following sections describe the popular approaches in PK/TK modeling.

The popular PK/TK modeling approaches are

1. Physiologically-based pharmacokinetic modeling
2. Classical compartmental analysis

   a. Linear compartmental models
   b. Non-linear compartmental models
   c. Compartmental models with one- or two-tissue binding
3. Non-Compartmental analysis
4. Population pharmacokinetics
5. Pharmacokinetic/pharmacodynamic modeling
6. Toxicokinetic/toxicodynamic modeling
7. Stochastic modeling
8. Neuronal network models

9.2 PHYSIOLOGICALLY BASED PHARMACOKINETICS MODELS (PBPK MODELS)

9.2.1 Description

The basic idea behind physiological modeling is to overlay xenobiotic data onto anatomical models encompassing all relevant organs/tissues of the body, and incorporate quantitative parameters of each anatomical region in the mathematical description of the compound's disposition (Bischoff et al., 1971; Bischoff, 1980, 1986). These parameters include blood flow, organ/tissue volumes, tissue affinity, enzymatic and transporter activity, plasma protein binding, membrane permeability, transport and elimination from an organ, and metabolism by the liver or other sites of metabolism. Therefore, based on the measurements of xenobiotics in blood/plasma taken from a peripheral vein, the kinetics of the compound at different dose levels administered through various routes of administration can be estimated for the impacts at various organs or tissues. Furthermore, since there are parallelisms in the anatomy of mammalian systems, such as having all organs between arterial blood and venous system or most organs having comparable fractions of the body weight, the physiological models of xenobiotics developed in experimental animals can be extrapolated to humans by changing the physiological parameters to those in a human (Dedrick et al., 1970; Harrison and Gibaldi, 1977; Nakashima et al., 1987; Chow, 1997; Rowland et al., 2004). The intended application of physiological models determines their complexity and there is no limit to the number of compartments. The PBPK models can be highly complex and comprehensive, mimicking the anatomical/regional geometry of the body and describing molecular and cellular events. A typical diagram of a semi-comprehensive model is presented in Figure 9.1. The models can be simplified by having lumped together various tissues. Lumping is the terminology used in constructing simplified PBPK models and refers to combining organs and tissues with similar kinetic behavior into one compartment (Figures 9.2 through 9.4). The simplified models, known as "hybrid-PBPK" (Kawahara et al., 1999; Gallo et al., 2004) or "minimal-PBPK" (Henthorn et al., 1992; Cao et al., 2012, 2013) are used to focus mainly on the disposition and uptake of specific organs of interest (Figure 9.4). The selection of model, expanded or simplified, depends on the goals and objectives of investigation; it is not an arbitrary decision and requires clear justifications for expansion or contraction of compartments.

The important features and applications of PBPK modeling are

- To provide a detailed and quantitative portrayal of xenobiotic disposition in the body of mammalian species and related risk assessment
- To improve the selection and optimization process of lead compound(s) in drug discovery and development
- To extrapolate more reliably the results from animals such as mice, rats, dogs and primates upward to man. This extrapolation also known as "scaling up" may also be applicable in certain cases in the opposite direction, that is, scaling down to lower species
- To predict *in vivo* data based on *in vitro* observations

- To assess human health risk, including cancer risk following exposure to toxic environmental chemicals/carcinogens through different routes of administration and estimate the internal dose
- To extrapolate from one route to another route of administration and calculating the tissue dose
- To predict the safe or low dose exposure, by high dose to low dose extrapolation in chemical carcinogen/toxic compounds
- To provide a useful framework for predicting the influence of PK/TK interactions from binary to more complex chemical mixtures
- To predict the time course of xenobiotic levels in different organs
- To facilitate simulation of drug disposition in disease states
- To predict fetal exposure from parental exposure to a xenobiotic

9.2.2 Model Development

To develop a physiological model for a xenobiotic, a diagram similar to Figure 9.1 is considered with relevant organs to describe the disposition of the compound. The number of organs and tissues included in the model is determined by the physicochemical characteristics of the compound such as lipid solubility, plasma protein binding, microsomal or hepatic intrinsic clearance, cell membrane permeability estimated from Caco-2 cells, degree of ionization of the compound and preferential binding by an organ, and physiological/biochemical parameters (Cowles et al., 1971; Rowland, 1985; Ritchel and Banerjee, 1986; Gallo et al., 1987; Brown et al., 1997, 1998; Corley et al., 2000; Blaauboer, 2003; Lipscomb and Poet, 2008). For example, if the compound is a highly lipid soluble drug, the adipose tissue is included in the model, and if the compound is a water-soluble the adipose tissue is excluded. Therefore, the *in vitro* and in silico data are important in the selection of an

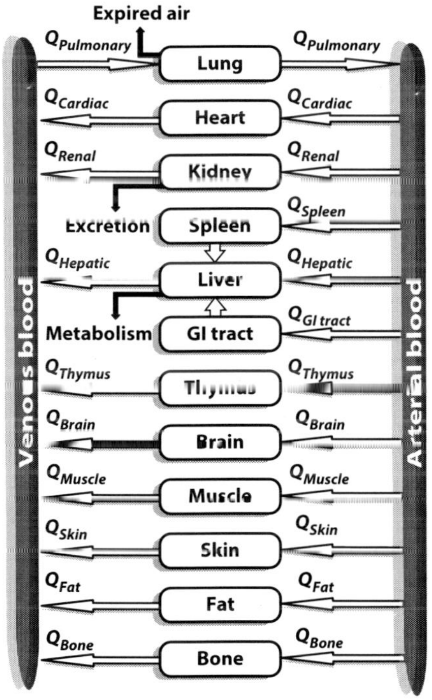

FIGURE 9.1 A typical diagram of a semi-comprehensive physiologically based pharmacokinetic (PBPK) model.

appropriate model. It should be noted that the tacit assumption in physiological modeling is that all included organs and tissues are well-stirred compartments.

Another important characteristic of physiological modeling is the presentation of each organ or region in terms of its sub-compartments. For example the distribution and uptake of a compound in a region can be defined not only in terms of capillaries concentration, but also in interstitial fluid and intracellular volume.

9.2.2.1 Flow-Limited (Perfusion-Limited) Models

When the compound is taken up rapidly by the interstitial and intracellular sub-compartments of an organ, and they are essentially in equilibrium with the concentration in the capillaries, the limiting factor of the uptake is blood flow, that is, the magnitude of uptake by the organ depends on how fast the compound reaches the organ through blood flow. Under this condition the uptake is considered blood flow-limited process and the organ is capable of not only taking up the compound rapidly but also holding large quantities over longer period of time. On the other hand, if the transfer of a drug between extracellular (i.e., blood and interstitial fluid) and intracellular fluids is limited by the permeation of the drug through the cell membrane, the process is considered membrane-limited or permeability-limited process (see also Chapter 3, Section 3.2.2.1). For volatile organic compounds (VOC), for example, a general seven-compartment flow-limited model has been proposed that consists of blood, adipose tissues, skin, kidney, liver, rapidly and slowly perfused tissues, and a gas exchange compartment (Mumtaz et al., 2012) (Figure 9.2). The example of non-volatile but lipophilic compounds is a six-compartment model for benzo(a)pyrene and dibenzo(def,p)chrysen (Anderson et al., 1987; Crowell et al., 2011), which includes blood, lung, adipose tissues, and liver, including other highly perfused organs and low perfused tissues (Figure 9.3). The diagram of minimal PBPK is presented in Figure 9.4.

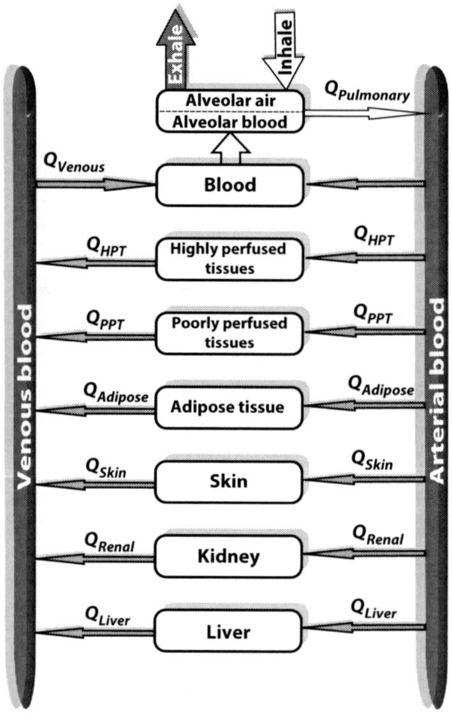

FIGURE 9.2 A physiologically based PK/TK model for volatile organic compounds with the simplified compartments of rapidly and slowly perfused tissues.

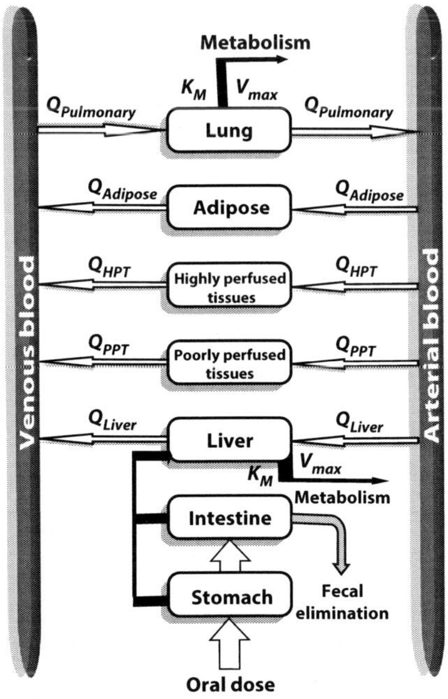

FIGURE 9.3 A physiologically based PK/TK model for non-volatile but lipophilic compounds administered orally with elimination only through biotransformation.

Similarly, seven-compartment models have also been developed for xenobiotics intended for therapeutic use (Hudacheck and Gustafson, 2013).

Once the relevant body regions have been identified in a comprehensive physiological model, the mass balances for blood and the organs can be written by considering the rates of input and the rates of output. For example, the following is the overall mass balance equation for a flow-limited model:

$$V_B \frac{dC_{Blood}}{dt} = \left[Q_{brain} \frac{C_{brain}}{R_{brain}} + Q_{heart} \frac{C_{heart}}{R_{heart}} + Q_{liver} \frac{C_{liver}}{R_{liver}} + Q_{kidney} \frac{C_{kidneys}}{R_{kidneys}} + Q_{fat} \frac{C_{fat}}{R_{fat}} + \cdots \right]$$
$$- \left[Q_{brain} + Q_{heart} + Q_{liver} + Q_{kineys} + Q_{fat} + \cdots \right] C_{blood} + Dose(t) \tag{9.1}$$

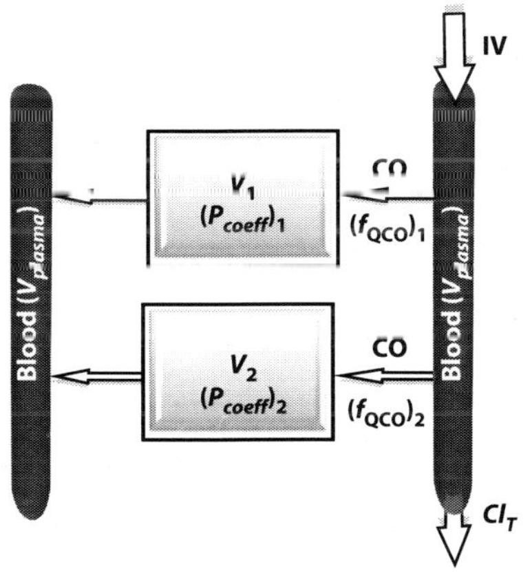

FIGURE 9.4 Diagram of a minimal physiologically based PK/TK model.

where (C) is the concentration, (Q) is the blood flow, (V) is the volume, and (R) is the distribution ratio of tissue/blood. The multiplication of Q (volume/time) by C (mass/volume) yields the units of rate in terms of mass/time. (R) has no units and the multiplication of V (volume) by dC/dt (mass/volume time) yields the units of mass/time defining the rate of change in amount or concentration in the blood. The above rate equation requires that the dose be included as a function of time.

The mass balance equations of organs with elimination are

$$\text{Liver:}\quad V_{liver}\frac{dC_{liver}}{dt} = Q_{liver}\left[C_{blood} - \frac{C_{liver}}{R_{liver}}\right] - Cl_m\frac{C_{liver}}{R_{liver}} \tag{9.2}$$

$$V_{liver}\frac{dC_{liver}}{dt} = Q_{liver}\left[C_{arterial} - C_{liver-venous}\right] - \frac{V_{max} \times C_{liver-venous}}{K_M + C_{liver-venous}} \tag{9.3}$$

$$\text{Kidney:}\quad V_{kidney}\frac{dC_{kidney}}{dt} = Q_{kidney}\left[C_{blood} - \frac{C_{kidney}}{R_{kidney}}\right] - Cl_r\frac{C_{kidney}}{R_{kidney}} \tag{9.4}$$

$$V_{kidney}\frac{dC_{kidney}}{dt} = Q_{kidney}\left[C_{arterial} - C_{kidney-venous}\right] - Cl_r\frac{C_{kidney-venous}}{R_{kidney}} \tag{9.5}$$

$$\text{Lung:}\quad V_{lung}\frac{dC_{lung}}{dt} = Q_{lung}\left[C_{blood} - \frac{C_{lung}}{R_{lung}}\right] - Cl_l\frac{C_{lung}}{R_{lung}} \tag{9.6}$$

where Cl_m, Cl_r, and Cl_l are metabolic, renal and lung clearances, respectively, with units if volume/time.

The differential equations for other organ or regions also follow the general concept of

The rate of change of amount in the organ = Rate of input – rate of output

That is,

$$V_{organ}\frac{dC_{organ}}{dt} = Q_{organ}\left[C_{blood} - \frac{C_{organ}}{R_{organ}}\right] \tag{9.7}$$

For highly perfused (*hpt*) and slowly perfused (*spt*) lumped compartments, the equations are

$$\text{Highly perfused tissues:}\quad V_{hpt}\frac{dC_{hpt}}{dt} = Q_{hpt} \times \left(C_{arterial} - C_{venous-hpt}\right) \tag{9.8}$$

$$\text{Slowly perfused tissues:}\quad V_{spt}\frac{dC_{spt}}{dt} = Q_{spt} \times \left(C_{arterial} - C_{venous-spt}\right) \tag{9.9}$$

For the minimal PBPK model with only two lumped compartments (Figure 9.4), the differential equations of the model are (Cao et al., 2012)

$$\frac{dCp}{dt} = \frac{Dose}{V_{plasma}} + (f_{Q_{CO}})_1 \times Q_{CO} \times \frac{C_1}{(P_{coeff1})_{tissue1} \times V_{plasma}} + (f_{Q_{CO}})_2 \times Q_{CO} \times \frac{C_2}{(P_{coeff2})_{tissue2} \times V_{plasma}} - Cp \times \frac{(f_{Q_{CO}})_1 \times Q_{CO} + (f_{Q_{CO}})_2 \times Q_{CO} + Cl_T}{V_{plasma}} \quad (9.10)$$

$$\frac{dC_1}{dt} = \left(Cp - \frac{C_1}{(P_{coeff1})_{tissue1}}\right) \times \frac{(f_{Q_{CO}})_1 \times Q_{CO}}{V_1} \quad (9.11)$$

$$\frac{dC_2}{dt} = \left(Cp - \frac{C_2}{(P_{coeff2})_{tissue2}}\right) \times \frac{(f_{Q_{CO}})_2 \times Q_{CO}}{V_2} \quad (9.12)$$

where, V_{plasma} is the volume of plasma, or blood; Cp is the concentration of compound in V_{plasma}; Q_{CO} is cardiac blood flow or plasma flow; $(f_{Q_{CO}})_1$ and $(f_{Q_{CO}})_2$ are fractions of cardiac output considered for compartment 1 and 2; $(P_{coeff1})_{tissue1}$ and $(P_{coeff2})_{tissue2}$ are partition coefficients of tissue compartment 1 and tissue compartment 2; Cl_T is the total body clearance; V_1 and V_2 are the volumes of tissue compartments 1 and 2. The assumptions of the model are

- $(f_{Q_{CO}})_1 + (f_{Q_{CO}})_2 \leq 1$, and (9.13)
- $V_1 + V_2 + V_{plasma}$ = Total volume of extracellular fluid (9.14)
- At time zero the concentration terms are equal to zero

The model can be expanded to include specific tissue compartments.

9.2.2.2 Permeability-Limited (Membrane-Limited) Models

The permeability-limited PBPK operates on a slower timescale and represents the disposition of a large number of xenobiotics (Thompson and Beard, 2011; Huang and Rowland, 2012). For permeability-limited PBPK models the compartments are subdivided into two sub-compartments and the differential equation of each organ is expressed in terms of two rate equations: (1) rate of change of xenobiotic concentration in extracellular fluid, and (2) rate of change of the concentration in intracellular compartment. This means that each compartment of a model like Figure 9.5 is subdivided into two compartments of extracellular and intracellular fluids as is shown in Figure 9.6.

The related differential equations for tissue blood compartment and cellular matrix for a permeability-limited model are

$$\text{Tissue blood: } V_{tissue1}\frac{dC_{tissue1}}{dt} - Q_t[C_{input} - C_{output}] - [PA]\frac{(C_{tissue1} - C_{tissue2})}{P_{t.b.coeff}} \quad (9.15)$$

Equation 9.15 can also be presented as

$$\frac{dC_1}{dt} = \frac{Q}{V_1}(C_{input} - C_{tissue1}) - \frac{PA}{V_1}\left(C_{tissue1} - \frac{C_{tissue2}}{P_{t.b.coeff}}\right) \quad (9.16)$$

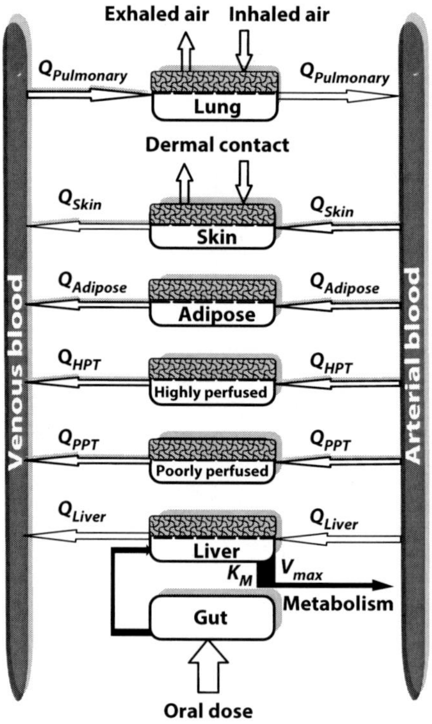

FIGURE 9.5 Permeability-limited physiologically based PK/TK model.

$$\text{Cellular Matrix: } V_{tissue2}\frac{dC_{tissue2}}{dt} = [PA] \times \frac{C_{tissue1} - C_{tissue2}}{P_{t.b.coeff}} \tag{9.17}$$

$$\frac{dC_{tissue2}}{dt} = \frac{PA}{V_2}\left(C_{tissue1} - \frac{C_{tissue2}}{P_{t.b.coeff}}\right) \tag{9.18}$$

where, Q is blood flow; C_{input} is the concentration of input or inflow; V_1 is the vascular volume, PA is the membrane permeability-area coefficient; $C_{tissue1}$ is the concentration in the vascular sub-compartment; $C_{tissue2}$ is the concentration in the extravascular sub-compartment; $(C_{tissue1} - C_{tissue2})$ is the concentration gradient across the membrane, and $P_{t.b.coeff}$ is the partition coefficient between $C_{tissue1}$ and $C_{tissue2}$.

Physiological models provide the versatility of identifying and evaluating potentially saturable processes like capacity-limited metabolism or active secretion.

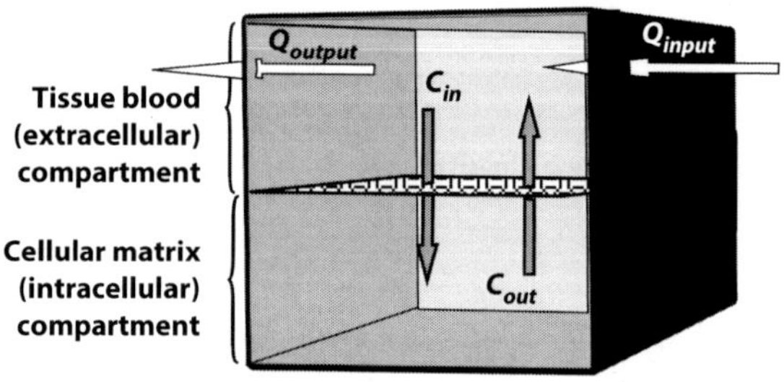

FIGURE 9.6 An isolated diagram of permeability-limited organ included in physiologically based PK/TK model (Figure 9.5) with subdivision of extracellular and intracellular fluids.

9.2.2.3 Variability of Physiological/Biochemical Key Parameters

Most of the estimated physiological parameters are based on allometric scaling functions. Various approaches have been used to establish the most realistic whole body PBPK models (Price et al., 2003; Kazama et al., 2003; Willman et al., 2003). For organ weight estimation, two of the recommended approaches are (Willmann et al., 2007; Upton et al., 2013):

$$(\text{Organ weight})_{individual} = \left(\frac{\text{Height}_{individual}}{\text{Height}_{mean}}\right)^{0.75} \times (\text{Organ weight})_{mean} \tag{9.19}$$

Where "mean" parameters refer to the mean values of a virtual individual based on data reported by the International Commission on Radiological Protection (ICRP, 2002; Willmann et al., 2007).

$$(\text{Organ weight})_{individual} = \left(\frac{(\text{Weight})_{individual}}{(\text{Weight})_{standard}}\right)^{1} \times (\text{Organ weight})_{standard} \tag{9.20}$$

The subscript "standard" refers to the values for a virtual standard man (Upton et al., 2012). Both "standard" and "mean" values were created based on the anthropometric reported data. The actual organ weight based on autopsy measurements have been reported to correlate better with the height of the body than body weight (de la Grandmaison et al., 2001). However, there are organs that are independent of height and weight, and organs that correlate well with the body weight, for example,

$$(\text{Brain weight})_{individual} \equiv (\text{Brain weight})_{mean} \equiv (\text{Brain weight})_{standard} \tag{9.21}$$

The assumption can be the brain weight of individual is equivalent to the brain weight of virtual men and because the skin weight proportional to body surface area (BSA), it correlates better with the body weight.

$$(\text{Skin weight})_{individual} = \left(\frac{(\text{Body weight})_{individual}}{(\text{Body weight})_{Reference}}\right)^{0.5} \times (\text{Skin weight})_{Reference} \tag{9.22}$$

The blood flow as a function of cardiac output (CO) is also estimated according to (Willmann et al., 2007)

$$(\text{CO})_{individual} = \left(\frac{\text{Height}_{individual}}{\text{Height}_{mean}}\right)^{0.75} \times (\text{CO})_{mean} \tag{9.23}$$

Equation 9.24 (Upton et al., 2012) is for the estimation of cardiac output for an individual within a population

$$(\text{CO})_{individual} = (\text{CO})_{standard}\left(\frac{(\text{Body weight})_{individual}}{(\text{Body weight})_{standard}}\right)^{0.41} \times \left(1.0309 + 0.0295\frac{(\text{Age})_{individual}}{(\text{Age})_{standard}} - 0.06033\left(\frac{(\text{Age})_{individual}}{(\text{Age})_{standard}}\right)^{2}\right) \tag{9.24}$$

There are differences, though not significant (Table 9.1) yet may contribute to the cumulative random errors when combined with differences in other estimates of physiological parameters such as the volume or weight of an organ.

TABLE 9.1
Comparison of Estimated Organ Blood Flow for "Mean" and "Standard" Men

Organ	Q (L/min) Mean (Caucasian Virtual Man) 73 kg, 5 ft 8in Male, BMI = 23.6 kg/m^2	Q (L/min) Standard (Virtual Man) 69 kg with 30% Fat and 40% Muscle
Lung	6.106	6.314
Brain	0.780	0.686
Kidneys	1.325	0.962
Heart	0.260	0.242
Liver	0.420	1.449 (Portal Drained Viscera + Artery)
Muscle	1.106	1.377
Adipose	0.325	1.035

Soruce: The "Mean" data are from Willmann, S. et al. 2007. *J Pharmacokinet Pharmacodyn* 34(3):401–31, and the "Standard" values are from Upton, R. N. et al. 2012. *J Pharmacokinet Pharmacodyn* 39:561–76.

9.2.3 Predictive Capability and Sensitivity Analysis

The analysis of experimental data according to physiological modeling is mostly based on simulation of the time course of drug in different anatomical regions. The simulation is achieved by substituting in the equations of a selected model, the physiological parameters of the organs (blood flow, weight, etc.), the input function of the experimental conditions and the physico-chemical parameters of the compound. When used in PBPK, the physiological parameters have usually been considered as exact values, ignoring uncertainties in the variability of the values and their impact on model prediction. The simulation provides the predicted concentrations of the drug. The observed data (i.e., the experimental values) are then judged based on their closeness to the predicted values. From the predicted and observed concentration–time profiles other pharmacokinetic parameters such as clearances, area under plasma concentration–time curve, half-life, maximum plasma concentration, etc. can be estimated. The fundamental issues in PBPK modeling, particularly when they are used in regulatory agencies, are the predictive capability of the model and its sensitivity analysis.

The predictive capability of the model is usually assessed by estimating the percent error of prediction (PE%) (Sheiner and Beal, 1981; Wu, 1995):

$$\text{PE\%} = \frac{(\text{Value})_{\text{predicted}} - (\text{Value})_{\text{observed}}}{(\text{Value})_{\text{observed}}} \tag{9.25}$$

The precision of the prediction is measured by the median absolute prediction error (MAPE%):

$$\text{MAPE\%} = \text{Median of } \left(|\text{PE\%}1|, |\text{PE\%}2|, |\text{PE\%}3|, \ldots, |\text{PE\%}n|\right) \tag{9.26}$$

The bias of the prediction is measured by the median prediction error (MPE%):

$$\text{MPE\%} = \text{Median } (\text{PE\%}1, \text{PE\%}2, \text{PE\%}3, \ldots, \text{PE\%}n) \tag{9.27}$$

An important attribute of PBPK modeling is its predictive potential, particularly for toxic dose or toxic metabolites reaching the target organs. The PBPK models incorporate significant numbers of biochemical, physiological, and physicochemical parameters related to linear and nonlinear biological processes. Each of these parameters carries some degree of error, and the interaction among them also contributes to the magnitude of the error. Contributing to the level of uncertainty of the model's output include dissimilarities in metabolism and nonlinear kinetics in the body and the impacts of the relationship between parameters of a model such as body mass, organ mass, etc. To assess the influences of these parameters, their interaction and the presence of nonlinearity on the output, the sensitivity analyses are employed. The sensitivity analysis allows the evaluation of the model output uncertainty to be attributed to the model's parameters, and present ways to establish consistency between the model structure and its output. There are various methodologies for sensitivity analysis that are grouped into "global" and "local" analysis (McNally et al., 2011). The global analyses assess the contribution of a parameter over the set of all possible input parameters (McNally et al., 2011). The local sensitivity analyses, used more often, evaluate sensitivities that are associated with a specific set of input parameters and are estimated by the normalized sensitivity coefficient (NSC) (Bois et al., 1991; Hetrick et al., 1991; Clewell et al., 1994; Loccisano et al., 2012). The local sensitivity analysis involves the adjustment of one parameter at a time while keeping the other parameters constant. The NSC is the percent change in the output of the model produced by a fixed 1% percent change in the parameter (Plowchalk et al., 1997), that is,

$$\mathrm{NSC} = \frac{(\mathrm{A} + \mathrm{B})/\mathrm{B}}{(\mathrm{C} - \mathrm{D})/\mathrm{D}} \tag{9.28}$$

where A is the model prediction with a 1% increase in parameter value, B is the model prediction with original parameter value, C is the parameter value increased by 1%, and D is the original parameter value. Generally, NSC values of less than 0.15 ($|\mathrm{NSC}| < 0.15$) are considered low sensitivity, the values between 0.15 and 0.5 ($0.15 < |\mathrm{NSC}| < 0.5$) moderate sensitivity and values greater than 0.5 ($|\mathrm{NSC}| > 0.5$) reflect high sensitivity. The positive values of NSC are indicative of direct relationship between output and parameter, and negative values represent an inverse relationship. The values greater than one ($|\mathrm{NSC}| > 1$) may represent the augmentation of error (Allen et al., 1996).

9.3 COMPARTMENTAL ANALYSIS—LINEAR PHARMACOKINETICS

The classical compartmental analysis, the commonly employed methodology for interpretation of *in vivo* and *in vitro* data, is different from physiological modeling. In this approach the body is assumed as a series of one, two, three or more compartments, which may or may not have anatomical significance. The ADME characteristics of a compound are considered the outcome of collective contribution of various physiological processes in the body. The assumption is that the compound after introduction into the body and appearance in the systemic circulation distribute to different organs/regions including the receptor site. Along with the overall distribution of the compound, certain physiological processes influence its concentration in blood or plasma and thus at the receptor site. Among these processes protein binding, storage in adipose tissues, metabolism by the liver, excretion by the kidney, interaction with receptor site, influence of efflux proteins, removal by other organs of elimination, and sometimes enterohepatic re-circulation are the most important factors (Chapter 1, Figure 1.1). The interaction of the compound with macromolecules such as plasma proteins, enzymes, and various tissues is mostly reversible and follow the law of mass action. Thus, the free amount in the blood or plasma can eventually reach to a quasi equilibrium with the amount bound to macromolecules and tissues. If the equilibrium between blood or plasma and tissues is achieved very rapidly, as is the case for highly perfused organs such as the liver, heart, and kidney, these organs are considered indistinguishable from the systemic circulation.

The PK/TK models of compartmental analysis are based on the mathematical description of the time course of xenobiotic concentration in biological samples, including blood, plasma, urine, bile, expired air, saliva, milk, and other biological samples. Depending upon the chemical nature of the compound, its behavior in the body and extent of its interaction with proteins of plasma, its concentration is measured as a total (i.e., bound and free drug in blood or plasma), or just free concentration in plasma. The molecules of xenobiotic bound to plasma proteins, blood cells, or peripheral tissues are not free for interaction with the receptor site, and no pharmacological response is expected from the bound molecules or molecules that dwell in muscle or adipose tissues during the distribution. Hence, it is realistic to consider the changes in the plasma free concentration equivalent to changes of free concentration at the receptor site which, according to the equilibrium equation (Equation 9.29) and the law of mass action, is proportional to the bound concentration to the receptor site and thus, pharmacological response.

$$C_{\text{plasma}} \propto (C)_{\text{Receptor site}} + R \underset{k_2}{\overset{k_1}{\rightleftarrows}} C - R \Rightarrow \text{Response} \tag{9.29}$$

Based on the above reasoning, the following are the assumptions of linear compartmental analysis:

- Only free drug interacts with its receptor site to provide the pharmacological response.
- The concentration of a free drug in the circulatory system is proportional to its concentration at its receptor site.
- In linear compartmental analysis all biological processes are governed by first order kinetics (Appendix A, Section A.4), thus the rate constants, half-lives, and clearances of biological processes remain constant during disposition of a compound.

The above assumptions facilitate the development of mathematical formulas of linear compartmental models, taking into consideration the physico-chemical characteristics of xenobiotics and its disposition profile in the body. As the systemic circulation is connected to peripheral tissues and organs, it is assumed that changes in free plasma concentration, or in some instances the total concentration in plasma, reflects changes in concentration of a drug in the peripheral compartments. Thus, the free plasma concentration–time curve can provide important information about the ways the body handles the drug. The basic approach in this quantification is to summarize the data of the free concentration–time curves and/or its metabolite(s) by mathematical equations or models. These mathematical relationships are based on conceiving the body as discrete compartments. This means that the body may be divided into one or more compartments without making any anatomical assumptions about the specific content of these compartments.

Contrary to the physiological models, the inputs and outputs of the compartments are not identified as blood flow or diffusion fluxes, but are defined in general forms of first-order rate constants. The compartments are assumed as homogeneous well mixed combinations of various organs or tissues that together handle a compound's disposition (Rescigno and Serge, 1966; Jacquez, 1972; Wagner, 1975; Gibaldi and Perrier, 1982). The highly perfused tissues, because of their rapid blood flow rate and quick equilibrium with blood, are pooled with systemic circulation and considered the central compartment. The central compartment is the major sampling compartment. To include other regions of the body that reach equilibrium with the concentration in systemic circulation at a slower rate, or when the compound is taken up preferentially by a particular organ or region, more complex models with two, three, or four compartments are used. Determining the optimum number of compartments for a xenobiotic is done by fitting the plasma concentration–time curve and statistical considerations such as Akaike information criteria (AIC) and/or Schwarz information criterion (also known as Bayesian information criteria—BIC) (see Appendix A). Depending on the disposition profile of a compound, the simultaneous curve fitting of the plasma concentration and cumulative urinary data; or plasma concentration and cumulative biliary data; or plasma concentration simultaneously fit with cumulative urinary and biliary data are considered more useful to achieve a more rational fit for data. All commercially available pharmacokinetic software packages have relevant statistical tests for goodness of fit to determine the optimum number of compartments in this type of modeling (Metzler, 1981).

9.3.1 Mathematical Description of Dose-Independent Compartmental Models

The mathematical description of a compartmental model is defined by a set of differential equations for concentration change in a compartment and the related integrated relationships. The descriptions of a few models are discussed here to review practical approaches for establishing the equations of the models.

9.3.1.1 Mathematical Relationships for an Orally Administered Compound—Time Course of the Amount at the Site of Absorption, in the Body and in Urine

An orally administered compound that follows the one-compartment model can be depicted as Figure 9.7, where compartment A_D represents the amount at the site of absorption at anytime t, compartment A is the amount of free drug in the systemic circulation, which includes highly perfused tissues at time t, and compartment A_e is the amount of the free compound eliminated unchanged from the body. Although Figure 9.7 shows three compartments, the model is a one-compartment model with respect to the central compartment. The number of compartments is usually determined by the compartments that represent the distributional regions in the body.

The rate of change in the central compartment is the difference between the rate of input (i.e., absorption) and the rate of output (i.e., elimination). The site of absorption has only an output rate, which is negative and represents the decline of the administered dose at the site of absorption with respect to time. The end compartment (A_e), on the other hand, with only the input rate is positive and corresponds to the accumulation of the compound with respect to time outside of the systemic circulation. The rates of input and output are defined as the rate constants of the physiological processes related to ADME multiplied by the amount of donor compartments at time t. When the processes are governed by first-order kinetics, the rate constants have units of "time^{-1}" (Appendix A) and the rate of change in the amount of xenobiotic for each compartment is the differential equation of the compartment. The rate equations, that is, differential equations, can be integrated by the Laplace transforms (Appendix A), to determine the measurable amount or concentration in any compartment as a function of time.

Site of absorption:

$$\text{Differential equation:} \quad \frac{dA_D}{dt} = -k_a A_D \tag{9.30}$$

$$\text{Laplace transform:} \quad \begin{aligned} s\tilde{A}_D - A_D^0 &= -k_u \tilde{A}_D \Rightarrow (s + k_u)\tilde{A}_D = A_D^0 \\ \therefore \tilde{A}_D &= \frac{A_D^0}{s + k_a} \end{aligned} \tag{9.31}$$

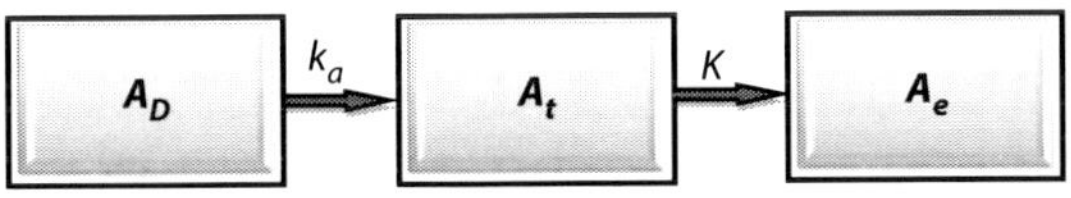

FIGURE 9.7 Diagram of a one-compartment model with respect to the amount in the systemic circulation (A_t); absorbable amount at the site of absorption (A_D); and the amount excreted in urine (A_e).

Central compartment:

$$\text{Differential equation:} \quad \frac{dA}{dt} = k_a A_D - KA \tag{9.32}$$

$$\text{Laplace transform:} \quad \begin{aligned} s\tilde{A} - 0 &= k_a A_D - K\tilde{A} \\ \therefore -k_a\tilde{A}_D &+ (s+K)\tilde{A} = 0 \end{aligned} \tag{9.33}$$

End compartment:

$$\text{Differential equation:} \quad \frac{dA_e}{dt} = KA \tag{9.34}$$

$$\text{Laplace transform:} \quad \begin{aligned} s\tilde{A}_e - 0 &= k\tilde{A} \\ \therefore -K\tilde{A} &+ s\tilde{A}_e = 0 \end{aligned} \tag{9.35}$$

The inverse Laplace transform of $\tilde{A}_D$ (Equation 9.31), the amount of xenobiotic at the site of absorption at time t is

$$A_D = A_D^0(e^{-k_a t}) \tag{9.36}$$

($\tilde{A}$) and ($\tilde{A}_e$) are the Laplace transform of A and A_e, respectively. The inverse Laplace transform of $\tilde{A}$, representing the amount in the body at time t, is determined from Equations 9.31 and 9.33 using a quotient of 2 × 2 determinants (Appendix A) and the Laplace table (Table A.1). At time zero when the dose is introduced at the site of absorption, the systemic circulation concentration is zero, therefore the initial condition for $\tilde{A}$ is zero

$$\tilde{A} = \frac{\begin{vmatrix} s+k_a & A_D^0 \\ -k_a & 0 \end{vmatrix}}{\begin{vmatrix} s+k_a & 0 \\ -k_a & s+K \end{vmatrix}} = \frac{A_D^0 \times k_a}{(s+k_a)(s+K)} \tag{9.37}$$

$$\therefore A = \frac{\left(A_D^0\right)(k_a)}{K - ka}(e^{-Kt} - e^{-k_a t}) \tag{9.38}$$

To determine the amount eliminated from the body per unit of time, A_e, three Laplace transforms (Equations 9.31, 9.33, and 9.35) are solved simultaneously with the help of a quotient of 3 × 3 determinants (Appendix A):

$$\tilde{A}_e = \frac{\begin{vmatrix} s+k_a & 0 & A_D^0 \\ -k_a & s+K & 0 \\ 0 & -K & 0 \end{vmatrix}}{\begin{vmatrix} s+k_a & 0 & 0 \\ -k_a & s+K & 0 \\ 0 & -K & s \end{vmatrix}} = \frac{(A_D^0)(K)(k_a)}{s(s+k_a)(s+K)} \tag{9.39}$$

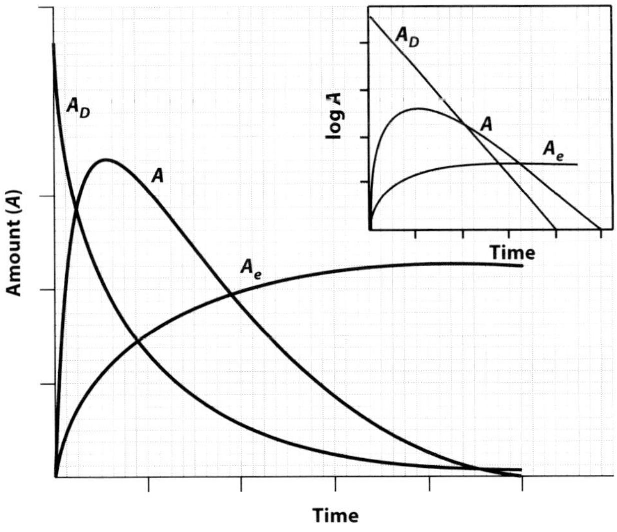

FIGURE 9.8 Time course of the amount at the site of absorption (A_D), in the systemic circulation (A), and cumulative amount excreted in urine (A_e) according to the model presented in Figure 9.7.

Taking the inverse Laplace transform of Equation 9.39 (using Table A.1—Appendix A) provides the following equation representing the time course of amount eliminated from the body at time (t):

$$A_e = A_D^0 K k_a \left[\frac{1}{k_a K} + \frac{K e^{-k_a t} - k_a e^{-Kt}}{k_a K (k_a - K)} \right]$$
$$\text{or,} \quad A_e = A_D^0 \left[1 + \frac{K e^{-k_a t}}{(k_a - K)} - \frac{k_a e^{-Kt}}{(k_a - K)} \right] \tag{9.40}$$

In the above integrated equations (Equations 9.36, 9.38, and 9.40), A_D^0 is the absorbable amount of administered dose at time zero, A_e is the total amount eliminated up to the time t, A_D represents the amount of the drug at the site of absorption at time t, A is the amount in the body at time t, and k_a and K are the rate constants of absorption and elimination. Plots of A_D, A, and A_e versus time are presented in Figure 9.8.

9.3.1.2 Mathematical Relationships for an Intravenously Administered Compound—Time Course of the Amount in the Body, Formation of Metabolite(s), and Elimination from the Body

A similar one-compartment model for a compound that also metabolizes in the body is depicted in Figure 9.9. The intravenous bolus dose is administered rapidly into compartment A. Thus, the initial condition for the central compartment is equal to dose. Compartment A_e is the cumulative amount eliminated unchanged via the urine at time t. Compartment A_m is the amount of metabolite(s) formed in the body at time t. Compartment A_{me} is the cumulative amount of metabolite(s) eliminated from the body at time t. The model is still one compartment with respect to the unchanged drug and one compartment with respect to metabolite(s).

Compartment A:

$$\tilde{A} = \frac{A_D^0}{(s + K)}, \quad A = A_D^0 e^{-Kt} \tag{9.41}$$

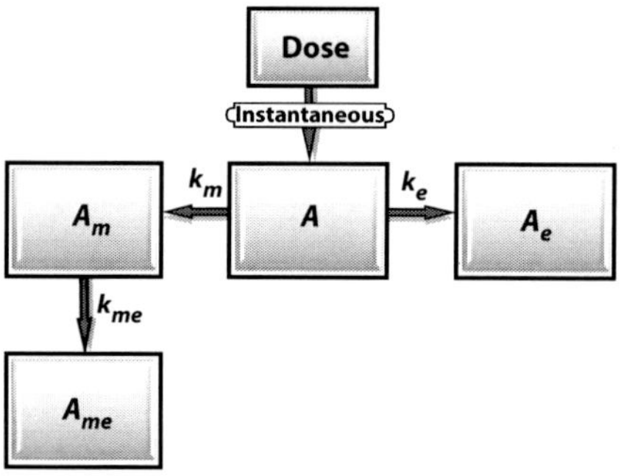

FIGURE 9.9 Diagram of a one-compartment open model with respect to unchanged xenobiotic and one compartment with respect to its metabolite, the formation and elimination of metabolite and excretion of unchanged compound are governed by first-order kinetics.

Compartment A_e:

$$\tilde{A}_e = \frac{k_e A_D^0}{s(s+K)}, \quad A_e = \frac{k_e A_D^0}{K}(1 - e^{-Kt}) \tag{9.42}$$

Compartment A_m:

$$\tilde{A}_m = \frac{k_m A_D^0}{(s+k_{me})(s+K)}, \quad A_m = \frac{k_m A_D^0}{K - k_{me}}(e^{-k_{me}t} - e^{-Kt}) \tag{9.43}$$

Compartment A_{me}:

$$\tilde{A}_{me} = \frac{k_{me} k_m A_D^0}{s(s+k_{me})(s+K)}, \quad A_{me} = \frac{k_m A_D^0}{K}\left[1 + \frac{1}{k_{me} - K}(Ke^{-k_{me}t} - k_{me}e^{-Kt})\right] \tag{9.44}$$

The rate constants include excretion rate constant (k_e), metabolic rate constant (k_m), and elimination rate constant of metabolite(s) (k_{me}). All rate constants are assumed first-order and the overall elimination rate constant, K, is equal to k_e plus k_m.

The input-disposition functions (Benet, 1972) can also be applied to determine the Laplace transforms of the compartments as presented in Table 9.2 (see also Section A.3 of Appendix A).

9.3.1.3 Mathematical Relationships for an Intravenously Administered Compound That Follows the Multicompartment Model: Use of Input-Disposition Function and General Partial Fraction Theorem—Central Compartment

Application of the input-disposition function for obtaining the integrated mathematical relationship of a multicompartment model, as depicted in Figure 9.10, is described below. The model consists of a central compartment and three peripheral compartments interdependent with the central compartment. All four compartments are considered driving-force compartments and eliminate into the outside environment, not necessarily the same environment.

The overall exit rate constants are

$$\text{Compartment 1: } E_1 = k_{10} + k_{12} + k_{13} + k_{14} \tag{9.45}$$

$$\text{Compartment 2: } E_2 = k_{20} + k_{21} \tag{9.46}$$

TABLE 9.2
Summary of Input-Disposition Functions of Model Depicted in Figure 9.9

Compartment	Input Function	Disposition Function	Laplace Transform	
A	A_D^0	$\frac{1}{s+K}$	$\tilde{A} = \frac{A_D^0}{s+K}$	
A_e	$k_e\tilde{A}$	$\frac{1}{s}$	$\tilde{A}_e = \frac{k_e\tilde{A}}{s}$	
			$\tilde{A}_e = \frac{k_e A_D^0}{s(s+K)}$	
A_m	$k_m\tilde{A}$	$\frac{1}{s+k_{me}}$	$\tilde{A}_m = \frac{k_m\tilde{A}}{s+k_{me}}$	$\tilde{A}_m = \frac{k_m A_D^0}{(s+k_{me})(s+K)}$
A_{me}	$k_{me}\tilde{A}_m$	$\frac{1}{s}$	$A_{me} = \frac{k_{me}\tilde{A}_m}{s}$	$\tilde{A}_{me} = \frac{k_{me}k_m A_D^0}{s(s+k_{me})(s+K)}$

$$\text{Compartment 3: } E_3 = k_{30} + k_{31} \tag{9.47}$$

$$\text{Compartment 4: } E_4 = k_{40} + k_{41} \tag{9.48}$$

The disposition function of the compartment 1 according to Equation A.33 (Appendix A) is as follows:

$$(Disp.)_{s,1} = \frac{(s+E_2)(s+E_3)(s+E_4)}{(s+E_1)(s+E_2)(s+E_3)(s+E_4) - k_{12}k_{21}(s+E_3)(s+E_4) - k_{13}k_{31}(s+E_2)(s+E_4) - k_{14}k_{41}(s+E_2)(s+E_3)} \tag{9.49}$$

Following the substitution of the miniature rate constants for E_1, E_2, E_3, and E_4 (Equations 9.45 through 9.48) in the denominator of Equation 9.49, implementing mathematical manipulation and

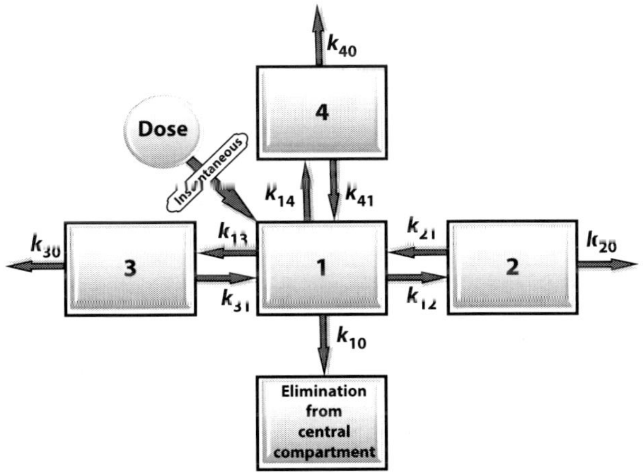

FIGURE 9.10 A typical multicompartment mammillary with a central compartment and three peripheral compartments interdependent with the central compartment. All four compartments are considered driving-force compartments and have elimination to the outside environment, not necessarily the same environment.

expressing α, β, γ, and δ as the hybrid rate constants of the new combination of miniature rate constants, the equation will change to the following disposition function:

$$(Disp.)_{s1} = \frac{(s+E_2)(s+E_3)(s+E_4)}{(s+\alpha)(s+\beta)(s+\gamma)(s+\delta)} \tag{9.50}$$

Multiplication of the disposition function (Equation 9.50) with any of the input functions yields the Laplace transform of the central compartment following that input. For example, the Laplace transform of Compartment 1 assuming intravenous bolus as the input is

$$\mathcal{L}(A_1) = \tilde{A} = Input \times (Disp.)_{s1} = \frac{Dose(s+E_2)(s+E_3)(s+E_4)}{(s+)(s+\beta)(s+\gamma)(s+\delta)} \tag{9.51}$$

The inverse Laplace transform of the central compartment (Equation 9.51), using the partial fraction theorem (Benet and Turi, 1971) (Appendix A, Section A3.3) is

$$\begin{aligned} A_1 = & \frac{Dose(E_2-\alpha)(E_3-\alpha)(E_4-\alpha)}{(\beta-\alpha)(\gamma-\alpha)(\delta-\alpha)}e^{-\alpha t} + \frac{Dose(E_2-\beta)(E_3-\beta)(E_4-\beta)}{(\alpha-\beta)(\gamma-\beta)(\delta-\beta)}e^{-\beta t} \\ & + \frac{Dose(E_2-\gamma)(E_3-\gamma)(E_4-\gamma)}{(\alpha-\gamma)(\beta-\gamma)(\delta-\gamma)}e^{-\gamma t} + \frac{Dose(E_2-\delta)(E_3-\delta)(E_4-\delta)}{(\alpha-\delta)(\beta-\delta)(\gamma-\delta)}e^{-\delta t} \end{aligned} \tag{9.52}$$

where, A_l is the amount in the central compartment at time t. The four exponential terms of the equation are created when s in the Laplace transform of the compartment is set equal to −α, −β, −γ, and −δ followed by multiplication with their exponential terms.

9.3.1.4 Mathematical Relationship for an Intravenously Administered Compound That Follows the Multicompartment Model: Use of Input-Disposition Function and General Partial Fraction Theorem—Peripheral Compartment

Once the Laplace transform of the central compartment is established, the Laplace transform of peripheral compartments can be obtained by using the method of substitution. For instance, for compartment 2 of Figure 9.10, the related equations are

$$\frac{dA_2}{dt} = k_{12}A_1 - k_{21}A_2 - k_{20}A_2 \tag{9.53}$$

Setting the overall exit rate constant as

$$E_2 = k_{21} + k_{20} \tag{9.54}$$

Equation 9.53 changes to

$$\frac{dA_2}{dt} = k_{12}A_1 - E_2A_2 \tag{9.55}$$

$$\mathcal{L}\left[\frac{dA_2}{dt}\right] = s\tilde{A}_2 - 0 = k_{12}\tilde{A}_1 - E_2\tilde{A}_2 \tag{9.56}$$

$$\tilde{A}_2 = \frac{k_{12}\tilde{A}_1}{(s + E_2)} \tag{9.57}$$

Substitution of $\tilde{A}_1$ as defined in Equation 9.51 into Equation 9.57 yields the Laplace transform of the second compartment as

$$\begin{aligned} \bar{A}_2 &= \frac{k_{12}[Dose(s + E_2)(s + E_3)(s + E_4)]}{(s + E_2)(s + \alpha)(s + \beta)(s + \gamma)(s + \delta)} \\ &= \frac{k_{12}[Dose(s + E_3)(s + E_4)]}{(s + \alpha)(s + \beta)(s + \gamma)(s + \delta)} \end{aligned} \tag{9.58}$$

Therefore, the integrated equation is

$$\begin{aligned} A_2 &= \frac{k_{12}[Dose(E_3 - \alpha)(E_4 - \alpha)]}{(\beta - \alpha)(\gamma - \alpha)(\delta - \alpha)} e^{-\alpha t} + \frac{k_{12}[Dose(E_3 - \beta)(E_4 - \beta)]}{(\alpha - \beta)(\gamma - \beta)(\delta - \beta)} e^{-\beta t} \\ &+ \frac{k_{12}[Dose(E_3 - \gamma)(E_4 - \gamma)]}{(\alpha - \gamma)(\beta - \gamma)(\delta - \gamma)} e^{-\gamma t} + \frac{k_{12}[Dose(E_3 - \delta)(E_4 - \delta)]}{(\alpha - \delta)(\beta - \delta)(\gamma - \delta)} e^{-\delta t} \end{aligned} \tag{9.59}$$

In the above relationships, the input function into the second compartment is k_{12} multiplied by the Laplace transform of the central compartment.

9.3.1.5 Multicompartmental Modeling of a Xenobiotic and Its Metabolite(s) Following Intravenous Administration

The schematic diagram of the model is presented in Figure 9.11. Compartment A_1 is the central compartment representing the amount in the systemic circulation and highly perfused tissues; compartment A_2 is the amount of parent compound in the peripheral compartment; compartment A_3 is the amount of metabolite(s) in the central compartment and A_4 is the amount of metabolite(s) associated with the peripheral compartment. The formation rate constant of the metabolite(s) is k_{13},

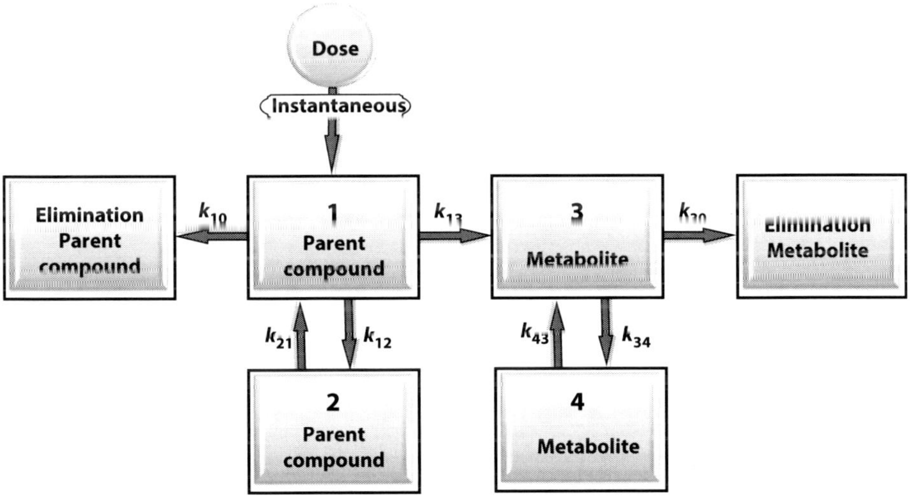

FIGURE 9.11 Diagram of a linear two-compartment model with respect to unchanged xenobiotic and two-compartment model with respect to its metabolite; the distribution, metabolism and elimination are governed by first-order kinetics.

the elimination rate constant of the metabolite(s) is k_{30}; the distribution rate constants of the parent compound between the central and peripheral compartment are k_{12} and k_{21} and the distribution rate constants of the metabolite(s) between compartment A_3 and A_4 are k_{34} and k_{43}. The differential equations of the compartments are

$$\frac{dA_1}{dt} = k_{21}A_2 - k_{12}A_1 - k_{10}A_1 - k_{13}A_1 \tag{9.60}$$

$$\frac{dA_2}{dt} = k_{12}A_1 - k_{21}A_2 \tag{9.61}$$

$$\frac{dA_3}{dt} = k_{13}A_1 + k_{43}A_4 - k_{34}A_3 - k_{30}A_3 \tag{9.62}$$

$$\frac{dA_4}{dt} = k_{34}A_3 - k_{43}A_4 \tag{9.63}$$

The integrated equations for the parent and metabolite(s) concentrations in the central compartment are

$$Cp = \frac{Dose}{V_1}\left(\frac{k_{21} - \alpha}{\beta - \alpha}e^{-\alpha t} + \frac{k_{21} - \beta}{\alpha - \beta}e^{-\beta t}\right) \tag{9.64}$$

where α and β are the hybrid rate constants representing:

$$\alpha + \beta = k_{13} + k_{12} + k_{21} \quad \text{and} \tag{9.65}$$

$$\alpha\beta = k_{13}k_{21} \tag{9.66}$$

$$\begin{aligned} C3 = \frac{k_{13}D}{V_3}\Bigg[& \frac{(k_{21} - \alpha)(k_{43} - \alpha)}{(\beta - \alpha)(\delta - \alpha)(\gamma - \alpha)}e^{-\alpha t} + \frac{(k_{21} - \beta)(k_{43} - \beta)}{(\alpha - \beta)(\delta - \beta)(\gamma - \beta)}e^{-\beta t} \\ & + \frac{(k_{21} - \delta)(k_{43} - \delta)}{(\alpha - \delta)(\beta - \delta)(\gamma - \delta)}e^{-\delta t} + \frac{(k_{21} - \gamma)(k_{43} - \gamma)}{(\alpha - \gamma)(\beta - \gamma)(\delta - \gamma)}e^{-\gamma t}\Bigg] \end{aligned} \tag{9.67}$$

The hybrid disposition rate constants of metabolite(s) correspond to the following relationships:

$$\delta + \gamma = k_{30} + k_{34} + k_{43} \tag{9.68}$$

$$\delta\gamma = k_{30}k_{43} \tag{9.69}$$

The complex models such as the one depicted in Figure 9.11 and defined by Equations 9.60 through 9.63 are usually calibrated to define sufficiently the experimental data. When the compartmental model sufficiently describes the data, it should be questioned whether the parameters/constants of the model can be determined unambiguously from the amount and quality of experimental data. This requires further evaluation of the model predictions, which are known as identifiability analysis. The methodology for identifiability analysis is well established and when the parameters of the model can be determined explicitly, the model is considered uniquely identifiable (Bellman and Åström, 1970; Cobelli and Mari, 1983; Chapman and Godfrey, 1987; Chappell et al., 1990; Dutta et al., 1996; van den Hof, 1998).

9.3.2 Dose-Dependent Compartmental Analysis

The differentiation between linear and nonlinear PK/TK is critical in compartmental analysis. As in physiological modeling, the nonlinear pharmacokinetics exists when a drug is absorbed or eliminated by mechanisms that are potentially saturable. The nonlinear absorption, metabolism, or excretion is governed by the Michaelis–Menten equation (Chapter 6, Section 6.6). It is also referred to as dose-dependent pharmacokinetics. The rate constants, half-lives, and clearances, which are constant in linear PK/TK, become variable and their values depend on the dose or plasma concentration.

The nonlinearity of PK/TK data is often recognizable from graphical analysis of the data. Figure 9.12 depicts the logarithm of plasma concentration versus time when the compound at high plasma levels exhibits the nonlinear or dose-dependent kinetics and at low concentrations shows dose-independent behavior (Gibaldi and Perrier, 1982). The plot of plasma concentration versus time, not logarithmic, exhibits a profile that resembles a hockey stick (Figure 9.13) (Wagner, 1975).

9.3.2.1 Compartmental Models with Michaelis–Menten Kinetics

The simplest model for consideration in this category is the one-compartment model with intravenous bolus injection and the assumption that elimination from the compartment is concentration-dependent, nonlinear, and governed by the Michaelis–Menten equation. The decline of the plasma concentration of xenobiotics is through the nonlinear rate of elimination from the compartment, that is,

$$\frac{dCp}{dt} = -\frac{V_{max} \times Cp}{K_M + Cp} \tag{9.70}$$

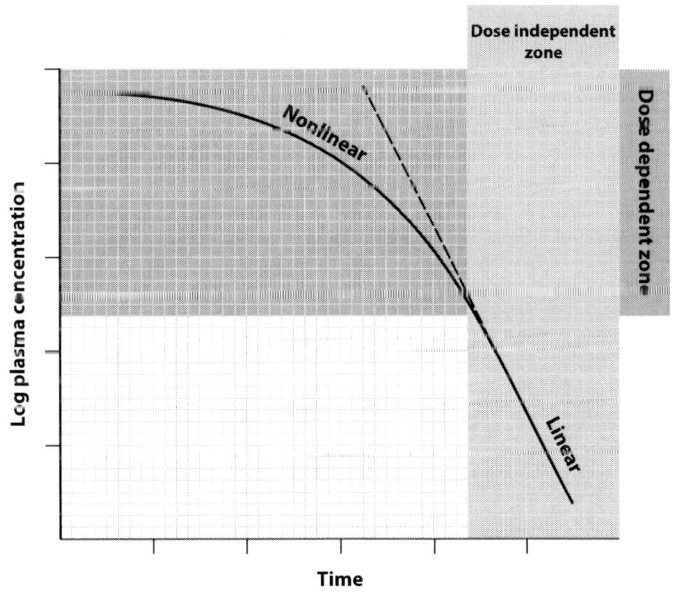

FIGURE 9.12 Plot of the logarithm of plasma concentration versus time for a xenobiotic that exhibits nonlinear dose-dependent kinetics at high doses and linear dose-independent at low concentrations.

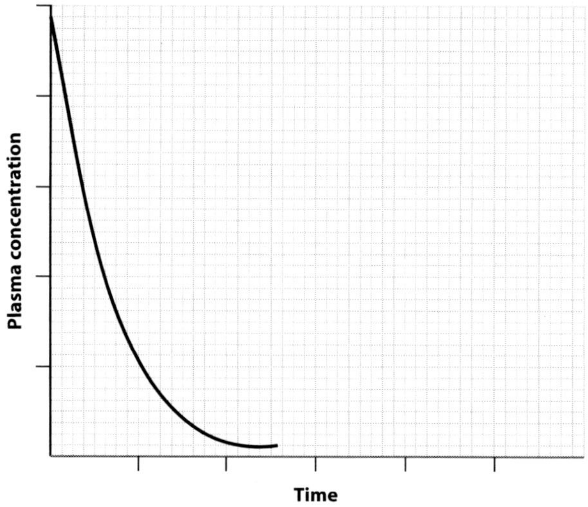

FIGURE 9.13 A typical hockey stick profile of plasma concentration–time plot of a xenobiotic that exhibits nonlinear PK/TK characteristics.

$$-dCp(K_M + Cp) = dt(V_{max} \times Cp) \tag{9.71}$$

$$-\frac{dCpK_M}{Cp} - dCp = V_{max}dt \tag{9.72}$$

Integration yields

$$-KM \times \ln C = V_{max}t + \text{Constant} \tag{9.73}$$

Evaluation of the constant at $t = 0$ is

$$\text{Constant} = Cp^0 - K_M \ln Cp^0 \tag{9.74}$$

Substitution of Equation 9.74 in Equation 9.73 and solving for the variables yields

$$\ln Cp = \frac{Cp^0 - Cp}{K_M} + \ln Cp^0 - \frac{V_{max}}{K_M} t \tag{9.75}$$

$$\text{or,} \quad Cp^0 - Cp + K_M \ln \frac{Cp^0}{Cp} = V_{max}t \tag{9.76}$$

$$t = \frac{K_M}{V_{max}} \ln \frac{Cp^0}{Cp} + \frac{Cp^0 - Cp}{V_{max}} \tag{9.77}$$

A modified version of the equation, having the initial condition of both variables, appropriate for curve fitting is (Gibaldi and Perrier, 1982)

$$t - t^0 = \frac{K_M}{V_{max}} \ln \frac{Cp^0}{Cp} + \frac{Cp^0 - Cp}{V_{max}} \tag{9.78}$$

To estimate preliminary values for V_{max} and K_M from plasma concentration–time data, the approaches described in Chapter 6, Section 6.6 can be employed. For example, Equation 9.70 can be modified to a difference equation (Equation 9.79) and by using the linear Lineweaver–Burk plot (Double Reciprocal), Hanes–Woolfe plot, or Eadie–Hofstee plot, the initial estimates of the parameters can be determined. In the following equations $Cp_{midpoint}$ corresponds to the midpoint plasma concentration that corresponds to the rate defined as $(\Delta Cp/\Delta t)$:

$$\frac{\Delta Cp}{\Delta t} = -\frac{V_{max} \times Cp}{K_M + Cp} \tag{9.79}$$

The linear versions of the equation are

$$\frac{1}{(\Delta Cp/\Delta t)} = \frac{K_M}{V_{max} Cp_{midpoint}} + \frac{1}{V_{max}} \tag{9.80}$$

$$\frac{Cp_{midpoint}}{\Delta Cp/\Delta t} = \frac{K_M}{V_{max}} + \frac{Cp_{midpoint}}{V_{max}} \tag{9.81}$$

$$\frac{\Delta Cp}{\Delta t} = V_{max} - \frac{(\Delta Cp/\Delta t)K_M}{Cp_{midpoint}} \tag{9.82}$$

When the elimination from the body is by multiple and parallel Michaelis–Menten processes, the rate of elimination is the pooled equation of all processes (Figure 9.14) (Wagner, 1975):

$$-\frac{dCp}{dt} = \frac{V_{max_1} Cp}{K_{M_1} + Cp} + \frac{V_{max_2} Cp}{K_{M_2} + Cp} + \frac{V_{max_3} Cp}{K_{M_3} + Cp} + \cdots\cdots + \frac{V_{max_n} Cp}{K_{M_n} + Cp} \tag{9.83}$$

Setting the boundary conditions: $at\ t^0, Cp = Cp^0$ and as $t \Rightarrow \infty, Cp \Rightarrow 0$ the dose-dependent parameters of Equation 9.83 can be estimated as (Wagner, 1975):

$$V_{max_n} = \frac{Cp^0 \left[\sum_{i=1}^{n} \left(V_{max_i}/(K_{M_i} + Cp^0)\right)\right]\left[\sum_{i=1}^{n} (V_{max_i}/K_{M_i})\right]}{\sum_{i=1}^{n} \left(V_{max_i}/K_{M_i}\right) - \sum_{i=1}^{n} \left(V_{max_i}/(K_{M_i} + Cp^0)\right)} \tag{9.84}$$

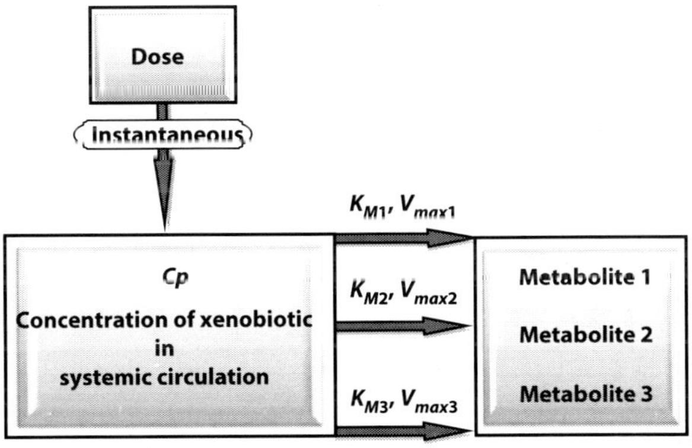

FIGURE 9.14 Diagram of a pharmacokinetic model with multiple and parallel nonlinear metabolism governed by the Michaelis–Menten equation.

$$K_{M_n} = \frac{Cp^0\left[\sum_{i=1}^{n}\left(V_{max_i}/(K_{M_i} + Cp^0)\right)\right]}{\sum_{i=1}^{n}\left(V_{max_1}/K_{M_i}\right) - \sum_{i=1}^{n}\left(V_{max_i}/K_{M_i} + Cp^0\right)} \tag{9.85}$$

At $Cp = Cp^0$, when plasma is the highest and greater than K_{M_n}, the dose-independent parameters are

$$V_{\max_n} = \sum_{i=1}^{n}(V_{max_i}) \tag{9.86}$$

$$K_{M_i} = \frac{\sum_{i=1}^{n}(V_{max_i})}{\sum_{i=1}^{n}\left(V_{max_i}/K_{M_i}\right)} \tag{9.87}$$

When the compound is eliminated from the body by simultaneous dose-independent excretion and dose-dependent metabolism, the overall rate of elimination is the combination of both processes

$$\frac{dC}{dt} = -\left(\frac{V_{max}Cp}{K_M + Cp} + k_e Cp\right) \tag{9.88}$$

$$\therefore \frac{dCp}{dt} = \frac{-k_e Cp(K_M + Cp) - V_{max}Cp}{K_M + Cp} \tag{9.89}$$

$$\text{or,} \quad \frac{dCp}{dt} = \frac{Cp(-k_e K_M - V_{max} - k_e Cp)}{K_M + Cp} \tag{9.90}$$

Similar approaches are used when the absorption of a compound follows first-order kinetics, but the elimination is governed by dose-dependent metabolism. The rate of concentration change in the body is defined as

$$\frac{dCp}{dt} = k_a C\bar{p}^0 - \frac{V_{max}Cp}{K_M + Cp} \tag{9.91}$$

where

$$C\bar{p}^0 = \frac{FD}{Vd}, \tag{9.92}$$

(F) is the fraction of dose absorbed and Vd is the apparent volume of distribution.

The rate equations of the two-compartment model with elimination from the central compartment after intravenous bolus dose are

$$\frac{dCp}{dt} = k_{21}C_2 - k_{12}Cp - \frac{V_{max}Cp}{K_M + Cp} \tag{9.93}$$

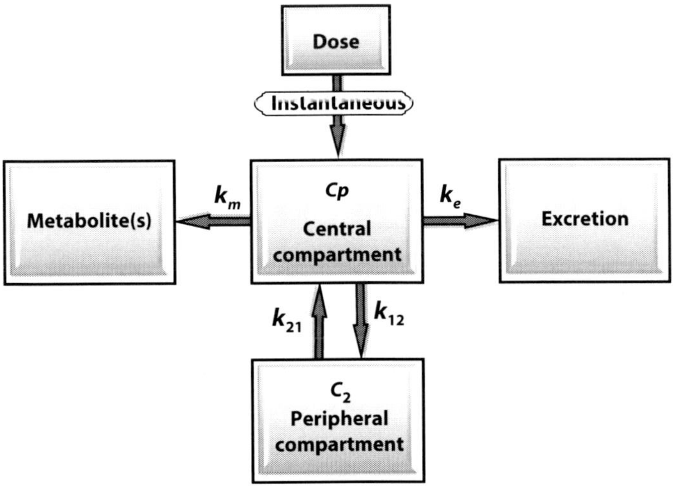

FIGURE 9.15 Diagram of a two-compartment model with parallel metabolism and excretion from the central compartment; under nonlinear conditions, the linear biotransformation rate constant (k_m) will change to K_M and V_{max}, the parameters of Michaelis–Menten equation.

$$\frac{dC_2}{dt} = k_{12}Cp - k_{21}C_2$$

The equation for the parallel dose-dependent and dose-independent eliminations from the central compartment (Figure 9.15) is

$$\frac{dCp}{dt} = k_{21}C_2 - \left(k_{12}Cp + \frac{V_{max}Cp}{K_M + Cp} + k_eCp\right) \tag{9.94}$$

9.4 NON-COMPARTMENTAL ANALYSIS BASED ON STATISTICAL MOMENT THEORY

The principle of non-compartmental analysis is based on the assumption that the passage of a compound through the body is a random process, (i.e., stochastic), and thus is governed by the principles of probability and can be defined by the probability function. In theory, a set of plasma concentration–time data is considered a statistical distribution. The equation of observations that defines the distribution is called the probability function, which is characterized by the moments of the function. The moments of the function can be estimated about the origin or about the mean. If $f(x)$ is the function, the moments of the function are defined by integrating the function with a suitable power of the independent variable over an interval of $[a,b]$:

$$\mu_r = \int_a^b x^r \cdot f(x)\,dx \tag{9.95}$$

When r, the power of x, is equal to zero:

$$\text{Zero moment:} \quad \mu_0 = \int_a^b f(x)\,dx$$

When $r = 1$:

$$\text{First moment: } \mu_1 = \int_a^b x f(x)dx$$

When $r = 2$:

$$\text{Second moment: } \mu_2 = \int_a^b x^2 f(x)dx$$

When $r = n$:

$$\text{n}^{\text{th}} \text{ moment: } \mu_n = \int_a^b x^n f(x)dx$$

Parallel to the above statistical equations of moments, the following equations are developed for the statistical moments of the plasma concentration–time curve (Cutler, 1978; Yamaoka et al., 1978; Benet and Galeazzi, 1979; Riegelman and Collier, 1980; Cawello et al., 1999). The PK/TK based on plasma concentration–time data and the related moments about the origin within the observation interval of $a = 0$, $b = \infty$ are

$$\mu_0 = \int_0^\infty Cpdt = \int_0^{t_n} Cpdt + \int_{t_n}^\infty Cpdt = AUC_0^\infty \quad (9.96)$$

$$\mu_1 = \int_0^\infty tCpdt = \int_0^{t_n} tCpdt + \int_{t_n}^\infty tCpdt = AUMC_0^\infty \quad (9.97)$$

$$\mu_2 = \int_0^\infty t^2Cpdt = \int_0^{t_n} t^2Cpdt + \int_{t_n}^\infty t^2Cpdt = AUM_2C_0^\infty \quad (9.98)$$

(AUC_0^∞) is the area under the plasma concentration–time curve or the area under the zero-moment curve. This parameter is calculated by a model-independent numerical integration, for example, linear trapezoidal, or log trapezoidal as described in Appendix A.

$(AUMC_0^\infty)$ is the area under the first-moment curve. The observed plasma concentrations are multiplied by the corresponding time points and plotted against the same time points. The method of estimation is the same as *AUC*. The terminal area of *AUMC* is calculated as

$$(\text{Terminal area})_{AUMC} = (Cp_n t_n / K) + (Cp / K^2).$$

$(AUM_2C_0^\infty)$ is the area under the second moment curve, which is constructed by multiplying each plasma concentration by the square of its time point and then plotted against the time point. The area under the second moment curve provides a variety of residences for a drug in the body.

Moments higher than $AUM_2C_0^\infty$ are used if the measures of skewness and kurtosis of the distribution need to be defined.

The advantages of non-compartmental analysis are the simplicity of the approach, estimating parameters without model assumptions and limitations, and practical use in drug development, clinical trials, and FDA approval. However, the methodology does not elucidate fully the organ specific function or the related physiological parameters and constants. The parameters are solely based on the observed data, and thus the analysis is sensitive to experimental protocol, such as frequency and duration of sampling.

9.4.1 Mean Residence Time (MRT) and Mean Input Time (MIPT)

Normalizing the first and second moments (Equations 9.97 and 9.98) with respect to the zero moment (Equation 9.96), yields the following PK/TK parameters:

$$MRT = \frac{\mu_1}{\mu_0} = \frac{AUMC_0^\infty}{AUC_0^\infty} \tag{9.99}$$

$$VRT = \frac{\mu_2}{\mu_0} = \frac{AUM_2C_0^\infty}{AUC_0^\infty} \tag{9.100}$$

where *MRT* is mean residence time (Benet, 1985; Cutler, 1987; Kasuya et al., 1987) and *VRT* is the variance of the residence time that may reflect the magnitude of the random error. For drugs that are given by bolus injection and follow the one-compartment model, Equation 9.99 can also be presented as

$$MRT = \frac{Cp\circ / K^2}{Cp\circ / K} = \frac{1}{K} = 1.44 \times T_{1/2} \tag{9.101}$$

Therefore, the mean residence time for drugs that follow the one-compartment model is equal to $1.44T_{1/2}$, which is also known as time constant, turnover time, transit time, or sojourn time (Shipley and Clark, 1972). The value of *MRT* is also estimated by dividing the area under the curve by the initial plasma concentration.

$$MRT = \frac{AUC_0^\infty}{Cp\circ} \tag{9.102}$$

For compounds that follow the two-compartment model (Equation 9.103), the MRT equation can be estimated as Equation 9.104.

$$Cp = ae^{-\alpha t} + be^{-\beta t} \tag{9.103}$$

$$MRT = \frac{(a/\alpha^2) + (b/\beta^2)}{(a/\alpha) + (b/\beta)} \tag{9.104}$$

When $\alpha \gg \beta$

$$MRT \cong \frac{1}{\beta} = 1.44 \times (T_{1/2})_\beta \tag{9.105}$$

If the system and data are considered linear and dose independent, the mean residence time would represent the time required for 63.3% of the dose to be eliminated via all routes of elimination. Since in linear systems *AUC* is equal to $Dose/Cl_T$, *MRT* can also be estimated as

$$MRT = \frac{AUMC \times Cl_T}{Dose} \tag{9.106}$$

When the input is not instantaneous like intravenous bolus injection, for example, if the input is zero-order infusion or first-order absorption, the mean residence time is equal to the mean residence time of bolus injection plus the mean input time (*MIPT*) such as mean infusion time or mean absorption time, that is,

$$MRT_{infusion} = MRT_{bolus} + \frac{T}{2} \tag{9.107}$$

where T represents the time of infusion and $0.5T$ is the mean infusion time (*MIT*). For oral administration

$$MRT_{oral} = MRT_{bolus} + \frac{1}{k_a} \tag{9.108}$$

where (k_a) is the absorption rate constant and $1/k_a$ represents the mean absorption time (*MAT*).

Therefore, *MAT* can be estimated as

$$MAT = MRT_{oral} - MRT_{bolus} = \frac{1}{k_a} = 1.44(T_{1/2})_{absorption} \tag{9.109}$$

When the absolute bioavailability is one or close to one, (*MAT*) represents the true mean absorption time (Cutler, 1978). If the bioavailability is low due to the pre-systemic metabolism or decomposition of the drug at the site of absorption, etc., the value of *MAT* is only an apparent value. In general, for the routes of administration such as oral route where bioavailability is a concern, the *MRT* can be estimated as

$$MRT_{oral} = MRT_{bolus} \times \frac{\left(AUC_0^\infty\right)_{oral.}}{F \times \left(AUC_0^\infty\right)_{bolus}} + MIPT \tag{9.110}$$

9.4.2 Total Body Clearance and Apparent Volume of Distribution

The total body clearance in non-compartmental analysis is estimated by the model-independent equation of

$$(Cl_T)_{bolus} = \frac{Dose}{AUC_0^\infty} \tag{9.111}$$

$$(Cl_T)_{oral} = \frac{F \times Dose}{AUC_0^\infty} \tag{9.112}$$

Equation 9.112 is also used for other routes of administration when bioavailability is a concern. Substituting AUC_0^∞ from Equation 9.99 into Equation 9.112 yields the following relationship for the total body clearance:

$$(Cl_T)_{bolus} = \frac{Dose \times MRT}{AUMC} \quad \text{or} \quad (Cl_T)_{oral} = \frac{FD \times MRT}{AUMC} \tag{9.113}$$

The apparent volume of distribution of the non-compartmental model (denoted as $(Vd)_{ss}$), is estimated from total body clearance and the mean residence time

$$(Vd)_{ss} = Cl_T \times MRT \tag{9.114}$$

Substitution of Equations 9.113 and 9.99 in Equation 9.114 yields the following relationships for the apparent volume of distribution (Benet and Galeazzi, 1979):

$$(Vd)_{ss} = \left(\frac{Dose}{AUC}\right)\left(\frac{AUMC}{AUC}\right) \tag{9.115}$$

$$(Vd)_{ss} = \frac{Dose \times AUMC}{AUC^2} \tag{9.116}$$

Equations 9.114 through 9.116 are used for xenobiotics that are given by bolus injection regardless of the dispositional profile.

Substitution of Equations 9.110 and 9.112 in Equation 9.114 provides the volume of distribution of a compound administered orally. In addition, the relationship applies to similar routes when the diffusion through a barrier can be considered a rate-limiting factor:

$$Vd_{ss} = \frac{F \times Dose}{\left(AUC_0^\infty\right)_{oral}}\left[MRT_{bolus} \times \frac{\left(AUC_0^\infty\right)_{oral}}{F \times \left(AUC_0^\infty\right)_{bolus}} + MIPT\right] \tag{9.117}$$

$$\therefore\ Vd_{ss} = \frac{Dose \times MRT_{bolus}}{AUC_{bolus}} + \frac{F \times Dose \times MIPT}{AUC_{oral}} \tag{9.118}$$

The volume of distribution for compounds given by short-term infusion is defined as

$$Vd_{ss} = \frac{A_{inf\ used} \times AUMC}{AUC^2} - \frac{A_{inf\ used} \times T}{2 \times AUC} \tag{9.119}$$

where $A_{infused}$ is the total amount administered by infusion, expressed as $A_{inf\ used} = k_0 \times T$, that is, the rate of infusion multiplied by the time of infusion. Therefore, Equation 9.110 is also presented as

$$Vd_{ss} = \frac{k_0 \times T \times AUMC}{AUC^2} - \frac{k_0 \times T^2}{2 \times AUC} \tag{9.120}$$

9.5 PK/PD AND TK/TD MODELING

9.5.1 Introduction

The quantitative evaluation of a xenobiotic interacting with its receptor site provides the pharmacological response known as pharmacodynamics (PD) and the toxic response known as toxicodynamics (TD). Similar to any dynamic processes, the initial and final conditions of response are the important parameters in PD/TD analysis. In other words, at the initial phase of the evaluation when the dose is zero, no response is anticipated. When the dose is optimum, the PD desired outcome is obtained, and when the dose is high and beyond the acceptable range, the TD response is expected.

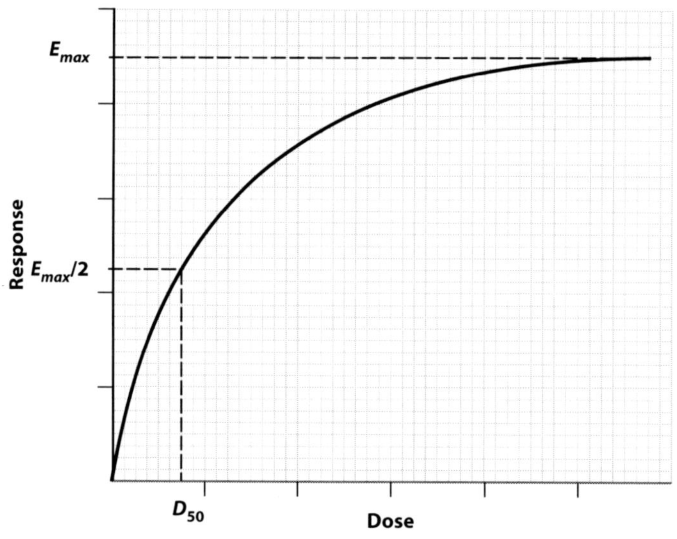

FIGURE 9.16 The hyperbolic dose-response plot.

The outcome, whether therapeutic or toxic, is a function of the amount or concentration of a xenobiotic and its affinity for interaction with its receptor. In most *in vitro* systems, for example, in cell culture, the concentration of xenobiotic can be kept constant and the interaction with its receptor site is direct and rather rapid during the controlled experimental conditions. Under *in vivo* conditions, however, the concentration of a xenobiotic at the receptor site or biophase is a function of time and the physiological processes of ADME. Therefore, the *in vivo* PD/TD observation is a function of the PK/TK profile of the compound.

In PD/TD analysis, the response is measured directly or indirectly. The examples of direct response are the measurable physiological/biochemical changes such as lowering blood pressure, reducing intraoccular pressure, change in the coagulation time of blood, binding to macromolecules, interaction with enzymes, etc. The examples of indirect response are the measuring specific biomarkers such as serum levels of creatine kinase MB, lactate dehydrogenase, troponin T and endothelin I in cardiomyopathy evaluation. In a dose-response relationship, for example, it can be observed where the dose is zero the response is zero, the magnitude of the dose increases and the magnitude of response increases until increasing the dose no longer increases the response. This maximum response is denoted as E_{max} and the dose that corresponds to half of E_{max} is known as D_{50} (Figure 9.16). The scenario is analogous to the Michaelis–Menten kinetics discussed in Chapter 7.

The plot of response versus logarithm of dose is sigmoidal as presented in Figure 9.17. The dose-response or dose-effect curves can also be presented as concentration-response or log concentration-response curve.

9.5.2 Xenobiotic–Receptor Interaction and the Law of Mass Action

Xenobiotics interact with their receptor site according to the following relationship:

$$[R] + [D] \rightleftharpoons [RD] \longrightarrow \text{Effect} \tag{9.121}$$

where $[R]$ is the concentration of free receptor, $[D]$ is the concentration of free compound in the environments of the receptor and $[RD]$ is the concentration of xenobiotic-receptor complex.

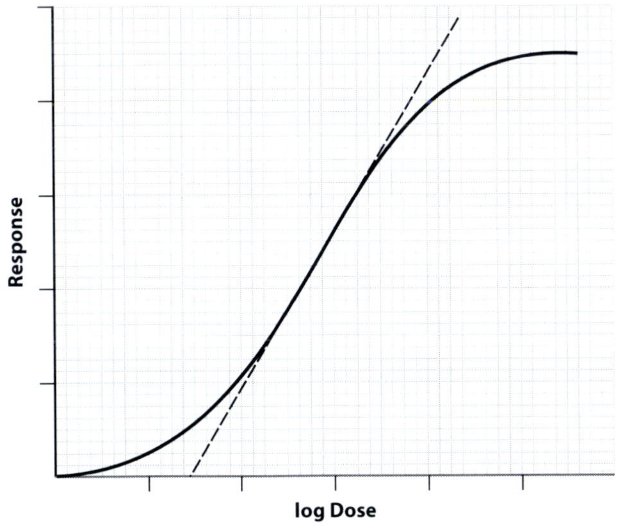

FIGURE 9.17 The sigmoidal curve of response versus logarithm of dose.

Equation 9.121 is a typical complexation reaction with the following assumptions:

1. $[D]$ and $[R]$ interact rather rapidly to form $[RD]$ complex
2. Only a single molecule of $[D]$ interacts with a single receptor site
3. $[R]$, $[D]$ and $[RD]$ are at equilibrium
4. The interaction is reversible

From the relationship presented in Equation 9.121, the association constant (K_A) and dissociation constant (K_D) are estimated as

$$K_A = \frac{[RD]}{[R][D]} \tag{9.122}$$

$$K_D = \frac{[R][D]}{[RD]} \tag{9.123}$$

The density of free receptor $[R]$ is the difference between total density $[R_T]$ (i.e., free and bound), and bound density $[RD]$

$$[R] = [R_T] - [RD] \tag{9.124}$$

Substitution of Equation 9.124 in Equation 9.123 yields

$$K_D = \frac{([R_T] - [RD]) \times [D]}{[RD]} = \frac{[R_T][D] - [RD][D]}{[RD]} \tag{9.125}$$

$$[RD](K_D + [D] = [R_T][D] \tag{9.126}$$

$$[RD] = \frac{[R_T][D]}{K_D + [D]} \tag{9.127}$$

Equation 9.127 indicates that $[RD]$ is directly proportional to $[D]$ at the receptor site. Equation 9.127 can also be presented in terms of plasma concentration:

$$[RCp] = \frac{[R_T][Cp]}{K_D + [Cp]} \tag{9.128}$$

An important assumption useful in defining Equations 9.127 and 9.128 is the "occupancy theory," which states the intensity of the response is proportional to the number of receptors occupied by the compound, $[RCp]$ (Ariens, 1964). Thus, the maximum response (E_{max}) occurs when all available receptor sites $[R_T]$ are occupied, in other words the maximum response is proportional to complete occupancy of receptor sites

$$\therefore [R_T] \propto E_{max}$$

Thus, Equation 9.128 can be written as

$$E = \frac{E_{max} \times [D]}{K_D + [D]} \tag{9.129}$$

Similar to the Michaelis–Menten equation (Chapter 7), Equation 9.129 can also be presented in the following linear version:

$$\log\left(\frac{E}{E_{max} - E}\right) = \log[D] - \log K_D \tag{9.130}$$

Thus, the plot of $(\log(E/(E_{max} - E))$ versus $[D]$ is a straight line with slope equal to one and y-intercept of $\log K_D$.

When effect is half-maximum, that is, $E = (E_{max}/2)$, the equilibrium constant K_D will be equal to the concentration that corresponds to 50% of maximum response (Figure 9.16), that is,

$$\frac{E_{max}}{2} = \frac{E_{max} \times [D]_{50}}{K_D + [D]_{50}} \tag{9.131}$$

$$\frac{E_{max}}{2} = \frac{E_{max} \times [Cp]_{50}}{K_D + [Cp]_{50}} \tag{9.132}$$

$$K_D = [D]_{50} = [Cp]_{50} = E_{50} \tag{9.133}$$

The dose-response relationship is used in toxicodynamics to develop an understanding of the integrated biological processes underlying a response, hazard identification, and safety assessment. The determination of the lowest dose causing adverse effects, or determination of threshold effect, or a NO-Observed-Adverse-Effect-Level (NOAEL) in the dose-response evaluation of lead compounds in drug discovery and development, has been associated with analytical uncertainty and biological variability. PK/PD or TK/TD modeling in association with toxicogenomics and population PK/TK analysis has reduced this uncertainty significantly.

9.5.3 Pharmacodynamic Models of Plasma Concentration and Response

When the response is measured as the outcome of a therapeutic dose or a known level of exposure, the magnitude of the outcome can be linked to plasma concentration through linear or nonlinear relationships.

9.5.3.1 Linear Pharmacodynamic Model

The dependent variable in this model is the response/effect, which is assumed to be directly proportional to plasma concentration, that is,

$$E = mCp + b \tag{9.134}$$

where (b), the y-intercept, is the response at time zero, that is, $b = E_0$. The model assumes a higher concentration provide a greater response. The maximum effect, E_{max}, cannot be estimated from the relationship (Equation 9.134). The model may only be useful for the correlation of response with the low concentrations when the receptor sites are not fully occupied (Figure 9.18).

9.5.3.2 Log-Linear Pharmacodynamic Model

This model is based on the assumption that the intensity of response is linearly dependent on the logarithm of plasma concentration (Levy, 1964, 1966)

$$E = m\log Cp + b \tag{9.135}$$

Equation 9.135 defines the midlinear segment of the sigmoidal curve presented in Figure 9.17. It follows that the response at time zero corresponds to the plasma concentration at time zero:

$$E_0 = m\log Cp^0 + b \tag{9.136}$$

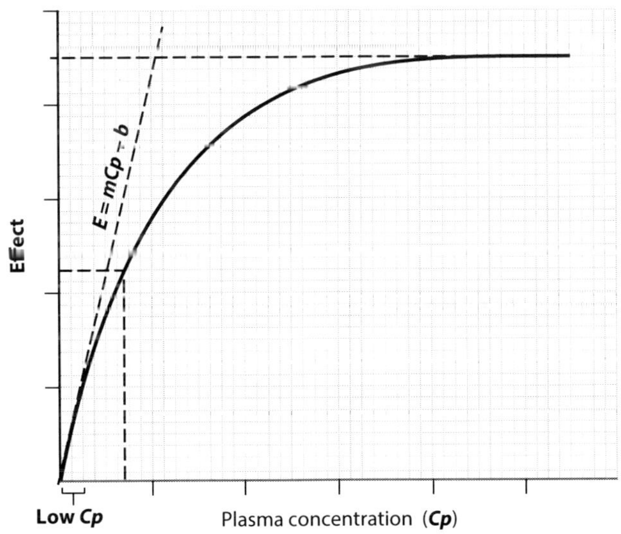

FIGURE 9.18 Depiction of the linear pharmacodynamic model according to Equation 9.134.

Solving for log Cp (Equation 9.135) and log Cp^0 (Equation 9.136) provides the following relationships:

$$\log Cp = \frac{E - b}{m} \tag{9.137}$$

$$\log Cp^0 = \frac{E_0 - b}{m} \tag{9.138}$$

Substitution of Equations 9.137 and 9.138 in a PK/TK model, for example, $\log Cp = \log Cp^0 - (Kt/2.303)$, yields

$$\frac{E - b}{m} = \frac{E_0 - b}{m} - \frac{Kt}{2.303} \tag{9.139}$$

$$\therefore E = E_0 - \frac{K \times t \times m}{2.303} \tag{9.140}$$

Equation 9.140, represents the zero-order time course of response as a function of $-(K/2.303)$, the slope of log plasma concentration versus time curve (log Cp – *time*), and m the slope of the response (E) versus the log Cp curve. In other words, plasma concentration declines exponentially and the intensity of response decreases linearly.

9.5.3.3 Nonlinear Hyperbolic E_{max} Model

Based on Equations 9.129, and 9.131 through 9.133, the relationship between the effect and plasma concentration can be presented as Equation 9.141 (Wagner, 1968)

$$E = \frac{E_{max} \times Cp}{E_{50} + Cp} \tag{9.141}$$

where E_{max} is the maximum response and correspond to total receptor binding (R_T) and E_{50} is the concentration of xenobiotic at half-maximum effect. Equation 9.141 indicates that the effect is a rectangular hyperbolic function of the free plasma level of the compound.

When the effect is considered a rectangular hyperbolic function of receptor-xenobiotic complex (RCp), the relationship between the effect (E) and the complex (RCp), using Equation 9.128, can be defined as

$$E = \frac{E_{max}[RCp]}{K_E + [RCp]} \tag{9.142}$$

where K_E correspond to the concentration of RCp that brings forth the half-maximum effect.

Combining Equation 9.128 and Equation 9.142 yields Equation 9.143, which shows that "when a hyperbolic agonist-receptor interaction feeds into a hyperbolic effector system, a hyperbolic curve results" (Black and Leff, 1983)

$$E = \frac{E_{max}[R_T]Cp}{K_D K_E + (R_T + K_E)Cp} \tag{9.143}$$

Analogous to other hyperbolic equations, for example, the Michaelis–Menten equation, at low concentrations when $E_{50} \gg Cp$, the relationship between E and Cp is linear with a slope of E_{max}/E_{50} and the equation of the line is: $E = (E_{max}/E_{50})\,Cp$.

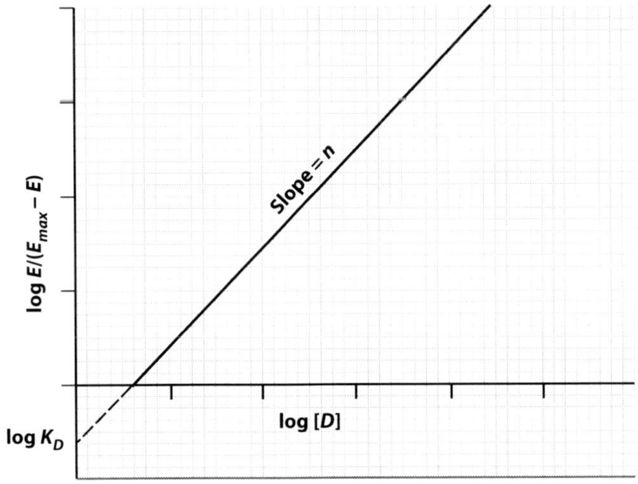

FIGURE 9.19 Plot of the logarithmic version of the model according to Equation 9.146.

At high concentrations of plasma is when $Cp \gg E_{50}$, $E = E_{max}$, which represents the plateau line of the hyperbolic curve parallel to the x-axis.

As presented in Figure 9.16, the rise in the magnitude of response versus concentration occurs at certain concentrations that are less than E_{50}. The slope of this gradual increase in response is determined as a function of ln*Cp* (i.e., $dE/d\ln Cp$) as follows.

Because $e^{\ln Cp} = Cp$, Equation 9.141 can also be written as $E = (E_{max} \times e^{\ln Cp})/(E_{50} + e^{\ln Cp})$. The slope representing change in response as a function of ln *Cp* is

$$\frac{dE}{d\ln Cp} = \frac{E_{max} \times Cp(E_{50} + Cp) - E_{max}Cp^2}{(E_{50} + Cp)^2} = \frac{E_{max} \times E_{50} \times Cp}{(E_{50} + Cp)^2} \tag{9.144}$$

By setting $Cp = E_{50}$, Equation 9.144 changes to Equation 9.145:

$$\frac{dE}{d\ln Cp} = \frac{E_{max}E_{50}^2}{(2E_{50})^2} = \frac{E_{max} \times E_{50}^2}{4E_{50}^2} = \frac{E_{max}}{4} \tag{9.145}$$

Thus, the slope of the initial rise in response versus ln *Cp* is a function of E_{max} and equal to 1/4 of the total occupancy of the receptor sites for a given compound.

Similar to Equation 9.130, the E_{max} model can also be presented in logarithmic form of Equation 9.146 (Figure 9.19):

$$\log\left(\frac{E}{E_{max} - E}\right) = \log Cp - \log E_{50} \tag{9.146}$$

9.5.3.4 Non-Hyperbolic Sigmoidal Model

When more than one molecule of xenobiotic binds to a receptor, the interaction can be presented as

$$[R] + [nD] \rightleftharpoons [R\ nD] \longrightarrow \text{Effect}[E] \tag{9.147}$$

The equilibrium constant, K_D, of the interaction is

$$K_D = \frac{[R][D]^n}{[D_nR]} \tag{9.148}$$

Mathematical manipulations similar to Equations 9.124 through 9.128, yield equations

$$[D_nR] = \frac{[R_T] \times [D]^n}{K_D + [D]^n} \tag{9.149}$$

$$E = \frac{E_{max} \times [D]^n}{K_D + [D]^n} = \frac{E_{max} \times Cp^n}{E_{50}^n + Cp^n} \tag{9.150}$$

Equation 9.150 represents the sigmoidal curve of response versus the concentration of the drug. It can also be used for the curve fitting purpose of dose-response curves that are not visibly sigmoidal. In that case, the value of n may no longer represent the number of reacting molecules with the receptor site, but rather an element to improve the curve fitting. The value of n influences the slope of linear segment of the curve.

Similar to Equation 9.145, the slope of rise in response as a function of $\ln Cp$ is defined as

$$\frac{dE}{d\ln Cp} = \frac{nE_{max}}{4} \tag{9.151}$$

Therefore, a larger n corresponds to a sharper slope and making the linear segment of the curve more perpendicular.

Equation 9.150 can also be changed into a linear relationship through the following steps:

$$\frac{E}{E_{max}} = \frac{[D]^n}{K_D + [D]^n} \tag{9.152}$$

$$[D]^n(E_{max} - E) = E \times K_D \tag{9.153}$$

$$K_D = \frac{[D]^n(E_{max} - E)}{E} \tag{9.154}$$

Taking a logarithm of both sides yields

$$\log K_D = n\log[D] + \log\frac{E_{max} - E}{E} \tag{9.155}$$

$$-\log\frac{E_{max} - E}{E} = -\log K_D + n\log D \tag{9.156}$$

Therefore,

$$\log\frac{E}{E_{max} - E} = n\log[D] - \log K_D \tag{9.157}$$

Thus, a plot of log ($E/E_{max} - E$) versus log [D] provides a linear relationship with a slope of n and a y-intercept of log K_D (Figure 9.19).

When $E = (E_{max}/2) \Rightarrow \log(E/(E_{max} - E)) = 0$

$$n \log[D] = \log K_D$$

$$\log K_D = n \log E_{50} \quad \text{or,} \quad K_D = E_{50}^n$$

Equation 9.158 defines effect (E) in terms of xenobiotic-receptor complex (RCp) for non-hyperbolic relationship, a modification of Equation 9.142:

$$E = \frac{E_{max}[RCp]^n}{K_E^n + [RCp]^n} \tag{9.158}$$

When $n = 1$, Equation 9.158 defines the rectangular hyperbola of Equation 9.142. Substituting Equation 9.128 into Equation 9.158 yields the following non-hyperbolic relationship for the effect in terms of plasma concentration (Black and Leff, 1983):

$$E = \frac{E_{max}[R_T]^n[Cp]^n}{K_E^n(K_D + [Cp])^n + [R_T]^n[Cp]^n} \tag{9.159}$$

9.5.4 Linking PK/TK Models to PD/TD Models

The delay in the pharmacological response following the administration of a compound indicates the response/effect is a function of the dispositional profile of the compound in the body. Linking the PK/TK of a compound to its PD/TD initiated by introducing the concept of biophase, and later on the hypothetical effect compartment linked to the PK/TK models (Segre, 1968; Wagner, 1968; Levy et al., 1969; Galeazzi et al., 1976; Dahlstrom et al., 1978; Kramer et al., 1979). In linking the effect to the disposition profile of a compound, if the fluctuation of response is in parallel with the rise and fall of plasma concentration, the linkage between the two models is direct and often as simple as the correlation between the measured plasma concentration and the magnitude of response. However, the accuracy of the direct correlation depends on the simultaneous measurements of plasma concentration and response. If the concentration of the drug at the effect compartment is proportional but not the same as plasma concentration, a separate compartment is needed to define the amount/concentration in the effect compartment.

Based on the assumption that concentration at the effect compartment is the same as the plasma concentration, the following relationships for one- and multi-compartment model can be developed. In the case of one compartment model, the receptor sites or biophase is assumed to be associated with the systemic circulation (Figure 9.20).

In the case of two-compartment model, the biophase can be associated with the central or peripheral compartment (Figure 9.21). This type of PK-PD modeling may be used to predict the response from the calculated or observed plasma concentration.

9.5.4.1 Linking the Nonlinear Hyperbolic E_{max} Model to PK/TK Models with Different Input

1. *PK/TK model:*

$$Cp = Cp^0 e^{-Kt} \tag{9.160}$$

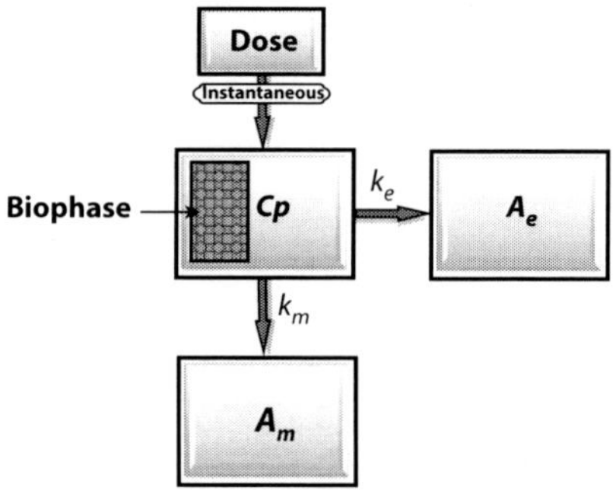

FIGURE 9.20 Schematic diagram of biophase associated with the systemic circulation of a one-compartment model.

Input: Intravenous bolus injection:

Cp = plasma concentration at time t, Cp^0 = initial plasma concentration and K = overall elimination rate constant.

Substituting the PK/TK model into the Equation 9.141 yields:

$$\text{Linked model:} \quad E = \frac{E_{max} \times Cp^0 e^{-Kt}}{E_{50} + Cp^0 e^{-Kt}} \tag{9.161}$$

Dividing the numerator and denominator of Equation 9.161 by Cp^0 yields:

$$E = \frac{E_{max} \times e^{-Kt}}{(E_{50}/Cp^0) + e^{-Kt}} = \frac{E_{max} \times f_b}{f_{b_{50}} + f_b} \tag{9.162}$$

where $f_{b_{50}}$ is the fraction of the administered dose in the body that provides an effect equal to 50% of E_{max} and represents the potency of the drug; the smaller $f_{b_{50}}$ represents a more potent drug.

2. *PK/TK model:*

$$Cp = \frac{k_0}{KVd}(1 - e^{-Kt}) \tag{9.163}$$

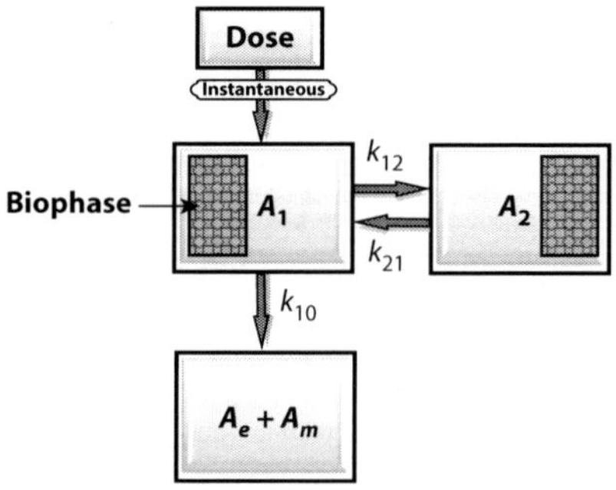

FIGURE 9.21 Schematic diagram of biophase associated with the central compartment, peripheral compartment, or both compartments of a two-compartment model.

Input: Zero-order input in systemic circulation:

k_0 = zero-order infusion, Vd is the volume of distribution

Substitution of the model in the Equation 9.141 yields the following linked models:
Linked model (before achieving steady-state level):

$$E = \frac{E_{max} \times Cp_{ss}(1 - e^{-Kt})}{E_{50} + Cp_{ss}(1 - e^{-Kt})} = \frac{E_{max}(1 - e^{-Kt})}{(E_{50}/Cp_{ss}) + (1 - e^{-Kt})} = \frac{E_{max}(f_{ss})}{f_{ss_{50}} + f_{ss}} \tag{9.164}$$

Linked model (at steady state):

$$E = \frac{E_{max}}{f_{ss_{50}} + 1} \tag{9.165}$$

where $(f_{ss_{50}})$ is the fraction of the steady-state plasma concentration that produces an effect equal to 50% of E_{max}. When E_{50} is equal to Cp_{ss} the magnitude of effect is equal to $E_{max}/2$.

3. *PK/TK model:*

$$Cp = \frac{FDk_a}{Vd(k_a - K)}(e^{-Kt} - e^{-k_a t}) \tag{9.166}$$

Input: First-order input (oral administration, intramuscular, inhalation, etc.)
Setting $\bar{Cp}^0 = (FDka/Vd(k_a - K))$ and substituting in Equation 9.141 yields

$$E = \frac{E_{max}\left[\bar{Cp}^0(e^{-Kt} - e^{-k_a t})\right]}{E_{50} + \bar{Cp}^0(e^{-Kt} - e^{-k_a t})} = \frac{E_{max}(e^{-Kt} - e^{-k_a t})}{(E_{50}/\bar{Cp}^0) + (e^{-Kt} - e^{-k_a t})} \tag{9.167}$$

where (k_a) is the first-order absorption rate constant. Equation 9.167 determines the time course of effect during the absorption process. When the absorption is complete, the term $e^{-k_a t} \Rightarrow 0$ and the equation is modified to

$$E = \frac{E_{max}e^{-Kt}}{(E_{50}/\bar{Cp}^0) + e^{-Kt}} \tag{9.168}$$

The response corresponding to the maximum plasma concentration $Cp_{max} = (FD/Vd)$ $(e^{-KT_{\max}})$ is defined as

$$E = \frac{E_{max} \times (FD/Vd)e^{-KT_{max}}}{E_{50} + (FD/Vd)e^{-KT_{max}}} = \frac{E_{max} \times FDe^{-KT_{max}}}{(E_{50} \times Vd) + FDe^{-KT_{max}}} \tag{9.169}$$

where (T_{max}), the time point of (Cp_{max}) is $T_{max} = (2.303/k_a - K)\log(k_a/K)$.
Based on Equation 9.169, it can be stated that the only time the maximum response is associated with the maximum plasma concentration is when $Cp_{max} \gg E_{50}$, and when $Cp_{max} = E_{50}$, the effect is half of the maximum response, $E = (E_{max}/2)$.

4. *PK/TK model:*

$$(Cp_{ss})_{peak} = \frac{Cp^0}{1 - e^{-Kt}} \quad \text{and} \quad (Cp_{ss})_{trough} = \frac{Cp^0 e^{-K\tau}}{1 - e^{-K\tau}} \tag{9.170}$$

Multiple dosing kinetics steady state peak and trough plasma concentrations:

Input: Multiple bolus injections, fixed dose and dosing interval after achieving steady state peak and trough levels

The fluctuation of response at steady state that corresponds to the peak and trough levels of plasma concentrations are

$$E_{peak} = \frac{E_{max} \times (Cp^0/1 - e^{-K\tau})}{E_{50} + (Cp^0/1 - e^{-K\tau})} = \frac{E_{max} \times Cp^0}{E_{50}(1 - e^{-K\tau}) + Cp^0} = \frac{E_{max} \times Cp^0}{E_{50}(f_{el})_\tau + Cp^0} \tag{9.171}$$

where, $(f_{el})_\tau = 1 - e^{-K\tau}$ is the fraction of the dose eliminated in one dosing interval at steady state and Cp^0 is the initial plasma concentration of a maintenance dose. The response corresponding to the trough level is

$$\begin{aligned} E_{trough} &= \frac{E_{max} \times (Cp^0 e^{-K\tau}/1 - e^{-K\tau})}{E_{50} + (Cp^0 e^{-K\tau}/1 - e^{-K\tau})} \\ &= \frac{E_{max} \times Cp^0}{E_{50}\left((1 - e^{-K\tau})/e^{-K\tau}\right) + Cp^0} = \frac{E_{max} \times Cp^0}{\left((E_{50}/e^{-K\tau}) - 1\right) + Cp^0} \end{aligned} \tag{9.172}$$

The average response at steady state can be determined by linking the E_{max} model with an average plasma concentration at steady state:

$$\begin{aligned} E_{ave} &= \frac{E_{max} \times (Cp^0/K\tau)}{E_{50} + Cp^0/K\tau)} = \frac{E_{max} \times Cp^0}{E_{50}(K\tau) + Cp^0} \\ &= \frac{E_{max} \times AUC}{E_{50}(\tau) + AUC} = \frac{E_{max} \times D}{E_{50}(Cl_T \times \tau) + D} \end{aligned} \tag{9.173}$$

For multiple-dosing kinetics with first order input, the E_{max} linked model is the same as Equations 9.171 through 9.173 with the stipulation that Cp^0 is $\bar{Cp}^0$ and the dose is equal to FD.

5. *PK/TK model*:

$$Cp = ae^{-\alpha t} + be^{-\beta t} \quad \text{(two-compartment model)} \tag{9.174}$$

Input: Single dose intravenous bolus–central compartment:

$$E = \frac{E_{max} \times \left(ae^{-\alpha t} + be^{-\beta t}\right)}{E_{50} + \left(ae^{-\alpha t} + be^{-\beta t}\right)} \tag{9.175}$$

The response in the post-distributive phase can be estimated by

$$E = \frac{E_{max} \times be^{-\beta t}}{E_{50} + be^{-\beta t}} \tag{9.176}$$

When the effect is associated with the amount of compound in the peripheral compartment, the relationship of the model can be presented as

$$E = \frac{E_{max} \times \left(\frac{k_{12}D}{\alpha - \beta}\left(e^{-\beta t} - e^{-\alpha}\right)\right)}{E_{50} + \left(\frac{k_{12}D}{\alpha - \beta}\left(e^{-\beta t} - e^{-\alpha}\right)\right)} \tag{9.177}$$

6. *PK/TK model*:

$$(Cp_{max})_{ss} = \frac{a}{1-e^{-\alpha\tau}} + \frac{b}{1-e^{-\beta\tau}} \quad \text{and} \quad (Cp_{min})_{ss} = \frac{ae^{-\alpha\tau}}{1-e^{-\alpha\tau}} + \frac{be^{-\beta\tau}}{1-e^{-\beta\tau}}$$

Multiple dosing kinetics, steady state peak and trough plasma concentrations:
Input: Multiple bolus injections, fixed dose and dosing interval
Equations for prediction of response at steady state peak and trough levels of a drug that follows two-compartment model are

$$E_{peak} = \frac{E_{max} \times (Cp_{max})_{ss}}{E_{50} + (Cp_{max})_{ss}} \quad \text{and} \quad E_{trough} = \frac{E_{max} \times (Cp_{min})_{ss}}{E_{50} + (Cp_{min})_{ss}} \tag{9.178}$$

9.5.4.2 Linking Non-Hyperbolic Sigmoidal Model to PK/TK Models with Different Input

The linking is the same as it was discussed for the E_{max} model. A few examples are

Model I:

$$E = \frac{E_{max} \times f_b^n}{f_{b50}^n + f_b^n} \tag{9.179}$$

Model II: Before achieving steady-state level:

$$E = \frac{E_{max} \times f_{ss}^n}{f_{ss50}^n + f_{ss}^n} \tag{9.180}$$

At steady state:

$$E = \frac{E_{max}}{f_{ss50}^n + 1} \tag{9.181}$$

Model III: During the absorption phase:

$$E = \frac{E_{max} \times \left(e^{-Kt} - e^{-k_a t}\right)^n}{\left(E_{50}/C\bar{p}^0\right)^n + \left(e^{-Kt} - e^{-k_a t}\right)^n} \tag{9.182}$$

In post-absorptive phase:

$$E = \frac{E_{max} \times (e^{-Kt})^n}{(E_{50}/C\bar{p}^0)^n + (e^{-Kt})^n} \tag{9.183}$$

At maximum plasma concentration:

$$E = \frac{E_{max} \times (FDe^{-KT_{max}})^n}{(E_{50} \times Vd)^n + (FDe^{-KT_{max}})^n} \tag{9.184}$$

Model IV:

$$E_{peak} = \frac{E_{max} \times (Cp^0)^n}{(E_{50}(f_{el})_\tau)^n + (Cp^0)^n} \quad \text{and} \quad E_{trough} = \frac{E_{max} \times (Cp^0)^n}{((E_{50}/e^{-K\tau}) - 1)^n + (Cp^0)^n} \tag{9.185}$$

$$\begin{aligned} E_{ave} &= \frac{E_{max} \times (Cp\circ)^n}{(E_{50} \times K\tau)^n + (Cp\circ)^n} = \frac{E_{max} \times (AUC)^n}{(E_{50} \times \tau)^n + (AUC)^n} \\ &= \frac{E_{max} \times D^n}{(E_{50} \times Cl_T \times \tau)^n + D^n} \end{aligned} \tag{9.186}$$

Model V:

$$E = \frac{E_{max} \times (ae^{-\alpha t} + be^{-\beta t})^n}{(E_{50})^n + (ae^{-\alpha t} + be^{-\beta t})^n} \tag{9.187}$$

In post-distributive phase:

$$E = \frac{E_{max} \times (be^{-\beta t})^n}{(E_{50})^n + (be^{-\beta t})^n} \tag{9.188}$$

Model VI:

$$E_{peak} = \frac{E_{max} \times (Cp_{max})^n_{ss}}{(E_{50})^n + (Cp_{max})^n_{ss}} \quad \text{and} \quad E_{trough} = \frac{E_{max} \times (Cp_{min})^n_{ss}}{(E_{50})^n + (Cp_{min})^n_{ss}} \tag{9.189}$$

9.5.5 PK/PD, TK/TD Modeling-Effect Compartment

The tacit assumption in the discussion of Section 9.5.4 was that the concentration of xenobiotic at the receptor site is the same as the plasma concentration. This means that the compound in receptor environments can reach a rapid equilibrium with plasma concentration. There are, however, receptor environments that do not reach rapid equilibrium with plasma, such as receptors in regions separated from plasma by a physiological barrier, or associated with the intracellular environment, or located in regions where blood perfusion is poor. For this type of receptor, the direct method of substitution may not be pertinent. One of the reasons is that the interaction of compounds with this type of receptor occurs after a lag time or a threshold of response, which the direct linking, as it was discussed in Section 9.5.4, cannot predict. A different approach based on determining the amount or concentration of xenobiotic at the receptor site has been developed, which considers effect as a separate compartment to represent the receptor environments with the following assumptions (Sheiner et al., 1979; Colburn, 1981; Fuseau and Sheiner, 1984; Hochhaus and Derendorf, 1995; Vora and Boroujerdi, 1996; Vaidyanathan and Boroujerdi, 2000) in Figures 9.22 and 9.23:

- The effect compartment is in contact with the systemic circulation.
- The transfer of the xenobiotic from the systemic circulation to the effect compartment is governed by linear processes.

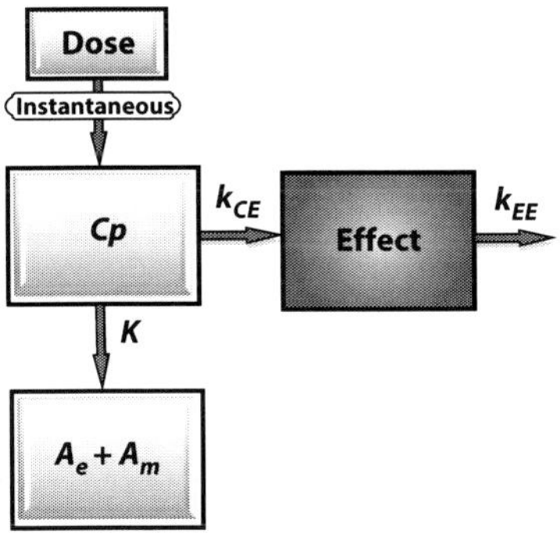

FIGURE 9.22 Schematic diagram of a one-compartment model separate effect compartment associated with the systemic circulation.

- The rate constant of transfer from the systemic circulation to the effect compartment and the amount in the effect compartment are considered negligible with respect to the overall disposition of compound in the body.
- At steady state the concentration of xenobiotic in the effect compartment is the same as the systemic circulation.
- The effect compartment has a first-order exit rate constant that is also small and considered a negligible value.

The rates of change in the amount of xenobiotic in the systemic circulation and the effect compartment, as depicted in Figure 9.22, following intravenous bolus administration are

$$\text{Central compartment: } \frac{dA}{dt} = -KA \tag{9.190}$$

$$\text{Effect compartment: } \frac{dA_E}{dt} = k_{CE}A - k_{EE}A_E \tag{9.191}$$

Solving the two differential equations by the Laplace transform (Appendix A) yields the following integrated equations:

$$(A)_t = A^0 e^{-Kt} \tag{9.192}$$

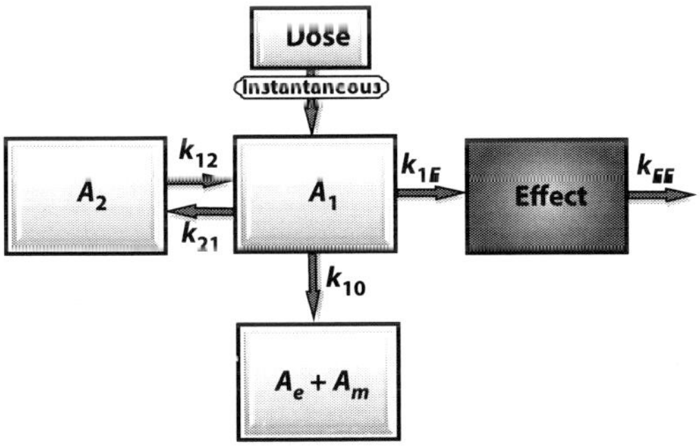

FIGURE 9.23 Schematic diagram of a two-compartment model with the effect compartment being associated with the central compartment.

$$(A_E)_t = \frac{k_{CE}A^0}{k_{EE} - K}(e^{-Kt} - e^{-k_{EE}t}) \tag{9.193}$$

where $(A_E)_t$ is the amount of the drug in the effect compartment at time t, K is the overall elimination rate constant ($K = k_e + k_m + k_{CE}$), k_{CE} and k_{EE} are the input and exit rate constants of the effect compartment, respectively, and A^0 is the administered dose.

Calculations of concentration in the effect compartment requires the knowledge of its apparent volume of distribution. Since at steady state the concentration of xenobiotic in the effect compartment is assumed to be approximately equal to plasma concentration, one approach is to determine a relevant proportionality constant between the apparent volume of distribution of compound in plasma and the volume of the effect compartment.

At steady state, the rate of input into the effect compartment is equal to the rate of output from the effect compartment, that is,

$$k_{CE} \times (Cp)_t \times Vd = k_{EE} \times (C_E)_t \times V_E \tag{9.194}$$

$$V_E = Vd\frac{k_{CE} \times (Cp)_t}{k_{EE} \times (C_E)_t}$$

Since $(Cp)_t \cong (C_E)_t$

$$V_E = Vd\frac{k_{CE}}{k_{EE}} \tag{9.195}$$

The apparent concentration of the effect can be estimated as

$$(C_E)_t = \frac{(A_E)_t}{V_E} \tag{9.196}$$

When the input in the central compartment is continuous and considered zero-order input, the rate equation of the effect compartment remains the same as Equation 9.191 and the rate equation of the central compartment is

$$\frac{dA}{dt} = k_0 - k_eA - k_mA - k_{CE}A = k_0 - KA \tag{9.197}$$

The integrated equations of the model before achieving the steady-state condition are

$$(Cp)_t = \frac{k_0}{K \times Vd}(1 - e^{-Kt}) \tag{9.198}$$

$$(A_E)_t = \frac{k_{CE}k_0}{K(k_{EE} - K)}(1 - e^{-Kt}) + \frac{k_{CE}k_0}{k_{EE}(K - k_{EE})}(1 - e^{-k_{EE}t}) \tag{9.199}$$

After stopping the input, the integrated equations are

$$(Cp)_t = \frac{k_0}{K \times Vd}(1 - e^{-Kt})e^{-Kt'} \tag{9.200}$$

$$(A_E)_t = \frac{k_{CE}k_0}{K(k_{EE} - K)}(1 - e^{-Kt})e^{-Kt'} + \frac{k_{CE}k_0}{k_{EE}(K - k_{EE})}(1 - e^{-k_{EE}t})e^{-k_{EE}t'} \tag{9.201}$$

After achieving the steady state

$$Cp_{ss} = \frac{k_0}{KVd} \tag{9.202}$$

$$(A_E)_t = \frac{k_{CE}k_0}{K(k_{EE} - K)} + \frac{k_{CE}k_0}{k_{EE}(K - k_{EE})} \tag{9.203}$$

For the first-order absorption through the GI tract, intramuscular, sublingual, inhalation, etc., the rate equation of the central compartment can be defined as Equation 9.204 and the rate equation of the effect compartment as Equation 9.191.

$$\frac{dA}{dt} = k_aFD - k_eA - k_mA - k_{CE}A = kaFD - KA \tag{9.204}$$

The integrated equations are

$$(Cp)_t = \frac{k_aFD}{Vd(k_a - K)}(e^{-Kt} - e^{-k_at}) \tag{9.205}$$

$$\begin{aligned}(A_E)_t = {} & \frac{k_{CE}k_aFD}{(K - k_a)(k_{EE} - k_a)}e^{-k_at} + \frac{k_{CE}k_aFD}{(k_a - K)(k_{EE} - K)}e^{-Kt} \\ & + \frac{k_{CE}k_aFD}{(k_a - k_{EE})(K - k_{EE})}e^{-k_{EE}t}\end{aligned} \tag{9.206}$$

The integrated equations of the effect compartment following a multiple-dosing regimen at steady state are

$$(A_E)_{peak} = \frac{k_{CE}D}{(k_{EE} - K)}\left(\frac{1}{1 - e^{-K\tau}} - \frac{1}{1 - e^{-k_{EE}\tau}}\right) \tag{9.207}$$

$$(A_E)_{trough} = \frac{k_{CE}D}{(k_{EE} - K)}\left(\frac{e^{-K\tau}}{1 - e^{-K\tau}} - \frac{e^{-k_{EE}\tau}}{1 - e^{-k_{EE}\tau}}\right) \tag{9.208}$$

In dealing with multiple compartment models, the rate equations of the effect compartment remain the same, but the integrated equation will be different. For example, for a two-compartment model depicted in Figure 9.23, the integrated equation of the effect compartment is

$$(A_E)_t = \frac{k_{CE}}{(k_{EE} - \alpha)}Ae^{-\alpha t} + \frac{k_{CE}}{(k_{EE} - \beta)}Be^{-\beta t} + \frac{k_{CE}D(k_{21} - k_{EE})}{(\alpha - k_{EE})(\beta - k_{EE})}e^{-k_{EE}t} \tag{9.209}$$

9.6 PHYSIOLOGICALLY BASED PK/PD OR TK/TD MODELING

Modeling of physiologically based PK/TK with the effect compartment is similar to the ones described in Section 9.5.5. The effect compartment is considered closely associated with a specific organ or tissue in the body. Depending upon the type of effect data, the rate constants or clearance exchange between the organ and the effect compartment is determined based on how the effect is perceived and measured. The effect can be a separate single compartment attached to the organ, or two separate compartments of free receptors and bound receptors with the exchange between the two compartments, while associated with the organ (Figure 9.24).

A hybrid approach to PBPK-PD/TK seems to be more practical when the association of the effect compartment with the organ is well established (Johnson et al., 2011) (Figure 9.25). The model is described as

$$\frac{dA_1}{dt} = \frac{Q}{V_2}A_2 - \frac{Q}{V_1}A_1 - \frac{Cl}{V1}A_1$$

$$\frac{dA_2}{dt} = \frac{Q}{V_1}A_1 - \frac{Q}{V_2}A_2$$

$$\frac{dA_3}{dt} = \left(\frac{Cl_3}{V_1}\right)A_1 - \left(\frac{Cl_3}{V_3}\right)A_3 - \left(\frac{Cl_4}{V_3}\right)(f_u \times A_3) + \left(\frac{Cl_4}{V_4}\right)(f_u \times A_4)$$

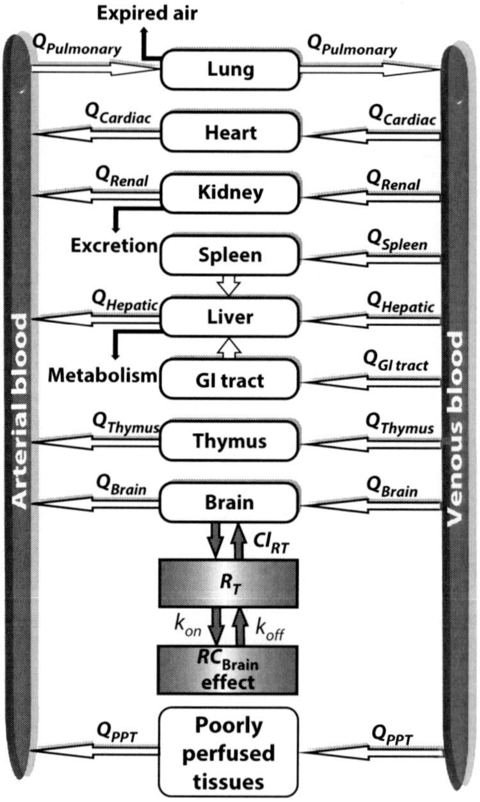

FIGURE 9.24 Physiologically based pharmacokinetic model with the effect compartment closely associated with the receptor density compartment of a specific organ or tissue in the body.

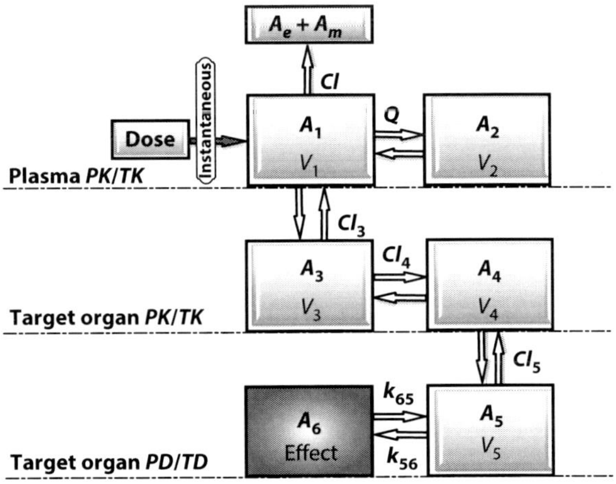

FIGURE 9.25 A modified version of PBPK with fewer compartments and focus on a target organ and associated PK/TK and PD/TD phases.

$$\frac{dA_4}{dt} = \left(\frac{Cl_4}{V_3}\right)(f_u \times A_3) - \left(\frac{Cl_4}{V_4}\right)(f_{u_5} \times A_4) - \left(\frac{Cl_5}{V_5}\right)(f_{u_5} \times A_4) + \left(\frac{Cl_5}{V_5}\right)(f_{u_5} \times A_5)$$

$$\frac{dA_5}{dt} = \left(\frac{Cl_5}{V_4}\right)(f_{u_5} \times A_4) - \left(\frac{Cl_5}{V_5}\right)(f_{u_5} \times A_5) - \left(k_{56} \times f_{u_5} \times A_5 \left(R_T - RC_5\right)\right) + k_{65}A_6$$

$$\frac{dA_6}{dt} = k_{56} \times f_{u_5} \times A_5 \left(R_T - RC_5\right) - k_{65}A_6$$

Compartment A_1 and A_2 are the amount in the systemic circulation and peripheral tissues, respectively; A_3 is the amount in the target organ's vascular compartment; A_4 is the amount of the target organ's extravascular compartment; A_5 is the amount in the receptor environments; A_6 is the amount bound to the receptor, or the effect compartment; V_1–V_5 are the volume of distribution of compartments 1 to five; Q is the blood flow and Cl_{1-5} are the compartmental clearance terms, f_u and f_{u_5} are xenobiotic free fraction in plasma and the target organ, respectively; R_t is the receptor density; and RC_5 is the concentration bound to the receptor.

9.7 HYSTERESIS CURVE

PK-PD models attempt to relate the plasma concentration to the intensity of response. When the response is plotted against the plasma concentration according to their time order, a loop with clockwise or counterclockwise direction can be obtained (Figure 9.26). The direction of the loop determines the latency of response. A counterclockwise direction is usually associated with the delay or threshold of response. The delay of response is indicative of a slow equilibration of the plasma concentration with the receptor environments or increases in the concentration of the receptor or its activity. Theoretically, failure of the hysteresis of effect versus the plasma concentration to collapse on plotting E versus concentration of the effect compartment C_E may indicate that the delay in the pharmacological response may be related to the upregulation of receptors. A clockwise direction occurs when the response to the initial concentration is rather highm but declines rapidly even when the concentration increases; a scenario that often is observed with the development of tolerance.

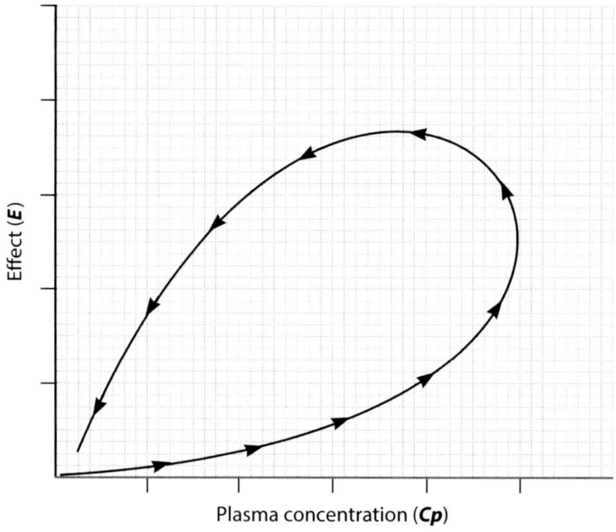

FIGURE 9.26 Counterclockwise hysteresis of effect versus plasma concentration.

REFERENCES

Allen, B. C., Covington, T. R., Clevwell, H. J. 1996. Investigation of the impact of pharmacokinetic variability and uncertainty on risks predicted with a pharmacokinetic model for chloroform. *Toxicology* 111:289–3.

Anderson, M. E., Clewell, H. J. III., Gargas, M. L., Smith, F. A., Reitz, R. H. 1987. Physiologically based pharmacokinetics and the risk assessment process for methylene chloride. *Toxicol Appl Pharmacol* 87:185–205.

Ariens, E. 1964. The mode of action of biologically active compounds. In *Molecular Pharmacology,* ed. G. De Stevens. New York: Academic Press.

Bellman, R., Åström, K. J. 1970. On structural identifiability. *Math Biosciences* 7 (3–4):329–39.

Benet, L. Z., Turi, J. S. 1971. Use of a general partial fraction theorem for obtaining inverse Laplace transforms in pharmaceutical analysis. *J Pharm Sci* 60:1593–4.

Benet, L. Z. 1972. General treatment of linear mammillary models with elimination from any compartment as used in pharmacokinetics. *J Pharm Sci* 61(4):536–41.

Benet, L. Z. 1985. Mean residence time in the body versus mean residence time in the central compartment. *J Pharmacokin Biopharm* 13:555–8.

Benet, L. Z., Galeazzi, R. L. 1979. Noncompartmental determination of the steady state volume of distribution. *J Pharm Sci* 68:1071–4.

Bischoff, K. B. 1980. Current applications of physiological pharmacokinetics. *Fed Proc* 39(7):2456–9.

Bischoff, K. B. 1986. Physiological pharmacokintics. *Bull Math Biol* 48(3/4):309–22.

Bischoff, K. B., Dedrick, R. L., Zaharko, D. S., Longstreth, J. A. 1971. Methotrexate pharmacokinetics. *J Pharm Sci* 60(8):1128–33.

Blaauboer, B. J. 2003. The integration of data on physicochemical properties, in vitro-derived toxicity data and physiologically based kinetic and dynamic as modeling a tool in hazard and risk assessment. A commentary. *Toxicol Lett* 138(1–2):161–71.

Black, J. W., Leff, P. 1983. Operational models of pharmacological agonism. *Proc R Soc Lond* B 220:141–62.

Boise, F. Y., Woodruff, T. J., Spear, R. C. 1991. Comparison of three physiologically based pharmacokinetic models for benzene disposition. *Toxicol Appl Pharmacol* 110:79–88.

Brown, E. A., Shelley, M. L., Fisher, J. W. 1998. A pharmacokinetic study of occupational and environmental benzene exposure with regard to gender. *Risk Anal* 18:205–13.

Brown, R. P., Delp, M. D., Lindstedt, S. L., Rhomberg, L. R., Beliles, R. P. 1997. Physiological parameter values for physiologically based pharmacokinetic models. *Toxicol Ind Health* 13: 407-84.

Cao, Y., Balthasar, J. P., Jusko, W. J. 2013. Second generation minimal physiologically-based pharmacokinetic model for monoclonal antibodies. *J Pharmacokinet Phrmacodyn* 40:597–607.

Cao, Y., Jusko, W. J. 2012. Applications of minimal physiologically-based pharmacoklinetic models. *J Pharmacokinet Pharmacodyn* 39:71–23.

Cawello, W., Brett, M., Weimann, H-J., Zimmerman, H., Pabst, G., Sierakowski, B., Gieschke, R., Baumann, A. 1999. *Parameters for Compartment-free Pharmacokinetics: Standardisation of Study Design, Data Analysis, and Reporting*. Aachen, Germany: Shaker Verlag.

Chapman, M. J., Godfrey, K. R. 1987. On structural equivalence and identifiability constraint ordering. In: *Identifiability of Parametric Model*, ed. E. Walter, 32–41. Tarrytown, NY: Pergamon.

Chappell, M. J., Godfrey, K. R., Vajda, S. 1990. Global identifiability of the parameters of nonlinear systems with specified inputs: A comparison of methods. *Math Biosciences* 102:41–73.

Chow, H. H. 1997. A physiologically based pharmacokinetic model of zidovudine (AZT) in the mouse: Model development and scale-up to human. *J Pharm Sci* 86(11):1223–8.

Clewell, H. J., III; Lee, T. S., Carpeter, R. L. 1994. Sensitivity of physiologically based pharmacokinetic models to variation in model parameters-methylene chloride. *Risk Anal* 14:521–31.

Cobelli, C., Mari, A. 1983. Validation of mathematical models of complex endocrine-metabolic systems. A case study on a model of glucose regulation. *Med Biol Eng Comp* 21(4):390–9.

Colburn, W. A. 1981. Simultaneous pharmacokinetic and pharmacodynamic modeling. *J Pharmacokinet Biopharm* 9(3):367–88.

Corley, R. A., Gordon, S. M., Wallace, L. A. 2000. Physiologically based pharmacokinetic modeling of the temperature-dependent dermal absorption of chloroform by humans following bath water exposure. *Toxicol Sci* 53:13–23.

Cowles, A. L., Borgstedt, H. H., Gillies, A. J. 1971. Tissue weights and rates of blow flow in man for the prediction of anesthetic uptake and distribution. *Anesthesiology* 35:523–6.

Crowell, S. R., Amin, S. G., Anderson, K. A., Krishnegowda, G., Sharma, A. K., Soelberg, J. J., Williams, D. E., Coreley, R. A. 2011. Preliminary physiologically based pharmacokinetic models for benzo(a)pyrene and dibenzo(def,p)chrysene in rhodents. *Toxicol Appl Pharmacol* 257:365–76.

Cutler, D. J. 1978. Theory of the mean absorption time, an adjunct to conventional bioavailability studies. *J Pharm Pharmacol* 30:476–8.

Cutler, D. J. 1987. Definition of mean residence times in pharmacokinetics. *Biopharm Drug Disp* 8:87–97.

Cutler, D. J. 1978. Theory of mean absorption time, an adjunct to conventional bioavailability studies. *J Pharm Pharmacol* 30:476–8.

Dahlstrom, B., Paalzow, L., Segre, G., Ågren, A. J. 1978. Relation between morphine pharmacokinetics and Analgesia. *J Pharmacokinet Biopharm* 6:41–53.

Dedrick, R. L., Bischoff, K. B., Zaharko, D. S. 1970. Interspecies correlation of plasma concentration history of methotrexate (NSC – 740). *Cancer Chemother Rep* 54:95–101.

de la Grandmaison, G. L., Clairand, I., Durigon, M. 2001. Organ weight in 684 adult autopsies: New table for a Caucasoid population. *Forensic Sci Int* 119(2):149–54.

Dutta, S., Mutsumoto, Y., Ebling, W. F. 1996. Is it possible to estimate the parameters of the sigmoid Emax model with truncated data typical of clinical studies? *J Pharm Sci* 85:232–9.

Fuseau, E., Sheiner, L. B. 1984. Simultaneous modelling of pharmacokinetics and pharmacodynamics with a nonparametric pharmacodynamic model. *Clin Pharmacol Ther* 35(6):733–41.

Galcazzi, R. L., Benet, L., Sheiner, L. 1976. Relationship between the pharmacokinetics and pharmacodynamics of procainamide. *Clin Pharmacol Ther* 20:278–89.

Gallo, J. M., Lam, F. C., Perrier, D. G. 1987. Area method for the estimation of partition coefficient for physiological pharmacokinetic model. *J Pharmacokin Biopharm* 15(3):271–80.

Gallo, J. M., Vicini, P., Orlansky, A., Li, S., Zhou, F., Ma, J., Pulfer, S., Bookman, M. A., Guo, P. 2004. Pharmacokinetic model–predicted anticancer drug concentrations in human tumors. *Clin Cancer Res* 10:8048–58.

Gibaldi, M., Perrier, D. 1982. *Pharmacokinetics 2nd Ed.*. New York: Marcel Dekker, Inc, 276–318.

Henthorn, T. K., Krejcie, T. C., Avram, M. J. 1992. The relationship between alfentanil distribution kinetics and cardiac output. *Clin Pharmacol Ther* 52(2):190–6.

Harrison, L. I., Gibaldi, M. 1977. Physiologically based pharmacokinetic model for digoxin disposition in dogs and its preliminary application to humans. *J Pharm Sci* 66(12):1679–83.

Hetrick, D. M., Jarabek, A. M., Travis, C. C. 1991. Sensitivity analysis for physiologically-based pharmacokinetic models. *J Pharmacokinet Biopharm* 19:1–20.

Hochhaus, G. Derendorf, H. 1995. Dose optimization based on pharmacokinetic-pharmacodynamic modeling. In: *Pharmacokinetic/Pharmacodynamic Correlation,* Chapter 4, eds. H. Derendorf, and G. Hochhaus, Boca Raton, FL: CRC Press.

Huang, S. M., Rowland, M. 2012. The role of physiologically based pharmacokinetic modeling in regulatory review. *Clin Pharmacol Ther* 91:542–9.

Hudacheck, S. F., Gustafson, D. L. 2013. Physiologically based pharmacokinetic model of lapatinib developed in mice and scaled to human. *J Pharmacokinet Pharmacodyn* 40:157–76.

ICRP. 2002. *Basic Anatomical and Physiological Data for Use in Radiological Protection: Reference Values*. Amsterdam, The Netherlands: Elsevier Science.

Jacquez, J. A. 1972. Compartmental analysis in biology and medicine: Kinetics of distribution of tracer-labeled materials. New York: Elsevier Publishing Company.

Johnson, M., Kozielska, M., Reddy, V. P., Vermeulen, A., Li, C., Grimwood, S., de Greef, R. Groothuis, G. M. M., Danhof, M., Proost, J. H. 2011. Mechanism-based pharmacokinetic–pharmacodynamic modeling of the dopamine D2 receptor occupancy of olanzapine in rats. *Pharm Res* 28:2490–504.

Kasuya, Y., Hirayama, H., Kubota, N., Pang, S. 1987. Interpretation and estimates of mean residence time with statistical moment theory. *Biopharm Drug Disp* 8:223–34.

Kawahara, M., Sakara, A., Miyashita, T., Tamai, I., Tsuji, A. 1999. Physiologically based pharmacokinetics of digoxin in mdr1a knockout mice. *J Pharm Sci* 88:1281–7.

Kazama, T., Morita, K., Ikeda, T., Kurita, T., Sato, S. 2003. Comparison of predicted induction dose with predetermined physiologic characteristics of patients and with pharmacokinetic models incorporating those characteristics as covariant. *Anesthesiology* 98:299–305.

Kramer, W. G., Kolibash, A. J., Lewis, R. P., Bathala, M. S., Visconti, J. A., Reuning, R. H. 1979. Pharmacokinetics of digoxin: Relationship between response intensity and predicted compartmental drug levels in man. *J Pharmacokinet Biopharm* 7:47–61.

Lipscomb, J. C., Poet, T. S. 2008. *in vitro* measurements of metabolism for application in pharmacokinetic modeling. *Pharmacol Ther* 118(1):82–103.

Loccisano, A. E., Campbell, J. L. Jr., Butenhoff, J. L., Anderson, M. E., Clewell H. J. III. 2012. Comparison and evaluation of pharmacokinetics of PFOA and PFOS in the adult rats using a physiologically based pharmacokinetic model. *Reprod Toxicol* 33(4):452–67.

Levy, G. 1964. Relationship between elimination rate of drugs and rate of decline of their pharmacological effect. *J Pharm Sci* 53:342–3.

Levy, G. 1966. Kinetics of pharmacologic effect. *Clin Pharm Therapeut* 7:362–72.

Levy, G., Gibaldi, M., Jusko, W. 1969. Multicompartment pharmacokinetic models and pharmacologic effects. *J Pharm Sci* 58:422–24.

McNally, K., Cotton, R., Loizou, G. D. 2011. A workflow for global sensitivity analysis of PBPK models. *Front Pharmacol* 2:31.

Metzler, C. M. 1981. Estimation of pharmacokinetic parameters: Statistical considerations. *Pharmacol Ther* 13:543.

Mumtaz, M. M., Ray, M., Crowell, S. R., Keys, D., Fisher, J., Ruiz, P. 2012. Translational research to develop a human PBPK models tool kit—Volatile organic compounds (VOCs). *J Toxicol Environ Health A* 75:6–24.

Nakashima, E., Yokogawa, K., Ichimura, F., Kurata, K., Kido, H., Yamaguchi, N., Yamana, T. 1987. A physiologically based pharmacokinetic model for biperiden in animals and its extrapolation to human. *Chem Pharm Bull* 35(2):718–25.

Plowchalk, D. R., Andersen, M. E., Bogdanffy, M. S. 1997. Physiologically based modeling of vinyl acetate uptake, metabolism, and intracellular pH changes in the rat nasal cavity. *Toxico Appl Pharmacol* 142:386–400.

Price, P. S., Connolly, R. B., Chaisson, C. F., Gross, E. A., Young, J. S., Mathis, E. T., Tedder, D. R. 2003. Modeling interindividual variation in physiological factors used in PBPK models of humans. *Crit Rev Toxicol* 33(5):469–503.

Rescigno, A., Serge, G. 1966. *Drug and Tracer Kinetics*. New York: Blaisdell.

Riegelman, S., Collier, P. 1980. The application of statistical moment theory to the evaluation of *in vivo* dissolution time and absorption time. *J Pharmacokinet Biopharm* 8:509–34.

Ritchel, W. A., Banerjee, P. S. 1986. Physiologic pharmacokinetic models: Application, limitations and outlook. *Meth Exp Clin Pharmacol* 8:603.

Rowland, M. 1985. Physiologic pharmacokinetic models and interanimal species scaling. *Pharmac Ther* 29:49–68.

Rowland, M., Balant, L., Peck, C. 2004. Physiologically based pharmacokinetics in drug development and regulatory science: A workshop report. *AAPS Pharm Sci* 6 (1) Article 6 (http://www.aapspharmsci.org).

Segre, G. 1968. Kinetics of interaction between drugs and biological system. *Il Far4maco Ed Sci* 23:907–18.

Sheiner, L. B., Beal, S. L. 1981. Some suggestions for measuring predictive performance. *J Pharmacokinet Biopharm* 9(4):503–12.

Sheiner, L. B., Stanski, D. R., Vozeh, S., Miller, R. D., Ham, J. 1979. Simultaneous modeling of pharmacokinetics and pharmacodynamics: Application to d-tubocurarine. *Clin Pharmaco Ther* 25(3):357–71.

Shipley, R. A., Clark, R. E. 1972. *Tracer methods for in vivo kinetics. Theory and Applications*. New York: Academic.

Thompson, M. D., Beard. D. A. 2011. Development of appropriate equations for physiologically-based pharmacokinetic modeling of permeability-limited and flow-limited transport. *J Pharmacokinet Pharmacodyn* 38:405–21.

Upton, R. N., Foster, D. J. R., Christrup, L. L., Dale, O., Moksnes, K., Popper, L. 2012. A physiologically-based recirculatory meta-model for nasal fentanyl in ma. *J Pharmacokinet Pharmacodyn* 39(5):561–76.

Vaidyanathan, S., Boroujerdi, M. 2000. Effect of Tamoxifen Pre-treatment on the pharmacokinetics, metabolism and cardiotoxicity of Doxorubicin in female rats. *Cancer Chemotherapy and Pharmacology* 46(3):185–92.

van den Hof, J. M. 1998. Structural identifiability of linear compartmental systems. *IEEE Transactions Autom Control* 43(6):800–18.

Vora, J., Boroujerdi, M. 1996. Pharmacokinetic-toxicodynamic relationship of adriamycin in rat. Prediction of a phenolic antioxidant mediated reduction in anthracycline cardiotoxicity. *J Pharm Pharmacol* 48(12):1264–69.

Wagner, J. G. 1968. Kinetics of pharmacological response, I. Proposed relationships between response and drug concentration in the intact animal and man. *J Theor Biol* 20:173–201.

Wagner, J. G. 1975. *Fundamentals of Clinical Pharmacokinetics*. 1st Edition. Hamilton, IL: Drug Intelligence Publication.

Willmann, S., Höhn, K., Edington, A., Sevestre, M., Solodenko, J., Weiss, W., Lippert, J., Schmitt, W. 2007. Development of a physiology-based whole-body population model for assessing the influence of individual variability on the pharmacokinetics of drugs. *J Pharmacokinet Pharmacodyn* 34(3):401–31.

Willmann, S., Lippert, J., Sevestre, M., Solodenko, J., Fois, F., Schmit, W. 2003. PK-Sim®: A physiologically based pharmacokinetic "whole body" model. *Biosilico* 1(4):121–4.

Wu, G. 1995. Calculating predictive performance: A user's note. *Pharm Res* 31(6):393–99.

Yamaoka, K., Nakagawa, T., Uno, T. 1978. Statistical moments in pharmacokinetics. *J Pharmacokinet Biopharm* 6:547–58.

10 PK/TK Analysis of Instantaneous Exposure to Xenobiotics (Single Bolus Injection)

10.1 INTRODUCTION

As discussed in Chapter 9, the PK/TK models describe in a quantitative manner some salient features of mammalian systems in handling xenobiotics. The focus of this chapter is to evaluate the properties of these models following the instantaneous input into the systemic circulation. The purpose is to describe the complex biological processes involved in handling the xenobiotics more explicitly in terms of model dependent parameters and constant. The instantaneous input (intravenous bolus injection) enables one to evaluate the disposition (distribution, metabolism, and excretion) of a compound without the complications associated with the passage across the boundary between the body and environment or complex barrier of the GI tract. This establishes the intravenous (IV) bolus data as reference for bioavailability evaluations and similar considerations.

The practical applications of compartmental analysis and non-compartmental analysis in establishing kinetic parameters and constants have been widely accepted. The detailed description and assumptions of the models are described in Chapter 9. For compartmental analysis, the body is assumed as a collection of pools and compartments. Each compartment is defined as a given amount of xenobiotic within a volume, which may represent the size of the compartment. Each compartment within its spatial assignment behaves kinetically as a well-mixed homogeneous pool. When a compartmental model is in contact with its external environment and has input (dose) and output (elimination) it is referred to as an open compartment. When the compartment has no exchange with its external environment, it is referred to as a closed compartment. All compartmental models used to describe the fate of exogenous compounds in the body are considered open models. The closed compartmental models are more relevant to endogenous compounds that may participate in, or are precursors for, the formation of other endogenous compounds in the body.

10.2 LINEAR ONE-COMPARTMENT OPEN MODEL

The one-compartment open model has the historical significance of being the model first considered for disposition of xenobiotics in the body. It has also the significance of offering applicable methods of estimating PK/TK parameters and constants in research and particularly in clinical settings. The model has had an instructive role in confirming the definition and meaning of PK/TK parameters, *in vivo* and *in vitro*.

The one-compartment model assumes that the input (intravenous dose) is mixed and exchanged very rapidly between the plasma and tissues, which behave collectively as a single well-mixed and homogeneous volume. Plasma, blood or serum, and urine are the reference samples of this well-mixed volume. Therefore, any change in the concentration or amount of compound in the reference samples reflects the equivalent changes in the compartment (Gibaldi and Perrier, 1982). Another assumption of the model is that the elimination (i.e., metabolism and excretion) is governed by first-order kinetics. The depiction of the model is presented in Figure 9.9 (Chapter 9).

The instantaneous input refers to an input without a definable rate, that is, at time zero the drug is outside of the compartment and at no time (i.e., at $t = 0$) is the compound in the body. A good example of instantaneous input is intravenous bolus injection. The first-order overall elimination rate constant K represents renal excretion, hepatic metabolism, and other significant elimination rate constants, that is,

$$K = k_e + k_m + k_n \tag{10.1}$$

In Equation 10.1, k_e is the first-order excretion (i.e., renal) rate constant; k_m is the first-order metabolic (i.e., hepatic) rate constant and k_n is the collective first-order rate constant of elimination, such as those for elimination through the sweat glands, milk, exhaled air, etc. The (k_n) value for most xenobiotics is considered negligible unless it is proven otherwise and Equation 10.1 is simplified to

$$K = k_e + k_m \tag{10.2}$$

The amount in the compartment at time (t), (A_t), is the product of free concentration in plasma at time (t), that is, (Cp_t), and the apparent volume of distribution Vd:

$$A_t = Cp_t \times Vd \tag{10.3}$$

The variables Cp_t and A_t are dependent variables on time and dose, whereas K and Vd are considered constant in linear PK/TK for a given compound.

The differential equation of the model, that is, the overall rate of elimination from the compartment is

$$\frac{dA_t}{dr} = -KA_t \tag{10.4}$$

According to Equation 10.4, the rate of elimination is a variable, the negative sign reflects the reduction of the injected dose in the body as a result of metabolism and excretion, and the rate constant of elimination K remains constant during the elimination process. The analogy can be made between the first-order elimination and a hydrodynamic model of two tanks of water with a drain outlet at the bottom (Figure 10.1) (Boroujerdi, 2002). The diameter of one opening is twice that of the other one. The flow of water from the tanks via the outlet per unit of time is analogous to the rate of elimination of a drug from the body. The flow rate of water from the tank depends on the volume of water and its hydrostatic pressure. As the volume of water declines, the pressure declines. This hydrostatic pressure of water is analogous to amount of a drug in the body. As the amount of drug in the body declines, so declines the rate of elimination. The diameter of the outlets that remain unchanged during the removal of water from the tanks is analogous to K. A smaller diameter tends to lower the flow rate, which is analogous to a reduction of K in disease states such as renal impairment and hepatic failure.

The important assumption of the model, as was generalized in Chapter 9, Section 9.5.5, is that only the free molecules of the injected compound in the compartment can interact with receptors to provide pharmacological response. In other words, the xenobiotic concentration of free molecules in plasma or serum is proportional to its concentration at the receptor site, which is in equilibrium with the receptor-bound concentration.

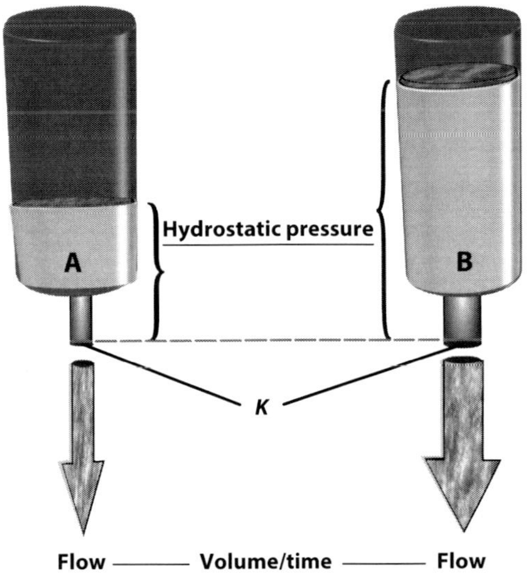

FIGURE 10.1 Illustration of a hydrodynamic model analogous to the first-order elimination from the body; the high level of fluid in the tank and larger opening of the outlet result in a greater exit flow rate.

The integrated forms of Equation 10.4 are

$$\ln A = \ln A^0 - Kt \tag{10.5}$$

$$\log A = \log A^0 - \frac{Kt}{2.303} \tag{10.6}$$

$$A = A^0 e^{-Kt} \tag{10.7}$$

In terms of plasma or serum free concentration of the compound

$$\ln Cp = \ln Cp^0 - Kt \tag{10.8}$$

$$\log Cp = \log Cp^0 - \frac{Kt}{2.303} \tag{10.9}$$

$$Cp = Cp^0 e^{-Kt} \tag{10.10}$$

where Cp and Cp^0 are the plasma free concentrations at time t and initial concentration at $t = 0$, respectively.

10.2.1 Half-Life of Elimination

A useful PK/TK constant of the model is the half-life of elimination. It is the time required for 50% of the compound at any initial level to be eliminated from the compartment, that is,

$$\text{At } t = T_{1/2},\ Cp_2 = \frac{Cp_1}{2},\ \text{or } A_2 = \frac{A_2}{2} \tag{10.11}$$

Therefore,

$$\log\frac{Cp_1}{2} = \log Cp_1 - \frac{K(T_{1/2})}{2.303} \tag{10.12}$$

$$KT_{1/2} = 2.303\log 2 \tag{10.13}$$

$$KT_{1/2} = 0.693 \tag{10.14}$$

$$\therefore T_{1/2} = \frac{0.693}{K}, \text{ or } T_{1/2} = \frac{\ln 2}{K}, \text{ or } T_{1/2} = \frac{2.303\log 2}{K} \tag{10.15}$$

$$\text{and } K = \frac{0.693}{T_{1/2}}, \text{ or } K = \frac{\ln 2}{T_{1/2}}, \text{ or } K = \frac{2.303\log 2}{T_{1/2}} \tag{10.16}$$

Equations 10.15 and 10.16 indicate that the half-life and rate constant of elimination for a compound that follows linear pharmacokinetics are constant, independent of the amount or concentration and inversely proportional to each other.

10.2.2 Time Constant

Occasionally it is more convenient to use the time constant rather than the half-life. The time constant is the reciprocal of the first-order rate constant. The relationship between the time constant and the half-life is as follows:

$$\text{Time Constant} = \frac{1}{K} \tag{10.17}$$

$$\frac{1}{K} = \frac{1}{0.693/T_{1/2}} = 1.44T_{1/2} \tag{10.18}$$

The time constant is also known as turnover time and in non-compartmental analysis of intravenous bolus injection, $1/K$ is referred to as mean residence time (MRT). The time constant or (MRT) represents the time for 63% of an intravenous bolus dose to be eliminated from the body. The significance of time constant (MRT) is further discussed at the end of this chapter in association with the non-compartmental analysis. The relationship between MRT and $T_{1/2}$ is

$$(MRT)_{i.v.bolus} = \frac{1}{K} = \frac{T_{1/2}}{0.693} \tag{10.19}$$

$$T_{1/2} = 0.693\ (MRT)_{i.v.bolus} \tag{10.20}$$

10.2.3 Apparent Volume of Distribution

A useful and often misunderstood constant of the linear compartmental model is the volume of distribution. It is identified as the apparent volume of distribution (Vd) to distinguish the value from the physiological volumes of distribution. This constant has no relationship with the total body water

(Chapter 3, Section 3.2.1), or other physiological volumes. It is only a constant of conversion for changing the concentration of a free xenobiotic in a compartment to its total amount, or vice versa. The definition is applicable even for multicompartmental models (Chiou, 1981). A large volume of distribution indicates that the concentration of measurable free compound in plasma is low because of its localization and association with regions that are not easily accessible or being a part of the sampling compartment.

The apparent volume of distribution is determined by dividing the amount by the plasma concentration:

$$Vd = \frac{A_t}{Cp_t} \tag{10.21}$$

At time zero, the amount of free compound in the body/compartment is equal to dose. The free concentration can also be estimated from the data. Thus, the apparent volume of distribution can be estimated as

$$Vd = \frac{\text{Dose}}{Cp^0} \tag{10.22}$$

10.2.4 Total Body Clearance

The total body clearance (Cl_T) is the rate of elimination of a xenobiotic from the body per unit of its concentration

$$Cl_T = \frac{\text{Rate}}{\text{Concentration}} = \frac{dA_t/dt}{Cp} = \frac{KA_t}{A_t/Vd} = K \times Vd \tag{10.23}$$

Therefore, total body clearance is the product of the apparent volume of distribution and the first-order overall elimination rate constant. It is also the sum of metabolic, renal, and residual clearances:

$$Cl_T = Vd \times K = Vd(k_e + k_m + k_n) = Cl_{renal} + Cl_{metabolic} + Cl_{residual} \tag{10.24}$$

10.2.5 Duration of Action

The duration of action is the time that plasma concentration remains within the therapeutic range of an agent. When the concentration falls below the minimum effective level, no effective response should be expected. For a single dose of intravenous bolus injection, the duration of action is estimated as

$$t_d = \frac{2.303}{K} \log \frac{Cp^0}{Cp_{MEC}} = 3.3T_{1/2} \log \frac{Cp^0}{Cp_{MEC}} \tag{10.25}$$

where (Cp_{MEC}) is the minimum effective concentration of therapeutic agent.

According to Equation 10.25, the duration of action is inversely proportional to K and directly proportional to the half-life of elimination. This indicates that the duration of action of a compound in patients with renal impairment or hepatic failure is longer than normal. The duration of action is not directly proportional to the administered dose, or initial concentration. For example, if the dose is doubled, Cp^0 will be doubled, but the duration of action will only increase by one half-life

$$(t_d)_1 = 3.3T_{1/2}\log\frac{Cp_1^0}{Cp_{MEC}} \tag{10.26}$$

$$(t_d)_2 = 3.3T_{1/2}\log\frac{2(Cp_1^0)}{Cp_{MEC}} \tag{10.27}$$

$$(t_d)_2 - (t_d)_1 = 3.3T_{1/2}\left(\log\frac{2Cp^0}{Cp_{MEC}} - \log\frac{Cp^0}{Cp_{MEC}}\right) = 1T_{1/2} \tag{10.28}$$

Thus, if the dose is quadrupled with respect to the first dose, the duration of action would increase by two half-lives.

10.2.6 Fraction of Dose in the Body at Time *t*

The fraction of the dose in the body at any time after the administration of the dose is estimated by normalizing Equation 10.10 with respect to the initial plasma concentration (Cp^0):

$$(f_b)_t = \frac{Cp}{Cp^0} = e^{-Kt} = e^{-\frac{0.693}{T_{1/2}}t} = (0.5)^{\frac{t}{T_{1/2}}} \tag{10.29}$$

Equation 10.29 indicates that

$$(f_b)_{t=0} = 1;\ (f_b)_{t=1T_{1/2}} = 0.5;\ (f_b)_{t=2T_{1/2}} = 0.25;\ (f_b)_{t=3.3T_{1/2}} = 0.1;\ (f_b)_{t=4.3T_{1/2}} = 0.05;$$
$$(f_b)_{t=6.6T_{1/2}} = 0.01;\ (f_b)_{t=7T_{1/2}} \cong 0$$

10.2.7 Fraction of Dose Eliminated by All Routes of Elimination at Time *t*

At time zero the fraction of the dose in the body is equal to one, as the dose eliminates exponentially, the sum of fractions remaining in the body and eliminated at time t is equal to 1. Therefore, the fraction of the dose eliminated from the body by all routes of elimination at time t is

$$(f_{el})_t = 1 - (f_b)_t = 1 - \frac{Cp}{Cp^0} = 1 - e^{-Kt} = 1 - e^{-\frac{0.693}{T_{1/2}}t} = 1 - (0.5)^{\frac{t}{T_{1/2}}} \tag{10.30}$$

10.2.8 Area under Plasma Concentration–Time Curve after IV Bolus

The area under plasma concentration–time curve (AUC) of an intravenously injected dose is a reference parameter in determining the bioavailability of the compound given by routes of administration other than intravenous injection. This parameter is also useful in calculating PK/TK parameters and constants of a model. (AUC) can be estimated by the model dependent method of integration or by model independent trapezoidal rule (Appendix A). The integration of Equation 10.10 provides the following relationship for AUC of the one-compartment model:

$$AUC = \int_0^\infty Cp^0 e^{-Kt} = Cp^0 \left.\frac{e^{-Kt}}{-K}\right|_0^\infty = \frac{Cp^0}{K} \tag{10.31}$$

(AUC) may also be calculated by using the trapezoidal rule

$$AUC\Big|_0^\infty = \sum_{i=0}^{n} \frac{Cp_i + Cp_{i+1}}{2}(t_{i+1} - t_i) + \frac{Cp_{last}}{K} \tag{10.32}$$

Multiplying the numerator and denominator of Equation 10.31 by the apparent volume of distribution yields the following relationship:

$$AUC = \frac{(Cp^0) \times Vd}{K \times Vd} = \frac{\text{Dose}}{Cl_T} \tag{10.33}$$

Equations 10.31 and 10.33 denote that in linear pharmacokinetics (i.e., dose-independent pharmacokinetics) the (AUC) is directly proportional to the initial plasma concentration or dose and inversely proportional to the overall elimination rate constant and clearance. This means that if the dose is doubled or if the clearance is declined by 50%, the area under the curve will be doubled (Figures 10.2 and 10.3).

The plot of *AUC* versus *Dose* is used to determine the dose dependency of the pharmacokinetics of a compound. A nonlinear relationship as presented in Figure 10.4 indicates that the rate of elimination of xenobiotic from the body is no longer first-order kinetics and the equations of linear pharmacokinetics as described in this chapter are not applicable. The nonlinear relationship between dose and *AUC* occurs when saturable processes govern the excretion and/or metabolism. As demonstrated in Figure 10.4, in a nonlinear relationship, if the dose is doubled, the *AUC* will increase disproportionately.

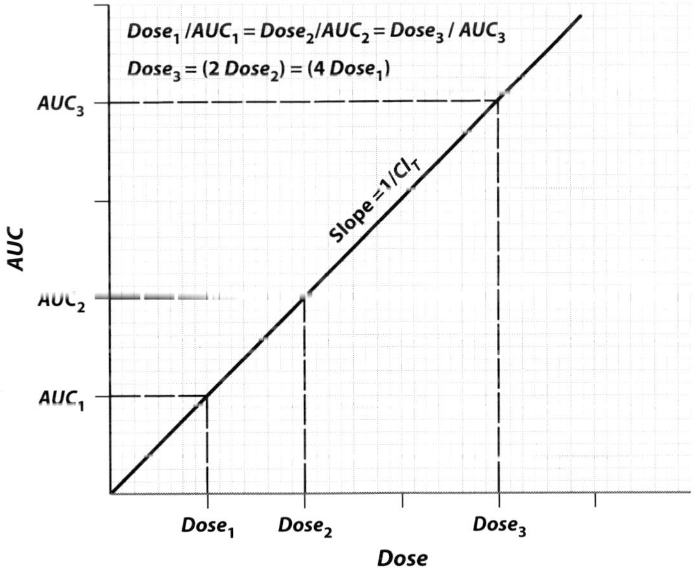

FIGURE 10.2 The linear relationship between the administered dose and the related area under the plasma concentration–time curve in dose-independent PK/TK.

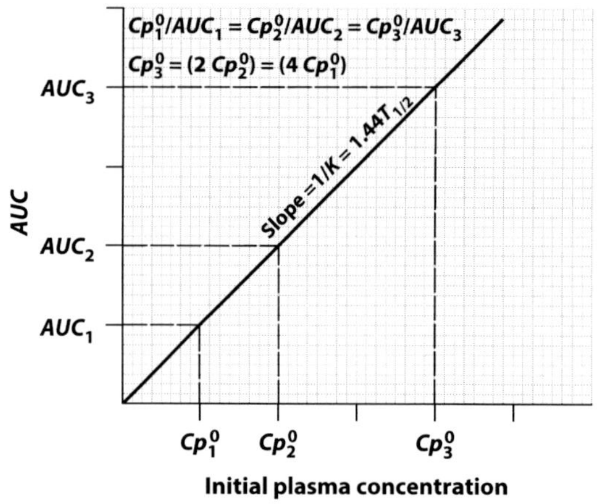

FIGURE 10.3 The linear relationship between initial plasma concentrations and the related *AUC* in dose-independent PK/TK.

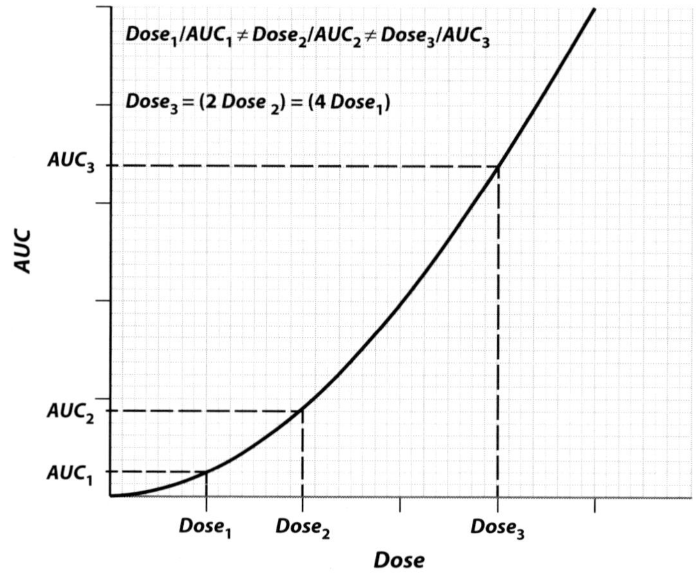

FIGURE 10.4 The linear relationship between initial plasma concentrations and the related *AUC* in dose-independent PK/TK.

10.3 LINEAR TWO-COMPARTMENT OPEN MODEL WITH ELIMINATION FROM THE CENTRAL COMPARTMENT

Among the multicompartmental models, the two-compartment model is a more realistic and practical model for quantitative PK/TK analysis of most xenobiotics, which also avoids the limitation of considering the body as a single compartment (Riegleman et al., 1968). The two-compartment model assumes the intravenously injected compound distributes in the central compartment and peripheral compartment in proportion to their blood perfusion and eliminates from the central compartment in proportion to its concentration. The "central" or "first" compartment represent the systemic circulation, extracellular fluid, and highly perfused organs/tissues such as the liver, kidney, lungs, etc. The highly perfused tissues achieve a rapid equilibrium with the systemic circulation and kinetically are indistinguishable from the systemic circulation (Rescigno and Seger, 1966; Wagner, 1975; Gibaldi et al., 1982). Thus, the compartment is assumed kinetically as a homogeneous compartment. The "peripheral" or "second" compartment represents tissues and regions of the body that have reduced blood perfusion and are less accessible. The examples of tissues that may be included in the peripheral compartment are adipose tissue, muscle, skin, etc.

On the bolus injection of a compound that follows the two-compartment model, the initial mixing in the central compartment is very rapid in relation to its distribution to the peripheral compartment and elimination via the central compartment. The rate of elimination (metabolism and excretion) is proportional to the concentration in the central compartment and the rate of distribution is proportional to the difference in concentrations between the two compartments.

10.3.1 Equations of the Model

The diagram of the model is depicted in Figure 10.5. The rate of change in the amount of xenobiotic in the central compartment depends on the rates of distribution and the elimination:

$$\frac{dA_1}{dt} = k_{21}A_2 - k_{12}A_1 - k_{10}A_1 \tag{10.34}$$

The input in the central compartment is instantaneous and has no rate. The rate of change in the amount of xenobiotic in the peripheral compartment is a function of distribution rates in and out of the compartment with no elimination from the peripheral compartment:

$$\frac{dA_2}{dt} = k_{12}A_1 - k_{21}A_2 \tag{10.35}$$

The variables A_1 and A_2 are the amount of xenobiotics at any time in the central and peripheral compartments, respectively. The rate constants of distribution are (k_{12}) and (k_{21}). The first subscript number represents the departure compartment and the second number represents the arrival compartment.

By using the Laplace method of integration (Chapter 9 and Appendix A) or method of input-disposition functions (Chapter 9 and Appendix A), the integrated equations of the model are defined as (Rescigno et al., 1966; Wagner, 1975, 1993; Gibaldi et al., 1982)

$$A_1 = \frac{D(\alpha - k_{21})}{\alpha - \beta}e^{-\alpha t} + \frac{D(k_{21} - \beta)}{\alpha - \beta}e^{-\beta t} \tag{10.36}$$

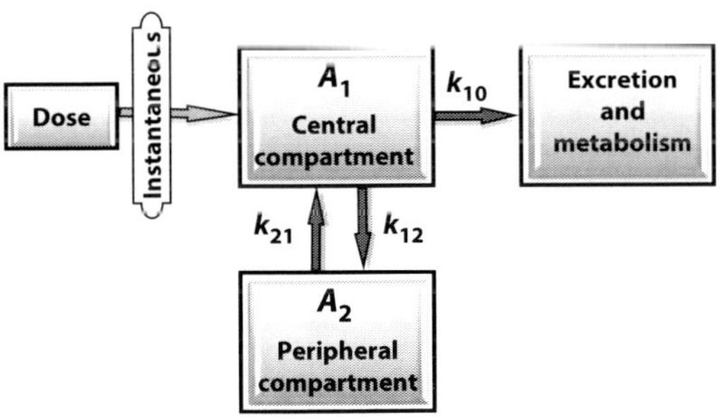

FIGURE 10.5 Diagram of a linear two-compartment model with excretion and metabolism from the central compartment.

$$A_2 = \frac{k_{12}D}{\alpha - \beta}(e^{-\beta t} - e^{-\alpha t}) \quad (10.37)$$

The first-order hybrid rate constants α and β, are the roots of quadratic equation generated during the integration. These rate constants are estimated in fitting the biexponential model to the plasma concentration versus time data. They are also estimated from the slope of the terminal portion of log plasma concentration–time curve (−β/2.303) and the slope of the residual line (−α/2.303).

Dividing Equation 10.36 by the apparent distribution volume of central compartment V_1 yields the equation of the model in terms of plasma concentration

$$Cp = \underbrace{\frac{D(\alpha - k_{21})}{V_1(\alpha - \beta)}}_{a} e^{-\alpha t} + \underbrace{\frac{D(k_{21} - \beta)}{V_1(\alpha - \beta)}}_{b} e^{-\beta t} \quad (10.38)$$

By substituting (*a*) and (*b*) for the coefficients of each exponential term, the biexponential equation of the model can be simplified to

$$Cp = ae^{-\alpha t} + be^{-\beta t} \quad (10.39)$$

The logarithm plot of plasma concentration versus time for a compound that follows the two-compartment model has two distinct phases (Figure 10.6). The early phase with steeper slope, referred to as the distributive phase, is formed by the rapid decline of plasma concentration in the central compartment. This rapid decline is the result of a number of physical and physiological processes that occur simultaneously and immediately after the injection. Among these processes are the rapid distribution and dilution of the drug into those fluids and tissues that make up the central compartment, transfer to the peripheral compartment and elimination from the central compartment. After a certain period of time that varies from compound to compound this initial phase comes to an end and the xenobiotic concentration in various tissues and fluids declines steadily. This slower and gradual decline appears as a second phase of a semilogarithmic plot of plasma concentration versus time curve and is referred to as the post-distributive phase or disposition phase. During this phase all physiological processes (excretion, metabolism, and distribution) contribute to the consistent decline of drug concentration in the central compartment, which is also a function of physicochemical characteristics of a drug and its interaction with biological molecules (Figure 10.6). Therefore, the rate constant calculated from the slope of the second phase of the curve β can only be identified as the disposition (i.e., distribution and elimination (metabolism and excretion)) rate constant and the related half-life as physiological, biological, or disposition half-life. This is one of the differences between the one- and two-compartment models.

Theoretically, extrapolation of the terminal portion of the curve to the *y*-axis gives the *y*-intercept, which is equal to the coefficient (*b*). This extrapolated line represents the biological processes that are involved in the disposition of a xenobiotic. Therefore, subtracting the data points of the extrapolated line that correspond to the early time points from the observed plasma concentrations of the distributive phase leads to a number of residual values that represent all other processes that, with disposition, influence the early sharp decline of plasma concentration. Plotting the residuals against the early time points forms a straight line with a slope of −α/2.303 and *y*-intercept of (*a*). Both α and β are first-order hybrid rate constants; α represents the rate constant of all physical and physiological processes other than the distribution and elimination that occurs immediately and simultaneously after the injection of a drug in the body during the distributive phase. The rate constant β represents the hybrid physiological process of distribution and elimination (i.e., disposition).

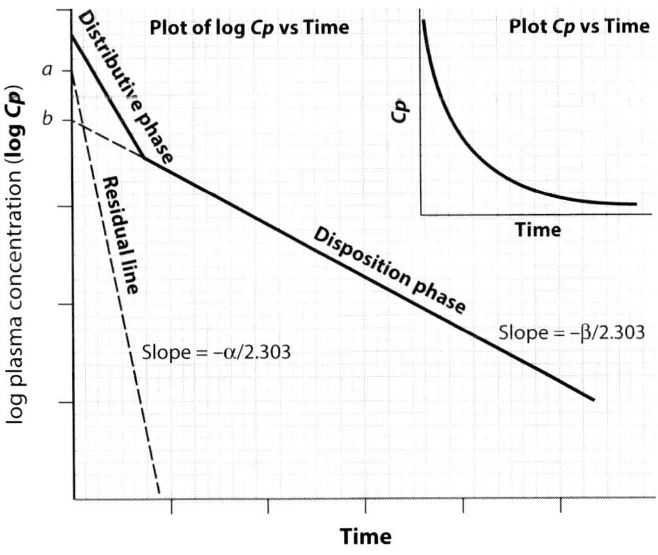

FIGURE 10.6 Diagram of plasma concentration–time profile of compounds that follow the two-compartment model.

The initial estimates of a, b, α, and β can be obtained by linear regression analysis of the extrapolated and residual lines; however, the best approach would be to use specialized pharmacokinetic software that are developed for the purpose of selecting and fitting the appropriate model to experimental data and estimating the related parameters and constants. The appropriate software would be the one that analyzes the entire plasma level time data (i.e., the distributive and post distributive phases of the curve together), using nonlinear regression analysis and the required statistical analysis/information criteria (see Appendix A).

10.3.2 Estimation of Initial Plasma Concentration and Apparent Volume of Distribution

The initial plasma concentration is estimated by setting t equal to zero in Equation 10.38

$$Cp^0 = a + b \tag{10.40}$$

Replacing the coefficients a and b with their formulas according to Equation 10.38 yields

$$Cp^0 = \frac{D(\alpha + k_{21})}{V_1(\alpha - \beta)} + \frac{D(k_{21} - \beta)}{V_1(\alpha - \beta)} = \frac{D}{V_1} \tag{10.41}$$

$$\therefore V_1 = \frac{D}{Cp^0} = \frac{D}{a + b} \tag{10.42}$$

Similar to the one-compartment model, the apparent volume of distribution of the central compartment is only a proportionality constant that is used to determine the amount in the central compartment (Gibaldi et al., 1969):

$$(A_1)_t = Cp_t \times V_1 \tag{10.43}$$

Equations 10.41 and 10.42 emphasize the importance of taking early plasma samples after the injection for the purpose of identifying the distributive phase more clearly. If the first blood sample is taken sometime after the end of the distributive phase, the profile of log Cp versus time constitutes only the disposition phase. This profile would be similar to that discussed for one-compartment models. Therefore, one mistakenly may analyze the data according to the one-compartment model, in which case some parameters such as initial plasma concentration would be underestimated.

10.3.3 Estimation of the Rate Constants of Distribution and Elimination

The first-order distribution rate constants are (k_{12}) and (k_{21}). The rate constant k_{21} representing the transfer rate constant from peripheral compartment to the central compartment, is determined from the a or b formula:

$$b = \frac{(a + b)(k_{21} - \beta)}{(\alpha - \beta)} \tag{10.44}$$

$$\therefore k_{21} = \frac{a\beta + b\alpha}{a + b} \tag{10.45}$$

Because the denominator of Equation 10.45 is the initial plasma concentration, a and b of the numerator can be normalized with respect to the initial plasma concentration as follows:

$$a_N = a/(a + b) \tag{10.46}$$

$$b_N = b/(a + b) \tag{10.47}$$

$$\therefore k_{21} = a_N\beta + b_N\alpha \tag{10.48}$$

where a_N and b_N are the normalized values of a and b.

To determine the equations of other miniature rate constants we need to review the disposition function of the model (Benet, 1972). According to the criteria set forth in Chapter 9 and Appendix A, the disposition function of the model is

$$(Disp)_s = \frac{s + k_{21}}{s^2 + (k_{12} + k_{21} + k_{10}) + k_{21}k_{10}} = \frac{s + E_2}{(s + \alpha)(s + \beta)} \tag{10.49}$$

The denominator of Equation 10.49 $(s + \alpha)(s + \beta)$ can be written as: $s^2 + s(\alpha + \beta) + \alpha\beta$, which corresponds to $s^2 + s(k_{12} + k_{21} + k_{10}) + k_{21}k_{10}$. The comparison yields two relationships

$$\alpha + \beta = k_{12} + k_{21} + k_{10} \tag{10.50}$$

$$\alpha\beta = k_{21}k_{10} \tag{10.51}$$

Therefore, the overall elimination rate constant k_{10} and the transfer rate constant from the central to the peripheral compartment can be determined by the equations

$$k_{10} = \frac{\alpha\beta}{k_{21}} \tag{10.52}$$

$$k_{12} = \alpha + \beta - k_{10} - k_{21} \tag{10.53}$$

10.3.4 Half-Lives of the Two-Compartment Model

Contrary to the one-compartment model with intravenous bolus injection that offers only one rate constant and one half-life of elimination, the rate constants of the two-compartment model enables us to estimate various half-lives.

10.3.4.1 Biological Half-Life

The biological half-life is also known as half-life of β, physiological half-life or half-life of disposition. It represents the combined effect of distribution and elimination and should not be mistaken with the half-life of elimination. It is estimated as

$$(T_{1/2})_\beta = \frac{0.693}{\beta} \tag{10.54}$$

In the two-compartment model, the fractions of the dose remaining in the body or eliminated depends on the biological half-life, that is,

$$(f_{el})_{t=1(T_{1/2})_\beta} = 0.50,\ (f_b)_{t=1(T_{1/2})_\beta} = 0.50;\ (f_{el})_{t=2(T_{1/2})_\beta} = 0.75,\ (f_b)_{t=2(T_{1/2})_\beta} = 0.25;$$
$$(f_{el})_{t=3.3(T_{1/2})_\beta} = 0.90,\ (f_b)_{t=3.3(T_{1/2})_\beta} = 0.10;\ (f_{el})_{t=4.3(T_{1/2})_\beta} = 0.95,\ (f_b)_{t=4.3(T_{1/2})_\beta} = 0.05;$$
$$(f_{el})_{t=6.6(T_{1/2})_\beta} = 0.99,\ (f_b)_{t=6.6(T_{1/2})_\beta} = 0.01$$

10.3.4.2 Elimination Half-Life

Elimination half-life or half-life of k_{10} represents the combined half-lives of metabolism, excretion, and other residual elimination processes:

$$(T_{1/2})_{elim} = \frac{0.693}{k_{10}} \tag{10.55}$$

where

$$k_{10} = k_e + k_m + k_n \tag{10.56}$$

10.3.4.3 Half-Life of α

The half-life of hybrid rate constant α represents physical and physiological processes other than distribution and elimination that occur immediately after the injection during the distributive phase. The distribution and elimination are excluded from its definition because its value is estimated from the residual line after subtraction of disposition phase from early distributive phase. This half-life is neither biological nor elimination half-life and is used only to calculate (α)

$$(T_{1/2})_\alpha = \frac{0.693}{\alpha} \tag{10.57}$$

10.3.4.4 Half-Life of k_{12}

This half-life denotes the half-life of xenobiotic transfer from the central to the peripheral compartment:

$$(T_{1/2})_{k_{12}} = \frac{0.693}{k_{12}} \tag{10.58}$$

10.3.4.5 Half-Life of k_{21}

This is the half-life of xenobiotic transfer from the peripheral to the central compartment:

$$(T_{1/2})_{k_{21}} = \frac{0.693}{k_{21}} \tag{10.59}$$

10.3.5 Area under the Curve, Volumes of Distribution, and Clearances

The area under plasma concentration–time curve of a compound that follows the two compartment model and is given intravenously can be estimated by the trapezoidal rule (Appendix A) or by the integration of Equation 10.39. The integration can be carried out from time zero to infinity or from zero to any time t, or the last the last time point of sampling:

$$AUC_0^\infty = \int_0^\infty ae^{-\alpha t} + be^{-\beta t} = \frac{a}{\alpha} + \frac{b}{\beta} \tag{10.60}$$

$$AUC_0^t = \int_0^t ae^{-\alpha t} + be^{-\beta t} = \frac{ae^{-\alpha t}}{-\alpha} + \frac{be^{-\beta t}}{-\beta} \tag{10.61}$$

Substitution of the a and b formulas in Equation 10.60 yields the following equation for AUC_0^∞

$$AUC_0^\infty = \frac{D(\alpha - k_{21})}{V_1(\alpha-\beta)\alpha} + \frac{D(k_{21}-\beta)}{V_1(\alpha-\beta)\beta} = \frac{\beta Cp^0(\alpha-k_{21}) + \alpha Cp^0(k_{21}-\beta)}{\alpha\beta(\alpha-\beta)} = \frac{(\alpha-\beta)k_{21}Cp^0}{(\alpha-\beta)\alpha\beta}$$

$$\therefore AUC_0^\infty = \frac{Cp^0}{k_{10}} \tag{10.62}$$

Equation 10.62 can also be used to calculate k_{10}

$$k_{10} = \frac{Cp^0}{AUC_0^\infty} = \frac{a+b}{(a/\alpha)+(b/\beta)} = \frac{1}{(a_N/\alpha)+(b_N/\beta)} \tag{10.63}$$

The volume of distribution, as it was discussed earlier in Section 10.2.3, is only a conversion factor that is used to change the concentration of xenobiotic in plasma to the total amount of drug in the body or vice versa. It has a very limited or no physiological significance and its magnitude depends on what particular model is used and what kinetic pool is considered. For a two-compartment model, the volumes of distribution include the volume of the central compartment, the volume of the peripheral compartment and the overall volume of distribution that encompasses both compartments.

The volume of distribution of the central compartment V_1 was discussed earlier and can be estimated by Equation 10.42.

To estimate the apparent overall volume of distribution of a compound that follows the two-compartment model, the following relationship known as $(Vd)_{area}$ or $(Vd)_{\beta}$ has been accepted as a practical formula. The formula is derived from the model independent relationship of Clearance = Dose/AUC:

$$(Vd)_{area} = \frac{D}{\beta \times AUC_0^{\infty}} = \frac{D}{\beta((a/\alpha) + (b/\beta))} \tag{10.64}$$

The apparent volume of distribution of the peripheral compartment is difficult to determine directly, mainly because the peripheral compartment of the model discussed here is not a sampling compartment, and the assumption of homogeneity, as it was assumed for the central compartment, is less applicable. The indirect determination of this volume term by subtracting the volume of the central compartment from the overall volume of distribution yields a value that is somewhat fictitious. As a result, any "concentration" term for the peripheral compartment would also be a fictitious parameter. Therefore, it is preferred to use the "amount" to characterize the time course of a drug in the peripheral compartment rather than "concentration".

The definition of the total body clearance here is the same as described in Chapter 4. The useful equations of estimating the total body clearance of a xenobiotic that follows the two-compartment model are

$$Cl_T = \frac{D}{AUC_0^{\infty}} \tag{10.65}$$

$$Cl_T = V_1 \times k_{10} \tag{10.66}$$

$$Cl_T = (Vd)_{area} \times \beta \tag{10.67}$$

Equations 10.66 and 10.67 also indicate that the rate of elimination from the body is the rate of elimination from the central compartment as long as no elimination takes place from the peripheral compartment. This point can be proven by setting the total amount of xenobiotic in the body equal to the combined amount in central compartment and peripheral compartment at any time t, that is, $A = A_1 + A_2$. Thus, the fraction of xenobiotic in the central compartment at any time would be

$$f_{central} = A_1/A \tag{10.68}$$

and the amount in the body and central compartment would be

$$A = A_1/f_{central} \tag{10.69}$$

$$A_1 - f_{central} \times A \tag{10.70}$$

Therefore, the overall rate of elimination from the body can be defined as Equation 10.71 and the relationship between β and k_{10} as Equation 10.72:

$$\frac{dA}{dt} = -\beta A = -k_{10}A_1 = -k_{10}(f_{central})A \tag{10.71}$$

and the relationship

$$\beta = k_{10}(f_{central}) = k_{10}\frac{A_1}{A} = k_{10}\frac{Cp \times V_1}{A} \tag{10.72}$$

Because for the model discussed here (Figure 10.5) the elimination occurs only from the central compartment, the total body clearance can also be expressed in terms of

$$\frac{\beta A}{Cp} = (Vd)_{area} \times \beta = k_{10} \times V_1 \tag{10.73}$$

By knowing the fraction of the dose eliminated from the body unchanged and fraction eliminated as metabolites, clearance terms such as renal and metabolic are estimated as

$$Cl_r = f_e Cl_T = \frac{A_e^\infty}{D} Cl_T = \frac{D - A_m^\infty}{D} Cl_T \tag{10.74}$$

$$Cl_m = f_m Cl_T = \frac{A_m^\infty}{D} Cl_T = \frac{D - A_e^\infty}{D} Cl_T \tag{10.75}$$

10.3.6 Time Course of Amount in the Peripheral Compartment

Equation 10.37, relating the amount of xenobiotic in the peripheral compartment at any time after the injection, is developed by the integration of differential Equation 10.35 and by using the method of input-disposition function or the method of substitution as described in Chapter 9 or Appendix A. The uptake magnitude by the peripheral compartment for a given compound depends on a number of factors such as lipid solubility of the compound, the extent of elimination from the central compartment, the degree of protein binding, and the type of interaction and accumulation in the tissues (Figure 10.7). It should be noted that in linear PK/TK, the slopes of terminal portion of log A_2 versus time and log A_1 versus time are the same and the lines are parallel (Figure 10.8).

The $(AUC)_{peripheral}$ (Figure 10.7), represents the extent of accumulation of xenobiotic in the peripheral compartment and can be estimated as

$$(AUC_0^\infty)_{peripheral} = \frac{k_{12}D}{\alpha - \beta}\left(\frac{1}{\beta} - \frac{1}{\alpha}\right) \tag{10.76}$$

The maximum amount in the peripheral compartment that corresponds to T_{max} of the curve (Figure 10.7) can be estimated by the following equations

$$(T_{max})_{peripheral} = \frac{2.303}{\alpha - \beta}\log\frac{\alpha}{\beta} \tag{10.77}$$

$$(A_{max})_{peripheral} = \frac{k_{12}D}{\alpha - \beta}(e^{-\beta T_{max}} - e^{-\alpha T_{max}}) \tag{10.78}$$

At T_{max} the rate of input from the central compartment into the peripheral compartment is equal to the rate of output from the peripheral compartment into the central compartment, which is indicative of the steady-state condition at that instant.

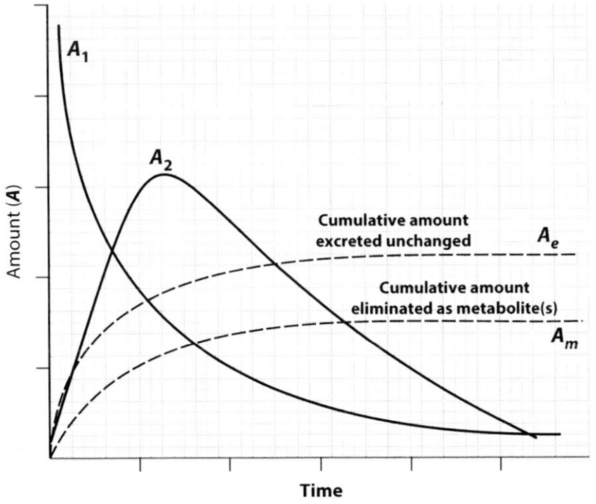

FIGURE 10.7 The time course of xenobiotic in the peripheral compartment in comparison to the central compartment following IV bolus administration.

Therefore,

$$\frac{dA_2}{dt} = 0 \qquad \therefore\ k_{12}A_1 = k_{21}A_2 \tag{10.79}$$

Based on this concept an overall volume of distribution is proposed that corresponds to the volume of distribution at the steady state, it is named $(VD)_{ss}$ and is derived according to the following assumptions and steps:

At T_{max}:

$$A_1 = Cp \times (V_1)_{ss} \tag{10.80}$$

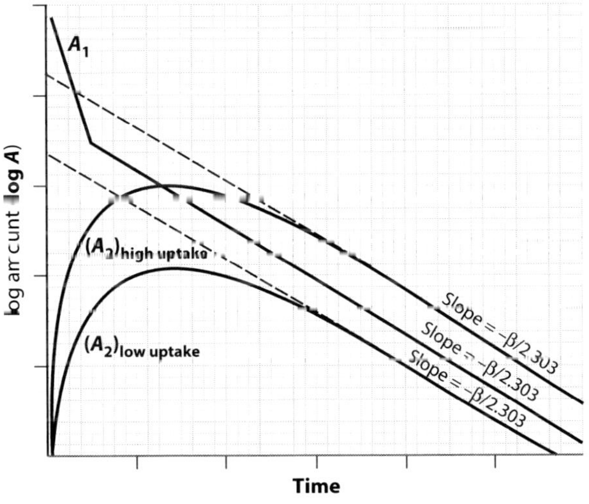

FIGURE 10.8 Semi-logarithmic plot of dose-independent time course of xenobiotics in the central and peripheral compartments with parallel terminal phase of similar slope.

$$A_2 = C_2 \times (V_2)_{ss} \tag{10.81}$$

$$(Vd)_{ss} = (V_1)_{ss} + (V_2)_{ss} \tag{10.82}$$

$$\begin{gathered} Cp = C_2 \\ k_{12}Cp(V_1)_{ss} = k_{21}C_2(V_2)_{ss} \end{gathered} \tag{10.83}$$

$$(V_2)_{ss} = \left(\frac{k_{12}}{k_{21}}\right)V_1 \tag{10.84}$$

Therefore,

$$(Vd)_{ss} = \left(\frac{k_{12}}{k_{21}}\right)V_1 + V_1 = \left(\frac{k_{12} + k_{21}}{k_{21}}\right)V_1 \tag{10.85}$$

It is important to note that $(Vd)_{ss}$ calculated by Equation 10.85 is different from Vd_{ss} of the non-compartmental analysis. Equation 10.85 may represent the volume of distribution only at the point where the amount of xenobiotic in the peripheral compartment reaches to its maximum.

10.4 LINEAR TWO-COMPARTMENT OPEN MODEL WITH BOLUS INJECTION IN THE CENTRAL COMPARTMENT AND ELIMINATION FROM PERIPHERAL COMPARTMENT

The diagram of the model is depicted in Figure 10.9. The differential equations of the model are

$$\frac{dA_1}{dt} = k_{21}A_2 - k_{12}A_1 \tag{10.86}$$

$$\frac{dA_2}{dt} = k_{12}A_1 - k_{21}A_2 - k_{20}A_2 \tag{10.87}$$

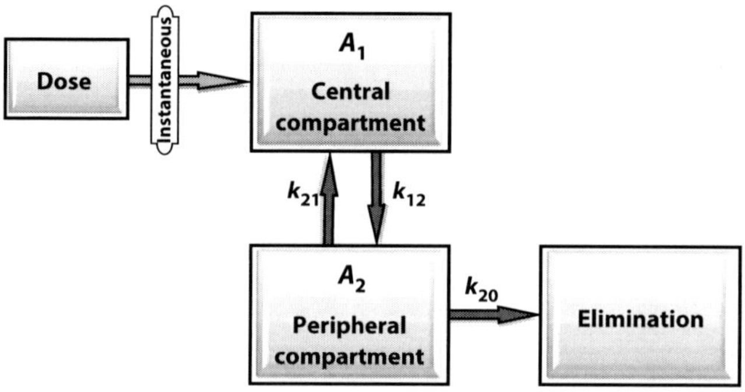

FIGURE 10.9 Schematic diagram of a linear two-compartment model with elimination from the peripheral compartment.

Using the Laplace transform or input disposition functions yields the following integrated equation of the model (Rowland and Riegelman, 1968; Wagner, 1993):

$$Cp = \left[\frac{D(k_{21} + k_{20} - \beta)}{V_1(\alpha - \beta)}\right] e^{-\beta t} + \left[\frac{D(k_{21} + k_{20} - \beta)}{V_1(\alpha - \beta)}\right] e^{-\alpha t} \tag{10.88}$$

Setting

$$a = \frac{D(k_{21} + k_{20} - \beta)}{V_1(\alpha - \beta)} \tag{10.89}$$

$$b = \frac{D(k_{21} + k_{20} - \alpha)}{V_1(\alpha - \beta)} \tag{10.90}$$

The simplified equation is the same as the model with elimination from the central compartment (Equation 10.39), and the miniature rate constants of elimination and distribution are estimated from the following relationships:

$$\alpha + \beta = k_{12} + k_{21} + k_{20} \tag{10.91}$$

$$\alpha\beta = k_{12}k_{20} \tag{10.92}$$

Equations 10.91 and 10.92 are different from Equations 10.50 and 10.51, respectively.

The distribution rate constant k_{12} can be estimated by Equation 10.93, if and only if the value of k_{20} is known:

$$k_{12} = \frac{\alpha\beta}{k_{20}} \tag{10.93}$$

A more practical approach is using the following relationships:

$$\phi = \frac{a\beta + b\alpha}{a + b} = k_{21} + k_{20} \tag{10.94}$$

Therefore,

$$k_{12} = \alpha + \beta - \phi \tag{10.95}$$

$$k_{20} = \frac{\alpha\beta}{k_{12}}, \text{ and} \tag{10.96}$$

$$k_{21} = \phi - k_{20} \tag{10.97}$$

The volume of distribution of the central compartment V_1, the overall volume of distribution of the compound, and the total body clearance is estimated the same as the model discussed in

Section 10.3 (Equations 10.42, 10.64, 10.67). If the elimination from the peripheral compartment is by means of metabolism only, the metabolic clearance can be estimated as

$$Cl_m = V_2 \times k_{20} \tag{10.98}$$

10.5 LINEAR THREE-COMPARTMENT OPEN MODEL WITH BOLUS INJECTION AND ELIMINATION FROM CENTRAL COMPARTMENT

The three-compartment model is often selected to define the disposition of compounds that, in addition to distribution and uptake by the central and peripheral compartments, are taken up preferentially by an organ or tissue with a longer residence time that is kinetically a distinguishable pool. The preferential uptake may or may not be obvious and predictable biologically, but it can be detected through the analysis of experimental data and curve fitting process. Furthermore, the third compartment may represent a deep compartment associated with different tissues but with similar functioning (Kaplan et al., 1973; Wagner, 1993).

A typical three-compartment model is depicted in Figure 10.10; its differential equation and the integrated equations, determined by the input-disposition functions, are as follows:

1. Central compartment (A_1) dispositional rate (i.e., rate of amount change):

$$\frac{dA_1}{dt} = k_{21}A_2 + k_{31}A_3 - k_{12}A_1 - k_{13}A_1 - k_{10}A_1 \tag{10.99}$$

2. Rate of change in the peripheral compartment (A_2):

$$\frac{dA_2}{dt} = k_{12}A_1 - k_{21}A_2 \tag{10.100}$$

3. Rate of quantity change in the peripheral compartment (A_3):

$$\frac{dA_3}{dt} = k_{13}A_1 - k_{31}A_3 \tag{10.101}$$

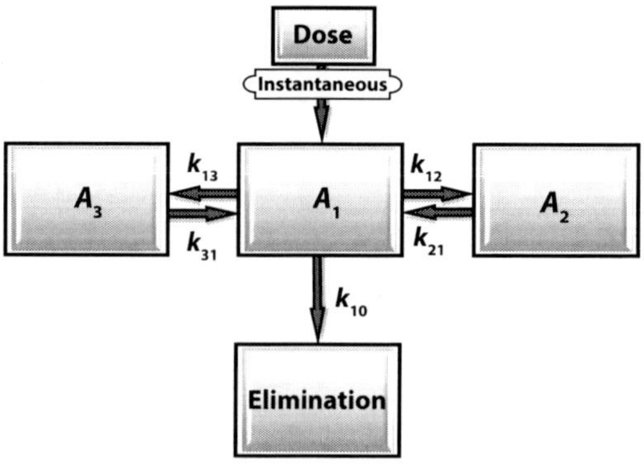

FIGURE 10.10 Schematic diagram of a linear three-compartment open model with elimination from the central compartment.

4. The integrated equation of the central compartment:

$$Cp = \frac{D(k_{21} - \alpha)(k_{31} - \alpha)}{V_1(\beta - \alpha)(\gamma - \alpha)} e^{-\alpha t} + \frac{D(k_{21} - \beta)(k_{31} - \beta)}{V_1(\alpha - \beta)(\gamma - \beta)} e^{-\beta t} + \frac{D(k_{21} - \gamma)(k_{31} - \gamma)}{V_1(\alpha - \gamma)(\beta - \gamma)} e^{-\gamma t} \quad (10.102)$$

By simplifying the coefficient of the exponential terms as a, b and c (Equations 10.103 through 10.105) and considering ($\alpha > \beta > \gamma$), Equation 10.102 can be written as Equation 10.106 with the following designations:

$$a = \frac{D(k_{21} - \alpha)(k_{31} - \alpha)}{V_1(\beta - \alpha)(\gamma - \alpha)} \quad (10.103)$$

$$b = \frac{D(k_{21} - \beta)(k_{31} - \beta)}{V_1(\alpha - \beta)(\gamma - \beta)} \quad (10.104)$$

$$c = \frac{D(k_{21} - \gamma)(k_{31} - \gamma)}{V_1(\alpha - \gamma)(\beta - \gamma)} \quad (10.105)$$

$$Cp = ae^{-\alpha t} + be^{-\beta t} + ce^{-\gamma t} \quad (10.106)$$

Therefore, the initial plasma concentration of a compound that follows the three-compartment model can be estimated by Equation 10.107 and the volume of distribution of the central compartment by Equation 10.108:

$$Cp^0 = a + b + c \quad (10.107)$$

$$V_1 = \frac{\text{Dose}}{a + b + c} \quad (10.108)$$

The hybrid rate constants of α, β, and γ correspond to the following combinations of the distribution and elimination rate miniature rate constants.

$$\alpha + \beta + \gamma = k_{10} + k_{12} + k_{13} + k_{21} + k_{31} \quad (10.109)$$

$$\alpha\beta + \alpha\gamma + \beta\gamma = k_{10}k_{21} + k_{13}k_{21} + k_{10}k_{21} + k_{21}k_{31} + k_{12}k_{31} \quad (10.110)$$

$$\alpha\beta\gamma = k_{10}k_{21}k_{31} \quad (10.111)$$

Therefore, the rate constants of distribution and elimination can be estimated as

- The transfer rate constant from peripheral compartment (A_2) to central compartment:

$$k_{21} = \beta + \frac{b(\gamma - \beta)(\beta - \alpha)}{(\beta - k_{13})(a + b + c)} \quad (10.112)$$

- The transfer rate constant from peripheral compartment (A_3) to central compartment:

$$k_{31} = \alpha + \frac{c(\gamma - \alpha)(\beta - \alpha)}{(k_{31} - \alpha)(a + b + c)} \quad (10.113)$$

- The overall elimination rate constant from central compartment:

$$k_{10} = \frac{\alpha\beta\gamma}{k_{21}k_{31}} \tag{10.114}$$

- The transfer rate constant from central compartment to peripheral compartment (A_2):

$$k_{12} = (\beta\gamma + \alpha\beta + \alpha\gamma) - k_{21}(\alpha + \beta + \gamma) - k_{10}k_{31} + \left(\frac{k_{21}^2}{k_{31} - k_{21}}\right) \tag{10.115}$$

- The transfer rate constant from central compartment to peripheral compartment (A_3)

$$k_{13} = \alpha + \beta + \gamma - (k_{10} - k_{12} + k_{21} + k_{31}) \tag{10.116}$$

The total body clearance is estimated according to the model independent Equation 10.65.

10.6 LINEAR THREE-COMPARTMENT OPEN MODEL WITH ELIMINATION FROM PERIPHERAL COMPARTMENT

The model is depicted in Figure 10.11. The differential equations of the model are

$$\frac{dA_1}{dt} = k_{21}A_2 + k_{31}A_3 - k_{13}A_1 - k_{12}A_1 \tag{10.117}$$

$$\frac{dA_2}{dt} = k_{12}A_1 - k_{21}A_2 - k_{20}A_2 \tag{10.118}$$

$$\frac{dA_3}{dt} = k_{13}A_1 - k_{31}A_3 \tag{10.119}$$

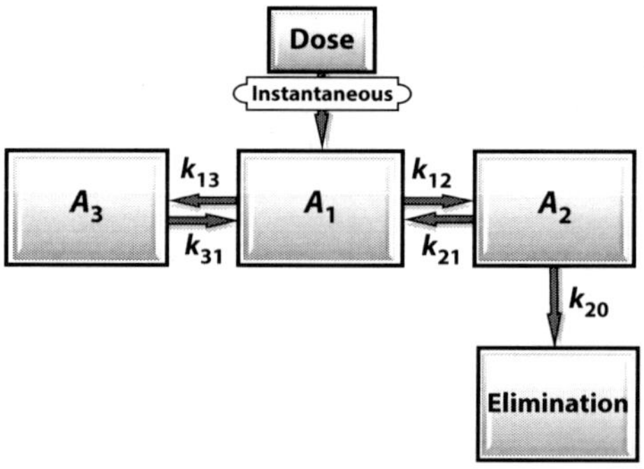

FIGURE 10.11 Schematic diagram of a linear three-compartment open model with elimination from a peripheral compartment.

The integrated equation by the use of input-disposition function and partial fraction theorem is

$$Cp = \frac{D(k_{20} + k_{21} - \alpha)(k_{31} - \alpha)}{V_1(\beta - \alpha)(\gamma - \alpha)} e^{-\alpha t} + \frac{D(k_{20} - k_{21} - \beta)(k_{31} - \beta)}{V_1(\alpha - \beta)(\gamma - \beta)} e^{-\beta t}$$
$$+ \frac{D(k_{20} - k_{21} - \gamma)(k_{31} - \gamma)}{V_1(\beta - \gamma)(\alpha - \gamma)} e^{-\gamma t} \quad (10.120)$$

The coefficient of exponential terms is different from the previous three-compartment model with elimination from the central compartment and can be abridged as

$$a = \frac{D(k_{20} + k_{21} - \alpha)(k_{31} - \alpha)}{V_1(\beta - \alpha)(\gamma - \alpha)} \quad (10.121)$$

$$b = \frac{D(k_{21} - k_{21} - \beta)(k_{31} - \beta)}{V_1(\alpha - \beta)(\gamma - \beta)} \quad (10.122)$$

$$c = \frac{D(k_{20} - k_{21} - \gamma)(k_{31} - \gamma)}{V_1(\beta - \gamma)(\alpha - \gamma)} \quad (10.123)$$

Therefore,

$$Cp = ae^{\alpha t} + be^{\beta t} + ce^{\gamma t} \quad (10.124)$$

The general equation is the same as the other three compartment models. However, the makeup of the coefficients and the formulas for estimation of miniature rate constants of distribution and elimination are different (Nagashima et al., 1968; Wagner, 1993). The development of formulas for estimation of transfer and elimination rate constants is more complex than the previous model. One approach is setting the following substitutions for the sake of brevity of mathematical manipulations of the constants and parameters, assuming ($\alpha > \beta > \gamma$):

$$\Phi_1 = \alpha + \beta + \gamma = k_{12} + k_{13} + k_{20} + k_{21} + k_{31} \quad (10.125)$$

$$\Phi_2 = k_{12}(k_{20} + k_{31}) + k_{20}(k_{13} + k_{31}) + k_{21}(k_{13} + k_{31}) \quad (10.126)$$

$$\Phi_3 = k_{12}k_{20}k_{31} \quad (10.127)$$

Normalizing the coefficients of the exponential terms with respect to the initial plasma concentration yields

$$X_1 = \frac{a}{a + b + c} = \frac{a}{Cp^0} \quad (10.128)$$

$$X_2 = \frac{b}{Cp^0} \quad (10.129)$$

$$X_3 = \frac{c}{Cp^0} \quad (10.130)$$

Setting the following secondary substitutions yields

$$m_1 = k_{12} + k_{13} = \Phi_1 - m_2 - k_{31} \tag{10.131}$$

$$m_2 = k_{21} + k_{20} = (P + S)/2 \tag{10.132}$$

$$m_3 = k_{31} = (P + S)/2 \tag{10.133}$$

$$P = m_2 + m_3 = X_1(\alpha - \gamma) + X_2(\alpha - \beta) + \gamma + \beta \tag{10.134}$$

$$S = m_2 m_3 = \left[\left\{ X_1(\alpha - \gamma) + X_2(\alpha - \beta) + \gamma - \beta \right\}^2 + 4X_2(\beta - \gamma)(\alpha - \beta) \right]^{1/2} \tag{10.135}$$

Therefore, the transfer rate constant from the central compartment to the peripheral compartment A_3 is

$$k_{13} = \frac{\Phi_3 - m_3\left(\Phi_2 - m_1 m_3 - m_2 m_3\right)}{m_3\left(m_3 - m_2\right)} \tag{10.136}$$

The transfer rate constant from the central compartment to the peripheral compartment A_2 and vice versa are

$$k_{12} = m_1 - k_{13} \tag{10.137}$$

$$k_{21} = \frac{m_1 m_2 + m_2 m_3 + m_3 m_1 - \Phi_2 - k_{13} m_3}{k_{12}} \tag{10.138}$$

The overall elimination rate constant can be estimated by

$$k_{20} = m_2 - k_{21} \tag{10.139}$$

The transfer rate constant from (A_3) to (A_1) as

$$k_{31} = m_3 = (P + S)/2 \tag{10.140}$$

Selection of multicompartmental models for any compound is mostly achieved through the use of curve filling and related goodness of fit based on statistical criteria such as Akaike Information Criterion (AIC) (Akaike, 1974) and other tests, and discrepancy between the observed and calculated parameters and constant, which are used for identifying the most appropriate model and accurate assessment of the PK/TK parameters and constants (see Appendix A, Comparing Models). A set of experimental data may be defined by more than one mathematical relationship. Thus, the uniqueness of the solution should also be confirmed. Commercially available specialized software for PK/TK analysis have these criteria in their curve fitting and numerical analysis of data.

10.7 COMPARING MODELS

The most commonly implemented model selection approaches are the Akaike Information Criterion (AIC) (Akaike, 1974) and Schwarz criterion, also known as Bayesian Information Criterion (BIC) (Schwarz, 1978). These two methods enable one to simultaneously compare multiple models, assess the selection uncertainty, and allow for the realistic estimation of the model's parameters and constants. In the absence of a perfect model, both AIC and BIC methods minimize the errors in estimation of parameters and constants by choosing the most realistic and appropriate PK/TK model. In comparing several models for a given set of data, the model with lowest AIC or BIC is considered the most relevant model among the models chosen to describe the data set. More discussion on AIC and BIC criteria and related equations are included in Appendix A, Section A.7.

REFERENCES

Akaike, H. 1974. A new look at the statistical model identification, *IEEE T Automat Contr* 19(6):716–23.

Benet, L. Z. 1972. General treatment of linear mammillary models with elimination from any compartment as used in pharmacokinetics. *J Pharm Sci* 61(4):536–41.

Boroujerdi, M. 2002. *Pharmacokinetics: Principles and Applications*, New York: McGraw Hill—Medical Publishing Division.

Chiou, W. L. 1981. The physiological significance of the apparent volume of distribution, Vd_{area} or Vd_{β}, in pharmacokinetic studies. *Res Commun Chem Pathol Pharmacol* 33(3):499–508.

Gibaldi, M., Perrier, D. 1982. *Pharmacokinetics*, 2nd Edition. New York: Marcel Dekker, Inc.

Gibaldi, M., Nagashima, R., Levy, G. 1969. Relationship between drug concentration in plasma or serum and amount of drug in the body. *J Pharm Sci* 58:193–7.

Kaplan, S. A., Jack, M. L., Alexander, K., Weinfeld, R. E. 1973. Pharmacokinetic profile of diazepam in man following single intravenous and oral and chronic oral administration. *J Pharm Sci* 62:1789–96.

Nagashima, R. N., Levy, G., O'Reilly, R. A. 1968. Comparative pharmacokinetics of coumarin anticoagulants IV. Application of a three compartment model to the analysis of dose-dependent kinetics of bishydroxycoumarin elimination. *J Pharm Sci* 57:1888–95.

Rescigno, A., Seger, G. 1966. *Drug and Tracer Kinetics*, Waltham, MA: Blaisdell Publishing Co.

Riegleman, S., Loo, J., Rowland, M. 1968. Shortcomings in pharmacokinetic analysis by conceiving the body to exhibit properties of a single compartment. *J Pharm Sci* 57:117–123.

Rowland, M., Riegelman, S. 1968. Pharmacokinetics of acetylsalicylic acid after intravenous administration in man. *J Pharm Sci* 57:1313–19.

Schwarz, G. E. 1978. Estimating the dimension of a model. *Ann Stat* 6(2): 461–4.

Wagner, J. G. 1975. *Fundamentals of Clinical Pharmacokinetics*, 1st Edition. Hamilton, IL. Drug Intelligence Publication, 57.

Wagner, J. G. 1993 *Pharmacokinetics for the Pharmaceutical Scientist*, Basel, Switzerland. Technomic Publishing Company, Inc.

11 PK/TK Analysis of Continuous Zero-Order Exposure to Xenobiotics (Intravenous Infusion)

11.1 INTRODUCTION

The zero-order input in this chapter refers to the administration of solution with known concentration of a compound into the systemic circulation via intravenous administration over a period of time. Thus, the input is not instantaneous, as was discussed in Chapter 10, and the constant rate of input is considered zero-order kinetics. The specialized infusions through routes of administration such as epidural or intrathecal are not a part of this discussion. Intravenous infusion is used commonly to treat illnesses that require long-term exposure and duration of action, or when the administration of bolus dose can be hazardous to the human subject or experimental animals. Often the long-term exposure to organic solvents that can absorb through inhalation and enter the systemic circulation without impediment can also be considered zero-order input. An intravenous infusion requires that the compound remains stable and soluble in the solution, and tolerable by the subject. The vehicles are usually solutions containing water of highest purity and sterility with dextrose, saline or other specialized additives and/or medications.

The intravenous (IV) infusion of therapeutic agents and fluids has become an important part of treatment in the hospitals, ambulatory infusion centers, home infusion therapies, home tele-infusion and telecare industries, etc. The IV infusion, also known as IV drip, is usually administered by infusion pump, which uses pressure to administer the solution at a constant rate, or smart infusion pump with software capability to prevent medication and/or programming errors.

The intravenous infusion is used to ensure a constant supply of the therapeutic agent into the body. The objective of long-term infusions is to accumulate the compound in the body and achieve a steady-state level within the therapeutic range in order to attain a consistent pharmacological response and extended therapeutic outcome. It is the method of choice for the administration of compounds with a narrow therapeutic range and short half-life. Short-term infusions are used to deliver compounds where their bolus injection can cause side effects such as precipitation or crystallization at the site of administration, speed shock, etc. The objective of the short infusion is not to achieve the steady-state level, but to deliver the compound carefully. An advantage of using intravenous infusion is also to circumvent fluctuations of plasma concentration when the compound is given by multiple dosing administrations. During the long-term infusion, because of achieving the steady-state level, a more accurate assessment of total body clearance can be achieved.

Examples of the application of continuous infusion are

- Total parenteral nutrition (TPN) or peripheral parenteral nutrition (PPN) therapy
- Antineoplastic therapy
- Pain management

- Chelation therapy (e.g., removal of high levels of aluminum from plasma of patient on dialysis)
- Chemotherapy with low therapeutic index drugs
- Infectious disease therapy
- Lifetime immunosuppressive therapy
- Hydration therapy
- Adjunctive therapies (such as simultaneous intravenous therapy with antiemetics, antidiarrheals, anticonvulsant, and anti-inflammatories in AIDS patients to control the symptoms of disease and/or the side effects of medications)
- Anesthesia
- Programmable implantable infusion pumps

Some of the relevant PK/TK models are discussed next.

11.2 COMPARTMENTAL ANALYSIS

11.2.1 Linear One-Compartment Model with Zero-Order Input and First-Order Elimination

The assumptions of the model are the same as discussed in Chapter 10. The difference is the administration of the dose, which is through a constant zero-order rate of input with units of mass/time, or volume/time for solutions with known concentration (Figure 11.1). The first-order rate of elimination depends on the amount or concentration of the infused compound in the body at a given time during infusion. Thus, at the beginning when the amount of the therapeutic agent in the compartment is low, the rate of elimination is less than the zero-order rate of input, that is, $KA < k_0$. As the infusion continues, the compound gradually accumulates in the body and the rate of elimination increases accordingly and eventually equals the zero-order rate of input ($KA = k_0$); that is, the amount introduced into the body per unit of time is equal to the total amount eliminated from the body by all routes of elimination per unit of time. Therefore, the amount or concentration of drug in the body remains constant. This plateau concentration is the steady-state plasma level and represents the maximum accumulation of drug in the body that can be achieved with the selected zero-order rate of input (Figure 11.2). In long-term infusions, the objective is to achieve the steady-state level within the therapeutic range for achieving a consistent pharmacodynamic outcome.

The steady-state level remains constant as long as the same rate of infusion is continued.

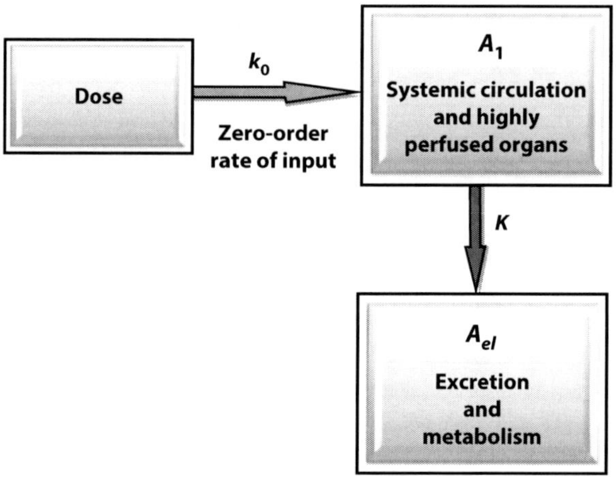

FIGURE 11.1 Schematic diagram of one-compartment model with zero-order input and first-order output.

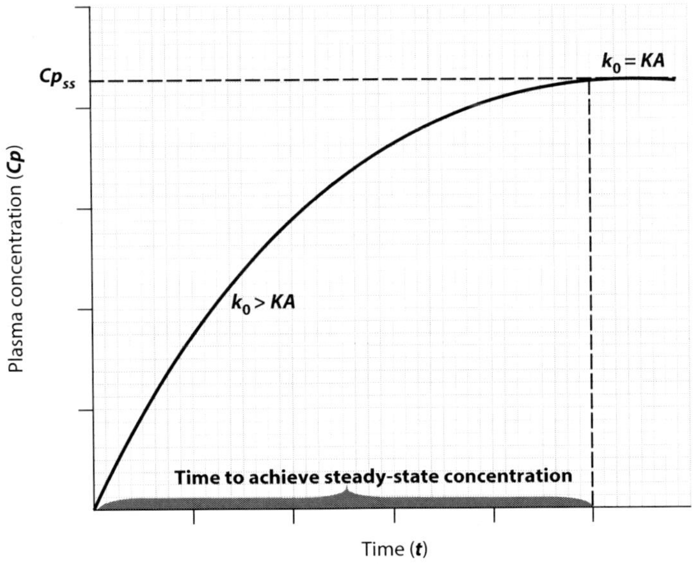

FIGURE 11.2 Plasma concentration–time profile of a zero-order input of a xenobiotic over an interval long enough to attain a steady-state concentration and the zero-order rate of input becomes equal to the first-order rate of elimination.

The following differential equation defines the rate of change (dA/dt) or rate of accumulation of the compound in the body:

$$\left(\frac{dA}{dt}\right)_t = k_0 - KA_t \tag{11.1}$$

The rate of elimination is

$$KA_t = Cp_t VdK = Cp_t Cl_T \tag{11.2}$$

Therefore,

$$\left(\frac{dA}{dt}\right)_t = k_0 - (Cl_T \times Cp_t) \tag{11.3}$$

where A_t and Cp_t are the amount in the body and plasma concentration at time t, respectively; Cl_T is the total body clearance.

At steady state,

$$\frac{dA}{dt} = 0$$
$$k_0 = KA_{ss} = Cl_T \times Cp_{ss} \tag{11.4}$$

$$\text{Therefore, } A_{ss} = \frac{k_0}{K} \text{ and} \tag{11.5}$$

$$Cp_{ss} = \frac{k_0}{Cl_T} \tag{11.6}$$

Integration of Equation 6.1 using the Laplace transform yields Equations 11.7 and 11.8 in terms of amount and concentration, respectively:

$$A_t = \frac{k_0}{K}\left(1 - e^{-Kt}\right) \quad (11.7)$$

$$Cp_t = \frac{k_0}{K \times Vd}\left(1 - e^{-Kt}\right) \quad (11.8)$$

As infusion proceeds ($t \Rightarrow \infty$, $e^{-Kt} \Rightarrow 0, Cp_t \Rightarrow Cp_{ss}$) Equation 11.8 changes to Equation 11.6. Substitution of Equation 11.6 into Equation 11.8 yields

$$Cp_t = Cp_{ss}\left(1 - e^{-Kt}\right) \Rightarrow \log\frac{Cp_{ss} - Cp_t}{Cp_{SS}} = -\frac{Kt}{2.303} \quad (11.9)$$

$$Cp_t = Cp_{ss}\left(1 - (1/2)^{\frac{t}{T_{1/2}}}\right) \quad (11.10)$$

Equation 11.6 is also used to estimate the total body clearance more accurately (Chiou et al., 1978; Rowland and Tozer, 1994)

$$Cl_T = \frac{k_0}{Cp_{ss}} \quad (11.11)$$

where k_0 is the infusion rate, that is, the total dose divided by the interval of infusion, or is equal to the desired steady-state concentration multiplied by the total body clearance of the compound:

$$k_0 = Cp_{ss} \times Cl_T \quad (11.12)$$

11.2.1.1 Required Time for Achieving Steady-State Plasma Concentration

According to Equation 11.10, if the compound is infused for one half-life ($t_{infusion} = T_{1/2}$) the plasma concentration will be at 50% of the targeted steady-state level, if the time of infusion is twice the half-life ($t = 2T_{1/2}$), the plasma concentration will be at 75% of the steady-state level. At $t = 3.3T_{1/2}$, $Cp = 90\%Cp_{ss}$; at $t = 4.3T_{1/2}$, $Cp = 95\%Cp_{ss}$; and at $t = 6.6T_{1/2}$, $Cp = 99\%Cp_{ss}$. The inverse of this concept is also useful, for example, if the ratio of Cp_t/Cp_{ss} is 0.75, the time of infusion is twice the half-life, that is,

$$f_{ss} = \frac{Cp_t}{Cp_{ss}} = \left(1 - e^{-Kt}\right) \quad (11.13)$$

In Equation 11.13, 1 represents 100% of the steady-state level, e^{-Kt} is the fraction left to reach to the steady state level, and $1 - e^{-Kt}$ (i.e., f_{ss}) is the fraction of steady state level (Figure 11.3).

In clinical practice after infusion of a therapeutic agent for about four to five half-lives it is assumed that the steady-state level has been achieved. This is a practical assumption because 95%–97% of a steady state level is close enough to be considered 100%. Mathematically, however, it takes about seven half-lives for Equation 11.13 to achieve approximately 99%–100% of a targeted steady-state level.

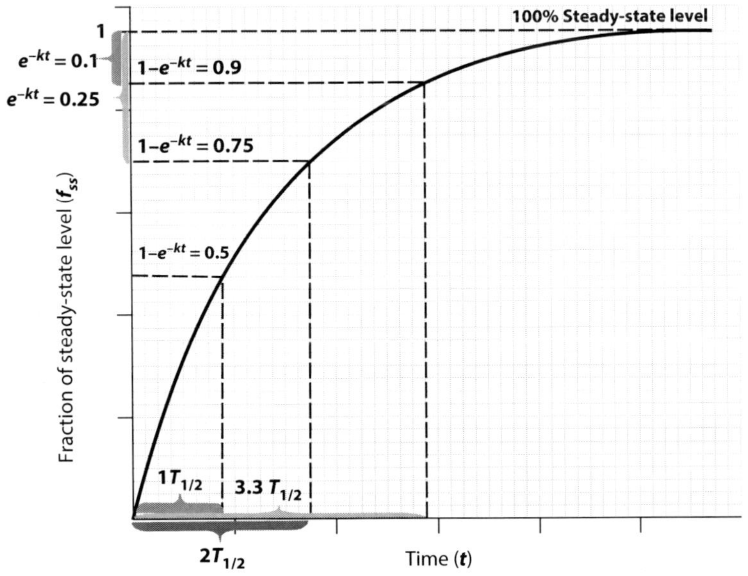

FIGURE 11.3 Plot of fraction of steady-state concentration versus time according to Equation 11.13.

If the duration of an infusion is less than seven half-lives of a compound, the plasma concentration at any time during the infusion is estimated by using Equation 11.9 or Equation 11.10. When the steady-state level is achieved, its magnitude depends on the rate of infusion; however, the time required to achieve any level of a steady-state with any rate of infusion depends only on the half-life of the therapeutic agent. In other words, by increasing or decreasing the rate of infusion, a different steady-state level is achieved, but it takes the same length of time to attain the steady-state levels (Figure 11.4).

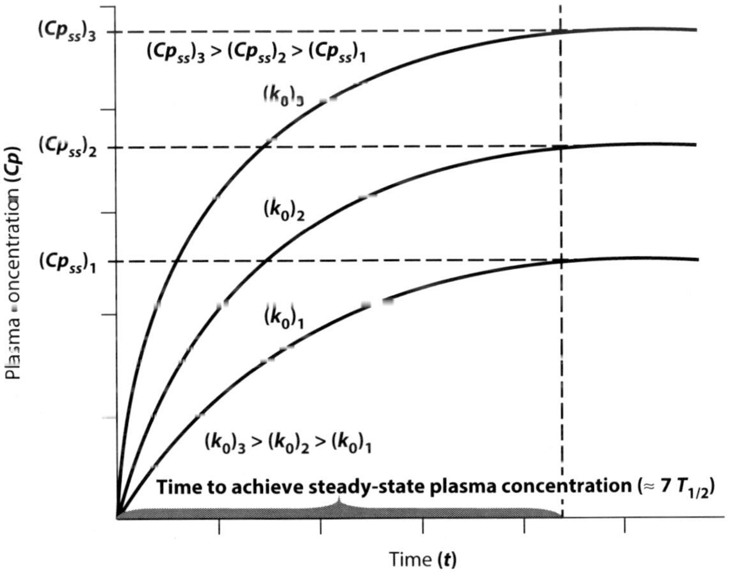

FIGURE 11.4 The steady-state plasma concentration is directly proportional to the zero-order rate of infusion; a higher rate of input results in a higher steady-state level; however, the time required for attaining the concentration is independent of the rate of input and remains the same for different input rates.

The following equation, derived from Equation 11.11, defines the relationship between a steady-state level and half-life more clearly:

$$Cp_{ss} = \frac{k_0}{K \times Vd} = \frac{1.44 \times T_{1/2} \times k_0}{Vd} \tag{11.14}$$

In Equation 11.14, as K or $T_{1/2}$ and Vd are constant, the steady-state plasma concentration will only be proportional to k_0. In disease states such as renal impairment or hepatic failure, however, when the half-life is longer than normal or K is smaller, if the treatment of a patient is continued at a rate that is recommended for patients with normal renal and hepatic functions, a higher steady-state level is achieved, which may require dosage adjustment. The time to reach to that steady state is directly proportional to the half-life of the therapeutic agent in the patient with renal or hepatic failure.

11.2.1.2 Administration of Loading Dose Followed by Zero-Order Infusion (Simultaneous Intravenous Bolus and Infusion, or Consecutive Fast and Slow Infusions)

One approach in achieving the steady-state level of a xenobiotic without delay is to give a loading dose, either as intravenous bolus dose (Wagner, 1974; Gibaldi and Perrier, 1982), or as fast intravenous infusion.

To determine the bolus loading dose, the desired steady-state level is set equal to the initial plasma concentration of the bolus dose, that is, $Cp_{ss} = Cp^0$. The loading dose would then be equal to:

$$D_L = Vd \times Cp_{ss} \tag{11.15}$$

The combination of the intravenous bolus and infusion can be presented as

$$Cp_{total} = Cp_{ss}e^{-Kt} + Cp_{ss}(1 - e^{-Kt}) = Cp_{ss} \tag{11.16}$$

Therefore, when a drug follows a one-compartment model, this combination provides an immediate steady-state level (Figure 11.5).

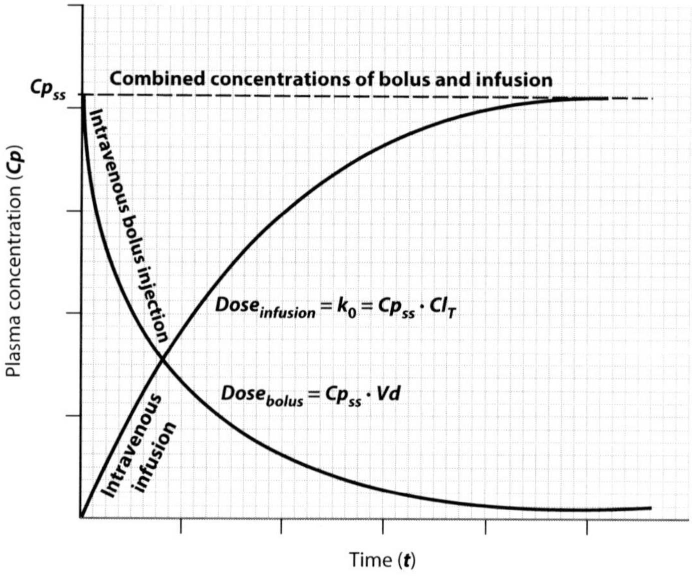

FIGURE 11.5 Plasma concentration–time profile of simultaneous intravenous bolus injection and zero-order input to achieve immediate steady-state concentration (the dashed line).

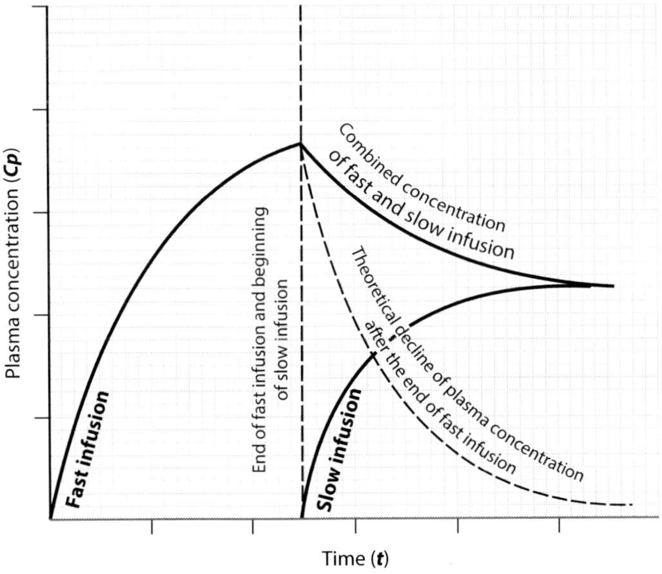

FIGURE 11.6 Theoretical profile of plasma concentration of a compound following consecutive administration of fast and slow zero-order input rate.

There are circumstances that giving a bolus-loading dose for reasons such as speed shock or precipitation, and/or crystallization of drug at the site of injection, etc., may not be appropriate. Under such circumstances the loading dose is administered by a fast infusion to achieve a desired accumulation of drug in the body and then followed by a slower infusion to maintain the plasma level. Because any change in the rate of infusion will have direct influence on the plasma concentration of the xenobiotic, terminating one infusion at one rate and initiating a second infusion with a different rate creates a temporary fluctuation in plasma concentration (Figure 11.6) (Boroujerdi, 2002).

To administer the loading dose as fast infusion, the following two parameters should be established initially:

- The time of fast infusion (t_{fast})
- Target concentration (Cp_t)

The apparent volume of distribution and the overall elimination rate constants are usually known for a compound. Thus, the fast infusion rate can be determined as

$$Cp_t = \frac{(k_0)_{fast}}{Cl_T}\left(1 - e^{-Kt_{fast}}\right) \quad (11.17)$$

$$(k_0)_{fast} = \frac{Cp \times Cl_T}{1 - e^{-Kt_{fast}}} = \frac{(\text{Rate of elimination})_{t_{fast}}}{(f_{ss})_{t_{fast}}} \quad (11.18)$$

The total plasma concentration following the fast and slow infusions without including the fluctuation of plasma is

$$(Cp_{total})_t = \frac{(k_0)_{fast}}{Cl_T}\left(1 - e^{-Kt_{fast}}\right) + \frac{(k_0)_{slow}}{Cl_T}\left(1 - e^{-Kt_{slow}}\right) \quad (11.19)$$

Including the fluctuation of plasma in the equation yields

$$\left(Cp_{total}\right)_t = \frac{\left(k_0\right)_{fast}}{Cl_T}\left(1 - e^{-Kt_{fast}}\right)e^{-Kt_{slow}} + \frac{\left(k_0\right)_{slow}}{Cl_T}\left(1 - e^{-Kt_{slow}}\right) \quad (11.20)$$

As t_{slow} increases, $e^{-Kt_{slow}}$ approaches zero and $(Cp_{total})_t$ will become the steady-state concentration of the slow infusion, that is,

$$\left(Cp_{total}\right)_{ss} = \frac{k_{0_{slow}}}{Cl_T} \quad (11.21)$$

11.2.1.3 Plasma Concentration after Termination of Infusion

After termination of infusion, whether the steady-state plasma level is achieved or not, the drug concentration of plasma, because of continuous first-order elimination process, declines exponentially similar to the decline of plasma concentration after bolus injection (Loo and Riegelman, 1970). The appropriate equations for estimation of plasma concentration after the end of infusion are (Gibaldi and Perrier, 1982; DeVane and Jusko, 1986)

$$Cp_{t'} = \left[Cp_{ss}(1 - e^{-Kt_{inf}})\right]e^{-Kt'} \quad (11.22)$$

$$Cp_{t'} = Cp_{end}e^{-Kt'} \quad (11.23)$$

where (t') is any time after the termination of infusion and t_{inf} is the period of infusion (Figure 11.7).

Similarly, if the infusion is discontinued after the steady-state level is achieved, the plasma concentration at any time t' after the termination of infusion can be calculated as

$$Cp_{t'} = Cp_{ss}e^{-Kt'} \quad (11.24)$$

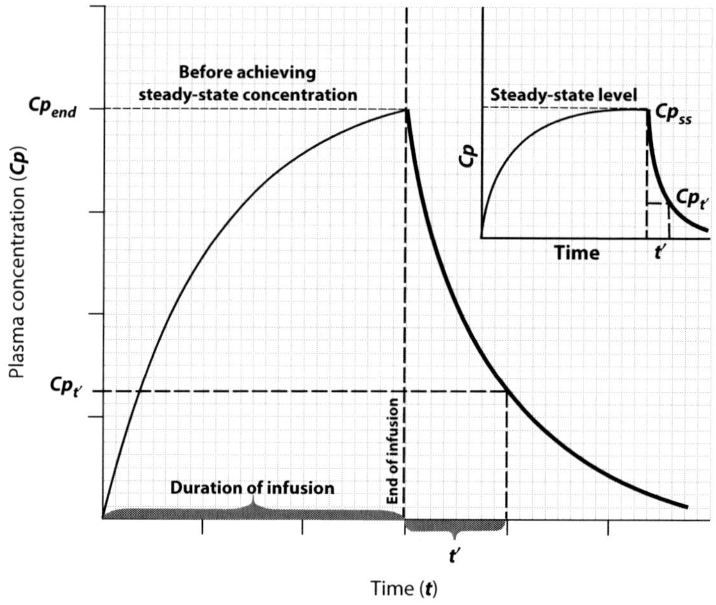

FIGURE 11.7 Profile of plasma concentration after stopping the infusion, before and after achieving the steady-state level.

11.2.1.4 Estimation of Duration of Action

Duration of action of a compound during an infusion is the time that the plasma concentration is above the minimum therapeutic range. If an infusion is initiated without a loading dose, the initial period when the concentration is rising to reach the minimum effective concentration, that is, time to the onset of action (t_{onset}), should be subtracted from the duration of infusion. In addition, after termination of infusion the plasma concentration, although declining, may remain above the minimum effective concentration for some time. This postinfusion period (t_{post}) should be added to the duration of action of infusion. The following stepwise calculations may be used to determine the duration of action of a therapeutic agent during infusion:

1. The minimum effective concentration in terms of fraction of steady-state plasma concentration is

$$\left(f_{ss}\right)_{MEC} = \frac{Cp_{MEC}}{Cp_{ss}} = 1 - e^{-Kt_{onset}} \tag{11.25}$$

2. The time to onset of action using Equation 7.8 (Chapter 7) is

$$t_{onset} = -\frac{\ln\left(1 - f_{ss}\right)}{K} \tag{11.26}$$

3. Subtraction of t_{onset} from the time of infusion yields the interval in which the concentration is above the minimum effective level while the infusion is running

$$\left(t_d\right)_{infusion} = t_{infusion} - t_{onset} \tag{11.27}$$

4. The time that, after the discontinuation of infusion, the plasma concentration remains above the minimum effective concentration is

$$t_{post} = \frac{2.303}{K} \log \frac{Cp_{ss}}{Cp_{MEC}} \tag{11.28}$$

or,

$$t_{post} = \frac{2.303}{K} \log \frac{Cp_{end}}{Cp_{MEC}} \tag{11.29}$$

5. The actual duration of action is

$$t_d = \left(t_d\right)_{infusion} + t_{post} \tag{11.30}$$

For infusions with an intravenous bolus-loading dose, the time to onset of action is zero ($t_{onset} = 0$).

11.2.2 Linear Two-Compartment Model with Zero-Order Input and First-Order Disposition

The description and assumptions of the two-compartment model are the same as discussed in Chapter 10. Briefly, the drug after entry into the systemic circulation distributes between central and peripheral compartments and simultaneously eliminates from the body (Krüger-Thiemer,

1968; O'Reilly et al., 1971; Gibaldi and Perrier, 1982; Wagner, 1993). All biological processes are governed by first-order kinetics, and pharmacokinetics of a drug is considered dose-independent and linear.

11.2.2.1 Equations of the Two-Compartment Model with Zero-Order Input into the Central Compartment and First-Order Output from the Central Compartment

As discussed in Chapters 9 and 10, the organs of elimination, the liver and kidneys, being highly perfused organs, are associated with the central compartment and metabolize and excrete the compound according to the principles of first-order kinetics, and the overall rate constant of elimination from the central compartment is $k_{10} = k_e + k_m$. The rate constants of distribution between the central and peripheral compartment and vice versa are also governed by first-order kinetics and the distribution is a function of concentration or amount in each compartment with different rate constant for the transfers (i.e., $k_{12} \neq k_{21}$).

The rate of change of the amount in the central compartment with zero-order input is

$$\frac{dA_1}{dt} = k_0 - k_{21}A_2 - \left(k_{10}A_1 + k_{12}A_1\right) \tag{11.31}$$

For the peripheral compartment the rate is

$$\frac{dA_2}{dt} = k_{12}A_1 - k_{21}A_2 \tag{11.32}$$

The disposition function of the central compartment is (Benet, 1972)

$$d(s)_1 = \frac{s + E_2}{\left(s + E_1\right)\left(s + E_2\right) - k_{12}k_{21}} = \frac{s + E_2}{(s + \alpha)(s + \beta)} \tag{11.33}$$

where

$E_1 = k_{10} + k_{21}$; $E_2 = k_{21}$; $\alpha + \beta = k_{12} + k_{21} + k_{10}$; and $\alpha\beta = k_{21}k_{10}$

The input function for constant rate of input (Appendix A) is

$$in_s = \frac{k_0\left(1 - e^{-bs}\right)}{s} \tag{11.34}$$

where (b) is the exposure time or the infusion time.

Thus, the Laplace transform and related integrated equation of the amount as a function of time t in the central compartment are

$$a_{(s)_1} = \frac{k_0\left(1 - e^{-bs}\right)\left(s + E_2\right)}{s(s + \alpha)(s + \beta)} \tag{11.35}$$

$$\left(A_1\right)_t = \frac{k_0\left(1 - e^{\alpha b}\right)\left(k_{21} - \alpha\right)}{\alpha(\beta - \alpha)}e^{-\alpha t} + \frac{k_0\left(\beta - k_{21}\right)\left(1 - e^{\beta b}\right)}{\beta(\alpha - \beta)}e^{-\beta t} \tag{11.36}$$

In the above equations, α and β represent the hybrid first-order rate constants and (b) is the time of infusion.

Dividing Equation 11.36 by the volume of distribution of the central compartment yields the equation of xenobiotic concentration in the central compartment as a function of the time:

$$\left(C_1\right)_t = \frac{k_0\left(1-e^{\alpha b}\right)\left(k_{21}-\alpha\right)}{V_1\alpha(\beta-\alpha)}e^{-\alpha t} + \frac{k_0\left(\beta-k_{21}\right)\left(1-e^{\beta b}\right)}{V_1\beta(\alpha-\beta)}e^{-\beta t} \tag{11.37}$$

As defined earlier, b is the time of infusion. Thus, during the infusion $b = t$ and the equation is modified to

$$\left(C_1\right)_t = \frac{k_0\left(k_{21}-\alpha\right)}{V_1\alpha(\beta-\alpha)}\left(e^{-\alpha t}-1\right) + \frac{k_0\left(\beta-k_{21}\right)}{V_1\beta(\alpha-\beta)}\left(e^{-\beta t}-1\right) \tag{11.38}$$

A different version of the Equation 11.38, given that $\alpha\beta = k_{21}k_{10}$, is Equation 11.39, which defines the plasma concentration during the infusion before achieving the steady state

$$\left(C_1\right)_t = \frac{k_0}{V_1 k_{10}}\left(1 + \frac{\beta-k_{10}}{\beta-\alpha}e^{-\alpha t} + \frac{k_{10}-\alpha}{\alpha-\beta}e^{-\beta t}\right) \tag{11.39}$$

When the time of infusion approaches seven biological half-lives (four to five half-lives clinically), $e^{-\alpha t}$ and $e^{-\beta t} \Rightarrow 0$ and Equation 11.39 changes to the steady-state plasma concentration:

$$Cp_{ss} = \frac{k_0}{V_1 k_{10}} \tag{11.40}$$

Since $(Vd)_{area} = \dfrac{Dose}{\beta \times AUC}$ and $AUC = \dfrac{Cp^0}{k_{10}}$, Equation 11.40 can also be presented as

$$Cp_{ss} = \frac{k_0}{(Vd)_{area} \times \beta} \tag{11.41}$$

Equation 11.37 is used to determine plasma concentrations after termination of infusion, that is,

$$(Cp)_{t'} = a'e^{-\alpha t'} + b'e^{-\beta t'} \tag{11.42}$$

where

$$a' = \frac{k_0\left(1-e^{\alpha b}\right)\left(k_{21}-\alpha\right)}{V_1\alpha(\beta-\alpha)} \tag{11.43}$$

$$b' = \frac{k_0\left(\beta-k_{21}\right)\left(1-e^{\beta b}\right)}{V_1\beta(\alpha-\beta)} \tag{11.44}$$

The plasma concentration–time profile of an intravenously infused drug that follows the two-compartment model is the same as Figure 11.2. The rate of input is higher than the rate of elimination ($k_0 > k_{10}A_1$) before achieving the steady state, and during the steady-state level $k_0 = Cp_{ss} \times (Vd)_{area} \times \beta$.

11.2.3 Simultaneous Bolus and Intravenous Infusion into the Central Compartment and Elimination from the Central Compartment—Two-Compartment Open Model

As discussed earlier, the immediate steady-state plasma concentration is achieved by simultaneous intravenous administration of bolus and infusion. The combination of bolus and infusion equations is

$$Cp_{total} = \left[\frac{D_{bolus}(\alpha - k_{21})}{V_1(\alpha - \beta)} e^{-\alpha t} + \frac{D_{bolus}(k_{21} - \beta)}{V_1(\alpha - \beta)} e^{-\beta t}\right] + \left[\frac{k_0}{V_1 k_{10}}\left(1 + \frac{\beta - k_{10}}{\beta - \alpha} e^{-\alpha t} + \frac{k_{10} - \alpha}{\alpha - \beta} e^{-\beta t}\right)\right] \tag{11.45}$$

Expanding Equation 11.45, taking the common denominator and simplifying the terms, yields

$$Cp_{total} = \frac{k_0}{V_1 k_{10}} + \frac{\alpha D_{bolus} - k_0}{V_1 k_{10}(\beta - \alpha)} e^{-\alpha t} + \frac{(\beta D_{bolus} - k_0)(\alpha - k_{10})}{V_1 k_{10}(\alpha - \beta)} e^{-\beta t} \tag{11.46}$$

For Cp_{total} to be equal to Cp_{ss}, the last two terms of Equation 11.46 must equal zero. This condition can only be achieved if $e^{-\alpha t}$ and $e^{-\beta t}$ are equal to zero (i.e., αt and βt are large values), or the numerator of each term is equal to zero. The conditions for the numerators to be equal to zero are not achievable because the only way they can be equal to zero is when $\alpha D_{bolus} = \beta D_{bolus} = k_0$, which is contrary to the assumptions of a two compartment model. Thus, the total plasma concentration can be equal to steady state plasma concentration ($Cp_{total} = Cp_{ss} = k_0/V_1 k_{10}$) if the exponential terms are equal to zero. Considering the plasma concentration–time profile of a compound that follows the two-compartment model, it should be expected that the combined administration may exhibit some fluctuation at the beginning of the co-administration, which is often due to the distributional phase of the loading dose (Figure 11.8).

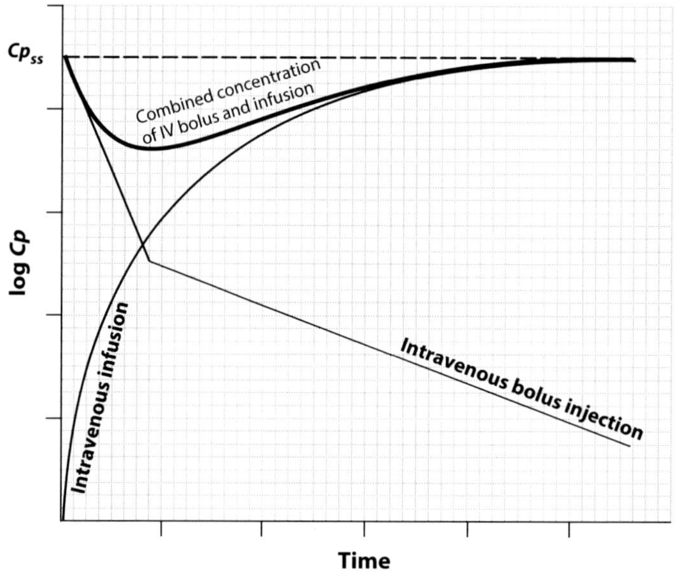

FIGURE 11.8 Profile of plasma concentration following simultaneous administration of an intravenous bolus injection and zero-order infusion of a compound that follows the two-compartment model.

11.2.4 Linear Two-Compartment Model with Two Consecutive Zero-Order Input and First-Order Elimination from the Central Compartment

When the loading dose to achieve a rapid steady-state level is an infusion with a faster rate than the maintenance infusion, the integrated equation of the model is different from combined bolus and infusion. The total amount or concentration in the central compartment during the second infusion is a function of the amount lingering in the body from the first infusion and the amount introduced by the second infusion (Wagner, 1993). During the first infusion, the plasma concentration at time t is estimated by Equation 11.47, which is the same as Equation 11.39 with a different arrangement of parameters and constants:

$$(C_1)_t = \frac{(k_0)_1}{V_1 k_{10}}\left[1 - \frac{k_{10} - \beta}{\alpha - \beta}e^{-\alpha t} - \frac{\alpha - k_{10}}{\alpha - \beta}e^{-\beta t}\right] \tag{11.47}$$

The integrated equation for the total concentration in the central compartment is

$$\begin{aligned} Cp_{total} = \frac{(k_0)_2}{V_1 k_{10}} & \left[\frac{(k_{21} - \alpha)\left((k_0)_2 - \alpha(I_1)_{t_1}\right) - \alpha k_{21}(I_2)_{t_1}}{\alpha(\alpha - \beta)V_1} e^{-\alpha(t_2 - t_1)}\right] \\ & - \left[\frac{(k_{21} - \beta)\left((k_0)_2 - \beta(I_1)_{t_1}\right) - \beta k_{21}(I_2)_{t_1}}{\beta(\alpha \quad \beta)V_1} e^{-\beta(t_2 - t_1)}\right] \end{aligned} \tag{11.48}$$

where

$$(I_1)_{t_1} = (k_0)_1\left[\frac{1}{k_{10}} + \frac{k_{21} - \alpha}{\alpha(\alpha - \beta)}e^{-\alpha t_1} - \frac{k_{21} - \beta}{\beta(\alpha - \beta)}e^{-\beta t_1}\right] \tag{11.49}$$

$$(I_2)_{t_1} = (k_0)_1 k_{12}\left[\frac{1}{k_{21}k_{10}} + \frac{e^{-\alpha t_2}}{\alpha(\alpha - \beta)} - \frac{e^{-\beta t_2}}{\beta(\alpha - \beta)}\right] \tag{11.50}$$

The parameters and constants of Equations 11.48 through 11.50 are: $(k_0)_1$ and $(k_0)_2$ are the first and second zero-order infusion rates; t_1 and t_2 are the lengths of first and second infusion; k_{10} is the first-order rate constant of elimination; k_{21} and k_{12} are first-order distribution rate constants, α and β are the first-order hybrid rate constants of the two-compartment model ($\alpha > \beta$), and V_1 is the volume of the central compartment.

The profile of combined plasma concentration of consecutive fast and slow infusions is depicted in Figure 11.6.

11.2.5 Three-Compartment Model with Zero-Order Input into the Central Compartment and First-Order Elimination from the Central Compartment

The assumptions and characteristics of the model (Figure 11.9) are the same as discussed in Chapter 10.

The differential and integrated equations defining the time course of infused xenobiotic in the central compartment during the infusion are defined in Equations 11.51 through 11.54

$$\frac{dA_1}{dt} = k_0 + k_{12}A_2 + k_{31}A_3 - A_1(k_{10} + k_{12} + k_{13}) \tag{11.51}$$

$$\frac{dA_2}{dt} = k_{12}A_1 - k_{21}A_2 \tag{11.52}$$

$$\frac{dA_3}{dt} = k_{13}A_1 - k_{31}A_3 \tag{11.53}$$

$$(C_1)_t = \frac{k_0}{V_1}\left[\frac{(k_{21}-\alpha)(k_{31}-\alpha)(e^{+\alpha b}-1)}{\alpha(\beta-\alpha)(\gamma-\beta)}e^{-\alpha t} + \frac{(k_{21}-\beta)(k_{31}-\beta)(e^{+\beta b}-1)}{\beta(\alpha-\beta)(\gamma-\beta)}e^{-\beta t} + \frac{(k_{21}-\gamma)(k_{31}-\gamma)(e^{+\gamma b}-1)}{\gamma(\alpha-\gamma)(\beta-\gamma)}e^{-\gamma t}\right] \tag{11.54}$$

where (b) is the length of infusion and plasma concentration is a function of time (t) during the infusion. The descriptions of other constants are the same as described earlier.

11.2.6 Three-Compartment Model with Zero-Order Input into the Central Compartment and First-Order Elimination from the Peripheral Compartment

The rate equations of the model are

$$\frac{dA_1}{dt} = k_0 + k_{21}A_2 + k_{31}A_3 - A_1(k_{12} + k_{13}) \tag{11.55}$$

$$\frac{dA_2}{dt} = k_{12}A_1 - A_2(k_{21} + k_{20}) \tag{11.56}$$

$$\frac{dA_3}{dt} = k_{13}A_1 - k_{31}A_3 \tag{11.57}$$

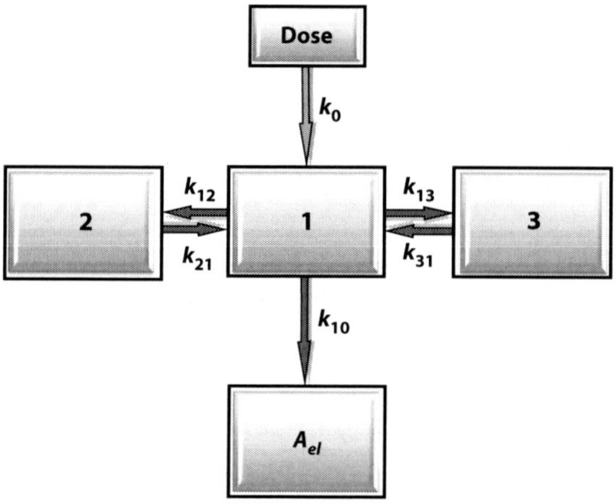

FIGURE 11.9 Diagram of a three-compartment model with zero-order input in the central compartment.

The integrated equation defining the time course of the infused xenobiotic in the central compartment is

$$(C_1)_t = \frac{k_0}{V_1}\left[\frac{(k_{20}+k_{21}-\alpha)(k_{31}-\beta)(1-e^{+\alpha b})}{-\alpha(\beta-\alpha)(\gamma-\alpha)}e^{-\alpha t} + \frac{(k_{20}+k_{21}-\beta)(k_{31}-\beta)(1-e^{+\beta b})}{-\beta(\alpha-\beta)(\gamma-\beta)}e^{-\beta t}\right.$$
$$\left. + \frac{(k_{20}+k_{21}-\gamma)(k_{31}-\gamma)(1+e^{+\gamma b})}{-\gamma(\alpha-\beta)(\gamma-\beta)}e^{-\gamma t}\right] \tag{11.58}$$

The definitions of parameters and constants are as defined before.

REFERENCES

Benet, L. Z. 1972. General treatment of linear mammillary models with elimination from any compartment as used in pharmacokinetics. *J Pharm Sci* 61(4):536–41.

Boroujerdi, M. 2002. *Pharmacokinetics: Principles and Applications*. New York: McGraw Hill-Medical Publishing Division, Chapter 7.

Chiou, W. L., Gadalla, M. A., Peng, G. W. 1978. Method for the rapid estimation of the total body drug clearance and adjustment of dosage regimen in patients during a constant-rate intravenous infusion. *J Pharmacokin Biopharm* 6(2):135–51.

DeVane, C. L., Jusko, W. J. 1986. Dosage regimen design. In *Pharmacokinetics: Theory and Methodology*, eds. M. Rowlan and G. Tucker, Oxford, UK: Pergamon Press, Chapter 10.

Gibaldi, M., Perrier, D. 1982. *Pharmacokinetics*, 2nd Edition. New York: Marcel Dekker, Inc, 63–65.

Krüger-Thiemer, E. 1968. Continuous intravenous infusion and multicompartment accumulation. *Eur J Pharmacol* 4:317–24.

Loo, J. C. L., Riegelman, S. 1970. Assessment of pharmacokinetic constants from postinfusion blood curves obtained after i.v. infusion. *J Pharm Sci* 59(1):53.

O'Reilly, R. A., Welling, P. G., Wagner, J. G. 1971. Pharmacokinetics of Warfarin following intravenous administration to man. *Thromb Diath Haemorrh* 25:178–86.

Rowland, M., Tozer, T. N. 1994. *Clinical Pharmacokinetics*, 3rd Edition. Media, PA: Williams and Wilkins, 66.

Wagner, J. G. 1974. A safe method of rapidly achieving plasma concentration plateaus. *Clin Pharmacol Ther* 16:691–700.

Wagner, J. 1993. *Pharmacokinetics for Pharmaceutical Scientist*. Lancaster, PA: Technomic Publishing Company, Inc.

12 PK/TK Analysis of Xenobiotics Absorbed via an Extravascular Route of Administration (Oral Administration)

12.1 INTRODUCTION

As discussed in Chapters 1, 2, and 7, the absorption of xenobiotics via an extravascular route of administration involves permeation through a physiological barrier before reaching the systemic circulation. The barriers can be as complex as the gastrointestinal wall (Chapter 8) or as simple as a capillary wall in the intramuscular, rectal, or sublingual route of administration (Chapters 1 and 2). The PK/TK profile of the compounds absorbed through extravascular routes involves all four biological processes of absorption, distribution, metabolism, and excretion. These processes occur simultaneously and not sequentially. The physicochemical characteristics of the xenobiotic and behavior of the body in dealing with the compound, influences the data which in return influences the selection of the PK/TK model. The general approaches in PK/TK modeling are discussed in Chapter 9. The focus of the current chapter is to evaluate the application and practicality of the models used in PK/TK analysis of compounds absorbed from an extravascular route of administration. Although the focus of discussion is on the oral administration, the models and parameters are applicable to the other extravacular route of administration, except for percutaneous absorption, which will be discussed in Chapter 13.

12.2 COMPARTMENTAL ANALYSIS

12.2.1 Linear One-Compartment Model with First-Order Input and First-Order Elimination

A diagram of the model for oral administration is presented in Figure 12.1. The assumptions of the model are (1) the dose is administered orally as a single dose at time zero; (2) the dose at the site of absorption is absorbed gradually into the systemic circulation, and declines exponentially with time; and (3) the driving force for absorption into the systemic circulation is the concentration gradient, and thus the absorption is governed by passive diffusion and follows the first-order kinetics. Therefore, the rate of absorption or input is defined as

$$\text{Rate of absorption} = \frac{dA_D}{dt} = k_a \times (A_D)_t \tag{12.1}$$

where (k_a) is the absorption rate constant and (A_D) is the absorbable amount of the dose at the site of absorption. As both processes of absorption and elimination follow first-order kinetics, their rates depend on the amount of the compound available for the process (Garrett, 1993, 1994). Thus,

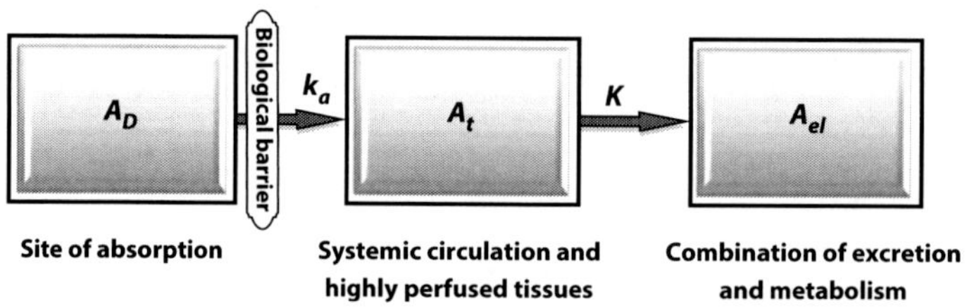

FIGURE 12.1 Linear one-compartment model with first-order input and first-order elimination; (A_D) represents the absorbable amount of xenobiotic at the site of absorption, (A_t) is the amount of unchanged xenobiotic in the body at a given time, (A_{el}) is the amount eliminated from the body by all routes of elimination.

the rates are variable as a function of the amount. The rate of absorption is higher at the early time points when the amount of xenobiotic is higher at the site of absorption. On the other hand, the rate of elimination (rate of output) at the early time points, because of the lesser amount in the body, is low but increases gradually as more of the administered dose is absorbed. It eventually reaches a time point when the rate of elimination becomes equal to the rate of absorption. The plasma concentration at that time point, known as T_{max}, reaches to its maximum, identified as Cp_{max}. The related rate of elimination at T_{max} is the product of total body clearance and maximum plasma concentration, which is equal to the rate of absorption at T_{max}:

$$Cl_T Cp_{max} = k_a (A_D)_{T_{max}} \tag{12.2}$$

After T_{max}, the amount at the site of absorption gradually declines and, thus, lessens the rate of absorption. In return, the amount in the body that has steadily increased causes the rate of elimination to become greater than the rate of absorption, until no more absorbable xenobiotic is left at the site of absorption. The combination of the changing rates of input and output influences the shape of the plasma concentration–time curve (Figure 12.2) to resemble a skewed bell-shaped curve. After

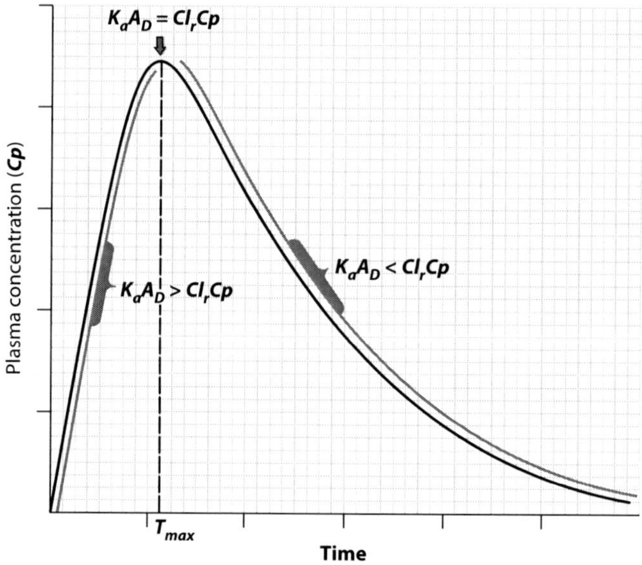

FIGURE 12.2 Profile of plasma concentration of an extravascularly administered xenobiotic; the changes in the rate of absorption and elimination contribute to the skewed bell-shaped curve of the profile.

the completion of absorption, the terminal portion of the plasma concentration–time curve is only a function of the elimination process.

Hence, the rate of change in the amount in the body, assuming the body behaves like a homogeneous compartment is

$$\frac{dA}{dt} = k_a(A_D)_t - K(A)_t \tag{12.3}$$

where $k_a(A_D)_t$ is the rate of absorption; $K(A)_t$ is the rate of elimination; $(A_D)_t$ is the absorbable amount at time t at the site of absorption; and $(A)_t$ is the amount in the body at time t.

At T_{max}, according to Equation 12.2, $Cl_T Cp_{max} = k_a(A_D)_{T_{max}}$ that is, the rate of absorption is equal to the rate of elimination, $k_a(A_D)_{T_{max}} = K(A)_{T_{max}}$. Therefore, at T_{max}, the rate of change of amount in the body is equal zero that is, $(dA/dt) = 0$.

The time for absorption of a compound from the site of absorption is approximately equal to $7 \times (T_{1/2})_{k_a}$, after this time, $k_a(A_D)_t = 0$ and the rate of change of amount in the body is $(dA/dt) = -K(A)_t$.

The integration of Equation 12.3 by the Laplace transform yields the following relationship:

$$(A)_t = \frac{(A_D^0)(k_a)}{K - k_a}(e^{-Kt} - e^{-k_a t}) \tag{12.4}$$

where $(A)_t$ is the amount of xenobiotic in the body at time t; A_D^0 represents the absorbable amount of the administered dose at the site of absorption at $t = 0$, which is the same as (FD). Where F corresponds to the fraction of the dose absorbed, a value that is equal or less than one $(0 \le F \le 1)$, it is also known as absolute bioavailability and $F \times$ Dose, (FD), representing the total amount absorbed or total absorbable amount of administered dose.

In terms of plasma concentration, Equation 12.4 is presented as

$$Cp_t = \frac{FDk_a}{Vd(k_a - K)}(e^{-Kt} - e^{-k_a t}) \tag{12.5}$$

Equations 12.4 or 12.5 represent the two biological processes of absorption and elimination that occur simultaneously during the absorption process:

$$Cp_t = \underbrace{\frac{FDk_a}{Vd(k_a - K)}e^{-Kt}}_{Elimination} - \underbrace{\frac{FDk_a}{Vd(k_a - K)}e^{-k_a t}}_{Absorption} \tag{12.6}$$

The plot of log Cp_t versus time according to Equation 12.5 is a skewed bell-shaped curve with a linear terminal portion, as is shown in Figure 12.3.

The slope of the linear terminal portion of Figure 12.3 is a function of the smallest rate constant. When $k_a > K$, the slope is equal to $-2.303/K$, majority of compounds absorbed from the GI tract fall into this category and follow Equation 12.5. However, when $K > k_a$, the slope equals $-ka/2.303$ and Equation 12.5 is modified to Equation 12.7:

$$Cp_t = \frac{FDk_a}{Vd(k_a - K)}(e^{-k_a t} - e^{-Kt}) \tag{12.7}$$

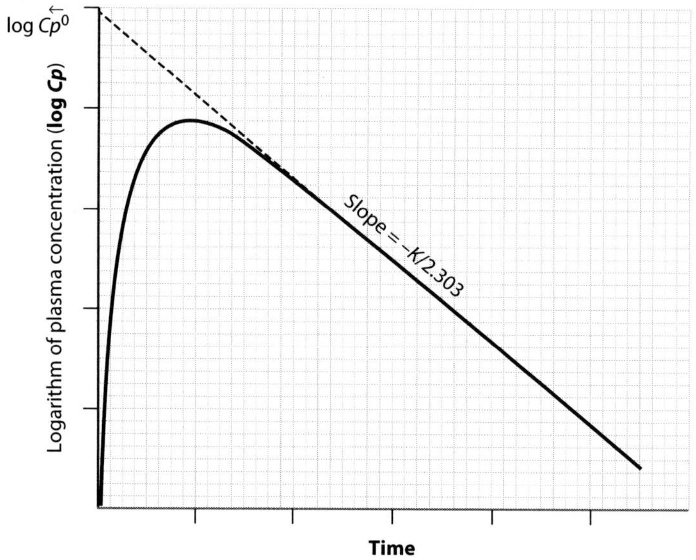

FIGURE 12.3 A typical plot of logarithm of plasma concentration versus time according to Equation 12.5 under the assumption that $k_a > K$.

Equation 12.7, representing the condition of $K > k_a$, is known as flip-flop kinetics and occurs when the rate of absorption is slower that the rate of elimination and exhibits absorption rate-limited elimination kinetics. The underlying mechanisms for such occurrences are most often related to the physicochemical characteristics of the compound, the formulation of therapeutic agent, for example, sustained-release dosage form, or species-related physiological/anatomical differences (Baggot, 1992). The comparative evaluation of flip-flop conditions ($K > k_a$) and the more common condition, $k_a > K$, is presented in Figure 12.4.

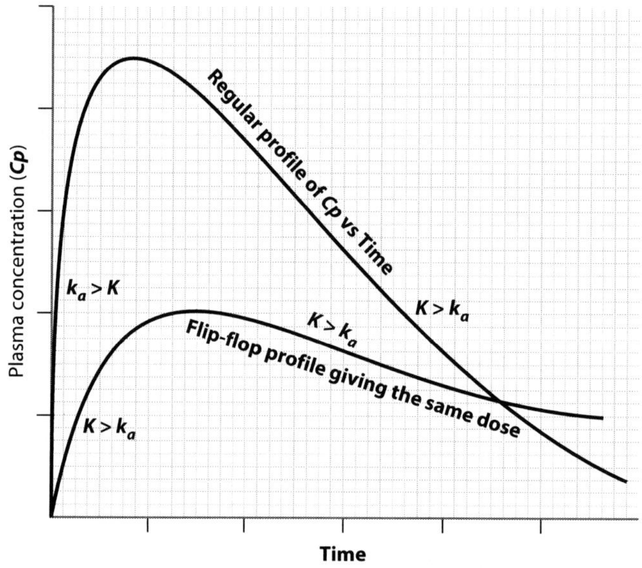

FIGURE 12.4 Comparison of plasma concentration–time profile of flip-flop condition ($K > k_a$) with that of the common condition ($k_a > K$).

Both Equations 12.5 and 12.7 generate the same bell-shaped curve. The flip-flop profile exhibits a smaller maximum plasma concentration and extended *AUC*. Identifying a flip-flop occurrence and managing the related data is an important consideration in bioavailability assessment and estimation of the *AUC*. It has been recommended that an extended and frequent sampling may aid in a more accurate evaluation of the related parameters and constants (Byron and Notari, 1976; Bredberg and Karlsson, 1991; Boxenbaum, 1998; Neelakantan and Veng-Pedersen, 2005; Yáñez et al., 2011). It should be noted that estimation of K from an intravenously administered dose can help identifying the flip-flop occurrence and differentiating between K and k_a (Figure 12.5).

When the two rate constants are equal ($K = k_a$), Equations 12.5 and 12.7 do not apply and it would be difficult to demarcate a terminal linear slope from the logarithmic plot. Under this condition the appropriate equation for plasma concentration is (Bialer, 1980)

$$Cp_t = \frac{\lambda FD}{Vd} te^{-\lambda t} \tag{12.8}$$

where $\lambda = K = k_a$ and the logarithmic form of the equation is

$$\log Cp_t = \log \frac{\lambda FD}{Vd} + \log t - \frac{\lambda t}{2.303} \tag{12.9}$$

When the absorption rate constant is significantly greater than the elimination rate constant ($k_a \gg K$), the absorption is fast and the plasma concentration–time profile resembles the intravenous bolus injection. Under this condition, $e^{-k_a t}$ in Equation 12.5 approaches zero, K is negligible in comparison to k_a, and the relationship is simplified to

$$Cp_t = \frac{FD \cancel{k_a}}{Vd(\cancel{k_a} - \cancel{K})}(\cancel{e^{-k_a t}} - e^{-Kt}) = \frac{FD}{Vd} e^{-Kt} \tag{12.10}$$

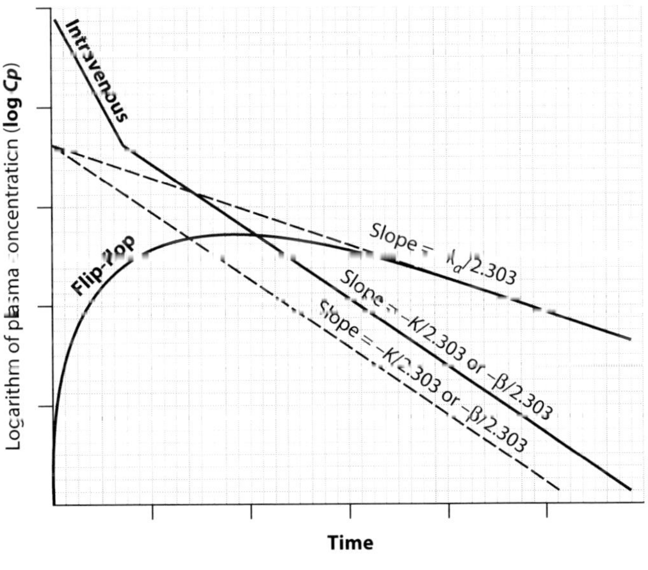

FIGURE 12.5 Identification of flip-flop occurrence and differentiation between K and k_a by comparison of overall elimination rate constant with the bolus injection of the same compound.

12.2.1.1 Calculation of Initial Estimates of K and k_a

The commercially available PK/TK software packages have made it possible to determine the values of K, k_a, and other parameters and constants of different models more accurately. The following are the basic principles and classical methods of determination of the initial estimates of the rate constants for elimination and absorption under the assumption that $k_a > K$.

12.2.1.1.1 Estimation of Overall Elimination Rate Constant (K)

For a set of ideal data with no random error, the plot of logarithm of plasma concentration versus time generates the bell-shape curve with a linear terminal portion. The slope of the linear terminal portion can be determined by using two data points in the slope relationship:

$$\text{Slope} = \frac{\log Cp_{n-1} - \log Cp_n}{t_{n-1} - t_n} = -\frac{K}{2.303} \tag{12.11}$$

Therefore,

$$K = 2.303 \times \text{Slope} \tag{12.12}$$

Extrapolation of the linear terminal portion of the curve with the slope calculated by Equation 12.11 to the *y-axis* provides the y-intercept of the extrapolated line. The extrapolated line represents the elimination component of Equation 12.5 (Figure 12.3). The y-intercept is equal to the coefficient of Equation 12.5, and for an ideal set of data can be estimated as

$$\bar{Cp}^0 = \frac{Cp_n}{e^{-Kt_n}} = \frac{FDk_a}{Vd(k_a - K)} \tag{12.13}$$

Therefore, the equation of the extrapolated line is

$$(\bar{Cp})_t = \bar{Cp}^0 e^{-Kt}, \quad \text{or} \quad \log(\bar{Cp})_t = \log \bar{Cp}^0 - \frac{Kt}{2.303} \tag{12.14}$$

When the plasma data on the linear terminal portion is scattered, a better initial estimate of K is determined by using regression analysis

$$\text{Slope} = \frac{\sum [(dt)(d \log Cp)]}{\sum dt^2} = -\frac{K}{2.303} \tag{12.15}$$

By substituting the numerical value of slope and mean values of the variables t and $\log(Cp)_t$ in equation $y = \bar{y} + m(x - \bar{x})$, $\bar{Cp}^0$ it can be estimated, that is,

$$\log(Cp)_t = \frac{\sum_{i=1}^{n} \log(Cp)_t}{n} + \left[\frac{\sum_{i=1}^{n} (dt)(d \log Cp)}{\sum_{i=1}^{n} (dt)^2}\left(t - \frac{\sum_{i=1}^{n} t}{n}\right)\right] \tag{12.16}$$

where n represents the number of data points located on the linear terminal portion of the curve (Curry, 1981). Substitution of the known values in Equation 12.16 gives rise to Equation 12.14.

The parameter $\bar{Cp}^0$ is a virtual value that cannot be measured experimentally. It is not the initial concentration of an orally administered dose. It can only be interpreted as the hypothetical y-intercept of an intravenous dose equal to FD.

12.2.1.1.2 Estimation of Absorption Rate Constant (k_a)

1. *Method of residuals*

This method is also known as the "feathering" or "peeling" method and involves the subtraction of Equation 12.14, the extrapolated line, from Equation 12.5. It yields Equation 12.17 that represents the absorption component of Equation 12.5, that is,

$$(\bar{Cp} - Cp)_t = \frac{FDk_a}{Vd(k_a - K)} e^{-k_a t}, \quad \text{or} \quad (\bar{Cp} - Cp)_t = \bar{Cp}^0 e^{-k_a t} \tag{12.17}$$

Plotting $\log(\bar{Cp} - Cp)_t$ versus time on the same graph on which that log Cp_t versus time was plotted gives the residual line with a slope of $-k_a/2.303$ (Figure 12.6). The method of residuals is applicable when there are enough early data points.

The manual calculation of data for the line of residuals is presented in Table 12.1.

Estimation of Lag-Time of Absorption: A y-intercept of the residual line greater than the y-intercept of the extrapolated line may indicate that the absorption starts after a delay, known as the lag-time of absorption. There are various reasons for the delay, for example, the physicochemical characteristics of a compound, such as being a weak base, and formulation factors, such as coated tablets, slow disintegration and dissolution and delayed-release dosage forms. A negative lag-time of absorption may represent an inadequate sampling and experimental design. The lag-time of absorption can be estimated graphically (Figure 12.7) or calculated by setting Equations 12.14 and 12.17 equal to each other and solving for lag time (Boroujerdi, 2002):

$$(\log \bar{Cp}^0)_{residual} - \frac{k_a t_l}{2.303} = (\log \bar{Cp}^0)_{extrapolated} - \frac{K t_l}{2.303}$$

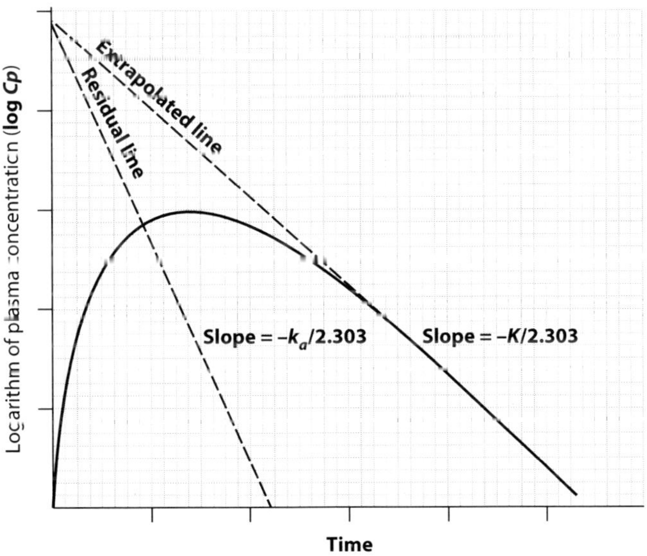

FIGURE 12.6 The conventional profile of logarithm of plasma concentration versus time and determination of extrapolated and residual lines, representing the elimination and absorption processes, respectively.

TABLE 12.1
Summary of Required Steps for Estimation of Residual Line

Time	$\bar{Cp} = \bar{Cp}^0 e^{-Kt}$	Cp	$\bar{Cp} - Cp$	$\log(\bar{Cp} - Cp)$
t_1	$\bar{Cp}_1$	Cp_1	$\bar{Cp}_1 - Cp_1$	$\log(\bar{Cp}_1 - Cp_1)$
t_2	$\bar{Cp}_2$	Cp_2	$\bar{Cp}_2 - Cp_2$	$\log(\bar{Cp}_2 - Cp_2)$
t_3	$\bar{Cp}_3$	Cp_3	$\bar{Cp} - Cp_3$	$\log(\bar{Cp}_3 - Cp_3)$
:	:	:	:	:

$$(\log \bar{Cp}^0)_{residual} - (\log \bar{Cp}^0)_{extrapolated} = t_l \left(\frac{k_a - K}{2.303} \right)$$

$$t_l = \frac{2.303}{k_a - K} \log \frac{(\bar{Cp}^0)_{residual}}{(\bar{Cp}^0)_{extrapolated}} \tag{12.18}$$

2. *Krüger-Thiemer method*
   The extravascular absorption rate constant of a compound can also be estimated from the y-intercept of the extrapolated line $\bar{Cp}^0$. The method requires that equal doses of a drug be given both intravenously and orally in a crossover design. The estimated Cp^0 of the bolus dose is then substituted for D/Vd in Equation 12.13 to develop Equation 12.20 for estimation of k_a:

$$\bar{Cp}^0 = \frac{F \times Cp^0 \times k_a}{(k_a - K)} \tag{12.19}$$

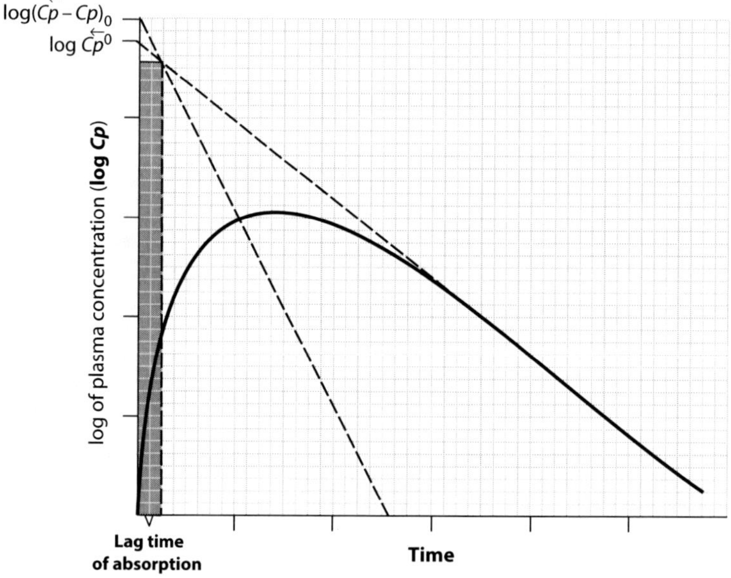

FIGURE 12.7 Graphical detection of lag time of absorption when the y-intercept of the residual line is greater than the y-intercept of extrapolated line.

$$\frac{k_a - K}{k_a} = \frac{FCp^0}{C\bar{p}^0}$$

$$\therefore k_a = \frac{K}{1 - (FCp^0 / C\bar{p}^0)} \tag{12.20}$$

This method is applicable when a compound can be administered both orally and intravenously.

3. *Wagner–Nelson method of using linear plot of percent remaining to be absorbed versus time*
   This method is based on the principle of mass balance for total amount absorbed at any time (Wagner and Nelson, 1964, 1970, 1975):

$$(A_D)_{abs} = A_t + A_{el} \tag{12.21}$$

where $(A_D)_{abs}$ is cumulative amount absorbed at time t, A_t = total amount in the body at time t, and A_{el} is the cumulative amount eliminated from the body by excretion and metabolism.

The differentiation of Equation 12.21 yields the rate equation (Equation 12.22), which defines the rate of absorption in terms of adding up the rate of amount change in the body and the rate of elimination:

$$\frac{dA_D}{dt} = \frac{dA_t}{dt} + \frac{dA_{el}}{dt} \tag{12.22}$$

Defining the rate equation in terms of concentration is achieved by substituting ($A_t = Cp_t \times Vd$) and (dA_{el}/dt = Rate of elimination $= KA_t = KVdCp_t = Cl_TCp$) in Equation 12.22:

$$\frac{dA_D}{dt} = Vd\frac{dCp}{dt} + KVdCp_t \tag{12.23}$$

Integration of Equation 12.23 yields

$$(A_D)_t = VdCp_t + KVd\int_0^t Cp_t dt = VdCp_t + KVd(AUC)_0^t \tag{12.24}$$

Dividing both sides of Equation 12.24 by Vd yields Equation 12.25, known as the Wagner–Nelson equation:

$$\frac{(A_D)_t}{Vd} = Cp_t + K\int_0^t Cp_t dt \tag{12.25}$$

Stepwise calculations of the Wagner–Nelson method for estimation of absorption rate constant is presented in Table 12.2. The total amount absorbed $(A_D)_\infty$ is obtained by integrating Equation 12.24 from zero to infinity:

$$(A_D)_\infty = (VdCp)_0^\infty + KVd\int_0^\infty Cp_t dt \tag{12.26}$$

TABLE 12.2
Stepwise Calculations of Wagner–Nelson Method for Estimation of Absorption Rate Constant

t	Cp	$AUC_{t_{n-1}}^{t_n}$	AUC_0^t	$\frac{(A_D)_{abs}}{Vd} = Cp + (K \times AUC_0^t)$	$\% \frac{(A_D)_{abs}}{Vd}$
t_1	Cp_1	$((0 + Cp_1)/2)(t_1 - 0) = AUC_1$	$AUC_0^1 = AUC_1$	$\frac{(A_{D1})_{abs}}{Vd} = Cp_1 + (K \times AUC_0^1)$	$\frac{(A_{D1})_{abs}}{Vd} \times \frac{100}{p}$
t_2	Cp_2	$((Cp_1 + Cp_2)/2)(t_2 - t_1) = AUC_2$	$AUC_0^2 = AUC_0^1 + AUC_2$	$\frac{(A_{D2})_{abs}}{Vd} = Cp_2 + (K \times AUC_0^2)$	$\frac{(A_{D2})_{abs}}{Vd} \times \frac{100}{p}$
t_3	Cp_3	$((Cp_2 + Cp_3)/2)(t_3 - t_2) = AUC_3$	$AUC_0^3 = AUC_0^2 + AUC_3$	$\frac{(A_{D3})_{abs}}{Vd} = Cp_3 + (K \times AUC_0^3)$	$\frac{(A_{D3})_{abs}}{Vd} \times \frac{100}{p}$
t_4	Cp_4	$((Cp_3 + Cp_4)/2)(t_4 - t_3) = AUC_4$	$AUC_0^4 = AUC_0^3 + AUC_4$	$\frac{(A_{D4})_{abs}}{Vd} = Cp_4 + (K \times AUC_0^4)$	$\frac{(A_{D4})_{abs}}{Vd} \times \frac{100}{p}$
t_5	Cp_5	$((Cp_4 + Cp_5)/2)(t_5 - t_4) = AUC_5$	$AUC_0^5 = AUC_0^4 + AUC_5$	$\frac{(A_{D5})_{abs}}{Vd} = Cp_5 + (K \times AUC_0^5)$	$\frac{(A_{D5})_{abs}}{Vd} \times \frac{100}{p}$
t_6	Cp_6	$((Cp_5 + Cp_6)/2)(t_6 - t_5) = AUC_6$	$AUC_0^6 = AUC_0^5 + AUC_6$	$\frac{(A_{D6})_{abs}}{Vd} = Cp_6 + (K \times AUC_0^6)$	$\frac{(A_{D6})_{abs}}{Vd} \times \frac{100}{p}$
⋮	⋮	⋮	⋮	⋮	
⋮	⋮	⋮	⋮	⋮	***Plateau***
t_n	Cp_n	$((Cp_{n-1} + Cp_n)/2)(t_n - t_{n-1}) = AUC_n$	AUC_0^n	$\frac{(A_{Dn})_{abs}}{Vd} = Cp_n + (K \times AUC_0^n)$	(p)

For a compound given through extravascular route of administration, the initial concentration ($Cp_{t=0}$) and infinity concentration ($Cp_{t=\infty}$) both equal zero, thus, the total amount of the dose that is ultimately absorbed is

$$(A_D)_\infty = KVd\int_0^\infty Cp_t dt = KVd(AUC)_0^\infty \tag{12.27}$$

Dividing Equation 12.24 by Equation 12.27 gives rise to the fraction of dose absorbed at time t

$$\frac{(A_D)_t}{(A_D)_\infty} = \frac{VdCp_t + KVd(AUC)_0^t}{KVd(AUC)_0^\infty} \tag{12.28}$$

For an extravascularly administered compound, the product of total body clearance (KVd) and area under the plasma concentration–time curve from time zero to infinity is equal to the absorbable amount of the administered dose, that is,

$$KVd(AUC)_0^\infty = FDose \tag{12.29}$$

Therefore, Equation 12.28 is the fraction of the absorbable amount of dose at time t, and by canceling the volume of distribution it is expressed in terms of concentration

$$\frac{(A_D)_t}{(A_D)_\infty} = \frac{Cp_t + K(AUC)_0^t}{K(AUC)_0^\infty} \tag{12.30}$$

Equations 12.28 or 12.30 can be expressed in terms of percent remaining to be absorbed, that is,

$$\text{Percent of absorbable amount of dose remaining to be absorbed} = \left(1 - \frac{(A_D)_t}{(A_D)_\infty}\right) \times 100 \qquad (12.31)$$

Equation 12.24 in terms of plasma concentration is defined as

$$\frac{(A_D)_{abs}}{Vd} = Cp_t + (K \times AUC_0^t) \qquad (12.32)$$

where the quantity of $(A_D)_{abs}/Vd$ is the amount absorbed per unit of volume of distribution. Plot of $(A_D)_{abs}/Vd$ versus time give rise to a hyperbolic curve representing the cumulative amount absorbed from time zero to infinity. The curve reaches a plateau level that represents the total amount absorbed at $t \Rightarrow \infty$ (Figure 12.8).

Dividing $((A_D)_{abs}/Vd)$ values on the linear segment of the curve by their average values on the plateau level results in fractional values of the points on the linear segment. These fractions after multiplying by 100 can also be plotted as percent amounts absorbed per unit of volume, that is, % $((A_D)_{abs}/Vd)$, or 100 – % $((A_D)_{abs}/Vd)$, which corresponds to the percent amount remaining to be absorbed per unit of volume of distribution. The stepwise calculations are presented in Table 12.2.

A plot of $\log(100 - \%((A_D)_{abs}/Vd))$ versus time provides a straight line with a slope of $-k_a/2.303$ (Figure 12.9).

The Wagner–Nelson method of the k_a estimation relies greatly on (K) and (AUC). An error in estimation of K or presence of truncation errors in AUC can significantly influence the calculated value of the absorption rate constant (Wang and Nedelman, 2002).

4. *Wagner–Nelson method of using amount excreted unchanged in urine*
   As discussed in Chapter 5 (Section 5.5.1), the rate of urinary excretion of the administered dose of a compound is estimated as (Wagner and Nelson, 1964)

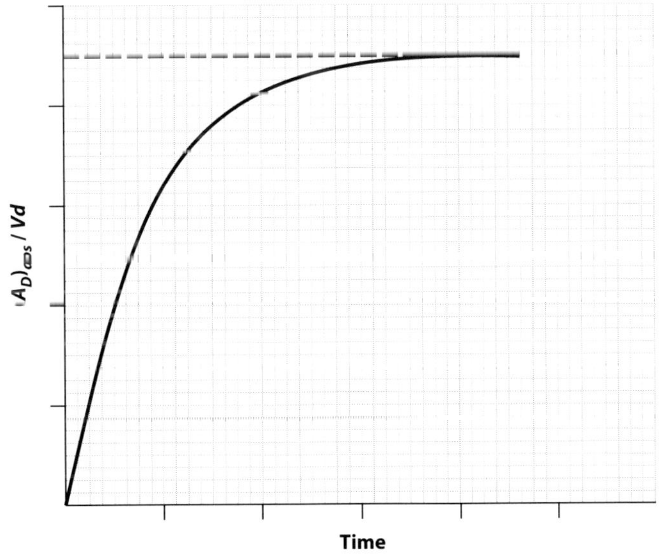

FIGURE 12.8 Theoretical plot of cumulative amount absorbed normalized with respect to the apparent volume of distribution versus time.

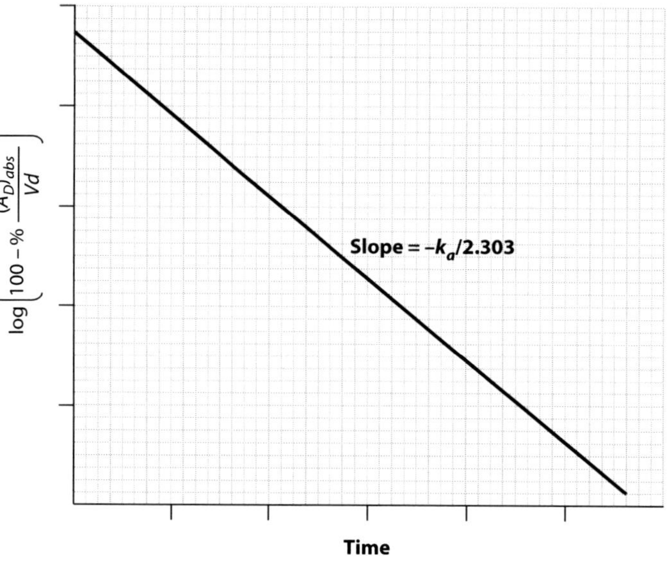

FIGURE 12.9 Plot of percentage remaining to be absorbed versus time.

$$\frac{dA_e}{dt} = k_e A_t \tag{12.33}$$

where (k_e) is the first-order excretion rate constant and (A_t) is the amount in the body at time t equal to $VdCp$, therefore

$$\frac{dA_e}{dt} = k_e VdCp = Cl_r Cp \tag{12.34}$$

Solving for Cp yields

$$Cp = \frac{dA_e/dt}{k_e Vd} \tag{12.35}$$

Substitution of Equation 12.35 in Equation 12.23 provides the absorption rate equation in terms of urinary data

$$\frac{d(A_D)_{abs}}{dt} = Vd\frac{d((dA_e/dt)/k_e Vd)}{dt} + KVd\frac{(dA_e/dt)}{k_e Vd} \tag{12.36}$$

$$\frac{d(A_D)_{abs}}{dt} = \frac{d(dA_e/dt)}{k_e dt} + \frac{K(dA_e/dt)}{k_e} \tag{12.37}$$

Upon integrating Equation 12.37 from zero to infinity, the total amount absorbed at time t can be estimated in terms of amount excreted in the urine

$$(A_D)_{abs} = \frac{1}{k_e}\left(\frac{dAe}{dt}\right) + \frac{K}{k_e}(A_e)_t \tag{12.38}$$

Substitution of Equation 12.33 ($(dA_e/dt) = k_e A_t$), in Equation 12.38 yields,

$$(A_D)_{abs} = \frac{1}{\cancel{k_e}}(\cancel{k_e} A_t) + \frac{K}{ke}(A_e)_t = A_t + \frac{K}{k_e}(A_e)_t \tag{12.39}$$

$$\text{At } t = \infty, \quad A_t \Rightarrow 0, \quad \text{and} \quad (A_D)_{abs} \Rightarrow (A_D)_\infty \tag{12.40}$$

Therefore,

$$(A_D)_\infty = \frac{K}{k_e}(A_e)_\infty \tag{12.41}$$

where $(A_D)_\infty$ is the total amount of drug absorbed or the absorbable amount, and $(A_e)_\infty$ is the total amount excreted unchanged that corresponds to the plateau level of cumulative curve of urinary excretion data (see Chapter 5, Section 5.5).

The proportion of Equations 12.38 through 12.41, gives the fraction of the absorbable amount of dose absorbed at time (t), that is,

$$\left(\frac{(A_D)_{abs}}{(A_D)_\infty}\right)_t = \frac{\frac{1}{\cancel{k_e}}(dAe/dt) + K/\cancel{k_e}(A_e)_t}{K/\cancel{k_e}(A_e)_\infty} = \frac{(dAe/dt) + K(A_e)_t}{K(A_e)_\infty} \tag{12.42}$$

$$\text{At } t = \infty, \quad \left(\frac{(A_D)_{abs}}{(A_D)_\infty}\right)_t \Rightarrow \left(\frac{(A_D)_{abs}}{(A_D)_\infty}\right)_\infty \Rightarrow 1 \tag{12.43}$$

Therefore, at time (t):

$$\text{Fraction of dose remaining to be absorbed} = 1 - \left(\frac{(A_D)_{abs}}{(A_D)_\infty}\right)_t \tag{12.44}$$

When the absorption process is linear and follows first-order kinetics, a plot of the logarithm of fraction or percent of absorbable amount of the administered dose remaining to be absorbed versus time will be a straight line with a slope of $-k_a/2.303$ (Wagner, 1975) (Figure 12.10).

12.2.1.2 Estimation of Peak Time (T_{max})

The three foremost parameters of the model used in the evaluation of bioavailability and bioequivalence are time of maximum plasma concentration, T_{max}, maximum plasma concentration, Cp_{max}, and extent of absorption or area under the plasma concentration–time curve, AUC. The peak time (T_{max}) is an indicator of how fast a compound absorbs and it is the tacit reflection of the rate of absorption. A short T_{max} suggests a fast rate of absorption and a long one signifies the slow absorption rate. The maximum plasma concentration Cp_{max} represents the highest concentration that can be achieved with a single dose. For therapeutic agents it is preferable to achieve this concentration within the therapeutic range. The parameters (Cp_{max}) and (T_{max}) profile the plasma concentration–time curve of an extravascularly administered dose by one point over the entire period of sampling and data collection. The area under the plasma-concentration curve AUC, on the other hand, is a parameter that

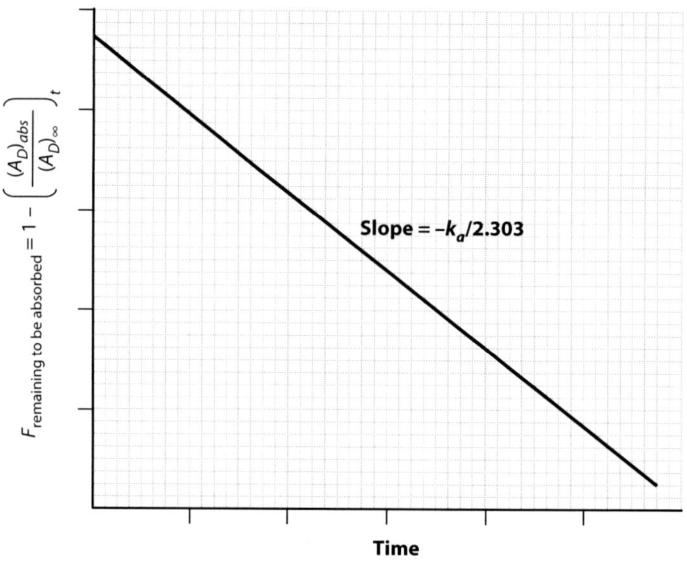

FIGURE 12.10 Treating the amount excreted unchanged in urine by the Wagner–Nelson method yields the same outcome as plasma concentration (exhibited in Figure 12.9) for estimation of the absorption rate constant.

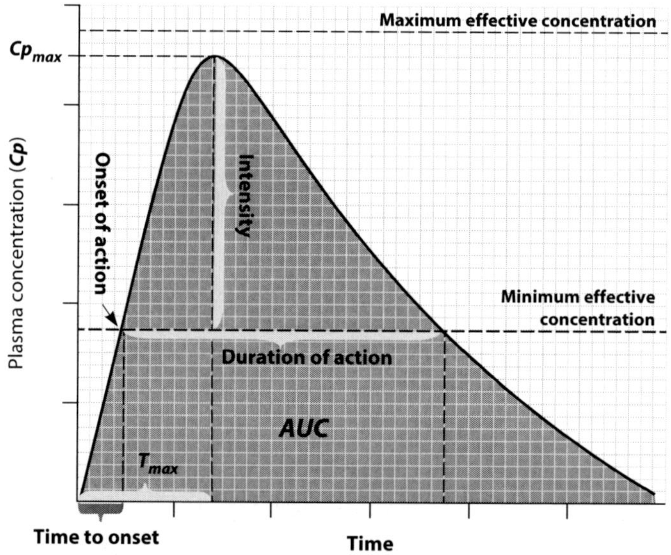

FIGURE 12.11 Principal parameters of plasma concentration–time profile of an extravascularly administered xenobiotic.

encompasses the entire sampling time and beyond, and represents the totality of absorption. These and other parameters of oral absorption are presented graphically in Figure 12.11.

The equation of T_{max} is derived from Equation 12.5, using the calculus principle of determining the maximum or minimum of a function by setting its first derivative equal to zero. As discussed earlier, at T_{max} the rate of absorption equals the rate of elimination, thus the rate of plasma concentration change in the body with respect to time at this particular time point is equal to zero. Thus, taking the first derivative of Equation 12.5, the integrated equation of the model, and setting it equal to zero should provide the T_{max} relationship:

$$\frac{dCp}{dt} = -K\bar{Cp}^0 e^{-Kt} + k_a \bar{Cp}^0 e^{-k_a t} = 0 \tag{12.45}$$

$$\therefore\ Ke^{-KT_{max}} = k_a e^{-k_a T_{max}} \tag{12.46}$$

Dividing both sides of Equation 12.46 by K, taking natural logarithm of both sides and solving for T_{max} yields

$$T_{max} = \frac{1}{k_a - K} \ln \frac{k_a}{K} \tag{12.47}$$

In terms of log base 10 is

$$T_{max} = \frac{2.303}{k_a - K} \log \frac{k_a}{K} \tag{12.48}$$

According to Equation 12.48, T_{max} is a function of the rate constants of absorption and elimination. Therefore, disease states, physiological factors, diet, and other xenobiotics that have an effect on the absorption and elimination processes of a compound may change the value of T_{max}. It is a dose-independent parameter.

The area under the plasma concentration–time curve from zero to T_{max} known as the partial AUC ($PAUC$), is considered a sensitive parameter in comparative evaluations of two compounds or dosage forms (Chen, 1992; Bois et al., 1994; Rostami-Hojgan et al., 1994; Endrenyi et al., 1998a,b; Chen et al., 2011).

12.2.1.3 Estimation of Peak Concentration (Cp_{max})

Contrary to T_{max}, Cp_{max} is a dose-dependent parameter that corresponds to the intensity of the pharmacological response. It is a parameter used in bioavailabilty and bioequivalence studies and the selection of most appropriate extravascular route of administration for a lead compound in drug discovery and development. Similar to T_{max}, Cp_{max} is an indirect measure of the absorption rate of a compound (Chen et al., 2001). The regulatory agencies favor the observed values of both parameters, and the non-compartmental analyses are often preferred in bioavailability assessment. This is mostly to avoid the biases associated with the assumptions of the models. However, the comparative evaluation of observed and calculated values of T_{max} and Cp_{max} may add confidence in the selected model and the related calculated parameters and constants.

The estimation of Cp_{max} based on the model described in Section 12.2.1 is as follows (Gibaldi and Perrier, 1982).

From Equation 12.46:

$$e^{-k_a t} = \frac{K}{k_a} e^{-KT_{max}} \tag{12.49}$$

Substitution of Equation 12.49 in Equation 12.7 yields

$$Cp_{max} = \frac{FDk_a}{Vd(k_a - K)} \left(e^{-KT_{max}} - \frac{K}{k_a} e^{-KT_{max}} \right) \tag{12.50}$$

$$Cp_{max} = \frac{FD}{Vd} (e^{-KT_{max}}) \tag{12.51}$$

Based on Equation 12.49, the maximum plasma concentration is a function of the total absorbable amount of dose, FD, and the fraction of dose in the body at T_{max}, that is,

$$(f_b)_{max} = e^{-KT_{max}}$$

Thus, the maximum amount of dose in the body at T_{max} can be estimated as

$$A_{max} = FD \times f_{bmax} \tag{12.52}$$

12.2.1.4 Estimation of Area Under Plasma Concentration–Time Curve (*AUC*)

The area under plasma concentration–time curve, AUC, is another important parameter of extravascularly administered compounds, in particular oral administration. It is considered a sensitive indicator of bioavailability and provides a collective understanding of the extent of exposure and magnitude of elimination. Theoretically, equal doses of a compound under similar conditions should provide equal AUC. In addition to being a key parameter in bioavailability and bioequivalence assessment, it is also an indicator of pharmacological response (Krzyzanski and Jusko, 1998). The (AUC) of exogenous compounds has the initial condition of zero, that is, ($AUC = 0$ at $t = 0$) and it can be estimated from zero concentration until the elimination is complete, or it can be estimated between the boundary of any two points. For endogenous compounds, however, the initial condition of zero does not exist and approaches are different from xenobiotics (Scheff et al., 2011).

The methods of AUC estimation, depending on area $0 \to t$ or $0 \to \infty$, are

1. $$AUC_0^\infty = \frac{FD}{Cl_t} \tag{12.53}$$

2. Using the trapezoidal rule (Appendix A):

$$AUC_0^t = \sum_{i=0}^{n} [(y_i + y_{i+1})/2](x_{i+1} - x_i) \tag{12.54}$$

$$AUC_0^\infty = \sum_{i=0}^{n} [(y_i + y_{i+1})/2](x_{i+1} - x_i) + AUC_n^\infty \tag{12.55}$$

3. Using the integration method:

$$AUC_0^\infty = \int_0^\infty C\bar{p}^0(e^{-Kt} - e^{-k_a t}) = C\bar{p}^0\left(\frac{1}{K} - \frac{1}{k_a}\right) \tag{12.56}$$

$$AUC_0^t = \int_0^t C\bar{p}^0(e^{-Kt} - e^{-k_a t}) = C\bar{p}^0\left(\frac{e^{-Kt}}{-K} + \frac{e^{-k_a t}}{k_a}\right)_0^t$$

$$= \frac{C\bar{p}^0}{K \times k_a}(Ke^{-k_a t} - k_a e^{-Kt})$$

The given time t in methods 2 and 3 can be the duration of sampling, or any time point such as T_{max}.

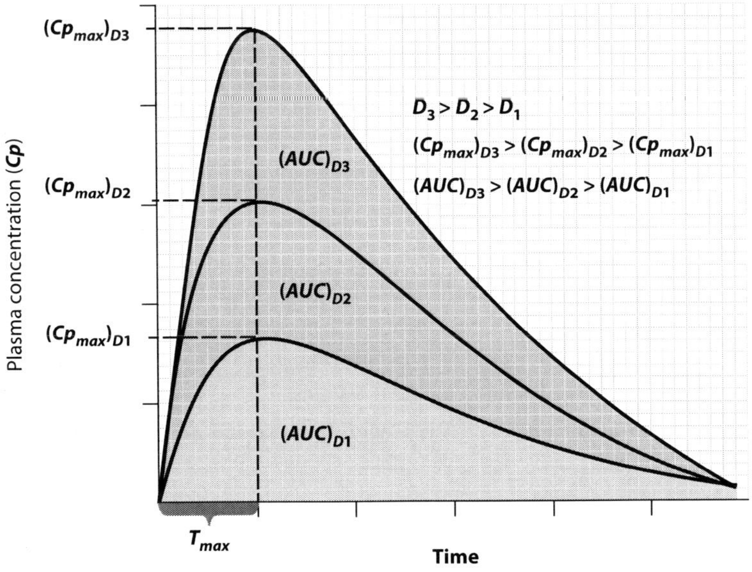

FIGURE 12.12 Illustration of dose dependency of AUC and Cp_{max} and dose independency of T_{max}.

The dose dependency of AUC and Cp_{max} and the dose independency of T_{max} are presented in Figure 12.12.

The influence of changing the rate of absorption on all three parameters of T_{max}, Cp_{max}, and AUC is presented in Figure 12.13. A scenario that is relevant to the selection of the most appropriate extravascular route of administration, or the dosage form for a therapeutic agent. If the total amount absorbed remains the same but the rate of absorption changes with each dose, the slower the rate of absorption, the longer T_{max} and lower Cp_{max}. The area under the curve depending on the extent of absorption, however, may or may not remain the same. Therefore, AUC, although considered a sensitive indicator of completeness of absorption, provides no information about the rate of absorption, T_{max} or Cp_{max}.

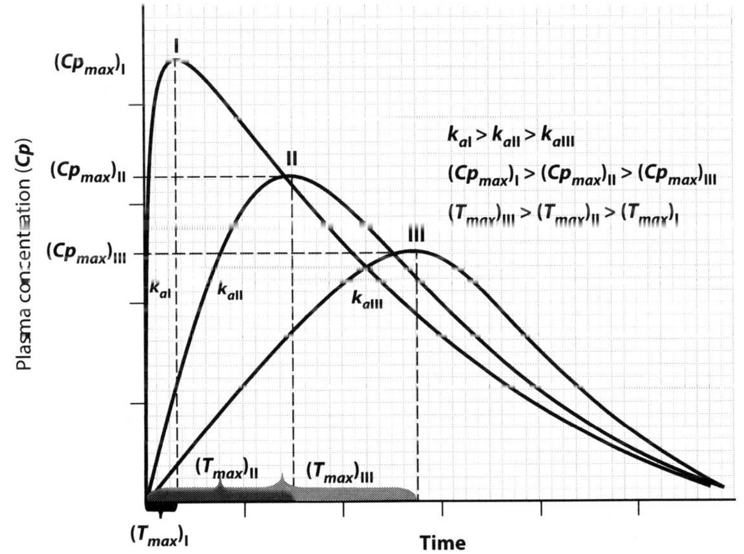

FIGURE 12.13 Illustration of the influence of rate of absorption on T_{max}, Cp_{max}, and AUC.

12.2.1.5 Estimation of Total Body Clearance and Apparent Volume of Distribution

The definitions of total body clearance and apparent volume of distribution are the same as discussed in Chapter 4. The following are practical methods of estimation of these constants within the context of one-compartment model. Using the model-independent Equation 12.53, the total body clearance is

$$Cl_T = \frac{F \times \text{Dose}}{AUC_0^\infty} \tag{12.57}$$

Using model-dependent equation, the clearance is

$$Cl_T = K \times Vd \tag{12.58}$$

Equation 12.58 requires the volume of distribution, which can be estimated by Equation 12.59, or 12.60

$$Vd = \frac{F \times \text{Dose}}{K \times AUC_0^\infty} \tag{12.59}$$

$$Vd = \frac{FD \times k_a}{C\bar{p}^0 (k_a - K)} \tag{12.60}$$

It should be noted that in most cases $C\bar{p}^0 \neq (Cp^0)_{i.v.Bolus}$ unless $F = 1$ and $k_a \gg K$. When the F value is not known, often the volume of distribution is reported as the following normalized value with respect to F:

$$\frac{Vd}{F} = \frac{D \times k_a}{C\bar{p}^0 (k_a - K)} \tag{12.61}$$

12.2.1.6 Fraction of Dose Absorbed (*F*)

The fraction of dose absorbed, also known as the absolute bioavailability, is the ratio of the area under the plasma concentration of an extravascularly administered dose to that of intravenous administration. It represents the absorbable fraction of the administered dose with the significance of formative data about the dosage form, a route of administration, or interaction of the compound with other compounds, therapeutic, environmental, or herbal/diet. It is also required for the calculation of the apparent volume of distribution or total body clearance (Equations 12.57, 12.59, and 12.60). The following are the methodologies used in estimation of this fraction:

1. Dose-independent approach, assuming the administered doses and total body clearances are the same:

$$F = \frac{(AUC)_{extravascular}}{(AUC)_{intravenous}} \tag{12.62}$$

   When the doses are different, the *AUC*s are normalized with respect to dose.

2. From *y*-intercept of the extrapolated line:

$$F = \frac{C\bar{p}^0 \times Vd \times (k_a - K)}{D \times k_a} \tag{12.63}$$

3. From Equation 12.57

$$F = \frac{Cl_T \times AUC_0^{\infty}}{D} \tag{12.64}$$

The F value, as the dose descriptive in the biopharmaceutics classification system (Chapter 8, Section 8.2.2.8), is estimated as

$$F = \frac{D - A_{dissolved} - A_{undissolved}}{D} \tag{12.65}$$

where D is the administered dose, $A_{dissolved}$ is the amount of dose dissolved at the site of absorption, and $A_{undissolved}$ is the amount of the dose remaining to be dissolved (Charkoftaki et al., 2012). Equation 12.65 is further simplified to

$$F = 1 - \left(\frac{r_p}{r_0}\right)^3 - \phi \tag{12.66}$$

where r_0 is the initial radius of drug particles; r_p and ϕ correspond to the time equal to the mean intestinal transit time (Oh et al., 1993; Yu et al., 1996; Rinaki et al., 2004; Charkoftaki et al., 2012).

12.2.1.7 Duration of Action

The equation of the extrapolated line can be used to estimate the duration of action of an orally administered dose, as long as it is corrected for the time to the onset of action (Figure 12.11). The appropriate equation is

$$t_d = \left(\frac{2.303}{K}\log\frac{\bar{Cp}^0}{Cp_{MEC}}\right) - t_{onset} \tag{12.67}$$

where (t_{onset}) is the time to the onset of action and approaches zero $t_{onset} \rightarrow 0$ when absorption is fast and the rate constant of absorption is much greater than the rate constant of elimination.

The duration of action of an extravasculary administered compound is the time required for the concentration of the extrapolated line that corresponds to the onset of action (point O in Figure 12.14) to decline to the minimum effective concentration (point M in Figure 12.14).

12.2.2 Linear Two-Compartment Model with First Order Input in the Central Compartment and First-Order Elimination from the Central Compartment

The profile of the plasma concentration–time curve of compounds that are given extravascularly and follow the two-compartment model is a reflection of four biological processes of absorption, distribution, metabolism, and excretion (Teorel, 1937; Rescigno and Segre, 1966). The plasma concentration–time profile for a two-compartment model is similar to the profile of the one-compartment model. The distinction between the two bell-shape profiles is often difficult to be made visually. However, depending on the values of the absorption rate constant in relation to the hybrid rate constants of the two-compartment model (α and β) and combined influence of ADME processes, the two-compartment model may exhibit a distributional phase around the Cp_{max}, which is known as distributional nose. Frequently, when it is possible, the drug is also given intravenously to establish with certainty whether the drug follows the two- or more compartment model.

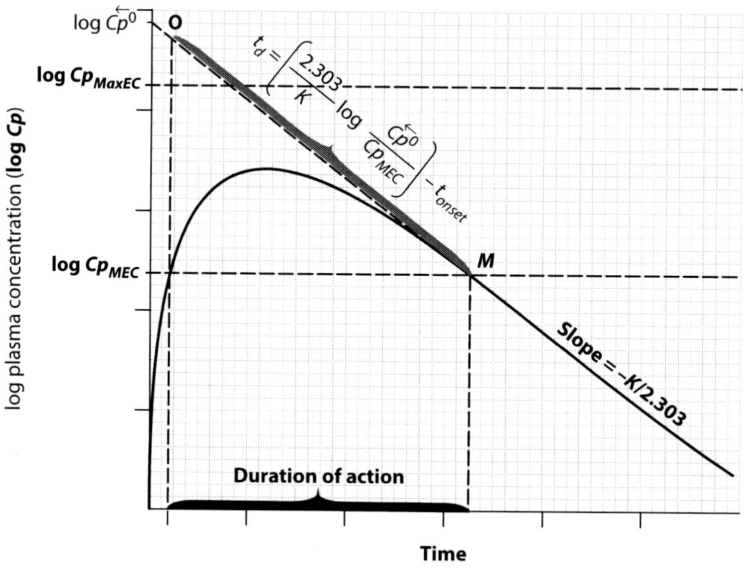

FIGURE 12.14 Illustration of duration of action for an extravascularly administered xenobiotic.

12.2.2.1 Equations of the Model

The diagram of the model is depicted in Figure 12.15. The model is comprised of the site of absorption, central compartment, and peripheral compartment. The central compartment, as discussed for intravenous administration, represents the systemic circulation and highly perfused tissues. The peripheral compartment is the less accessible or slow equilibrating tissues and organs. The equations of the model and the related parameters and constants represent the overall outcome of the combined biological processes at the site of absorption, central and peripheral compartments.

The rate equations of the model, representing the change in each compartment per units of time, are

Site of absorption:

$$\frac{dA_D}{dt} = -k_a A_D \tag{12.68}$$

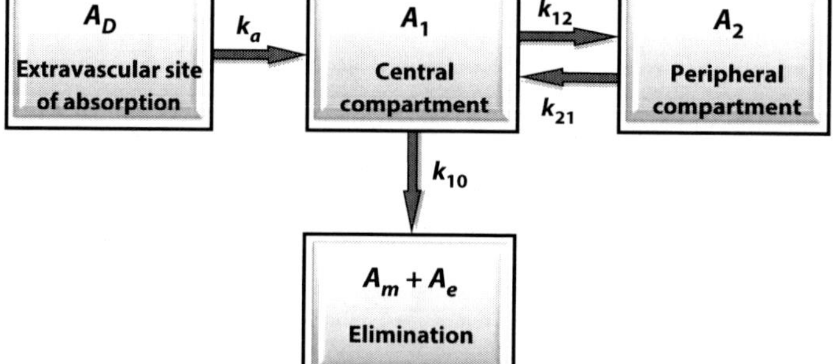

FIGURE 12.15 Diagram of a two-compartment model with first-order absorption, first-order distribution, and first-order elimination from the central compartment.

Central compartment:

$$\frac{dA_1}{dt} = k_a A_D + k_{21} A_2 - k_{12} A_1 - k_{10} A_1 \tag{12.69}$$

Peripheral compartment:

$$\frac{dA_2}{dt} = k_{12} A_1 - k_{21} A_2 \tag{12.70}$$

The input function of the model is (Appendix A):

$$\text{Input} = \frac{k_a FD}{s + k_a} \tag{12.71}$$

The disposition function of the central compartment (Appendix A) is

$$\begin{aligned}(Disp)_{s1} &= \frac{s + E_2}{(s + E_1)(s + E_2) - k_{12}k_{21}} \\ &= \frac{s + k_{21}}{s^2 + s(k_{21} + k_{10} + k_{12}) + k_{21}k_{10}} = \frac{s + E_2}{(s + \alpha)(s + \beta)}\end{aligned} \tag{12.72}$$

where E_1 and E_2 are the sum of the exit rate constants of the central and peripheral compartments, respectively. Multiplying the input with the disposition function yields the Laplace transform of the central compartment:

$$L(A_1) = \frac{k_a FD(s + k_{21})}{(s + k_a)(s + \alpha)(s + \beta)} \tag{12.73}$$

The integrated form, defining the amount in the central compartment as a function of time, is

$$A_1 = \frac{k_a FD(k_{21} - k_a)}{(\alpha - k_a)(\beta - k_a)} e^{-k_a t} + \frac{k_a FD(k_{21} - \alpha)}{(k_a - \alpha)(\beta - \alpha)} e^{-\alpha t} + \frac{k_a FD(k_{21} - \beta)}{(k_a - \beta)(\alpha - \beta)} e^{-\beta t} \tag{12.74}$$

where F is the absolute bioavailability, D is the administered dose, k_a is the first-order absorption rate constant, k_{21} is the first-order distribution rate constant from peripheral compartment to the central compartment, α and β are the first-order hybrid rate constants estimated from the slope of distributive and post-distributive phases. Dividing Equation 12.74 by the apparent volume of distribution of the central compartment yields Equation 12.75, which by denoting the coefficients of exponential terms as a^*, b^*, and c^* is further simplified to Equation 12.74.

$$Cp = \frac{k_a FD(k_{21} - k_a)}{V_1(\alpha - k_a)(\beta - k_a)} e^{-k_a t} + \frac{k_a FD(k_{21} - \alpha)}{V_1(k_a - \alpha)(\beta - \alpha)} e^{-\alpha t} + \frac{k_a FD(k_{21} - \beta)}{V_1(k_a - \beta)(\alpha - \beta)} e^{-\beta t} \tag{12.75}$$

$$Cp = a^* e^{-\alpha t} + b^* e^{-\beta t} + c^* e^{-k_a t} \tag{12.76}$$

where

$$c^* = \frac{k_a FD(k_{21} - k_a)}{V_1(\alpha - k_a)(\beta - k_a)}, \tag{12.77}$$

$$a^* = \frac{k_a FD(k_{21} - \alpha)}{V_1(k_a - \alpha)(\beta - \alpha)} \tag{12.78}$$

and

$$b^* = \frac{k_a FD(k_{21} - \beta)}{V_1(k_a - \beta)(\alpha - \beta)} \tag{12.79}$$

12.2.2.2 Interpretation of k_a, α, and β

As indicated earlier, depending on the frequency of plasma sampling and comparative magnitude of α and k_a, the plot of log plasma concentration versus time of an orally administered compound that follows the two-compartment model may exhibit a typical maximum peak or distributional nose, which is the combine influence of distribution and absorption processes at the early time points (Figure 12.16). The only constant that can be identified unambiguously from Equation 12.76 and Figure 12.16 is the smallest hybrid rate constants β, or the disposition rate constant, that corresponds to the slope of the terminal phase of the logCp – time curve, that is, $-\beta/2.303$. The extrapolation of the linear segment will then provide the y-intercept b^*.

Theoretically, the other two rate constants, namely α and k_a, should be estimated from the residuals. The equation of the first residual curve is determined by subtraction of equation of extrapolated line from Equation 12.76:

$$(Cp - \bar{Cp})_{residual\text{-}1} = (a^* e^{-\alpha t} + b^* e^{-\beta t} + c^* e^{-k_a t}) - (b^* e^{-\beta t}) \tag{12.80}$$

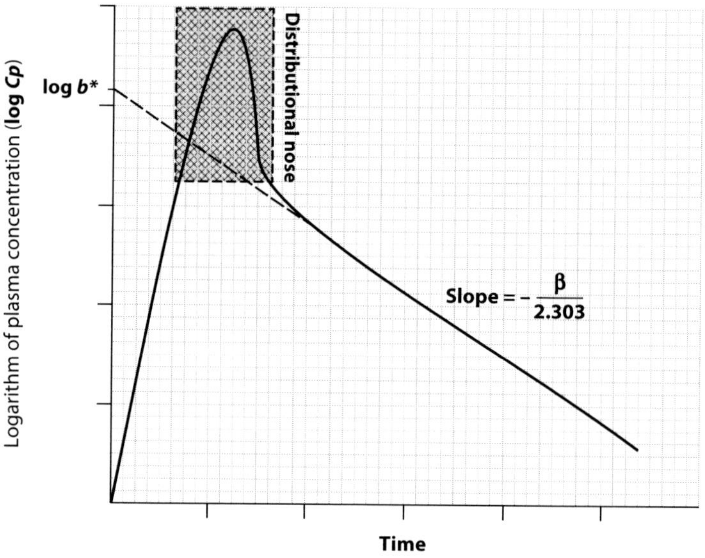

FIGURE 12.16 Plot of log plasma concentration versus time of an extravascularly administered compound that exhibits the distributional nose of a two-compartmental, which is created because of the combined influence of distribution and absorption processes at the initial phase of absorption.

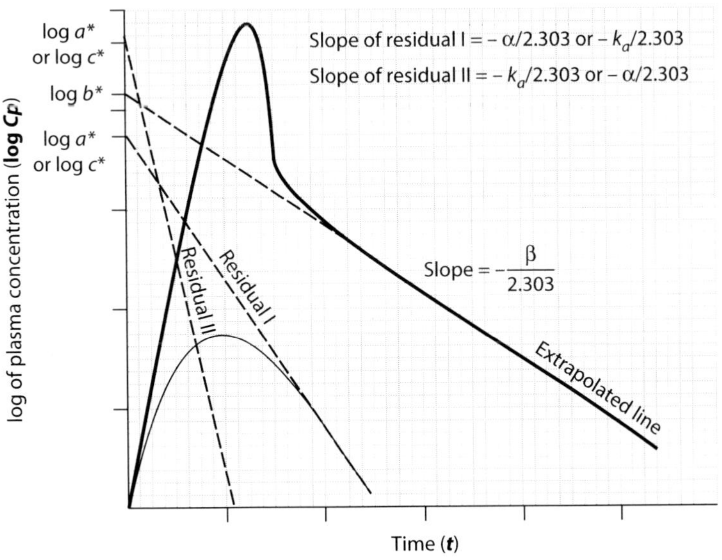

FIGURE 12.17 Plot of positive residuals according to Equations 12.81 through 12.83.

$$(Cp \quad \bar{Cp})_{residual\text{-}1} = a^* e^{-\alpha t} + c^* e^{-k_a t} \tag{12.81}$$

Equation 12.81 represents the positive residual values, which when plotted as $\log(Cp - \bar{Cp})$ versus time exhibits profile of a parabolic curve, similar to the one-compartment model.

The slope of the terminal linear portion of the rectangular parabolic curve is equal to $-\alpha/2.303$ (if $\alpha < k_a$) or equal to $-k_a/2.303$ (if $\alpha > k_a$). The y-intercept of the extrapolated line is either a^* or c^* depending on which rate constant (α or k_a) is smaller. The residual line of the residual parabolic curve (Equation 12.81) is calculated by subtracting the equation of the extrapolated line of the residual curve from Equation 12.81, that is,

When $\alpha < k_a$

$$\left[(Cp - \bar{Cp}) - \overleftarrow{Cp}\right] = (a^* e^{-\alpha t} + c^* e^{k_a t}) - a^* e^{-\alpha t} = c^* e^{-k_a t} \tag{12.82}$$

When $\alpha > k_a$

$$\left[(Cp - \bar{Cp}) - \overleftarrow{Cp}\right] = (a^* e^{-\alpha t} + c^* e^{k_a t}) - c^* e^{-k_a t} = a^* e^{-\alpha t} \tag{12.83}$$

Depending on which rate constant is the largest, the slope of the second residual line (Equation 12.82 or 12.83) is either $-\alpha/2.303$ or $-k_a/2.303$, and the y-intercept of the line is the coefficient associated with the largest rate constant (Figure 12.17).

When α and k_a are similar (i.e., $\alpha \cong k_a$) the log plasma concentration–time curve is similar to the profile of one-compartment model with first-order input (Ronfeld and Benet, 1977).

12.2.2.3 Parameters and Constants of the Model

12.2.2.3.1 Initial Plasma Concentration

Setting $t = 0$ in Equation 12.76, the initial plasma concentration can be estimated as:

$$Cp^0 = a^* + b^* + c^* \tag{12.84}$$

12.2.2.3.2 Area under Plasma Concentration–Time Curve

The integration of Equation 12.76 yields

$$AUC_0^\infty = \frac{a^*}{\alpha} + \frac{b^*}{\beta} + \frac{c^*}{k_a} \tag{12.85}$$

Substitution of the equations for a^*, b^* and c^* (Equations 12.77 through 12.79) yields

$$AUC_0^\infty = \frac{k_a FD(k_{21} - \alpha)}{\alpha V_1 (k_a - \alpha)(\beta - \alpha)} + \frac{k_a FD(k_{21} - \beta)}{\beta V_1 (k_a - \beta)(\alpha - \beta)} + \frac{k_a FD(k_{21} - k_a)}{k_a V_1 (\alpha - k_a)(\beta - k_a)} \tag{12.86}$$

Mathematical manipulation of Equation 12.86 yields Equation 12.88.

$$AUC_0^\infty = \frac{k_a FD}{k_a V_1 \alpha\beta}\left[\frac{\alpha\beta(k_{21} - k_a)(\alpha - \beta) + k_a\beta(k_{21} - \alpha)(\beta - k_a) - k_a\alpha(k_{21} - \beta)(\alpha - k_a)}{(\alpha - k_a)(\beta - k_a)(\alpha - \beta)}\right] \tag{12.87}$$

$$\therefore AUC_0^\infty = \frac{FD}{V_1\alpha\beta}\left[\frac{k_{21}\left(k_a^2\alpha - \alpha^2 k_a - \alpha\beta^2 + \alpha^2\beta + k_a\beta^2 - k_a^2\beta\right)}{\left(k_a^2\alpha - \alpha^2 k_a - \alpha\beta^2 + \alpha^2\beta + k_a\beta^2 - k_a^2\beta\right)}\right] = \frac{FDk_{21}}{V_1\alpha\beta} \tag{12.88}$$

For the two-compartment model $\alpha\beta = k_{21}k_{10}$, the area under the plasma concentration–time curve can also be estimated as

$$AUC_0^\infty = \frac{FD}{Cl_T} \tag{12.89}$$

$$AUC_0^\infty = \frac{FD}{V_1 k_{10}} \tag{12.90}$$

$$AUC_0^\infty = \frac{FD}{Vd_\beta \beta} \tag{12.91}$$

The model-independent trapezoidal rule can always be used to determine the area under the curve from $0 \rightarrow \infty$ or from $0 \rightarrow t$. The estimated terminal area for AUC_0^∞ is achieved by dividing the last plasma concentration by β.

12.2.2.3.3 Apparent Volumes of Distribution

The overall volume of distribution, Vd_β, and the volume of the central compartment are determined from Equations 12.90 and 12.91:

$$Vd_\beta = \frac{FD}{\beta \times AUC_0^\infty} \tag{12.92}$$

$$V_1 = \frac{FD}{k_{10} \times AUC_0^\infty} \tag{12.93}$$

12.2.2.3.4 Total Amount Eliminated between $0 \rightarrow t$ and $0 \rightarrow \infty$

The differential equation defining the total amount of drug eliminated from the body is

$$\frac{dA_{el}}{dt} = k_{10}A_1 \tag{12.94}$$

The integration of Equation 12.94 between $0 \rightarrow t$ gives

$$A_{el} = \int_0^t k_{10}A_1 dt = k_{10}V_1 \int_0^t Cpdt \tag{12.95}$$

Therefore, the total amount eliminated between $0 \rightarrow t$ is

$$A_{el} = k_{10}V_1 AUC_0^t = Cl_T AUC_0^t \tag{12.96}$$

When t approaches infinity $A_{el} = FD$ and the total amount eliminated between $0 \rightarrow \infty$ is equal to the total amount absorbed:

$$FD = k_{10}V_1 AUC_0^\infty = Cl_T AUC_0^\infty \tag{12.97}$$

Because $A_{el} = A_e + A_m$, the total amount excreted unchanged between $0 \rightarrow t$ and $0 \rightarrow \infty$ can be estimated as

$$A_e = f_e\left(k_{10}V_1 AUC_0^t\right) = Cl_r \times AUC_0^t \tag{12.98}$$

$$A_e^\infty = f_e\left(k_{10}V_1 AUC_0^\infty\right) = Cl_r \times AUC_0^\infty = f_e FD \tag{12.99}$$

where f_e is the fraction of absorbed dose excreted unchanged and Cl_r is the renal clearance. The fraction of absorbed dose excreted unchanged and eliminated as metabolites are

$$f_e = \frac{A_e^\infty}{FD} \tag{12.100}$$

$$f_m = \frac{FD - A_e^\infty}{FD} \tag{12.101}$$

The total amount of the absorbed dose eliminated as metabolites at time t and ∞ can be estimated as

$$(A_m)_t = f_m\left(k_{10}V_1 AUC_0^t\right) = Cl_m \times AUC_0^t \tag{12.102}$$

$$A_m^\infty = f_m\left(k_{10}V_1 AUC_0^\infty\right) = Cl_m \times AUC_0^\infty = f_m FD \tag{12.103}$$

12.2.2.3.5 Estimation of Distribution and Elimination Rate Constants

The rate constants of distribution and elimination are estimated according to methodology discussed in Chapter 10, Section 10.3.3.

12.2.2.4 Estimation of First-Order Absorption Rate Constant—Loo–Riegelman Method

This method is based on the mass balance equation of the amount or concentration between the site of absorption, the central and peripheral compartments, and the total amount eliminated from the body (Loo and Riegelman, 1968). The method requires a prior knowledge of the distribution rate constants, k_{12} and k_{21}, and also the overall elimination rate constant k_{10}. The application of the method requires a good understanding of the principles and the assumptions of the equations. There are some similarities between this method and the Wagner–Nelson method (Chapter 12, Section 12.2.1.1). The Wagner–Nelson method is favored for the one-compartment model, although it can also be used for the two-compartment model (Wagner, 1974). The Loo–Riegelman method is preferred for the two-compartment model.

The Loo–Riegelman (Loo and Riegelman, 1968) method starts with the mass balance equation at time t:

$$(A_{total})_t = (A_1)_t + (A_2)_t + (A_e)_t \tag{12.104}$$

where $(A_{total})_t$ is the total amount absorbed at time t; $(A_1)_t$ and $(A_2)_t$ are the amount in the central and peripheral compartments at time t, respectively; and $(A_e)_t$ is the amount eliminated from the body at time t.

The total amount of a drug eliminated at time t and ∞ (Equations 12.96 and 12.97) is

$$(A_{el})_t = k_{10}V_1AUC$$

$$A_{el}^{\infty} = k_{10}V_1AUC_0^{\infty} = FD$$

Therefore, the total percentage absorbed at time t is

$$\frac{(A_{total})_t}{FD} \times 100 = \%(A_1)_t + \%(A_2)_t + \left[\frac{k_{10}V_1AUC_0^t}{k_{10}V_1AUC_0^{\infty}}\right] \times 100 \tag{12.105}$$

Equation 12.105 defines the amount eliminated from the body in terms of the volume of the central compartment and the area under the plasma concentration–time curve from $0 \to t$ and $0 \to \infty$. Here, a prior knowledge of k_{10} is required.

The second term of Equation 12.105 is the amount in the peripheral compartment, which for an orally administered dose is unknown, but by using the following equations and methodology it can be expressed in terms of the amount in the central compartment, that is, the sampling compartment.

The amount in the central compartment between two time points of t_n and t_{n-1}, where $t_n - t_{n-1} = \Delta t$ is

$$(A_1)_{t_n} = (A_1)_{t_{n-1}} + \left(\frac{\Delta A_1}{\Delta t}\right)(\Delta t) \tag{12.106}$$

The rate of the amount changed in the peripheral compartment is the same as described before:

$$\frac{dA_2}{dt} = k_{12}(A_1)_t - k_{21}(A_2)_t$$

Substituting Equation 12.106 for A_l in the above equation yields

$$\frac{(dA_2)_{t_n}}{dt} = k_{12}\left[(A_1)_{t_{n-1}} + \frac{\Delta A_1}{\Delta t}(\Delta t)\right] - k_{21}(A_2)_{t_n} \tag{12.107}$$

$$\frac{(dA_2)_{t_n}}{dt} = k_{12}(A_1)_{t_{n-1}} + k_{12}\left[\frac{\Delta A_1}{\Delta t}\right]\Delta t - k_{21}(A_2)_{t_n} \tag{12.108}$$

Using the Laplace transforms from Appendix A, Table A.1 yields the integrated form of Equation 12.108 as follows:

$$L(1) = 1/s$$

$$L(t) = 1/s^2$$

$$L(A_1) = \tilde{A}_1$$

$$L(A_2) = \tilde{A}_2$$

$$L\left(\frac{dA_2}{dt}\right) = s(\tilde{A}_2)_{t_n} - (\tilde{A}_2)_{t_{n-1}}$$

$(\tilde{A}_1)_{t_{n-1}}$ is considered the initial condition of the interval Δt.

Therefore, the Laplace transform of Equation 12.108 is

$$s\left(\tilde{A}_2\right)_{t_n} - \left(\tilde{A}_2\right)_{t_{n-1}} = \frac{k_{12}\left(\tilde{A}_1\right)_{t_{n-1}}}{s} + \frac{k_{12}(\Delta A_1/\Delta t)\Delta t}{s^2} - k_{21}\left(\tilde{A}_2\right)_{t_n} \tag{12.109}$$

Rearranging Equation 12.109 to solve for $(\tilde{A}_2)_{t_n}$ yields

$$\left(\tilde{A}_2\right)_{t_n} = \frac{\left(\tilde{A}_2\right)_{t_{n-1}}}{s + k_{21}} + \frac{k_{12}\left(\tilde{A}_1\right)_{t_{n-1}}}{s(s + k_{21})} + \frac{k_{12}(\Delta A_1/\Delta t)\Delta t}{s^2(s + k_{21})} \tag{12.110}$$

The inverse Laplace transform of Equation 12.110 is

$$(A_2)_{t_n} = \underbrace{(A_2)_{t_{n-1}}e^{-k_{21}\Delta t}}_{\text{1st}} + \underbrace{\frac{k_{12}}{k_{21}}(A_1)_{t_{n-1}}[1 - e^{-k_{21}\Delta t}]}_{\text{2nd}} + \underbrace{\frac{k_{12}}{k_{21}}\Delta A_1 - \frac{k_{12}}{k_{21}^2} \times \frac{\Delta A_1}{\Delta t}[1 - e^{-k_{21}\Delta t}]}_{\text{3rd term}} \tag{12.111}$$

To simplify Equation 12.111, Taylor series expansion is applied to the exponential function of the third term, that is,

$$e^{-k_{21}\Delta t} \cong 1 - k_{21}\Delta t + \frac{k_{21}^2(\Delta t)^2}{2} \tag{12.112}$$

Substituting Equation 12.112 into Equation 12.111 yields

$$(A_2)_{t_n} = \underbrace{(A_2)_{t_{n-1}} e^{-k_{21}\Delta t}}_{\text{1st}} + \underbrace{\frac{k_{12}}{k_{21}}(A_1)_{t_{n-1}}[1 - e^{-k_{21}\Delta t}]}_{\text{2nd}} + \underbrace{\frac{k_{12}}{k_{21}}\Delta A_1 - \frac{k_{12}}{k_{21}} \times \frac{\Delta A_1}{\Delta t}\left[1 - k_{21}\Delta t + \frac{k_{21}^2(\Delta t)^2}{2}\right]}_{\text{3rd term}} \tag{12.113}$$

Further simplification of the third term of Equation 12.113 yields the following relationship that represents $(A_2)t_n$ term in Equations 12.104 and 12.105:

$$(A_2)_{t_n} = (A_2)_{t_{n-1}} e^{-k_{21}\Delta t} + \frac{k_{12}}{k_{21}}(A_1)_{t_{n-1}}[1 - e^{-k_{21}\Delta t}] + \frac{k_{12}}{2}\Delta A_1 \Delta t \tag{12.114}$$

Therefore,

$$A_{total} = \underbrace{(A_1)_{t_n}}_{A_1} + \underbrace{(A_2)_{t_{n-1}} + \frac{k_{12}}{k_{21}}(A_1)_{t_{n-1}}[1 - e^{-k_{21}\Delta t}] + \frac{k_{12}}{2}\Delta A_1 \Delta t}_{A_2} + \underbrace{k_{10}V_1 AUC_0^t}_{A_{el}} \tag{12.115}$$

Dividing Equation 12.115 by the volume of distribution yields Equation 12.116, which defines the total concentration of the compound:

$$C_{total} = (Cp)_{t_n} + (C_2)_{t_{n-1}} + \frac{k_{12}}{k_{21}}(Cp)_{t_{n-1}}[1 - e^{-k_{21}\Delta t}] + \frac{k_{12}}{2}\Delta Cp_1 \Delta t + k_{10} AUC_0^t \tag{12.116}$$

The use of the two-term Taylor expansion in simplifying the calculation of the absorption rate constant by the Loo–Riegelman method has been demonstrated to introduce errors in the calculation of absorption rate constant (Boxenbaum and Kaplan, 1975; Zeng et al., 1983). Equation 12.116 is expressed in terms of plasma concentration, except for $(C_2)t_{n-1}$, which is estimated by Equation 12.117. Equation 12.117 is the same as Equation 12.114 only in terms of the concentration of the second compartment. The starting point for the calculation of concentration of the second compartment is $(C_2)t_{n-1} = 0$.

$$(C_2)_{t_n} = (C_2)_{t_{n-1}} e^{-k_{21}\Delta t} + \frac{k_{12}}{k_{21}}(Cp)_{t_{n-1}}[1 - e^{-k_{21}\Delta t}] + \frac{k_{12}}{2}\Delta Cp \Delta t \tag{12.117}$$

Using Equation 12.115 the expression of "%absorbed" (Equation 12.105), can be defined as

$$\frac{A_{total}}{FD} \times 100 = \%(A_1)_{t_n} + \%\left[(A_2)_{t_{n-1}} + \frac{k_{12}}{k_{21}}(A_1)_{t_{n-1}}[1 - e^{-k_{21}\Delta t}] + \frac{k_{12}}{2}\Delta A_1 \Delta t\right] + \frac{k_{10}V_1 AUC_0^t}{k_{10}V_1 AUC_0^\infty} \times 100 \tag{12.118}$$

The plot of the logarithm percentage remaining to be absorbed, that is, $\log[100 - ((A_{total}/FD) \times 100)]$ versus time should be linear with a slope of $-k_a/2.303$.

The numerical example of the Loo–Riegelman method is presented in Tables 12.3 through 12.6. The data presented in Table 12.3 are the plasma concentrations of a compound that follows two compartment model given orally and intravenously following administration of equal doses.

The biexponential equation summarizing the intravenous data is determined as $Cp = 120e^{-4.145t} + 91.41e^{-0.4145t}$. The calculated rate constants were estimated as $k_{21} = 2.0275$ h^{-1}, $k_{10} = 0.8474$ h^{-1} and $k_{12} = 1.6846$ h^{-1}. The stepwise calculation of the absorption rate constant according to Equations 12.116 and 12.117 and as presented in Tables 12.4 through 12.6 generated the independent variable of $\log(100 - \%(A_{total}/FD))$. The linear plot of this variable versus time provided the slope of $-k_a/2.303 = -0.94$ h^{-1}, and $k_a = 2.165$ h^{-1}.

TABLE 12.3
Data Related to Application

Time (h)	*Cp* Oral (mg/L)	*Cp* IV (mg/L)
0.10	1.000	166.979
0.20	12.000	136.516
0.30	28.000	115.326
0.40	46.000	100.306
0.60	56.000	81.262
0.75	59.700	72.345
1.00	60.000	62.293
2.00	40.000	39.929
3.00	26.200	26.361
4.00	17.500	17.415
5.00	11.500	11.506
6.00	7.580	7.602
10.00	1.400	1.448
14.00	0.268	0.276

TABLE 12.4
Calculations of Absorption Rate Constant Using the Loo–Riegelman Method

t_n	t_{n-1}	Δt	$(Cp)t_n$	$(Cp)t_{n-1}$	ΔCp	$\frac{k_{12}}{2}\Delta Cp\Delta t$	$\frac{k_{12}}{k_{21}}(Cp)_{t_{n-1}}[1 - e^{-k_{21}\Delta t}]$
0.10	0.0	0.10	1.000	0	1.000	0.0842	0.0000
0.20	0.10	0.10	12.000	1.000	11.000	0.9265	0.1525
0.30	0.10	0.10	28.000	12.000	16.000	1.3477	1.8298
0.40	0.10	0.10	46.000	28.000	18.000	1.5161	4.2694
0.60	0.10	0.20	56.000	46.000	10.000	1.6846	12.7410
0.75	0.20	0.15	59.700	56.000	3.700	0.4675	12.2014
1.00	0.15	0.25	60.000	59.700	0.300	0.0632	19.7235
2.00	0.25	1.00	40.000	60.000	−20.000	−16.8460	43.2887
3.00	1.00	1.00	26.200	40.000	−13.800	−11.6237	28.8592
4.00	1.00	1.00	17.500	26.200	−8.700	−7.3280	18.9027
5.00	1.00	1.00	11.500	17.500	−6.000	−5.0538	12.6259
6.00	1.00	1.00	7.580	11.500	−3.920	−3.3018	8.2970
10.00	1.00	4.00	1.400	7.580	−6.180	−20.8217	6.2961
14.00	4.00	4.00	0.268	1.400	−1.132	−3.8139	1.1629

TABLE 12.5
Calculations of Absorption Rate Constant Using the Loo–Riegelman Method

$(C_2)_{t_{n-1}}e^{-k_{21}\Delta t}$	$(C_2 t_n)$	$AUC_{t_{n-1}}^{t_n}$	$AUC_0^{t_n}$	$k_{10} \times AUC_0^{t_n}$	C_{total}
0	0.0842	0.050	0.050	0.042	1.127
0.0687	1.1477	0.650	0.700	0.593	13.741
0.9370	4.1145	2.000	2.700	2.288	34.402
3.3594	9.1449	3.700	6.400	5.423	60.568
7.4667	21.8923	10.200	16.600	14.067	91.959
14.5943	27.2632	8.677	25.277	21.420	108.383
20.1139	39.9006	14.962	40.239	34.099	133.999
24.0351	50.4778	50.000	90.239	76.469	166.946
6.6461	23.8816	33.100	123.339	104.517	154.599
3.1443	14.7190	21.850	145.189	123.033	155.252
1.9379	9.5100	14.500	159.689	135.320	156.330
1.2521	6.2473	9.540	169.229	143.405	157.232
0.8225	−13.7031	17.960	187.189	158.624	146.321
−4.1180	−6.7690	3.336	190.525	161.451	154.950

TABLE 12.6
Calculations of Absorption Rate Constant Using the Loo–Riegelman Method

$\frac{A_{total}}{FD} \times 100$	$100 - \% \frac{A_{total}}{FD}$	$\log\left(100 - \% \frac{A_{total}}{FD}\right)$
0.722	99.278	1.99685
8.808	91.192	1.95996
22.053	77.947	1.89180
38.826	61.174	1.78657
58.948	41.052	1.61333
69.476	30.524	1.48464
85.897	14.103	1.14931
107.017		
99.102		
99.521		
100.212		
100.790		
93.795		
99.327		

12.2.3 Linear Two-Compartment Model with First-Order Input in the Peripheral Compartment and First-Order Elimination from the Peripheral Compartment

The diagram of the model is depicted in Figure 12.18. The differential equations of the model are

$$\text{Central compartment:} \quad \frac{dA_1}{dt} = k_{21}A_2 - k_{12}A_1 \tag{12.119}$$

$$\text{Peripheral compartment:} \quad \frac{dA_2}{dt} = k_a FD + k_{12}A_1 - k_{21}A_2 - k_{20}A_2 \tag{12.120}$$

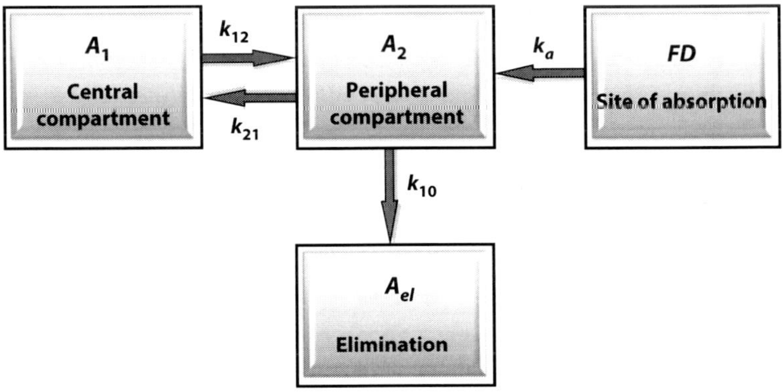

FIGURE 12.18 Diagram of a two-compartment model with first-order input into the peripheral compartment and elimination from the peripheral compartment.

The integrated equation of the time course of the compound in the central compartment, the sampling compartment, is

$$Cp_t = \frac{FDk_a}{V_1}\left[\frac{k_{21}+k_{20}-\alpha}{(k_a-\alpha)(\beta-\alpha)}e^{-\alpha t} + \frac{k_{21}+k_{20}-\beta}{(k_a-\beta)(\alpha-\beta)}e^{-\beta t} + \frac{k_{21}+k_{20}-k_a}{(\alpha-k_a)(\beta-k_a)}e^{-k_a t}\right] \tag{12.121}$$

The abbreviate equation is

$$Cp_t = a^* e^{-\alpha t} + b^* e^{-\beta t} + c^* e^{-k_a t} \tag{12.122}$$

where

$$a^* = \frac{FDk_a(k_{21}+k_{20}-\alpha)}{V_1(k_a-\alpha)(\beta-\alpha)} \tag{12.123}$$

$$b^* = \frac{FDk_a(k_{21}+k_{20}-\beta)}{V_1(k_a-\beta)(\alpha-\beta)} \tag{12.124}$$

$$c^* = \frac{FDk_a(k_{21}+k_{20}-k_a)}{V_1(\alpha-k_a)(\beta-k_a)} \tag{12.125}$$

The distribution and elimination rate constants are estimated using the coefficients of the exponential as follows:

$$k_{20}+k_{21} = \frac{a^*\beta k_a + b^*\alpha k_a + c^*\alpha\beta}{a^*(k_a-\alpha)+\beta(k_a-\beta)} \tag{12.126}$$

$$k_{12} = \alpha+\beta-(k_{20}+k_{21}) \tag{12.127}$$

$$k_{20} = \frac{\alpha\beta}{k_{12}} \tag{12.128}$$

$$k_{21} = (k_{20} + k_{21}) - k_{20} \tag{12.129}$$

The apparent volume of distribution of the central compartment can be estimated as

$$\frac{V_1}{F} = \frac{(k_a D/a^*)(k_{21} + k_{20} - \alpha)}{(k_a - \alpha)(\beta - \alpha)} \tag{12.130}$$

$$V_1 = \frac{(k_a FD \, / \, a^*)(k_{21} + k_{20} - \alpha)}{(k_a - \alpha)(\beta - \alpha)} \tag{12.131}$$

The plasma concentration–time curve for this model exhibits a more defined distributional nose.

12.2.4 Linear Three-Compartment Model with First-Order Input in the Central Compartment and First-Order Elimination from the Central Compartment

The diagram of the model is depicted in Figure 12.19. The assumptions of the model are the same as discussed before for intravenous administration. The central compartment represents the systemic circulation and highly perfused tissues/organs. It receives the absorbable dose subsequent to absorption from an extravascular route of administration. The number of peripheral compartments are determined either based on the physiological/biochemical circumstances confirmed by the curve fitting data, or the number is determined solely through the curve fitting process (Wagner, 1988; Wagner et al., 1990).

The differential equations of the model are

$$\frac{dA_1}{dt} = k_a FD + k_{21}A_2 + k_{31}A_3 - A_1(k_{12} + k_{13} + k_{10}) \tag{12.132}$$

$$\frac{dA_3}{dt} = k_{13}A_1 - k_{31}A_3 \tag{12.133}$$

$$\frac{dA_2}{dt} = k_{12}A_1 - k_{21}A_2 \tag{12.134}$$

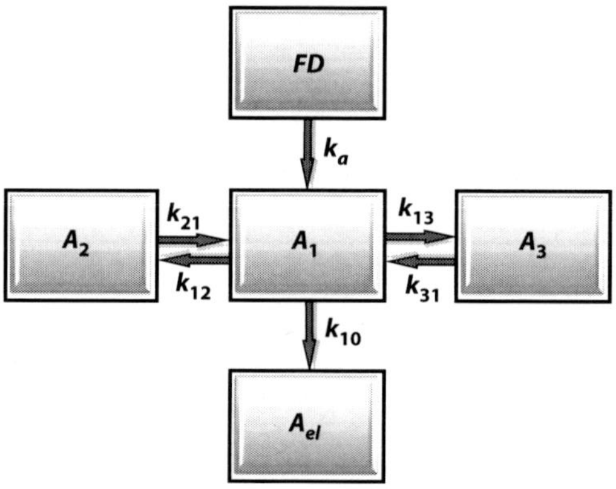

FIGURE 12.19 Diagram of a three-compartment model with first-order input into the central compartment and first-order elimination from the peripheral compartment.

The integrated equation representing the time course of the administered absorbed dose in the systemic circulation is

$$Cp = \frac{FDk_a}{V_1}\left[\frac{(k_{21}-\alpha)(k_{31}-\alpha)}{(\beta-\alpha)(\gamma-\alpha)(k_a-\alpha)}e^{-\alpha t} + \frac{(k_{21}-\beta)(k_{31}-\beta)}{(\alpha-\beta)(\gamma-\beta)(k_a-\beta)}e^{-\beta t} + \frac{(k_{21}-\gamma)(k_{31}-\gamma)}{(\alpha-\gamma)(\beta-\gamma)(k_a-\gamma)}e^{-\gamma t} + \frac{(k_{21}-k_a)(k_{31}-k_a)}{(\alpha-k_a)(\beta-k_a)(\gamma-k_a)}e^{-k_a t}\right] \quad (12.135)$$

The abbreviated version of Equation 12.135 is

$$Cp = a^* e^{-\alpha t} + b^* e^{-\beta t} + c^* e^{-\gamma t} + d^* e^{-k_a t}$$

where

$$a^* = \frac{FDk_a(k_{21}-\alpha)(k_{31}-\alpha)}{V_1(\beta-\alpha)(\gamma-\alpha)(k_a-\alpha)}$$

$$b^* = \frac{FDk_a(k_{21}-\beta)(k_{31}-\beta)}{V_1(\alpha-\beta)(\gamma-\beta)(k_a-\beta)}$$

$$c^* = \frac{FDk_a(k_{21}-\gamma)(k_{31}-\gamma)}{V_1(\alpha-\gamma)(\beta-\gamma)(k_a-\gamma)}$$

$$d^* = \frac{FDk_a(k_{21}-k_a)(k_{31}-k_a)}{V_1(\alpha-k_a)(\beta-k_a)(\gamma-k_a)}$$

The calculation of the absorption rate constant is exactly by the Loo–Riegelman method (Loo and Riegelman, 1968; Wagner, 1983; Proost, 1985).

REFERENCES

Baggot, J. D. 1992. Review bioavailability and bioequivalence of veterinary drug dosage forms, with particular reference to horses. An overview. *J Vet Pharmacol Ther* 15(2):160–73.

Bialer, M. 1980. A simple method for determining whether absorption and elimination rate constants are equal in the one-compartment open model with first-order input. *J Pharmacokinet Biopharm* 8(1):111–13.

Bois, F. Y., Tozer, T. N., Hauck, W. W., Chen, M. L., Patnaik, R., Williams, R. L. 1994. Bioequivalence performance of several measures of rate of absorption. *Pharm Res* 11:966–74.

Boroujerdi, M. 2002. *Pharmacokinetics: Principles and Applications*, Chapter 8. New York: McGraw Hill-Medical Publishing Division.

Boxenbaum, H. 1998. Pharmacokinetics tricks and traps: Flip-flop models. *J Pharm Pharm Sci* 1(3):90–1.

Boxenbaum, H. G., Kaplan, S. A. 1975. Potential source of error in absorption rate calculations. *J Pharmacokin Biopharm* 3(4):257–64.

Bredberg, U., Karlsson, M. O. 1991. *In vivo* evaluation of the semi-simultaneous method for bioavailability estimation using controlled intravenous infusion as an "extravascular" route of administration. *Biopharm Drug Dispos* 12(8):583–97.

Byron, P. R., Notari, R. E. 1976. Critical analysis of "flip-flop" phenomenon in two-compartment pharmacokinetic model. *J Pharm Sci* 65(8):1140–4.

Charkoftaki, G., Dokoumetzidis, A., Valsami, G., Macheras, P. 2012. Elucidating the role of dose in the biopharmaceutics classification of drugs: The concepts of critical dose, effective *in vivo* solubility, and dose-dependent BCS. *Pharm Res.* DOI 10.1007/s11095-012-0815-4.

Chen, M. L. 1992. An alternative approach for assessment of rate of absorption in bioequivalence studies. *Pharm Res* 9:1380–5

Chen, M-L., Barbara Davit, B., Lionberger, R., Wahba, Z., Ahn, H-Y., Lawrence X., Yu, L. X. 2011. Using partial area for evaluation of bioavailability and bioequivalence. *Pharm Res* 28:1939–47.

Chen, M. L., Lesko, L. J., Williams, R. L. 2001. Measures of exposure versus measures of rate and extent of absorption. *Clin Pharmacokinet* 40:565–72.

Curry, S. H. 1981. Theoretical considerations in calculation of terminal phase half-times following oral doses, illustrated with model data. *Biopharm Drug Disp* 2: 115–21.

Endrenyi, L., Csizmadia, F., Tothfalusi, L., Balch, A. H., Chen, M. L. 1998a. The duration of measuring partial AUCs for the assessment of bioequivalence. *Pharm Res* 15:399–404.

Endrenyi, L., Csizmadia, F., Tothfalusi, L., Chen, M. L. 1998b. Metrics comparing simulated early concentration profiles for the determination of bioequivalence. *Pharm Res* 15:1292–9.

Garrett, E. R. 1993. Simplified methods for the evaluation of the parameters of the time course of plasma concentration in the one-compartment body model with first-order invasion and first-order drug elimination including methods for ascertaining when such rate constants are equal. *J Pharmacokinet Biopharm* 21(6):689–734.

Garrett, E. R. 1994. The Bateman function revisited: A critical reevaluation of the quantitative expressions to characterize concentrations in the one compartment body model as a function of time with first-order invasion and first-order elimination. *J Pharmacokinet Biopharm* 22(2):103–28.

Gibaldi, M., Perrier, D. 1982. *Pharmacokinetics*, 2nd Edition. New York: Marcel Dekker, Inc., 63–65.

Krzyzanski, W., Jusko, W. J. 1998. Integrated functions for four basic models of indirect pharmacodynamic response. *J Pharm Sci* 87(1):67–72.

Loo, J. C. K., Riegelman, S. 1968. New method for calculating the intrinsic absorption rate of drugs. *J Pharm Sci* 57:918–28.

Neelakantan, S., Veng-Pedersen, P. 2005. Determination of drug absorption rate in time-variant disposition by direct deconvolution using β clearance correction and end-constrained non-parametric regression. *Biopharm Drug Dispos* 26(8):353–70.

Oh, D. M., Curl, R. L., Amidon, G. L. 1993. Estimating the fraction dose absorbed from suspensions of poorly soluble compounds in humans: A mathematical model. *Pharm Res* 10:264–70.

Proost, J. H. 1985. Wagner's exact Loo-Riegelman equation: The need for a criterion to choose between the linear and logarithmic trapezoidal rule. *J Pharm Sci* 74:793–4.

Rescigno, A., Segre, G. 1966. *Drugs and Tracer Kinetics*, Waltham, MA: Blaisdell Publishing Company.

Rinaki, E., Dokoumetzidis, A., Valsami, G., Panos Macheras, P. 2004. Identification of biowaivers among Class II drugs: Theoretical justification and practical examples. *Pharm Res* 21(9):1567–72.

Ronfeld, R. A., Benet, L. Z. 1977. Interpretation of plasma concentration-time curves after oral dosing. *J Pharm Sci* 66(2):178–80.

Rostami-Hodjegan A., Jackson P.R., Tucker G.T. 1994. Sensitivity of indirect metrics for assessing "rate" in bioequivalence studies—Moving the "goalposts" or changing the "game". *J Pharm Sci* 83:1554–7.

Scheff, J. D., Almon, R. R., DuBois, D. C., Jusko, W. J., Androulakis, I. P. 2011. Assessment of pharmacologic area under the curve when baselines are variable. *Pharm Res* 28:1081–9.

Teorell, T. 1937. Kinetics of distribution of substances administered to the body I. The extravascular mode of administration. *Arch Int Pharmacodyn* 57:205–25.

Wagner, J. G. 1970. "Absorption rate constant" calculated according to the one-compartment model with first-order absorption: Implications in *in vivo – in vitro* correlation. *J Pharm Sci* 59:1049.

Wagner, J. G. 1974. Application of the Wagner-Nelson absorption method to the two-compartment model. *J Pharmacokin Biopharm* 2(6):469–86.

Wagner, J. G. 1975. *Fundamentals of Clinical Pharmacokinetics*. Hamilton, IL: Drug Intelligence Publications, Inc. 174–7.

Wagner, J. G. 1983. Pharmacokinetic absorption plots from oral data alone or oral/intravenous data and an exact Loo-Riegelman equation. *J Pharm Sci* 72:838–42.

Wagner, J. G. 1988. Types of mean residence times. *Biopharm Drug Dispos* 9:41–57.

Wagner, J. G., Ganes, D. A., Midha, K. K., Gonzalez-Younes, I., Sackellares, J. C., Olsen, L. D., Affrime, M. B., Patrick, J. E. 1990. Stepwise determination of multicompartment disposition and absorption parameters from extravascular concentration-time data. Application to mesoridazine, flurbiprofen, flunarizine, labetanol and diazepam. *J Pharmacokin Biopharm* 19:413–55.

Wagner, J. G., Nelson, E. 1964. Kinetic analysis of blood levels and urinary excretion in the absorptive phase after single doses of drug. *J Pharm Sci* 53:1392–403.

Wang, Y., Nedelman, J. 2002. Bias in the Wagner–Nelson estimate of the fraction of drug absorbed. *Pharm Res* 19(4):470–6.

Yáñez, J. A., Remsberg, C. M., Sayre, C. L., Forrest, M. L. 2011. Flip-flop pharmacokinetics—Delivering a reversal of disposition: Challenges and opportunities during drug development. *Ther Deliv* 2(5):643–72.

Yu, L. X., Crison, J. R., Amidon, G. L. 1996. Compartmental transit and dispersion model analysis of small intestinal transit flow in humans. *Int J Pharm* 140:111–18.

Zeng, Y. L., Akkermans, A. A. M. D., Breimer, D. D. 1983. Factors affecting the error in the Loo-Riegelman method for estimating the rate of drug absorption. *Arzneim Forsch* 33(1)5:757–60.

13 PK/TK of Xenobiotics Absorbed through the Skin (Transdermal Absorption)

13.1 INTRODUCTION

The main barrier to percutaneous absorption of xenobiotics is the stratum corneum (the Latin words for the outer horny layer of skin) that covers the epidermis layer, the precursor layer to the stratum corneum. The layers under the epidermis include the dermis, and the subcutaneous region that covers the muscle. The hair follicles, sweat glands, and two types of gland—apocrine and eccerine—are the appendages of the skin. The passage of xenobiotic is mostly through the epidermis and the widely distributed hair follicle and eccrine glands on the surface of the body.

13.1.1 Stratum Corneum

The stratum corneum, the horny layer of the skin, is the outer membrane of the epidermis. It is formed from 15–25 layers of biochemically inactive tissues originated from living keratinocytes (Odland, 1983) with estimated weight density of 1.3 to 1.4 gm/cm^2 (Scheuplein, 1978; Scheuplein and Bronaugh, 1983). The stratum corneum structurally has two distinct protein and lipoidal phases and plays an important part in making the skin a remarkable physiological barrier to the penetration of xenobiotics. This important physiological role is related to its anatomical structure, which is created by impermeable densely packed corneocytes, separated by lipid bilayers, thus minimizing the transdermal absorption of both hydrophobic and hydrophilic compounds. The diffusion of xenobiotics through the stratum corneum is dependent on diffusion through the lipid $(D_{coeff})_{lipid}$, diffusion through the protein $(D_{coeff})_{protein}$, and the partition coefficient of the compound P_{coeff} between the protein and lipid environments. Using the aforementioned parameters, the diffusivity of the stratum corneum (σ) can be estimated as

$$\sigma = \left(P_{coeff}\right) \times \left(\frac{\left(D_{coeff}\right)_{lipid}}{\left(D_{coeff}\right)_{protein}}\right) \tag{13.1}$$

The lipid layers of the stratum corneum are considered the main avenue for transdermal absorption. Various mathematical models have been proposed treating the stratum corneum as the brick-and-mortar models, organized as a fully aligned or fully staggered, or partially staggered two-dimensional structure (Michaels et al., 1975; Elias, 1983; Tojo, 1987; Lieckfeld and Lee, 1992; Heisig et al., 1996; Charalambopoulou et al., 2000; Frasch and Barbero, 2003), which all define the permeation through the stratum corneum according to the assumptions of the model and hexagonal shape of the corneocytes. The applicability and validity of the models are based on how well they predict the observed data and the ease of parameter estimation (Figure 13.1).

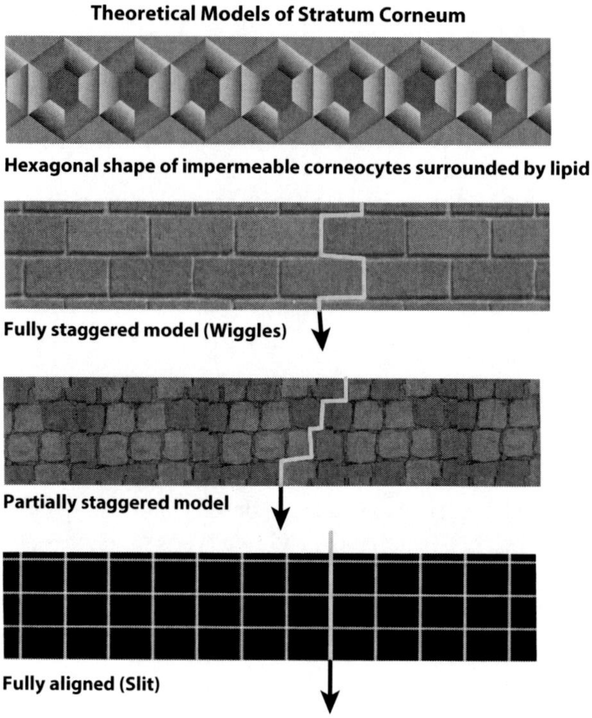

FIGURE 13.1 Structural models/analogies attributed to the permeation of xenobiotics through the stratum corneum.

13.1.2 Epidermis

When a compound crosses the stratum corneum, it encounters the viable epidermis layer, also known as the malpighian layer: a moist environment, with no capillaries, but with living cells that facilitate the transfer of xenobiotics to the lower layers (Sceuplein, 1976; Flynn, 1985). The viable epidermis poses a significant barrier to transdermal diffusion even in the absence of the stratum corneum (Andrews et al., 2013). It has a variable thickness of about 150–220 μm and holds two primary cells types: keratinocytes (in ascending order: stratum basale, stratum spinosom, stratum granulosum, stratum lucidum, and stratum corneum) and non-keratinocytes cells such as Markel cells, melanocytes, and Langerhans cells (Figure 13.2). The basal cells continuously generate new cells to be moved toward the outer membrane to replace the stratum corneum layer that are continually removed from the skin surface. The turnover rate in human and pig is about 30 days and in rodents it is much faster (Weinstein, 1966; Halprin, 1972; Bergstresser and Taylor, 1977). Carcinogenic agents/elements can cause irreparable damage to basal cells and trigger skin cancer, or squamous cell carcinomas. Melanocytes and Markel cells are derivatives of the neural crest and axone, respectively, and are located in the basal region of the epidermis. Langerhans cells are derived from bone marrow and play an important role in the skin immune response.

The epidermal living cells are joined together firmly and form another barrier to the penetration of xenobiotics. After crossing the stratum corneum, xenobiotics must permeate through the lopoidal cell membranes. For ions and polar compounds this is not a thermodynamically favorable action and the lipophilic compounds after permeating through the cell membranes face the aqueous cytoplasmic environment (Flynn, 1985). Thus, regardless of differences in physicochemical characteristics of xenobiotics, passing through the live epidermis layer of the skin is not an easy task for xenobiotics (Flynn, 1985; Masters et al., 1997).

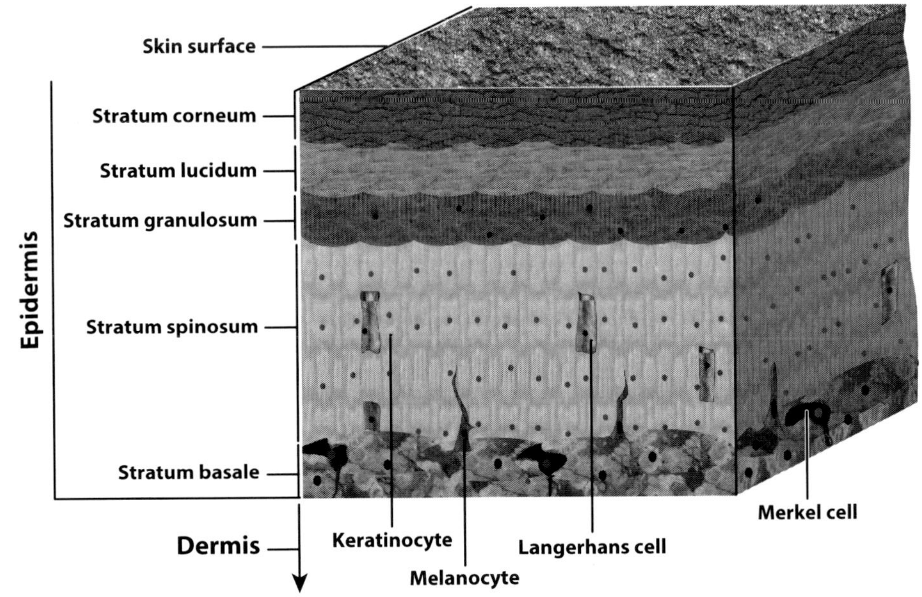

FIGURE 13.2 Layer of epidermis.

13.1.3 Dermis and Adnexial Appendages

The next layer of the skin is the dermal region that is connected to the epidermis through a thin layer called the basement membrane. The dermis has two regions; one region is attached to the epidermis called the papillary region, and the other is in the lower layer of the skin, a more vascular region known as the reticular region (James et al., 2005). The dermis is formed by a network of elastic fibers, a collagen and extrafibrilator matrix, and a watery gel environment called ground substance (Achterberg et al., 2014). The dermis holds a large number of functional elements of skin such as capillary network including anastomoses, touch and heat receptors, pilosebaceous units that refer collectively to hair follicles and sebaceous gland, exocrine and apocrine sweat glands, and the lymphatic network (Figure 13.3).

Apocrine glands open onto hair follicle and are found in the armpits and around genitals and contain chemicals known as pheromones. Eccrine glands that arise directly from the subcutaneous region to the skin surface are not connected to hair follicles, they open directly onto the skin surface and are found over the entire body surface area except lips, clitoris, labia minora, and external ear canal. The fluid secreted is sweat, which contains water, salt, urea, ammonia, and uric acid, which evaporates, providing a cooling sensation. The gaps in the interlocking lattice of the dermis are large and have little or no ability to filter molecules smaller than their openings. Thus, the permeation through the dermis is relatively facile without any significant molecular selectivity.

The absorption of xenobiotics through the skin occurs by two major routes: transepidermal and transappendageal. Transepidermal is the diffusion through or between the epidermis cells and transappendageal is the passage via the appendages such as hair follicles or sebaceous glands. The diffusion through the transappendageal route is more efficient, but the openings of hair follicles and sweat glands constitute only 0.6% of the skin surface area. Thus, the transepidermal diffusion seems play a more prominent role in transdermal absorption.

In general, skin is the largest organ that interfaces with the external environment, but has a highly restrictive permeability barrier. Xenobiotics must overcome the stratum corneum and associated layers underneath to reach the systemic circulation. This organ performs countless functions, which independently or collectively can influence the absorption of xenobiotics. By acting as an

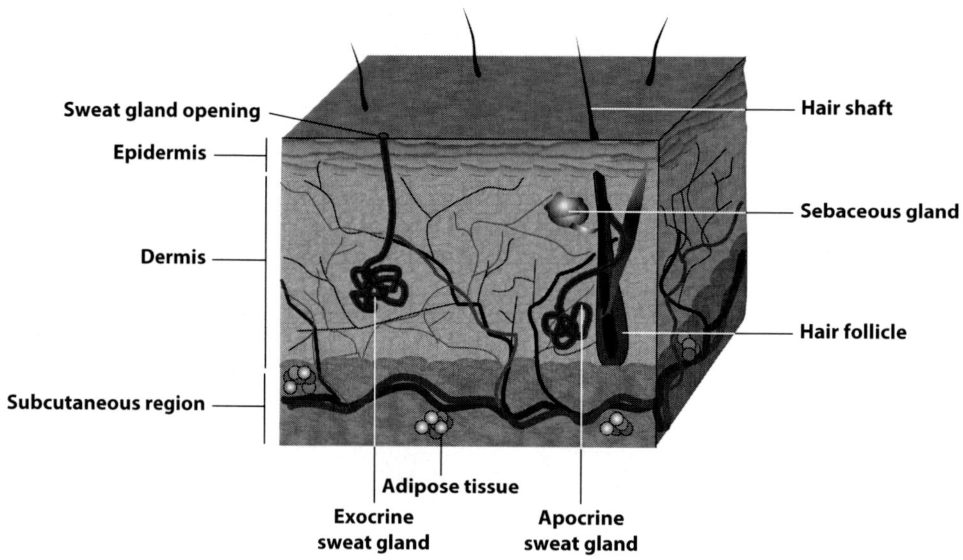

FIGURE 13.3 Structural components of skin.

insulator containing fat cells, blood vessels, and sweat glands, it regulates body temperature; it acts as a relatively impermeable membrane to keep fluid, electrolytes, and other components inside the body; it accounts for about 16% of total body weight, and provides a large surface area with variable thickness and characteristics.

13.2 FACTORS INFLUENCING THE TRANSDERMAL ABSORPTION OF XENOBIOTICS

The physicochemical factors of the compound such as diffusion of the xenobiotics in the formulation applied to the skin, or any type of moieties that are in contact with the skin can act as a rate-limiting factor or facilitator for the absorption of compounds through the skin and influence the rate and extent of transdermal absorption. The xenobiotic-related factors include the diffusion coefficient of drug through the skin; concentration of the compound in terms of thermodynamic activity; solubility in the formulation, pKa of the compound and pH of formulation; particle and molecular size; and partition coefficient of the compound.

Formulation factors affect not only the physical state of compounds applied to the skin, but also the physical characteristics of the skin. They include concentration of water in terms of thermodynamic activity of water in the formulation; viscosity of the formulation; the presence of alcohol, ether or other organic solvent; the presence of a surfactant (e.g., presence of 5% sodium lauryl sulfate, or 0.50% polysorbate 80) to make skin more permeable; the presence of penetration enhancers such as dimethylsulfoxide (DMSO), dimethylformamide (DMF), dimethyl acetamide (DMA) and Azone; hydrogenated soybean phospholipid (Nishihata et al., 1988); alcohols with long carbon chains (C8-C14); cyclicmonoterpenes and n-octanol (Parsaee et al., 2002; Ho et al., 1994); nonionic surfactants (Iwasa et al., 1991); and propylene glycol and isopropyl myristate (Ho et al., 1994; Santoyo et al., 1995).

The cutaneous metabolism is also a major factor in transdermal absorption of xenobiotics at the site of administration. The expression of multiple CYP450 enzymes such as CYP1A1, CYP1B1, CYP2B6, CYP2E1, CP3A4, and CYP3A5 have been found in keratinocyte culture; in addition, the presence of multidrug resistance-associated transport proteins 1 and 3–6 and MDR-1 have also been identified (Randolph et al., 1998; Baron et al., 2001).

The environmental factors that can influence the transdermal absorption of xenobiotics include temperature of the environment, which influence the blood flow and the state of hydration of skin, thus the permeability; relative humidity, which influences the state of hydration of the skin; and air current, which changes the state of hydration of skin by evaporation its water content.

13.3 MATHEMATICAL MODEL FOR PREDICTION OF SKIN CONCENTRATION

The interests in quantitative transdermal absorption extend over the transdermal absorption of therapeutic agents, environmental pollutant, and occupational chemical exposure, etc. The amount of a xenobiotic applied or deposited on the skin varies by the activity, lipophilicity, and other physicochemical characteristics of the compound, including the rate of deposition or application (Kissel et al., 1996). The interpretation of the data is often complex and may not follow a unified approach.

13.3.1 General Approaches Summarizing the Permeation and Biotransformation of Xenobiotics

The diffusion models are mostly based on Fick's second law of diffusion, which is obtained from Fick's first law by introducing the idea of conservation of mass to eliminate the flux J (Jost, 1952; Crank, 1975; Scheuplein, 1967; Scheuplein and Blank, 1973; Hada et al., 2005; Sugibayashi et al., 2010).

The sequence of events of xenobiotic permeation through the skin after exposure to the surface of skin is

1. Availability of the compound on the surface of the skin. For the *in vitro* systems this availability is the release from the device or applied layer on the exposed surface of the skin
2. Partitioning on skin surface
3. Diffusion and binding in stratum corneum
4. Permeation through the stratum corneum
5. Partition in viable epidermis
6. Diffusion through the viable epidermis and dermis
7. Metabolism in dermis and viable epidermis
8. Permeation through the capillaries and uptake by the systemic circulation
9. Residual uptake, if any, by the subcutaneous fatty layer

A general relationship summarizing the permeation and biotransformation of xenobiotic is presented in Equation 13.2 (Tojo, 1988)

$$\{1 + B(C)\}\frac{\partial C}{\partial t} = \frac{1}{r}\frac{\partial}{\partial r}\left(D_{coeff_r} r\frac{\partial C}{\partial r}\right) - \frac{\partial}{\partial x}\left(D_{coeff_x}\frac{\partial C}{\partial x}\right) - u_x\frac{\partial C}{\partial x} - F(C) - M(C) \quad (13.2)$$

where $B(C)$ is the binding term expressed by $p/(1 + qC)^2$ in the stratum corneum; p and q are identified as the model parameters based on the Langmuir isotherm; $F(C)$ is the facilitated transport through the stratum corneum in the cases where electrophoretic transport is applied; and $M(C)$ is the biotransformation defined by the Michaelis–Menten equation. The first term of the right hand side of Equation 13.2, $1/r\partial/\partial r(D_{coeff_r}, r\partial C/\partial r)$ stands for the radial diffusion; the second term, $\partial/\partial x\left(D_{coeff_x}\partial C/\partial x\right)$, stands for longitudinal diffusion; and the third term, $u_x\partial C/\partial x$, stands for convective transport.

A common approach for predicting dermal exposure and absorption is based on the assumption that a finite dose of xenobiotic is deposited on and absorbed by the skin. The application or exposure is uniform over the exposed skin surface area and the absorption is slow such that it can be differentiated from the application. The related relationship on the sequential processes of application followed by the absorption is

$$A_{abs} = Area(\text{cm}^2) \times A_{deposited}(\text{mg/cm}^2) \times F_{weight} \times F_{abs} \tag{13.3}$$

where A_{abs} is the amount absorbed; *Area* (cm^2) is the surface area available for contact; $A_{deposited}$ (mg/cm^2) is the amount deposited/applied on the skin; F_{weight} is the weight fraction of the compound in the mixture; and F_{abs} is the fraction of applied dose absorbed through the skin during the time of exposure.

13.3.2 Skin-Perm Model

The permeation coefficient through the stratum corneum for lipophilic compound using the lipid fraction of the stratum corneum can be estimated by the following empirical equation (Fehrenbacher and ten Berge, 2000):

$$\log P_{Stratum} = -1.326 + \left(0.6097 \times \log P_{coeff}\right) - \left(0.1786 \times MW^{0.5}\right) \tag{13.4}$$

where $\log P_{Stratum}$ is the permeation coefficient through the stratum corneum; P_{coeff} is partition coefficient of the compound in octanol-water system and MW is its molecular weight.

The empirical equations for estimating the permeation coefficient of the protein fraction of the stratum corneum and the aqueous layer of epidermis are

$$P_{protein}(\text{cm/h}) = \frac{0.0001519}{\sqrt{MW}} \tag{13.5}$$

$$P_{aqueous}(\text{cm/h}) = \frac{2.5}{\sqrt{MW}} \tag{13.6}$$

The permeation coefficient of skin from aqueous solution or from clear liquid is

$$P_{skin-water} = \frac{1}{\dfrac{1}{P_{Stratum} + P_{protein}} + \dfrac{1}{P_{aqueous}}} \tag{13.7}$$

The product of the permeation coefficient of the skin by the concentration of xenobiotic in the solution (mg/cm^3) yields the permeation rate in $\text{mg/cm}^2\text{/h}$ (Lotens and Wammes, 1993; Wilschut et al., 1995).

For absorption of gases or vapors, the aqueous-air partition coefficient is determined first (Equation 13.8) followed by multiplication with the skin-water permeation coefficient (Equation 13.7). The aqueous-air partition coefficient is calculated as

$$P_{coeff_{aq.air}} = \frac{R \times T \times W_{sb}}{Vp \times MW} \tag{13.8}$$

where (R) is the gas constant (8.314 Joule/mol/degree Kelvin, or Newton meter/mol/Kelvin) and T is the temperature in Kelvin. The product of $P_{coeff_{aq.air}}$ with $P_{skin-water}$ yields the permeation coefficient for the interaction of gaseous compound and skin:

$$P_{skin-air} = P_{coeffaq.air}\left(\frac{1}{\frac{1}{P_{Stratum} + P_{protein}} + \frac{1}{P_{aqueous}}}\right) \tag{13.9}$$

For gasses that absorb through the skin rapidly, the rate-limiting factor is the diffusion from the air to the skin. The stagnant air between the skin surface and the environment, *d*, is assumed to be 3 cm (Lotens et al., 1993):

$$D_{air} = 360\left(\sqrt{\frac{76}{MW}}\right) \tag{13.10}$$

$$P_{air} = \frac{D_{air}}{d} \tag{13.11}$$

$$P_{skin-air-air} = \frac{1}{\frac{1}{P_{skin-air}} + \frac{1}{P_{air}}} \tag{13.12}$$

where D_{air} is the air diffusion coefficient in cm^2/h; P_{air} is the permeation of the air layer in contact with the skin; and $P_{skin-air-air}$ is the permeation coefficient of the skin when the diffusion through the air is the rate-limiting factor for the absorption of gaseous compound. The multiplication of the permeation coefficient of the skin with the concentration of gaseous compound in the air (mg/cm^3) provides the rate of permeation.

The skin-perm model provides an estimation of the skin permeation of vapors. The advantage of the model is to evaluate the skin permeation of gaseous or vapors from the permeation coefficient of aqueous solutions.

13.3.3 One-Layered Diffusion Model

When the skin surface comes in contact with a xenobiotic, the absorption follows the passive diffusion consistent with the compounds activity gradient. According to the Fick's first law, the flux, *J*, of a substance through a one-layer barrier perpendicular to the direction of diffusion is directly proportional to the concentration gradient, that is,

$$J = -D_{coeff}\left(\frac{dC}{dx}\right) \tag{13.13}$$

where (*J*) is the quantity of a compound diffusing per unit time through a unit area of the barrier, *C* is the concentration and *x* is the distance. The quantity of the compound crossing the barrier at *x* in time δt is $J(\delta t)$ whereas the quantity leaving through the barrier at $(x + \delta x)$ at the same time is $(J + \partial J/\partial x\ \delta x)\delta t$. Thus, the net gain in the quantity of diffusing compound can be expressed in terms of the difference between *x* and δx, that is,

$$\delta C\delta x = J\delta t - \left(J + \frac{\partial J}{\partial x}\right)\delta t = -\frac{\partial J}{\partial x}\delta x\delta t \tag{13.14}$$

The second law defines the diffusion of the compound in the barrier as

$$\frac{\partial C}{\partial t} = D_{coeff} \frac{\partial^2 C}{\partial x^2} \tag{13.15}$$

This can be integrated as the following finite system:

$$C = X(x)T(t) = \left[A \sin \lambda x + B \cos \lambda x\right] e^{-\lambda^2 D_{coeff} t} \tag{13.16}$$

Under the assumption of homogenous one-layer membrane and sink condition at time t and distance x, the initial conditions at time $t = 0$ is as follows:

$$f(x) = C^0 \quad \text{for} \quad 0 < x < h$$
$$f(x) = 0 \quad \text{for} \quad h < x < L$$

The solution for this particular case is

$$C = C^0 \left(\frac{h}{L} + \frac{2}{\pi} \sum_{1}^{\infty} \frac{1}{n} e^{-(n\pi/L)^2 D_{coeff} t} \cos \frac{n\pi x}{L} \sin \frac{n\pi h}{L} \right) \tag{13.17}$$

It may be necessary to evaluate D_{coeff} as a function of time of the total solute, which has diffused across the boundary. This quantity, A_{TD}, is given by

$$A_{TD} = -\int_0^t D_{coeff} \left(\frac{\partial C}{\partial x} \right)_{x=h} dt \tag{13.18}$$

From Equation 13.17, $(\partial C/\partial x)_{x=h}$ can be determined as

$$\frac{\partial C}{\partial x} = -C^0 \frac{2}{L} \sum_{1}^{\infty} e^{-D_{coeff} t \left(\frac{n\pi h}{L} \right)^2} \sin^2 \frac{n\pi h}{L} \tag{13.19}$$

Substitution of Equation 13.19 into Equation 13.18 yields

$$M = C^0 \frac{1}{2\pi^2} \sum_{1}^{\infty} \frac{1}{n^2} \sin^2 \frac{n\pi h}{L} \left[1 - \exp\left\{ -\frac{D_{coeff} n^2 \pi^2 t}{L^2} \right\} \right] \tag{13.20}$$

The parameter A_{TD}, the transdermally absorbed amount, can be evaluated for various ratios of h/L. For instance, when $h/L = 1/2$, Equation 13.20 reduces to Equation 13.21:

$$A_{TD} = \frac{C^0 L}{4} \left[1 - \frac{8}{\pi^2} \sum_{m=1}^{\infty} \frac{1}{(2m-1)^2} \exp\left\{ -\frac{D_{coef} (2m-1)^2 \pi^2 t}{L^2} \right\} \right] \tag{13.21}$$

where $m = 0, 1, 2, 3, \ldots$

The average membrane concentration of xenobiotic is estimated as

$$C_{ave} = \frac{P_{coeff}C}{2}\left\{1 - \frac{8}{\pi^2}\sum_{m=1}^{\infty}\frac{1}{(2m-1)^2}\exp\left(-\frac{D_{coeff}(2m-1)^2\pi^2 t}{L^2}\right)\right\} \tag{13.22}$$

The steady state, that is at $t = \infty$,

$$\frac{8}{\pi^2}\sum_{m=1}^{\infty}\frac{1}{(2m-1)^2}\exp\left(-\frac{D_{coeff}(2m-1)^2\pi^2 t}{L^2}\right) \Rightarrow 0, \text{ and} \tag{13.23}$$

$$C_{ss} = \frac{P_{coeff}C}{2} \tag{13.24}$$

Equations 13.22 and 13.24 indicate that the average or steady-state concentration is directly proportional to the partition coefficient of xenobiotics.

13.3.4 Two-Layered Diffusion Model

This model resembles the anatomical features of the skin more closely. It takes into consideration the concentration of xenobiotic on the surface of skin, in the stratum corneum, and the epidermis and dermis. The boundary conditions of the model between the stratum corneum and epidermis-dermis layers include (Sugibayashi et al., 2010):

$$C_{ED} - \left(P_{coeff}\right)_{Stratum}\left(P_{coeff}\right)_{ED} C_{Stratum}$$

and

$$\left(D_{coeff}\right)_{Stratum}\frac{dC_{Stratum}}{dx} - D_{ED}\frac{dC_{ED}}{dx}$$

where ED stands for viable epidermis and dermis layer, P is the partition coefficient, (D_{coeff} and D_{ED}) are the diffusion coefficient and C represents the concentration of the xenobiotic in different layer. The total permeability coefficient of the whole skin includes the permeability coefficient of the stratum corneum and that of viable epidermis-dermis layer (Ghanem et al., 1992):

$$P_T - \frac{1}{\frac{1}{P_{Stratum}} + \frac{1}{P_{ED}}} \tag{13.25}$$

The reciprocal of the permeability coefficient is referred to as the permeability resistance coefficient. Thus, the total permeability of the skin (P_T) is the reciprocal of permeability resistances of stratum corneum ($(P_R)_{Stratum}$) and viable epidermis-dermis layers ($(P_R)_{ED}$).

$$P_T = \frac{1}{\left(P_R\right)_{Stratum} + \left(P_R\right)_{ED}} \tag{13.26}$$

The permeability coefficient of the stratum corneum is a function of its transcellular and paracellular permeations, that is,

$$P_{Stratum} = P_{trans} + P_{para} \tag{13.27}$$

The amount of xenobiotic absorbed per unit of area of stratum corneum is defined as

$$A_{Stratum} = \frac{\left(P_{coeff}\right)_{Stratum} \times C \times L_{Stratum}\left(1 + \frac{\left(P_R\right)_{Stratum}}{\left(P_R\right)_T}\right)}{2} \tag{13.28}$$

The amount of compound per unit area of viable epidermis and dermis layer is

$$A_{ED} = \frac{\left(P_{coeff}\right)_{ED} \times C \times L_{ED} \times \frac{\left(P_R\right)_{ED}}{\left(P_R\right)_T}}{2} \tag{13.29}$$

where $L_{Stratum}$ and L_{ED} are the thickness of the stratum corneum and epidermis-dermis layer, respectively; $(P_R)_T$ is the total permeability resistance of the skin, that is, $1/P_T$; and C is the concentration of the xenobiotic on the surface of skin.

Merging Equations 13.28 and 13.29 yields Equation 13.30, which defines the total amount of xenobiotic per unit area of skin (Sugibayashi et al., 2010):

$$A_T = \frac{C}{2}\left\{\left(P_{coeff}\right)_{Stratum} \times L_{Stratum}\left(1 + \frac{\left(P_R\right)_{Stratum}}{\left(P_R\right)_T}\right) + \left(P_{coeff}\right)_{ED} \times L_{ED} \times \frac{\left(P_R\right)_{ED}}{\left(P_R\right)_T}\right\} \tag{13.30}$$

Dividing Equation 13.30 by total thickness of the skin yields the average concentration of the compound in the skin:

$$C_{ave} = \frac{C}{2L_T}\left\{\left(P_{coeff}\right)_{Stratum} \times L_{Stratum}\left(1 + \frac{\left(P_R\right)_{Stratum}}{\left(P_R\right)_T}\right) + \left(P_{coeff}\right)_{ED} \times L_{ED} \times \frac{\left(P_R\right)_{ED}}{\left(P_R\right)_T}\right\} \tag{13.31}$$

where L_T is the total thickness of the skin, that is, $L_T = L_{Stratum} + L_{ED}$.

The average concentration (Equation 13.31) in terms of the permeability is defined as

$$C_{ave} = \frac{C}{2L_T}\left\{\left(P_{coeff}\right)_{Stratum} \times L_{Stratum}\left(1 + \frac{P_T}{P_{Stratum}}\right) + \left(P_{coeff}\right)_{ED} \times L_{ED} \times \frac{P_T}{P_{ED}}\right\} \tag{13.32}$$

13.4 PK/TK OF PERCUTANEOUS ABSORPTION

13.4.1 Estimation of PK/TK Parameters and Constants of Skin Penetration from Urinary Data

The *in vivo* disposition kinetics of a transdermally absorbed compound into the systemic circulation is often complex to estimate, mostly because of the unknown magnitude of the absorbed

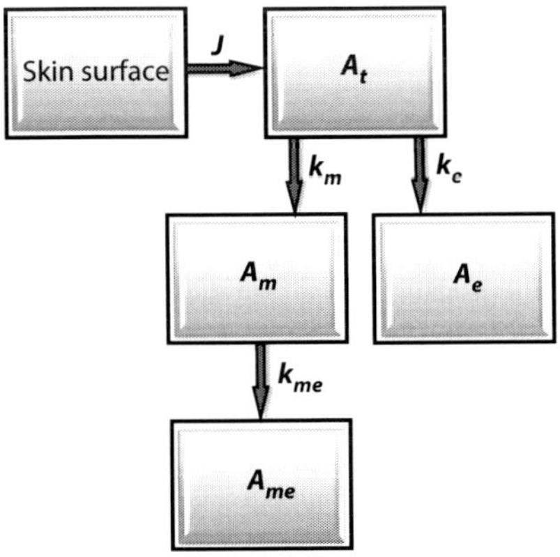

FIGURE 13.4 Diagram of a one-compartment model used to estimate PK/TK parameters and constants of skin penetration from urinary data.

dose. The compound can be rubbed off or spreads out on the skin, or skin can be in contact with a large volume of a reservoir of the applied dose (Wurster and Kramer, 1961) Regardless of type of application/exposure, the urinary data can be used to estimate the magnitude of absorbed dose. One approach is based on a single compartment model with exit rate constant for excretion and metabolism (Cooper and Berner, 1985).

The general differential equation of the model is similar to the one-compartment model with absorption from an extravascular site of administration (Figure 13.4) with the following modifications:

$$\frac{dA}{dt} = \left(Area_{skin}\right)J - KA_t \tag{13.33}$$

The amount in the body or the compartment, A and J, is the flux across the skin at its inner layer. The rate constant of elimination from the compartment is K and $Area_{skin}$ is the area of skin application/exposed and dA/dt is the rate of change of amount in the body with respect to time.

The rate constant of elimination represents the excretion of the unchanged compound and elimination of its metabolites, that is,

$$K = k_e + k_m + k_{nr} \tag{13.34}$$

The rate of change of the amount of metabolite or unchanged in the body with respect to time are

$$\begin{aligned} \frac{dA_m}{dt} &= k_m A - k_{me} A_m \\ \frac{dA_e}{dt} &= k_e A_t \end{aligned} \tag{13.35}$$

where k_m is the constant of formation of metabolite; k_{me} is the rate constant of elimination of metabolite; A_m is the amount of metabolite, and k_e is the excretion rate constants.

13.4.1.1 Urinary Data and PK/TK Relationships of Infinite Dose

In the case of infinite dose, one approach is to rely on the urinary excretion and metabolism of the compound. The total rate of elimination from the body based on the assumptions of this model (i.e., using a one-compartment model, assuming that all biological processes follow first-order kinetics, and the effective dose for systemic effect is the amount absorbed in the systemic circulation) is

$$\text{Total rate of elimination} = KA = k_e A_t + k_{me} A_m \tag{13.36}$$

The total amount of the applied dose eliminated unchanged and as a metabolite in the urine is based on the following integrated equations:

$$A_e = k_e \int_0^t A_t dt \tag{13.37}$$

$$A_{me} = k_{me} \int_0^t A_{m_t} dt \tag{13.38}$$

The asymptotic limit with an infinite dose yields (Cooper, 1976; Cooper et al., 1985)

$$A_e^{Area} = \lim_{t \gg t_{lag}^{Ae}} A_e = (Area)_{skin} \times J_{ss} \times \left(\frac{k_e}{K}\right) \times \left(t - t_{lag}^{A_e}\right) \tag{13.39}$$

$$A_{me}^{Area} = \lim_{t \gg t_{lag}^{me}} = (Area)_{skin} \times J_{ss} \times \left(\frac{k_m}{K}\right) \times \left(t - t_{lag}^{A_{me}}\right) \tag{13.40}$$

where (J_{ss}) is the steady-state flux. The lag time of excretion and elimination of metabolite are

$$t_{lag}^{A_e} = \frac{1}{k_{D_{coeff}}} + \frac{1}{K} \tag{13.41}$$

$$t_{lag}^{A_{me}} = \frac{1}{k_{D_{coeff}}} + \frac{1}{k_{me}} + \frac{1}{K} \tag{13.42}$$

Therefore, the lag time of the system is the sum of lag time of excretion and lag time of elimination of metabolite(s). The missing fractional sum from the model is the amount eliminated through non-renal elimination including biliary excretion and elimination of metabolites.

In Equations 13.41 and 13.42, $k_{D_{coeff}}$ is

$$k_{D_{coeff}} = \frac{6D_{coeff}}{l^2} \tag{13.43}$$

where D_{coeff} and l^2 are the diffusion coefficient and thickness of stratum corneum, respectively.

Combining Equations 13.39 and 13.40 yields Equation 13.44, representing the total amount absorbed through the exposed area of skin into the systemic circulation that eliminates from the body by excretion and metabolism at steady-state flux:

$$(A_t)_{Area} = A_e^{Area} + A_{me}^{Area} = (Area)_{skin} \times J_{ss} \times \left(\frac{k_e + k_m}{K}\right) \times \left[t - \left(\frac{k_e}{k_e + k_m}\right) t_{lag}^{A_e} - \left(\frac{k_m}{k_e + k_m}\right) t_{lag}^{A_{me}}\right] \tag{13.44}$$

Equation 13.44 is based on the assumption that the urinary excretion and metabolism reach a maximum, and if this maximum is sufficiently high enough before declining with time the steady-state flux is achieved. At the steady state, the total elimination of the transdermally absorbed compound is $(Area)_{skin} \times J_{ss}$ times the fraction excreted unchanged and eliminated as metabolite.

13.4.1.2 Urinary Data and PK/TK Relationships of Finite Dose

The finite dose implies that a known volume of dose is in contact with the skin surface and the solvent or the vehicle of the xenobiotic does not evaporate. The model is the same as discussed before, except the flux is determined by Equation 13.45:

$$V_{dose} \times \frac{dC_{dose}}{dt} = Area \times D_{coeff} \times \left(\frac{\partial C_{skin}}{\partial x} \right)_{x=0} \tag{13.45}$$

where V_{dose} is the volume of the dose, dC_{dose}/dt is the rate of change of dose concentration with respect to time, and $(\partial C_{skin}/\partial x)_{x=0}$ is the partial differential equation of the concentration migration in the skin with the initial condition of $x = 0$. The partial differential equation of concentration change in the skin with respect to time is

$$\frac{\partial C_{skin}}{\partial t} = D_{coeff} \frac{\partial^2 C_{skin}}{\partial x^2} \tag{13.46}$$

The initial conditions of Equation 13.46 are

$$C_{skin}(x,0) = 0 \quad C_{skin}(l,t) = 0 \quad C_{skin}(0,t) = P_{coeff} C_{dose}(t) \tag{13.47}$$

The partition coefficient, P_{coeff}, is the partition coefficient of xenobiotic between the dose and skin, C_{skin}/C_{dose}. The assumptions of the model include (1) the rate of diffusion in the applied volume of the dose on the surface of skin is much faster than the diffusion in the skin, and (2) the systemic circulation in the viable epidermis and dermis presents a sink condition.

The solution of Equations 13.45 and 13.46 based on the boundary conditions of Equation 13.47 is defined (Carslaw and Jaeger, 1959) as follows:

$$J = -D_{coeff} \left(\frac{\partial C_{skin}}{\partial x} \right)_{x=l} = 2J_{ss} \sum_{n=1}^{\infty} f_n e^{-\gamma_n t} \tag{13.48}$$

where f_n and γ_n are

$$f_n = \frac{\alpha_n^2}{\cos\alpha_n \left(\beta + \beta^2 + \alpha_n^2 \right)} \tag{13.49}$$

$$\gamma_n = \frac{\alpha_n^2 D_{coeff}}{l^2} = \frac{\alpha_n^2}{6} k_{D_{coeff}} \tag{13.50}$$

where α_n is the root of $(\alpha \tan \alpha = \beta)$; and β in Equations 13.49 and 13.50 is defined as (Carslaw et al., 1959)

$$\beta = \frac{P_{coeff} \times l_{skin}}{l_{dose}} \tag{13.51}$$

Other parameters of Equations 13.48 through 13.51 are $l_{dose} = V_{dose}/Area$, which is referred to as the thickness of the applied dose; J_{ss} is the steady-state flux that is dependent on the concentration.

Substitution of Equation 13.48 into Equation 13.33 followed by the integration yields Equation 13.52, which defines the amount in the body, at a given time with a finite dose (Cooper et al., 1985):

$$\left(A_t\right)_{Area} = 2\left(Area \times J_{ss}\right)\sum_{n=1}^{\infty}\frac{f_n}{\left(K - \gamma_n\right)}\left(e^{-\gamma_n t} - e^{-Kt}\right) \tag{13.52}$$

The total amount of the metabolite in the compartment can then be presented as

$$A_m = 2\left(Area \times J_{ss}\right)\sum_{n=1}^{\infty}\frac{k_m f_n}{K - \gamma_n}\left[\frac{e^{-\gamma_n t} - e^{-k_{me}t}}{\left(k_{me} - \gamma_n\right)} - \frac{e^{-Kt} - e^{-k_{me}t}}{\left(k_{me} - K\right)}\right] \tag{13.53}$$

If the rate constant of elimination of metabolites is greater than the overall elimination rate constant and the parameter γ_n, Equation 13.53 reduces to Equation 13.54 and the total rate of elimination can be defined as Equation 13.55

$$\lim_{\substack{k_{me} \gg \gamma_n \\ k_{me} \gg K}} \left(A_m\right)_t = \frac{k_m}{k_{me}} A_t \tag{13.54}$$

The total rate of elimination is

$$\frac{k_e + k_m}{K}\left(Area \times J_{ss}\right) = k_e A_t + k_{me} A_m \tag{13.55}$$

where the total fraction of absorbed dose eliminated in the urine is $k_e + k_m/K$.

13.4.1.3 PK/TK Model for Simultaneous Loss from the Skin Surface and Absorption through the Skin

The model is a simple one-compartment model with respect to the systemic circulation and takes into consideration the amount on the surface of the skin as a compartment with zero-order or first-order loss from the surface (Guy and Hadgraft, 1982; 1984). The release from the skin compartment is into the central compartment, which includes systemic circulation and highly perfused tissues. The schematic diagram of the model is presented in Figure 13.5. For the purpose of simplicity, the elimination from the central compartment (systemic circulation) and highly perfused tissues are not included in the mathematics of the model (Guy et al., 1984).

13.4.1.3.1 PK/TK Model with Zero-Order Loss from the Skin Surface

The differential equations of the model are

$$\frac{dC_{surface}}{dt} = -k_0 - k_1 C_{surface} \tag{13.56}$$

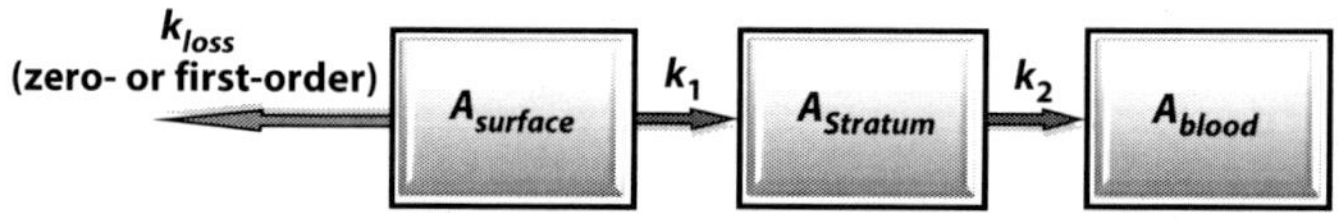

FIGURE 13.5 Diagram of a general model with first-order or zero-order loss from skin surface.

$$\frac{dC_{Stratum}}{dt} = \left(\frac{V_{surface}}{V_{Stratum}}\right) k_1 C_{surface} - k_2 C_{Stratum} \tag{13.57}$$

$$\frac{dC_{blood}}{dt} = \left(\frac{V_{Stratum}}{Vd}\right)\left(k_2 C_{Stratum}\right) \tag{13.58}$$

where $dC_{surface}/dt$ is the rate of concentration change of the applied dose on the surface of skin; k_0 is the zero-order rate constant of loss from the surface; k_1 is the first-order transfer rate constant from the surface to the stratum corneum; $C_{surface}$ is the concentration of the applied dose on the surface; $dC_{Stratum}/dt$ is the rate of change of concentration in stratum corneum; $V_{surface}$, $V_{Stratum}$, and Vd are the respective volumes of compartments; $C_{Stratum}$ is the concentration in the stratum corneum; k_2 is the first-order transfer rate constant from stratum corneum to the systemic circulation; and dC_{blood}/dt is the first-order rate of entry into the systemic circulation.

The integration of the differential equations using the Laplace transform yields the following relationship:

$$\left(A_{surface}\right)_N = \left(1 + \frac{k_0}{k_1}\right) e^{-k_1 t} - \frac{k_0}{k_1} \tag{13.59}$$

$$\left(\frac{V_{Stratum}}{V_{surface}}\right)\left(A_{Stratum}\right)_N - \frac{k_1}{k_2 - k_1}\left(e^{-k_1 t} - e^{-k_2 t}\right) - \frac{k_0}{k_2}\left[1 + \frac{1}{k_2 - k_1}\left(k_1 e^{-k_2 t} - k_2 e^{-k_1 t}\right)\right] \tag{13.60}$$

$$\begin{aligned}\left(\frac{Vd}{V_{surface}}\right)\left(A_{blood-input}\right)_N &= 1 + \frac{1}{k_2 - k_1}\left(k_1 e^{-k_2 t} - k_2 e^{-k_1 t}\right) \\ &- k_1 k_2 k_0 \left[\frac{t}{k_1 k_2} - \frac{k_1 + k_2}{k_1^2 k_2^2} + \frac{e^{-k_1 t}}{k_1^2 \left(k_1 - k_2\right)} + \frac{e^{-k_2 t}}{k_2^2 \left(k_1 - k_2\right)}\right]\end{aligned} \tag{13.61}$$

where $(A_{surface})_N$, $(A_{Stratum})_N$, and $(A_{blood\ input})_N$ are normalized amounts on the surface, stratum corneum, and blood with respect to the initial amount on the skin at time zero. The values are also normalized with respect to the ratio of the volume of distribution of receiver/donor compartments. Equations 13.59 through 13.61 are valid as long as the xenobiotic is present on the surface of the skin. The time required for the total dose on the surface to be lost or absorbed into the stratum corneum ($t_{duration}$) is

$$t_{duration} = \frac{1}{k_1} \ln\left(\frac{k_1}{k_0} + 1\right) \tag{13.62}$$

13.4.1.3.2 PK/TK Model with First-Order Loss from the Skin Surface

The differential equations of the model (according to the schematic diagram of Figure 13.5) for stratum and blood compartments are the same as those described for zero-order loss from the surface. The differential equation of change in concentration of xenobiotic on the surface, however, is different as described below

$$\frac{dC_{surface}}{dt} = -\left(k_{loss} + k_1\right) C_{surface} \tag{13.63}$$

where k_{loss} represents the first-order rate constant for the surface loss.

The integrated equations of the model using the Laplace transform are

$$\left(A_{surface}\right)_N = e^{-\left(k_{loss}+k_1\right)t} \tag{13.64}$$

$$\left(\frac{V_{Stratum}}{V_{surface}}\right)\left(A_{Stratum}\right)_N = \frac{k_1}{k_2-\left(k_{loss}+k_1\right)}\left(e^{-\left(k_{loss}+k_1\right)t}-e^{-k_2t}\right) \tag{13.65}$$

$$\left(\frac{Vd}{V_{surface}}\right)\left(A_{blood-input}\right)_N = \frac{k_1k_2}{\left(k_{loss}+k_1\right)k_2}\left[1+\frac{k_2e^{-\left(k_{loss}+k_1\right)t}-\left(k_{loss}+k_1\right)e^{-k_2t}}{\left(k_{loss}+k_1\right)-k_2}\right] \tag{13.66}$$

For the first-order loss, because Equations 13.64 through 13.66 are exponential, the determination of duration of presence of xenobiotic on the surface, similar to what was described for zero-order loss (Equation 13.62), is not feasible. The value can be estimated, however, as the time required for a certain fraction of the amount on the surface to disappear. For example, the time required for 90% of xenobiotic on the surface to disappear by first-order loss and first-order absorption into the stratum corneum is

$$\left(t_{duration}\right)_{90\%} = \frac{\ln(10)}{k_{loss}+k_1} \tag{13.67}$$

13.4.1.4 PK/TK Model for Xenobiotic Disposition Kinetics in Skin and Plasma and the Reservoir Effect

The focus of the PK/TK model considered in the previous section (Section 13.4.1.3) was related to the role of the surface concentration and the parallel removal from the surface. Thus, the blood compartment was considered only a receiver compartment without the exit rates. In this section, the model as depicted in Figure 13.6, which in addition to including the elimination from the systemic circulation takes into consideration the preferential binding of xenobiotic to stratum corneum (Guy et al., 1984). The binding is identified for the compounds that are known to exhibit a reservoir effect (McKenzie and Stoughton, 1962; Vickers, 1963; Stoughton, 1965; Carr and Wieland, 1966; Winkelmann, 1969).

The differential equations of the model are

$$\frac{dC_{surface}}{dt} = -k_1C_{surface} \tag{13.68}$$

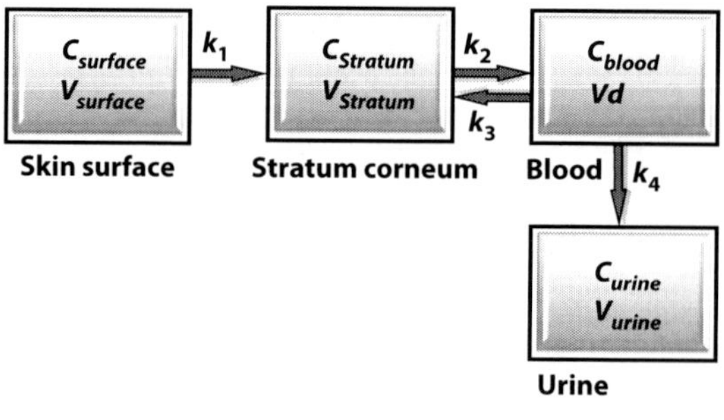

FIGURE 13.6 Diagram of compartmental model assuming the stratum corneum exhibits a reservoir effect and is in exchange with the central compartment.

$$\frac{dC_{Stratum}}{dt} = \frac{V_{surface}}{V_{Stratum}} k_1 C_{surface} - k_2 C_{Stratum} + \frac{Vd}{V_{Stratum}} k_3 C_{blood} \tag{13.69}$$

$$\frac{dC_{blood}}{dt} = \frac{V_{Stratum}}{Vd} k_2 C_{Stratum} - \left(k_3 + k_4\right) C_{blood} \tag{13.70}$$

$$\frac{dC_{urine}}{dt} = \frac{Vd}{V_{urine}} k_4 C_{blood} \tag{13.71}$$

The integrated form of Equation 13.68 is a simple monoexponetial equation that represents the time course of concentration change on the surface of the skin. The integrated equations for the stratum corneum and blood compartment are presented below. The concentration terms similar to the previous section are normalized with respect to the initial concentration:

$$\left(C_{Stratum}\right)_N = \frac{V_{surface}}{V_{Stratum}} k_1 \left[\frac{\left(k_3 + k_4 + k_1\right)e^{-k_1 t}}{\left(k_1 - \alpha\right)\left(k_1 - \beta\right)} + \frac{\left(k_3 + k_4 - \alpha\right)e^{-\alpha t}}{(\alpha - \beta)\left(\alpha - k_1\right)} + \frac{\left(k_3 + k_4 - \beta\right)e^{-\beta t}}{\left(\beta - k_1\right)(\beta - \alpha)}\right] \tag{13.72}$$

$$\left(C_{blood}\right)_N = \frac{V_{surface}}{Vd} k_1 k_2 \left[\frac{e^{-k_1 t}}{\left(k_1 - \alpha\right)\left(k_1 - \beta\right)} + \frac{e^{-\alpha t}}{(\alpha - \beta)\left(\alpha - k_1\right)} + \frac{e^{-\beta t}}{\left(\beta - k_1\right)(\beta - \alpha)}\right] \tag{13.73}$$

where $(C_{Stratum})_N$ and $(C_{blood})_N$ are the normalized concentration of the stratum corneum and blood, that is,

$$\left(C_{Stratum}\right)_N = \frac{C_{Stratum}}{C_0}, \text{ and} \tag{13.74}$$

$$\left(C_{blood}\right)_N = \frac{C_{blood}}{C_0} \tag{13.75}$$

where the initial concentration on the surface of the skin at time zero is C_0; the rate constant of permeation from the surface to the stratum corneum is k_1, estimated from the rate of disappearance from the surface; the rate constant of penetration into the skin is k_2, which includes permeation through the viable dermis; the relative affinity of the compound for the stratum corneum is k_3, it is considered in this model larger than k_2; the rate constant of elimination from the blood compartment is k_4; the concentrations at each compartment is identified as C of that compartment; and the volume of each compartment is represented by V.

The hybrid rate constants of α and β, the roots of the quadratic equation generated during the integration by the Laplace transform (Appendix A), are equal to

$$\alpha \times \beta = k_2 k_4 \tag{13.76}$$

$$\alpha + \beta = k_2 + k_3 + k_4 \tag{13.77}$$

13.4.1.5 PK/TK Model for Percutaneous Absorption through the Viable Epidermis with Parallel Penetration via Appendages

In addition to permeation through the layers of the stratum corneum, viable epidermis and dermis, and capillaries, some xenobiotics may reach the systemic circulation via transappendageal pathways of permeation (Wallace and Barnett, 1978). Figures 13.7 through 13.9 are the diagrams of proposed models for the combined penetration into the systemic circulation.

The model presented in Figure 13.7, compartment 1 is the constant amount of compound in solution or suspension applied on the surface of the skin. Compartment 2 is considered the skin barrier as a single homogeneous layer. Compartment 3 is the receiver compartment with a negligible exit rate constant of k_{32}. This makes compartment 3 compliable with the sink condition (Wallace et al., 1978). The differential equations of the model based on the assumption of each compartment are

$$\frac{dA_1}{dt} = 0 \tag{13.78}$$

$$\frac{dA_2}{dt} = k_{12}A_1 - (k_{21} + k_{23})A_2 \tag{13.79}$$

$$\frac{dA_3}{dt} = k_{23}A_2 \tag{13.80}$$

The integration of Equation 13.80 yields Equation 13.81, which is the amount in compartment 3 as a function of time. The model is suitable for *in vitro* assessment of percutaneous absorption of xenobiotics with compartment 3 as the sampling compartment.

$$(A_3)_t = \frac{A_1 k_{12} k_{23}}{k_{21} + k_{23}}\left(t - \frac{1}{k_{21} + k_{23}} + \frac{e^{-(k_{21}+k_{23})t}}{k_{21} + k_{23}}\right) \tag{13.81}$$

At steady state, that is, when $t \Rightarrow \infty$ and $e^{-(k_{21}+k_{23})t} \Rightarrow 0$, Equation 13.81 reduces to the straight line of Equation 13.82:

$$(A_3)_{ss} = \frac{A_1 k_{12} k_{23}}{k_{21} + k_{23}}\left(t - \frac{1}{k_{21} + k_{23}}\right) \tag{13.82}$$

The slope of the line is the diffusion rate or flux $J = \dfrac{A_1 k_{12} k_{23}}{k_{21} + k_{23}}$ (13.83)

The y-intercept is the lag time of absorption: $t_{lag} = \dfrac{1}{k_{21} + k_{23}}$ (13.84)

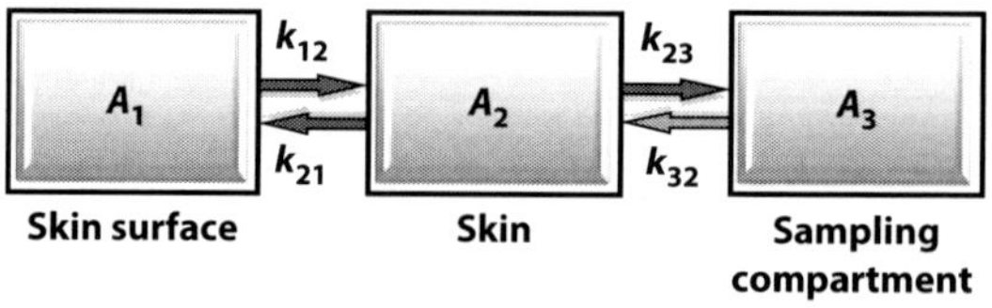

FIGURE 13.7 Diagram of a three-compartment model for skin, considering the skin barrier as a single homogeneous layer and a receiver/sampling compartment that complies with the sink condition.

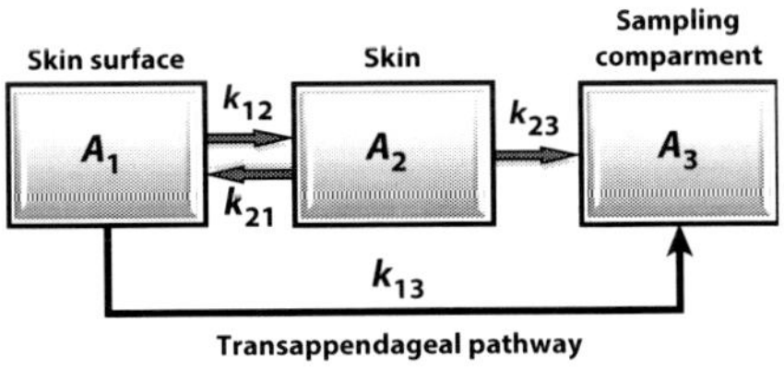

FIGURE 13.8 Diagram of a compartmental model for parallel route of penetration through the skin and also direct entry into the receiver compartment through transappendageal pathway.

The rate of transfer from compartment 1 to compartment 2 is A_1k_{12} and the total amount in compartment 2 at steady state is the ratio of the rate to the exit rate constants:

$$(A)_{ss} = \frac{A_1k_{12}}{k_{21} + k_{23}} \tag{13.85}$$

The model presented in Figure 13.8 characterizes the parallel routes through the skin layers and also direct entry via the transappendageal pathway into compartment 3. The differential equation of compartment 3, representing the rates of input into compartment 3 is

$$\frac{dA_3}{dt} = k_{13}A_1 + k_{23}A_2 \tag{13.86}$$

The integration of Equation 13.86 yields the amount of the compound in the sampling compartment as a function of time as presented in Equation 13.87 (Wallace et al., 1978)

$$A_3 = A_1k_{13}t + \left(\frac{A_1k_{12}k_{23}}{k_{21} + k_{23}}\right)\left(t - \frac{1}{k_{21} + k_{23}} + \frac{e^{-(k_{21}+k_{23})t}}{k_{21} + k_{23}}\right) \tag{13.87}$$

The diffusion rate of the model (flux) at steady state includes the bypass passageway to compartment 3

$$J_{ss} = A_1\left(\frac{k_{13} + k_{12}k_{23}}{k_{21} + k_{23}}\right) \tag{13.88}$$

The lag time of absorption in this model due the bypass passageway is shorter than the previous model and is equal to

$$t_{lag} = \frac{1}{k_{21} + k_{23}} \quad \frac{k_{13}}{(k_{21} + k_{23})k_{13} + k_{12}k_{23}} \tag{13.89}$$

The amount in compartment 2 remains the same as described in previous model.

The model presented in Figure 13.9 is a more complex model based on permeation through different tissue compartments of the skin in combination with the transappendageal absorption. The skin layer is represented by three compartments (Barnett and Locko, 1977; Wallace et al., 1978). Compartment 1 represents the skin surface; compartments 2 and 3 are the tissue compartments representing the stratum corneum, and viable epidermis and dermis layers. The transappendageal

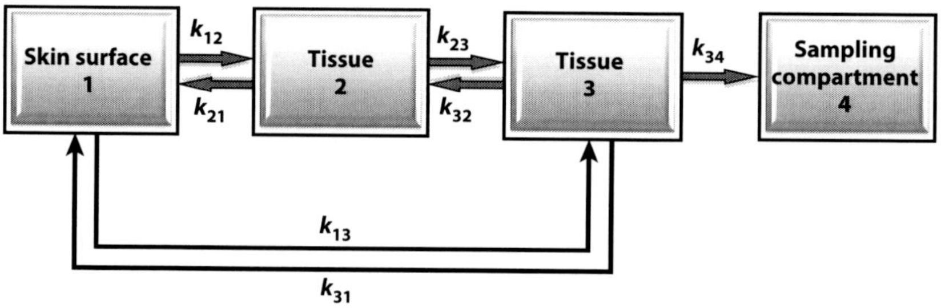

FIGURE 13.9 Diagram of a more complex compartmental model based on permeation through different layers of skin in combination with the transappendageal absorption and exchange.

transfer is in compartment 3 and is considered in both directions, an occurrence that may exist only for a few compounds. The combined entry into compartment 3 makes it a reservoir compartment with input into the sampling compartment. The complex nature of the model lessens its practicality.

13.4.1.6 Diffusion–Diffusion Model and Statistical Moments for Percutaneous Absorption

The diffusion–diffusion model is based on the assumption that the applied dose on the surface of the skin and the skin itself are two simple diffusion layers with a thickness of $x_{surface}$ and x_{skin}, respectively (Kubota and Ishizaki, 1986). The $x_{surface}$ value is equal to the real thickness of the layer applied to the skin multiplied by the bioavailability of the applied xenobiotics. The volume of the applied dose on the surface and that of the skin are defined as $(Area \times x_{surface})$ and $(Area \times x_{skin})$, and the diffusion coefficient of each volume is $(D_{coeff})_{surface}$ and $(D_{coeff})_{skin}$. The normalized diffusion coefficient, a rate constant similar to the first-order absorption rate constant, is estimated as

$$k_{diffusion} = \frac{\left(D_{coeff}\right)_{skin}}{\left(x_{skin}\right)^2} \tag{13.90}$$

The surface and skin concentrations terms are $C_{surface}$ and C_{skin}, respectively.

The fluxes of the xenobiotic on the surface and through the skin are defined as

$$J_{surface} = -\left(D_{coeff}\right)_{surface} \times Area \times \frac{\partial C_{surface}}{\partial x_{surface}} \tag{13.91}$$

$$J_{skin} = -\left(D_{coeff}\right)_{skin} \times Area \times \frac{\partial C_{skin}}{\partial x_{skin}} \tag{13.92}$$

The partial differential equations according the Fick's second law for the surface and the skin are

$$\frac{\partial C_{surface}}{\partial t} = \left(D_{coeff}\right)_{surface} \frac{\partial^2 C_{surface}}{\partial x_{surface}^2} \tag{13.93}$$

$$\frac{\partial C_{skin}}{\partial t} = \left(D_{coeff}\right)_{skin} \frac{\partial^2 C_{skin}}{\partial x_{skin}^2} \tag{13.94}$$

The boundary conditions are

$$\frac{\partial C_{surface}}{\partial t}\left(-x_{surface},t\right) = 0 \tag{13.95}$$

$$C_{surface}(0,\ t)P_{coeff} = C(0,\ t) \tag{13.96}$$

$$J_{surface}(0,\ t) = J_{skin}(0,\ t) \tag{13.97}$$

$$J_{skin} = \left(x_{skin},\ t\right) = Cl_{capillary}C\left(x_{skin},\ t\right) \tag{13.98}$$

The initial conditions are

$$C_{surface}\left(x_{surface},0\right) = C_0 \tag{13.99}$$

$$C_{skin}\left(x_{skin},\ 0\right) = 0 \tag{13.100}$$

where $Cl_{capillary}$ is the clearance of the compound from the dermis into the capillaries and the product of $Cl_{capillary} \times C_{skin}$ is the rate of removal of the compound from the dermis into the blood flow in the capillaries.

According to the diffusion–diffusion model and statistical moments, the mean residence time for the surface and skin are (Kubota and Ishizaki, 1986, 1985)

$$\left(MRT\right)_{surface} = \frac{x_{surface}}{x_{diffusion}}\left(\frac{1}{k_{diffusion}}\right) + \frac{1}{3} \times \frac{x_{surface}^2}{\left(D_{coeff}\right)_{surface}} \tag{13.101}$$

The term $1/3 \times x_{surface}^2/\left(D_{coeff}\right)_{surface}$ in Equation 13.101 is related to the diffusion process through the layer applied on the surface of the skin and is not a function of the diffusion through the skin. When $k_{diffusion}$ and $k_{clearance}$ are equal to zero, the MRT and VRT equations are

$$\left(MRT\right)_{surface} = \frac{1}{3} \times \frac{x_{surface}^2}{\left(D_{coeff}\right)_{surface}} \tag{13.102}$$

$$(MRT)_{skin} = \frac{1}{2k_{diffusion}} + \frac{1}{k_{clearance}} \tag{13.103}$$

where $x_{diffusion}$ is the apparent diffusion length of the skin, $k_{diffusion}$ is the normalized diffusion coefficient, and $k_{clearance}$ is the normalized skin-capillary boundary clearance defined as

$$k_{clearance} = \frac{Cl_{capillary}}{V_{skin}} \tag{13.104}$$

The parameter V_{skin} is the volume of the skin.

The variance of mean residence times (Yamaoka et al., 1978; Kubota et al., 1985) is

$$(VRT)_{surface} = \frac{2}{45} \times \frac{x_{surface}^4}{\left(D_{coeff}\right)_{surface}^2} + \frac{2x_{surface}}{x_{diffusion}}\left(\frac{1}{3k_{diffusion}^2} + \frac{1}{k_{diffusion}k_{clearance}} + \frac{1}{k_{clearance}^2}\right) + \left(MRT_{skin}\right)^2 \tag{13.105}$$

When $k_{diffusion}$ and $k_{clearance}$ are equal to zero:

$$(VRT)_{surface} = \frac{7}{45} \times \frac{x_{surface}^4}{\left(D_{coeff}^2\right)_{surface}} \tag{13.106}$$

$$(VRT)_{skin} = \frac{1}{6k_{diffusion}^2} + \frac{2}{3} \times \frac{1}{k_{diffusion}k_{clearance}} + \frac{1}{k_{clearance}^2} \tag{13.107}$$

The overall mean residence time and related variance of the surface and the skin combined is

$$(MRT)_{overall} = (MRT)_{surface} + (MRT)_{skin} + MRT_{blood} \tag{13.108}$$

$$(VRT)_{overall} = (VRT)_{surface} + (VRT)_{skin} + (VRT)_{blood} \tag{13.109}$$

The (MRT_{blood}) and (VRT_{blood}) are estimated from the plasma concentration–time curve.

13.4.1.7 Physiological Modeling of Percutaneous Absorption of Xenobiotics

A typical physiological modeling of drug absorption and disposition in different skin layers, muscle and systemic circulation is presented in Figure 13.10. The parameters of the model include the concentration of absorbed xenobiotic in different regions of skin; clearance from each physiological region, blood flow rates and the physiological volumes (Singh and Roberts, 1993). The assumptions of the model are (1) pseudo equilibrium between physiological compartments and their blood flow, (2) horizontal diffusion under the surface is negligible, and (3) all compartments associated with the vertical diffusion behave as well-stirred compartments with no elimination.

The general rate equations of the model are described as (Singh et al., 1993)

$$V_{surface}\frac{dC_{surface}}{dt} = \left(Cl_{s\leftrightarrow d}\right)C_{surface}^0 e^{-(k_{s\rightarrow d})t} \tag{13.110}$$

$$\begin{aligned}(V_u)_{dermis}\frac{d(C_u)_{dermis}}{dt} &= \left(Cl_{s\leftrightarrow d} \times C_{surface}^0 e^{-(k_{s\leftrightarrow d})t}\right) + \left(Q_{dermis} \times (f_u)_{blood} \times C_{blood}\right) \\ &\quad - \left(Q_{dermis} \times (f_u)_{dermis} \times C_{dermis}\right) + Cl_{d\leftrightarrow sc}\left((f_u)_{sc} \times C_{sc} - (f_u)_{dermis} \times C_{dermis}\right)\end{aligned} \tag{13.111}$$

where $C_{surface}^0$ is the initial concentration on the surface, $k_{s\leftrightarrow d}$ is the transfer rate constant of surface to dermis layer, $(V_u)_{dermis}$ is the apparent volume of distribution of unbound xenobiotic in dermis, $(f_u)_{dermis}$ is the fraction of unbound compound in dermis, $(C_u)_{dermis}$ is the concentration of the unbound compound in dermis, $Cl_{s\rightarrow d}$ is the clearance from the surface to the dermis layer, Q_{dermis} is the blood flow to dermis, $(f_u)_{blood}$ is the fraction of unbound compound in the blood, C_{dermis} is the

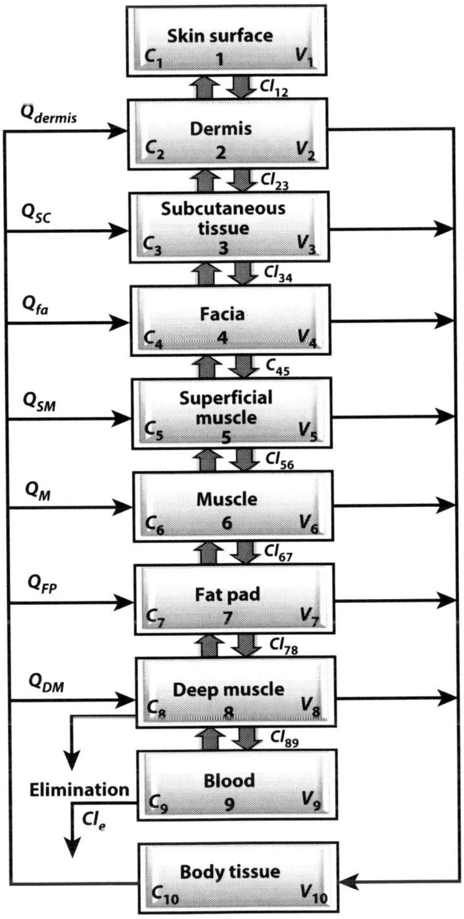

FIGURE 13.10 Diagram of a typical physiologically based PK/TK model with xenobiotic absorption and disposition in different skin layers, muscle and systemic circulation.

concentration in dermis, $Cl_{d\to sc}$ is the clearance from dermis to subcutaneous layer, and $(f_u)_{sc}$ and C_{sc} are the fraction of the unbound compound and concentration in subcutaneous layer, respectively.

When $\left(f_u\right)_{dermis} = \left(f_u\right)_{sc}$ and $\left(f_u\right)_{blood}/\left(f_u\right)_{dermis} = \left(P_{coeff}\right)_{db}$ where, $\left(P_{coeff}\right)_{db}$ is the partition coefficient between dermis and blood, Equation 13.111 reduces to Equation 13.112:

$$\left(V_u\right)_{dermis}\frac{dC_{dermis}}{dt} = \left\{\left(Cl_{s\leftrightarrow d} \times C^0_{surface}e^{-(k_{s\leftrightarrow d})t}\right)\left(f_u\right)_{dermis} + \left(Q_{dermis} \times C_{blood} \times \left(P_{coeff}\right)_{db}\right) - \left(Q_{dermis} \times C_{dermis}\right) + Cl_{d\leftrightarrow sc}\left(C_{sc} - C_{dermis}\right)\right\}\left(f_u\right)_{dermis} \quad (13.112)$$

Integration of Equation 13.112 by the Laplace transform and rearranging the constants and parameter yields Equation 13.113 for the area under the zero-moment curve (Singh et al., 1993):

$$(AUC)_{dermis} = \frac{\left(Cl_{s\leftrightarrow d} \times (AUC)_{surface} \times \left(f_u\right)_{dermis}\right) + \left(Q_{dermis} \times (AUC)_{blood} \times \left(P_{coeff}\right)_{db}\right) + \left(Cl_{d\leftrightarrow sc} \times (AUC)_{sc}\right)}{Q_{dermis} + Cl_{d\leftrightarrow sc}} \quad (13.113)$$

The equation of the mean residence time of dermis is also defined as:

$$(MRT)_{dermis} = \frac{\begin{array}{c}(MRT_{int})_{dermis} + \left[Cl_{s\leftrightarrow d} \times (f_u)_{dermis}\left(AUC_{surface}/AUC_{dermis}\right)MRT_{surface}\right] + Q_{dermis} \\ \times \left(P_{coeff}\right)_{db} \times \left(AUC_{blood}/AUC_{deremis}\right)MRT_{blood} + Cl_{d\leftrightarrow sc}\left(AUC_{sc}/AUC_{dermis}\right)MRT_{sc}\end{array}}{Q_{dermis} + Cl_{d\leftrightarrow sc}} \tag{13.114}$$

where $(MRT_{int})_{dermis}$ is the intrinsic mean residence time of the compound in dermis and is estimated as

$$(MRT_{\text{int}})_{dermis} = \frac{V_{dermis}}{(f_u)_{dermis}\left(Q_{dermis} + Cl_{d\leftrightarrow sc}\right)} \tag{13.115}$$

13.4.1.8 Six-Compartment Intradermal Disposition Kinetics of Xenobiotics with Contralateral Compartments

The six-compartment model is proposed to refine further the physiological factors influencing the intradermal permeation of xenobiotics (Higaki et al., 2002; Jepps et al., 2013). The diagram of the model is depicted in Figure 13.11. The model is considered dose independent with the following linear differential equations (Higaki et al., 2002):

1. Surface (applied dose):

$$V_{surface}\frac{dC_{surface}}{dt} = -Cl_{surface\rightarrow vs} \times C_{surface} \tag{13.116}$$

2. Viable skin (*vs*):

$$V_{vs}\frac{dC_{vs}}{dt} = \left(Cl_{surface-vs} \times C_{surface}\right) + \left(Cl_{muscle-vs} \times C_{muscle}\right) + \left(Cl_{plasma-vs} \times Cp\right) - \left(Cl_{vs-muscle} + Cl_{vs-plasma}\right)C_{vs} \tag{13.117}$$

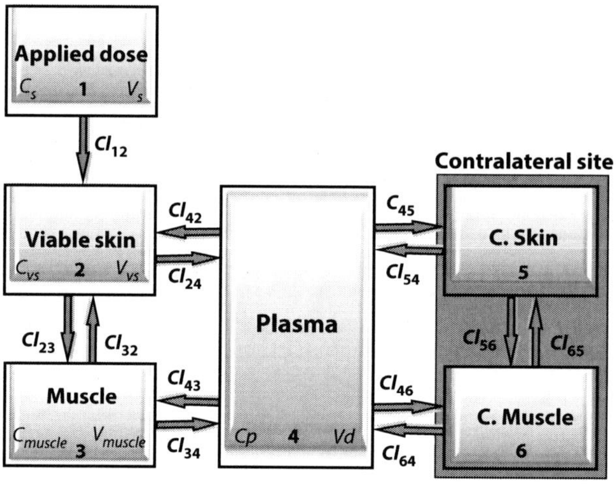

FIGURE 13.11 Diagram of a six-compartmental model for intradermal permeation and disposition of xenobiotics with contralateral compartments.

3. Muscle:

$$V_{muscle}\frac{dC_{muscle}}{dt} = \left(Cl_{vs-muscle} \times C_{vs}\right) + \left(Cl_{plasma-muscle} \times Cp\right) - \left(Cl_{muscle-vs} + Cl_{muscle-plasma} + Cl_{muscle}\right) \times C_{muscle} \quad (13.118)$$

4. Contralateral viable skin (*cvs*):

$$V_{cvs}\frac{dC_{cvs}}{dt} = \left(Cl_{muscle-vs} \times \frac{V_{c-muscle}}{V_{muscle}} \times C_{c-muscle}\right) + \left(Cl_{plasma-vs} \times \frac{V_{cvs}}{V_{vs}} Cp\right) - \left(Cl_{vs-muscle} + Cl_{vs-plasma}\right) \times \frac{V_{cvs}}{V_{vs}} \times C_{cvs} \quad (13.119)$$

5. Contralateral muscle (*c–muscle*):

$$V_{c-muscle}\frac{dC_{m}}{dt} = \left(Cl_{vs-muscle} \times \frac{V_{cvs}}{V_{vs}} \times C_{cvs}\right) + \left(Cl_{plasma-muscle} \times \frac{V_{c-muscle}}{V_{muscle}} \times Cp\right) - \left(Cl_{muscle-vs} + Cl_{muscle-plasma} + Cl_{muscle}\right) \times \frac{V_{c-muscle}}{V_{muscle}} \times C_{c-muscle}$$

$$Vd\frac{dCp}{dt} = Cl_{vs-plasma} \times \left(C_{vs} + \frac{V_{cvs}}{V_{vs}} \times C_{cvs}\right) + \left(C_{muscle} + \frac{V_{c-muscle}}{V_{muscle}} \times C_{c-muscle}\right) - \left(Cl_{plasma-vs} + Cl_{plasma-vs}\frac{V_{cvs}}{V_{vs}} + Cl_{plasma-muscle}\frac{V_{c-muscle}}{V_{muscle}} + Cl_{total}\right) \times Cp \quad (13.120)$$

6. Plasma

$$Vd\frac{dCp}{dt} = Cl_{vs-plasma} \times \left(C_{vs} + \frac{V_{cvs}}{V_{vs}} \times C_{cvs}\right) + Cl_{muscle-plasma}\left(C_{muscle} + \frac{V_{c-muscle}}{V_{muscle}} \times C_{c-muscle}\right) - \left(Cl_{plasma-vs} + Cl_{plasma-muscle} + Cl_{plasma-vs}\frac{V_{cvs}}{V_{vs}} + Cl_{plasma-muscle}\frac{V_{c-muscle}}{V_{muscle}} + Cl_{total}\right) \times Cp \quad (13.121)$$

Clearance terms with a hyphenated subscript represent clearance from the first compartment of the subscript to the second compartment, for example, ($Cl_{muscle-plasma}$) identifies the clearance from muscle to plasma. The following are to identify the clearance terms of Figure 13.11.

$Cl_{12} = Cl_{surface \to skin}$
$Cl_{23} = Cl_{vs \to muscle}$
$Cl_{32} = Cl_{muscle \to vs}$
$Cl_{24} = Cl_{vs \to plasma}$
$Cl_{42} = Cl_{plasma \to vs}$
$Cl_{34} = Cl_{muscle \to plasma}$
$Cl_{43} = Cl_{plasma \to muscle}$
$Cl_{45} = Cl_{plasma \to cvs}$
$Cl_{54} = Cl_{cvs \to plasma}$
$Cl_{56} = Cl_{cvs \to c-muscle}$
$Cl_{65} = Cl_{c-muscle \to cvs}$
$Cl_{46} = Cl_{plasma \to c-muscle}$
$Cl_{64} = Cl_{c-muscle \to plasma}$

The model has been used successfully to fit the data from ten compounds with different physicochemical characteristics (Higaki et al., 2002).

REFERENCES

Achterberg, V. F., Buscemi, L., Diekmann, H., Smith-Clerc, J., Schwengler, H., Meister, J. J., Wenk, H., Gallinat, S., Hinz, B. 2014. The nano-scale mechanical properties of the extracellular matrix regulate dermal fibroblast function. *J Invest Dermatol* 134:1862–72.

Andrews, S. N., Jeong, E., Prausnitz, M. R. 2013. Transdermal delivery of molecules is limited by full epidermis, not just stratum corneum. *Pharm Res* 30:1099–1109.

Barnett, G., Locko, V. 1977. Transport across epithelia: A kinetic evaluation. *Biochem Biophys Acta* 464:276–86.

Baron, J. M., Höller, D., Schiffer, R., Frankenberg, S., Neis, M., Merk, H. F., Jugert, F. K. 2001. Expression of multiple cytochrome p450 enzymes and multidrug resistance-associated transport proteins in human skin keratinocytes. *J Invest Dermatol* 116:541–8.

Bergstresser, P. R., Taylor, J. R. 1977. Epidermal turnover time a new examination. *Br J Dermatol* 96:503–9.

Carr, R. D., Wieland, R. G. 1966. Corticosteroid reservoir in the stratum corneum. *Arch Dermatol* 94(1):81–4.

Carslaw, H. S., Jaeger, J. C. 1959. *Conduction of Heat in Solids*, 128. Oxford: Clarendon Press.

Charalambopoulou, G. Ch., Karamertzanis, P., Kikkinides, E. S., Stubos, A. K., Kanellopoulos, N. K., Papaioannou, A. Th. 2000. A study on structural and diffusion properties of porcine stratum corneum based on very small angle neutron scattering data. *Pharm Res* 17(9):1085–91.

Cooper, E. R. 1976. Pharmacokinetics of skin penetration. *J Pharm Sci* 65(9):1396–7.

Cooper, E. R., Berner, B. 1985. Finite dose pharmacokinetics of skin penetration. *J Pharm Sci* 74(10):1100–2.

Crank, J. 1975. *The Mathmatics of Diffusion*, New York, NY: Oxford University Press Inc.

Elias, P. M. 1983. Epidermal lipids, barrier function, and desquamation. *J Invest Dermatol* 80:44–9.

Fehrenbacher, M. C., ten Berge, W. F. 2000. Dermal exposure modeling in Charles. *Mathematical Models for Estimating Occupational Exposure to Chemicals*, ed B. Keil, 65–74. Falls Church, VA: *AIHA* (A publication of the American Industrial Hygiene Association).

Frasch, H. F., Barbero, A. M. 2003. Steady-state flux and lag time in the stratum corneum lipid pathway: Results from finite element models. *J Pharm Sci* 92(11):2196–207.

Flynn, G. L. 1985. Mechanism of percutaneous absorption from physiochemical evidence. In *Percutaneous Absorption*, eds. Bronaugh, R. L. and Maibach, H. I. New York: Marcel Dekker, Inc.

Ghanem A. H., Mahmoud H., Higuchi W. I., Liu P. 1992. The effects of ethanol on the transport of lipophilic and polar permeants across hairless mouse skin: Methods/validation of a novel approach. *Int J Pharm* 78:137–56.

Guy, R. H., Hadgraft, J. 1982. A pharmacokinetic model for percutaneous absorption. *Internat J Pharmaceut* 11:119–29.

Guy, R. H., Hadgraft, J. 1984. Percutaneous absorption kinetics of topically applied agents liable to surface loss. *J Soc Cosmet Chem* 45:103–13.

Guy, R. H., Hadgraft, J. 1984. Prediction of drug disposition kinetics in skin and plasma following topical administration. *J Pharm Sci* 73(7):883–87.

Hada, N., Hasegawa, T., Takahashi, H., Ishibashi, T., Sugibayashi, K. 2005. Cultured skin loaded with tetracycline HCl and chloramphenicol as dermal delivery system: Mathematical evaluation of the cultured skin containing antibiotics. *J Control Release* 108:341–50.

Halprin, K. M. 1972. Epidermal turnover time a new examination. *Br J Dermatol* 96:503–9.

Heisig, M., Lieckfeldt, R., Wittum, G., Mazurkevich, G., Lee, G. 1996. Non steady-state descriptions of drug permeation through stratum corneum. I. The biphasic brick-and-mortar model. *Pharm Res* 13:421–426.

Higaki, K., Asai, M., Suyama, T., Nakayama, K., Ogawara, K-I., Kimura, T. 2002. Estimation of intradermal disposition kinetics of drugs: II. Factors determining penetration of drugs from viable skin to muscular layer. *J Int Pharm* 239:129–41.

Ho, H. O., Huang, F. C., Sokolaski, T. D., Sheu, M. T. 1994. The influence of cosolvents on the in-vitro percutaneous penetration of diclofenac sodium from a gel system. *J Pharm Pharmacol* 46:636–42.

Iwasa, A., Irimoto, K., Kasai, S., Okuyama, H., Nagai, H. 1991. Effect of nonionic surfactants on percutaneous absorption of diclofenac sodium. *Yakuzaigaku* 51:16–21.

James, W. D., Berger, T., Elston, D. 2005. *Andrews Diseases of the Skin: Clinical Dermatology*, 10th Edition. 1, 11–12. Philadelphia, PA: Saunders Elsevier.

Jepps, O. G., Danci, Y., Anissimov, Y. G., Roberts, M. S. 2013. Modeling the human skin barrier-toward a better understanding of dermal absorption. *Adv Drug Deliv Rev* 65:152–68.

Jost, W. 1952. *Diffusion in Solids, Liquids, Gases*. New York: Academic Press Inc.

Kissel, J., Richter, K., Fenske, R. 1996. Field measurements of dermal soil loading attributes to various activities: Implications for exposure assessment. *Risk Analysis* 16:116–125.

Kubota, K., Ishizaki, T. 1986. A diffusion–diffusion model for percutaneous absorption. *J Pharmacokin Biopharm* 14(4):409–39.

Kubota, K., Ishizaki, T. 1985. A theoretical consideration of percutaneous drug absorption. *J Pharmacokin Biopharm* 13:55–71.
Lieckfeldt, R., Lee, G. 1992. Use of a model lipid matrix to demonstrate the dependence of the stratum corneum's barrier properties on its internal geometry. *J Contr Rel* 20:183–94
Lotens, W. A., Wammes, L. J. A. 1993. Vapor transfer in two-layer clothing due to diffusion and ventilation. *Ergonomics* 36(10):1223–40.
Masters, B. R., So, P. T., Gratton, E. 1997. Multiphoton excitation fluorescence microscopy and spectroscopy of *in vivo* human skin. *Biophys J* 72:2405–12.
McKenzie, A. W., Stoughton, R. B. 1962. Method for comparing percutaneous absorption of steroids. *Arch Dermatol* 86(5):608–10.
Michaels, A. S., Chandrasekaran, S. K., Shaw, J. E. 1975. Drug permeation through human skin: Theory and *in vitro* experimental measurement. *AIChE J* 21:985–96.
Nishihata, T., Kamada, A., Sakai, K., Takahashi, K., Matsumoto, K., Shinozaki, K., Tabata, Y., Keigami, M., Miyagi, T., Tatsumi, N. 1988. Percutaneous absorption of diclofenac in rats and humans: Aqueous gel formulation. *Int J Pharm* 46:1–7.
Odland, G. F. 1983. Structure of skin. In *Biochemistry and Physiology of the Skin*, ed. L. A. Goldsmith, vol. I. Oxford: Oxford University Press.
Parsaee S., Sarbolouki M. N., Parnianpour, M. 2002. *In vitro* release of diclofenac diethylammonium from lipid-based formulations. *Int J Pharm* 241:185–90.
Randolph, G. J., Beaulieu, S., Pope, M., Sugawara, I., Hoffman, L., Steinman, R. M., Muller, W. A. 1998. A physiologic function for p-glycoprotein (MDR-1) during the migration of dendritic cells from skin via afferent lymphatic vessels. *Proc Natl Acad Sci USA* 95:6924–29.
Santoyo, S, Arellano, A, Ygartua, P, Martin, C. 1995. Penetration enhancer effects on the *in vitro* percutaneous absorption of piroxicam through rat skin. *Int J Pharm* 117:219–24.
Scheuplein, R. J., Blank, I. H. 1973. Mechanism of percutaneous absorption. IV. Penetration of nonelectrolytes (alcohols) from aqueous solutions and from pure liquids. *J Invest Dermatol* 60:286–96.
Scheuplein, R. J. and Bronaugh, R.L. 1983. Percutaneous absorption. In Goldsmith, L. A. (ed.) *Biochemistry and Physiology of the Skin* V. II. New York: Oxford University Press.
Scheuplein, R. J. 1967. Mechanism of percutaneous absorption. II. Transient diffusion and the relative importance of various routes of skin penetration. *J Invest Derm* 48:79.
Scheuplein, R. J. 1976. Percutaneous absorption after twenty-five years: Or old wine in new wineskins. *J Invest Dermatol* 67:31–8.
Scheuplein, R. 1978. Skin as a barrier. In *The Physiology and Pathophysiology of Skin*, vol. 5. New York: Academic Press.
Singh, P., Roberts, M. S. 1993. Dermal and underlying tissue pharmacokinetics of salicylic after topical application. *J Pharmacokin Biopharm* 21(4):337–73.
Stoughton, R. B. 1965. Dimethylsulfoxide (DMSO) induction of a steroid reservoir in human skin. *Arch Dermatol* 91(6):657–60.
Sugibayashi, K., Todo, H., Oshizaka, T., Owada, Y. 2010. Mathematical model to predict skin concentration of drugs: Toward utilization of silicone membrane to predict skin concentration of drugs as an animal testing alternative. *Pharm Res* 27(1):134–42.
Tojo, K.1987. Random brick model for drug transport across stratum corneum. *J Pharm Sci* 76:889–91.
Tojo, K. 1988. Concentration profile in plasma after transdermal drug delivery. *Inter J Pharm* 43(3):201–5.
Vickers, C. F. 1963. Existence of reservoir in the stratum corneum. Experimental proof. *Arch Dermatol* 88:20–3.
Wallace, S. M., Barnett, G. 1978. Pharmacokinetic analysis of percutaneous absorption: Evidence of parallel penetration pathways for methotrexate. *J Pharmacokin Biopharm* 6(4):315–325.
Weinstein, G. D. 1966. Comparison turnover time of keratinous protein fractions in swine and human epidermis. In *Swine in Biomedical Research*, eds. L. K. Bustad, R. O. McClellan, M. P. Burns, 287–97. Richland, WA: Pacific Northwest Laboratory.
Wilschut, A., ten Berge, W. F., Robinson, P. J., McKone, T. E. 1995. Estimating skin permeation. The validation of five mathematical skin permeation models. *Chemosphere* 30:1275–96.
Winkelmann, R. K. 1969. The relationship of structure of the epidermis to percutaneous absorption. *Br J Dermatol* 81(4):11–22.
Wurster, D. E., Kramer, S. F. 1961. Investigation of some factors influencing percutaneous absorption. *J Pharm Sci* 50(4):288–93.
Yamaoka, K., Kakagawa, T., Uno, T. 1978. Statistical moments the pharmacokinetics. *J Pharmacokin Biopharm* 6:547–58.

14 PK/TK Analysis of Multiple Dosing

14.1 INTRODUCTION

The outcome of exposure to xenobiotics, whether it is as regimented as multiple dosing kinetics of therapeutic agents following the PK/TK models (Chapters 10 through 12), or as unsystematic as random exposure to environmental pollutants following the stochastic models, is the accumulation of the compounds in the body. In the case of regimented exposure, the purpose is to achieve a consistent pharmacological response for a period of time longer than duration of action of a single dose (Krüger-Thiemer, 1966, 1969; Krüger-Thiemer and Bünger, 1965). Similar to uninterrupted zero-order input (Chapter 11), which resulted in the accumulation of the infused compound in the body, multiple dosing administration also accumulates the compound in the body following the administration of a fixed dose and dosing interval (Van Rossum and Tomey, 1968). Unlike the continuous infusion, however, the plasma concentration of administered compound fluctuates after each dose reaches a maximum concentration, or peak level, and then declines to a minimum concentration or trough level (Wagner et al., 1965; Gibaldi and Perrier, 1982). In this type of regimented exposure, the dose is usually kept constant and a dosing interval (τ) is selected as a fixed interval approximately equal to the half-life of the compound. This means that during each dosing interval approximately 50% of amount in the body eliminates and 50% remains in the body. Thus, peak and trough levels of each dose will be higher than previous doses. Since the dose and dosing interval are kept constant, after a finite time the administered dose will become equal to the amount eliminated from the body. This is when the steady-state fluctuations (i.e., peak and trough) are achieved and the amount accumulated in the body is at its anticipated maximum level of the designed dosing regimen (Buell et al., 1969; Wagner, 1975; Gibaldi, 1982). A well-designed dosing regimen for therapeutic agents maintains the fluctuation at steady state within the therapeutic range (Levy, 1974). The concept of attaining and maintaining the steady state is similar to intravenous infusion, which is the rate of input equal to the rate of output. In a multiple dosing steady state, the fixed administered dose is equal to the amount eliminated from the body in a dosing interval. The fixed dose that is given on a regular basis in a fixed dosing interval is known as the maintenance dose.

14.2 KINETICS OF MULTIPLE INTRAVENOUS BOLUS INJECTIONS—ONE-COMPARTMENT MODEL

In general, the kinetic descriptions of multiple dosing are based on a geometric series (Appendix A). The summary of plasma levels in terms of peak and trough levels following a regimen of fixed intravenous dose and dosing intervals is presented in Table 14.1.

In Table 14.1, A^0, A_{max}, and A_{min} are the initial, maximum, and minimum amounts in the body, respectively. The last row of Table 14.1 is the geometric series of the multiple dosing functions of maximum concentrations. As the number of doses increases the number of exponential terms increases, the equation becomes longer, and using them for the purpose of calculating the

TABLE 14.1
Summary of Peak and Trough Levels in Multiple Intravenous Injections and the Related Geometric Series

τ	*Dose*	A_{max}	A_{min}
τ_1	D_1	$(A_{max})_1 = A^0$	$(A_{max})_1 = A^0\, e^{-K\tau}$
τ_2	D_2	$(A_{max})_2 = A^0\,(1 - e^{-K\tau})$	$(A_{min})_2 = A^0(e^{-K\tau} + e^{-K\tau})$
τ_3	D_3	$(A_{max})_3 = A^0\,(1 + e^{-K\tau} + e^{-2K\tau})$	$(A_{min})_3 = A^0(e^{-K\tau} + e^{-2K\tau} + e^{-3K\tau})$
τ_4	D_4	$(A_{max})_4 = A^0\,(1 + e^{-K\tau} + e^{-2K\tau} + e^{-3K\tau})$	$(A_{min})_4 = A^0(e^{-K\tau} + e^{-2K\tau} + e^{-3K\tau} + e^{-4K\tau})$
$\vdots$	$\vdots$	$\vdots \quad \vdots \quad \vdots$	$\vdots \quad \vdots \quad \vdots$
τ_n	D_n	$(A_{max})_4 = A^0\,(1 + e^{-K\tau} + e^{-2K\tau} + e^{-3K\tau} + e^{-4K\tau} + \ldots + e^{-(n-1)K\tau})$	$(A_{max})_4 = A^0\,(e^{-K\tau} + e^{-2K\tau} + e^{-3K\tau} + e^{-4K\tau} + \ldots + e^{-(n-1)K\tau} + e^{-nK\tau})$
		Geometric Series: $S_n = 1 + e^{-K\tau} + e^{-2K\tau} + e^{-3K\tau} + e^{-4K\tau} + \ldots + e^{-(n-1)K\tau}$	Geometric Series: $S_n e^{-K\tau} = e^{-K\tau} + e^{-2K\tau} + e^{-3K\tau} + e^{-4K\tau} + \ldots + e^{-(n-1)K\tau} + e^{-nK\tau}$

concentration of the compound in plasma is cumbersome. To make the calculations more practical, the solution of the geometric series for the *n*th dose is determined as follows:

The difference between peak and trough levels of the *n*th dose is

$$(A_{max})_n - (A_{min})_n = A^0(S_n) - A^0(S_n)e^{-K\tau} = A^0(S_n)(1 - e^{-K\tau}) \tag{14.1}$$

where S_n is the geometric series of *nth* dose as identified in Table 14.1.

Also, the difference between $(A_{min})_n$ and $(A_{max})_n$ of the last two entries of Table 14.1 is

$$(A_{\min})_n - (A_{\max})_n = A^0(\cancel{e^{-Kt}} + \cancel{e^{-2Kt}} + \cancel{e^{-3Kt}} + \cancel{e^{-4Kt}} + \cancel{e^{-5Kt}} + \ldots + \cancel{e^{(n-1)Kt}} + e^{-nKt})$$
$$-A^0(1 + \cancel{e^{-Kt}} + \cancel{e^{-2Kt}} + \cancel{e^{-3Kt}} + \cancel{e^{-4Kt}} + \ldots + \cancel{e^{(n-1)Kt}}) = e^{-nK\tau} - 1$$

Therefore,

$$(A_{max})_n - (A_{min})_n = 1 - e^{-nK\tau} \tag{14.2}$$

Setting Equation 14.1 equal to Equation 14.2 and solving for S_n yields

$$S_n = \frac{1 - e^{-nK\tau}}{1 - e^{-K\tau}} \tag{14.3}$$

Equation 14.3 summarizes all the exponential terms of $(A_{max})_n$ in Table 14.1 and represents the solution of geometric series S_n of multiple dosing after *n*th dose.

As $n \Rightarrow \infty$, the accumulation reaches the expected maximum level of steady state, $e^{-nK\tau} \Rightarrow 0$, and Equation 14.3 changes to Equation 14.4, which represents the solution of the geometric series at steady state.

$$(S_n)_{ss} = \frac{1}{1 - e^{-K\tau}} \tag{14.4}$$

14.2.1 Equations of Peak and Trough Levels

The peak and trough levels of the first dose are

$$(Cp_{max})_1 = Cp^0 \tag{14.5}$$

$$(Cp_{min})_1 = (Cp_{max})_1 e^{-\tau K} = Cp^0 e^{-K\tau} \tag{14.6}$$

$$(Cp_t)_1 = (Cp_{max})_1 e^{-Kt} = Cp^0 e^{-Kt} \tag{14.7}$$

Equation 14.7 represents plasma concentration at any time t between peak and trough levels of the first dose ($t < \tau$).

Multiplying Equations 14.5 through 14.7 by Equation 14.3, yields peak, trough, and any t concentration before achieving the steady-state level (Figure 14.1) (Gibaldi and Perrier, 1982):

$$(Cp_{max})_n = Cp^0 \left(\frac{1 - e^{-nK\tau}}{1 - e^{-K\tau}} \right) \tag{14.8}$$

$$(Cp_{min})_n = Cp^0 \left(\frac{1 - e^{-nK\tau}}{1 - e^{-K\tau}} \right) e^{-K\tau} = (Cp_{max})_n e^{-K\tau} \tag{14.9}$$

$$(Cp_t)_n = Cp^0 \left(\frac{1 - e^{-nK\tau}}{1 - e^{-K\tau}} \right) e^{-Kt} = (Cp_{max})_n e^{-Kt} \tag{14.10}$$

Multiplying Equations 14.5 through 14.7 by Equation 14.4 generates equations of steady-state levels

$$(Cp_{max})_{ss} = Cp^0 \left(\frac{1}{1 - e^{-K\tau}} \right) \tag{14.11}$$

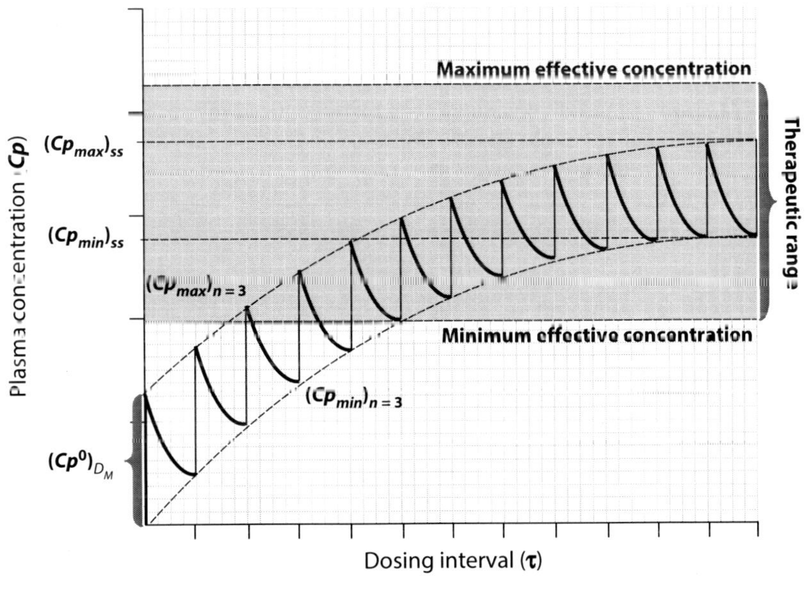

FIGURE 14.1 Plasma concentration–time profile of multiple intravenous bolus injections of the same dose given every fixed dosing interval.

$$(Cp_{min})_{ss} = Cp^0\left(\frac{e^{-K\tau}}{1-e^{-K\tau}}\right) = (Cp_{max})_{ss}e^{-K\tau} \tag{14.12}$$

$$(Cp_t)_{ss} = Cp\circ\left(\frac{e^{-Kt}}{1-e^{-K\tau}}\right) = (Cp_{max})_{ss}e^{-Kt} \tag{14.13}$$

In Equations 14.8 through 14.13, the initial plasma concentration, Cp^0, is the initial plasma concentration of the maintenance dose, that is, D_M/Vd, which also correspond to the difference between the steady-state peak and trough levels (Figure 14.1).

14.2.2 Time Required to Achieve Steady-State Plasma Levels

In multiple-dosing kinetics, contrary to zero-order input (Chapter 11) where the amount and time of input are continuous, the interval of time consists of several dosing intervals, and is not expressed as continuous t, but rather as $n\tau$. The equation of $n\tau$, or the time needed to achieve any fraction of steady-state levels is determined by comparing the numerators of Equations 14.3 and 14.4. In both equations "1" corresponds to 100% of the steady-state level, "$e^{-nK\tau}$" is the fraction left to reach the steady-state levels and "$1 - e^{-nK\tau}$" of Equation 14.3 represents the fraction of steady-state levels:

$$f_{ss} = 1 - e^{-nK\tau} \tag{14.14}$$

Solving for $n\tau$ by taking the natural logarithm of Equation 14.14 yields

$$n\tau = -\frac{\ln(1-f_{ss})}{K}, \text{ or} \tag{14.15}$$

$$n\tau = -3.3T_{1/2}\log(1-f_{ss}) \tag{14.16}$$

According to Equation 14.17 and analogous to zero-order input (Chapter 11), the time required for achieving any fraction of steady state is directly proportional to half-life, or indirectly proportional to the overall elimination rate constant. For example, to achieve 50% of steady-state levels

$$f_{ss} = 0.50 \qquad n\tau = -3.3T_{1/2}\log(0.50) = 1T_{1/2}$$
$$f_{ss} = 0.75 \qquad n\tau = -3.3T_{1/2}\log(0.25) = 2T_{1/2}$$
$$f_{ss} = 0.90 \qquad n\tau = -3.3T_{1/2}\log(0.10) = 3.3T_{1/2}$$
$$f_{ss} = 0.95 \qquad n\tau = -3.3T_{1/2}\log(0.05) = 4.3T_{1/2}$$
$$f_{ss} = 0.99 \qquad n\tau = -3.3T_{1/2}\log(0.01) = 6.6T_{1/2}$$

14.2.3 Average Steady-State Plasma Concentration

The average plasma concentration at steady state is a parameter based on the definition of steady state. At steady state the rate of input is equal to the rate of output. In the case of zero-order input (Chapter 11) the rate of input is continuous whereas in multiple dosing it is sequential and the rate is defined as a maintenance dose given every dosing interval. The rate of output in zero-order input at steady state, although first-order, remains constant because of a constant steady-state plasma level.

In multiple dosing, the rate of output changes during a dosing interval and declines from a maximum rate to a minimum because of the change in the plasma level. To define a single rate of output and input during a dosing interval at steady state, similar to zero-order input, the average rates are calculated as follows:

$$\text{Average rate of input} = \text{Maintenance dose} \div \text{Dosing interval} = \frac{D_M}{\tau} \tag{14.17}$$

$$\text{Average rate of output} = Cl_T \times (Cp_{ave})_{ss} = K \times (A_{ave})_{ss} \tag{14.18}$$

where $(Cp_{ave})_{ss}$ and $(A_{ave})_{ss}$ are average plasma concentrations and average amounts in the body at steady state, respectively.

Setting the rate of input equal to the rate of output and solving for $(Cp_{ave})_{ss}$ yields the following relationship, which can be presented in various forms (Levy, 1974):

$$(Cp_{ave})_{ss} = \frac{D_M}{Cl_t \times \tau} \tag{14.19}$$

$$(Cp_{ave})_{ss} = \frac{D_M}{K \times Vd \times \tau} \tag{14.20}$$

$$(Cp_{ave})_{ss} = \frac{1.44 \times T_{1/2} \times D_M}{Vd \times \tau} \tag{14.21}$$

$$(Cp_{ave})_{ss} = \frac{Cp^0}{K \times \tau} \tag{14.22}$$

$$(Cp_{ave})_{ss} = \frac{(AUC)_\tau}{\tau} \tag{14.23}$$

The definition of the average steady-state plasma concentration based on Equations 14.19 through 14.23 can be either "the ratio of average rate of input divided by clearance" or "the area under plasma concentration of a maintenance dose at steady state divided by the dosing interval." Equations 14.19 through 14.23 are used in designing dosing regimen, adjusting the dosing regimen for patients with renal failure, and used in therapeutic drug monitoring. The shortcoming of the equations is that they provide no information about the fluctuation of peak and trough levels within the therapeutic range. The magnitude of the fluctuation, however, can be estimated from D_M/Vd.

14.2.4 Loading Dose and Maintenance Dose

To reduce the time required to achieve the steady-state fluctuations, a loading dose (D_L) can be administered at the start of multiple dosing regimen. This one-time dose should be equal to the maximum amount in the body at steady state and provide an initial concentration equal to $(Cp_{max})_{ss}$; in other words, the loading dose is the product of maximum plasma concentration at steady state and the volume of distribution (Perrier and Gibaldi, 1973)

$$D_L = (Cp_{max})_{ss} \times Vd = (A_{max})_{ss} \tag{14.24}$$

The maintenance dose (D_M) then equals the portion of the loading dose that leaves the body by all routes of elimination at steady state during the first dosing interval after the administration of D_L:

$$(f_{el})_{\tau_{ss}} = 1 - e^{-K\tau} \tag{14.25}$$

$$D_M = D_L(1 - e^{-K\tau}) \tag{14.26}$$

The maintenance dose can also be defined as multiplication of steady-state fluctuation and the volume of distribution. The fluctuation is the difference between the peak and trough levels, that is,

$$D_M = ((Cp_{max})_{ss} - (Cp_{min})_{ss}) \times Vd \tag{14.27}$$

It can also be estimated from Equation 14.19

$$D_M = (Cp_{ave})_{ss} \times Cl_T \times \tau \tag{14.28}$$

Equation 14.28 indicates that the maintenance dose is the average rate of elimination at steady state ($(Cp_{ave})_{ss} \times Cl_T$) in one dosing interval multiplied by the dosing interval.

Thus, in a multiple-dosing regimen if a loading dose is followed by a maintenance dose at every fixed dosing interval, the steady-state fluctuations can be achieved immediately and maintained as long as the administration of the maintenance dose continues on a fixed dose schedule (Figure 14.2).

Using Equation 14.26, the relationship between the loading dose and maintenance dose is described as

$$D_L = \frac{D_M}{1 - e^{-K\tau}} = \frac{D_M}{(f_{el})_{\tau_{ss}}} \tag{14.29}$$

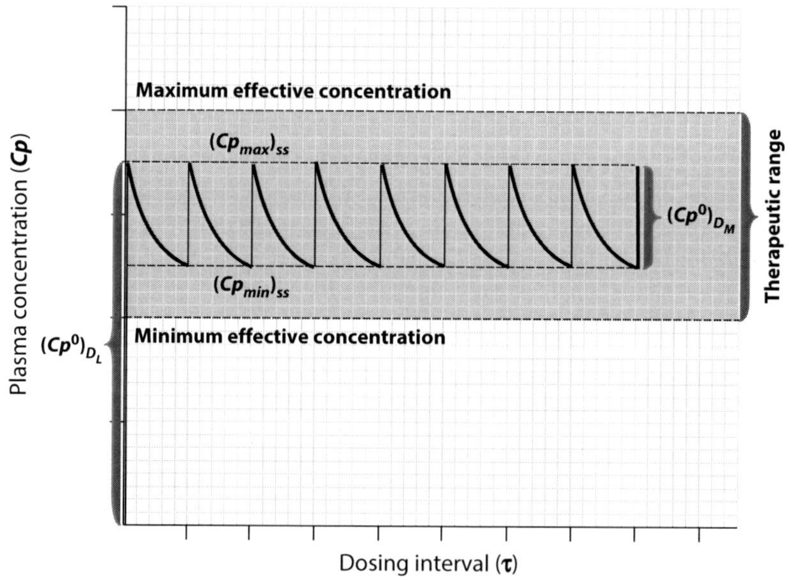

FIGURE 14.2 Profile of fluctuation of plasma concentration following administration of loading dose and seven maintenance doses.

When the half-life is very short in comparison to the dosing interval ($\tau \gg T_{1/2}$) the fraction eliminated from the body in one dosing interval approaches one, that is, $(f_{el})_{\tau_{ss}} \Rightarrow 1$, representing approximately 100% elimination and $D_L = D_M$. An example would be a compound with a wide range effect and very short half-life, say less than 30 minutes, given every 4 hours.

When the dosing interval is equal to half-life ($\tau = T_{1/2}$), the fraction of loading dose eliminated in the first dosing interval is half of the loading dose, that is, $D_L = 2D_M$, or $D_M = D_L/2$.

When $\tau \ll T_{1/2}$, $(f_{el})_{\tau_{ss}} = 1 - e^{\approx 0} \cong 1 - 1 \cong 0$, the denominator of Equation 14.29 approaches zero and the relationship is undefined. An example is giving a compound with a half-life of 7 days, once a day. A practical approach is to set the loading dose equal to the average amount at steady state and define the relationship between D_L and D_M according to Equation 14.30:

$$D_L = (A_{ave})_{ss} = \frac{D_M}{K\tau} \tag{14.30}$$

14.2.5 Extent of Accumulation

The degree of accumulation in the body at steady state is defined by a number known as the accumulation ratio, which is the ratio of the steady-state peak level to that of the first dose, or the ratio of the steady-state trough level to the trough level of the first dose (Equation 14.31). The first dose here is referred to the maintenance dose according to the profile of Figure 14.1. The ratio indicates how many folds have accumulated in the body at peak or trough levels of steady state with respect to the first dose for a given dosing regimen. It also corresponds to the ratio of D_L/D_M, which is the reciprocal of the fraction of the dose eliminated from the body by all routes of elimination during a dosing interval at steady state. For example, if the ratio is equal to 5, it is indicative of five times more accumulation at steady state with respect to the first dose, or $(Cp_{max})_{ss} = 5 \times (Cp_{max})_1$ or $(Cp_{min})_{ss} = 5 \times (Cp_{min})_1$.

$$R = \frac{(Cp_{max})_{ss}}{(Cp_{max})_1} = \frac{(Cp_{max})_{ss}}{(Cp^0)_{D_M}} = \frac{(Cp_{min})_{ss}}{(Cp_{min})_1} = \frac{D_L}{D_M} = \frac{1}{1 - e^{-K\tau}} = \frac{1}{(f_{el})_\tau} \tag{14.31}$$

The extent of accumulation can also be determined by using the area under the plasma concentration–time curve (Colburn, 1983), where $R = (AUC_0^\infty)_1 \div (AUC_0^\tau)_1$ and $R = (AUC_0^\tau)_{ss} \div (AUC_0^\tau)_1$.

14.2.6 Decline of Plasma Concentration after the Last Dose

After the last dose, whether steady state is achieved or not, the plasma concentration declines exponentially until the total accumulated amount is completely removed from the body. The knowledge of plasma concentration after the last dose of the first compound may help to determine when the second treatment with a different compound should be initiated. The following equations can be used to estimate the plasma concentration after the last dose of a multiple dosing regimen of a compound that follows the one-compartment model (Figure 14.3).

$$Cp_{t'} = Cp_{max}e^{-Kt'} \quad \text{or,} \quad Cp_{t'} = Cp_{min}e^{-K(t'-\tau)} \tag{14.32}$$

14.2.7 Design of a Dosing Regimen

The purpose of designing a dosing regimen is to determine the loading dose, maintenance dose, and dosing interval for a given compound based on the PK/TK information. The following are a few of the practical approaches:

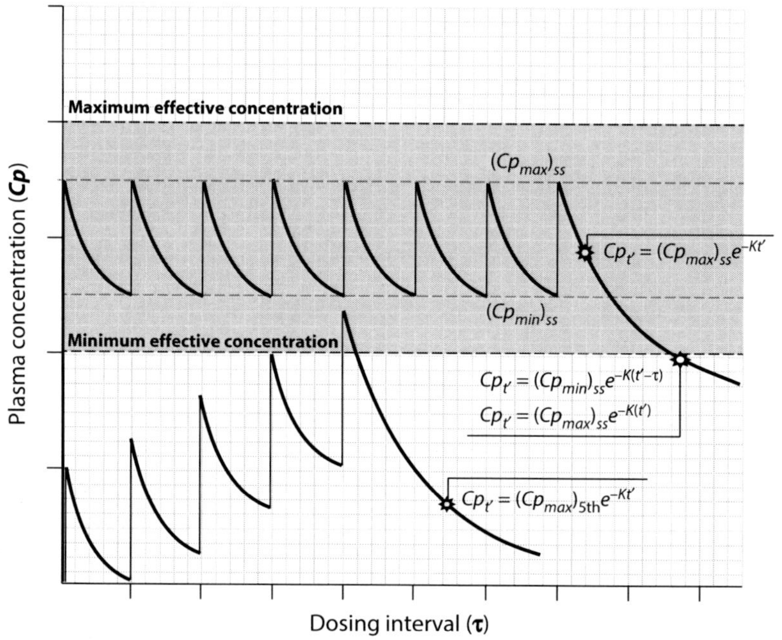

FIGURE 14.3 Profile of plasma concentration decline after the last dose of a multiple dosing regimen before and after achieving steady-state levels.

14.2.7.1 Dosing Regimen Based on a Target Concentration

The target concentration is set equal to the average steady-state plasma concentration; a dosing interval is selected based on the half-life of the compound, and the convenience of administration and the maintenance dose is estimated as follows:

$$D_M = (Cp_{ave})_{ss} \times Vd \times \tau \times K \tag{14.33}$$

If the dosing interval is set equal to half-life $D_L = 2D_M$. If the dosing interval is different from the half-life, the loading dose is determined according to Equation 14.29. For compounds with a very short half-life, the dosing interval is usually set equal to four hours for the convenience of administration and $D_L = D_M$.

It would be advisable to verify whether the peak and trough levels of the dosing regimen are within the acceptable range. This is particularly important for therapeutic agents with narrow therapeutic range. A quick check is

$$(Cp_{max})_{ss} = \frac{D_L}{Vd} \tag{14.34}$$

$$(Cp_{min})_{ss} = (Cp_{max})_{ss} - \frac{D_M}{Vd} \tag{14.35}$$

14.2.7.2 Dosing Regimen Based on Steady-State Peak and Trough Levels

The fluctuation of plasma concentrations at steady state is set equal to the initial plasma concentration of the maintenance dose (Boroujerdi, 2002)

$$\text{Fluctuation} = (Cp_{max})_{ss} - (Cp_{min})_{ss} = (Cp^0)_{D_M} \tag{14.36}$$

Therefore,

$$D_M = (Cp_{max} - Cp_{min})_{ss} \times Vd \tag{14.37}$$

A trough level equal to half of the peak level indicates that the dosing interval is equal to the half-life. The loading dose is then equal to twice of the maintenance dose. If the trough level is not half of the peak level, the dosing interval can be estimated as

$$\tau = \frac{2.303}{K}\log\frac{(Cp_{max})_{ss}}{(Cp_{min})_{ss}} \tag{14.38}$$

14.2.7.3 Dosing Regimen Based on Minimum Steady-State Plasma Concentration

When the minimum steady-state plasma concentration is known, the dosing interval is set preferentially equal to the half-life and the maintenance dose is determined using Equation 14.12:

$$D_M = \frac{(Cp_{min})_{ss} \times Vd \times (1 - e^{-K\tau})}{e^{-K\tau}} \tag{14.39}$$

Depending on the dosing interval, the loading dose is calculated as described previously (Boroujerdi, 2002).

14.3 KINETICS OF MULTIPLE ORAL ADMINISTRATION

The principle of kinetics of oral administration is the same as that for multiple intravenous injections (Section 14.2). The main difference is the influence of the absorption process on the plasma concentration of the administered compound and the related geometric series. Similar to the elimination process, absorption of multiple dosing can also be defined in terms of a geometric series. For multiple dosing kinetics of oral administration there are two biological processes of absorption and elimination, two geometric series and two solutions, one for each process. The solutions of the geometric series before achieving the steady state are

$$\text{Elimination} : (S_n)_K = \frac{1 - e^{-nK\tau}}{1 - e^{-K\tau}} \tag{14.40}$$

$$\text{Absorption} : (S_n)_{k_a} = \frac{1 - e^{-nk_a\tau}}{1 - e^{-k_a\tau}} \tag{14.41}$$

At steady state, Equations 14.40 and 14.41 are modified to

$$(\text{Elimination})_{ss} : (S_n)_K = \frac{1}{1 - e^{-K\tau}} \tag{14.42}$$

$$(\text{Absorption})_{ss} : (S_n)_{k_a} = \frac{1}{1 - e^{-k_a\tau}} \tag{14.43}$$

The profile of plasma concentration following multiple oral dosing is presented in Figure 14.4. The only noticeable difference is the bell-shaped characteristic of oral doses.

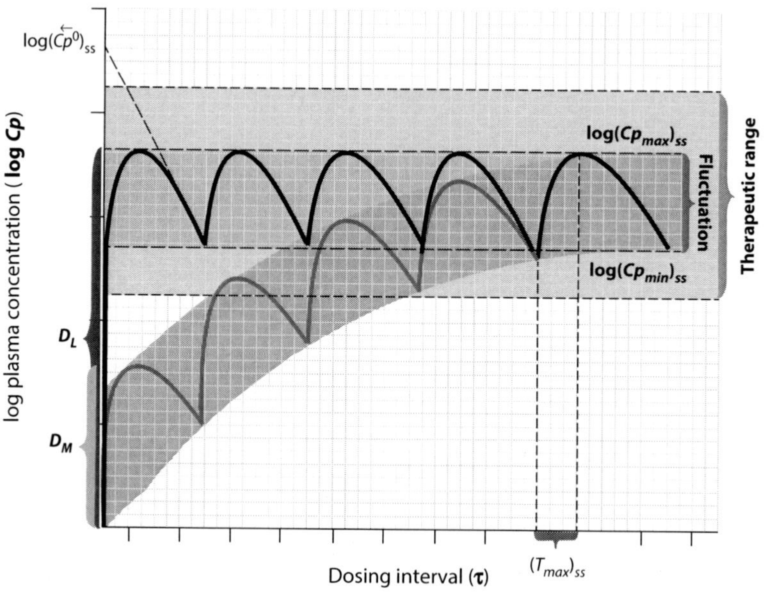

FIGURE 14.4 Steady-state plasma concentration–time profile of multiple oral dosing regimen following administration of a loading dose and five maintenance doses in comparison with gradual accumulation of the compound in the body toward the steady-state fluctuation by giving only the maintenance doses.

14.3.1 Peak, Trough, and Average Plasma Concentrations before and after Achieving Steady State

Multiplying the equation of the first dose by the solutions of the geometric series yields the trough levels of multiple oral administrations.

The equation of the trough level of the first dose is

$$(Cp_{min})_1 = \overleftarrow{Cp}^0(e^{-K\tau} - e^{-k_a\tau}) \tag{14.44}$$

The equation of the trough level after nth dose before achieving steady state is (Gibaldi and Perrier, 1982)

$$(Cp_{min})_n = \overleftarrow{Cp}^0\left[\left(\frac{1-e^{-nK\tau}}{1-e^{-K\tau}}\right)e^{-K\tau} - \left(\frac{1-e^{-nk_a\tau}}{1-e^{-k_a\tau}}\right)e^{-k_a\tau}\right] \tag{14.45}$$

The plasma concentration at any time before achieving the steady state is

$$(Cp_{min})_n = \overleftarrow{Cp}^0\left[\left(\frac{1-e^{-nK\tau}}{1-e^{-K\tau}}\right)e^{-Kt} - \left(\frac{1-e^{-nk_a\tau}}{1-e^{-k_a\tau}}\right)e^{-k_at}\right] \tag{14.46}$$

where n is the number of doses; τ is the dosing interval, and t is the time after the administration of nth dose.

Equations 14.47 and 14.48 define the trough level and plasma concentration at any time during the steady state

$$(Cp_{min})_{ss} = \overleftarrow{Cp}^0\left[\frac{e^{-K\tau}}{1-e^{-K\tau}} - \frac{e^{-k_a\tau}}{1-e^{-k_a\tau}}\right] \tag{14.47}$$

$$(Cp_{min})_{ss} = C\bar{p}^0 \left[\frac{e^{-Kt}}{1 - e^{-K\tau}} - \frac{e^{-k_a t}}{1 - e^{-k_a \tau}} \right] \tag{14.48}$$

When the absorption process is fast and the dosing interval is equal or greater than $7(T_{1/2})_{k_a}$, the exponential term representing the absorption approaches zero ($e^{-k_a\tau} \Rightarrow 0$) and the trough level will only be a function of the elimination process (Equations 14.49 and 14.50). In other words, the maintenance dose is given when the absorption process is complete. In clinical practice it can be assumed that the amount of dose left at the site of absorption is insignificant after 4–5 half-lives of absorption.

$$(Cp_{min})_n = C\bar{p}^0 \left(\frac{1 - e^{-nK\tau}}{1 - e^{-K\tau}} \right) e^{-K\tau} \tag{14.49}$$

$$(Cp_{min})_{ss} = C\bar{p}^0 \left(\frac{e^{-K\tau}}{1 - e^{-K\tau}} \right) \tag{14.50}$$

Estimation of the peak levels depends on the value of T_{max}. As the T_{max} of steady-state levels remains constant and reproducible, it is considered a more consistent value than the T_{max} before achieving the steady state. By taking the first derivative of Equation 14.48, the equation of T_{max} for multiple oral dosing at steady state can be developed as follows:

The first derivative of Equation 14.48 is

$$\frac{d(Cp_t)_{ss}}{dt} = C\bar{p}^0 \left[\frac{-Ke^{-Kt}}{1 - e^{-K\tau}} + \frac{k_a e^{-k_a t}}{1 - e^{-k_a \tau}} \right] \tag{14.51}$$

Setting Equation 10.42 equal to zero and solving for T_{max} we obtain

$$\frac{e^{-KT_{max}}}{e^{-k_a T_{max}}} = \frac{k_a(1 - e^{-K\tau})}{K(1 - e^{-k_a\tau})} \tag{14.52}$$

$$(T_{max})_{ss} = \frac{2.303}{k_a - K} \log \left(\frac{k_a(1 - e^{-K\tau})}{K(1 - e^{-k_a\tau})} \right) \tag{14.53}$$

Therefore, multiplying the equation of maximum plasma concentration after the first dose (Chapter 12, Equation 12.50) by Equation 14.42 and substituting $(T_{max})_{ss}$ for t yields the equation for the maximum plasma concentration at steady state (Gibaldi and Perrier, 1982)

$$(Cp_{max})_{ss} = \frac{FD}{Vd} \left(\frac{e^{-K(T_{max})_{ss}}}{1 - e^{-K\tau}} \right) \tag{14.54}$$

The average plasma concentration at steady state for multiple oral dosing is similar to Equations 14.19 through 14.23, that is,

$$(Cp_{ave})_{ss} = \frac{FD}{Cl_t \times \tau} = \frac{FD}{K \times Vd \times \tau} = \frac{1.44 \times T_{1/2} \times FD}{Vd \times \tau} = \frac{(AUC)_\tau}{\tau} \tag{14.55}$$

14.3.2 Extent of Accumulation

The extent of accumulation at a steady state is determined as described earlier by Equation 14.31 for intravenous administration:

When $\tau < 7(T_{1/2})_{k_a}$:

$$R = \frac{(Cp_{min})_{ss}}{(Cp_{min})_1} = \frac{1}{(1 - e^{-K\tau})(1 - e^{-k_a\tau})} \tag{14.56}$$

When $\tau > 7(T_{1/2})_{k_a}$:

$$R = \frac{1}{(1 - e^{-K\tau})} \tag{14.57}$$

14.3.3 Loading Dose, Maintenance Dose, and Designing a Dosing Regimen

The procedure for designing a regimen for multiple oral dosing is the same as described for multiple intravenous injections, with the exception of including the bioavailability factor, F, in the calculation of D_M or D_L (Slattery et al., 1980). For example, Equation 14.33 changes to:

$$D_M = \frac{(Cp_{ave})_{ss} \times Vd \times \tau \times K}{F} \tag{14.58}$$

The relationship between the loading dose and the maintenance dose, assuming F values are equal, can be described as:

$$D_L = \frac{D_M}{(1 - e^{-K\tau})(1 - e^{-k_a\tau})} \quad (\text{When } \tau < 7(T_{1/2})_{k_a}) \tag{14.59}$$

or,

$$D_L = \frac{D_M}{1 - e^{-K\tau}} \quad (\text{When } \tau > 7(T_{1/2})_{k_a}) \tag{14.60}$$

The time required for reaching any fraction of steady state during multiple oral dosing is calculated by Equations 14.15 through 14.16.

14.4 EFFECT OF CHANGING DOSE, DOSING INTERVAL, AND HALF-LIFE ON ACCUMULATION IN THE BODY AND FLUCTUATION OF PLASMA CONCENTRATION

The magnitude of the fluctuation of plasma concentration is the difference between the peak and trough levels; for intravenous multiple administration it is equal to the initial plasma concentration of a single maintenance dose. The measure of accumulation in the body is the average amount at steady-state levels. For a given dosing regimen, if the dosing interval is kept constant and the dose is increased, the following parameters increase: the peak and trough levels before and after achieving the steady-state levels, the average plasma concentration at steady state, the accumulation in the

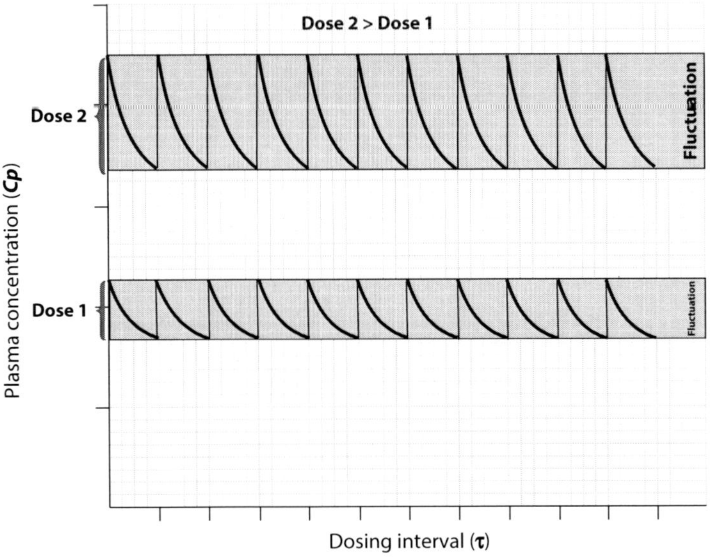

FIGURE 14.5 Influence of changing dose on fluctuation and steady-state levels of plasma concentration when dosing interval is kept the same.

body, and the fluctuation of plasma concentration. Decreasing the dose would reduce the steady-state levels, plasma fluctuation, and drug accumulation in the body (Figure 14.5).

If the dose is kept constant and the dosing interval is altered, increasing the dosing interval would increase the fluctuation, lower the average steady-state level, and reduce the accumulation in the body. If the dosing interval is shortened, a reduced dosing interval would increase the accumulation, reduce the fluctuation, and increase the steady-state levels (Figure 14.6).

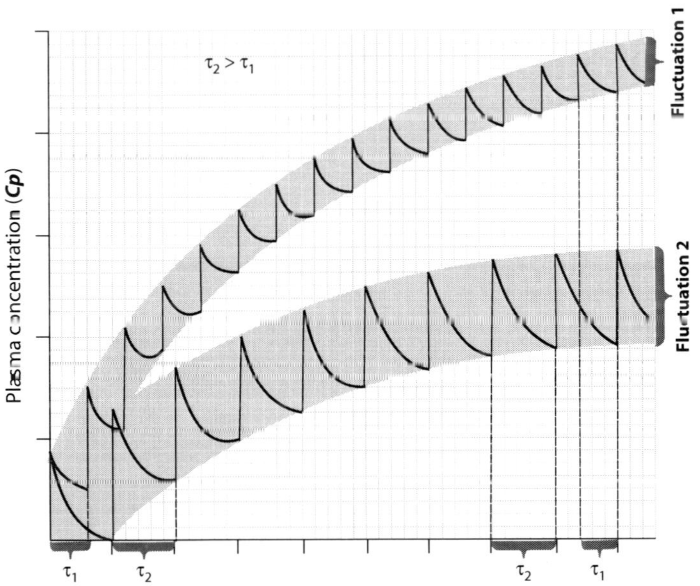

FIGURE 14.6 Influence of changing dosing interval on fluctuation and steady-state levels of plasma concentration when doses are kept the same.

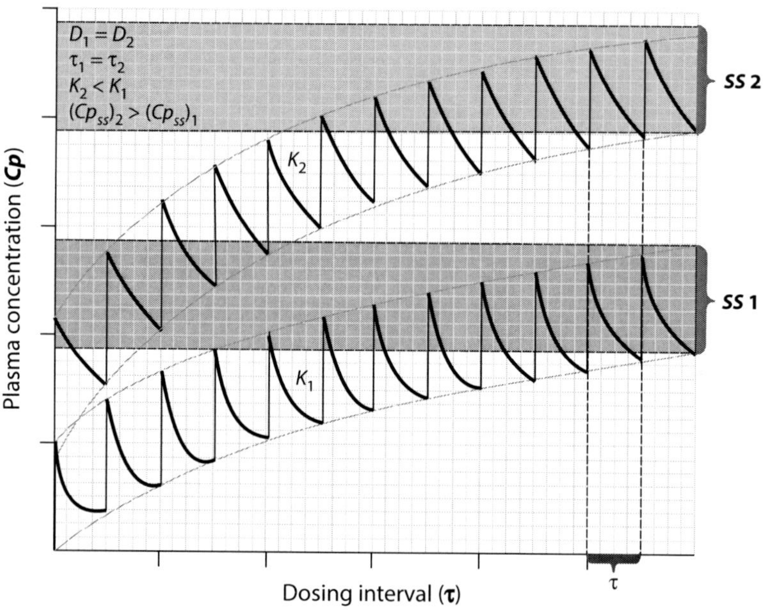

FIGURE 14.7 Influence of change in the overall elimination rate constant (or half-life) due to disease state (renal and/or hepatic failure) on the steady-state plasma concentrations.

In disease states such as renal impairment or hepatic failure the half-life is longer than normal for a given compound. Recommending the dose and dosing interval that are designed for normal renal or hepatic function would increase the steady-state levels and accumulation in the body, and adjustment of dosage regimen is required (Figure 14.7).

14.5 EFFECT OF IRREGULAR DOSING INTERVAL ON THE PLASMA CONCENTRATION OF MULTIPLE-DOSING REGIMEN

The irregular change in the dosing interval influences the fluctuation, accumulation, and steady-state levels. The issue may not be crucial for therapeutic agents with very wide therapeutic ranges. However, it should be considered important for drugs with low therapeutic index and a narrow therapeutic range. Figure 14.8 depicts the plasma concentration fluctuations when a compound with a half-life of 6 hours and narrow therapeutic range is administered on a regular dosing interval of 6-12-6-12 (q.i.d. around the clock or q6h) versus the plasma concentration of the same dose given on an irregular dosing interval 8-12-4-8 (so called 4 tablets a day). The latter schedule would create inconsistency in accumulation in the body and reduce the plasma levels below the minimum effective concentration. However, when the therapeutic range is wide, as shown in Figure 14.8, the inconsistency may not be of any concern since the lowest concentration still remains within the therapeutic range and both schedules are considered reasonable.

14.6 MULTIPLE DOSING KINETICS—TWO-COMPARTMENT MODEL

The equations of multiple dosing kinetics of a compound that follows the two-compartment model are also based on solving the geometric series for each exponential term of the single dose equation. The multiple-dosing relationships are then derived by multiplying each exponential term of the single dose equation by the multiple dosing functions, which are the solution of the geometric series. The fast distributive phase and slow disposition phase, characteristics of the plasma

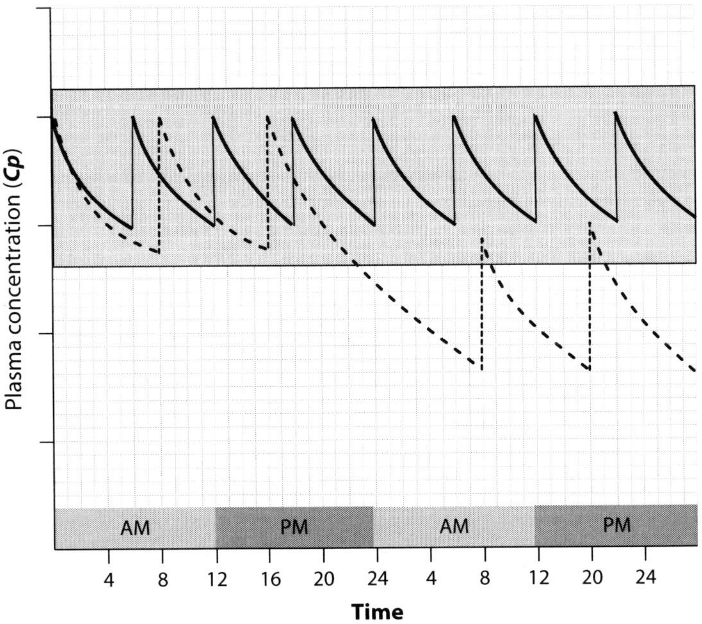

FIGURE 14.8 Comparison of plasma-level fluctuations between regular around the clock dosing versus irregular dosing regimen.

concentration–time profile of the two-compartment model, repeats itself after giving each dose of multiple-dosing regimen (Figure 14.9). The hybrid rate constants of α and β remain essentially the same as the first dose; however, the coefficients *a* and *b* change after each dose before achieving the steady-state levels. During the steady state *a* and *b* remain relatively the same as long as the dosing interval is kept constant.

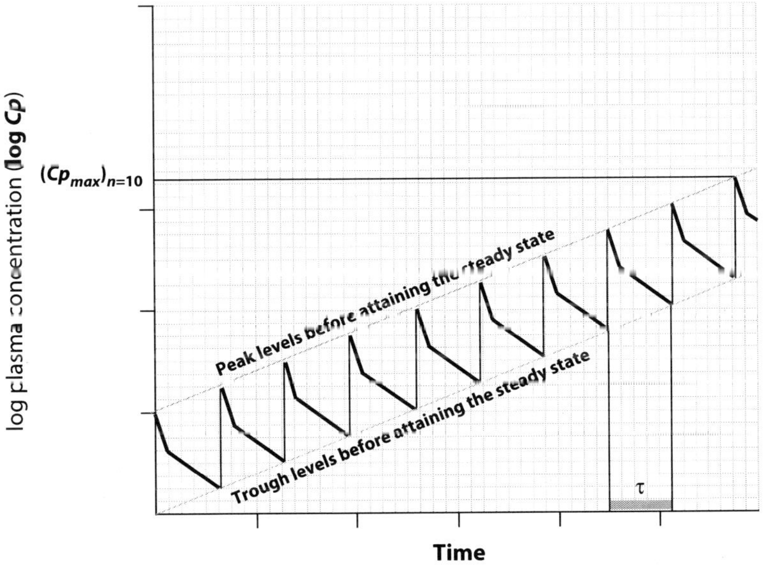

FIGURE 14.9 Profile of plasma concentrations of a compound that follows the two-compartment open model given on a multiple dosing regimen.

14.6.1 Peak, Trough, and Average Plasma Concentrations before and after Achieving Steady State

Solutions of the geometric series of α and β for concentrations before and during steady-state levels are similar to what was discussed for the one-compartment model, that is,

Before achieving steady state

$$S_n(e^{-\alpha\tau})_n = \frac{1 - e^{-n\alpha\tau}}{1 - e^{-\alpha\tau}} \quad \text{and} \quad S_n(e^{-\beta\tau})_n = \frac{1 - e^{-n\beta\tau}}{1 - e^{-\beta\tau}} \tag{14.61}$$

For steady state to be reached without giving a loading dose, a number of maintenance doses should be administered. This means that n in Equation 14.61 increases, $e^{-n\alpha\tau} \Rightarrow 0$ and $e^{-n\beta\tau} \Rightarrow 0$, and the numerators approach one

$$S_n(e^{-\alpha\tau})_{ss} = \frac{1}{1 - e^{-\alpha\tau}} \quad \text{and} \quad S_n(e^{-\beta\tau})_{ss} = \frac{1}{1 - e^{-\beta\tau}} \tag{14.62}$$

The peak and trough levels of the first dose are

$$(Cp_{max})_1 = Cp^0 = a + b \tag{14.63}$$

$$(Cp_{min})_1 = ae^{-\alpha\tau} + be^{-\beta\tau} \tag{14.64}$$

The equations of peak and trough levels before achieving steady state can be developed by multiplying Equations 14.63 and 14.6 by Equation 14.61:

$$(Cp_{max})_n = a\left(\frac{1 - e^{-n\alpha\tau}}{1 - e^{-n\alpha\tau}}\right) + b\left(\frac{1 - e^{-n\beta\tau}}{1 - e^{\beta\tau}}\right) \tag{14.65}$$

$$(Cp_{min})_n = a\left(\frac{1 - e^{-n\alpha\tau}}{1 - e^{\alpha\tau}}\right)e^{-\alpha\tau} + b\left(\frac{1 - e^{-n\beta\tau}}{1 - e^{-\beta\tau}}\right)e^{-\beta\tau} \tag{14.66}$$

It is important to recognize that when the dosing interval is equal or greater than seven half-lives of α (i.e., $\tau \geq (T_{1/2})_\alpha$), $e^{-\alpha\tau} \Rightarrow 0$, $e^{-n\alpha\tau} \Rightarrow 0$ and Equations 14.65 and 14.66 will change to

$$(Cp_{max})_n = a + b\left(\frac{1 - e^{-n\beta\tau}}{1 - e^{-\beta\tau}}\right) \tag{14.67}$$

$$(Cp_{min})_n = b\left(\frac{1 - e^{-n\beta\tau}}{1 - e^{-\beta\tau}}\right)e^{-\beta\tau} \tag{14.68}$$

To calculate plasma concentration at any time between the peak and trough levels of any dose before steady state, the following relationship can be applied:

$$(Cp_t)_n = a\left(\frac{1 - e^{-n\alpha\tau}}{1 - e^{\alpha\tau}}\right)e^{-\alpha t} + b\left(\frac{1 - e^{-n\beta\tau}}{1 - e^{-\beta\tau}}\right)e^{-\beta t} \tag{14.69}$$

When $t \geq 7(T_{1/2})_\alpha$, Equation 14.69 is modified to

$$(Cpt)_n = b\left(\frac{1 - e^{-n\beta\tau}}{1 - e^{-\beta\tau}}\right)e^{-\beta t} \tag{14.70}$$

A similar approach is used to develop equations of peak and trough levels after achieving the steady-state fluctuation, that is,

$$(Cp_{max})_{ss} = \frac{a}{1 - e^{-\alpha\tau}} + \frac{b}{1 - e^{\beta\tau}} \tag{14.71}$$

$$(Cp_{min})_{ss} = \frac{ae^{-\alpha\tau}}{1 - e^{-\alpha\tau}} + \frac{be^{-\beta\tau}}{1 - e^{-\beta\tau}} \tag{14.72}$$

When $\tau \geq 7(T_{1/2})_\alpha$ Equations 14.71 and 14.72 convert to

$$(Cp_{max})_{ss} = a + \frac{b}{1 - e^{\beta\tau}} \tag{14.73}$$

$$(Cp_{min})_{ss} = \frac{be^{-\beta\tau}}{1 - e^{-\beta\tau}} \tag{14.74}$$

Plasma concentrations between the peak and trough levels of any dose during steady state can be estimated by

$$(Cp_t)_{ss} = \frac{ae^{-\alpha t}}{1 - e^{-\alpha\tau}} + \frac{be^{-\beta t}}{1 - e^{-\beta\tau}} \tag{14.75}$$

When $\tau \geq (7T_{1/2})_\alpha$

$$(Cp_t)_{ss} = \frac{be^{-\beta t}}{1 - e^{-\beta\tau}} \tag{14.76}$$

The average steady-state plasma concentration is also calculated similarly to that for the one-compartment model with the same definition (Levy, 1974, 1975; Van Rossum, 1968)

$$(Cp_{ave})_{ss} = \frac{D}{Cl_T \times \tau} \tag{14.77}$$

$$(Cp_{ave})_{ss} = \frac{AUC}{\tau} \tag{14.78}$$

$$(Cp_{ave})_{ss} = \frac{D}{\beta \times Vd_{area} \times \tau} \tag{14.79}$$

$$(Cp_{ave})_{ss} = \frac{D}{k_{10} \times V_1 \times \tau} \tag{14.80}$$

$$(Cp_{ave})_{ss} = \frac{Cp\circ}{k_{10} \times \tau} \tag{14.81}$$

$$(Cp_{ave})_{ss} = \frac{1.44 \times (T_{1/2})_{biol} \times D}{Vd_{area} \times \tau} \tag{14.82}$$

14.6.2 Time Required to Achieve Steady-State Plasma Levels

Analogous to Equations 14.14 through 14.16, the time required to attain steady-state fluctuations is estimated in terms of $n\tau$, that is,

The fraction of steady-state levels is defined as

$$f_{ss} = 1 - e^{-n\beta\tau} \tag{14.83}$$

$$n\tau = -3.3(T_{1/2})_\beta \log(1 - f_{ss}) \tag{14.84}$$

According to Equation 14.83, the time required to achieve any fraction of steady state is dependent solely on the biological half-life (Boroujerdi, 2002):

To achieve:	$f_{ss} = 0.5$	$n\tau = 1(T_{1/2})_\beta$
	$f_{ss} = 0.75$	$n\tau = 2(T_{1/2})_\beta$
	$f_{ss} = 0.90$	$n\tau = 3.3(T_{1/2})_\beta$
	$f_{ss} = 0.95$	$n\tau = 4.3(T_{1/2})_\beta$
	$f_{ss} = 0.99$	$n\tau = 6.6(T_{1/2})_\beta$

14.6.3 Fraction of Steady State, Accumulation Index, and Relationship between Loading Dose and Maintenance Dose

Equation 14.82 represents the fraction of steady state. It can be used to estimate the trough levels before achieving the steady state without relying on the values of a and α, and is solely based on the steady-state trough levels.

The accumulation index and the relationship between the loading dose and maintenance dose are similar to equations of the one-compartment model. The difference is replacing K with the disposition rate constant β.

$$R = \frac{(Cp_{min})_{ss}}{(Cp_{min})_1} = \frac{1}{1 - e^{-\beta\tau}} \tag{14.85}$$

$$D_L = \frac{D_M}{1 - e^{-\beta\tau}} = \frac{D_M}{1 - (1/2)^{\frac{(T_{1/2})_{biol}}{\tau}}} \tag{14.86}$$

The combination of loading dose and maintenance dose provides immediate steady-state levels (Figure 14.10).

Equation 14.86 indicates that for compounds with a very short biological half-life, no loading dose is necessary, that is, $e^{-\beta\tau} \Rightarrow 0$ and $D_L = D_M$. By setting the dosing interval equal to biological half-life, the loading dose will be twice the maintenance dose, that is, $\tau = (T_{1/2})_\beta \Rightarrow D_L = 2D_M$. For compounds with a very long half-life $e^{-\beta\tau} \Rightarrow e^0 \Rightarrow 1$ the denominator of the equation approaches

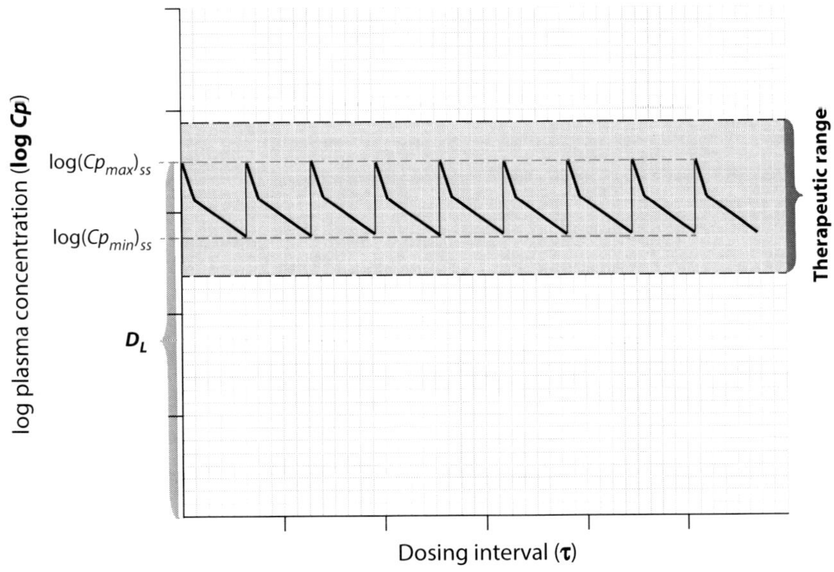

FIGURE 14.10 Profile of plasma concentrations following administration of a loading dose and eight maintenance doses of a compound that follows the two-compartment model.

zero, making the relationship undefined. The loading dose under this condition is set equal to the average amount in the body at steady state, that is,

$$D_L = \frac{D_M}{\beta\tau} \tag{14.87}$$

14.6.4 Decline of Plasma Concentration after the Last Dose

After the last dose is given, regardless of whether the steady-state levels are achieved, the plasma concentration declines biexponentially according to the two-compartment model. When $t' < \tau$, Equations 14.69 through 14.70 or 14.75 through 14.76 are used to estimate plasma concentration before or after achieving the steady state, respectively. However, when $t' > \tau$ the trough level can be used as the initial concentration of a simple monoexponential decline with the slope of $-\beta/2.303$:

$$Cp_{t'} = Cp_{min}e^{-\beta(t'-\tau)} \tag{14.88}$$

$$\log Cp_{t'} = \log Cp_{min} - \frac{\beta(t'-\tau)}{2.303} \tag{14.89}$$

Using Cp_{max} as the initial plasma concentration of the decline after the last dose would add error to the estimation of when $Cp_{t'}$ (Figure 12.3).

14.6.5 The Concept of Half-Life in Multiple Dosing Kinetics of Multicompartmental Model

In the design of a dosing regimen, setting the dosing interval equal to the half-life necessitates an accurate estimate of this constant. An accurate estimate of the constant becomes more important when the pharmacokinetics of the compound follows the multicompartment model. In the one-compartment model the amount in the systemic circulation approximates the amount in the body, whereas in a multicompartment model the half-life estimated from the central compartment concentration may not reflect the true elimination half-life of the amount in the body. One approach in dealing with this dilemma is an approach identified as the "operational multiple dosing half-life" (Sahin and Benet, 2008), which requires weighing the individual half-life by the fractional area under the plasma concentration curve, that is,

$$\frac{1}{(T_{1/2})_{MD}} = \sum_{i=1}^{n} \frac{f_{AUC,i}}{(T_{1/2})_i} \tag{14.90}$$

where $(T_{1/2})_{MD}$ is the multiple dosing half-life, $f_{AUC,i}$ is the fraction of the area under the curve corresponding to each half life estimated as

$$f_{AUC,i} = \frac{L_i/\lambda_i}{\sum_{i=1}^{n} L_i/\lambda_i} \tag{14.91}$$

where λ_i and L_i represent the exponential terms of the multicompartment model and the coefficients of the exponential terms of the systemic concentration–time curve; and n is the number of exponential terms.

Therefore, the multiple dosing half-life for the amount in the body, $(T_{1/2})_{Amt}$, can be characterized as

$$(T_{1/2})_{Amt} = \sum_{i=1}^{n} f_{AUC,i} \times (T_{1/2})_i \tag{14.92}$$

A different approach for setting a suitable dosing interval is the use of the sum of mean residence time of the central compartment and the mean residence time of the absorption site multiplied by a factor (Wagner, 1987). For the two-compartment model with IV bolus administration, the factor was determined as 0.75 (Wagner, 1987). The value of a dosing interval estimated by this approach will provide a ratio of peak/trough levels equal to 2, and the variation of steady-state levels remain within ±33%. For the intravenous administration, the MRT of the absorption site is equal to zero. The half-life of the amount in the body based on this approach is estimated as $0.75 \times MRT$. Whereas the half-life calculated from the mean residence time of the non-compartmental model is

$$(T_{1/2})_{Amt} = 0.693 \times MRT \tag{14.93}$$

$$(T_{1/2})_{Amt} = 0.693 \times \frac{Vd_{ss}}{Cl_T} \tag{14.94}$$

Another approach in estimating the half-life of the amount in the body for the design of a dosing regimen is the concept of "effective half-life" (EHL) (Kwan et al., 1984; Boxenbaum and Battle, 1995), which is based on the accumulation ratio, R, of the compound in the body estimated as

$$R = \frac{(AUC_{ss})_{0\to\tau}}{(AUC_{single\ dose})_{0\to\tau}} \quad \text{or;} \tag{14.95}$$

$$R = \frac{(AUC_{single\ dose})_{0\to\infty}}{(AUC_{single\ dose})_{0\to\tau}} \tag{14.96}$$

where $(AUC_{ss})_{0\to\tau}$ is the area under plasma concentration curve at steady state during a dosing interval; $(AUC_{single\ dose})_{0\to\tau}$ and $(AUC_{single\ dose})_{0\to\infty}$ are the areas under a single dose during a time equal to the dosing interval τ and from time zero to infinity, respectively.

The effective half-life is then estimated as a function of dosing interval

$$(T_{1/2})_{effective} = \frac{0.693 \times \tau}{\ln[(R-1)/R]} \tag{14.97}$$

14.7 MULTIPLE INTRAVENOUS INFUSIONS

The multiple dosing equations of a compound that is administered by intravenous infusion are the same as described for the bolus or first-order absorption. The equation of single dose intravenous infusion, as discussed in Chapter 11 (Equation 11.8), is

$$Cp_t = \frac{k_0}{K \times Vd}(1 - e^{-Kt}) \tag{14.98}$$

Multiplication of Equation 14.98 by the multiple dosing functions yields the following relationship before achieving the steady state:

$$(Cp)_n = \frac{k_0}{K \times Vd}(1 - e^{-Kt})\left(\frac{1 - e^{-nK\tau}}{1 - e^{-K\tau}}\right) \tag{14.99}$$

where (t) is the time of infusion, n is the number of infusions and τ is the dosing interval. The plasma level at time (t), that is, the concentration at the end of nth infusion, is the peak level before the next dose and its trough level is the concentration at time $(\tau - t)$ after the end of nth infusion, that is,

$$(Cp)_n = \frac{k_0}{K \times Vd}(1 - e^{-Kt})\left(\frac{1 - e^{-nK\tau}}{1 - e^{-K\tau}}\right)e^{-K(\tau - t)} \tag{14.100}$$

As the number of infusions increases, $n\uparrow$ and $e^{-nK\tau} \Rightarrow 0$, a steady-state fluctuation can be achieved with a peak level that corresponds to the concentration at the end of one infusion and a trough level that corresponds to the plasma level before the start of next infusion. The related equations after achieving the steady state are given in Equations 14.101 through 14.103.

The single infusion steady-state equation (Chapter 11, Equation 11.6) is

$$Cp_{ss} = \frac{k_0}{Cl_T} \tag{14.101}$$

The multiple dosing relationships for peak and trough levels at steady state are

$$\text{Peak}: (Cp_{t_{infusion}})_{ss} = \frac{k_0}{KVd(1 - e^{-K\tau})} = \frac{Cp_{ss}}{1 - e^{-K\tau}} \tag{14.102}$$

$$\text{Trough} : (Cp_{t'})_{ss} = \frac{Cp_{ss}e^{-Kt'}}{1 - e^{-K\tau}} \tag{14.103}$$

The volume of distribution of the compound administered by a multiple dosing infusion is estimated by the following equation (Sawchuk et al., 1977):

$$Vd = \frac{k_0(1 - e^{-Kt})}{K(Cp_{max} - Cp_{predose}e^{-Kt})}$$

where ($Cp_{predose}$) is the concentration before the start of the succeeding infusion.

REFERENCES

Boroujerdi, M. 2002. *Pharmacokinetics: Principles and Applications*, New York: McGraw Hill-Medical Publishing Division.

Boxenbaum, H. G., Battle, M. 1995. Effective half-life in clinical pharmacology. *J Clin Pharmacol* 35:763–6.

Buell, J., Jelliffe, R., Kalaba, R., Sridhar, R. 1969. Modern control theory and optimal drug regimens, I: The plateau effect. *Math Biosci* 5(3–4):285–96.

Colburn, W. A. 1983. Estimating the accumulation of drugs. *J Pharm Sci* 72(7):833–4.

Gibaldi, M., Perrier, D. 1982. *Pharmacokinetics*, 2nd Edition. New York: Marcel Dekker, Inc.

Krüger-Thiemer, E. 1966. Formal theory of drug dosage regimens, I. *J Theo Biol* 13:212–35

Krüger-Thiemer, E. 1969. Formal theory of drug dosage regimens, II. The exact plateau effect. *J Theo Biol* 23(2):169–90.

Krüger-Thiemer, E., Bünger, P. 1965. The role of the therapeutic regimen, Part II. *Chemotherapia* 10:129–44.

Kwan, K. C., Bohidar, N. R., Hwang, S. S. 1984. Estimation of an effective half-life. In *Pharmacokinetics: A Modern View*, eds. L. Z. Benet, G. Levy and B. Ferraiolo, 147–162. New York: Plenum.

Levy, G. 1974. Pharmacokinetic control and clinical interpretation of steady-state blood levels of drugs. *Clin Pharmacol Ther* 16:130–4.

Levy, G. 1975. Salicylate pharmacokinetics in the human neonate. In *Basic and Therapeutic Aspects of Prenatal Pharmacology*, eds. P. L. Morselli, S. Garattini and F. Sereni. New York: Raven Press.

Perrier, D., Gibaldi, M. 1973. Relationship between plasma or serum drug concentration and amount of drug in the body at steady state upon multiple dosing. *J Pharmacokin Biopharm* 1:17–19.

Sawchuk, R. J., Zaske, D.E., Cipolle, R. J., Wargin, W. A. 1977. Kinetic model for gentamicin dosing with the use of individual patient parameters. *Clin Pharmacol Ther* 21:M362–5.

Sahin, S., Benet, L. Z. 2008. The Operational multiple dosing half-life: A key to defining drug accumulation in patients and to designing extended release dosage forms. *Pharm Res* 25(12):2869–77.

Slattery, J. T., Gibaldi, M., Koup, J. R. 1980. Prediction of maintenance dose required to attain a desired drug concentration at steady state from a single determination of concentration after an initial dose. *Clin Pharmacokin* 5:377–85.

Van Rossum, J. M., Tomey, A. H. M. 1968. Rate of accumulation and plateau plasma concentration of drugs after chronic medication. *J Pharm Pharmacol* 20:390–1.

Wagner, J. G., Northam, J. I., Always, C. D., Carpenter, O. S. 1965. Blood levels of drug at equilibrium state after multiple dosing. *Nature* 207:1301–2.

Wagner, J. G. 1975. *Clinical Pharmacokinetics*, Hamilton, IL: Drug Intelligence Publications Inc.

Wagner, J. G. 1987. Dosage intervals based on mean residence times. *J Pharm Sci* 76:35–38.

15 Bioavailability, Bioequivalence, and Biosimilar Evaluations

15.1 INTRODUCTION

A critical role of bioavailability (BA) and bioequivalence (BE) evaluations is in drug discovery and development of small-molecule therapeutic agents. The focus of evaluation is the performance of a new drug entity or comparative evaluation of more than one compound. BA assessment is based on PK parameters and constants obtained from quantitative analysis of blood or plasma concentration of active compound. The PK measurements of systemic exposure such as *AUC* or Cp_{max} and other parameters define the BA and BE (Chen et al., 2001a,b). The factors discussed in Chapter 8 can influence the absorption, and thus the bioavailability of a compound. Factors such as such as gastric emptying rate, small intestinal transit time, blood flow rate, intestinal and hepatic first-pass effect, influx and efflux proteins, physicochemical characteristics of the compound, formulation factors associated with dosage forms, factors that influence drug dissolution (Amidon et al., 1995), and inactive ingredients can influence the absorption and bioavailability of a compound. Furthermore, variation in splanchnic blood flow and biliary secretion may influence the first-pass effect. The magnitude of bile salts can affect the solubility of the lipophilic compounds (Fleisher et al., 1999; Karalis et al., 2008), species differences, intra-individual variability (McGilveray et al., 1990), and specific site-dependency of absorption, The presence of food in the GI tract (Olanoff et al., 1986; Gupta and Benet, 1990) or other compounds will always remain sources of variability in BA measurements (Griffin, 1981; Walter-Sack, 1987).

The regulatory guidelines related to bioavailability and bioequivalence evaluations are described in the appropriate sections of Title 21 of the Code of Federal Regulations. Additional publications of the U.S. Department of Health and Human Services Food and Drug Administration (FDA), include the Guidance for Industry—Bioavailability and Bioequivalence Studies Submitted in NDAs or INDs—General Considerations (FDA Guidance, 2014) prepared by the Office of Communications Center for Drug Evaluation and Research (CDER); Center for Drug Evaluation and Research (CDER) and ICH related guidelines provide valuable details of the BA and BE evaluation. Other FDA core guidance on bioavailability and bioequivalence are

- U.S. Food and Drug Administration, Title 21 Code of Federal Regulations (CFR) Part 320, Office of Federal Register, National Archives and Records Administration (2014)
- Food and Drug Administration Modernization Act, Public Law No. 105-115, 111 Stat. 2296, 1997 (http://www.fda.gov/opacom/7modact.html)
- U.S. Food and Drug Administration, Center for Drug Evaluation and Research and Center for Biologics Evaluation and Research
- Guidance for Industry—Providing Clinical Evidence of Effectiveness for Human Drugs and Biological Products. Office of Training and Communications, Division of Communications Management, Drug Information Branch, HFD-210, Rockville, Maryland 20857, May 1998
- U.S. Food and Drug Administration, Title 21 CFR 21 CFR 314.50(d) (1)–(6), Office of Federal Register, National Archives and Records Administration (2001)
- U.S. Food and Drug Administration, Title 21 CFR 314.94, Office of Federal Register, National Archives and Records Administration (2001)

- Waiver of *in vivo* bioavailability and bioequivalence studies for immediate release, solid oral dosage forms based on a biopharmaceutics classification system (published August 2000)
- Bioavailability and bioequivalence studies for orally administered drug products—General considerations (published October 2000)
- Statistical approaches to establishing bioequivalence (published January 2001)
- Bioanalytical method validation for human studies (published May 2001)
- Food-effect bioavailability and bioequivalence studies (draft published October 1997)
- Topical dermatological drug product NDAs and ANDAs—bioavailability, bioequivalence, *in vitro* release, and associated studies
- Bioavailability and bioequivalence studies for nasal aerosols and nasal sprays for local actions

15.1.1 Terminology and Definitions

A few of the key definitions are noted here to signify the applications of pharmacokinetic parameters and constants in drug approval and regulatory considerations:

15.1.1.1 Bioavailability

> This term means the rate and extent to which active ingredient or active moiety is absorbed from a drug product and becomes available at the site of action. For drug products that are not intended to be absorbed into the bloodstream, bioavailability may be assessed by measurements intended to reflect the rate and extent to which the active ingredient or active moiety becomes available at the site of action (21 CRF 320.1).

The rate represents how fast a drug is absorbed from an extravascularly administered dosage form and the extent represents how much from the active ingredient of the dosage form reaches the systemic circulation.

15.1.1.2 Pharmaceutical Equivalents

The FDA defines pharmaceutical equivalents as drug products that contain the same active ingredient(s), are of the same dosage form route of administration and are identical in strength or concentration.

In other words, two or more drug products with identical active ingredient(s) (i.e., the same salt or ester of the same therapeutic moiety that have the same quality, purity, identity. and strength according to USP or other standards) are considered pharmaceutically equivalents. This indicates that pharmaceutical equivalents can be different in their inactive ingredients (excipients) including preservatives, coloring and flavoring agents. They can also be different in release mechanism, expiration date, packaging and labeling (with some limitation) and shape.

15.1.1.3 Pharmaceutical Alternatives

According to FDA drug products are considered a pharmaceutical alternative if they contain the same therapeutic moiety, but are different salts, esters, or complexes of that moiety, or are different dosage forms or strengths.

Therefore, in contrast to pharmaceutical equivalents, pharmaceutical alternatives have different salts of the active ingredient(s) such as hydrochloride, sulfate, phosphate, etc. They can also be of different strengths in different dosage forms. Thus, controlled-release capsules and standard-release tablets are pharmaceutical alternatives.

15.1.1.4 Bioequivalence

> The absence of a significant difference in the rate and extent to which the active ingredient or active moiety in pharmaceutical equivalents, or pharmaceutical alternatives, become available at the site of

drug action when administered at the same molar dose under similar conditions in an appropriately designed study. (21 CRF 320.1)

Therefore, pharmaceutical equivalents or pharmaceutical alternatives that have the same bioavailability under the same experimental protocol are considered bioequivalent. Section 505(j) (7) (B) of the Federal Food, Drug, and Cosmetic Act describes the circumstances under which two formulations are considered bioequivalent drug products as when:

The rate and extent of absorption of the test drug do not show a significant difference from the rate and extent of absorption of the reference drug when administered at the same molar dose of the therapeutic ingredient under similar experimental conditions in either a single dose or multiple doses; or the extent of absorption of the test drug does not show a significant difference from the extent of absorption of the reference drug when administered at the same molar dose of the therapeutic ingredient under similar experimental conditions in either a single dose or multiple doses and the difference from the reference drug in the rate of absorption of the drug is intentional, is reflected in its proposed labeling, is not essential to the attainment of effective body drug concentrations on chronic use, and is considered medically insignificant for the drug.

The above definitions are intended for drug products that are absorbed into the systemic circulation.

15.1.1.5 Therapeutic Equivalents

The FDA Orange Book that provides the list of approved generic products for substitution purposes, define the therapeutic equivalent as

Drug products are considered to be therapeutic equivalents only if they are pharmaceutically equivalent and if they can be expected to have the same clinical effect and safety profile when administered to patients under the conditions specified in the labeling.

15.1.1.6 Absolute and Relative Bioavailability

As noted earlier, a low bioavailability can be indicative of a number of events such as hepatic first-pass effect, pre-systemic metabolism by intestinal CYP450, influence of P-glycoprotein, simultaneous removal of drug from the site of absorption by chemical degradation, instability of a drug in the GI tract, etc. The intravenously injected drugs bypass the absorption related hindrances and are introduced completely in the systemic circulation; their bioavailability is considered 100% that is, fraction reaching the systemic circulation is equal to one ($F = 1$). If the bioavailability of an oral dosage form, or any extravascularly administered dosage form, is compared with the intravenous injection, it is called absolute bioavailability (Figure 16.1). It represents the fraction of dose absorbed from an oral dosage form divided by 1 (i.e., F_{oral}/F_{iv}, or $F_{oral}/1$). Thus, the absolute bioavailability cannot exceed one. When $F_{oral} = 1$, the drug is presumed fully absorbed and totally bioavailable, however, the absorption rate of the product may or may not be fast. Achieving an absolute bioavailability equal to 1 may not be attainable for a number of extravacular routes including oral administration.

The relative bioavailability is referred to the comparison of a new formulation of a drug with its reference dosage form using the same or a different route of administration (Colburn and Welling 1986). The intravenous injection usually is not included in this comparison. The relative bioavailability also applies to the comparison of effect(s) of different conditions such as fasting, food, exercise, drug interaction, etc. on the similar or different dosage forms intended for the same or different routes of administration (Colburn et al., 1985). In contrast to absolute bioavailability, relative bioavailability can exceed one. A relative bioavailability greater than one is indicative of the active ingredient being more absorbed from the test formulation and the test having a higher bioavailability. In addition, if the influence of a condition is under investigation, a relative bioavailability of greater than one indicates that the condition enhances the absorption of the therapeutic agent (Weidekamm et al., 1998). In the comparison of two similar dosage forms, if relative bioavailability is less than one, it indicates that the absorption from the test formulation is not comparable to that of the reference, or the condition under investigation hinders the absorption of the drug (Cassidy et al., 1999; Marathe et al., 1998).

15.2 PEAK EXPOSURE, TOTAL EXPOSURE, AND EARLY EXPOSURE

As discussed earlier, the BA and BE assessments are based on the rate and extent of absorption. The extent of absorption is measured by the *AUC* of the plasma concentration–time curve during the sampling time, that is, AUC_0^t, or to time infinity, AUC_0^∞. The rate of absorption is commonly defined by Cp_{max} and T_{max}, which are not the direct measures of the rate. They are estimated from the observed data and, if calculated, a relevant PK model must be used (see Chapter 12). Both Cp_{max} and AUC_0^∞ or AUC_0^t are labeled by the FDA as the peak exposure and total exposure, respectively. The guidance also recommends the estimation of early exposure by determining the partial *AUC* between time zero and the population median T_{max}, contingent upon having at least two plasma measurements before the T_{max} for more accurate estimation of the partial area representing the early exposure. The early exposure has been shown to be more pertinent than Cp_{max} in expressing the absorption rate differences in relative bioavailability studies (Bois et al., 1994; Endrenyi et al., 1998a,b; Chen, 1992; Rostami-Hodjegan et al., 1994; Macheras et al., 1994). The rate constant of absorption, a model-dependent rate constant, is less commonly employed in BA and BE assessment (Chen et al., 2001).

The preferred method for estimation of total or partial exposure is the trapezoidal rule (Appendix A).

The following sections are model-independent approaches for estimation of absolute and relative bioavailability. These approaches are applicable when all biological processes follow first-order kinetics and they are valid only for dose-independent pharmacokinetics.

15.2.1 Estimation of Absolute Bioavailability from Plasma Data: Single Dose

Using the following model-independent equations of oral administration and intravenous injection:

Intravenous administration:

$$Cl_T = \frac{Dose}{AUC} \therefore \frac{Cl_T \times AUC}{Dose} = 1 \tag{15.1}$$

Oral administration:

$$Cl_T = \frac{F \times Dose}{AUC} \therefore \frac{Cl_T \times AUC}{Dose} = F \tag{15.2}$$

Therefore, the absolute bioavailability can be estimated as

$$\frac{F}{1} = F_{absolute} = \frac{(Cl_T)_{oral} \times (AUC)_{oral} \times (Dose)_{i.v.}}{(Cl_T)_{i.v.} \times (AUC)_{i.v.} \times (Dose)_{oral}} \tag{15.3}$$

Because in linear pharmacokinetics clearance is considered a constant and independent of the amount of dug in the body or route of administration, theoretically $(Cl_T)_{oral}$ is equal to $(Cl_T)_{iv}$ and if equal doses are administered orally and intravenously, Equation 15.3 is simplified to

$$F_{absolute} = \frac{(AUC)_{oral}}{(AUC)_{i.v.}} \tag{15.4}$$

15.2.2 Estimation of Absolute Bioavailability from Total Amount Eliminated from the Body: Single Dose

For the assessment of bioavailability and bioequivalence, primarily plasma concentration–time data is recommended (FDA, 1977). The use of urinary data is viewed as an alternative source of information. Contingent upon the feasibility of the measurements and sensitivity of analytical methodology, the following relationship can be employed to estimate the bioavailability of a compound from total amount of unchanged and/or metabolites in the urine:

Intravenous administration:

$$Dose = A_e^\infty + A_m^\infty \tag{15.5}$$

Oral administration:

$$F \times Dose = A_e^\infty + A_m^\infty \tag{15.6}$$

Setting Equation 15.6 equal to Equation 15.5 and solving for $F_{absolute}$ yields

$$F_{absolute} = \frac{(A_e^\infty + A_m^\infty)_{oral} \times (Dose)_{i.v.}}{(A_e^\infty + A_m^\infty)_{i.v.} \times (Dose)_{oral}} \tag{15.7}$$

Equation 15.6 also yields

$$F_{absolute} = \frac{A_e^\infty + A_m^\infty}{Dose} \tag{15.8}$$

It is also feasible to estimate the bioavailability from the unchanged compound in the urine using the following relationships:

Intravenous administration:

$$A_e^\infty = f_e \times Dose \tag{15.9}$$

Oral administration:

$$A_e^\infty = f_e \times F \times Dose \tag{15.10}$$

Therefore, using Equations 15.9 and 15.10 yields the following relationship for the absolute

$$F_{absolute} = \frac{(A_e^\infty)_{oral} \times (f_e Dose)_{i.v.}}{(A_e^\infty)_{i.v.} \times (f_e Dose)_{oral}} \tag{15.11}$$

15.2.3 Estimation of Relative Bioavailability from Plasma Data: Single Dose

Similar to the absolute bioavailability, the related equations are similar to the absolute bioavailability

$$\frac{F_{test}}{F_{reference}} = \frac{(Cl_T)_{test} \times (AUC)_{test} \times (Dose)_{reference}}{(Cl_T)_{reference} \times (AUC)_{reference} \times (Dose)_{test}} \tag{15.12}$$

15.2.4 Estimation of Relative Bioavailability from Total Amount Eliminated from the Body: Single Dose

The equations are similar to those used to estimate absolute bioavailability from the total amount eliminated from the body:

Test:

$$F_{test} \times Dose = A_e^{\infty} + A_m^{\infty} \tag{15.13}$$

Reference:

$$F_{reference} \times Dose = A_e^{\infty} + A_m^{\infty} \tag{15.14}$$

Therefore,

$$\frac{F_{test}}{F_{reference}} = \frac{(A_e^{\infty} + A_m^{\infty})_{test} \times (Dose)_{reference}}{(A_e^{\infty} + A_m^{\infty})_{reference} \times (Dose)_{test}} \tag{15.15}$$

and

$$\frac{F_{test}}{F_{reference}} = \frac{(A_e^{\infty})_{test} \times (f_e Dose)_{reference.}}{(A_e^{\infty})_{reference} \times (f_e Dose)_{test}} \tag{15.16}$$

15.3 BIOAVAILABILITY AND FIRST-PASS METABOLISM

A number of factors can influence the bioavailability and reduce the amount of a compound reaching the systemic circulation. The possible contributing factors include first-pass intestinal and hepatic metabolism, chemical degradation at the site of absorption, inadequate permeability through the GI tract wall, and poor performing dosage form. Thus, the absolute bioavailability of an orally administered dosage form can be considered the product of various fractions that outlast the barriers of pre-disposition (Pang and Gillette, 1978; Colburn, 1979; Cassidy and Houston, 1980; Kwan, 1997)

$$F_{absolute} = F_{abs} \times (F_{escaped})_{IFPE} \times (F_{escaped})_{HFPE} \tag{15.17}$$

where F_{abs} is the fraction at the site of absorption; $(F_{escaped})_{IFPE}$ and $(F_{escaped})_{HFPE}$ are the fractions that escape the intestinal first-pass effect (IFPE) and hepatic first-pass effect (HFPE), respectively. The magnitude of the fraction that escapes first-pass metabolism depends on the rate and extent of first-pass metabolism. A conceptual approach for estimating the fraction that escapes the hepatic first-pass metabolism is

$$(A_{liver})_{input} = Q \times AUC_{input} \tag{15.18}$$

$$(A_{liver})_{output} = Q \times AUC_{output} \tag{15.19}$$

$$A_m = Cl_m \times AUC_{input} \tag{15.20}$$

where the amount of compound entering the liver is $(A_{liver})_{input}$, the amount escaping the first-pass metabolism is $(A_{liver})_{output}$, the blood flow is Q, and the amount of first-pass metabolism is A_m.

Therefore, the fraction that is metabolized by the liver is the ratio of the metabolic clearance to the blood flow of the liver

$$F_m = \frac{Cl_m \times (AUC)_{input}}{Q_{liver} \times (AUC)_{input}} = \frac{Cl_m}{Q_{liver}} \quad (15.21)$$

The fraction that escapes the hepatic metabolism and represents the bioavailability of the drug is estimated as

$$(F_{escaped})_{HFPE} = 1 - F_m \quad (15.22)$$

In experimental animal studies, when dosing into and sampling from the portal vein and duodenal region are feasible, the following equations differentiate between the gut and hepatic first-pass effect of the orally administered compound (Lee et al., 2001):

Fraction that escapes hepatic first-pass effect:

$$1 - F_m = 1 - (ER)_{liver} = \frac{(AUC)_{portal}}{(AUC)_{iv}} \times \frac{D_{iv}}{D_{portal}} \quad (15.23)$$

Fraction that escapes intestinal first-pass effect:

$$1 - (ER)_{GI} = \frac{1}{F_{abs}} \times \frac{(AUC)_{ID}}{(AUC)_{portal}} \times \frac{D_{portal}}{D_{ID}} \quad (15.24)$$

where (D_{ID}) stands for intra-duodenal dose, and ER is the extraction ratio.

The amount absorbed, not injected, into portal vein can be estimated as Equations 15.25 through 15.26:

$$(A_{abs})_{portal} = D_{ID} \times F_{abs} \times (1 - (ER)_{GI}) \quad (15.25)$$

$$(A_{abs})_{portal} = \int_{t_1}^{t_2} Q_{portal} \times (C_{portal} - Cp) \quad (15.26)$$

Therefore, the fraction that escapes the GI first-pass effect can also be estimated as (Lee et al., 2001; Tam-Zaman et al., 2004)

$$1 - (ER)_{GI} = \frac{1}{D_{ID}} \times \frac{1}{F_{abs}} \times \int_{t_1}^{t_2} Q_{portal}(C_{portal} - Cp) \quad (15.27)$$

15.4 CONFIRMATION OF LINEARITY OF RELATIVE AND ABSOLUTE BIOAVAILABILITY DURING MULTIPLE DOSING REGIMEN

To confirm that the bioavailability of a compound remains the same during a multiple dosing regimen, the following simple relationships may be used to evaluate the dose independency of the compound during a multiple dosing regimen at steady state:

$$\%PTF = \frac{(Cp_{max})_{ss} - (Cp_{min})_{ss}}{(Cp_{ave})_{ss}} \times 100 \quad (15.28)$$

$$\%Change = \frac{(Cp_{max})_{ss} - (Cp_{min})_{ss}}{(Cp_{min})_{ss}} \times 100 \tag{15.29}$$

where *PTF* is referring to peak-trough fluctuation.

For certain drugs the single dose study may not be predictive of the steady-state peak and trough levels. The common reason is that for certain drugs the linear pharmacokinetic characteristics observed after a single-dose administration may change during multiple dosing to dose-dependent or nonlinear pharmacokinetics. Thus, prediction of parameters of multiple dosing of such a drug based on the constants and parameters of the single dose administration is unsafe, and the data from a multiple dosing study are necessary to provide a complete bioavailability evaluation.

A different approach is to compare the area under the plasma concentration–time curve of the single dose from zero to infinity, with the area under plasma concentration of a maintenance dose during one dosing interval at steady state. For drugs that follow linear pharmacokinetics these two areas should be the same.

15.5 BIOEQUIVALENCE STUDY

Two or more drug products are often compared in a bioequivalence study. The objective of such a study is to determine whether a test drug product is equivalent to that of an existing reference product *in vivo* (FDA Guidance, 1997, 1999a,b). The reference product is an approved and marketed product. The test product is alleged to be interchangeable with reference product in clinical practice. This means that the test product is claimed to be therapeutically equivalent to the reference product. The reference product is often referred to as the "innovator product," which indicates the product is authorized by the FDA to be marketed based on its documented safety, efficacy, and quality. If the reference is a brand name product, the manufacturer is the patent holder of the product until the expiration of the patent.

According to the definition, the equivalence testing is referred to "*The absence of a significant difference in the rate and extent to which the active ingredient or active moiety in pharmaceutical equivalents, or pharmaceutical alternatives, become available at the site of drug action when administered at the same molar dose under similar conditions in an appropriately designed study*" (21 CFR part 320 and § 314.70). The phrase "absence of a significant difference" is referred to testing a null hypothesis in a randomized, often open-label, parallel or crossover design. The crossover design is usually preferred and it requires the administration of the test and reference to the same group of subjects separated by a washout period. The washout period is determined based on the half-life or mean residence time of the active ingredient and it is selected to ensure complete removal of the first dose (e.g., ≈ 7 half-lives) before the administration of the dose from the other product. When it is not feasible or possible to carry out a cross over trial, the parallel design is employed. The examples of a bioequivalence trial with parallel design include drugs that are highly toxic, or compounds that have a long mean residence time. An in-depth review of the relevant statistical analysis and experimental design can be found in statistical methodology books and references (Dunnett and Gent, 1977; Selwyn et al., 1981; Hauck and Anderson, 1984; Anderson and Hauck, 1990; Hauschke et al., 1990; Phillips, 1990; Steinijans and Hauschke, 1990; Steinijans et al., 1992; Farolfi et al., 1999; Hauck et al., 2000; Midha et al., 2005). In addition, other sources such as the following governmental resources should be consulted:

- Statistical Approaches Establishing Bioequivalence (USFDA, 2001)
- Bioavailability and Bioequivalence Studies Submitted in NDAs or INDs—General Considerations (USFDA, 2014)
- ICH Guidance Documents—Food and Drug Administration (FDA, 2013)
- Biostatistical Methodology in Clinical Trials (EMEA, European Medicines Agency, 1993)
- BA/BE for HVDs/HVDPs: Concept Paper (EMEA, 2006)

The null hypothesis of the practical approaches in bioequivalence evaluation is briefly discussed here.

Similar to the null hypothesis for student t-test and ANOVA (i.e., H_0: $\mu_{Test} = \mu_{Ref}$ and H_1: $\mu_{Test} \neq \mu_{Ref}$) where μ_{Test} and μ_{Ref} are population means for test and reference products, for the bioequivalence trial the hypothesis is

$$H_0: \left|\mu_{Test} - \mu_{Ref}\right| > \Delta \;\Rightarrow\; \text{bioinequivalence}$$

$$H_1: \left|\mu_{Test} - \mu_{Ref}\right| \leq \Delta \;\Rightarrow\; \text{bioequivalence}$$

Often the distribution of pharmacokinetic parameters such as AUC and Cp_{max} are positively skewed and exhibit heterogeneity of variances. Under this condition, a logarithmic transformation is usually considered to achieve a relatively homogeneous variance (Chow et al., 1991). Under the multiplicative models, the null hypothesis for the bioequivalence trial is

$$H_0: \frac{\mu_{Test}}{\mu_{Ref}} < \Delta_1, \quad \text{or} \quad \frac{\mu_{Test}}{\mu_{Ref}} > \Delta_2 \Rightarrow \text{bioinequivalence}$$

$$H_1: \Delta_1 \leq \frac{\mu_{Test}}{\mu_{Ref}} \leq \Delta_2 \Rightarrow \text{bioequivalence}$$

The recommended values of Δ_1 and Δ_2 by the FDA are 0.8 and 1.25, respectively. The two numbers reflect the symmetrical normal distribution of ±0.223 around the central tendency of zero that corresponds to 100% equivalent. In other words ln 0.8 = −0.2331 and ln 1.25 = +0.2231 falls on either side of the central tendency of ln 1 = 0.

The above hypothesis become additive after logarithmic transformation:

$$H_0: \ln\mu_{Test} - \ln\mu_{Ref} < \ln\Delta_1 \quad \text{or} \quad \ln\mu_{Test} - \ln\mu_{Ref} > \ln\Delta_2 \Rightarrow \text{bioinequivalence}$$

$$H_1: \ln\Delta_1 \leq \ln\mu_{Test} - \ln\mu_{Ref} \leq \ln\Delta_2 \Rightarrow \text{bioequivalence}$$

15.5.1 Required PK Parameters and Other Necessities

The required data for the testing of a null hypothesis are the PK parameters and constants obtained from accessible biological samples representing a summary of the comparative data on absorption and disposition of both test and reference products. The measurements of the active compound and/or its metabolite(s) must be conducted under the Good Laboratory Practice (GLP) guidelines. The subjects should be 18 years old or older and from different age groups and racial groups with a balance of men and women, unless the drug is used in a specific gender group. If the drugs are intended for children, a similar diversity should be followed in recruitment of young patients, if it is feasible. At least 12 to 18 blood/plasma samples should be drawn at appropriate time points. The measurements should include a control (pre-dose) measurement, early-time point samplings during the absorptive phase, and enough measurements during elimination phase. The sampling should span over at least three half-lifes of elimination. Criteria for data deletion due to vomiting or influence of food should be established and validated in advance.

The parameters of single dose study include:

- The area under the plasma concentration–time curve from time zero to the last measurement time point (total exposure): (AUC_0^t)

- The area under the plasma concentration–time curve from time zero to infinity (total exposure): AUC_0^∞
- The area under the early exposure curve: $AUC_0^{T_{max}}$
- Maximum plasma concentration of single dose administration (peak exposure): Cp_{max}
- Time to maximum plasma concentration: T_{max}
- Mean residence time: MRT
- Mean absorption time: MAT
- The rate constant of elimination based on the terminal segment of the curve.
- Volume of distribution: Vd_{ss} (estimated by non-compartmental analysis)
- Total body clearance: Cl_T (estimated from model-independent equation $Cl_T = (Dose/AUC)$)

The parameters of the multiple dosing study include:

- The area under the plasma concentration–time curve of one dosing interval at steady state: AUC_{ss}^{τ}
- Peak level at steady state: $(Cp_{max})_{ss}$
- Trough level at steady state: $(Cp_{min})_{ss}$
- Average steady state plasma concentration: $(Cp_{ave})_{ss}$
- Time to maximum plasma concentration in one dosing interval at steady state: $(T_{max})_{ss}$
- Calculated percent of fluctuation at steady state: $\%Fluctuation = \frac{(Cp_{max})_{ss} - (Cp_{min})_{ss}}{(Cp_{ave})_{ss}}$

All the parameters and constants of single and multiple dosing are based on model-independent approaches, such as reading directly from the observed data or estimating by non-compartmental analysis.

15.5.2 Statistical Analysis of PK Data

For a conventional two-treatment comparison of test and reference using two-period, two-sequence randomized crossover design, the statistical analysis of the PK data include analysis of variance (ANOVA) with the source of variability among the groups, among the subjects in groups, and among periods of study and treatment. The "80/20 Power Rule" is used for sample size determination in the planning stage of a trial (Diletti et al., 1992). The rule requires that the sample size be large enough to provide an 80% probability of detecting a 20% differences between the mean bioavailability characteristics of the two products. Except for the T_{max}, the concentration and concentration-dependent parameters and constants are based on logarithmic transformation. The parametric and/or non-parametric 90% confidence intervals of all parameters should fall within the window of 80% to 125%. For compounds with a narrow therapeutic range, the window is closer, for example, 90% to 110%.

15.5.3 Required PD Data

The PD parameters are used in bioequivalence studies when the measurement of plasma concentration or other biological samples are not feasible or relevant and/or when the measurements of the active compound and/or its metabolite(s) do not correspond to therapeutic response. The therapeutic response must be relevant and determined under the Good Clinical Practice (GCP) guidelines. The response must be gauged by a validated dose-response relationship and the placebo effect should be taken under consideration. The experimental design is the crossover design unless it is necessary to use the parallel design. The assessment of the outcome and related null hypothesis are the same as described earlier.

15.6 BIOSIMILARS (BIOSIMILARITY AND INTERCHANGEABILITY)

Therapeutic biopharmaceutical agents are typically complex protein molecules produced in living systems. The examples are: monoclonal antibodies, streptokinase, interleukin (IL)-2, plasminogen activator, granulocyte colony-stimulating factor, alpha and gamma interferon, erythropoietin, and many more emerging biologic moieties (De Lorenzo and D'Alessio, 2009; Roovers et al., 2007; Alley et al., 2010; Bell et al., 2011; Horton et al., 2012). Contrary to the chemically synthesized small molecules, the "generic" form of a protein drug is not structurally identical to the reference product. The term "generic" does not even apply to the protein drugs with a similar structure to the reference product. Thus, the similar products, not identical, are considered biologically similar or biosimilar products. Because minor modifications in the structure of the protein drugs can modify their efficacy, potency, safety (e.g., immune response) and purity, the FDA has established guidance for scientific considerations in demonstrating biosimilarity to a reference product, which can be found at http://www.fda.gov/downloads/Drugs/GuidanceComplianceRegulatoryInformation/Guidances/UCM291128.pdf.

The international agencies such as WHO (World Health Organization), EMA (European Medicines Agency), and other agencies in different countries have similar policies related to biosimilars. Other helpful materials include the USP-NF chapters related to potency of biologic medicine (USP. USP-NF. http://www.usp.org/usp-nf.Sep2013), USP has also agreed to work with NIBSC (National Institute for Biological Standards and Control) and other governmental laboratories to assure that the its biologic reference materials are in accordance with national and international materials (Williams et al., 2014).

The *in vitro* and *in vivo* PK analyses of therapeutic biologics are challenging and often inconsistent. Few investigations have shed light on the complex behavior of these compounds in the body (Faggioni, 1992; Vugmeyster et al., 2011; Vugmeyster et al., 2012; Fronton et al., 2014). For example, the PK analysis of monoclonal antibodies has been investigated and various PBPK or a classical compartmental model or other mechanistic or dynamic models have been proposed, yet no clear consensus on the disposition profile of the compound has emerged (Keizer et al., 2010; Vugmeyster et al., 2012; Xiao, 2012; Jones et al., 2013; Fronton et al., 2014). The challenges associated with PK analysis of therapeutic biologics in comparison to the small synthetic molecules include (1) physicochemical characteristics and properties of these large molecules, their size, shape and complex molecular charge and different stability kinetics; (2) they are administered most often through intravenous, subcutaneous or nasal route, and each route manifest inter- and intra-individual differences; (3) their distribution profile depends on their size, shape and charge, and binding to the target site; (4) their elimination is through proteolysis and processes such as nonspecific endocytosis, Fc receptor-mediated clearance (Roopenian and Akilesh, 2007; Keizer et al., 2010; Kuo and Aveson, 2011), and target mediated clearance; (5) their target-binding and target-mediated clearance are capacity-limited, saturable, and kinetically nonlinear; (6) the subject variability, species differences in PK and PD profile, immunogenicity and off-target effects also contribute to the complexity of PK/PD analysis and human/species extrapolation; and (7) the absence of validated *in vitro* systems for absorption or metabolism studies of biologics adds to the complexity of PK/PD challenges.

Concluding from the current language of the regulations, a biosimilar product can be defined as a therapeutic biologic agent that has similar *in vitro* and *in vivo* biological properties and characteristics, and similar clinical outcomes in efficacy, safety, and immunogenicity to the reference product. Interchangeability refers to the biosimilars that can be substituted for the reference products.

REFERENCES

Alley, S. C., Okeley, N. M., Senter, P. D. 2010. Antibody-drug conjugates: Targeted drug delivery for cancer. *Curr Opin Chem Biol* 14:529–37.

Amidon, G. L., Lennernas, H., Shah, V. P., Crison, J. R 1995. A theoretical basis for a biopharmaceutic drug classification: the correlation of *in vitro* drug product dissolution and *in vivo* bioavailability. *Pharm Res* 12:413–20.

Anderson, S., Hauck, W. W. 1990. Consideration of individual bioequivalence. *J Pharmacokin Biopharm* 18(3):259–73.

Bell, D. A., Hooper, A. J., Burnett, J. R. 2011. Mipomersen, an antisense apolipoprotein B synthesis inhibitor. *Expert Opin Investig Drugs* 20:265–72.

Bois, F. Y., Tozer, T. N., Hauck, W. W., Chen, M. L., Patnaik, R., Williams, R. L. 1994. Bioequivalence performance of several measures of rate of absorption. *Pharm Res* 11:966–74.

Cassidy, M. K., Houston, J. B. 1980. *In vivo* assessment of extrahepatic conjugation metabolism in first pass effects using the model compound phenol. *J Pharm Pharmac* 32:57–9.

Cassidy, J., Twelves, C., Cameron, D., Steward, W., O'Byrne, K., Jordell, D., Banken, I. et al. 1999. Bioequivalence of two tablet formulations of capecitabine and exploration of age, gender, body surface area, and creatinine clearance as factors influencing systemic exposure in cancer patients. *Cancer Chemother Pharmacol* 44:453–60.

Chen, M. L. 1992. An alternative approach for assessment of rate of absorption in bioequivalence studies. *Pharm Res* 9:1380–5.

Chen, M. L., Lesko, L. J., Williams, R. L. 2001a. Measures of exposure versus measures of rate and extent of absorption. *Clin Pharmacokinet* 40:565–72.

Chen, M. L., Shah, V., Patnaik, R., Adams, W., Hussain, A., Conner, D., Mehta, M. et al. 2001b. Bioavailability and bioequivalence: An FDA regulatory overview. *Pharm Res* 18(12):1645–50.

Chow, S. C., Peace, K. E., Shao, J. 1991. Assessment of bioequivalence using a multiplicative model. *J Biopharm Stat* 1(2):193–203.

Colburn, W. 1979. A pharmacokinetic model to differentiate preabsorption, gut epithelial, and hepatic first-pass metabolism. *J Pharmacokin Biopharm* 4:407–15.

Colburn, W. A., Gibson, D. M., Rodriguez, L. C., Bugge, C. J. L., Blumenthal, H. P. 1985. Effect of meals on the kinetics of etretinate. *J Clin Pharmacol* 25:583–89.

Colburn, W. A., Welling, P. G. 1986. Relative bioavailability: What reference? *J Pharm Sci* 75(9):921–22.

De Lorenzo, C., D'Alessio, G. 2009. Human anti-ErbB2 immunoagents–immunoRNases and compact antibodies. *FEBS J* 276:1527–35.

Diletti, E., Hauschke, D., Steinijous, V. W. 1992. Sample size determination: Extended tables for the multiplicative model and bioequivalence ranges of 0.9 to 1.11 and 0.7 to 1.43. *Int J Clin Pharmacol Ther Tox* 30(Suppl. 1):S59–62.

Dunnett, C. W., Gent, M. 1977. Significance testing to establish equivalence between treatments with special reference to data in the form of 2 × 2 tables. *Biometrics* 33:593–602.

Endrenyi, L., Csizmadia, F., Tothfalusi, L., Balch, A. H., Chen, M. L. 1998a. The duration of measuring partial AUCs for the assessment of bioequivalence. *Pharm Res* 15:399–404.

Endrenyi, L., Csizmadia, F., Tothfalusi, L., Chen, M. L. 1998b. Metrics comparing simulated early concentration profiles for the determination of bioequivalence. *Pharm Res* 15:1292–9.

Faggioni, R. 1992. *Protein Pharmacokinetics and Metabolism*. New York: Plenum Press.

Farolfi, M., Powers, J. D., Rescigno, A. 1999. On the determination of bioequivalence. *Pharmacol Res* 39:1–4.

FDA Center for Drug Evaluation and Research: Guidance for Industry 1997. Extended Release Oral Dosage Forms: Development, Evaluation, and Application of *in Vitro/in Vivo* Correlation. Rockville, MD.

FDA Center for Drug Evaluation and Research: Guidance for Industry 1999a. Average, Population and Individual Approaches to Establishin Bioequivalence, Rockville, MD.

FDA Center for Drug Evaluation and Research: Guidance for Industry 1999b. BA and BE Studies for Orally Administered Drug Products—General Consideration, Rockville, MD.

FDA – CDER. 2014. Bioavailability and Bioequivalence—Studies Submitted in NDAs or INDs http://www.fda.gov/Drugs/GuidanceComplianceRegulatoryInformation/Guidances/default.htm http://www.fda.gov/downloads/Drugs/GuidanceComplianceRegulatoryInformation/Guidances/UCM389370.pdf

Fleisher, D., Li, C., Zhou, Y., Pao, L. H, A. Karim, A. 1999. Drug, meal and formulation interactions influencing drug absorption after oral administration clinical implications. *Clin Pharmacokinet* 36(3):233–54.

Food and Drug Administration. 1977. Bioavailability and bioequivalence requirements. *Fed Regist* 42:1624–53.

Fronton, L., Pilari, S., Huisinga, W. 2014. Monoclinal antibody disposition: A simplified PBPK model and its implications for the derivation and interpretation of classical compartment model. *J Phatmacokinet Pharmacodyn* 41(2):87–107.

Griffin, J. 1981. Drug interactions occurring during absorption from the gastrointestinal tract. *Pharmacol Ther* 15:79–8.

Gupta, S., Benet, L. 1990. High-fat meals increase the clearance of cyclosporine. *Pharm Res* 7:46–8.

Hauck, W. W., Andeson, S. 1984. A new statistical procedure for testing equivalence in two-group comparative bioavailability trials. *J Pharmacokin Bioharm* 12(1):83–91.

Hauck, W. W., Hyslop, T., Chen, M-L., Patnaik, R., Williams, R. L., and the FDA Population/Individual Bioequivalence Working Group. 2000. Subject-by-formulation interaction in bioequivalence: Conceptual and statistical issues. *Pharm Res* 17(4):375–84.

Hauschke, D., Steinijans, V. W., Diletti, E. 1990. A distribution-free procedure for the statistical analysis of bioequivalence studies. *Int J Clin Pharmacol Ther Tox* 28(2):72–8.

Horton, S., Walsh, C., Emery, P. 2012. Certolizumab pegol for the treatment of rheumatoid arthritis. *Expert Opin Biol Ther* 12:235–49.

Jones, H., Mayawala, K., Poulin, P. 2013. Dose selection based on physiologically based pharmacokinetic (PBPK) approaches. *AAPS J* 15(2):377–87.

Karalis, V., Macheras, P., Van Peer, A., Shah, V. P. 2008. Bioavailability and bioequivalence: Focus on physiological factors and variability. *Pharm Res* 25(8):1956–62.

Keizer, R. J., Huitema, A. D. R., Schellens, J. H. M., Beijnen, J. H. 2010. Clinical pharmacokinetics of therapeutic monoclonal antibodies. *Clin Pharmacokinet* 49:493–507.

Kuo, T. T., Aveson, V. G. 2011. Neonatal Fc receptor and IgG-based therapeutics. *mAbs* 3:422–30.

Kwan, K. C. 1997. Oral bioavailability and first-pass effects. *Drug Metab Dispos* 25(12):1329–36

Lee, Y-H., Perry, B. A., Lee, H-S., Kunta, J. R., Sutyak, J. P., Patrick J. Sinko, P. J. 2001. Differentiation of gut and hepatic first-pass effect of drugs: 1. Studies of verapamil in ported dogs. *Pharm Res* 18(12):1721–28.

Macheras, P., Symillides, M., Reppas, C. 1994. The cutoff time point of the partial area method for assessment of rate of absorption in bioequivalence studies. *Pharm Res* 11:831–4.

Marathe, P. H., Greene, D. S., Kollia, G. D., Barbhaiya, R. H. 1998. Evaluation of the effect of food on the pharmacokinetics of avitriptan. *Biopharm Drug Disp* 19:381–94.

McGilveray, I. J., Midha, K. K., Skelly, J. P., S., Dighe, S, Doluisio, J. T., French, I. W., A., Karim, A., Burford, R. 1990. Consensus report from "Bio International '89": Issues in the evaluation of bioavailability data. *J Pharm Sci* 79(10):945–46.

Midha, K. K., Rawson, M. J., Hubbard, J. W. 2005. The bioequivalence of highly variable drugs and drug products. *Int J Clin Pharmacol Ther* 43(10):485–98.

Olanoff, L., Walle, T., Cowart, T., Walle, W., Oexmann, M., Conradi. E. 1986. Food effects on propranolol systemic and oral clearance: Support for a blood flow hypothesis. *Clin Pharmacol Ther* 40:408–14.

Pang, K. S., Gillette, J. R.1978. Theoretical relationships between area under the curve and route of administration of drugs and their precursors for evaluating sites and pathways of metabolism. *J Pharm Sci* 67:703–4.

Phillips, K. F. 1990. Power of the two one-sided tests procedure in bioequivalence. *J Pharmacokin Biopharm* 18:137–44.

Roopenian, D. C., Akilesh, S. 2007. FcRn: The neonatal Fc receptor comes of age. *Nat Rev Immunol* 7:715–25.

Roovers, R. C., van Dongen, G. A., van Bergen en Henegouwen, P. M. 2007. Nanobodies in therapeutic applications. *Curr Opin Mol Ther* 9:327–35.

Rostami-Hodjegan, A., Jackson, P. R., Tucker, G. T. 1994. Sensitivity of indirect metrics for assessing "rate" in bioequivalence studies—Moving the "goalposts" or changing the "game". *J Pharm Sci* 83:1554–7.

Selwyn, M. R., Dempster, A. P., Hall, N. R. 1981. A Bayesian approach to bioequivalence for the 2 × 2 changeover design. *Biometrics* 37:11–21.

Steinijans, V. W., Hauschke, D. 1990. Update on the statistical analysis of bioequivalence studies. *Int J Clin Pharmacol Ther Tox* 28(3):105–110.

Steinijans, V. W., Hauschke, D., Jonkman, J. H. G. 1992. Controversies in bioequivalence studies. *Clin Pharmacokinet* 22(4):247–53.

Tam-Zaman, N., Tam, Y. K., Tawfik, S., Wiltshire, H. 2004. Factors responsible for the variability of saquinavir absorption: Studies using an instrumented dog model. *Pharm Res* 21(3):436–42.

U.S. Department of Health and Human Services, Food and Drug Administration Center for Drug Evaluation and Research (CDER). 2014. Guidance for Industry, Bioavailability and Bioequivalence Studies Submitted in NDAs or INDs—General Considerations. http://www.fda.gov/downloads/drugs/guidancecomplianceregulatoryinformation/guidances/ucm389370.pdf. Title 21—Food and drugs, Chapter I—Food and Drug Administration. Department of health and Human Services, Subchapter D—Drugs for Human Use,

Bioavailability and Bioequivalence, updated 2014, part 320, Subpart A and B. http://www.accessdata.fda.gov/scripts/cdrh/cfdocs/cfcfr/CFRSearch.cfm?CFRPart=320.

U.S. Department of Health and Human Services, Food and Drug Administration, Center for Drug Evaluation and research (CDER). 2001. Guidance for Industry, Statistical Approaches to Establishing Bioequivalence. http://www.fda.gov/downloads/Drugs/Guidances/ucm070244.pdf

U.S. Food and Drug Administration, Title 21 CFR 314.94, Office of Federal Register. 2001. National Archives and Records Administration.

Vugmeyster, Y., Szklut, P., Wensel, D., Ross, J., Xu, X., Awwad, M., Gill, D., Tchistiakov, L. and Warner, G. 2011. Complex pharmacokinetics of a humanized antibody against human amyloid beta peptide, anti-abeta Ab2, in nonclinical species. *Pharm Res* 28:1696–706.

Vugmeyster, Y., Xu, X., Theil, F. P., Khawli, L., Leach, M. W. 2012. Pharmacokinetics and toxicology of therapeutic proteins: Advances and challenges. *World J Biol Chem* 3:73–92.

Walter-Sack. I. 1987. The influence of nutrition on the systemic availability of drugs. Part I: Drug absorption. *Klin Wochenschr* 65:927–35.

Weidekamm, E., Rüsing, G., Caplain, H., Sörgel, F., Crevoisier, C. 1998. Lack of bioequivalence of a generic mefloquine tablet with the standard product. *Eur J Clin Pharmacol* 54:615–19.

Williams, R. L., Bristow, A. F., Hauck, W. W., Srini Srinivasan, V., Tina Morris, T., Atouf, F., Ambrose, M. et al. 2014. Role of public standards in the safety and efficacy of biologic medicines. *AAPS J* 16(3):516–21.

Xiao, J. J. 2012. Pharmacokinetic models for FcRn-mediated IgG disposition. *J Biomed Biotechnol* 2012:282989.

16 Preclinical PK/TK and Clinical Trials

16.1 INTRODUCTION

The PK/TK analysis is an essential part of assessing drug safety in the course of drug discovery and development through preclinical investigations, clinical trials and the post-marketing surveillance of a new drug entity. The preclinical TK analysis is associated with the quantitative toxicity assessment of a new synthetic compound immediately after the initial phase of discovery. The analysis provides information and guidance in planning to optimize the dose and dosing regimen for human. The merged notion of PK/TK analysis is clearly defined and expressed in the ICH guidelines S3A as "*... the generation of pharmacokinetic data, either as an integral component of the conduct of non-clinical toxicity studies or in specialty designed supportive studies, in order to assess systemic exposure.*"

Many of the failures of drug candidates in drug discovery and development are related to their undesirable PK/TK properties. For example too long or too short $T_{1/2}$; poor absorption; extensive hepatic first-pass metabolism; Pgp-CYP3A4 coordinated intestinal metabolism; and nonlinear characteristics of lead compound following single or multiple dosing administration. The required preclinical toxicokinetic and pharmacokinetic assessment include: (1) the *in vitro* metabolic studies and *in vitro* characterization of metabolite(s) when the measured metabolites are greater than 10% of the total drug-related exposure, which include hepatocytes, microsomal and/or cytosolic incubations; (2) repeated-dose toxicity, known as chronic toxicity studies; and (3) pharmacokinetics of the lead compound in experimental animals. The preclinical PK/TK *in vivo* and *in vitro* evaluations must be done before the clinical trials.

16.2 PRECLINICAL PK/TK

16.2.1 Estimation of the First Dose in Humans

The determination of the first dose in human is an important step in the preclinical phase of drug discovery and development (DDD) and will remain a challenging aspect of the process (Lowe et al., 2007; Huang et al., 2008). The dose prediction for human from preclinical PK/PD/TK data and biopharmaceutical properties is a multifaceted task. The challenge is to estimate an efficacious dose and dosing regimen for the first time in human (FTIH) based on the interspecies extrapolation of PK/PD data (Miller et al., 2005) and preclinical data such as physicochemical characteristics of the lead compound; its biopharmaceutical factors such as solubility, type of dosage form, dissolution of the solid dosage form; and all physiological factors of GI tract including the intestinal and hepatic first-pass metabolism of oral dosage form and other factors that influence the permeability and absorption of the lead compound (Amidon et al., 1995; Lipinski, 2001; Li et al., 2005; Custodio et al., 2008). Various attempts have been made to predict the human PK/PD constants and parameters from preclinical data (Wajima et al., 2004; De Buck et al., 2007; Fura et al., 2008). The projection of PK parameters and constants such as the steady-state volume of distribution (Oie and Tozer, 1979; Obach et al., 1997, 2008; Mahmood, 2005; Hosea et al., 2009), bioavailability (Fura et al., 2008; Sinha et al., 2008; Hosea et al., 2009), clearance (Oie and Tozer, 1979; Obach et al., 1997, 2008; Mahmood and Yuan, 1999; Stoner et al., 2004; Mahmood, 2005, 2006; Tang and

Mayersohn, 2005; Hosea et al., 2009), etc. from preclinical data are mostly based on interspecies scaling using physiological modeling (Luttringer et al., 2003; Parrott et al., 2005; Jones et al., 2006; De Buck et al., 2007). The combined PK/PD/TK evaluation at the early phase of drug development is essential because of the development of a realistic safe dose for FTIH and also toxicokinetic and risk identification. The old concept of LD_{50}, the lethal dose required to kill 50% of experimental animals, is no longer considered a helpful measure of toxicity, and its use has been discontinued. The following are other measures which are recommended for the estimation of the first dose:

1. NTEL dose (No Toxic Effect Level dose): The largest dose that produces no toxic response in the most sensitive experimental animals.
2. NOAEL dose (No Observed Adverse Effect Level dose): The largest dose with no undesirable side effects such as tissue toxicity, weight loss, or seizures.
3. MTD (Maximum Tolerated Dose): The maximum tolerated dose with no unwanted side effect in multiple dosing.
4. NOEL dose (No Observed Effect dose): The dose that produces the threshold of pharmacological or toxicological response.
5. Microdose trial.

For exploratory clinical trials conducted at the initial part of the Phase I clinical trial, the ICH guideline (M3(R2), 2009, 2013) has recommended the use of microdose trials, which may involve human subject, healthy volunteers or patients from selected populations. The microdose regimen is supposed to produce limited exposure and is not intended for the measurement of any clinical outcomes. The related biological samples, however, can be used to investigate PK/TK or PD/TD parameters and constants at a low dose, single administration and multiple-dosing regimen. There are five approaches for the starting dose:

First approach: Single microdose trial, the subjects receive 100 μg/subject in one injection. This is the dose that can be used to target receptor interaction or disposition profile of the ligand in a PET ligands study.

Second approach: Multiple microdose trial, the subjects receive five administrations of the 100 μg/subject for a total of 500 μg/subject.

Third approach: Single sub-therapeutic administration of a dose yielding up to one half of the NOAEL exposure.

Fourth approach: A dosing regimen for a maximum of 14 days to attain the concentration within the therapeutic range. It is not intended to evaluate the therapeutic outcome of MTD.

Fifth approach: A dosing regimen for a maximum of 14 days such that it achieves the concentration within the therapeutic range and it does not exceed duration of dosing in non-rodent preclinical evaluation. The regimen is not intended to assess the pharmacological response of MTD.

16.2.2 PK/TK Preclinical Requirements

The preclinical metabolic and pharmacokinetic data should include the data from two rodent species (usually the rat and mouse) and a non-rodent species (usually the dog). If nonlinear PK or dose dependency is observed in metabolic and pharmacokinetic or toxicity studies with one species, the same range of doses should be used in metabolic and pharmacokinetic studies with other species. The PK data are considered more effective if metabolism and pharmacokinetic studies are conducted in both sexes of young adult animals of the same species and strain. Commonly reported biological samples and data are from blood (RBCs, plasma, and serum), urine, bile, and feces. In addition, a few representative organ and tissue samples should be taken, such as liver, kidney, fat, and suspected target organs for uptake studies. Sampling times depend on the compound being tested and the selected route of administration. The number of blood samples should be taken frequently such that different phases of the concentration–time profile can be defined with certainty. Time spacing of samples will depend on the rates of uptake and elimination of the compounds. It is often recommended to take blood and tissue samples in a "power of 2" series, that is, 2, 4, 8, 16, and 30 minutes follows by 1, 2, 4, 8, and 16 hours; or seven time points using "power of 3," that is, 3, 9, 30 minutes and 1, 3, 9, 24 hours. The sampling schedule for an oral dosing experiment can be established as 15, 30 minutes and 1, 2, 4, 8, 24, 48, and 72 hours. When the absorption is rapid more frequent initial sampling is recommended and the measurement of concentration at the late time points depends on the sensitivity of the analytical methodology and clearance of the compound. The required supportive data for the *in vivo* PK/TK analysis include the *in vitro* metabolic analysis, related kinetic characteristics and significant metabolic pathways and the intrinsic clearances. In addition, the *in vitro* measurements of protein binding, transport across the cell membranes (e.g., data generated from Caco2 cells), and conjugate formation should be identified.

The integration of in vitro, *in vivo* and in silico data generated during discovery and preclinical phase of development has been facilitated by using PK modeling software such as GastroPlus (Lukacova et al., 2009; Parrott et al., 2009); STELLA (Shono et al., 2009, 2010, 2011) and Simcyp (Shaffer et al., 2012; Vieira et al., 2012); (ADMET) Predictor™ (Agoram et al., 2001); and other similar marketed tools.

16.2.2.1 Safety Pharmacology and Toxicity Testing

Other required data for the PK/TK analysis of a lead compound are the toxicity testing and evaluation. The focus of preclinical safety and toxicological evaluations are related to the structural and biochemical consequences of administration of a new small drug entity. The testing relies on the histopathological evaluation of organs, physiological and functional observations. An important part of the preclinical evaluation is the safety pharmacology data and observation.

Safety pharmacology is the assessment of adverse effects of drugs on the central and peripheral nervous system, for example, seizure; adverse effect on cardiovascular system for example, hypertension, hypotension and arrhythmia; adverse effect on respiratory system asthma and bronchoconstriction; adverse effect on the renal glomerular filtration rate; and adverse effect on GI tract and its function (Pugsley et al., 2008). The ICH guideline S7A (Guidance for Industry S7A) defines the safety pharmacology as "*those studies that investigate the potential undesirable pharmacodynamic effects of a substance on physiological functions in relationship to exposure in the therapeutic range and above*." The primary objectives of safety pharmacology are indentifying potential pharmacodynamic risk to human; investigating the mechanisms of risk posed by the lead compounds; correlating the PD response with peak drug levels; using the information to determine low-observed-effect level (LOEL) and maximum dose with no observed-effect level (NOEL); and assuring human safety in advance of first-time-in-human exposure (FTIH).

Acute toxicity studies deal with toxicity testing after single administration of a lead compound. The main purpose is to determine the degree of toxicity quantitatively and qualitatively (Robinson et al., 2008). Other purposes are to determine the level of exposure for chronic studies and to determine the onset and disappearance of toxic response, if it disappears. The acute toxicity study usually involves at least two mammalian animal species, normally rats (Sprague-Dawley, Fisher or Wistar) and mice or hamster using two different routes of administration, one is the route intended for human administration and the second preferred route is intravenous administration. The dose should be high enough to induce significant toxicity, its pH should be 7.4 with physiological osmolarity and its volume cannot exceed 10 mL/kg b wt. Necropsy should be carried out on live and dead animals. Body weight, food and water consumption, urine and feces analysis, cardiovascular functions, hematological parameters, histological data, respiration and other visible changes must be recorded.

Chronic toxicity studies are carried out to identify the target organs of toxicity and estimation of dose levels for clinical trials. The initial study is carried out on rodents between 2 weeks and 6 months and non-rodent between 2 weeks and 9 months in duration. The experimental protocol for systemic drugs usually involves one rodent (typically rat) and one non-rodent (e.g., dog). However, the selection of the animals is based on the kinetics and metabolic characteristics of the lead compound in the selected animals in comparison to humans. A lead compound may exhibit substrate specificity for an enzyme of CYP450 in the animal, which may not present in human. The strains of animals used in toxicity testing (acute and chronic) are: Swiss mice, NMR1 mice, Wistar rats, Sprague Dawley rats, Beagle dogs, guinea pigs, Himalayan rabbits, New Zealand white rabbit, Cynomolgus monkeys, Rhesus monkeys, baboons, and mini pigs. The FDA recommendations for the selection of the most appropriate species are (1) similarity in ADME of the lead compound between species, (2) species should be predictive for a known pharmacological class, (3) the *in vitro* metabolic profile should be similar, (4) the species should have similar biochemistry and physiology to man, and (5) have similar pharmacology/receptor binding properties to man. Other considerations in chronic toxicity study are standardization of time and daily administration of the dose; larger non-rodent species should be fasted before the administration of the compound; the behavioral, hematological with a complete clinical chemistry data must be determined so that the symptoms of an illness are not confused with the symptoms of toxicity. At the end all sacrificed animals go under macroscopic evaluation, including weight of each organ and histopathological evaluation. For anticancer drugs, the NCI recommendations are, single and daily for 5 days lethality studies in mice, single and daily for 5 days toxicity studies in dog and single and daily for 5 days toxicity study in rodents (optional).

The recommended toxicity screening assays include (1) ATP measurement using high-throughput methodology and viable cells in 96-well or 384-well plates with the lead compound at various concentrations to establish the dose–response curve; (2) release of the liver-specific cytosolic enzymes alanine aminotransferase (ALT) and aspartate aminotransferase (AST) are commonly used as the endpoints for the evaluation of cytotoxicity; (3) macromolecular synthesis is an assay used to measure the dividing cells by measuring incorporation of labeled precursors (^{3}H or ^{14}C–labelled thymidine, uridine, or leucine) in DNA, RNA; (4) Glutathione measurement which involves the measurement of GSA in exposed cells in comparison to non-treated cells; and (5) stem cells measurement.

16.2.2.2 Metabolic Evaluations in Preclinical Phase

Metabolic processes influence parameters that are relevant to the safety and efficacy of drugs, and the *in vivo* and *in vitro* metabolic data are important in the selection of viable drug candidates and evaluation of their bioavailability, systemic clearance and toxicity. Identification of biotransformation pathways that might produce toxicity, and prediction of drug–drug interactions that might lead to alterations in the pharmacokinetic profiles of co-administered drug are a few challenges in preclinical phase for PK/TK study. The design and conduct of drug metabolism studies and interpretation of results must take into consideration the advances in pharmacogenomics, pharmacogenetics and transporters. In discovery phase, metabolic data provides a basis for choosing chemical structures suited for lead compounds. In preclinical phase, metabolic data aids in the development of clinical plans with regard to human drug exposures and safety and throughout the DDD process, metabolic data is needed for PK/PD analysis of the lead compounds.

Drug metabolism methodologies employed during different stages of discovery and development can be vastly different, primarily because of the different needs and endpoints. In discovery, the primary purpose is to screen large numbers of compounds in order to select ideal candidates for development, hence require technologies with high throughout capabilities. The development studies require more in-depth analyses of a single compound, employing methods that have been thoroughly validated, from a good laboratory practice (GLP) perspective.

The metabolic liabilities of lead compounds include formation of reactive metabolite which may exert adverse effects by inactivating the same drug-metabolizing enzyme that catalyzes its formation. Thus, it may raise the exposure/concentration of either the parent compound or other compounds that rely on the same enzyme system for metabolism and clearance and the possibility of binding to other proteins, DNA, RNA causing genotoxicity and irreparable damage. The reactive metabolites may interact with GSH covalently and lower GSH levels; a potential source of oxidative stress for the cell that may lead to various adverse outcomes. Furthermore, the reactive metabolites by covalent binding to proteins in the cell may cause apoptosis or necrosis. At the organ level, the formation of reactive metabolites in the liver and non-hepatic metabolism in the kidneys, adrenal glands, and skin may manifest as acute organ toxicity, changing the clearance of the compound by the organs of elimination. A different aspect of the interaction of metabolites of lead compound is their non-covalent off-target binding, which may result in unexpected and often problematic pharmacological response. The importance of preclinical metabolism studies is evident in cases when a parent lead compound is precluded from the advancement into development phase, merely because it acts as inhibitor of CYP subfamilies, in particular CYP3A.

In drug discovery phase, depending on the physiological relevance, time and cost, various *in vitro, in silico, in situ*, isolated organs or *in vivo* methodologies are employed (see Chapter 6). In the preclinical phase of drug development, metabolic studies require more in-depth analysis and the employed *in vitro* and *in vivo* methodologies must be thoroughly validated under the GLP guidelines.

As discussed in Chapter 6, the *in vitro* methodologies can be divided into two major groups of cellular fractions and organ fraction. The cellular fractions derived from the endoplasmic reticulum of liver include liver microsomes, cytosol, and S-9 fraction. They all require cofactors such as NAD, NADPH, FAD, and FAM during incubation with the lead compound. The application of cellular fractions in preclinical phase includes the high throughput investigation of metabolism of lead compounds; investigation of the effect of inducers or inhibitors on metabolism of the compounds; drug-drug interaction; metabolic activation and deactivation; identification of metabolites; mutagenicity tests, comparative evaluation of a compounds metabolism in different species. The major disadvantage of the cellular fractions is the lack cellular control therefore caution is exercised in extrapolation of the metabolic data from these systems to *in vivo* condition.

The organ fractions are the isolated intact perfused liver, liver slices and intact hepatocytes. Among the organ fractions, the isolated intact perfused liver, due to the intact tissue and vascular systems, is a good model for metabolic investigation of drugs. However, the methodology is expensive, somewhat time consuming and the organ cannot be maintained viable over a long period of time. The liver slices are less desirable unless they are precision-cut (Chapter 6, Section 6.4.3), even with the precision cut slices, many cells in the system are damaged and there is limited availability of oxygen. Hepatocytes offer a novel tool in metabolic investigation of lead compound in preclinical phase (see also Chapter 6, Section 6.4.2). They are multitask parenchymal cells of the liver that are responsible for metabolism of xenobiotics. Because the integrity of the cells is maintained, the metabolism profile mimics the *in vivo* profile more closely.

The *in vivo* assessment of metabolism of a lead compound is based on the measurement of metabolites in urine, blood, bile and portal vein blood. Generally, the concentration of metabolites in plasma for most drugs is below the sensitivity of the detection and urine is usually contains mostly Phase II metabolites. Bile duct cannulation and portal vein cannulation are useful technique but invasive for human. Substituting feces for bile duct cannulation is neither accurate nor practical for the measurement of metabolites. Animal models are often used for determination of metabolic profile of a drug in the portal vein and bile. Comparison of drug concentration of hepatoportal vein before entering into the liver with blood levels of systemic circulation provides important information on the liver first-pass effect and absorption of the lead compounds.

16.3 PK/TK AND CLINICAL TRIALS

The general considerations and all technical requirements for clinical trials and the process of clinical development of pharmaceuticals for human use are established in ICH Guidelines (ICH-E8, 1997). The ICH Guideline for Good Clinical Practice (GCP, E6 (R1)) defines the Clinical Trial/Study as

> Any investigation in human subjects intended to discover or verify the clinical, pharmacological and/or other pharmacodynamic effects of an investigational product(s), and/or to identify any adverse reactions

to an investigational product(s), and/or to study absorption, distribution, metabolism, and excretion of an investigational product(s) with the object of ascertaining its safety and/or efficacy. The terms clinical trial and clinical study are synonymous.

Clinical study of a lead compound starts with Phase I clinical trial with the objective of evaluating the PK/TK and safety of drug candidate for the first time in human by giving single and multiple ascending doses. Phase I study has two phases of Phase I-*a* and I-*b*.

16.3.1 Phase I-*a* Clinical Trial

During this clinical trial the tolerability, PK/TK/PD and safety of the lead compound is determined. It usually involves 40–60 healthy subjects in a randomized double blind, placebo-controlled experimental design with escalating dose for about 6 months study. The single ascending dose protocol involves giving the lowest dose to the first group of subjects. If the safety and tolerability are acceptable, the dose will be increased for the second group, and if the conditions of safety and tolerability are met, the dose will be increased for the third group and so on.

16.3.2 Phase I-*b* Clinical Trial

This trial is also designed to evaluate safety, tolerability, adverse effects and PK/TK/PD of the lead compound. The same number of healthy subject (≈40–60) is used in this study and experimental design is the same. The difference is giving the escalating dose by multiple dosing regimens.

The dose escalation of I-*a* and I-*b* for a lead compound with low toxicity is based on exponential regimen of 1D, 2D, 4D, 8D, 16D and for highly toxic compound is D, 2D, 4D, 6D, 8D, 10D, etc. The PK/TK/PD evaluations of Phase I are first time in human and include complete investigation of a lead compound's ADME, its absolute bioavailability and bioequivalence study, if it is a new formulation. Further investigations of Phase I includes drug-drug interaction studies; geriatric PK and safety studies; ethnic population PK and pharmacogenomics; and hepatic and renal impairment PK and safety. The design of escalating multiple dosing regimen of I-*b* is based on PK analysis of I-*a* phase of study.

Phase II clinical trial is the crucial phase in drug discovery and development with main objective of "Proof of Concept" and efficacy of the lead compound. Phase II can also be divided into Phases of II-*a* and II-*b*.

16.3.3 Phase II-*a* Clinical Trial

This trial is an exploratory, safety and efficacy study of the lead compound. The decision to continue with the lead compound as a viable therapeutic agent is made based on the results of this study. The human subjects are patients for whom the lead compound is intended. The study is based on placebo controlled randomized double blind, or open label, and preferentially crossover design using 50–200 patients. The intended dose and regimen are based on the data from Phase I clinical trial.

16.3.4 Phase II-*b* Clinical Trial

This trial is carried out on ≈200–500 patients. The objectives are confirmation and optimization of the therapeutic dose for patient population, followed by the safety, efficacy and tolerability of the dosing regimen.

The main purpose of the Phase II clinical trial (II-*a* and II-*b*) is the proof of concept (PoC), that is, to determine whether the perceived hypothesis for the lead compound's therapeutic outcome and receptor interaction is credible, logical, or consistent in clinical reality of the disease state, and whether the differentiation points can be established for the lead compound with the marketed products.

16.3.5 Phase III Clinical Trial

This study is to confirm the efficacy and safety of the lead compound in a larger and diverse population of patients. The study design is placebo controlled randomized double blind crossover design (preferred) in 500–1000 patients in multicenter under supervision of qualified and approved medical staff. Patients are selected from different sex, age, and ethnicity. The dose for the study is the confirmed dosing regimen of Phase II clinical trial, and the clinical data are used to confirm the dose and dosing regimen for efficacy and safety of the finalized dosing regimen in comparison to control group or existing therapy.

16.3.6 Phase IV Clinical Trial

This phase is an obligatory post-marketing surveillance study in a large diverse population (>10,000) of patients. The objective of this study is to evaluate the adverse effects, drug-drug interaction, pharmacogenetic variances, etc., using the recommended therapeutic dose/dosing regimen in a parallel experimental design.

All clinical studies and use of human subjects are expected to be carried out according to the ethical requirements of the declaration of Helsinki (WMA 2008), the ICH (E6) guidelines, and good clinical practice (GCP).

16.4 DOSE AND DOSING REGIMEN ADJUSTMENT FOR PATIENTS WITH VARIABLE RENAL IMPAIRMENTS IN CLINICAL TRIALS

If the lead compound is intended for population of patients with variable renal insufficiencies, for example, geriatric patients, one approach is to adjust the dose and dosing regimen of the Phase I clinical trial according to the estimated glomerular filtration rate of patients with renal impairment (Levy, 1977). The following sections are the proposed approaches for the adjustment of dosage regimen. The methodology includes first the estimation of the overall elimination rate constant, or the half-life of a lead compound, in patient population with renal insufficiency and second adjusting the dosing regimen according to the calculated overall elimination rate constant (Chiou and Hsu, 1975; Slatlery et al., 1980; DeVane and Jusko, 1982; Ritschel, 1983; Burton et al., 1985; Van Dalen et al., 1986).

16.4.1 Estimation of the Overall Elimination Rate Constant or Half-Life of a Therapeutic Agent in Patients with Renal Failure Based on the Estimated Glomerular Filtration Rate

Based on the assumption that the metabolic rate constant of patients with renal impairment remains unaffected, the overall elimination rate constant of drugs can be presented as

$$\bar{K} = \bar{k}_e + k_m \text{ (One-compartment model)} \tag{16.1}$$

$$\bar{k}_{10} = \bar{k}_e + k_m \text{ (Two-compartment model)} \tag{16.2}$$

where $\bar{k}_e$ is the reduced rate constant of urinary excretion.

The assumption of unaltered metabolic rate constant is true only for therapeutic agents that maintain first-order metabolism, and the increase in plasma free concentration, because of renal impairment, has no influence on the hepatic Phase I and/or Phase II metabolism, that is, liver function remains the same and the plasma concentration continues less than Michaelis–Menten constant ($C_p \ll K_M$). The adjustment of dose and/or dosing regimen reduces the plasma concentration to the normal level.

The first step to adjust the dose and dosing regimen is to estimate the overall elimination rate constant of a compound in patients with renal impairment according to their glomerular filtration rate, which can be estimated using creatinine clearance (Chapter 5, Section 5.4). The estimated GFR is then used to adjust the normal value of K or half-life to determine $\bar{K}$, the overall elimination rate constant of patient with renal impairment.

16.4.1.1 Compounds That Are Eliminated Entirely by the Renal Route

For drugs that are eliminated entirely by renal route of elimination and glomerular filtration

$$K = k_e \tag{16.3}$$

Rate of elimination is equal to the rate of excretion, that is,

$$\frac{dA}{dt} = \frac{dA_e}{dt} = -KA = -k_eA = Cl_TCp = Cl_rCp \tag{16.4}$$

Therefore, $Cl_T = Cl_r$

$$K = \frac{dA_e/dt}{A} = \frac{dA_e/dt}{Cp \times Vd} = \frac{Cl_T}{Vd}, \text{ and} \tag{16.5}$$

$$Cl_T = \frac{dA_e/dt}{Cp} \tag{16.6}$$

16.4.1.1.1 Wagner's Method (Wagner, 1975)

The ratio of normal total body clearance to normal creatinine clearance is

$$R = \frac{Cl_T}{Ccr} \tag{16.7}$$

$$\therefore\ Cl_T = R \times Ccr \tag{16.8}$$

The excretion rate of the compound is then expressed as

$$\frac{dA_e}{dt} = R \times Ccr \times Cp = Cl_T \times Cp = K \times Vd \times Cp \tag{16.9}$$

Therefore the normal overall elimination rate can be defined as

$$K = \frac{R}{Vd}Ccr \tag{16.10}$$

Equation 16.10 indicates that for compounds that are entirely eliminated through renal excretion, a plot of overall elimination rate constant, K, versus creatinine clearance Ccr yields a straight line through the origin with slope of R/Vd (Figure 16.1). Thus the product of slope and creatinine clearance of patient with renal impairment yields the overall elimination rate constant of subject with renal impairment, $\bar{K}$.

$$\bar{K} = \left(\frac{R}{Vd}\right) \times \overline{Ccr} \tag{16.11}$$

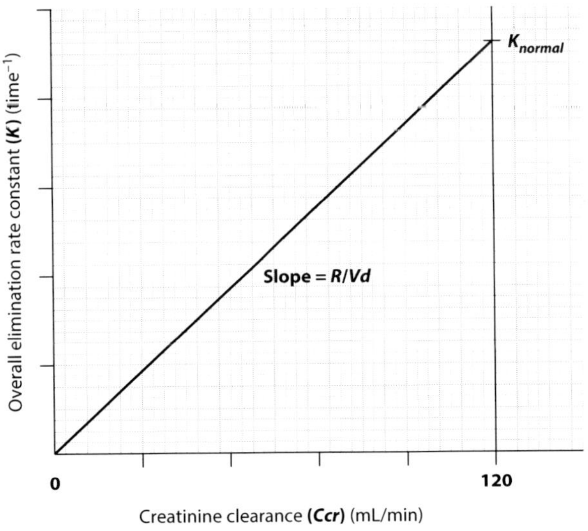

FIGURE 16.1 Plot of overall elimination rate constant versus creatinine clearance according to Equations 16.10 and 16.11.

16.4.1.1.2 Dettli's Method (Dettli, 1977)

In this method k_e is assumed to be proportional to glomerular filtration rate or *Ccr*, that is,

$$k_e = K = aCcr \tag{16.12}$$

Similar to Wagner method a plot of *K* versus *Ccr* yields a straight line with slope of "*a*," the proportionality constant, according to the following equations (Figure 16.2):

$$a = K/Ccr \tag{16.13}$$

$$a = \bar{K}/\bar{C}cr \tag{16.14}$$

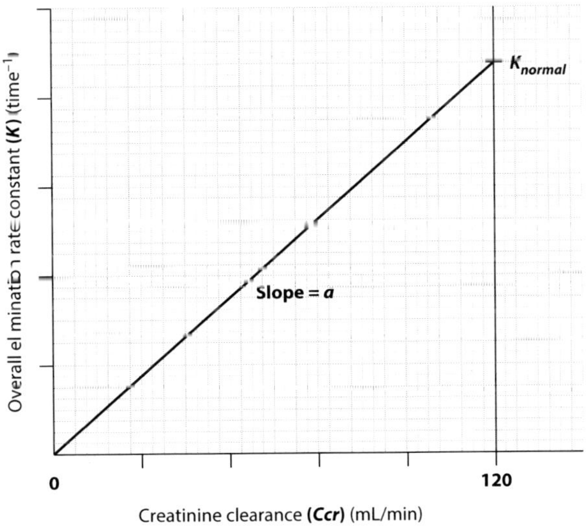

FIGURE 16.2 Plot of the overall elimination rate constant versus creatinine clearance according to Equations 16.12 through 16.14.

Setting Equation 16.14 equal to 16.13 and solving for $\bar{K}$ yields

$$\bar{K} = K\frac{\bar{Ccr}}{Ccr} \tag{16.15}$$

16.4.1.1.3 Giusti's Method (Giusti and Hayton, 1973)

According to this method the excretion rate constant of patients with renal failure can be defined in terms of normal excretion rate constant with the ratio of $\bar{Ccr}/Ccr$ as the correction factor:

$$\bar{K} = \bar{k}_e = k_e\frac{\bar{Ccr}}{Ccr} \tag{16.16}$$

Dividing both sides of Equation 16.6 by K yields

$$\frac{\bar{K}}{K} = f_e\frac{\bar{Ccr}}{Ccr} \tag{16.17}$$

Since for the first group $f_e = 1$, the equation will be the same as 14.28.

16.4.1.2 Compounds That Are Eliminated Entirely by the Non-Renal Route

The overall elimination rate constant is independent of renal function (i.e., $k_e = 0$ and $K = k_m$), depends on hepatic/other organs metabolism, activity of metabolic enzyme systems and concentration of the compound in the systemic circulation. Therefore, unless there is a change in the linearity of elimination and pharmacokinetics of the compound, no adjustment is needed. For these drugs the plot of K versus Ccr is parallel to the x-axis.

16.4.1.3 Compounds That Are Eliminated Entirely by Renal and Non-Renal Routes of Elimination

Most therapeutic agents are eliminated by renal excretion and metabolism. Adding the first-order metabolic rate constant to Equations 16.11, 16.15, and 16.16, or 16.17 yields the following relationships for estimation of the overall elimination rate constant in patient with renal impairment:

16.4.1.3.1 Wagner's Method (Wagner, 1975)

$$\bar{K} = k_m + \left(\frac{R}{Vd}\right) \times \bar{Ccr} \tag{16.18}$$

16.4.1.3.2 Dettli's Method (Dettli, 1977)

$$K = aCcr + k_m \tag{16.19}$$

$$\therefore a = \frac{K - k_m}{Ccr} \tag{16.20}$$

The slope of the line, a, is calculated by the equation $(y_2 - y_1)/(x_2 - x_1)$ where $(y_2 - y_1) = K - k_m$ and $(x_2 - x_1) = Ccr = 120$ mL/min.

Substituting Equations 16.20 in 16.19 and solving for $\bar{K}$ yields

$$\bar{K} = \frac{K - k_m}{Ccr}\bar{C}cr + k_m \tag{16.21}$$

Equations 16.18 and 16.21 represent a similar plot as is shown in Figure 16.3, which indicates that when creatinine clearance is equal to zero, for example, end-stage renal impairment, the y-intercept is equal to k_m, the metabolic rate constant. When the clearance of creatinine is normal, that is, $Ccr \cong 120$ mL/min, the overall elimination rate constant represents the normal value (i.e., $K = 0.693/T_{1/2}$). Therefore, as the renal failure progresses and $\bar{C}cr$ decreases, the overall elimination rate constant (K or k_{10}) also decreases whereas the half-life, $\bar{T}_{1/2}$, increases.

16.4.1.3.3 *Giusti's Method (Giusti and Hayton, 1973)*

Modification of Equation 16.16 to include the metabolic rate constant yields

$$\bar{K} = k_e\frac{\bar{C}cr}{Ccr} + k_m \tag{16.22}$$

Dividing Equation 16.22 by K changes the equation into a relationship in terms of fraction of dose excreted unchanged

$$\frac{\bar{K}}{K} = f_e\frac{\bar{C}cr}{Ccr} + 1 - f_e \tag{16.23}$$

Since $k_e/K = f_e$ and $k_m/K = f_m = 1 - f_e$, Equation 16.23 can be written as Equation 16.24 and modified to Equation 16.26

$$\frac{\bar{K}}{K} = 1 - f_e + f_e\frac{\bar{C}cr}{Ccr} \tag{16.24}$$

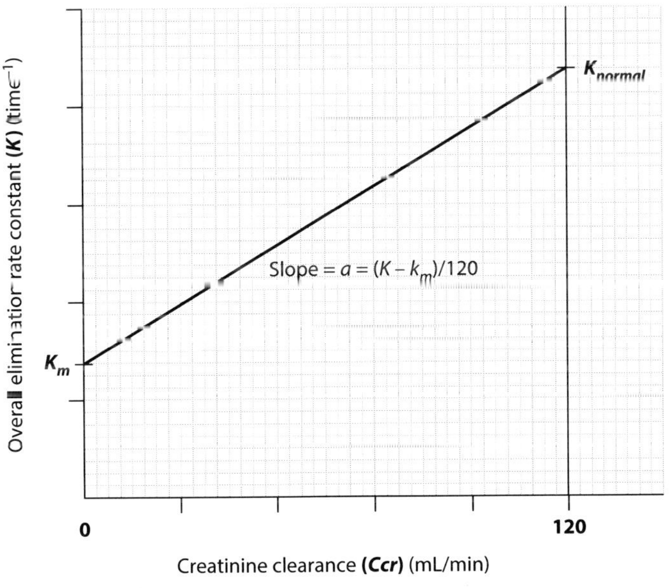

FIGURE 16.3 Plot of the overall elimination rate constant versus creatinine clearance according to Equations 16.18 and 16.21.

$$\therefore \frac{\bar{K}}{K} = 1 - \left(f_e \left(1 - \frac{\bar{C}cr}{Ccr} \right) \right) \tag{16.25}$$

$$\bar{K} = K\left[1 - \left(f_e \left(1 - \frac{\bar{C}cr}{Ccr} \right) \right) \right] \tag{16.26}$$

The methods discussed here basically provide the same answer for the overall elimination rate constant of patient with renal insufficiency. The cautionary notes about the methods of estimation of $\bar{K}$ based on creatinine clearance are (1) the methods may provide a realistic estimate of $\bar{K}$ only when the serum creatinine concentration is at steady state and (2) the pharmacokinetics of the compound under investigations remains linear and dose independent, during the adjusted therapy for the patients.

16.4.2 Adjustment of Dosing Regimen for Patients with Renal Insufficiency Based on the Estimated Overall Elimination Rate Constant

For patients with mild or moderate renal impairment, because the rate of output is reduced significantly, the rate of input should be adjusted accordingly. Thus, the objective of adjusting a dosing regimen is to modify dose and dosing interval such that the accumulation of the lead compound in the body of patients with renal impairment at steady state ($\bar{A}_{ss}$) equals that of the recommended level for patients with normal renal function (A_{ss}), that is,

$$\bar{A}_{ss} = A_{ss} \tag{16.27}$$

$$(\bar{A}_{ave})_{ss} = (A_{ave})_{ss} \tag{16.28}$$

Using the equation of average plasma concentration (Chapter 14, Equations 14.19 through 14.23) for oral absorption assuming bioavailability remains the same, Equation 14.39 can be expressed as

$$\frac{F\bar{D}}{\bar{K} \times \bar{\tau}} = \frac{FD}{K \times \tau} \tag{16.29}$$

Therefore,

$$\bar{D} = \frac{D \times \bar{K} \times \bar{\tau}}{K \times \tau} \tag{16.30}$$

For intravenous infusion:

$$\frac{\bar{k}_0}{\bar{K}} = \frac{k_0}{K} \tag{16.31}$$

$$\frac{\bar{k}_0}{C\bar{l}_T} = \frac{k_0}{Cl_T} \tag{16.32}$$

$$\bar{k}_0 = k_0 \frac{\bar{K}}{K} = k_0 \frac{\bar{Cl}_T}{Cl_T} \tag{16.33}$$

For intravenous bolus one-compartment model:

$$\frac{\bar{D}}{\bar{K} \times \bar{\tau}} = \frac{D}{K \times \tau} \tag{16.34}$$

$$\bar{D} = \frac{D \times \bar{K} \times \bar{\tau}}{K \times \tau} \tag{16.35}$$

For intravenous bolus injection two-compartment model:

$$\frac{\bar{D}}{\bar{k}_{10} \bar{\tau}} = \frac{D}{k_{10} \tau} \tag{16.36}$$

$$\bar{D} = \frac{D \bar{k}_{10} \bar{\tau}}{k_{10} \tau} \tag{16.37}$$

If the dosing interval is kept the same as the healthy subjects and the dose is changed, Equations 16.30 or 16.35 are simplified to

$$\{\bar{\tau} = \tau\} \Rightarrow \bar{D} = D \frac{\bar{K}}{K} \tag{16.38}$$

If the dose is kept the same as standard and the dosing interval is changed:

$$\{\bar{D} = D\} \Rightarrow \bar{\tau} = \tau \frac{K}{\bar{K}} \tag{16.39}$$

If both the dose and dosing interval are changed, Equations 16.30 or 16.35 will have two unknowns. One of the unknowns, either the dose or dosing interval, is then set to a convenient value and the equation can then be used to solve for the second unknown. For example, if the standard dose is given every 6 hours around the clock (q.i.d.), the dosing interval can be set equal to 8 hours and the adjusted dose can be calculated based on three times around the clock:

$$\{\bar{\tau} = \lambda\} \Rightarrow \bar{D} = \frac{D \times \bar{K} \times \lambda}{K \times \tau} \tag{16.40}$$

If the dose can be set different from the standard regimen, the dosing interval can be estimated as

$$\{\bar{D} = dose\} \Rightarrow \bar{\tau} = \frac{dose \times K \times \tau}{D \times \bar{K}} \tag{16.41}$$

The assumptions here are the fraction of dose absorbed and the volume of distribution of patient with renal insufficiency are the same as patients with normal renal function. If the volume of distribution changes in renal impairment, one option would be to adjust the dosing interval according to the area under plasma concentration curve at steady state and then adjust the dose according to total body clearance:

$$\{C\bar{p}_{ave} = Cp_{ave}\} \Rightarrow \frac{A\bar{U}C}{\bar{\tau}} = \frac{AUC}{\tau} \quad (16.42)$$

$$\bar{\tau} = \tau \frac{(A\bar{U}C)_\tau}{(AUC)_\tau} \quad (16.43)$$

$$\bar{D} = \frac{\bar{\tau} \times D \times C\bar{l}_T}{\tau \times Cl_T} \quad (16.44)$$

A different approach for adjusting dosing regimen is to modify the regimen based on the normal average amount of accumulation in the body at steady state. The approach is also known as the accumulation ratio method and is accomplished according to the following steps:

Estimation of the normal accumulated average amount for the body of patients with normal renal function:

$$(A_{ave})_{ss} = \frac{F \times D}{K \times \tau} = \frac{1.44 \times T_{1/2} \times F \times D}{\tau} \quad (16.45)$$

Adjusting the dosing regimen with respect to the normal accumulation of the compound in the body by changing the dose or dosing interval or both as follows:

a. Solving for a new dose by keeping the dosing interval the same:

$$\bar{D} = \frac{(A_{ave})_{ss} \times \bar{K} \times \tau}{F} = \frac{(A_{ave})_{ss} \times \tau}{1.44 \times \bar{T}_{1/2} \times F} \quad (16.46)$$

b. Solving for a new dosing interval by keeping the dose the same:

$$\bar{\tau} = \frac{F \times D}{K \times (A_{ave})_{ss}} = \frac{1.44 \times \bar{T}_{1/2} \times F \times D}{(A_{ave})_{ss}} \quad (16.47)$$

c. Change both the dose and dosing interval:

$$\bar{D} = \frac{(A_{ave})_{ss} \times \bar{K} \times \bar{\tau}}{F} = \frac{(A_{ave})_{ss} \times \bar{\tau}}{1.44 \times \bar{T}_{1/2} \times F} \quad (16.48)$$

As discussed earlier, Equation 16.48 has two unknowns, one unknown can be set equal to an acceptable value.

A similar approach can also be used to adjust the infusion rate:

$$A_{ss} = \frac{k_0}{K} = 1.44 \times T_{1/2} \times k_0 \quad (16.49)$$

$$\bar{k}_0 = \frac{A_{ss}}{1.44 \times \bar{T}_{1/2}} \quad (16.50)$$

16.4.3 Dosing Regimen Adjustment with Respect to Peak and Trough Levels

The equations of peak or trough levels (Chapter 14, Equations 14.11 through 14.13) are also used to adjust the standard dosing regimen by calculating first the normal maximum amount of the compound in the body:

$$(A_{max})_{ss} = \frac{D}{1 - e^{-K\tau}} = \frac{D}{(f_{el})_\tau} \quad (16.51)$$

Followed by estimation of the dose by keeping dosing interval the same:

$$\bar{D} = (A_{max})_{ss}(1 - e^{-\bar{K}\tau}) \quad (16.52)$$

Or, calculation of the dosing interval by giving the same normal dose:

$$\bar{\tau} = \frac{2.303}{\bar{K}} \log \frac{(A_{max})_{ss}}{(A_{max})_{ss} - D} \quad (16.53)$$

Either Equations 16.52 or 16.53 can be used to change both the dosing interval and the dose ($\bar{\tau}$ and $\bar{D}$) depending on which of the two unknowns is set equal to a known value.

The trough level can also be used for the adjustments by estimating the normal minimum amount:

$$(A_{min})_{ss} = \frac{D \times e^{-K\tau}}{1 - e^{-K\tau}} \quad (16.54)$$

Estimation of the dose by keeping the dosing interval the same:

$$\bar{D} = \frac{(A_{min})_{ss}(1 - e^{-\bar{K}\tau})}{e^{-\bar{K}\tau}} \quad (16.55)$$

Estimation of the dosing interval by giving the standard dose:

$$\bar{\tau} = \frac{2.303}{\bar{K}} \log \frac{D - (A_{min})_{ss}}{(A_{min})_{ss}} \quad (16.56)$$

Equations 16.55 and 16.56 can also be used to determine both $\bar{\tau}$ and $\bar{D}$.

REFERENCES

Agoram, B., Woltosz, W. S., Bolger, M. B. 2001. Predicting the impact of physiological and biochemical processes on oral drug bioavailability. *Adv Drug Delivery Rev* 50:S41–67.

Amidon, G. L., Lennernaes, H., Shah, V. P., Crison, J. R. 1995. A theoretical basis for a biopharmaceutic drug classification: the correlation of *in vitro* drug product dissolution and *in vivo* bioavailability. *Pharm Res* 12:413–20.

Burton, M. E., Vasko, M. R., Brater, D. C. 1985. Comparison of drug dosing methods. *Clin Pharmacokin* 10(1):1–37.

Chiou, W. L., Hsu, F. H. 1975. A new simple and rapid method to monitor the renal function based on pharmacokinetic consideration of endogenous creatinine. *Res Commun Chem Pathol Pharmacol* 10(2):315–30.

Custodio, J. M., Wu, C.-Y., Benet, L. Z. 2008. Predicting drug disposition, absorption/elimination/transporter interplay and the role of food on drug absorption. *Adv Drug Deliv Rev* 60:717–33.

De Buck, S. S., Sinha, V. K., Fenu, L. A., Nijsen, M. J., Mackie, C. E., Gilissen, R. A. H. J. 2007. Prediction of human pharmacokinetics using physiologically based modeling: A retrospective analysis of 26 clinically tested drugs. *Drug Metab Dispos* 35:1766–80.

Dettli, L. 1977. Elimination kinetics and dosage adjustment of drugs in patients with kidney disease, In: *Progress in Pharmacology*, Vol 1, eds. H. Grobecker, New York: Gustav Fisher Verlag.

DeVane, C. L., Jusko, W. J. 1982. Dosage regimen design. *Pharmc Ther* 17(2):143–63.

Fura, A., Vyas, V., Humphreys, W., Chimalokonda, A., Rodrigues, D. 2008. Prediction of human oral pharmacokinetics using nonclinical data: examples involving four proprietary compounds. *Biopharm Drug Dispos* 29:455–68.

Giusti, D. L., Hayton, W. L. 1973. Dosage regimen adjustments in renal impairment. *Drug Intel Clin Pharm* 7:382–7.

Guidance for Industry S7A Safety Pharmacology Studies for Human Pharmaceuticals—ICH. 2001. http://www.fda.gov/downloads/Drugs/GuidanceComplianceRegulatoryInformation/Guidances/ucm074959.pdf

Guidance for Industry—M3(R2) Nonclinical Safety Studies for the Conduct of Human Clinical Trials and Marketing Authorization for Pharmaceuticals–Questions and Answers (R2) ICH. 2013. http://www.fda.gov/downloads/Drugs/GuidanceComplianceRegulatoryInformation/Guidances/UCM292340.pdf

ICH Guidelines E8–General Considerations for Clinical Trials. 1997. http://www.ich.org/fileadmin/Public_Web_Site/ICH_Products/Guidelines/Efficacy/E8/Step4/E8_Guideline.pdf

ICH guideline M3(R2) on non-clinical safety studies for the conduct of human clinical trials and marketing authorisation for pharmaceuticals EMA. 2009. http://www.ema.europa.eu/docs/en_GB/document_library/Scientific_guideline/2009/09/WC500002720.pdf

Hosea, N., Collard, W. T., Cole, S., Maurer, T. S., Fang, R. X., Jones, H., Kakar, S. M. et al. 2009. Prediction of human pharmacokinetics from preclinical information: comparative accuracy of quantitative prediction approaches. *J Clin Pharm* 49(5):513–33.

Huang, C., Zheng, M., Yang, Z., Rodrigues, A. D., Marathe, P. 2008. Projection of exposure and efficacious dose prior to first-in-human studies: How successful have we been? *Pharm Res* 25:713–26.

Jones, H. M., Parrott, N., Jorga, K., Lave, T. 2006. A novel strategy for physiologically based predictions of human pharmacokinetics. *Clin Pharmacokinet* 45:511–42.

Levy, G. 1977. Pharmacokinetics in renal disease. *Am J Med* 62:461–5.

Li, S., He, H., Parthiban, L. J., Yin, H., Serajuddin, A. T. M. 2005. IV–IVC considerations in the development of immediate-release oral dosage form. *J Pharm Sci* 94:1396–417.

Lipinski, C. A. 2001. Drug-like properties and the causes of poor solubility and poor permeability. *J Pharmacol Toxicol Methods* 44:235–49.

Lowe, P. J., Hijazi, Y., Luttringer, O., Yin, H., Sarangapani, R., Howard, D. 2007. On the anticipation of the human dose in first-in-man trials from preclinical and prior clinical information in early drug development. *Xenobiotica* 37:1331–54.

Lukacova, V., Woltosz, W. S., Bolger, M. B. 2009. Prediction of modified release pharmacokinetics and pharmacodynamics from in vitro, immediate release, and intravenous data. *AAPS J* 11(2):323–34.

Luttringer, O., Theil, F-P., Poulin, P., Schmitt-Hoffmann, A. H., Guentert, T. W., Lave, T. 2003. Physiologically based pharmacokinetic (PBPK) modeling of disposition of epiroprim in humans. *J Pharm Sci* 92:1990–2007.

Mahmood, I. 2005. Interspecies scaling of drugs cleared by the kidneys and the bile. In: *Interspecies Pharmacokinetic Scaling: Principles and Application of Allometric Scaling*, ed. I. Mahmood, Rockville, MD: Pine House Publishers.

Mahmood, I. 2006. Prediction of human drug clearance from animal data: Application of the rule of exponents and 'fu corrected intercept method' (FCIM). *J Pharm Sci* 95:1810–21.

Mahmood, I., Yuan, R. 1999. A comparative study of allometric scaling with plasma concentrations predicted by species-invariant time methods. *Biopharm Drug Dispos* 20:137–44.

Miller, R., Ewy, W., Corrigan, B. W., Ouellet, D., Hermann, D., Kowalski, K. G., Lockwood, P. et al. 2005. How modeling and simultion have enhanced decision making in new drug development. *J Pharmacokinet Pharmacodyn* 32:185–97.

Note for Guidance on Toxicokinetics: The assessment of Systemic Exposure in Toxicity Studies—ICH S3A. 1994. http://www.ich.org/fileadmin/Public_Web_Site/ICH_Products/Guidelines/Safety/S3A/Step4/S3A_Guideline.pdf

Obach, R. S., Baxter, J. G., Liston, T. E., Silber, B. M., Jones, B. C., Macintyre, F., Rance, D. J., Wastall, P. 1997. The prediction of human pharmacokinetic parameters from preclinical and *in vitro* metabolism data. *J Pharmacol Exp Ther* 283:46–58.

Obach, R. S., Lombardo, F., Waters, N. J. 2008. Trend analysis of a database of intravenous pharmacokinetic parameters in humans for 670 drug compounds. *Drug Metab Dispos* 36:1385–405.

Oie, S., Tozer, T. N. 1979. Effect of altered plasma protein binding on apparent volume of distribution. *J Pharm Sci* 68:1203–5.

Parrott, N., Jones, H., Paquereau, N., Lave, T. 2005. Application of full physiological models for pharmaceutical drug candidate selection and extrapolation of pharmacokinetics to man. *Basic Clin Pharmacol Toxicol* 96:193–9.

Parrott, N., Lukacova, V., Fraczkiewicz, G., Bolger, M. B. 2009. Predicting pharmacokinetics of drugs using physiologically based modeling—Application to food effects. *AAPS J* 11(1):45–53.

Pugsley, M. K., Authier, S., Curtis, M. J. 2008. Principles of safety pharmacology. *Br J Pharmacol* 154(7):1382–99.

Ritschel, W. A. 1983. A simple method for dosage regimen adjustment. *Meth Find Exptl Clin Pharmacol* 5(6):407–12.

Robinson, S., Delongeas, J. L., Donald, E., Dreher, D., Festag, M., Kervyn, S., Lampo, A. et al. 2008. A European pharmaceutical company initiative challenging the regulatory requirement for acute toxicity studies in pharmaceutical drug development. *Regul Toxicol Pharmacol* 50:345–52.

Shaffer, C. L., Scialis, R. J., Rong, H. J., Obach, R. S. 2012. Using Simcyp to project human oral pharmacokinetic variability in early drug research to mitigate mechanism-based adverse events. *Biopharm Drug Dispos* 33(2):72–84.

Shono, Y., Jantratid, E., Dressman, J. B. 2011. Precipitation in the small intestine may play a more important role in the *in vivo* performance of poorly soluble weak bases in the fasted state: case example nelfinavir. *Eur J Pharm Biopharm* 79(2):349–56.

Shono, Y., Jantratid, E., Janssen, N., Kesisoglou, F., Mao, Y., Vertzoni, M., Reppas, C., Dressman, J. B. 2009. Prediction of food effects on the absorption of celecoxib based on biorelevant dissolution testing coupled with physiologically based pharmacokinetic modeling. *Eur J Pharm Biopharm* 73(1):107–14.

Shono, Y., Jantratid, E., Kesisoglou, F., Reppas, C., Dressman, J. B. 2010. Forecasting *in vivo* oral absorption and food effect of micronized and nanosized aprepitant formulations in humans. *Eur J Pharm Biopharm* 76(1):95–104.

Sinha, V. K., De Buck, S. S., Fenu, L. A., Smit, J. W., Nijsen, M., Gilissen, A. H. J., VanPeer, A., Lavrijsen, K., Mackie, C. E. 2008. Predicting oral clearance in humans: How close can we get with allometry? *Clin Pharmacokinet* 47:35–45.

Slatlery, J. T., Gibaldi, M., Koup, J. R. 1980. Prediction of maintenance dose required to attain a desired drug concentration at steady state from a single determination of concentration after an initial dose. *Clin Pharmacokin* 5(4):377–85.

Stoner, C. L., Cleton, A., Johnson, K., Oh, D-M., Hallak, H., Brodfuehrer, J., Surendran, N., Han, H. K. 2004 Integrated oral bioavailability projection using *in vitro* screening data as a selection tool in drug discovery. *Int J Pharm* 269(1):241–9.

Tang, H., Mayersohn, M. 2005. A novel model for prediction of human drug clearance by allometric scaling. *Drug Metab Dispos* 33:1297–303.

Van Dalen, R., Vree, T. B., Baars, A. M., Termond, E. 1986. Dosage adjustment for ceftazidine in patients with impaired renal function. *Eur J Clin Pharmacol* 30(5): 597–605.

Vieira, M. L. T., Zhao, P., Berglund, E. G., Reynolds, K. S., Zhang, L., Lesko, L. J., Huang, S.M. 2012. Predicting drug interaction potential with a physiologically based pharmacokinetic model. A case study of telithromycin, a time-dependent CYP3A inhibitor. *Clin Pharmacol Ther* 91(4):700–8.

Wagner, J. G. 1975. *Fundamentals of Clinical Pharmacokinetics*. Hamilton, IL: Drug Intelligence Publications, Inc.

Wajima, T., Yano, Y., Fukumura, K., Oguma, T. 2004. Prediction of human pharmacokinetic profile in animal scale up based on normalizing time course profiles. *J Pharm Sci* 93:1890–900.

World Medical Association (WMA), Declaration of Helsinki. 1964. Adapted by the 18th WMA General Assembly—Handbook of WMA Policy.

17 Population Pharmacokinetics/ Toxicokinetics

17.1 INTRODUCTION

A definition of population pharmacokinetics (PPK) is "the study of the sources and correlates of variability in drug concentrations among individuals who are the target patient population receiving clinically relevant doses of a drug of interest" (Aarons, 1991; FDA Guidance for Industry, 1999). The goals of PPK studies are mainly to obtain relevant pharmacokinetic parameters in a group of patients that represent the target population, and measuring the variability between the subjects and linking it to various demographical, pathophysiological, genetic, and environmental factors. The ultimate aim is to optimize different aspects of therapy; for example, designing effective dosing regimen for a patient population that suffers from organ failure or population of patients that require combination therapy and are vulnerable to drug-drug interaction, etc.

A typical example of PPK study is the Phase III clinical trial (Chapter 16) or Phase IV post-marketing surveillance where large sets of data are gathered from large representative groups of target population at different medical centers with non-uniform and sparse data collection with inter- and intra-subject variability (Sheiner, 1984; FDA Guidance for Industry, 1999). Many PPK study design factors can influence the study outcomes and related interpretations. For example, design factors such as sampling strategy (Ette et al., 1995a,b), number of observations (Ette et al., 1995a,b), sampling time recording (Sun et al., 1996), and study compliance (Girard et al., 1996) may influence the outcome of a PPK investigation. Furthermore, the PPK study outcomes are influenced by factors such as

- Demographic characteristics of the population (e.g., age; gender; race; body weight or surface area; etc.)
- Type of illness
- Environmental factors such as smoking, meals and diet
- Genetic factors such as CYP2D6, CYP2C19, polymorphism
- Physiological and pathological factors for example, hepatic or renal impairment; multiple disease states (co-morbidity)
- Drug–drug interaction
- Circadian variation
- Formulation of medication
- Other factors can modify the PK profile of a compound and related parameters and constants

There are various approaches for analysis of population pharmacokinetics data. Some approaches like the Naïve Pooled-Data approach (discussed in this chapter) relies on the isolated PK analysis of an individual subject overlooking the inter- and intra-individual variability. In other approaches, for example, the nonlinear mixed-effects model, the PK analysis is an integrated component of a broader analysis, which takes into consideration some of the factors discussed above.

Nowadays most PPK analyses use the latter approach, which considers the relationship between physiology (normal and disease states) and pharmacokinetics/pharmacodynamics of a compound in population of patients. It assesses the extent of inter-individual variability in the population and its influence on PK analysis. It identifies the factors (i.e., demographic, pathophysiological, environmental, or drug-related origin, etc.) that influence the PK behavior of a drug. It is considered an enabling tool for pharmacokinetic analysis of sparsely sampled data, particularly when there are limitation in collecting the data, for example, AIDS patients (Pfister et al., 2003), critical care patients (Georges et al., 2009), cancer patients (Sugiyama et al., 2010) and neonates (Urien et al., 2011; Shellhaas et al., 2013), to name a few. It is believed that the combination of heterogeneous data enhances the power of analysis in determining the linearity or nonlinearity of the data and achieves precision in analysis. It analyzes the residual inter-individual variability and estimates the magnitude of unexplained variability. It allows combining diverse sets of data from various sources (e.g., intensely and sparsely plasma, serum, blood data from assorted sources) for analysis.

Population analyses of PK and PD (PPK and PPD) data in support of new drug applications are often discussed and encouraged by the regulatory agencies. Various FDA and ICH guidelines have sections on PPK and helpful guidance for the NDA submission and approval process of a new drug entity. Examples are ICH E5 guidelines (ICH E5, 1998) related to the influence of race factors on the PK assessment of a population; FDA guidelines on the influence of gender on PPK (FDA Guideline, 1993); ICH E7 guidelines (ICH E7, 1993) related to the influence of age and PK in geriatrics; ICH E11 guidelines (ICH E11, 2000) pertaining to the evaluation of pediatric PPK; ICH E7 (ICH E7, 1993) and FDA guidance on pharmacokinetics in patients with impaired hepatic function (FDA Guideline, 2003); and FDA guidance on *in vivo* drug interaction studies (FDA Guidance, 2006), etc.

17.2 FIXED EFFECT AND RANDOM EFFECT PARAMETERS

The emphasis in evaluation of PK/PD parameters by PPK analysis is the elucidation of the dose-concentration–response relationship of a drug based on inter- and intra-individual variability. In analysis of variability, two types of parameters are taken into consideration (Sheiner and Beal, 1983; Whiting et al., 1986; Karlsson and Sheiner, 1993): fixed effect and random effect.

17.2.1 Fixed Effect Parameters

Fixed effect parameters are referred to the average values of PK parameters in a patient population, and/or average relationship between the PK parameters and the measurable biological factors, for example, biomarkers. Thus, the characteristics of the population are reflected by the average values with no covariates.

17.2.2 Random Effect Parameters

Random effect parameters quantify the random variability of the PPK data, and represent the influence of inter-individual variability; inter-occasion variability (i.e., the random differences in an individual between different occasions); and residual variability (i.e., the unexplained variability often referred to as intra-individual or within subject variability).

17.3 LINEAR AND NONLINEAR MIXED-EFFECTS MODELS

The mixed-effects models are used when the investigation of population pharmacokinetics of a compound is desired and the data, that is, dependent and independent variables, from all subjects in the population are used simultaneously. The required elements for developing a population pharmacokinetic mixed-effect model are: (1) the data from the population; (2) structural model based on the plasma concentration of the drug in the population, described by an appropriate model (see Chapter 10 and Chapter 16 [Section 16.7]); (3) statistical model to define random variability, such as between-individual, between-occasion, residuals, etc., variability around the structural model; (4) covariate model to include important covariates that are related to the properties of the drug, for example, for highly lipophilic and metabolized drug covariates such as hepatic CYP isozymes, measurable genetic factors, weight, etc. are included; and (5) a modeling software capable of bringing the data from all elements together.

Similar to other statistical models, mixed-effects models describe the relationship between a response variable (e.g., PK parameters) and covariates that have been measured in conjunction with the response. The challenge of PPK analysis in using these models is the identification of the most important covariate and its interaction with other covariates of the model (Ribbing and Jonsson, 2004). For example, the influence of body weight on the clearance of a drug, which in turn affects the plasma concentration, is a relevant covariate. However, selecting the body weight as the only covariate defeats the purpose of PPK analysis and increases the bias in parameter estimation. The significance of using these models for population analysis is in the relationship between PK parameters and covariates and between the covariates, for example, body weight with gender, ethnicity, etc. An appropriate PPK model with relevant covariates can add to the knowledge of drug efficacy and optimum therapy.

Required data for PPK analysis include: (1) the plasma/serum/whole blood concentration data, optimized or sparse, determined by a validated analytical methodology; (2) dose and dosing regimen; (3) route of administration and formulation of the drug; (4) measurements of pharmacological response(s); (5) data related to the selected covariates, for example, disease state(s), concurrent administration of other therapeutic agents, demographical data, environmental factors, etc.; (6) accurate reporting of days and times of sampling and data collection; and (7) report of any side effects and toxicological response.

The mixed-effects models can be identified as linear or nonlinear mixed-effects models.

17.3.1 Linear Mixed-Effects Model (LME)

The models that describe the response variable as a linear function of both the fixed and random effects plus unit error term is called linear mixed-effects model (LME) (Laird and Ware, 1982). These models represent the relationship between a response and independent covariates with coefficients that can change with respect to one or more covariates. The standard form of a linear mixed-effects model is

$$y = M_F\phi + M_r\beta + \varepsilon \tag{17.1}$$

where (y) is the response, a vector of n observations; M_F is the fixed-effects matrix; ϕ is the fixed-effects vector; M_r is the random effects matrix; β is the random effects vector; and ε is the observation error vector. Thus, $M_F\alpha$ is the fixed effects part of the equation; $M_r\beta$ is the random component of the deviation from the population mean; and ε is the error term.

17.3.2 Nonlinear Mixed-Effects Model (NLME)

When the response is expressed as a nonlinear function of both fixed and random effects (mixed effects) plus unit error term, it is identified as nonlinear mixed-effects model (NLME) (Sheiner and Beal, 1980). In other words, the drug concentration or dependent variable is a nonlinear function of the model parameters and independent variable. The nonlinear mixed-effects model is a widely accepted PPK model. The detection and characterization of nonlinear processes, whether related to capacity-limited metabolism, absorption or excretion, are important for the safe and effective use of a therapeutic agent (Ludden, 1991).

In the NLME model, the PK model for estimation of PK parameters is embedded in a statistical model. The assumptions of the statistical model should be determined in terms of covariates as they relate to the expected variability within and among individuals.

For more extensive mathematical and statistical steps of the nonlinear mixed-effects model, relevant references should be consulted (Davidian and Giltinan, 1995, 2003; Bonate, 2005; Nedelman, 2005; Laffont and Concordet, 2011).

The initial basic setup is summarized below (Davidian, 2009) for a population of N individuals and a number of measurements for each individual:

$$N = \text{Number of individuals of } i = 1, \ldots, N \tag{17.2}$$

$$n_i = \text{Number of measurements for individual } i \tag{17.3}$$

$$\text{PK/PD or TK/TD measurements of the outcome per individual } i = y_{i1}, y_{i2}, y_{i3}, \ldots y_{in_i} \tag{17.4}$$

PK/TK measurements are referred to as plasma concentration or other biological samples and PD/TD measurements are the measurable response or measurements of relevant biomarkers.

The PK/PD or TK/TD measurements are taken at time points $t_{i1}, t_{i2}, t_{i3}, \ldots, t_{in_i}$.

Thus, the measurement for ith individual at t_{ij} is y_{ij}, where $j = 1, 2, 3, \ldots, n_i$.

The measurement-time relationship requires the input (i.e., dose), which is also "within-individual covariates." Setting it equal to Θ_i, it can be defined as:

For single oral dose (first-order absorption, i.e., compounds administered by an extravascular route):

$$\Theta_i = D_i \tag{17.5}$$

For zero-order input:

$$\Theta_i = D_i \text{ infused over the time of infusion } (t_{inf})\text{, i.e., } k_0$$

The "among-individual covariates" refer to demographical, environmental and pathophysiological covariates identified as Ξ_i, which theoretically do not change with time and remain constant during the observation and sampling.

The collected data from N individuals of the population can be defined as

$$Y_i, X_i \quad \text{where } i = 1, \ldots, N$$

$$Y_i = (y_{i1}, \ldots, y_{in_i}) \tag{17.6}$$

$$X_i = (\Theta_i, \Xi_i) \tag{17.7}$$

Therefore, the measurements of individual subject are

$$y_{ij} = m(t_{ij}, \Theta_i, \theta_i) + \varepsilon_{ij}, \quad j = 1, \ldots, n_i \tag{17.8}$$

where θ corresponds to the key pharmacokinetic parameters, for example, for a one-compartment model with first-order input there are three parameters (i.e., $r = 3$) and they correspond to the absorption rate constant, clearance, and the volume of distribution.

$$\theta_i = (k_{a_i}, Vd_i, Cl_i) = (\theta_{i1}, \theta_{i2}, \theta_{i3}) \tag{17.9}$$

Therefore, at the individual-level the one-compartment model with first-order input with $\Theta_i = D_i$ and $r = 3$ can be defined as

$$m(t_i, \Theta_i, \theta_i) = \frac{FD_i k_{a_i}}{Vd_i \left(k_{a_i} - Cl/Vd_i\right)} \left(e^{-\left(\frac{Cl_i}{Vd_i}\right)t} - e^{-k_{a_i}t} \right) \tag{17.10}$$

$$\varepsilon_{ij} = y_{ij} - m(t_{ij}, \Theta_i, \theta_i) \tag{17.11}$$

With conditional expectation (E) of

$$E(\varepsilon_{ij} | \Theta_i, \theta_i) = 0 \tag{17.12}$$

At the population level, the individual parameters is defined as

$$\theta_i = d(\Xi_i, \beta, b_i) \tag{17.13}$$

where

$$i = i, \ldots, N, (r \times 1); \tag{17.14}$$

(d) is r-dimensional function of (θ_i) and (Ξ_i) in terms of β and b_i, which correspond to fixed effects ($p \times 1$) and random effects ($q \times 1$), respectively.

Thus, the individual parameters at the population level are functions of "among-individual covariates," fixed and random effects. For example if the "among-individual covariates" are age, weight, and creatinine clearance

$$\Xi_i = (wt_i, age_i, Ccr_i), \tag{17.15}$$

$$b_i = (b_{i1}, b_{i2}, b_{i3}) \quad \text{for } (q = 3) \text{ and} \tag{17.16}$$

$$\beta - (\beta_1, \ldots, \beta_7) \quad \text{for } (p = 7) \tag{17.17}$$

Therefore,

$$k_{a_i} = \theta_{i1} = d_1(\Xi_i, \beta, b_i) = e^{\beta_1 + b_{i1}} \tag{17.18}$$

$$Vd_i = \theta_{i2} = d_2(\Xi_i, \beta, b_i) = e^{\beta_2 + \beta_4 wt_i + b_{i2}} \tag{17.19}$$

$$Cl_i = \theta_{i3} = d_3(\Xi_i, \beta, b_i) = e^{\beta_3 + \beta_5 wt_i + \beta_6 Ccr_i + \beta_7 age_i + b_{i3}} \tag{17.20}$$

Taking a logarithm of Equations 17.18 through 17.20 yields

$$\log k_{a_i} = \beta_1 + b_{i1} \tag{17.21}$$

$$\log Vd_i = \beta_2 + \beta_4 wt_i + b_{i2} \tag{17.22}$$

$$\log Cl_i \beta_3 + \beta_5 wt_i + \beta_6 cr_i + \beta_7 age_i + b_{i3} \tag{17.23}$$

17.4 PLME MODEL

A partially linear mixed-effects (PLME) model, also known as a semiparametric mixed-effects model, has also been proposed. The main emphasis of the model is to use a set of spline basis functions that are treated as the random effect for the purpose of neutralizing the influence of one variable, for example, neutralizing time and measuring the effect of another covariate on the response directly (Hardle et al., 2001; Ruppert et al., 2003; Bonate, 2005). The advantages of PLME are (1) the use of spline basis function removes the need for having a model that takes account of time, and (2) the model is based on the linear mixed-effects model and it is faster to compute. However, there are some limitations associated with the approach; for example, the model assumes the observations are independent and fails to consider the "within the subject" correlation. The utility of PLME models in PPK and PPD are evolving and further comparative analyses are required (Bonate, 2005).

17.5 NAÏVE-POOLED DATA APPROACH (NPD)

The naïve-pooled and two-stage approach had commonplace prior to the introduction of population-based models (Sheiner and Beal, 1980). The naïve-pooled data analysis consisted of pooling all data points as though they belong to one individual and using least-square fitting to determine the parameters of the data set. Although the convenience of parameter estimation may be considered an advantage, nonetheless the methodology ignores the inter-individual variability and may generate realistic estimates of parameters if the subject variability is very small; a scenario that can only be achieved in the lab with small experimental animals of the same strain. In general, the naïve-pooled approach is not considered an acceptable methodology for data analysis in humans. However, there have been reports indicating that the methodology can accurately estimate the mean parameters for a number of drugs used in anesthesia (Shafer et al., 1990; Gustafsson et al., 1992; Dyke et al., 1993).

17.5.1 Naïve Average Data Approach (NAD)

In this approach the mean value of the data for each time point of a sample collection is calculated first. Thus, all collected samples will change to one set of mean values. A model is then fitted to the mean values to determine the parameters of the population. This approach also ignores the subject variability and between occasion variability, and may lead to biased estimation of the population parameters.

17.6 TWO-STAGE APPROACH

This approach, still in use, is generally described as standard, global, or the iterative two-stage method.

17.6.1 Standard Two-Stage (STS) Approach

The first stage of the STS approach involves the calculations of model parameters for each subject individually. In the second stage, the parameters of the population are identified as the arithmetic or geometric mean plus variance/covariance of all individually calculated parameters. The method does not take into consideration the variability and reliability of the individual estimates, and has been shown to overestimate the parameters variance/covariance (Davidian and Giltinan, 1993). However, it is used commonly for PK parameter estimation in experimental animals.

17.6.2 Global Two-Stage (GTS) Approach

Similar to the standard method, the parameters are estimated for each individual first, and then iteratively optimized to estimate the population mean and variance/covariance of each parameter

(Steimer et al., 1984). Qualitatively, the global method is preferred over the standard, but in dealing with large populations, it suffers from the same shortcomings as the standard approach.

17.6.3 Iterative Two-Stage (ITS) Approach

The important feature of this approach is that the information from the sample mean and variance/covariance of the population are used as prior knowledge, in the context of Bayesian estimation, in quantifying the individual parameters from the data (Steimer et al., 1984). The method estimates the parameters more accurately, but the calculated data are not significantly different from the GTS method.

In all two-stage approaches, accurate and precise estimates of model parameters are essential in the first stage. This would require multiple and appropriately timed blood samples, which may not be feasible in some patient population, an inherent drawback of two-stage approaches.

17.7 BAYESIAN APPROACH

This approach is computationally more demanding and its solution is achieved through the application of Markov chain Monte Carlo methods (Racine-Poon, 1985; Gelfand and Smith, 1990; Wakefield et al., 1997). Briefly, the calculated parameters are viewed as random variables with a distribution attached to each parameter in order to express the uncertainty with their values. The distribution is known as the prior distribution, which represents the prior knowledge about the parameter of interest. The prior distribution of the parameters across the population and the calculated data from an individual are used in estimating an individual's parameters (Dokoumetzidis and Aarons, 2005). Prior distributions are incorporated in Bayesian analysis using Bayes' Rule (Dempster et al., 1977). The estimation of mean and variance/covariance requires the estimates of the priors for the parameters.

17.8 TOOLS OF COMPUTATIONAL STATISTICS FOR PPK

The commonly used software in PPK analysis include NONMEM 7 (Beal, 1994; Bauer, 2010); and the new version of Phenix® NLME™ with QRPEM (Quasi-Random Parametric Expectation Maximization), a method for likelihood expectation maximization (EM). It is considered an improvement in population pharmacokinetic analysis of a nonlinear mixed-effects model (Leary et al., 2011); WinBUGS (BUGS stands for Bayesian inference Using Gibbs Sampling) is the software for Bayesian analysis using a Markov chain Monte Carlo method (Lunn et al., 2009); and SimBiology for MATLAB.

There have been certain critiques about these programs. For example, the programs such as NONMEM (Sheiner and Beal, 1982) and SAS macro NLINMIX use first order Taylor series expansion (Galecki, 1998; Galecki et al., 2004). The series expansion, also known as first-order method, is considered the limitation of the program when dealing with nonlinear random effect parameters or nonlinear residual errors (Ette et al., 2007). The First-Order method of NLME has been reported to have potential in generating bias estimates (Sheiner and Beal, 1981; Beal, 1984; Sheiner, 1984; Grasela et al., 1986; Racine et al., 1986; Karlsson and Sheiner, 1993; Ette et al., 1994; 1995a,b). For one- and two-compartment models the bias has been shown to occur when the inter-subject variability is very high (White et al., 1991; Ette et al., 1998). The SAS macro NLINMIX use the alternative first-order expansion for both the fixed effects and random effects and the calculated parameters are similar to NONMEM. In general, the majority of these programs estimate PK parameters based on using maximum likelihood estimation.

REFERENCES

Aarons, L. 1991. Population pharmacokinetics: Theory and practice. *Br J Clin Pharmacol* 32(6):669–70.
Bauer, R. J. 2010. Introduction to NONMEM7, ftp://nonmem.iconplc.com/Public/nonmem712/intro712.pdf

Bauer, R. J. 2011. ICON Development Solutions. *NONMEM Users Guide: Introduction to NONMEM 7.2.0.* Ellicott City, Maryland. ftp://nonmem.iconplc.com/Public/nonmem720/guides/nm720.pdf

Beal, S. L. 1984. Population pharmacokinetic data and parameter estimation based on their first two statistical moments. *Drug Metab Rev* 15:173–93.

Bonate, P. L. 2005. Covariate detection in population pharmacokinetics using partially linear mixed effects models. *Pharm Res* 22(4):541–49.

Bonate, P. L. 2005. Recommended reading in population pharmacokinetic pharmacodynamics. *AAPS J* 7(2):E363–73.

Bulitta, J. B., Bingölbali, A., Shin, B. S., Landersdorfer, C. B. 2011. Development of a new pre- and post-processing tool (SADAPT-TRAN) for nonlinear mixed-effects modeling in S-ADAPT. *AAPS J* 13(2):201–11.

Chan, P. L., Jacqmin, P., Lavielle, M., McFadyen, L., Weatherley, B. 2011. The use of the SAEM algorithm in MONOLIX software for estimation of population pharmacokinetic-pharmacodynamic-viral dynamics parameters of maraviroc in asymptomatic HIV subjects. *J Pharmacokinet Pharmacodyn* 38(1):41–61.

Davidian, M. 2009. Non-linear mixed-effects models. In *Longitudinal Data Analysis*, eds. G. Fitzmaurice, M. Davidian, G. Verbeke, G. Molenberghs, 107–141. Boca Raton, FL: Chapman & Hall/CRC Press.

Davidian, M., Giltinan, D. M. 1993. Some general estimation methods for nonlinear mixed-effects models. *J Biopharm Stat* 3(1):23–55.

Davidian, M., Giltinan, D. 1995. *Nonlinear Models for Repeated Measurements Data.* New York: Chapman and Hall.

Davidian, M., Giltinan, D. M. 1995. *Nonlinear Models for Repeated Measurement Data.* Boca Raton, FL: Chapman & Hall/CRC Press.

Davidian, M., Giltinan, D. M. 2003. Nonlinear models for repeated measurement data: An overview and update. *J Agr Biol Env Stat* 8, 387–419.

Dempster, A. P., Laired, N. M., Rubin, D. B. 1977. Maximum likelihood from incomplete data via EM algorithm. *J R StatSoc B* 39:1–38.

Dokoumetzidis, A., Aarons, L. 2005. Propagation of population pharmacokinetic information using a Bayesian approach: Comparison with meta-analysis *J Pharacokinet Phrmacodyn* 32(3–4):401–18.

Dyck, J. B., Maze, M., Haack, C., Azarnoff, D. L., Vuorilehto, L., Shafer, S. L. 1993. Computer-controlled infusion of intravenous dexmedetomidine hydrochloride in adult human volunteers. *Anesthesiology* 78(5):821–8.

Ette, E. I., Kelman, A. W., Howie, C. A., Whiting, B. 1994. Influence of ineranimal variability on the estimation of population pharmacokinetic parameters in preclinical studies. *Clin Res Reg Affairs* 11:121–39.

Ette, E. I., Howie, C. A., Kelman, A. W., Whiting, B. 1995a. Experimental design and efficient parameter estimation in preclinical pharmacokinetics studies. *Pharm Res* 12:729–37.

Ette, E. I., Kelman, A. W., Howie, C. A., Whiting, B. 1995b. Analysis of animal pharmacokinetic data: Performance of the one point per animal design. *J Pharmacokinet Biopharm* 23:551–66.

Ette, E. I., Sun, H., Ludden, T. M. 1998. Balanced designs and longitudinal population pharmacokinetic studies. *J Clin Pharmacol* 38:417–23.

Ette, E. I., Williams, P. J., Ahmad, A. 2007. Population pharmacokinetic estimation methods. In *Pharmacometrics: The Science of Quantitative Pharmacology*, eds. E. I. Etteand P. J. Williams, Hoboken, NJ: John Wiley & Sons, Inc, Chapter 10.

FDA Final Guidance on population pharmacokinetics. 1999. http://www.fda.gov/downloads/Drugs/GuidanceComplianceRegulatoryInformation/Guidances/ucm072137.pdf; http://www.fda.gov/downloads/ScienceResearch/SpecialTopics/WomensHealthResearcUCM133184.pdf

FDA Final Guidance for study and evaluation of gender differences. 1993. http://www.fda.gov/downloads/RegulatoryInformation/Guidances/UCM126835.pdf

FDA/ICH Final Guideline on clinical investigation of medicinal products in the pediatric population (E11). 2000. http://www.fda.gov/downloads/Drugs/GuidanceComplianceRegulatoryInformation/Guidances/ucm072114.pdf

FDA Final Guidance on pharmacokinetics in patients with impaired hepatic function. 2003. http://www.fda.gov/downloads/Drugs/GuidanceComplianceRegulatoryInformation/Guidances/ucm072123.pdf

FDA Draft Guidance on *in vivo* drug interaction studies. 2006. http://www.fda.gov/OHRMS/DOCKETS/98fr/06d-0344-gdl0001.pdf; http://www.fda.gov/downloads/Drugs/GuidanceComplianceRegulatoryInformation/Guidances/ucm292362.pdf

Galecki, A. T. 1998. NLMEM: A new SAS/IML macro for hierarchical nonlinear models. *Comput Methods Programs Biomed* 5(3):207–16.

Galecki, A. T., Wolfinger, R. D., Linares, O. A., Smith, M. J., Halter, J. B. 2004. Ordinary differential equation PK/PD models using the SAS macro NLINMIX. *J Biophar Stat* 14(2):483–503.

Gelfand, A. E., Smith, A. F. M. 1990. Sampling-based approaches to calculating marginal densities. *J Am Stat Assoc* 85:398–409.

Georges, B., Conil, J. M., Seguin, T., Ruiz, S., Minville, V., Cougot, P., Decun, J. F. et al. 2009. Population pharmacokinetics of ceftazidime in intensive care unit patients: Influence of glomerular filtration rate, mechanical ventilation, and reason for admission. *Antimicrob Agents Chemother* 53(10):4483–9.

Girard, P., Sheiner, L. B., Kastrissios, H., Blaschke, T. F. 1996. Do we need full compliance data for population pharmacokinetic analysis? *J Pharmacokinet Biopharm* 24:265–82.

Grasela, Jr., T. H., Antal, E. J., Townsend, R. J., Smith, R. B. 1986. An evaluation of population pharmacokinetics in therapeutic trials. Part I. Comparison of methodologies. *Clin Pharmacol Ther* 39:605–12.

Gustafsson, L. L., Ebling, W. F., Osaki, E., Harapat, S., Stanski, D. R., Shafer, S. L. 1992. Plasma concentration clamping in the rat using a computer-controlled infusion pump. *Pharm Res* 9(6):800–7.

Hardle, W., Liang, H., Gao, J. 2001. *Partially Linear Models*, Rockville, MD: Springer Verlag.

ICH Final Guideline on special populations: Geriatrics (E7). 1993. http://www.ich.org/LOB/media/MEDIA483.pdf; http://www.fda.gov/downloads/Drugs/GuidanceComplianceRegulatoryInformation/Guidances/UCM189544.pdf

ICH Final Guideline on ethnic factors in the acceptability of foreign clinical data (E5). Released September 1998. http://www.ema.europa.eu/docs/en_GB/document_library/Scientific_guideline/2009/09/WC500002842.pdf

Karlsson, M. O., Sheiner, L. B. 1993. The importance of modeling interoccasion variability in population pharmacokinetic analysis. *J Pharmacokinet Biopharm* 21(6):735–50.

Laffont, C. M., Concordet, D. 2011. A new exact test for the evaluation of population pharmacokinetic and/or pharmacodynamic models using random projections. *Pharm Res* 28:1948–62.

Laird, N. M., Ware, J. H. 1982. Random-effect models for longitudinal data. *Biometrics* 38:963–74.

Leary, R., Dunlavey, M., Chittenden, J., Brett Matzuka, B., Guzy, S. 2011. QRPEM A New Standard of Accuracy, Precision, and Efficiency in NLME Population PK/PD Methods. http://www.pharsight.com/library/sign_in/whiteQRPEM.pdf

Ludden, T. M. 1991. Nonlinear pharmacokinetics. Clinical Implications. *Clin Pharmacokinet* 20:429–46.

Lunn, D., Spiegelhalter, D., Thomas, A., Best, N. 2009. The BUGS project: Evolution, critique and future directions. *Stat Med* 28(25):3049–67.

Nedelman, J. R. 2005. On some "disadvantages" of the population approach. *AAPS J* 7(2):E374–82.

Pfister, M., Labbé, L., Hammer, S. M., John Mellors, J., Bennett, K. K. Rosenkranz, S., Sheiner, L. B., and the AIDS Clinical Trial Group Protocol 398 Investigators. 2003. Population pharmacokinetics and pharmacodynamics of efavirenz, nelfinavir, and indinavir. *Antimicrob Agents Chemother* 47(1):130–7.

Phoenix® NLME™—Nonlinear Mixed Effects Modeling, Phoenix NLME Software Review. 2011. Phoenix; Pharsight. http://www.certara.com/products/pkpd/phx-nlme/

Racine, A., Grieve, A. P., Fluhler, H., Smith, A. F. M. 1986. Bayesian methods in practice: Experiences in pharmaceutical industry. *Appl Stat* 35:1–38.

Racine-Poon, A. 1985. A Bayesion approach to nonlinear random effect models. *Biometrics* 41:1015–23.

Ribbing, J., Jonsson, E. N. 2004. Power, selection bias and predictive performance of the population pharmacokinetic covariate model. *J Pharmacokinet Pharmacodyn* 31(2):109–34.

Ruppert, D., Wand, M. P., Carroll, R. J. 2003. *Semi Parametric Regression*, Cambridge, UK: Cambridge University Press.

Shafer, S. L., Varvel, J. R., Aziz, N., Scott, J. C. 1990. Pharmacokinetics of fentanyl administered by computer-controlled infusion pump. *Anesthesiology* 73(6):1091–102.

Sheiner, L. B. 1984. The population approach to pharmacokinetic data analysis: Rationale and standard data analysis methods. *Drug Metab Rev* 15(1–2):153–71.

Sheiner, L. B., Beal, S. L. 1980. The NONMEM system. *Am Stat* 34:118–19.

Sheiner, L. B., Beal, S. L. 1981. Evaluation of methods for estimating population pharmacokinetic parameters. I. Biexponential model and experimental pharmacokinetic data. *J Pharmacokinet Biopharm* 9:635–51.

Sheiner, L. B., Beal, S. L. 1982. Bayesian individualization of pharmacokinetics: Simple implementation and comparison with non-Bayesian methods. *J Pharm Sci* 71:1344–8.

Sheiner, L. B., Beal, S. L. 1983. Evaluation of methods for estimating population pharmacokinetic parameters III. Monoexponential model: Clinical pharmacokinetic data. *J Pharmacokinet Biopharm* 11:303–19.

Shellhaas, R. A., Ng, C. M., Dillon, C. H., Barks, J. D., Bhatt-Mehta, V. 2013. Population pharmacokinetics of phenobarbital in infants with neonatal encephalopathy treated with therapeutic hypothermia. *Pediatr Crit Care Med* 14(2):194–202.

Steimer, J. L., Golmard, J. L., Boisvieux, J. F. 1984. Alternative approaches to estimation of population pharmacokinetic parameters: Comparison with the nonlinear mixed-effect model. *Drug Metab Rev* 15:265–92.

Sugiyama, E., Kaniwa, N., Kim, S. R., Hasegawa, R., Saito, Y., Ueno, H., Okusaka, T. et al. 2010. Population pharmacokinetics of gemcitabine and its metabolite in Japanese cancer patients: Impact of genetic polymorphisms. *Clin Pharmacokinet* 49(8):549–58.

Sun, H., Ette, E. I., Lundden, T. M. 1996. On the recording of sample times and parameter estimation from repeated measures pharmacokinetic data. *J Pharmacokinet Biopharm* 24:637–50.

Urien, S., Firtion, G., Anderson, S. T., Hirt, D., Solas, C., Peytavin, G., Faye, A. et al. 2011. Lopinavir/ritonavir population pharmacokinetics in neonates and infants. *Br J Clin Pharmacol* 71(6):956–60.

Wakefield, J. C., Walker, S. G. 1997. Bayesian nonparametric population models: Formulation and comparison with likelihood approaches. *J Pharmacokinet Biopharm* 25:235–53.

White, D. B., Walawander, C. A., Tung, Y., Grasela, T. H. 1991. An evaluation of point and interval estimates in population pharmacokinetics using NONMEM analyses. *J Pharmacokinet Biopharm* 19:87–112.

Whiting, B., Kelman, A. W., Grevel, J. 1986. Population pharmacokinetics: Theory and clinical application. *Clin Pharmacokinet* (11):387–401.

18 Quantitative Extrapolation Interspecies Scaling/Low Dose Extrapolation

18.1 INTRODUCTION

The concept of quantitative extrapolation encompasses various fundamental areas of PK/TK analyses. Quantitative extrapolations are used commonly in different areas of contemporary research and evaluations; for example, environmental health sciences, drug discovery, and development and regulatory decisions, to name a few. The discipline of toxicology and the field of drug discovery and development can be considered sciences of extrapolation. Both subjects rely on the information and knowledge obtained from cell culture, *in vitro* systems, experimental animals, and a relatively small number of humans. Even when there are dissimilarities, it is assumed that the non-human systems can reasonably predict a better understanding of safety considerations in humans. The knowledge gained from extrapolation of PK/TK or PD/TD data, in general, are used to improve the understanding of xenobiotics behavior in the body and the assessment of their risk for human. Predictions of the *in vivo* PK/TK characteristics of xenobiotics from *in vitro* studies is another important qualitative and often quantitative extrapolation, which are carried out on a routine basis.

The focus of this chapter is on two areas of important extrapolation in PK/TK analysis, namely interspecies scaling and low dose extrapolation. The two selected topics are equally important and inter-reliant.

18.2 INTERSPECIES SCALING IN MAMMALS

The reliance on experimental animals is a cornerstone of preclinical phase of drug discovery and development and considered the essential component of toxicological evaluation and risk assessment. The experimental animals are commonly considered a predictive model for humans and the interspecies scaling is an important part of PK/TK analysis. All regulatory agencies, agricultural, food, chemical, environmental, and pharmaceutical/biotechnology industries rely on the extrapolated data from experimental animals to make decisions on the safety of products, and the data are used for approval and marketing decisions. The application of interspecies scaling is based on the principles of similitude and dimensional analysis (McMahon and Bonner, 1983; Peters, 1983; Calder, 1984).

There are many similarities in the anatomy and physiology of land mammalian species from a 10 g mouse of 3–4 week-old to a 6 ton African elephant. The blood flows in the same direction; the anatomical geometry is similar; the weight of each organ is a similar fraction of body weight, except skin and skeleton; cellular structure is the same; and the general biochemistry and physiology for the most part are similar. This remarkable similarity has allowed interspecies scaling of physiological parameters and constants such as longevity, blood volume and flow, heart rate, and PK/TK parameters.

There are two approaches to interspecies PK/TK scaling, an allometric or empirical approach and a physiologic approach. The allometric methodologies are straightforward, but frequently inaccurate and predict only the average value of parameters. In addition, they require significant data and the prediction is only for the parent compounds. The physiologic extrapolations are based on

PBPK modeling as discussed in Chapter 9 and other relevant chapters. The PBPK approach takes into consideration the full PK/TK profile of a compound and associated variability and uncertainty, including the formation of metabolites. It is a major undertaking and tends to be costly.

18.2.1 Allometric Approach

The historical research by a number of scientists established the following empirical allometric relationship, which defines the relationship between a physiological parameter and body weight (Rubner, 1883; Kleiber, 1932, 1947; Adolph, 1949; Davidson et al., 1986; Voisin et al., 1990):

$$Y = aW^b \tag{18.1}$$

where (Y) is the biological function or PK/TK parameter of interest, (W) is body weight, (a) and (b) are the coefficient and exponent of the relationship, respectively, which are considered species-independent constants. The geometric mean of b values is 0.82 (Table 18.1, the selected data are from Adolph, 1949; as presented in Mordenti, 1986).

TABLE 18.1
Allometric Equations of Physiological Parameters as a Function of Body Weight

Physiological Parameter (Y)	Units	Allometric Equation
	Physiological Rates	
Intake of water	mL/h	$0.01W^{0.88}$
Urine output	mL/h	$0.0064W^{0.82}$
Ventilation rate	mL/h	$120W^{0.74}$
	Oxygen Consumption	
Basal	mL STP/h	$3.80W^{0.734}$
Liver slice	mL STP/h	$3.3W^{0.77}$
	Clearance	
Urea	mL/h	$1.59W^{0.72}$
Inulin	mL/h	$1.74W^{0.77}$
Creatinine	mL/h	$4.20W^{0.69}$
Hippurate	mL/h	$5.4W^{0.8}$
	Physiological Period	
Heartbeat duration	h	$1.19 \times 10^{-5}\ W^{0.27}$
Breath duration	h	$4.7 \times 10^{-5}\ W^{0.28}$
	Organ Weight	
Kidneys	g	$0.0212W^{0.85}$
Brain	g	$0.081W^{0.7}$
Heart	g	$6.6 \times 10^{-3}\ W^{0.98}$
Lung	g	$0.0124W^{0.99}$
Liver	g	$0.082W^{0.87}$
Thyroids	g	$2.2 \times 10^{-4}\ W^{0.80}$
Adrenal	g	$1.1 \times 10^{-3}\ W^{0.92}$
Pituitary	g	$1.3 \times 10^{-4}\ W^{0.76}$
Stomach and intestine	g	$0.112W^{0.94}$
Blood	g	$0.055W^{0.99}$

Taking logarithm of both sides of Equation 18.1 yields

$$\log Y = \log a + b\log W \tag{18.2}$$

Therefore, a plot of the parameter versus the body weight on a log-log scale should generate a straight line with the slope of b and y-intercept of $\log a$ (Figure 18.1).

The allometric exponent b is an indicator of the relationship between the parameter of interest Y, and the body weight W (Chappel and Mordenti, 1989). For instance, when the exponent is greater than zero ($b > 0$), the parameter Y decreases as the body weight increases; the analogy is the influence of body weight on the heartbeat. When the exponent is equal to zero ($b = 0$), the parameter Y is independent of the weight, a relevant example is the body temperature, which is independent of the weight. When $b = 1$, the increase in parameter Y is proportional to the body weight, for example, blood volume and body weight. When $b > 1$, the increase in parameter Y is faster than W (Mordenti et al., 1991).

A feature of the allometric relationship is to define one parameter; for example, Y_1 in terms of another parameter Y_2 (Adolph, 1949), that is,

$$Y_1 = a_1 W^{b_1} \Rightarrow \log Y_1 = \log a_1 + b_1 \log W \tag{18.3}$$

$$Y_2 = a_2 W^{b_2} \Rightarrow \log Y_2 = \log a_2 + b_2 \log W \tag{18.4}$$

Therefore,

$$\log W = \frac{\log Y_1 - \log a_1}{b_1} \quad \text{and} \quad \log W = \frac{\log Y_2 - \log a_2}{b_2} \tag{18.5}$$

$$\frac{\log Y_1 - \log a_1}{b_1} = \frac{\log Y_2 - \log a_2}{b_2} \tag{18.6}$$

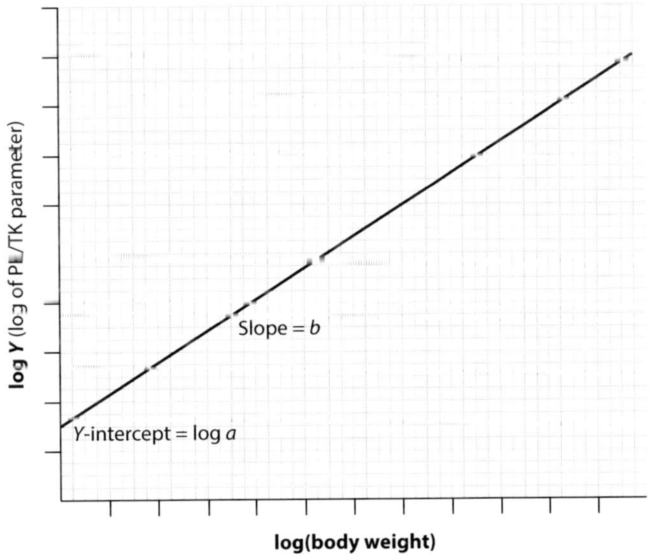

FIGURE 18.1 Plot of logarithm of PK/TK parameter or physiological parameter versus logarithm of body weight according to Equation 18.2.

Solving for Y_1 in terms of Y_2 yields

$$\log Y_1 = \log a_1 + \frac{b_1}{b_2}(\log Y_2 - \log a_2) \tag{18.7}$$

$$Y_1 = a_1 \left(\frac{Y_2}{a_2}\right)^{\frac{b_1}{b_2}} \tag{18.8}$$

The quotients of similar parameters can also provide useful information. For example, the ratio of renal blood flow to cardiac output yields

$$\frac{\text{Renal blood flow (mL/min)}}{\text{Cardiac output (mL/min)}} = \frac{43.06W^{0.77}}{166W^{0.79}} = 0.259 \tag{18.9}$$

This indicates that the renal blood flow is approximately 26% of cardiac output in mammalian species (Mordenti, 1986).

It should be noted that the sample size, methods of measurement of the parameter of interest and the variability of body weight contribute to the variation of empirical estimates of a and b. The theoretical and statistical values of allometric approaches have been questioned and criticized for shortcomings such as the sudden changes of slope in the allometry line; significant errors in extrapolation to extremes of biological parameters such as body weight, deviations of cardiac output, and deviation of cardiac energetic, to name a few (Gunther et al., 1966; White et al., 1968; Loiselle and Gibbs, 1979; Yates, 1979; Smith, 1980; Heusner, 1984).

18.2.1.1 Allometric Approach and Chronological Time

The concept of time is a critical parameter in allometric interspecies extrapolation. The small animals have a rapid heartbeat, faster anabolism and catabolism, and shorter life. The element of time associated with any of these processes is the chronological time. It is often assumed that one human year is equivalent to seven dog years. This assumption is based on the chronological life span of dog and human. In other words, a dog ages 7.14% of its life span per year, whereas human ages 7.14% of its life in seven years (Ings, 1990). Thus, theoretically if time-dependent biological processes are measured according to each species biological clock, animals exhibit to have the same rate.

The pioneering work on allometric scaling and the concept of invariant time was first introduced when the disposition of methotrexate was evaluated in five different mammalian species following intravenous administration of the compound (Dedrick et al., 1970). The transformation of chronological time to biological time, known as the Dedrick time equivalent model, is achieved when the dependent variable on the y-axis is normalized by dividing plasma concentrations by dose (mg/kg) and body weight (kg), and 0.25 (see Equation 18.12) represents a constant for the conversion from chronological time to biological time as the independent variable on the x-axis.

The following approach known as the Elementary Dedrick plot is expressed as

$$y\text{-}axis = \frac{Concentration}{Dose\,(\text{mg/kg})/W(\text{kg})} \tag{18.10}$$

$$x\text{-}axis = \frac{Time}{W^{1-b}} \tag{18.11}$$

where (b) is the exponent of clearance.

When considering the clearance and the volume of distribution, the plot is expressed as (Mahmood and Balian, 1999).

$$y\text{-}axis = \frac{Concentration}{Dose/W^c} \tag{18.12}$$

$$x\text{-}axis = \frac{Time}{W^{c-b}} \tag{18.13}$$

where b and c are exponents of clearance and volume of distribution, respectively.

In the Elementary Dedrick Plot, the interspecies superimposability occurs only when $y = 1$ (Boxenbaum, 1984). When $y \neq 1$ then both the intercept and slope will be species specific.

The chronological time is an essential component of PK/TK analysis in interspecies scaling. The concept of time in physical relativity, psychological relativity, and biological relativity has been philosophically discussed and various conceptual ideas have been presented (Boxenbaum, 1986). For example, the allometric mammalian heartbeat time (*HBT*) is characterized by Equation 18.14 (Gunther and DeLa Barra, 1966; Stahl, 1965; Boxenbaum, 1982):

$$HBT = 0.0428W^{0.28} \tag{18.14}$$

For a 30 g mouse, the heartbeat time or cardiochron is 0.111 seconds and for a 70 kg human is 0.973 seconds; W is body weight in grams.

The breath time or the pulmonary cycle time (*PCT*) identified as pneumatochron is

$$PCT = 1.169W^{0.28} \tag{18.15}$$

The breath time for the mouse and the human are 0.438 and 3.841 seconds, respectively. Dividing *PCT* by *HBT* is approximately 4, that is

$$\left(\frac{PCT}{HBT}\right)_{mouse} = \frac{0.438}{0.111} = 3.946 \tag{18.16}$$

$$\left(\frac{PCT}{HBT}\right)_{human} = \frac{3.841}{0.973} = 3.947 \tag{18.17}$$

Thus, every species has four heartbeats per breath time. From comparison of allometric exponents for the duration of periodic phenomena, it can be concluded that its value remains relatively constant around 0.25, 0.28, etc. Based on this observation, the concept of biological time is then expressed as Equation 18.18 (Boxenbaum, 1986).

$$t_{biological} = aW^{0.25} \tag{18.18}$$

where (a) is a constant.

The allometric approach is used to extrapolate the PK/TK parameters and constants of small experimental animals to human. The smaller animals with short lifespan clear xenobiotics at a faster rate per unit of body weight than the larger animals with longer lifetime. The lifespan of animals correlates well with their size and body weight. In general, the rate of ADME processes depends on the biological clock of the animals. In extrapolation of pharmacokinetic parameters and constants, each species has its own unique pharmacokinetic clock. Based on the concept of pharmacokinetic time, several units of time have been proposed (Boxenbaum and Ronfeld, 1983), which are discussed below. For example, "Kallynochron" is the unit of time that all species clear the same volume of plasma per body weight (kg). It is defined as

$$Kallynochron = \frac{t}{W^{1-b}} \tag{18.19}$$

where (b) is the exponent from the allometric equation for clearance.

"Apolysichron" is another PK/TK time that refers to the combination of clearance and the apparent volume of distribution. In one apolysichron, species eliminate the same fraction of a compound from their bodies and clear the same volume of plasma per kg body weight

$$Apolysichron = \frac{t}{W^{b'-b}} \tag{18.20}$$

There are two additional PK/TK time units similar to apolysichrons called "dienetichrons" and "syndesichrons" except that these units take into consideration the maximum life potential (*MLP*) and brain weight (*BW*) (see also Chapter 4, Section 4.2.2).

The *MLP* can be estimated by the following empirical equation (Sacher, 1959):

$$MLP(years) = 185.4 \times (BW)^{0.636} \times W^{-0.225} \tag{18.21}$$

where both brain weight (*BW*) and body weight (*W*) are in kilograms.

The application of allometric approach for extrapolation of PK/TK parameters such as total body clearance is discussed in Chapter 4, Section 4.4.2.1.

18.2.1.2 Application of Allometric in Converting Animal Dose to Human Dose

An application of the allometric methodology is scaling up the animal dose to human dose. The approach is similar to Equations 18.3 through 18.8, that is,

$$\frac{(Dose)_{human}}{(W_{human})^b} = \frac{(Dose)_{animal}}{(W_{animal})^b} \tag{18.22}$$

Dividing both sides of Equation 18.22 by W_{human} and rearranging the equation to determine mg dose per kg of human body weight yields

$$\frac{(Dose)_{human}}{W_{human}} = \frac{(Dose)_{animal}}{W_{human}} \times \left(\frac{W_{human}}{W_{animal}}\right)^b \tag{18.23}$$

Multiplying the numerator and denominator of the right side of Equation 18.23 by W_{animal} yields

$$\frac{(Dose)_{human}}{W_{human}} = \frac{(Dose)_{animal}}{W_{animal}} \times \frac{(W_{human})^b \times W_{animal}}{(W_{animal})^b \times W_{human}} \tag{18.24}$$

$$\frac{(Dose)_{human}}{W_{human}} = \frac{(Dose)_{animal}}{W_{animal}} \times \frac{(W_{animal})^{1-b}}{(W_{human})^{1-b}} = \frac{(Dose)_{animal}}{W_{animal}} \times \left(\frac{W_{animal}}{W_{human}}\right)^{1-b} \tag{18.25}$$

18.3 TOXICOGENOMICS

The prediction of human health hazards based on animal data is limited to the differences between human and experimental animal responses to xenobiotics. This includes the types of different responses the two species may have toward various dose levels. Thus, it is important to validate animal models for the regulatory processes including environmental sciences and drug discovery and development.

Toxicogenomics, a subdiscipline of genomics, contributes significantly to the selection of the appropriate animal model. Furthermore, the discipline has improved the understanding of the following areas:

- The knowledge of toxic mechanisms
- The understanding of *in vitro* and *in vivo* systems
- Rapid screening of xenobiotic toxicity
- Selection of the lead compounds in drug discovery
- The understanding between the genetic variability and response to xenobiotic exposure
- Prediction of differences between experimental animal and human responses to chemicals

These aspects collectively have increased the efficiency of predictive toxicology and more accurate and insightful knowledge of the safety profile of a xenobiotic, which in turn has improved the efficiency of the extrapolation of animal data. In addition its role in interspecies extrapolation, toxicogemics plays an essential role in identifying the variability of response among humans to xenobitics. Xenobiotics induce differential gene expressions, which can be defensive, adaptive, or repairing, by up or down regulation. Humans show different basal and inducible patterns of gene expression when exposed to xenobiotics. Often the exposure to xenobiotics may lead to the induction of genes that are not associated with their mechanism of toxicity, but the responses from these unrelated genes, although do not characterize a particular toxicity, can indicate the stress response from exposure to both chemical and physical stress, including UV light (Rieger and Chu, 2004). The inclusion of relevant genetic parameters in the allometric equations to improve their predictability is yet to be established.

18.4 LOW DOSE EXTRAPOLATION

The objective of this section is not to discuss the quantitative risk assessment, which is beyond the scope of this book, rather to explore the involvement of PK/TK in the low dose extrapolation. The challenges of estimating toxic effects of xenobiotics at low dose exposure have received considerable attention from toxicologists, statisticians, and kineticists over the past decades. The low dose extrapolation is one of the key issues in human health risk assessment and it deals with the uncertainty of extrapolation of data beyond the observations and measurement of the response (Figure 18.2). Fitting a number of appropriate mathematical models to experimental animal data is an approach used to determine the carcinogenicity of a compound and to predict the response at dose levels that are well below the sensitivity of the assays or experimental dose levels. A different scenario of dose extrapolation involves the extrapolation from high dose to low dose. Often, in attempting the interspecies extrapolation of the data from experimental animals to human, the data must be generated from a dose level that is extrapolated from a high dose to a low dose, appropriate for the objectives of the experimental protocol. The National Research Council (NRC, 2009), in its 2009 report "Science and Decisions: Advancing Risk Assessment" indicated that the exposure-response for cancer low dose extrapolation is driven by stochastic events and does not incorporate information on human sensitivity, but the non-cancer toxicity exposure is driven by the individual sensitivity and is not stochastic. There are many different mathematical models for estimating the dose exposure. Most models express the probability of response as a function of dose.

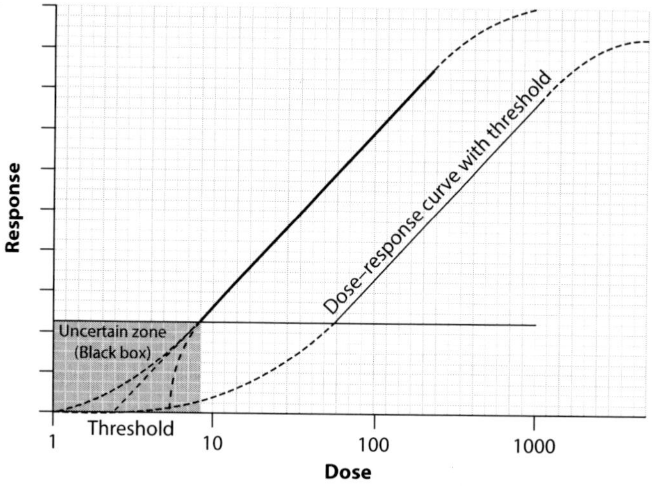

FIGURE 18.2 The uncertainty of extrapolation beyond the limit of observation and prediction of response.

All models start with the experimental evaluation of response at different dose levels in one or more species of experimental animals that are considered relevant to human risk assessment:

$$(\mathrm{Pr})_{response} = f(Dose) \tag{18.26}$$

The models differ only in the choice of function *f*.

The models reflect two types of threshold and non-threshold scenarios. The threshold models assume that below a threshold, which can be the magnitude of exposure, a dose, or concentration, there is no adverse health effect and there is a minimally effective dose. The non-threshold models assume there is no minimally acceptable dose that is, any nonzero dosage is considered unsafe, as is the intention of Delaney clause for carcinogenic response. (The Delaney clause was a 1958 amendment to the Food, Drugs, and Cosmetic Act of 1938 sponsored by Jim Delaney, the New York Congressman who verbalized the amendment as: "the Secretary of the FDA shall not approve for use in food any chemical additive found to induce cancer in man, or, after tests, found to induce cancer in animals.")

The types of models used in the extrapolation are

The probit model (Finney, 1952; Gad and Weil, 1986)
The logit model (Gad and Weil, 1986)
The one-hit model (Hoel et al., 1975; Gad and Weil, 1986; Bailar et al., 1988)
The gamma multi-hit model (Cornfield et al., 1979; Gad and Weil, 1986)
The Armitage–Doll multi-stage model (Armitage and Doll, 1961; Gad and Weil, 1986)
A simplified statistico-pharmacokinetic model (Cornfield, 1977; Gad and Weil, 1986)
The Weibull model (Gad and Weil, 1986)
The log-probit model (Gad and Weil, 1986)
The steady-state model (Gad and Weil, 1986)
The Gompertz function (Boxenbaum et al., 1988)
The modified Hill equation (Egorin et al., 1986, 1987)

All of these models except one are statistical in nature and do not rely on physiologic, biochemical or PK/TK properties of a xenobiotic. The discussion here will be limited to a few of the above models.

18.4.1 The Probit Model

The assumptions of the model is that the logarithm of tolerances have a normal distribution with mean and standard deviation of μ and σ, respectively. The proportion of subjects responding to dose that is, *P*(*Dose*) is

$$P(Dose) = \Phi[(\log Dose - \mu)/\sigma] = \Phi(\alpha + \beta \log Dose) \tag{18.27}$$

In Equation 18.27, $\Phi(x)$ is the standard normal integral from $-\infty$ to x; $\alpha = \mu/\sigma$ and $\beta = 1/\sigma$. The dose-response curve corresponding to the equation has *P*(*Dose*) near zero if *Dose* is close to zero and as *Dose* increases, *P*(*Dose*) approaches one. A plot of a probit dose-response is a sigmoidal with the quantity β as the slope of the probit line, where the equation line is

$$Y = \Phi^{-1}\left[P(Dose)\right] = \alpha + \beta \log Dose \tag{18.28}$$

18.4.2 The Logit Model

This model also generates a sigmoidal dose-response curve symmetric about 50% of the response:

$$P(Dose) = \frac{1}{1 + e^{-(\alpha+\beta \log Dose)}} \tag{18.29}$$

Similar to the probit model, as *Dose* deceases *P*(*Dose*) also decreases and approaches zero more slowly than the probit model, that is,

$$\lim_{Dose \to 0}\left[P(Dose)/(Dose)^{\beta}\right] = \text{constant} \tag{18.30}$$

In general, the logit model estimates a virtual safe dose, lower than the probit model.

18.4.3 The One-Hit Model

This model is essentially a linear dose-response model with the assumption that only one hit, corresponding to one genetic change, is needed to transform a normal cell into a cancerous cell. The relationship of the model is

$$P(Dose) = 1 - e^{-\lambda Dose} \tag{18.31}$$

where λ is a rate constant of change of the dose-response curve at $Dose = 0$; λ *Dose* is the expected number of hits at dose level (*Dose*). Being a linear relationship, the dose-response curve of the model is linear. Thus, any dose of carcinogen presents a risk of cancer and only when $Dose = 0$, $P(Dose) = 0$. Although the model tacitly refers to a biochemical interaction, it is considered a phenomenological model rather than a biochemical one. The probability of the phenomenon that is, a normal cell transformation into a cancerous cell, according to this model follows Poisson probability distribution. The model is viewed by some as simplistic and conservative.

18.4.4 The Gamma Multi-Hit Model

This model is an extension of the one-hit model and assumes that if a number of hits (identified as k hits) are required to induce cancer, then the probability of generating cancer as a function of exposure to a *Dose* is

$$P(Dose) = 1 - \sum_{i=0}^{k-1} \frac{(\lambda Dose)^i e^{-\lambda d}}{i!} \approx \frac{(\lambda Dose)^k}{k!} \tag{18.32}$$

When $k = 1$, Equation 18.32 changes to Equation 18.31, which is indicative of the linear low dose region of the plot. For $k > 1$ it is convex and for $k < 1$ it is concave.

For small $\lambda(Dose)^k$, the model can be modified to

$$P(Dose) = \lambda(Dose)^k, \text{or} \tag{18.33}$$

$$\log P(Dose) = \log \lambda + k \log Dose \tag{18.34}$$

Therefore, k is the slope of the straight line generated from plotting *P*(*Dose*) versus log*Dose*. Equation 18.32 describes a dose response curve that follows gamma distribution with k as the shape parameter.

18.4.5 The Armitage–Doll Multi-Stage Model

The assumption of the model is that cancer originates from a malignant cell, formed through a multi-staged mutation (Armitage and Doll, 1961). Thus, the effect of the agent is considered additive at those stages. The related probability function is

$$P(Dose) = 1 - e^{-\left[\sum_{i=0}^{\infty} \alpha_i (Dose)^i\right]} \quad \text{for} \quad \alpha_i \geq 0 \tag{18.35}$$

The model is also expressed as

$$P(Dose) = 1 - e^{-(\beta_i + \gamma_i (Dose))(\beta_{i+1} + \gamma_{i+1} (Dose))........(\beta_k + \gamma_k (Dose))} \tag{18.36}$$

where $\beta_i \geq 0$, $\gamma_i \geq 0$ and k is the number of mutational stages.

In a simpler format, Equation 18.36 generates Equation 18.37 that corresponds to the general Equation 18.35, that is,

$$P(Dose) = 1 - e^{-\alpha_0 + \alpha_1 (Dose) + \alpha_2 (Dose)^2 + ...+ \gamma_k (Dose)^k} \quad \text{for} \quad \alpha_i \geq 0 \tag{18.37}$$

The model also assumes that each stage of mutation follows a Poisson distribution.

18.4.6 Statistico-Pharmacokinetic Model

This model takes into consideration the simultaneous and reversible activation and inactivation of the agent in the body. The probability of a response is assumed to be linearly proportional to the activated complex. If the total amount of the compound in the body is equal to S and of deactivating agent is T, and the ratios of the back and forth rate constants of reactions governing the activation step are K, and deactivation step by K^*, the model can be expressed as

$$P(Dose) = \frac{Dose - S\left[P(Dose)\right] - y}{Dose - S\left[P(Dose)\right] - y + K} \quad \text{for } Dose > T \tag{18.38}$$

The parameter y in Equation 18.38 is

$$y = \frac{K\left[P(Dose)\right]T}{K\left[P(D)\right] + K^*\left[1 - P(D)\right]} \tag{18.39}$$

For $Dose < T$:

$$P(Dose) \cong \frac{Dose}{S + K\left(1 + \dfrac{T}{K^*}\right)} \tag{18.40}$$

At dose levels of $D < T$, the dose-response curve closely resembles linear plot. When $K^* = 0$, the dose-response curve has a threshold at $Dose = T$ and for $K^* > 0$ the dose-response curve resembles a hockey stick. The constants S, T, K, and K^* are considered species-dependent kinetic constants.

The few models discussed in this chapter for low dose extrapolation may also have other applications in biological sciences. Although they have no firm biological basis, their main purpose in low dose extrapolation is to establish an acceptable balance between the risk and benefit of xenobiotics.

REFERENCES

Adolph, E. F. 1949. Quantitative relations in the physiological constitutions of mammals. *Science* 109:579–85.

Armitage, P., Doll, R. 1961. Stochastic models for carcinogenesis. In *Proceedings of the Fourth Berkeley Symposium VI,* ed. J. Neyman, Berkeley, CA: University of California Press.

Bailar, J. C. III, Crouch, E, A., Shaikh, R., Spiegelman, D. 1988. One-hit models of carcinogenesis: Conservative or not? *Risk Anal* 8(4):485–97.

Boxenbaum, H. 1982. Interspecies scaling, allometry, physiological time, and the ground plan of pharmacokinetics. *J Pharmacokin Biopharm* 10(2):201–27.

Boxenbaum, H., Ronfeld, R. 1983. Interspecies pharmacokinetic scaling and the Dedrick plots. *Am J Physiol* 245:R768–74.

Boxenbaum, H. 1984. Interspecies pharmacokinetics scaling and the evolutionary-comparative paradigm. *Drug Metab Rev* 15(5–6):1071–121.

Boxenbaum, H. 1986. Time concept in physics, biology, and pharmacokinetics. *J Pharm Sci* 75(11):1053–62.

Boxenbaum, H., Neafsey, P. J., Fournier, D. J. 1988. Hormesis, Gompertz functions and risk assessment. *Drug Metab Rev* 19:195–29.

Calder III, W. A. 1984. *Size Function, and Life History,* Cambridge, MA: Harvard University Press.

Chappel, W. R., Mordenti, J. 1989. The use of interspecies scaling in toxicokinetics. In *Toxicokinetics and New Drug Development*, eds. A. Yacobi, J. P. Skelly and V. K. Batra, New York: Pergamon Press 42–96.

Cornfield, J. 1977. Carcinogenic risk assessment. *Science* 198:693–99.

Cornfield, J., Carlborg, F. W., Van Ryzin, J. 1979. Setting tolerance on the basis of mathematical treatment of dose-response data extrapolated to low doses. In *Proc of the First Internat Toxicol Congress on Toxicology,* NY: New York Academic Press.

Davidson, I. W. F., Parker J. C., Beliles, R. P. 1986. Biological basis for extrapolation across mammalian species. *Regul Toxicol Pharmacol* 6:211–37.

Dedrick, R. L., Bischoff, K. B., Zaharko, D. S. 1970. Interspecies correlation of plasma concentration history of methotrsxate (NSC-740). *Cancer Chemother Rep* 54:95–101.

Egorin, M. J., Conley, B. A., Forrest, A., Zuhowski, E. G., Sinibaldi, V., Van Echo, D. A. 1987. Phase I study and pharmacokinetics of menogaril (NSC 269148) in patients with hepatic dysfunction. *Cancer Res* 47:6104–10.

Egorin, M. J., Van Echo, D. A., Whitacare, M. Y., Forrest, A., Sigman, L. M., English, K. L., Aisner, J. 1986. Human pharmacokinetics, excretion and metabolism of the anthracycline antibiotic menogaril 97-OMEN, NSC 269148) and their correlation with clinical toxicities. *Cancer Res* 46:1513–20.

Finney, D. J. 1952. *Statistical Method in Biological Assay*, New York: Hafner Publishing Co.

Gad, S., Weil, C. S. 1986. *Statistics and Experimental Design for Toxicologist*, Caldwell, NJ: Telford Press.

Gunther, B., DeLa Barra, L. 1966. Physiometry of the mammalian circulatory system. *Acta Physiol Lat Am* 16:32–42.

Heusner, A. A. 1984. Biological similitude: Statistical and functional relationships in comparative physiology. *Am J Physiol* 246:R839–45.

Hoel, D. G., Galor, D. W., Kirschstein, R. L., Saffiotti, V., Schneiderman, M. A. 1975. Estimation of risks of irreversible delayed toxicity. *J Toxicol Environ Health* 1:133–51.

Ings, R. M. J. 1990. Interspecies scaling and comparisons in drug development and toxicokinetics. *Xenobiotica* 20(11):1201–31.

Kleiber, M. 1932. Body size and metabolism. *Hilgardia* 6:315–53.

Kleiber, M. 1947. Body size and metabolic rate. *Physiol Rev* 27:511–41.

Loiselle, D. S., Gibbs, C. L. 1979. Species differences in cardiac energetics. *Am J Physiol* 237(1):H90–8.

Mahmood, I., Balian, J. D. 1999. The pharmacokinetic principles behind scaling from preclinical results to phase I protocol. *Clin Pharmacokin* 36(1):1–11.

McMahon, T. A., Bonner, J. T. 1983. *On Size and Life*. New York: Scientific American Library, 69–110.

Mordenti, J. 1986. Man versus beast: Pharmacokinetic scaling in mammals. *J Pharm Sci* 75(11):1028–40.

Mordenti, J., Chen, S. A., Moore, J. A., Ferraiolo, B. L., Green, J. D. 1991. Interspecies scaling of clearance and volume of distribution data for five therapeutic proteins. *Pharm Res* 8(11):1351–59.

National Research Council: Science and Decisions: *Advancing Risk Assessment,* Washington DC: The National Press, 2009.

Peters, R. H. 1983. *The Ecological Implications of Body Size,* Cambridge, MA: Cambridge University Press.

Rieger, K., Chu, G. 2004. Portrait of transcriptional responses to ultraviolet light and ionizing radiation in human cells. *Nucl Acids Res* 32(16):4786–803.

Rubner, M. 1883. Ueber den einfluss der korpergrosse auf stoff und kraft wechsel. *Z Biol* 1919:535–62.

Sacher, G. 1959. Relation of lifespan to brain weight and body weight in mammals. In eds. G. E. W. Wolstenholme and M. O'Conner, Churchill, London CIBA Foundation colloquia on aging., 115–33.

Smith, R. J. 1980. Rethinking allometry. *J Theor Biol* 87:97–111.

Stahl, W. R. 1965. Organ weights in primates and other mammals. *Science* 150:1039–41.

Voisin, E. M., Ruthsatz, M. Collins, J. M., Hoyle, P. C. 1990. Extrapolation of animal toxicity to humans: Interspecies comparisons in drug development. *Regul Toxicol Pharmacol* 12:107–16.

White, L., Haines, H., Adam, T. 1968. Cardiac output related to body weights in small mammals. *Comp Biochem Physiol* 27:559–65.

Yates, F. E. 1979. Comparative physiology: Compared to what? *Am J Physiol* 237:R1–2.

Appendix A: Relevant Mathematical Concepts

A.1 THE LAPLACE TRANSFORM METHOD OF INTEGRATION

The idea behind the Laplace transform is that every solution of any differential equation has a Laplace transform and the derivatives of the function can also be expressed in terms of the Laplace transform. This means that any complex differential equation can be converted into a linear algebraic equation and the resulting equation is then solved by simpler methods. The final solution of the algebraic conversion is subjected to an inverse transformation, which may often be facilitated by a table of Laplace transforms, yielding the desired solution of the differential equation. To learn more about the basic properties and the uniqueness theorem of the Laplace transform, the textbooks of ordinary differential equation should be consulted (Brauer and Nohel, 1973). The following discussion represents the application of Laplace transform in PK/TK modeling.

Briefly, assume our function is $y = f(t)$. The Laplace transform of this function, represented by the notation $\mathbf{L}[f(t)]$, can be obtained by multiplying the function, $f(t)$, by $e^{-st}\,dt$ and integrating from time zero to ∞. Thus, the Laplace transform of the function $f(t)$ is

$$\mathbf{L}[f(t)] = \int_0^\infty e^{-st} f(t) \tag{A.1}$$

where "s" is the Laplace operator and the function is converted from the real variable "t" into a function of complex parameter "s."

As an example, consider the function of $f(t) = 1$. The Laplace transform of the function is

$$\mathbf{L}[f(t)] = \mathbf{L}(1) = \int_0^\infty e^{-st}(1)\,dt = \left.\frac{e^{-st}}{s}\right|_0^\infty = \frac{1}{s} \tag{A.2}$$

Therefore, the Laplace transform of 1 is $1/s$ and the inverse Laplace transform of $1/s$ is 1, that is,

$$\mathbf{L}^{-1}\left(\frac{1}{s}\right) = 1 \tag{A.3}$$

From the above transformation it can also be concluded that the Laplace transform of any constant A is A/s and the inverse Laplace transform of A/s is A. For example, $\mathbf{L}(5) = 5/s$ and $\mathbf{L}^{-1}(5/s) = 5$.

The Laplace transforms of exponential functions such as $f(t) = e^{kt}$ and $f(t) = e^{-kt}$ are

$$\mathbf{L}(e^{kt}) = \int_0^\infty e^{-st}e^{kt}\,dt = \int_0^\infty e^{(k-s)t} = \left.\frac{e^{(k-s)t}}{k-s}\right|_0^\infty = \frac{1}{s-k} \tag{A.4}$$

$$\mathbf{L}(e^{-kt}) = \int_0^\infty e^{-st}e^{-kt}dt = \int_0^\infty e^{-(s+k)t}dt = \left.\frac{e^{-(s+k)t}}{-(s+k)}\right|_0^\infty = \frac{1}{s+k} \tag{A.5}$$

Accordingly, the inverse Laplace transforms are

$$\mathbf{L}^{-1}\left[\frac{1}{s-k}\right] = e^{kt} \tag{A.6}$$

$$\mathbf{L}^{-1}\left[\frac{1}{s+k}\right] = e^{-kt} \tag{A.7}$$

Table A.1 contains a short list of Laplace transforms that are useful in the development of integrated equations of pharmacokinetics. In Table A.1, A and B are constants; k, k_1, k_2 and k_3 are rate constants in their general form, they are not related to any specific rate constant and are included under the assumption that $k \neq k_1 \neq k_2 \neq k_3$; m is a power constant and s is the Laplace parameter.

Among the theorems of the Laplace transformation the following are the most useful in PK/TK analysis:

1.

$$\text{If} \quad \mathbf{L}[f_1(t)] = F_1(s) \quad \text{and} \quad \mathbf{L}[f_2(t)] = F_2(s), \quad \text{then} \tag{A.8}$$

$$\mathbf{L}[f_1(t)] + \mathbf{L}[f_2(t)] = F_1(s) + F_2(s) \tag{A.9}$$

2. If a is a constant, then

$$\mathbf{L}[af(t)] = a\mathbf{L}[f(t)] = aF(s) \tag{A.10}$$

3. If a and b are constants, then

$$af_1(t) + bf_2(t) \text{ has Laplace transform } aF_1(s) + bF_2(s). \tag{A.11}$$

4. The Laplace transform of derivative dy/dx is the Laplace transform of y minus the value of the function at $t = 0$, that is,

$$\mathbf{L}\left[\frac{dy}{dx}\right] = s\tilde{y} - y_0 \tag{A.12}$$

where $\tilde{y}$ is the Laplace transform of y.

A.2 DETERMINANTS AND THE CRAMER'S RULE

In solving the pharmacokinetic systems of linear differential equations with constant coefficients using the Laplace transform, the coefficients of transformed functions and the constant terms of initial conditions are important in solving the equations, as briefly described here.

TABLE A.1
Relevant Laplace Transforms

Function	Laplace Transform
1	$\frac{1}{s}$
A	$\frac{A}{s}$
t	$\frac{1}{s^2}$
At	$\frac{A}{s^2}$
t^m	$\frac{m!}{s^{m+1}}$
e^{kt}	$\frac{1}{s-k}$
e^{-kt}	$\frac{1}{s+k}$
Ae^{kt}	$\frac{A}{s-k}$
Ae^{-kt}	$\frac{A}{s+k}$
Ate^{-kt}	$\frac{A}{(s+k)^2}$
$\frac{A}{k}(1 - e^{-kt})$	$\frac{A}{s(s+k)}$
$At^m e^{-kt}$	$\frac{A}{(s+k)^{m+1}}$
$\frac{A}{k_1} e^{-\left(\frac{k_2}{k_1}\right)t}$	$\frac{A}{k_1 s + k_2}$
$\left(\frac{A}{k}\right)t - \frac{A(1 - e^{-kt})}{k^2}$	$\frac{A}{s^2(s+k)}$
$Ae^{-kt} - \frac{B}{k}(1 - e^{-kt})$	$\frac{(As - B)}{s(s+k)}$
$\frac{(B - Ak_1)e^{-k_1 t} - (B - Ak_2 e^{-k_2 t})}{k_2 - k_1}$	$\frac{As + B}{(s+k_1)(s-k_2)}$
$\frac{A}{k_2 - k_1}(e^{-k_1 t} - e^{-k_2 t})$	$\frac{A}{(s+k_1)(s+k_2)}$
$\frac{1}{k_2 k_1} + \frac{k_2 e^{-k_1 t} - k_1 e^{-k_2 t}}{k_1 k_2 (k_1 - k_2)}$	$\frac{1}{s(s+k_1)(s+k_2)}$

(Continued)

TABLE A.1 (*Continued*)
Relevant Laplace Transforms

Function	Laplace Transform
$\frac{A}{k_1 - k_2}\left(\frac{1 - e^{-k_2 t}}{k_2} - \frac{1 - e^{-k_1 t}}{k_1}\right)$	$\frac{A}{s(s + k_1)(s + k_2)}$
$A\left(\frac{1}{k_1 k_2} + \frac{e^{-k_1}}{k_1(k_1 - k_2)} - \frac{e^{-k_2 t}}{k_2(k_1 + k_2)}\right)$	$\frac{A}{s(s + k_1)(s + k_2)}$
$\frac{A}{k}t - A(1 - e^{-kt})$	$\frac{A}{s^2(s + k)}$
$\frac{B}{k_1 k_2} - \frac{(Ak_1 - B)e^{-k_1 t}}{k_1(k_1 - k_2)} + \frac{(Ak_2 - B)e^{-k_2 t}}{k_2(k_1 - k_2)}$	$\frac{As + B}{s(s + k_1)(s + k_2)}$
$\frac{(A - k_1)e^{-k_1 t}}{(k_2 - k_1)(k_3 - k_1)} + \frac{(A - k_2)e^{-k_2 t}}{(k_1 - k_2)(k_3 - k_2)} + \frac{(A - k_3)e^{-k_3 t}}{(k_1 - k_3)(k_2 - k_3)}$	$\frac{s + A}{(s + k_1)(s + k_2)(s + k_3)}$

Any two- or three-variable linear equation can be written in the form of

$$ax + by = c \quad (A.13)$$

$$ax + by + cz = Q \quad (A.14)$$

Where a, b, and c are the constant coefficients and Q is also a constant. In solving the PK/TK system of linear differential equations by the Laplace transform, one frequently encounters systems of simultaneous equations of the form

$$\begin{aligned} ax + by &= C \\ dx + ey &= F \end{aligned} \quad (A.15)$$

or,

$$\begin{aligned} ax + by + cz &= Q \\ dx + ey + fz &= R \\ gx + hy + iz &= s \end{aligned} \quad (A.16)$$

and so on.

The use of Cramer's rule and determinants can make the process of solving the simultaneous set of differential equations simple. This is only one application of the rule and discussion on various applications and theorems of determinants are beyond the scope of this section.

The first step in solving the system of equations with three variables (Equation A.16) according to Cramer's rule is to establish the determinant (D) of the system by using the coefficients of the variables that are unknown:

$$D = \begin{pmatrix} a & b & c \\ d & e & f \\ g & h & i \end{pmatrix} \quad (A.17)$$

The solution of this determinant is

$$D = aei + bfg + cdh - ceg - bdi - afh \tag{A.18}$$

The signs of Equation A.18 are decided according to the following rule:

$$Sign = (-1)^{\text{No of Rows + No of Columns}}$$

That is,

$$\begin{vmatrix} +11 & -12 & +13 \\ -21 & +22 & -13 \\ +31 & -32 & +33 \end{vmatrix}$$

where the first number is the row number and the second is the column number. Therefore, position 32, for example, refers to the third row and second column.

According to Cramer's rule, if the solution of determinant D is not zero, then the equations have a unique solution and each variable can be expressed as a fraction of two determinants. The numerator of the fraction is expressed by replacing the column of coefficients of the variable under evaluation with the constants of the system. For example, for x in Equation A.16:

$$x = \frac{\begin{pmatrix} Q & b & c \\ R & e & f \\ S & h & i \end{pmatrix}}{\begin{pmatrix} a & b & c \\ d & e & f \\ g & h & i \end{pmatrix}} = \frac{Qei + bf\,S + Rhc - Sec - Rbi - Qf\,h}{aei + bfg + chd - gec - dbi - a\,fh} \tag{A.19}$$

A similar arrangement is repeated for each of the unknowns. For instance, for y the solution is as follows:

$$y = \frac{\begin{pmatrix} a & Q & c \\ d & R & f \\ g & S & i \end{pmatrix}}{\begin{pmatrix} a & b & c \\ d & e & f \\ g & h & i \end{pmatrix}} = \frac{aRi + Q\,fg + dSc - cRg - Qdi - f\,Sa}{aei + bfg + chd - gec - dbi - af\,h} \tag{A.20}$$

Similar calculations are used for the z variable. The unknowns of a set of simultaneous equations with two variables can also be determined in a similar way. For example, the solution for the variables in Equation A.15 is

$$x = \frac{\begin{pmatrix} C & b \\ F & e \end{pmatrix}}{\begin{pmatrix} a & b \\ d & e \end{pmatrix}} = \frac{Ce - bf}{ae - bd} \tag{A.21}$$

$$y = \frac{\begin{pmatrix} a & C \\ d & F \end{pmatrix}}{\begin{pmatrix} a & b \\ d & e \end{pmatrix}} = \frac{aF - dC}{ae - bd} \tag{A.22}$$

A.3 INPUT AND DISPOSITION FUNCTIONS

In compartmental analysis, the Laplace transform of linear differential equations for any compartment can be developed by multiplying its input function with its disposition function (Benet, 1972). This convolution method facilitates the development of Laplace transforms of differential equations of compartmental models significantly.

A.3.1 Input Functions

The input functions via the major routes of administration into the body are

Intravenous bolus injection:

$$Input = Dose = A_D^0 \tag{A.23}$$

Intravenous infusion:

$$f(t) = k_0 \tag{A.24}$$

(k_0 is the zero-order rate of infusion)

$$\mathbf{L}[f(t)] = \int_0^{t_{infusion}} k_0 e^{-st}\,dt = -\frac{k_0}{s} e^{-st_{infusion}} + \frac{k_0}{s} \tag{A.25}$$

When the infusion starts at $t = 0$ and continues for a time long enough such that

$$t_{infusion} \Rightarrow \infty$$

$$\therefore\ Input = \mathbf{L}[k_0] = \frac{k_0}{s} \tag{A.26}$$

When the infusion starts at $t = 0$ and continues for a time $t < \infty$

$$\therefore\ Input = \mathbf{L}[k_0] = \frac{k_0}{s}(1 - e^{-st_{infusion}}) \tag{A.27}$$

When the infusion starts at $t \neq 0$

$$Input = \mathbf{L}[f(t)] = \int_a^{t_{infusion}} k_0 e^{-st}\,dt = \frac{k_0(e^{-as} - e^{-st_{infusion}})}{s} \tag{A.28}$$

where $t_{infusion}$ is the time of infusion; a is a time different from zero.

- Oral absorption:

$$f(t) = FDk_a e^{-k_a t} \tag{A.29}$$

(FD is the amount of drug absorbed and k_a is the first-order absorption rate constant).

$$\mathbf{L}[f(t)] = FDk_a \int_0^{\infty} e^{-kat} e^{-st} = FDk_a \int_0^{\infty} e^{-(ka+s)t} \tag{A.30}$$

$$Input = \frac{FDk_a}{s + k_a} \tag{A.31}$$

- The input function from any driving-force compartment (DFC) into other compartments is defined as

$$(\text{Exit rate constant})_{\text{DFC}} \times (\text{Laplace transform})_{\text{DFC}} \tag{A.32}$$

A similar approach can be taken for development of other input functions.

A.3.2 Disposition Function

In more complex multicompartmental linear models when all peripheral compartments are interdependent with the central compartment, the disposition function of the central compartment is determined by the following equation (Benet, 1972):

$$(Disp.)_s = \frac{\prod_{i=2}^{n}(s + E_i)}{\prod_{i=1}^{n}(s + E_i) - \sum_{j=2}^{n}\left[k_{1j}k_{j1}\prod_{\substack{m=2\\ m\neq j}}^{n}(s + E_m)\right]} \tag{A.33}$$

where $(Disp.)_s$ is the disposition function of the central compartment as a function of Laplace operator, s; n is the number of the driving force compartment; E_i is the overall exit rate constant from the ith compartment; k_{1j} is the first-order rate constant from the central compartment to the jth compartment; k_{j1} is the first-order rate constant from the jth compartment to the central compartment; Σ and Π are the symbols for summation and continued product.

A.3.3 Determination of Inverse Laplace Transform by Method of Partial Fraction Theorem

This method is used when the denominator of the Laplace transform of a compartment (i.e., multiplication of input and disposition functions) has no repeated terms and contains more "s" terms than the numerator. Under these conditions, the inverse Laplace transform can be estimated by substituting the hybrid rate constants for each "s" term in numerator and denominator, followed by multiplication with the related exponential function (Benet and Turi, 1971). See Chapter 9, Section 9.3, Equations 9.51 and 9.52.

A.4 TRAPEZOIDAL RULE

Many functions in PK/TK are integrable. The integrability of the linear differential equations is a straightforward concept. For example, the integral of the "rate" or "slope" relationship, that is, a simple differential equation, is the equation of the line or curve and integral of the line or curve equation is an area under the line or curve. Integration of these functions is referred to as anti-differentiation or reversing and the following symbol is used to identify the operation:

$$\int f(x)dx = F(x) + C \tag{A.34}$$

When the interval of integration is not defined (Equation A.34), the operation is called the indefinite integration and $F(x)$ is the indefinite integral. The definite integral, on the other hand, is the magnitude of an integral over a well-defined interval. Mathematically, if $F(x)$ is the integral of $f(x)$ between an upper limit of $x = b$ and lower limit of $x = a$ (i.e., $a \leq x \leq b$), the integration is defined as

$$\int_a^b f(x)dx = F(b) - F(a) \tag{A.35}$$

An example of a definite integral is the area under the plasma concentration–time curve of a compound between time of administration ($a = 0$) and 6 hours after the administration ($b = 6\ h$). An example of indefinite integral is when the area is between the time of administration ($a = 0$) and the time that the dose is completely eliminated from the body ($b \cong \infty$). Both the upper and lower limits in the definite integral are real numbers. Therefore, the definite integral is a real number whereas an indefinite integral is a function.

The trapezoid rule is a practical method for numerical approximation of definite integrals. The method is based on dividing the interval of observation (i.e., $a \leq x \leq b$) into subintervals that correspond to the time of sampling and measurements. The area between the curve and each subinterval on the x-axis is a quadrilateral with one pair of parallel lines and one pair of unparallel lines forming a trapezoid with an area equal to

$$Area = base\left(\frac{y_1 + y_2}{2}\right) \tag{A.36}$$

where y_1 and y_2 correspond to plasma concentration or logarithm of plasma concentration at two adjacent time points and the base is the subinterval of the x axis that corresponds to the time interval between the two y values (Figure A.1). The total area under the curve that corresponds to the indefinite integral of plasma concentration–time curve or log plasma concentration–time curve is then approximated by adding up all trapezoidal areas. The terminal area is estimated by dividing the last plasma data point by the overall elimination rate constant. The following Equations A.37 through A.41 are the summary relationship for area calculations of Cp–time or logCp–time curve from time zero to time t or infinity:

$$AUC_0^{\infty} = \sum_{i=0}^{n}[(y_i + y_{i+1})/2](x_{i+1} - x_i) + AUC_n^{\infty} \tag{A.37}$$

$$AUC_0^{t} = \sum_{i=0}^{n}\left(\frac{Cp_i + Cp_{i+1}}{2}\right)(t_{i+1} - t_i) \tag{A.38}$$

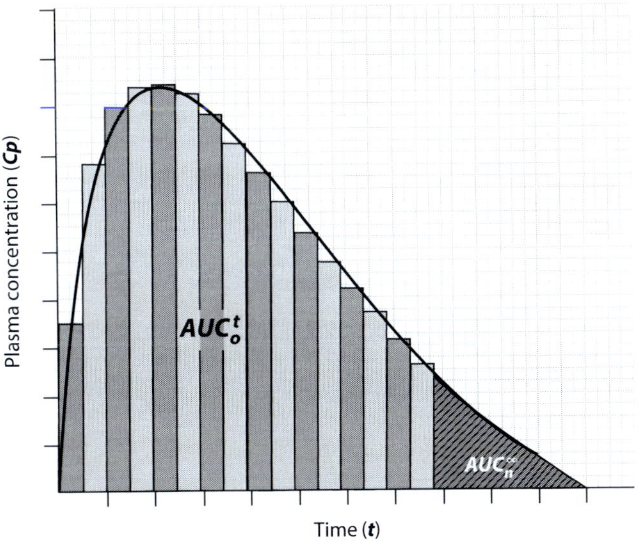

FIGURE A.1 A graphical presentation of trapezoidal rule for estimation of area under plasma concentration–time curve AUC_0^t, AUC_t^∞ and AUC_0^∞.

$$AUC_0^\infty = \sum_{i=0}^{n}\left(\frac{Cp_i + Cp_{i+1}}{2}\right)(t_{i+1} - t_i) + \frac{Cp_n}{K} \tag{A.39}$$

$$AUC_0^t = \sum_{i=0}^{n}\left(\frac{\log Cp_i + \log Cp_{i+1}}{2}\right)(t_{i+1} - t_i) \tag{A.40}$$

$$AUC_0^\infty = \sum_{i=0}^{n}\left(\frac{\log Cp_i + \log Cp_{i+1}}{2}\right)(t_{i+1} - t_i) + \frac{Cp_n}{K} \tag{A.41}$$

where n is the total number of observations, i identifies each measurement, $i = 0$ indicates that the interval starts from $a = 0$ and the initial values of x and y should be included, AUC_n^∞ represents the terminal area between the last data point and infinity, estimated as the last y value divided by the rate constant estimated from the slop of the terminal phase of the curve.

A.5 GEOMETRIC SERIES

A geometric series is used in connection with multiple-dosing kinetics. A geometric series is a series for which the ratio of each two consecutive terms is a constant function of the summation, that is,

$$S_n(x) = \frac{1 - (x)^{n+1}}{1 - (x)} \tag{A.42}$$

Therefore, for a function like $Cp = Cp^0 e^{-Kt}$, the geometric series is

$$S_n(x) = \frac{1 - e^{-n(K)\tau}}{1 - e^{-K\tau}} \tag{A.43}$$

As $n \Rightarrow \infty$, $e^{-n(K)\tau} \Rightarrow 0$ and the geometric series changes to

$$S_n(x) = \frac{1}{1 - e^{-K\tau}} \tag{A.44}$$

The application of geometric series is presented in Chapter 13.

A.6 KINETICS AND RATE EQUATIONS

The activity of a chemical or biological system is defined either dynamically or kinetically. The dynamic principles often deal with the quantitative and qualitative evaluations of transformation between an initial condition and its intended end condition. The evaluations can only determine the dynamics of the system or process in terms of receiving an input and providing a response, similar to the principles of thermodynamics, pharmacodynamics, or toxicodynamics. Kinetics, on the other hand, is the study of time dependency of the transformation from the initial condition to the intended end condition in terms of associated rates or velocities. In PK/TK, the biological processes are explored in terms of their rate of mass fluxes, and rapidity of distribution, biotransformation, and elimination. The mathematical expressions are the same as that used in chemical kinetics.

A.6.1 First-Order Kinetics

When the rate of a process is proportional to the first power of concentration, amount, or any measurable quantity under evaluation, the process is considered first-order kinetics, which can be summarized as

$$Rate \propto [A] \tag{A.45}$$

An equal sign and a proportionality constant, K, can replace the proportionality sign and change Equation A.45 into a linear differential equation

$$\frac{dA}{dt} = \pm KA \tag{A.46}$$

where A is the amount and K is the first-order rate constant. The plus sign symbolizes a positive outcome for the process such as "formation" or "appearance" and the minus sign represents a negative outcome such as "decomposition" or "disappearance," etc. The differential equation is dA/dt. The integration of Equation A.46 yields Equations A.49, A.51, and A.52

$$\int_0^\infty \frac{dA}{A} = \pm K \int_0^\infty dt \tag{A.47}$$

$$\ln A = \pm Kt + \text{constant} \tag{A.48}$$

$$\text{at } t = 0, \quad \text{constant} = \ln A_{t=0} = \ln A^0$$

$$\ln A = \ln A^0 \pm Kt \tag{A.49}$$

$$\ln(A/A^0) = \pm Kt \tag{A.50}$$

$$\ln A = 2.303 \log A \tag{A.51}$$

$$\log A = \log A^0 \pm \frac{Kt}{2.303} \tag{A.52}$$

$$\log(A/A^0) = \pm\frac{Kt}{2.303} \tag{A.53}$$

Raising Equation A.49 to the power of exponential ($e^{\ln A} = e^{\ln A^0} e^{\pm Kt}$) yields

$$A = A^0 e^{-Kt} \tag{A.54}$$

Equation A.45 establishes that the rate of a first-order process dA/dt is a variable and changes with amount or concentration. Its units are mass/time and the units of the rate constant K are "time^{-1}". Furthermore, Equation A.49 implies that plot of lnA versus time is linear with a slope of $\pm K$ and y intercept of $\ln A^0$. Equations A.50 and A.53 indicate that plots of ln (A/A^0) or log (A/A^0) versus time will pass through the origin with a slope $\pm K$ or $\pm K/2.303$, respectively; and the time required for a first-order process to advance to a certain level of completion, for example, 50% complete or 90% percent complete, is independent of the amount. This property is used to estimate the half life of the process under consideration. The half-life $T_{1/2}$ is the time for the process to be half-completed, or it is the time required for one-half of the initial amount or concentration of a drug to be used in a process, that is,

$$\text{At } t = T_{1/2}; \quad A = A^0/2$$

$$\ln\frac{A^0}{2} = \ln A^0 \pm \frac{KT_{1/2}}{2.303}$$

$$T_{1/2} = \frac{\ln 2}{K} \tag{A.55}$$

or

$$\log A^0 - \log\frac{A^0}{2} = \log\frac{A^0/2}{A^0} = \log 2 = 0.301$$

$$T_{1/2} = \frac{0.693}{K} \tag{A.56}$$

and

$$K = 0.693/T_{1/2} \tag{A.57}$$

For the processes that follow first-order kinetics, the half-life and K are constant.

A.6.2 Zero-Order Kinetics

Contrary to the first-order kinetics where the rate is a variable and directly proportional to the amount or concentration of the compound under consideration, the zero-order rate is constant, that is, the process occurs at a constant rate and the rate is independent of the concentration or amount. Therefore, the rate is controlled by factors other than the compound's concentration; for example, saturation of an enzyme when the rate of metabolism is at maximum, or the infusion rate of intravenously administered therapeutic agents, which can be changed through the infusion pump, etc. The rate equation of a zero-order process is

$$\frac{dA}{dt} = \pm k_0 \tag{A.58}$$

where k_0 is the zero-order rate or rate constant, which has the units of mass/time. The integration of Equation A.58 yields

$$A = A_0 \pm k_0 t \tag{A.59}$$

where the amount at time t is A and the initial amount is A_0. A plot of A versus t on a linear graph is a straight line with y-intercept = A^0 and slope = k_0. The half-life of a zero-order process is not constant and directly proportional to the 50% of the initial amount or concentration of the drug and is inversely proportional to the zero-order rate, that is,

$$A_0 - \frac{A_0}{2} = k_0 T_{1/2} \tag{A.60}$$

$$T_{1/2} = \frac{A_0}{2k_0} \tag{A.61}$$

A.7 COMPARING MODELS: AKAIKE AND SCHWARZ CRITERIA

In fitting models with different levels of complexity to a set of data, a better fit is usually achieved with a model with more parameters. However, a more complex model may not be the most relevant one; in other words, a model with fewer parameters is to be preferred to one with more. When comparing several conceivable models to select the most appropriate one, it is necessary to compensate for improvement of fit due to increased model complexity. The Akaike information criterion (AIC) (Akaike, 1974) and Schwarz criterion, also known as Bayesian information criterion (BIC) (Schwarz, 1978), are useful for comparing compartmental models with any types of input.

The AIC formula is remarkably a simple one:

$$AIC = 2K - 2\log(L) \tag{A.62}$$

where (K) is the number of predictors (parameters to be estimated) and (L) is the maximum likelihood value. The ($2K$) part of the formula controls the number of predictors and the ($-2\log(L)$) part of the equation is a large positive value. A model with the lowest AIC value is the preferred one.

The BIC formula is also simple to use

$$BIC = -2\ln L + K\ln(n) \tag{A.63}$$

where L is the maximum likelihood value, K is the number of predictors and n is the number of data points. When data are limited, BIC is usually preferred to AIC.

REFERENCES

Akaike, H. 1974. A new look at the statistical model identification. *IEEE T Automat Contr* 19(6): 716–23.

Benet, L. Z. 1972. General treatment of linear mammillary models with elimination from any compartment as used in pharmacokinetics. *J Pharm Sci* 61(4):536–41.

Benet, L. Z., Turi, J. S. 1971. Use of a general partial fraction theorem for obtaining inverse Laplace transforms in pharmaceutical analysis. *J Pharm Sci* 60:1593–4.

Brauer, F., Nohel, J. A. 1973. *Ordinary Differential Equations*. Menlo Park, CA: W. A. Benjamin.

Schwarz, G. E. 1978. Estimating the dimension of a model. *Ann Stat* 6(2):461–4.

Index

B

C

D

E

F

G

H

M

N

O

Q

R

S

T

U

V

W